AF528558

GEOGRAPHIC AND OCEANOGRAPHIC RESEARCH IN INDIAN WATERS.

By R. B. SEYMOUR SEWELL, C.I.E., Sc.D., F.R.S., F.R.A.S.B., Lt.-Col., I.M.S.

Geographic and Oceanographic Research in Indian Waters

R B Seymour Sewell

2023

Daya Publishing House®

A Division of

Astral International Pvt. Ltd.

New Delhi – 110 002

ISBN: 9789354615047 EBOOK

Publisher's Note:

Every possible effort has been made to ensure that the information contained in this book is accurate at the time of going to press, and the publisher and author cannot accept responsibility for any errors or omissions, however caused. No responsibility for loss or damage occasioned to any person acting, or refraining from action, as a result of the material in this publication can be accepted by the editor, the publisher or the author. The Publisher is not associated with any product or vendor mentioned in the book. The contents of this work are intended to further general scientific research, understanding and discussion only. Readers should consult with a specialist where appropriate.

Every effort has been made to trace the owners of copyright material used in this book, if any. The author and the publisher will be grateful for any omission brought to their notice for acknowledgement in the future editions of the book.

Published by : **Daya Publishing House®**
A Division of
Astral International Pvt. Ltd.
– ISO 9001:2015 Certified Company
4736/23, Ansari Road, Darya Ganj
New Delhi-110 002
Ph. 011-43549197, 23278134
E-mail: info@astralint.com
Website: www.astralint.com

CONTENTS.

PREFACE.

In this volume I have attempted to bring together the results that have been obtained from a Study of Oceanography, using the term in its widest sense to include not only the hydrographic and meteorological conditions present but also the topography of the sea-floor and the character of its deposits, as well as the formation and structure of the coral reefs and islands, in the area that has been surveyed by the Marine Survey of India, and which, therefore, may be termed Indian Waters.

The Marine Survey of India was inaugurated in 1875 and from its inception up to the year 1926 an Officer of the Indian Medical Service was attached to it in the dual capacity of Medical Officer to the Survey Ship 'Investigator' and Marine Biologist, and in this latter capacity this officer was, at first unofficially and later officially, one of the staff of the Zoological Section of the Indian Museum, Calcutta, and of its successor, the Zoological Survey of India. At first and for many years the attention of a succession of Surgeon-Naturalists was concentrated on the investigation of the deep-sea Fauna ; but with our gradually increasing knowledge of the Zoology of the ocean deeps, and the development of the science of Oceanography, it became increasingly important to investigate the conditions under which the animals were living, not only as regards the topography of the sea-floor and the character of its deposits but also the physical and chemical character of the sea water at all depths and the consequent movement of the deep-water currents.

A study of the bottom deposits was commenced by the late Lieut.-Colonel A. Alcock, F.R.S., but it was not until 1910 that a systematic investigation of the hydrography of this area was commenced and even then it was only possible for the Surgeon-Naturalist to undertake investigations regarding the temperature and salinity of the water, while the study of such important features as the hydrogen-ion concentration and the amount of oxygen, nitrate or phosphate, etc., in solution has had to be left to future investigators.

I had originally intended in a final chapter to attempt to correlate the hydrographic results that had been obtained with the distribution of the fauna and especially with the occurrence, or otherwise, of some of the planktonic organisms in the various regions ; but the severance of my connection with the Marine Survey of India in 1925, when I was appointed Director of the Zoological Survey of India, and my departure from India in 1933 has rendered this impossible.

In conclusion I should like to take this opportunity of placing on record my very great indebtedness to all my colleagues, both officers and men, of the Marine Survey of India, who not only by their cheery companionship made life on board extremely pleasant but who ungrudgingly devoted many hours to what was to them the monotonous business of trawling or of taking hydrographic observations, and of acknowledging the great kindness of the Royal Asiatic Society of Bengal in undertaking the publication of this series of papers.

R. B. SEYMOUR SEWELL.

The Zoological Laboratory, Cambridge.
Royal Asiatic Society of Bengal, Calcutta.
January, 1938.

LIST OF PLATES.

quite outside the waters themselves. Changes of temperature and in the rate of evaporation during the changing seasons give rise to a rhythmical variation in the conditions under which animals live that in Indian waters has a period of approximately six months, the temperature and salinity of the upper layers of the waters having two maxima in May and September respectively. Again, the rotation of the earth and the influence of the winds of the monsoon seasons give rise to internal ' seiches ' that produce oscillating changes in salinity and temperature of the water at different levels, having a period that may extend over days and weeks in accordance with the size and shape of the sea-basins; while tidal influences may produce a rise and fall in salinity twice in every twenty-four hours. It even seems probable that in tropical waters the daily variation of barometric pressure may be accompanied by rhythmical changes in the salinity of the sea-water at the surface.

In consequence of all this continual change, isolated observations such as those taken by the ' Challenger,' ' Valdivia,' etc., give us merely a picture of the conditions existing at the actual moment when the observations were made and furnish no guide as to what may be the state of affairs even a few hours later.

The effect of these changes on the endemic fauna must be considerable, and in consequence the study of the physical conditions that exist in any given area forms an important, if subsidiary, branch of the study of the bionomics of that area.

I. THE GEOGRAPHY OF THE ANDAMAN SEA BASIN.

The Andaman Sea is the name given to that comparatively small and partially isolated portion of the Indian Ocean that lies enclosed between the coast of Burma and the Malay Peninsula on the one hand, and the chain of the Andaman and Nicobar Islands and Sumatra on the other. This area has long been recognised as one that is specially rich in its fauna. I have in the following paper attempted to gather together into a more or less connected narrative the available data regarding the origin and evolution of this basin.

A study of the contours of the sea bottom shows clearly that the western boundary of the sea is a lofty submarine mountain chain, of which only the topmost peaks appear above the surface, while on either side the sea-floor slopes steeply downwards. At certain places the ridge is interrupted, and we thus get channels of various depths that permit of a free circulation of the more superficial waters between the Andaman Sea and the Bay of Bengal, while to the south there is a comparatively narrow, shallow channel through the straits of Malacca communicating with the enclosed seas of the Malay Archipelago. The channels that now communicate across the ridge with the Bay of Bengal are numerous and vary considerably in depth. There are three main channels, whose respective depths increase as we proceed from north to south. These channels are as follows:—

(1) The Preparis channel, divided into north and south portions by the island of the same name, in which the depth of water is a little over 100 fathoms.

(2) Ten Degree channel, situated between the Andaman and Nicobar Islands and having a depth of approximately 400 fathoms, and

(3) Great channel, lying between Great Nicobar Island and Sumatra, in which the depth is probably 750 odd fathoms, but is certainly not greater than 800 fathoms, as is clearly indicated by the temperature of the deeper waters of the basin.

The mountain chain that forms the western boundary of the basin and extends from Preparis island for a distance of approximately seven hundred miles to join the mountain chain of Sumatra, Java and the Malay Archipelago, is part of the great

Alpine-Himalayan system. Professor Gregory (1912, p. 202) tracing this great system from the west remarks "the main line continues eastward through the Himalaya until the resistance of the Chinese plateau has forced it southward, and part of it now lies beneath the Bay of Bengal; it re-appears in Sumatra and continues through Java and across the Malay Archipelago." Recently Gregory (1923, pp. 161, 162) has somewhat modified his views, and he now considers that "the direct prolongation of the Himalayan axis crosses Southern China, and that the Burmese Malay arcs form a loop to the south comparable to the Persian loop in Western Asia and the Apennine loop in Europe," while the Andaman Sea and the gulf of Martaban together form one of several "subsidences between the direct prolongation of the Himalayan line and its southern loop."

The work of Rink (1847), Hochstetter (1866), Ball (1870), Oldham (1885), and Tipper (1911) has shown that the geological structure of the Andaman and Nicobar Islands has many points of resemblance with that of the Arakan range in Upper Burma to the north and the islands of Sumatra and Java to the south, and that geologically these islands are part of a continuous chain stretching from north to south along the western boundary of the Andaman Sea.

Rink (1847, Trans. 1870) first put forward the view that this mountain chain was elevated to its present height from the bottom of the ocean. He describes the chain of islands as consisting "partly of those stratified deposits which occupied the level bottom of the sea before their appearance, partly of plutonic rocks which pierced through the former and came to the surface through that upheaval. . . . . . To the two above-mentioned formations has to be added a third, which is the result of the chemical and mechanical destruction of the plutonic masses; it is local and consists of clays and pebbles derived from the underlying strata. On some of the islands this formation has reduced the high and precipitous configuration of the plutonic mountain and produced an undulating hilly land; it is, however, evidently a submarine deposit and has been upheaved by a movement later than, and independent of, that which caused the principal upheaval of the islands." Hochstetter (1866, Translated 1870) has stated that the same geological formation can be traced further southward through Sumatra and Java, and that its appearance in all these localities is due to the same upheaval, the line of which runs through the whole archipelago.

Throughout the whole length of the Andaman-Nicobar ridge, two series of deposits, apart from the plutonic rocks, have been recognised, namely,

(1) The conglomerates of the North and Middle Islands, and the sandstones of South Andaman Island and of the Nicobar Islands and

(2) The clays and clay-marl of the Nicobars and Ritchie's Archipelago.

Hochstetter regarded these as different facies of the same deposit, but Oldham and Tipper consider that they are distinct, and, moreover, were laid down at different times, the clay deposits being the younger.

In the early part of 1924 I was able, thanks to the kindness of Col. Ferrar, O.B.E., Chief Commissioner of the Andamans and Nicobars, to make a tour round South and Little Andaman Islands and Ritchie's Archipelago in company with Mr. Gee, of the

Plate IA. Sandstone cliffs, west side of Little Andaman Island.

Plate IB. The western entrance to Nankauri Harbour.

Geological Survey of India. We were fortunate enough to discover on one of the islands of the archipelago, Outram island, a bed of fossiliferous sandstone, in which were the remains of many Gastropod and Lamellibranch shells and the carapace of a crab belonging apparently to the genus *Philyra*. The only previous record, that I have been able to lay my hand on, of the occurrence of a species of this genus in fossil form is that given by Reuss, (1859, pp. 66–69 and p. 82) of the fossilised remains of *Leucosia* and *Philyra* in a deposit in the East Indies, the age of which he gives as "uncertain tertiary or quaternary." Whatever the age of the deposit on Outram island may be, it appears indubitably to have been laid down in shallow water, probably of not more than 50 fathoms depth. A very similar sandstone was found on the west side of little Andaman Island. In both cases the deposit was markedly stratified (*vide* plate 1 A).

Many of the islands along the Andaman-Nicobar chain, both in Ritchie's Archipelago to the east of the Andamans, and Car Nicobar, Camorta, Trinkat and the greater portion of Nankauri island in the northern and central groups of the Nicobars consist largely of a clay formation that is characterised by being non-calcareous and by containing Radiolarian skeletons and silicious sponge-spicules. Certain small differences are found in the deposits in different islands; for instance,

(1) in the Nicobars these clay beds contain intercalated bands of conglomerate and sandstone, which are not found in the Andaman group,

(2) again, in the Nicobars the deposits occur in association with serpentine rocks, which is not the case in Ritchie's Archipelago and

(3) In the island of Car Nicobar these clays do contain a certain small percentage of calcium carbonate.

These clay-marls, to which the name "Archipelago series" has been given, have been pronounced by Ehrenburg (1850, p. 476) to be identical with similar deposits found in Barbadoes, even the species of *Polycistina* present in the two deposits being for the most part identical. So far as I know the only deposit, that is being formed at the present day and about whose process of formation we have definite knowledge, that presents the same characteristics as these clay deposits is Radiolarian ooze, and it has been accepted by several writers that we have in this area and in Barbadoes true deep-sea deposits that have become elevated above sea-level [*vide* Ehrenburg (1876, p. 117), Nicholson and Lydekker (1889, p. 33) and Nicholson (1893, p. 56)]. In the case of the Barbadoes deposit the view has been put forward that during the process of elevation the beds were protected from being washed away by wave action owing to the formation of a coral crust over them, but no such coral crust has been found in the Andamans and Nicobars.

If we accept Ehrenburg's view regarding this deposit, one must conclude that since its formation it has become elevated from a considerable depth (at least 2,000 fathoms) below the sea-level. Both Tipper and Oldham agree in regarding the clay deposits in the Andamans and those in the Nicobars as geologically identical and, moreover, have arrived at the conclusion that the deposit has been formed by the breaking down of the older serpentine rocks. If this latter view is, as I believe, correct, it follows that the resemblance to Radiolarian ooze is a spurious one and is due to the absence of calcium carbonate. Böggild (1916) has shown that a deficiency of calcium carbonate is a characteristic feature of the bottom deposits in enclosed sea-basins, and in a subsequent paper I shall indicate that a similar deficiency is found in the sub-marine deposits near the mouths of many of the larger rivers of India. It seems probable that the non-calcareous clays were formed, as Oldham and Tipper suggest, by the breaking down of older rocks, and that deposition took place in comparatively shallow water in an estuarine or brackish-water area.

In many places these clay beds are still *in situ* and present the same relationships to the older plutonic rocks that existed when they were first formed, and it seems clear that they have been elevated from some depth below the sea-surface to their present height *after* the subsidence of the basin had been completed.

Judging from the fossils embedded in the two series of deposits there must then have been more than one phase of upheaval along this line. The first upheaval apparently occurred in the Tertiary Period or even earlier. Oldham (1885, p. 143.), summing up the situation, remarks that "the Port Blair series" (consisting principally of sandstones and interbedded slaty shales) " is probably of early tertiary or possibly late-cretaceous age . . . and, by tracing them southwards, we find that the rocks of the Archipelago series are probably of miocene age or even newer." As regards this latter point Richthofen (1862, p. 327.), Jenkins (1864, p. 71) and Hochstetter (1866, p. 220.) are all agreed in placing the formation of the clay deposits in the Miocene epoch.[1]

According to Suess (1904, pp. 451–456) Upper Burma is divided into three zones, which can be traced southwards as follows:—

1. *The Western Zone.*—This is characterised by the line of hills forming the Arakan Range. This hill range as we follow it southwards passes down to Cape Negrais, and at the present day its continuation can still be followed, passing through Thalia Reef, Preparis island, and the Cocos islands, and then on through the Andamans and Nicobars to join the western hill-range of Sumatra.

2. *The Central Zone.*—This consists of a wide belt, comprising the valleys of the Irrawaddi and Sitang Rivers, as far as the eastern bank of the latter. This wide plain includes certain volcanoes and other evidences of volcanic action; and according to Suess, it is the southern continuation of this plain that now forms the whole bed of the Andaman Sea.

3. *The Eastern Zone.*—This comprises a further range of hills, that is continued southwards and forms the limestone islands of the Mergui Archipelago.

According to this view the Andaman Sea must be regarded as a recent formation caused by the subsidence of a large portion of the western and central zones, and the increase in width of this area, in comparison with the breadth of the Irrawaddi and

---

[1] Gee (1925, p. 37) in a preliminary report on his investigations remarks, " The main Archipelago Group, therefore, appears to be of Mid-Tertiary age. It is probable that the two southern islands—Sir Hugh Rose Island and Neil Island—are composed of more recent strata since the echinoid, *Maretia*, obtained from them varies only slightly from the living species, *M. planulata* now existing in the Andaman seas."

Sitang river basins to the north, suggests that the Andaman-Nicobar ridge has drifted towards the west away from the mainland, and has thus formed a pronounced curve with its apex in the region of Little Andaman Island.

Cotter (1918, p. 412 *et seq*) has, however, put forward the view that the Arakan range already existed at the commencement of the Tertiary period as a narrow peninsula, and that at this stage the Irrawaddi basin was already occupied by a shallow sea. Throughout the Tertiary epoch the Arakan ridge was steadily rising, while the sea basin to the east of it was slowly subsiding, though simultaneously its northern extremity was silting up. In Pleistocene times, however, this subsidence of the sea bed ceased and was replaced by a process of elevation.

The question at once arises, at what period of the world's history did this extensive movement of subsidence take place? In order to determine this point, it is necessary to examine in detail the conditions that are found to exist to-day in the mountain chains that form the enclosing walls of the basin.

In tracing the continuation of a line of folding of the earth's crust one would naturally expect to find certain variations in height in different areas, causing culminations and depressions along the length of the ridge. These may be due to a variety of causes, among which one may mention an alteration of pitch of the axis, and a variation in the rate of weathering and denudation in different regions. A study of the heights, to which the different peaks rise in the Andaman-Nicobar ridge and its continuation north and south, shows an extremely interesting continuous variation. I give below a list of the more important heights in series from north to south.

	Name of peak.	Height above sea-level.
Arakan Range, Burma	Chypotong	8,386 feet.
	Chinsikiatong	6,012 ,,
	Mengootsan	3,600 ,,
Andaman-Nicobar Ridge	North Andaman	2,902 ,,
	Middle Andaman	1,678 ,,
	Little Andaman	600 ,,
	Car Nicobar	200 ,,
	Batti Malv	150 ,,
	Chaura	343 ,,
	Katchal	835 ,,
	Little Nicobar	1,428 ,,
	Great Nicobar	2,705 ,,
Sumatra	Batu Mukurah	6,371 ,,
	Sinobong	12,140 ,,

As we follow the range southward we have a gradual fall and a subsequent rise in the level of the ridge, the difference between the highest peaks at each end, which rise to well over 8,000 feet, and the lowest of the peaks in the middle, namely, the island of Batti Malv, which is only 150 feet above sea-level, is at the least some 8,000 feet or approximately 1,340 fathoms, and it would appear probable that there has been, since the first formation of this mountain chain, a secondary depression that has lowered the level of the range in this area by approximately this amount.

Molengraaf (1921, p. 114. and 1922, p. 335) has called attention to the fact that a similar depression, though on a smaller scale, exists further to the south in the continuation of the chain through the Malay Archipelago to the east of Timor Island, though on Timor Island itself there is incontestible proof of a post-tertiary elevation in the shape of a succession of coral-reef terraces, that have been raised high above the present sea-level.

The view that the Andaman and Nicobar Islands have ever been directly connected with the mainland is open to very considerable objections. Kurz (1868, p. 18.) certainly held this view and even went still further in postulating a direct connection with India itself, for in describing the flora of the Andamans he remarks, "we may presume that a great number of species, which originally grew on the countries now submerged between these Islands, Burmah and Hindoostan, have been repelled by the advancing sea and the vegetation became thus comparatively richer in generic types as the area grew smaller." A study of the fauna, however, indicates that, if any direct connection ever existed between these islands and the mainland, it had become completely broken before the evolution of the Mammalia took place, for no indigenous land mammals occur on any of these Islands, with the possible exceptions of a species of *Paradoxurus* (*P. grayi* var. *tylleri*) and an insectivore (*Crocidura perotetti*) which have been captured in the Andamans, but which so far as we know do not occur in any of the Nicobar Islands. As regards the Batrachia only three species have been discovered, namely, *Bufo melanostictus*, that seems to be of almost universal distribution and occurs on many of the remote islands of the northern portion of the Indian Ocean, and *Rana limnocharis*, both of which occur in the Andamans and Nicobars, and, thirdly, *Rana nicobariensis* which, as the name implies, has been found only in the Nicobars. A fauna such as this certainly indicates the improbability of any connection with the main continent; and Boden Kloss (1903, p. 323) from a study of the mammalian and avian fauna reached the conclusion that these islands have never been in any way connected with the mainland. Since the Mammalia made their appearance towards the close of the Triassic period and the Andaman-Nicobar ridge did not apparently arise from the sea-bottom before the late Cretaceous epoch, it is certainly difficult to see how any connection with the mainland can reasonably be postulated, though I understand that geologists believe that originally the whole chain of islands was continuous, presumably above sea-level, but that, as Gregory (1923. p. 160) expresses it; "The continuity of the original mountains has been broken by wide gaps which, in the western part of the Burmese-Malay Arcs, for example, are larger than the remnants."

It is not improbable that originally the area occupied by the Andaman basin was considerably shallower than it is at the present day and partook more of the nature of a large brackish-water lake, into which flowed the Irrawaddi, Salween and the other rivers of Southern Burma, and out of which there were three channels, viz.:— one between the Burma coast and the Andamans in the north, a central exit between the Andamans and Nicobars, and a southern one between the Nicobars and Sumatra. In the middle of the Tertiary period this submerged area extended considerably further to the north than it does to-day and in Eocene and Miocene times extended well into Upper Burma. Theobold (1870) pointed out that practically the whole of the deltaic area has been formed by a process of elevation and that throughout it marine deposits, or as he terms them 'the older alluvium,' form the great bulk of the deltaic lands. With an elevation of this area the northernmost opening, the Preparis channel, must have tended to become considerably less deep, whereas on the other hand, the subsidence of the central part of the basin, which is indicated by the alteration of pitch of the axis of the ridge and to which I have referred above, has caused a deepening of the more southerly channels, though here the deepening may have been partially counteracted or even altogether abolished by a subsequent further elevation of the ridge. With a reduced depth of water the conditions inside the basin would be much more nearly those of a brackish-water lake, and it is interesting to note that at the present day, so Dr. Annandale informed me, the fresh-water fauna of the Andamans and Nicobars is essentially one of a brackish-water type (*vide* also, Annandale and Hora, 1925, p. 35).

Up to the present time our knowledge of the floor of the Andaman Sea has been incomplete, and even now there are large areas where few or no soundings have been taken. Thanks to the kindness of the Hydrographer, The Admiralty, Whitehall and the Officer in charge, Marine Survey of India, I have been able to collate all the recent soundings taken in this region and on them I have based the accompanying sketch-map (Chart I). So far as is at present possible, I have indicated the contour lines of the floor of the sea at intervals of 100 fathoms, but it must be clearly understood that the positions of these contours are in certain areas largely theoretical owing to insufficient data; but enough soundings have now been taken in this region to render the main features of the map reliable.

The first feature to which I would call attention is the clear indication that the Andaman Sea is not a single basin, as seems to have been accepted hitherto. The greater part of the sea is occupied by a large basin, the greatest depth of water in which is at least 2000 fathoms, but in the north-west region are two much smaller but quite independent basins, in each of which the depth is over 1000 fathoms.

Another noticeable feature of the map is the extraordinarily complicated arrangement of the contours of the Andaman-Nicobar ridge. These contours show clearly that this mountain chain is composite in character. The portion of the chain that is represented above sea-level by the Andaman Islands, can, as I have already indicated, be followed northwards to join the Arakan hill range; but commencing on the east coast of Little Andaman Island is a second well-marked range, that runs in a north-easterly direction and separates the main basin of the Andaman Sea on its south and east side from the other two small basins lying to the north and west of it.

This range of hills appears on the surface as "Flat rock," which is just awash, in the middle of Invisible Bank on the east side of the south end of South Andaman Island, and again about half-way along its length as Barren Island. Lying still further to the north-east and continuing the line of the ridge are two more peaks which are, however, entirely submarine, and connected with, but lying somewhat

it is not a 'deep' in the strict oceanographic sense, since its depth is only 2,400 fathoms. Carpenter's Ridge attains its greatest height near its southern end in approximately lat. 6° N. and gradually subsides again towards the south.

Molengraaf (1912, p. 273 *et seq.*) has shown that the region to the south and east of the Nicobars, that now constitutes the East Indian Archipelago, has within comparatively recent times been inundated by the sea, owing to a general rise in the sea-level. He has put forward the view, based on a considerable mass of evidence, that during the end of the Pliocene or at the commencement of the Pleistocene age the present Sunda Sea was probably occupied by a stretch of low-lying land or a group of islands, and that subsequently at the beginning of the ice-age the sea-level sank, so that a great area of land was exposed, connecting the present islands of Sumatra, Java and Borneo. At the close of the glacial period the sea-level rose again to its present height and all the low-lying country became submerged to form a great shelf-sea. Daly (1910, p. 399 *et seq.*) had already utilized this rise in sea-level to explain the formation of many of the existing coral-atols and barrier-reefs. As Professor Sollas has recently pointed out (Nature, 1923, 332-4), this alteration of the sea-level during the Quaternary epoch was not a simple one, but so far as the evidence at our disposal goes, seems to have partaken of the nature of a more or less regular oscillation, and this change in level has left indisputable evidence in various parts of the world. We should therefore expect to find some evidence of a change in sea-level in the Andaman-Nicobar region, but here the matter appears to be complicated by earth movements. Rink (1870, p. 122) in his account of a visit to the island of Bompoka in the central group of the Nicobars, remarks, "We reached a precipitous cliff, which to my great astonishment did not show plutonic rock, but only coral limestone. The coral formation formed terraces to a height of 60 to 80 feet. I had not as yet found upon the islands such clear proof of the more recent upheavement of the ground." Similar evidence is to be seen in the island of Katchal, where there is a cave in weather-worn coral limestone at a height of 200 feet above sea-level [*vide* Anderson (1903, pp. 163-4)]. We have here, doubtless, evidence of the same process of elevation that has given rise to the terraced coral-reefs found in Timor and the neighbouring islands, though the degree of elevation is very much less.

It seems probable that in the portion of the ridge forming the central group of the Nicobars this process of upheaval is still slowly continuing and the same seems to be the case further to the north in the Arakan district. Theobald (1870, p. 26-7) states that "towards Cape Negrais no prominent signs of elevation present themselves, but as we approach Gwa we find gradually proof of a somewhat recent rise of the coast, in the shape of coral banks raised above the present limits of its growth and in the presence, a few feet below the surface in the plains now removed from the shore, of shelly sand and shells of species living on the coast." During my visits to Nankauri Harbour in 1921–23, I have in one or two localities come across beds of reef-coral that now form part of the shore of one or other of the islands, and are raised considerably above high-water level. Since these beds must have originated at some depth below sea-level, it seems clear that elevation has quite recently taken place in this area. One must, however, bear in mind that an apparent elevation may be due to an actual fall of sea-level. Throughout the Tropical belt, in widely scattered regions, evidence has for many years past been accumulating that tends to show that in comparatively recent times there has been an apparent elevation of land. Fryer (1910, p. 266) in Aldabra and the neighbouring islands of the south-west portion of the Indian Ocean; Bourne (1888, p. 440) in Diego Garcia; Stanley Gardiner (1903, vol. 1, p. 36 *et seq.*) in the Maldives; Foote (1883, p, 70) in Rameswaram Island at the south end of India; Hadden (1889, p. 587) in the Torres Straits; and Saville Kent (1893, p. 130) in the great Barrier Reef of Australia, have all called attention to evidence of alteration in the relative sea and land levels, giving rise to an apparent rise in the level of the land. I have myself seen evidence of a similar change in the coral reefs of Mergui and in the Nicobar Islands, and in the sandstone cliffs on the west side of little Andaman Island As regards this latter region, there is at the foot of the southern point of these cliffs a well-marked terrace, at its widest from 8 to 10 feet across, just below the present high-water mark. Above this terrace, at a height of about 5 feet above high-water mark is a second terrace. If, as I believe, both these terraces have been formed at one time or other by wave-erosion, they indicate an alteration of level of some 6 feet. Further evidence of a change in the respective sea- and land-levels is to be seen, (1) at the south-west corner of Rutland Island in the Andaman Group, where there is a raised pebble beach, situated some 5-6 feet above the top-level of the present beach, and (2) at the S.-W. end of the westernmost of the Twin Islands, off the west side of Rutland Island, where again there is a raised beach of shingle conglomerate, about 6 feet above high-water mark. Running in a south-westerly direction from this Island is a reef, the rocks of which are composed of the same coarse-grained conglomerate. This condition indicates that there has been an elevation of about 5 to 6 feet and that erosion has since been taking place, and is probably continuing, if the condition of the jungle at this end of the Island is any guide.

Such wide-spread evidence of an apparent rise of land-level is most easily accounted for by a fall in sea-level of a corresponding amount. I am not, however, concerned here with the possible cause of such a fall of sea-level, and would merely point out that it would result in producing an apparent rise of land-level.

Kurz (1868), however, has put forward the view that the Andamans are sinking. He bases his opinion on certain local changes in the vegetation and the presence of stumps of trees in a mangrove swamp bordering the Middle Strait. Oldham (1885, p. 143) has accepted Kurz's view, although he describes how in 1840 Helfer sailed through Homfray's Strait, which in 1858 was found by the Andaman Committee to be unnavigable except for small boats, and furthermore admits that the Andaman midden that he carefully examined showed that the land had "not appreciably altered its level relatively to the sea" during "a period that must be measured by centuries, if not by tens of centuries." Suess declines to accept Kurz's view and I cannot but think that what he saw was due entirely to some local change—possibly the banking up of the sea in the strait during a strong north-easterly gale. I have myself in the Nicobars seen stumps of palm trees standing in the mud between tide marks about 20 yards out from the present beach in a small bay in Nankauri Harbour, but in view of all the other evidence to the contrary I cannot accept this as proof that these islands are sinking.

It is interesting to note that surrounding the islands of the Andaman-Nicobar ridge there is a typical "continental shelf" extending outwards from the shore. For years past there has been a great divergence of opinion as to the origin of this shelf. By many oceanographers, among whom are Spencer, Hull, Nansen, Richard and A. Milne-Edwardes, this shelf is regarded as an integral part of the land itself; and the continental edge, where the shelf joins the continental slope, is regarded by them as the old sea-beach, its submerged position to-day being attributed either to subsidence of the land or to elevation of the sea-level. Others, however, among whom are the late Sir John Murray and Dr. J. Y. Buchanan, regard the shelf as being a new formation that has been built up of debris and detritus derived from the continental surface and deposited around its margin. Probably both views are equally correct and in some regions the shelf is a 'drowned' portion of the continent, while in other areas it is merely the refuse heap of the land.

In view of the great changes in level that have taken place in the various regions of the Andaman-Nicobar ridge the only possible explanation of the presence of a continental shelf here seems to be that it has arisen as a comparatively recent formation, partly from erosion along the sea-coast, partly from the deposition of debris and detritus and probably largely from the outward growth of coral banks around the shores of the islands. Owing to the presence of deep channels that cut through the ridge the continental shelf is not a continuous one. All the Andaman Islands, the Cocos Islands and the outlying North and South Sentinel Islands, are enclosed within a common shelf. The central group of the Nicobars seems to be enclosed by another which probably also extends northwards round Car Nicobar though the soundings taken in this region are not sufficiently numerous to enable one to be absolutely certain of this.[1] And a third shelf encloses Great and Little Nicobar Islands.

As is only to be expected around islands such as these, situated so near to the equator, coral flourishes and forms reefs that extend outwards, often for very considerable distances from the shore, though the rate of growth and the extent of coral formation is usually considerably less on the east side of a great ocean, such as the Indian and Pacific, than it is on the west. In the case of the Pacific Ocean the western side is a maze of coral islands and reefs, while the east side is practically bare of such. The same is true to a certain extent of the Indian Ocean, but here local conditions have enabled considerable reefs to form wherever oceanic conditions

---

[1] A number of soundings taken recently by the R.I.M.S. "*Investigator*" show that the island of Katchal is separated off from the other islands of this group by a channel that has throughout its length a depth of water of well over 100 fathoms.

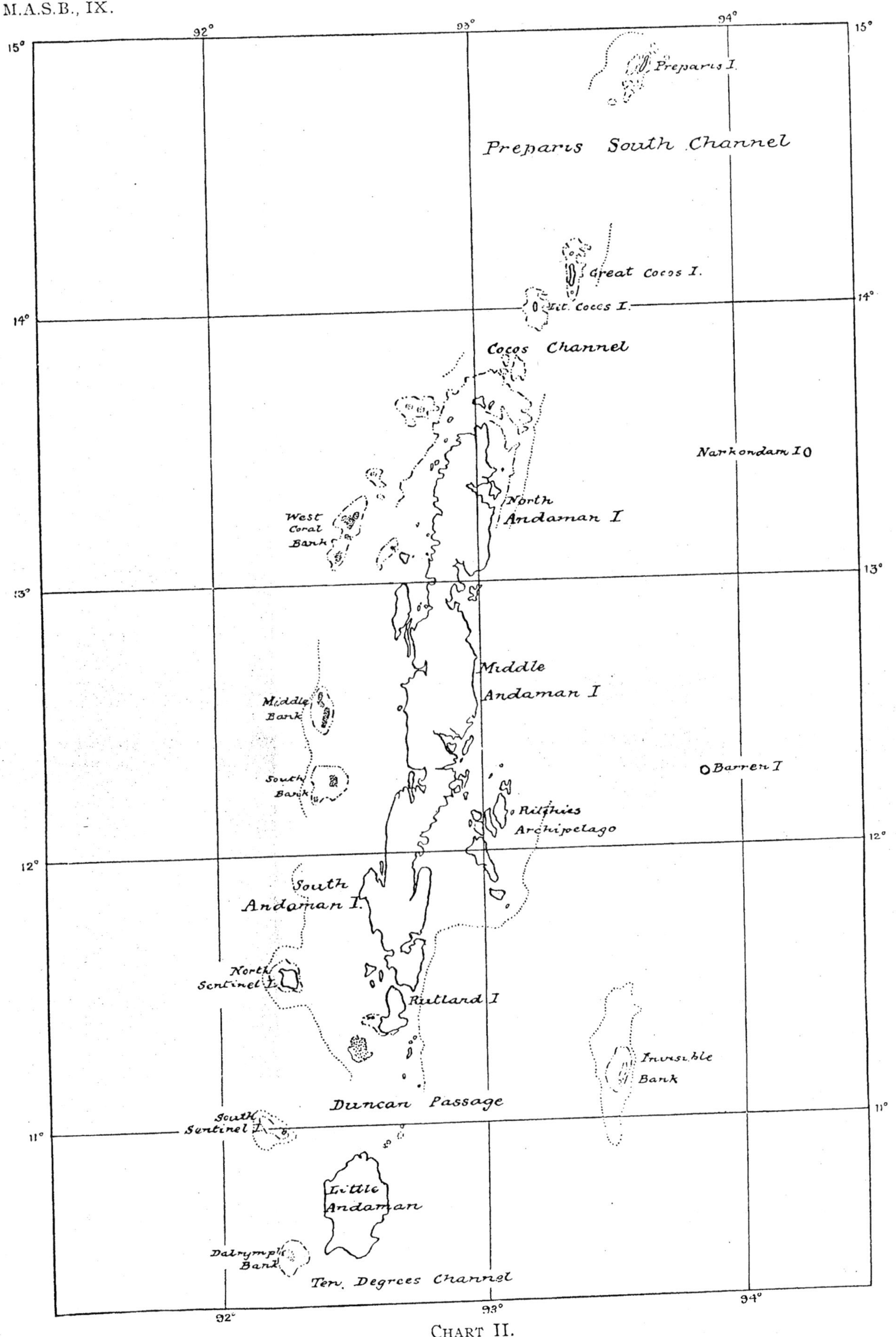

CHART II.

The coral banks on the west of the Andamans. Banks with less than 10 fathoms of water over them are indicated by dots.

can be maintained. The south-west Monsoon causes a flow of oceanic water that sets in towards the west side of the Andaman-Nicobar ridge and the Arakan coast of Burma and in consequence we find coral reefs extending well up the Burma coast till the lowered salinity of the sea-water, and the deposition of mud and silt due to the outflow of the Brahmaputra River prohibit their existence. Throughout the greater part of the year the reefs on one side or other of the ridge are being pounded by heavy seas. From November to January the north-east Monsoon causes high seas to break over the reefs on the eastern side and during the south-west Monsoon from June to September the western shores are subjected to the same process; while in April and October cyclones are particularly prevalent in these regions. As a result fragments of coral are continually being broken off and a building up of the reef-flat takes place, while simultaneously a certain amount of coral detritus is being swept out to sea and settles down to the bottom, where it helps to form a basis on which further generations of corals and nullipores arise and so extend the reef further seawards.

Around the group of the Andaman Islands the continental shelf is very much wider on the western than on the eastern side and along the western edge of the shelf is a series of coral banks and islands. Passing from north to south we have North, West, Middle and South Coral Banks, then come North and South Sentinel Islands, and finally there is Dalrymple Bank at the south end of the series (*vide* Chart II). Both North and South Sentinel Islands, though they attain a considerably higher level, as their name implies, than the banks, which are covered by about six fathoms of water, are of coral formation. Alcock (1902, p. 82), who visited south Sentinel Island in 1889, remarks that "the island seems to be nothing more than a coral reef raised a few feet above high-water mark, for loose coral-rock crops up everywhere, and even in its very centre we found masses of coral whose specific structure was still quite plainly recognisable." In 1922 I was able to visit both these islands and I can corroborate Alcock's statement. From what I saw of North Sentinel Island, that too seems to be undoubtedly of coral formation, and in places along the coast are cliffs composed of coral limestone that rise up sheer for 20–30 feet. Oldham (1885, p. 145, Note) remarks—"the two Sentinel Islands have been described to me as being composed of coral; this, as I found from experience in the Nicobar Islands, almost certainly refers to the fantastical weathering of the limestones of the Archipelago series," but there is no fantastical weathering about either of these islands and Oldham's informant was perfectly correct in his statement.

This remarkable chain of coral banks and islands is separated from the coast by a distance of approximately fourteen miles and a depth of water averaging some forty fathoms. It is possible that the increased growth of coral in this region, in comparison with what we find on the eastern side of these islands, may be to some extent due to the influx of oceanic water during the south-west monsoon and the consequent abundance of food supply, and Daly (1910) would explain the formation of such a reef and the channel intervening between it and the mainland as due to a rise in sea-level during pleistocene times. It seems to me that we have another and much more important factor in its production and one that simultaneously explains certain other physical features of the group.

Brouwer (1919) has pointed out that during the development of an anticlinal fold the axis may shift its position so as to cause an asymmetrical development of the two sides of the ridge and a difference in the lateral slopes, and Molengraaf (1921, p. 115.) has shown that such alterations have taken place in several of the islands of the East Indian Archipelago, such as Roti, Sumba and Jamdena. In a text figure, which I have taken the liberty of reproducing, he shows how such a change in the axis will affect a cross-section of a ridge and simultaneously cause an alteration in the position and character of the coral reef growing round its shores. Tipper (1911, p. 16) has remarked regarding the Andaman group, that its geological structure is "that of an anticline, thrust or folded over to the west," and he has called attention to the different slopes on the two sides of the islands, "a section across the Andamans

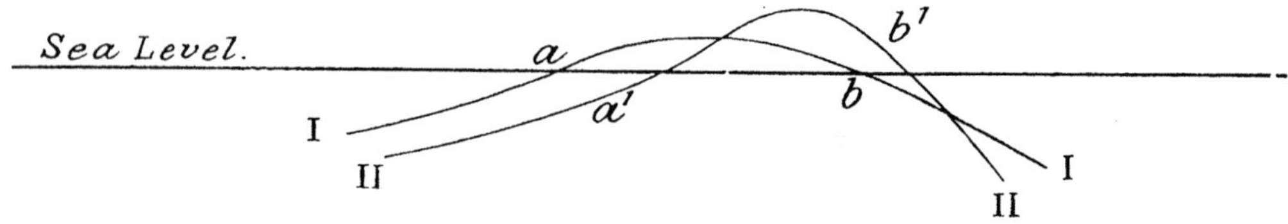

TEXT-FIG. 1.—Successive stages of development of an island on an anticline axis. Fringing reefs formed at "a" and "b" during Stage I would be found at "á" and "b'" in a later Stage II. The fringing reef at "a" (Stage I) would have been submerged and a barrier reef have developed at "á" (Stage II), whereas the fringing reef "b" (Stage I) would have been elevated and been found at "b'" (Stage II). After Molengraaf.

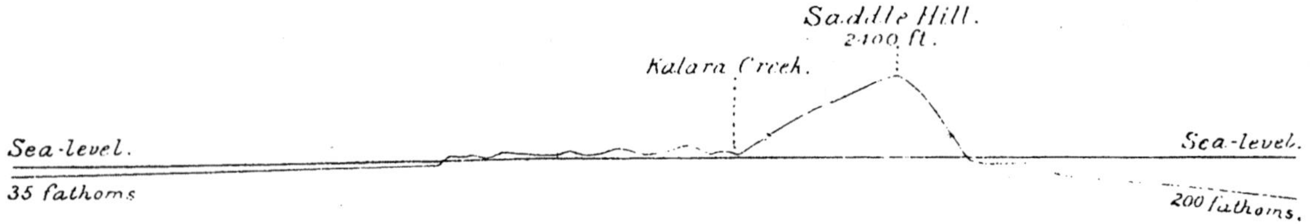

TEXT-FIG. 2.—Section across the North Andaman Island through Saddle Hill, showing the steeper eastward slope continued as a submarine feature after Tipper.

shows a long gradual slope from west to east, culminating at some point near the east coast, a steep eastward slope continued under the sea until about midway between the Andamans and Barren Island." He admits that this steep easterly slope is not what one would expect in an anticline thrust or folded to the west, and explains it by supposing that a fault has occurred on the east, and that it was along this fault that Barren Island and Narcondam arose. A comparison of the section given by Tipper across the Andaman ridge, which I reproduce, and Stage II of Molengraaf's figure reveals a degree of similarity that is very striking, and it seems to me that we have in the shifting of the axis of the ridge from west to east an explanation of many of the physical and geological features of this area. Such a shift would of itself produce the altered slopes on the two sides of the ridge and would, as Molengraaf points out, give rise "to the development of a barrier-reef at one coast of an island, whereas at the opposite coast simultaneously fringing reefs are elevated above sea-level."

Plate 3. A raised coral beach at the south end of Henry Lawrence Island, Ritchie's Archipelago.

The view that this chain of coral banks on the west of the Andamans might be a barrier reef was, I believe, first put forward by Alcock (1902, p. 49), who remarks—"The series of coral banks which flank the western shore of the Andamans may perhaps represent a discontinuous barrier reef." Anderson (1902, p. 1) agrees with this view and remarks—"Extending along the west coast of the Andamans, at an average distance of 14 miles, is a veritable barrier reef, composed of disconnected and mostly submerged coral banks, separated from one another by wide deep channels." I have no doubt that this view is the correct one, and that we have here a definite barrier reef extending from Lat. 13° 40′N., for a distance of approximately two hundred miles, to Lat. 10° 26′N. As in other barrier reefs we have a channel of some 40 fathoms depth between it and the coast line, and on the sea-ward side the ocean bed drops steeply down to a depth of 250–300 fathoms.

If the view I have expressed above, namely, that the cause of the formation of this barrier reef is a shifting of the axis of the anticline, we ought to find on the east side of the ridge a series of raised coral beaches extending along much the same area. Oldham (1885, p. 144-5) has recorded a large series of raised coral beaches all along the east side of the Andamans, extending from Outram and Laurence Islands in Ritchie's Archipelago, passing through Havelock Island, and along the shores of south Andaman Island, where it "can be seen forming a terrace from a few yards to over half a mile in width in almost every bay" (*vide* Pl. 3). I have been unable to find any account of the geological formation in Little Andaman, and therefore can only contribute a few brief notes that I made at the time I visited the island in 1922, and again in 1924. "On the east side is a deep indentation known as Hut Bay. Here the shore at the south end consists of rock interspersed with coral and with a well-marked fringing reef; further to the north the beach is sandy. Above the beach the surface of the land is composed largely of coral and coral debris. The bed of a small stream, which I followed for some distance, is composed entirely of boulders of coral limestone, and everywhere throughout the jungle one sees fragments of coral." It would seem, therefore, that here too the island is composed on its south-east side, of a raised coral reef. Still further to the south, on the other side of Ten-degree Channel, the island of Car Nicobar consists, to a large extent, of a raised coral formation, and here again this elevated coral reef is situated on the east, where it forms a wide semicircle enclosing the clay-deposits that form the central and western parts of the island. It would appear, therefore, that we have ample corroboration of the view that the changes, that have taken place in this region of the Andaman-Nicobar ridge, are due to a shifting of the axis of the anticline from west to east, and it is interesting to see how this change seems to be limited to that portion of the ridge where its westerly curvature attains its maximum, and where the line which to the north runs south by west, now alters its course and runs south by east.

Beyond the edge of the continental shelf on the east side of the Andamans the bottom slopes steeply downwards to 200 fathoms, which we may regard as the foot of the talus slope derived from the surrounding coral reefs; and beyond this the descent becomes much more gradual, and we have indications of a series of ridges and valleys alternating with each other as we follow the line of the ridge from north to south. These valleys can be traced downwards as far as the 1,100 fathom line. If this great range of hills were above sea-level, these valleys would be important catchment areas and would form the beds of large rivers, and it is interesting to note that, if we follow the line of each across the continental shelf on the east side of the Andaman Islands, it is found to correspond with what is now a large inlet or harbour on the east coast, namely, Port Cornwallis, Stewart Sound and Port Blair. Though it does not necessarily follow, it is not improbable that these gullies are the submerged beds of old rivers that originally drained the eastern side of the range, and that the connecting link between the present sea-level and the 200 fathom contour has become obliterated by the origin and formation of the continental shelf, which, as I have already pointed out, must have arisen since the subsidence of the ridge took place.

On the extreme southern end of the Andaman sea-basin, the contours indicate the presence of a wide gently-sloping valley that runs roughly in a north-west direction and can be traced down to a depth of 1,400 fathoms. Molengraaf (1921, p. 102 *et seq.*) has called attention to the presence of "drowned" rivers in the East Indian Archipelago, and it seems probable that in the southern end of the Andaman Sea and what is now the Straits of Malacca, we have a further drowned river-bed. Certainly the general trend of the lower reaches of the rivers in the north end of Sumatra and the corresponding coast-line of the Malay Archipelago is towards the north-west, and they probably originally were tributaries of one large river-system that has since become submerged. Incidentally, the depth to which we can trace these valleys and gullies, agrees very fairly closely with the depth to which the whole region, as I have already pointed out, has probably subsided since its original upheaval, namely some 1,300 to 1,400 fathoms. It thus appears probable that the subsidence of the basin, or at any rate of the deeper portion lying to the east of the volcanic ridge, occurred at sometime subsequent to the elevation of the mountain chains, probably in the Pleiocene or early Pleistocene period, and it is an interesting speculation as to how far subsidence may have been influenced by the upheaval of the volcanic portion of the range having caused a blocking of the exits of the shallow water basin, and the consequent accumulation of water over an area that was already in a state of unstable equilibrium.

On the north and east sides of the Andaman Sea the submarine contours are much more regular than on the west. Extending around the northern shore is a well-marked continental shelf, the edge of which corresponds closely with the 100-fathom line, and which is continuous on its western side with the shelf off the Arakan coast but is separated from that of the Andaman group by Preparis

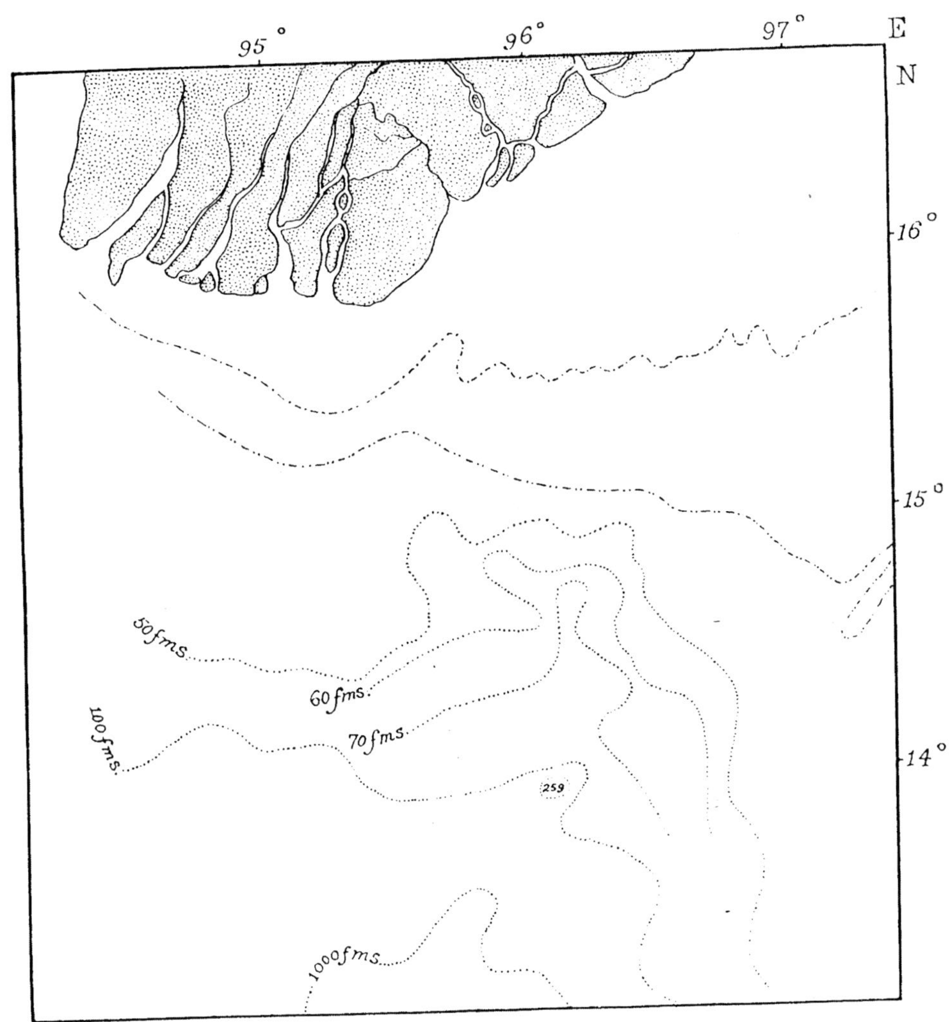

CHART III.
The submarine gully off the delta of the Irrawaddi River (from Admiralty Chart No. 70).

Channel. Along the Arakan coast in the Bay of Bengal, this shelf is comparatively narrow and its edge lies at a distance of approximately thirty to thirty-five miles from the coast-line, but along the southern shores of the Irrawaddi delta and in the Gulf of Martaban, the shelf increases enormously in width and in most parts is at least a hundred miles wide : indeed, the whole area of the Andaman Sea lying north of a line drawn between Preparis Island and the mouth of Tavoy River has a depth of less than 100 fathoms.

There are two features in this wide shelf to which I would draw attention. Near its western edge the sea-floor shows two distinct elevations, namely Thallia Reef and Preparis Island : both lie in the line connecting Cape Negrais, the southern extremity of the Arakan range, with the Andaman Islands, and they are undoubtedly pinnacles of the same great mountain chain. The second feature is of exactly the opposite nature. A study of the contour lines shows that, commencing opposite the Krishna Shoal, where the sea-coast bends sharply towards the north-east to form the western shore of the Gulf of Martaban, and running an S-shaped course across the shelf, is a well-marked gully (Chart III). In comparison with the deep submarine gullies, each known as a "swatch," which are present off the Indus and Ganges rivers, this off the Irrawaddi is a small and shallow one, for in the middle of its course it has a width of ten miles and a depth of only 120 feet. Immediately opposite the end of the gully, at the 100 fathom contour, is a small deep hole, where a sounding of 259 fathoms has been obtained, and it is interesting to note that beyond this again the contours down to the 1,300-fathom level show a well-marked bay running towards the north-east that might be taken to indicate a possible continuation of the gully down to this depth. Molengraaf (1921, p. 101, Fig. 2, and 1922, p. 277 *et seq.*) has shown that in the East Indian Archipelago, it is possible to trace on the floor of the Sunda Sea the submarine bed of a large river, that had its tributaries in Borneo and Sumatra, and ran eastwards into what is now the China Sea. This river excavated its channel when the Sunda lands were above sea-level, but owing to the rise of the sea in Pleistocene times it has now become submerged. It is possible that we have in this gully off the Irrawaddi delta a similar formation, but it is difficult to see how one can reconcile its presence, assuming that it is the old river-bed, with the conditions that exist at the present day. I have pointed out that the formation of a continental shelf on the east side of the Andaman ridge has obliterated all traces of the old valleys in that region down to a depth of 200 fathoms, and one would certainly expect that the deposition of the enormous quantities of mud and silt brought down by the Irrawaddi would have done the same in this region even if such a gully had previously existed. But as I have already mentioned, in Tertiary times the Andaman Sea extended much further to the north and the area that is now occupied by the delta was marine. Since this is so, it must surely be admitted that the continental shelf in this region is a new formation and has arisen partly by deposition of land detritus and partly by elevation of the sea-bottom, and in consequence, it is impossible to regard the gully as a drowned part of the old Irrawaddi river-bed. The part of the channel, which crosses the continental shelf must be a new formation due to the deposition of silt and mud along certain definite lines, and due to equally definite causes, the most important of which are the effect of mixture between the silt-laden river-water and the more saline waters of the sea, causing a rapid precipitation of the suspended mud, and the inflowing current through Preparis channel that gives to the outflowing river water its easterly trend. On the other hand, if we admit that the subsidence of the Andaman Sea basin, as indicated by the rise and fall of the mountain crest and by the presence of other gullies along the Andaman chain, has been so great as this, the deep portion of the gully from the edge of the shelf down to a depth of some 1,300 fathoms, may be another 'drowned' gully; but if so, its position on the east side of the volcanic chain indicates that it belongs to the Salween river system.

From the Gulf of Martaban southwards to Victoria Point the coast of Southern Burma is a complicated one and is fringed by a succession of islands, commencing with the Moscos Islands in Lat. 14° N. and continuing southwards through the whole

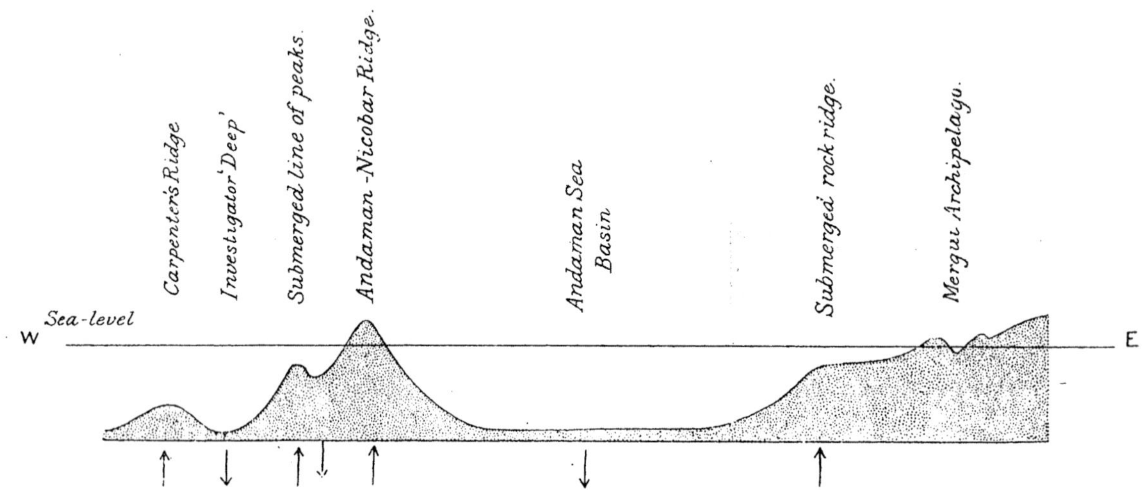

TEXT-FIG. 3.—Schematic representation of the Tertiary crustal movements in the Nicobar area.

length of the Mergui Archipelago to the islands of the Malay coast. The submarine contours lie fairly close together in the northern region but as we trace them southwards they become more and more widely separated so that the slope of the sea-floor, instead of being steep, is at the south end a gentle gradient, and in consequence the deep-water basin is reduced to a comparatively narrow gully. As we trace the continental shelf sea-wards in this region, we find that it is dotted over along its whole length with numerous islands, but slopes gradually downwards to the 40–50 fathom line, and there we get a small though fairly steep drop down to a second shelf that at the southern end of the sea extends westwards for a very considerable distance. This second and deeper shelf is due, as I shall show later when dealing with the nature of the sea-bottom, to a submerged rock ridge that can be traced along almost the whole length of the coast.

Molengraaf (1921, p. 108) has put forward the view that the deep-sea basins and the adjoining elevated islands in the East Indian Archipelago were formed simultaneously and are due to a folding of the earth's crust along certain definite lines. The elongate shape of the various islands and the presence of deep troughs, separating them from one another, he explains on the grounds that they are respectively the exposed tops of anticlinal folds, between which lie the synclinal depressions. We have already seen that the various mountain chains that together make up the complicated facies of the Andaman-Nicobar ridge can be traced southwards into Sumatra and the region beyond, and we should, therefore, be able to distinguish the same sequence of folds and depressions in this region that Molengraaf has been able to detect in the more southern area.

In a section running from W.S.W. to E.N.E. across the island of Sumatra, Molengraaf recognises two major folds, the tops of which now form the Mentawei Islands and Sumatra respectively; in his schematic representation (*vide* Molengraaf, 1921, p. 109, fig. 4) he shows, however, a third though much smaller fold lying further westward in the floor of the Indian ocean. All these folds can plainly be recognised in the region of the Andaman-Nicobar ridge, though here the fold that forms the Mentawei Islands does not make its appearance above sea-level.

I have shown these folds in the figure above, and a comparison with Molengraaf's figure at once reveals the similarity between the two areas. To make this more clear I have given the corresponding synclinal and anticlinal folds in the following table.

		Section across the Nicobar Islands. W.E.	Section across Sumatra. W.S.W.—E.N.E.
1.		Bay of Bengal	Indian Ocean.
2.	1st geoanticline ..	Carpenter's ridge	Raised area in floor of Indian ocean.
3.	1st geosyncline ..	"Investigator Deep" ..	Sunda trough.
4.	2nd geoanticline	Series of submerged peaks to west of the Andaman-Nicobar ridge.	Range of coastal islands girdling the west coast of Sumatra, including the Mentawei Islands.
5.	2nd geosyncline ..	Irregular depressed region lying to the west of the Andaman-Nicobar ridge between it and the submarine peaks.	Mentawei trough and corresponding trough shaped depths.
6.	3rd geoanticline ..	Andaman-Nicobar ridge ..	Non-volcanic and volcanic mountain ranges of Sumatra.
7.	3rd geosyncline ..	Deep basin of the Andaman Sea	Tertiary terrain, folded in late Tertiary and early Pleistocene times.
8.	.. ..	Rocky submarine hill-chain lying parallal to Burma coast and shelf area to east of it.	Stable Sunda land including the Sunda shelf.

Just as in the Malay arc a further complication of this system of ridges is introduced by the separation of the outer non-volcanic range, including the islands of Sunda, Timor and Buru, and known as the Timor-Ceram arc, from the volcanic range, which includes the islands of Bali, Alor and Wetter, and the formation between them of an extra deep-water area, the Wetter trough, so we find an exactly comparable condition in the northern end of the ridge where the line of the Andaman-Arakan range on the outer side of the chain becomes separated from the volcanic ridge of Barren Island and Narcondam, with the formation between them of the two small deep-water basins in the north-west area of the Andaman Sea.

Molengraaf has pointed out that at the commencement of this folding "the sea-basins were much shallower than they are to-day and that the islands were lower and smaller," so that, the view that I have expressed above that the Andaman basin was formerly much shallower and therefore partook more of the nature of a brackish water basin than of an open sea, is at least in agreement with what we know regarding the folding of the crustal regions of the earth in this locality, and the continuation of this folding even at the present time is indicated by the evidence that, at any rate, the Nicobars and the deltaic region of the Arakan peninsula are still rising.

The evidence at our disposal tends then to show that the Andaman sea-basin made its first appearance about the commencement of the Tertiary Epoch when the great Alpine-Himalayan system began to arise. Simultaneous with the formation of the parallel ranges of the Burmo-Malayan arc, the westernmost of which appears to have arisen in the Andaman-Nicobar section from the sea-bottom, there was formed a shallow-water basin into which all the great rivers of Southern Burma and the streams on the east side of the Andaman-Nicobar region poured their effluent and out of which the water escaped through channels between the various islands into the Bay of Bengal. This elevation of the mountain chain steadily progressed till the height of the ridge attained its maximum of about 9,000 feet above sea-level. At the close of the Tertiary Epoch extensive subsidence occurred throughout the whole area of the basin, but more especially on the east of the Nicobar portion of the ridge; the amount of the subsidence being at least 1,300 fathoms, and this state of affairs persisted for at least a sufficient length of time to enable thick deposits of clay to be laid down in the shallow waters around the small but still persisting islands of the chain. Finally, in still more recent times, elevation of the ridge and possibly a further sinking of the trough has occurred so that the clay deposits have been raised to the height of several hundred feet above sea-level. Finally, by continued subsidence or by subaerial denudation, or both, a shallow channel was opened up during the Pleistocene period along the valley between the Andaman-Nicobar ridge and its southern continuation on the west and the Malayan ridge on the east. This now forms the Straits of Malacca and indirectly connects the Andaman Sea with the Pacific Ocean. These various channels have permitted the entry into the basin of the rich shallow-water fauna of both Indian and Pacific Oceans, whereas the deep fauna must have been derived either from ancestors capable of living in moderate depths of less than 800-900 fathoms, who had already succeeded in establishing themselves in the Bay of Bengal, or else by recent migration of shallow-water forms downwards into the deep waters of the basin.

In conclusion I must express my thanks to Dr. Coggin Brown, of the Geological Survey of India, who has very kindly read through the above pages and has given me much valuable advice and references to recent papers with which I was unacquainted.

BIBLIOGRAPHY.

1. Alcock, A., 1902 .. .. "A Naturalist in Indian Seas." London.

2. Annandale, N., and Hora, S. L., 1925 .. .. .. "The Freshwater Fish from the Andaman Islands." *Rec. Ind. Mus.* Vol. XXVII, p. 33. Calcutta.

3. Anderson, A. R. S., 1902 .. "Andaman Islands, Zoology," Port Blair.

4. " " 1903 .. "Diary of a visit to the Nicobars for Census purposes." *Census of India*, 1901, Vol. III.

5. Ball, V., 1870 .. .. (*a*) "Brief notes on the Geology and on the fauna in the neighbourhood of Nancowry harbour, Nicobar Islands." *Journ. Asiatic Soc., Bengal*, Vol. XXXIX, p. 25. Calcutta.
(*b*) "Notes on the Geology of the vicinity of Port Blair, Andaman Islands." *Journ. Asiatic Soc., Bengal*, Vol. XXXIX, p. 231. Calcutta.

6. Böggild, O. D., 1916 .. .. "Meeresgrundproben der Siboga Expedition." *Siboga Expedition, Monograph*, LXV. Leiden.

7. Bourne, G. C., 1888 .. .. "The Atoll of Diego Garcia and the Coral formation of the Indian Ocean." *Proc. Royal Soc.*, Vol. XLIII, p. 440. London.

8. Brouwer, H. A., 1918 .. .. "On Reef-caps." *Proc. Royal Acad. Amsterdam*, Vol. XXI, p. 816. Amsterdam.

9. Burton, R. C., 1913 .. .. "Volcanic rocks of Teng-Yueh District." *Rec. Geol. Surv. India*, Vol. XLIII, p. 206. Calcutta.

10. Cotter, G. de P., 1918 .. .. "The Geotectonics of the Tertiary Irrawaddi Basin." *Jour. Asiatic Soc., Bengal*, (new series), Vol. XIV, p. 409. Calcutta.

11. Daly, R. A., 1910 .. .. "Pleistocene Glaciation and the Coral Reef Problem." *Amer. Journ. of Science*, Vol. XXX, p. 297. Connecticut.

12. Ehrenburg, 1850 .. .. "On an extensive rock-formation of Siliceous Polycystina from the Nicobar Islands." *Berlin Monatsbericht*, 1850, p. 476–478. [*Abstract by T. R. J. in Quar. Journ. Geol. Soc., London*, Vol. VII, Pt. 2, p. 118. 1851.]

13. " 1876 .. .. "Fortsetzung der Mikrogeologischen studien als Gesammt Ubersicht der Mikroscopischen Palaontologie gleichartig analysirter Gebirgsarten der Erde, mit specialler Ruck-sicht auf den Polycystinen-Mergal bei Barbados.' *Abhandl der Konig. Akad. der Wissenschaften zu Berlin.* Vol for 1875, p. 1.

14. Foote, B., 1883 .. .. "Geology of Madura and Tinnevelly Districts." *Mem. Geol. Surv. India*, Vol. XX. Calcutta.

15. Fryer, J. C. F., 1910 .. .. "The South-west Indian Ocean. *The Geographical Journal*, Vol. XXXVII, No. 3 London.

16. Gardiner, S., 1903 .. .. "The Fauna and Geography of the Maldive and Laccadive Archipelagoes." Cambridge.

17. Gee, E. R., 1925 .. .. "General Report for 1924" (by Dr. Pascoe, Director, Geological Survey of India.)
Rec. Geol. Surv. India; Vol. LVIII, Pt. I. Calcutta.

18. Gregory, J. W., 1912 .. .. "The Making of the Earth." London.

19. ,, ,, 1923 .. .. "The Alps of Chinese Tibet and their Geographical Relations."
The Geographical Journal, Vol. LXI, No. 3, p. 153. London.

20. Haddon, A. C., 1889 .. .. "Notes on the Geology of Torres Straits." Rep. British Association. London.

21. Hochstetter, 1866 .. .. "Beitrage zur Geologie and Physickalischen Geographie der Nicobar Inseln. Geologischen Beobachtungen Reise der Oesterreichischan Fregatte Novara um die Erde in Jahre 1857–59. Geologische Theil iii, pp. 85–112. Wien.
[*Translated. Selections Records Govt. Ind.*, Vol. LXXVII, pp. 208–229. 1870.]

22. Jenkins, 1864 .. .. "On some Tertiary Mollusca from Mount Sela in the Island of Java."
Quart. Journ. Geol. Soc., London, Vol. XX, p. 71. London.

23. Kloss, Boden, 1903 .. .. "In the Andamans and Nicobars." London.

24. Kurz, 1868 .. .. "Report on the Vegetation of the Andamans." Calcutta.

25. Lyell, 1872.. .. .. "Principles of Geology."

26. Medlicott and Blandford, 1879 .. "A Manual of the Geology of India." Calcutta.

27. Molengraaf, G. A. F., 1921 .. "Modern Deep Sea Research in the East Indian Archipelago."
The Geographical Journal, Vol. LVII, No. 2, p. 95. London.

28. ,, ,, 1922 .. "Geologie" in 'De Zeeën van Nederlandsch Oost-Indië.' Uitgegeven door het Koninklijk Nederlandsch Aardrigkskundig Genootschap. Leiden.

29. Nicholson and Lydekker, 1889 .. "Manual of Palaeontology." London and Edinburgh.

30. Nicholson, H. A., 1893 .. "Recent progress in Palaeontology."
Trans. Edin. Geol. Soc.; Vol. VI, p. 53. Edinburgh.

31. Oldham, R. D., 1885 .. "Note on the Geology of the Andaman Islands."
Rec. Geol. Surv. India, Vol. XVIII, Pt. 3, pp. 135–145. Calcutta.

32. ,, ,, ,, .. "The Geology of India." 2nd Edition. Calcutta.

33. Pascoe, P. H., 1912 .. .. "The Oilfields of Burma."
Mem. Geol. Surv. India, Vol. XL. Pt. 1. Calcutta.

34. Richthofen, 1862 .. .. "Bericht über einer Ausfluq in Java."
Zeitschrift für deutschen geol. gesellschaft, Vol. XIV, p. 327. Berlin.

35. Rink, P. H., 1847, p. 327 .. "Die Nikobar Inseln. Eine Geographische Skizze, mitspecieller Berücksichtigung der Geognosie." Kopenhagen.
Translated, Selections Rec. Govt. Ind., Vol. LXXVII, pp. 105–153, 1870.

36. Saville-Kent, 1893 .. .. "The Great Barrier Reef of Australia." London.

37. Sewell, R. B. S., 1922 .. .. "A Survey season in the Nicobar Islands."
Journal, Bombay Nat. Hist. Soc., Vol. XXVIII, No. 4, p. 570. Bombay.

38. Sollas, W. J., 1923 .. .. "Man and the Ice Age."
Nature, p. 332. London.

39. Suess. ,, 1904 .. .. "The Face of the Earth." Authorised English Translation. Clarendon Press. Oxford.

40. Theobald, W., Jnr, 1870 .. "On the alluvial deposits in the Irrawadi, more particularly as contrasted with those of the Ganges."
Rec. Geol. Surv. India, Vol. III, Pt. I., p. 17. Calcutta.

41. Tipper, G. H., 1911 .. .. "The Geological formation of the Andamans with references to the Nicobars."
Mem. Geol. Surv. India, Vol. XXXV, Pt. 4. Calcutta.

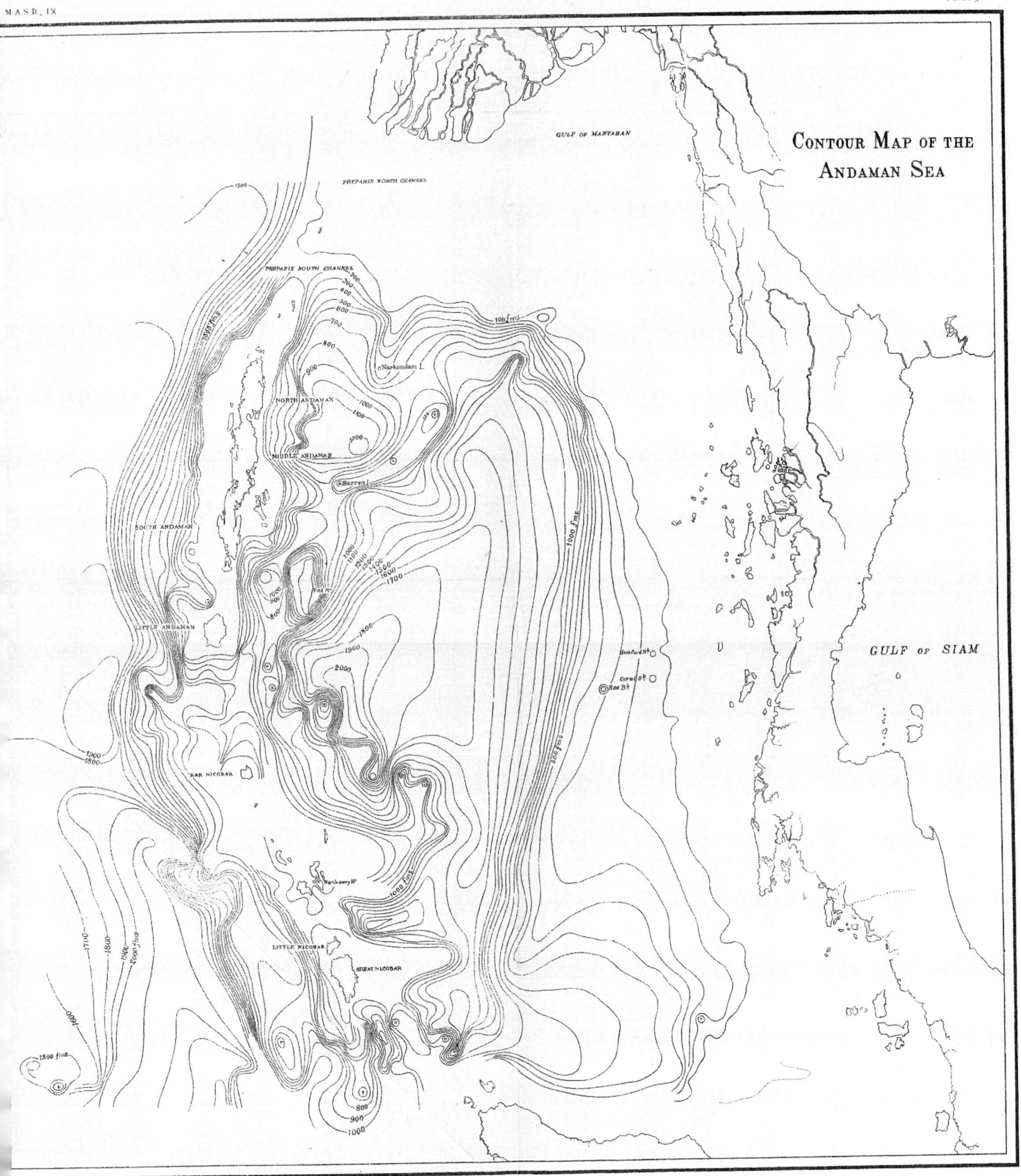

CHART I.

Map of the Andaman Sea basin showing the contours of the bottom. Submarine peaks are marked ⊕.
The coast lines are taken from Admiralty Chart No. 70.

GEOGRAPHIC AND OCEANOGRAPHIC RESEARCH IN INDIAN WATERS.

By R. B. SEYMOUR SEWELL, M.A., F.A.S.B., F.L.S., F.Z.S., Major, I.M.S.,
Surgeon-Naturalist to the Marine Survey of India and Director, Zoological Survey of India.

II.

A STUDY OF THE NATURE OF THE SEA-BED AND OF THE DEEP-SEA DEPOSITS OF THE ANDAMAN SEA AND BAY OF BENGAL.

In attempting a general survey of the nature of the sea-bed in Indian waters, one is at the outset faced with the difficulty of accurately discriminating between various types of deposit. Since the publication of the report on the deep-sea deposits obtained by the "Challenger" (Murray and Renard, 1891.) it has been the custom to classify samples of the sea-bottom as of Terrigenous or Pelagic origin, and, in the case of the former group, deposits are in the main denoted by their colour and consistency, as for example, Blue mud, Green sand, etc. In the greater number of bottom samples obtained by the R.I.M.S. "Investigator" the characters of the deposits are such that it is not infrequently impossible to assign them to any particular types and in many cases one is doubtful whether a sample should be classed as a Pelagic deposit or a Terrigenous one. In consequence it has been the custom to classify deposits by their colour and consistency, but on a more elaborate system than that adopted for the "Challenger" samples; thus according to the consistency of the deposit it may be termed a mud, an ooze or a clay, and the colour variations include not only the Green, Blue and Red of the "Challenger" scheme, but also Grey, Brown or Ochre. In such a classification the colour sense and delicacy of touch of the observer will play a leading part and hence the method can hardly be called a scientific one, but under the circumstances it was probably the best that could be devised and, since the functions of the Surgeon-Naturalist are primarily biological, we have all, with one exception, left the matter at that and have made no further attempts to discriminate between the samples by such means as chemical analysis or microscopical examination.

Alcock, however, during his tenure of office as Surgeon-Naturalist from 1888 to 1892, carried out both chemical and microscopical examinations of many of the samples of bottom deposits that were obtained during this period and although, owing to the conditions under which he was working, his analyses may not have attained such a high degree of accuracy as is now-a-days aimed at, the results that he obtained are so interesting that they deserve to be rescued from oblivion, and I take this opportunity of publishing, with his consent, those that refer to the areas under discussion. The method he employed consists of treating weighed quantities of the dried deposit with (*a*) dilute hydrochloric acid and (*b*) dilute acetic acid noting the amount dissolved by each; he also recorded the character of the ash after incineration, and the presence of organisms, such as Diatoms, Foraminifera, Radiolaria, etc., and in many cases he even gives the percentage of Foraminifera present in the whole deposit.

The amount of the deposit dissolved in acetic acid serves as a rough guide to the quantity of calcium carbonate present in a sample, and throughout this paper, when referring to the percentage of calcium carbonate in any sample, it is this figure that I have given.

In most, if not all, the published charts of the deposits of the ocean bed that of the Andaman Sea and Bay of Bengal is shown as consisting of terrigenous deposit throughout the whole area, and in the main this is correct. Murray and Renard (1891, p. 233.) state that "The Blue muds surround nearly all coasts and fill nearly all enclosed areas, like the Mediterranean and even the Arctic Ocean." One would, therefore, expect in a basin such as the Andaman Sea, which is largely land-locked and into which a number of large rivers, notably the Irrawaddi, Sitang, Salween, Tavoy and Tenasserim, pour their silt-laden waters, that the greater part of the sea floor below the 100 fathom line would be composed of mud; but in these waters records of Blue mud are rare, and its place is taken by mud of a distinctly brown tinge, though as regards other characters there seems to be little difference between the two types. Even as regards this typically terrigenous deposit the distribution is, however, somewhat peculiar and in certain regions the bottom deposits show marked differences that seem worthy of special mention.

A reference to the accompanying chart (Chart IV.) shows that in the Andaman Sea the area occupied by brown mud is in the main confined to the south-east part of the basin. At the north end, beyond Lat. 13° N., this type of deposit appears to be limited to the continental shelf, and in the region of the Irrawaddi delta its area of deposition is even more restricted, for here it appears to be found only in inshore waters in depths of less than 50 fathoms. South of the line Lat. 9° N. however this deposit appears to be widespread and covers the whole sea-floor even down to depths as great as 2,000 fathoms. This expanse of mud is not, however, a continuous one. Commencing opposite the mouth of the Gulf of Martaban in Lat. 13°50′ N.: Long 96°20′E. at the 100 fathom line and running southward through the whole length of the basin is a very remarkable line of rock, which I have indicated in the chart in red. This rock-ridge closely follows the curve of the present coast-line and lies roughly at a distance from it of 90 to 100 miles. It is apparently not continuous, though further soundings may reveal its presence in areas where we have as yet no record of its occurrence, for in certain regions it seems to disappear, only, however, to reappear again and in more or less the same line. The depth at which the ridge lies varies in different parts of its length. Commencing at its north end and tracing it southward we find that to the west of Tavoy River it sinks from 77 fathoms at the north end to 165 fathoms at the south, while opposite Mergui it lies at a depth of 190 fathoms. The ridge in this part of its course appears to be sinking and a little to the south of Mergui it disappears. It reappears again, however, to the west of Domal Island, this time at a depth of 106 fathoms and as we still trace it southward it sinks to 159 fathoms and again disappears, though the line of its continuation is almost certainly indicated by two banks, Heckford Bank and Coral Bank, that rise from a depth of over 100 fathoms to within 7 and 11 fathoms of the surface respectively. Possibly Roe Bank, that lies some thirty miles to the west and rises from about 250 fathoms up to 8 fathoms, represents an outlying spur of the same chain. The ridge now trends to the westward and the last actual trace that we get of this portion of it is in Lat. 8°52′ N.: Long. 96°29′ E., where rock has been recorded at a depth of 228 fathoms; the trend of the contour lines of the sea floor in this region seem to indicate, however, a considerable further extension towards the south-west. At a distance of 50 to 60 miles west of Kopah Inlet at a depth of 136 fathoms we again find a rock bottom making its appearance. This second rock ridge at first trends slightly towards the east, to a point 25 miles west of Pulo Rajai, where it lies at a depth of 100 fathoms. It then bends westward and after a big interruption reappears in an unnamed shoal, that carries only 37 fathoms of water, while all around there are depths of over 200 fathoms. Finally a rock bottom again reappears, at a depth of 196 fathoms, forty miles north of Diamond Point in Sumatra.

It seems to me a reasonable assumption to make that we have here a double range of hills; or possibly a single range that has been fractured across the middle of its length, the southern ends of each portion having been bent to the west. Probably these rock ridges, like the neighbouring islands and the coast of Burma, consist of limestone, but, since their orginal upheaval, they have become almost buried beneath the huge deposit of mud that is being continually brought down by the rivers and precipitated over the sea-floor. On the east or land-ward side of the hill-range this deposit of mud appears to have completely filled whatever depression originally existed between it and the hill range that forms the islands of the Mergui Archipelago, so that we now have a more or less flat plain, that slopes from the shore line down to approximately the 100 fathom line and might be mistaken for a broad continental shelf, as I have already pointed out when dealing with the contours of the Andaman Sea basin.

This hill range is probably one of the anticlinal folds of the Eastern Zone of Burma and probably became submerged when the depression of the central part of the Andaman Sea took place. Assuming this to be the case, we have here a hill-range extending for some five hundred miles, which has become completely submerged and whose presence is now only indicated on the surface by four coral banks, which, had the ridge been situated in the western part of an ocean instead of on the east side of an enclosed basin, would almost certainly have given rise to true coral atolls—a speculation that is not devoid of interest in view of the statements put forward by Murray, Agassiz and others, when criticising Darwin's theory of the origin of atolls, that it is impossible to conceive of a mountain chain being submerged without some at least of its peaks still showing above sea-level as true islands.

Turning now to the north-west and westerly region of the Andaman Sea basin we find, as I have mentioned already, that Brown mud is limited, opposite the Irrawaddi delta, to the inshore area of the continental shelf and its place as a deep-sea deposit is in the main taken by Green mud or Green ooze, also a terrigenous deposit but one that is characterised by the presence of glauconite, to which the colour is due. Murray and Renard (1891, p. 236.) state that Green mud is "almost always developed along bold and exposed coasts where no very large rivers pour their detrital matters into the sea," and further (*Loc. cit*: p. 237) that " along coasts where Green muds are laid down pelagic conditions appear to approach much nearer to the shores than where Blue muds prevail." In view of this latter statement it is interesting to note that in several places where this Green mud deposit begins to approach the Brown mud, in the region of lat. 14° N. long. 96 E. and along Barragua Flat off the Irrawaddi delta, we have records of " Grey ooze," which indicate a tendency in the deposit to approximate to Blue mud.

It is obvious that this deposit of Green mud is not derived from the deltaic region of the Irrawaddi, where the whole of the coast is low lying with wide mud-flats extending sea-wards. It is possible that a certain proportion may be derived from erosion of the coastal region of the Andaman Islands, but the main source seems to be the coast of Arakan in the Bay of Bengal. Along this coast, extending as far north as Lat. 19°40′, Green mud constitues nearly the only type of deposit even down to depths of over 1,500 fathoms. Throughout the greater part of the year from January to October a surface current is sweeping in from the Bay of Bengal through Preparis Channel into the Andaman Sea, and there can be little doubt that this will carry in with it quantities of detritus and silt derived from the Burma coast to the north of the channel, and will ultimately deposit it on the floor of the Andaman Sea [For the direction of the currents in this area I must refer the reader to the Admiralty Current Charts.].

I have in the preceding paper called attention to the volcanic ridge that runs diagonally in a N.E. direction across the floor of the Andaman Sea commencing opposite South Andaman Island. All along this ridge we have records of the occurrence of Pteropod ooze. Its occurrence in this area is probably due to several causes. The inflowing currents from the Bay of Bengal will carry in with them a number of these small pelagic animals and it is probable that, when this water comes into contact and mixes with the much less saline waters in the neighbourhood of the Irrawaddi and Salween river area, they are for the most part killed off and their shells settle down to the bottom. When once on the bottom the solution of the shells and their disappearance is probably less rapid along this ridge than it would be in deeper water. Murray and Renard (1891, p. 266) in discussing the occurrence of this type of deposit show that it is formed in areas of moderate depth in which there is a relatively small quantity of land debris; and since the bulk at any rate of the land debris in this region is probably derived from the Arakan Coast of Burma, the greater part of it must have already been deposited before the line of this ridge is reached.

Pteropod ooze has also been recorded by the "Valdivia" to the south-west of Great Nicobar Island in Lat. 7° N.; 93° 30 E. (approx.) at a depth of 296 fathoms and Alcock (1898, p. 45,) has also stated that patches of this deposit occur on the west side of the Andamans. I have, however, been unable to find any record of its occurrence in this latter region and, owing to the large growth of coral around these islands and especially along the barrier reef, the main deposit in this locality is coral sand and mud—a type of deposit that is usually in "Investigator" records known as Grey mud or ooze.

Along the floor of Ten-degree Channel and again in the neighbourhood of Great Channel the sea-bed is in many cases recorded as consisting of 'sand', even in depths as great as 1660 fathoms. A comparison of the chart of the bottom-deposits (Chart IV) and the contour-map of the Andaman Sea (Chart I) shows that in several instances this type of sea-bottom is found in close proximity to an outlying spur of the great submarine mountain-chain and often near one of the submarine peaks, and it seems probable that in at least some of these instances the sand is of volcanic origin.

Before leaving the area of the Andaman Sea there is one observation of the sea-bottom that is worthy of special mention and that is the occurrence in Lat. 13°17′ N.; Long. 93°07′ E. of a number of phosphatic nodules. They were dredged in a depth of 90 fathoms, and have been examined and reported on by Tipper (1911, p. 16.) who states that they resemble similar phosphatic nodules obtained from other parts of the world. It is interesting to note that this deposit occurs in close association with deposits of Green mud, and Murray (1910, p. 372.) states that "it is in areas where warm and cold currents meet at the surface that glauconite and phosphatic deposits are found at the bottom."

Unfortunately our knowledge of the bottom-deposits of the Bay of Bengal is by no means as complete as one could wish, but sufficient is known to enable us to divide this great basin into three distinct areas.

Commencing at the northern end of the Bay where the combined branches of the Ganges and Brahmaputra rivers have given rise to the Gangetic Delta and the Sunderbunds, we find that the bottom-deposit is, like that of the southern end of the Andaman basin, Brown mud. As we trace this deposit eastwards around the head of the bay we find that it extends as far as the deltaic area at the mouths of the Kaladan and Lemro rivers and there becomes replaced by Green mud which, as we have already seen, forms the deposit along the Arakan coast of Burma. On the western side of the bay a deposit of Brown mud, interspersed with occasional patches of Green mud, can be traced all along the east coast of the great Indian Peninsula as far as the north end of Ceylon, where again it is replaced by Green mud.

Opposite the delta of the Ganges and Brahmaputra rivers lies the deep Gangetic "Swatch of no ground," a deep gully that has a length of approximately seventy miles, and a depth that gradually alters from 60 fathoms at its shore end to 454 fathoms about the middle of its length, beyond which it again sinks to 595 fathoms at its mouth. On either hand are extensive mud flats, and a reference to Chart V shows that each of these flats is, along its shore side, traversed by mud banks, between which run the mouths of the main branches of the rivers and of the creeks between the various islands of the Sunderbunds. Around the head of the Swatch these two mud-flats become continuous, so that the Swatch may be said to divide what would otherwise be a continuous expanse into eastern and western portions. The main junction of the Ganges and Brahmaputra occurs at Goalundo and a line drawn south from this point follows roughly the course of the Haringhata volcanic matter in this line of soundings has been derived either from the same area or from the volcanic region in the neighbourhood of the Sumatran coast, and has been swept westward and finally deposited on the floor of the mouth of the Bay of Bengal by the equatorial current that runs from east to west.

On three occasions the trawl has bought up large pieces of pumice from the sea-bed in the middle of the Bay of Bengal. The positions where these samples were obtained are as follows:—

Lat. N.	Long. E.	Depth in fathoms.
9·°34·'00";	85·°43·'15"	1997.
11·58·00;	88·52·17	1748.
12·20·00;	85·08·00	1803.

The pumice was almost certainly derived from the great explosion of Krakatoa in 1880 and had drifted up into the Bay of Bengal, gradually becoming water-logged and sinking to the bottom.

A study of the amount of calcium carbonate contained in bottom deposits from various parts of the Bay of Bengal and Andaman sea, reveals several very interesting points.

Among Alcock's analyses there are unfortunately only four from the Andaman Sea [1]; viz:—

	Position. N.	Position. E.	Deposit in fathoms (or metres in brackets).	Type of Deposit.	Percentage of Calcium Carbonate.
1.	12.°34.'00";	93.°22.'40"	1130 (2068)	Dark mud	trace.
2.	12. 54. 00;	93. 32. 30	1159 (2121)	Dark mud	trace.
3.	12. 59. 00;	93. 23. 10	683 (1250)	Blue mud	24·0.
4.	13. 36. 00;	93. 18. 00	538 (985)	Olive mud	46·6.

The positions, at which the samples were obtained, all lie close to the Andaman Islands and in the neighbourhood of Preparis Channel. They cannot therefore be taken as being typical of the deposits in the basin, but a comparison of Nos. 3 and 4 indicates a rapid increase in the percentage of calcium carbonate in deposits, from the same depth, as one approaches Preparis Channel and the Bay of Bengal.

We are fortunate in having a number of analyses of samples from various positions in the Bay of Bengal itself. At first sight the results that Alcock obtained in this area exhibit an extraordinary range of variation; but a reference to a chart and a study of the various positions from which the samples were obtained reveal that we have two very definite and distinct series corresponding exactly with what we have already noted regarding the type of deposit in the two regions of the Bay,

---

[1] Since this paper was written a bottom-trawl has been carried out in the Andaman Sea, in 8°32′ N; 94°10′ E at a depth of 1,260 fathoms. The mud brought up by the trawl has the brown colour that is so constant a character of the bottom-deposits in this region. Dr. W. A. K. Christie, of the Geological Survey of India, has very kindly estimated the amount of calcium carbonate that is contained in a sample of the mud. The result obtained is surprisingly small, being only one per cent. The specimen contains a few *Globigerina* shells, probably sufficient to account for the whole of the carbonate present, but the chief characteristic of the sample, according to Mr. G. H. Tipper, of the Geological Survey, lies in the large number of Radiolarian skeletons and sponge spicules. The analysis agrees extremely well with the two most southerly samples in Alcock's series, Nos. 1 and 2. As already noted in my previous paper, the trawl brought up evidence in the shape of volcanic rocks, that the area had recently been in eruption.

namely (*a*) those from the south and east portion, and (*b*) those from the north and west. Taking the Southern area first, we find that the average percentage of calcium carbonate in the deposit is 45 4, so that it would be classified according to Thoulet's scheme as "calcareous" (*vide* Richard, 1907, p. 69). So high is the percentage in the majority of the samples that the application of the term "Globigerina ooze" is fully justified. In the table below I give the results of these analyses and the positions from which the deposits were obtained :—

Position. N. ° ′ ″	E. ° ′ ″	Depths in fathoms (or metres in brackets).	Character of Deposit.	Percentage of Calcium Carb.
7.25.00 ;	88.16.00	.. 2035 (3724)	.. Glob. ooze ..	.. 75.0
7.38.00 ;	89.48.00	.. 1651 (3021)	.. Glob. ooze ..	.. 95.0
7.42.00 ;	90.10.00	.. 1572 (2877)	.. Glob. ooze ..	.. 75.0
7.46.00 ;	90.25.00	.. 1623 (2970)	.. Glob. ooze ..	.. 75.0
9.08.00 ;	91.30.00	.. 1732 (3170)	.. Dark mud ..	.. 50.0
9.34.00 ;	85.43.15	.. 1997 (3655)	.. Grey ooze ..	.. 0.5
10.16.30 ;	88.13.00	.. 1860 (3404)	.. Grey ooze ..	.. 50.0
11.07.00 ;	89.11.00	.. 1773 (3245)	.. Glob. ooze ..	.. 50.0
11.11.00 ;	90.06.00	.. 1735 (3175)	.. Glob. ooze ..	.. 75.0
11.58.00 ;	88.52.17	.. 1748 (3199)	.. Glob. ooze ..	.. 50.0
12.05.00 ;	88.40.15	.. 1745 (3193)	.. Grey ooze ..	.. 10.0
12.11.00 ;	87.13.00	.. 1763 (3226)	.. Grey ooze ..	.. 30.0
12.18.00 ;	86.13.00	.. 1797 (3289)	.. Grey ooze ..	.. 30.0
12.20.00 ;	85.08.00	.. 1803 (3299)	.. Glob. ooze with pumice ..	20.0
12.21.30 ;	92.03.30	.. 780 (1427)	.. Glob. ooze ..	.. 75.0
12.24.00 ;	90.08.00	.. 1669 (3054)	.. Glob. ooze ..	.. 50.0
12.50.00 ;	90.52.00	.. 1644 (3009)	.. Grey ooze over clay ..	12.0
12.51.00 ;	88.53.00	.. 1674 (3063)	.. Glob. ooze ..	.. 75.0
13.07.00 ;	91.17.00	.. 1632 (2987)	.. Grey ooze over clay ..	15.0
13.16.00 ;	88.01.30	.. 1676 (3067)	.. Buff ooze over Blue clay ..	50.0
13.18.00 ;	91.38.30	.. 1672 (3060)	.. Ooze, Clay ..	.. 30.0
13.29.30 ;	91.59.00	.. 1637 (2996)	.. Brown mud ..	.. 31.5
13.34.00 ;	86.54.10	.. 1672 (3060)	.. Buff ooze over Brown clay	16.7
14.14.00 ;	85.58.00	.. 1654 (3027)	.. Grey ooze ..	.. 50.0

The "Valdivia" during her cruise carried out a few soundings in the region of the mouth of the Bay of Bengal and the percentages of calcium carbonate in these examples. as given by Murray and Philippi (1908), are as follows :—

N. ° ′ ″	E. ° ′ ″	Depth in Metres.	Character of deposit.	Percentage of Calcium Carb.
6.54.00 ;	93.28.48	.. 296	.. Pteropod ooze ..	.. 38·0
6.56.18 ;	93.32.42	.. 362	.. Pteropod ooze ..	.. 36·0
7.57.54 ,	91.47.12	.. 3974	.. Blue mud ..	.. 10·0
7.43.12 ;	88.44.54	.. 3692	.. Globigerina ooze	.. 47·5

So that these results agree well with Alcock's determinations.

Turning now to the deposits in the north and west regions of the Bay, there are in Alcock's records twenty-seven analyses, and I give the details below :—

The influx of river water into the Bay of Bengal and the Andaman Sea will still further tend to reduce the percentage of calcium carbonate in bottom deposits by its solvent action. Murray and Irvine (1890, p. 93) have pointed out that organic matter is brought down in large quantities by rivers and this, owing to its decomposition in the sea, sets up a chemical change that results in the formation of calcium sulphate. As they remark, "It is probable that the quantity of sulphate of lime in solution in the ocean is limited only by the amount of organic decomposition that takes place in ocean waters." In this respect there must be a marked resemblance between the conditions existing in the Andaman Sea and along the north and west shores of the Bay of Bengal, owing to the size and number of the rivers that flow into the sea in these two regions; and it is therefore not surprising that the character of the bottom deposits in the two areas should closely resemble each other at any rate in their shallower depths.

Böggild himself (1916, p. 12) realized that by taking the average percentage of calcium carbonate in all samples from the East Indian Archipelago he was not giving a true picture of the actual state of affairs; in consequence he further subdivided his samples into two series, those taken in the neighbourhood of large islands and those obtained away from such land areas. The composition of these two series of bottom samples shows a marked difference in the calcium carbonate content, especially in the shallower depths; the bottom deposits around large islands having a much smaller percentage. This, as Böggild remarks, is undoubtedly due to the greatly increased amount of terrigenous deposit in the samples taken in the neighbourhood of large land areas, and he might also have added that in the vicinity of large land areas and particularly in the tropical belt where such islands are clad in dense vegetation, the amount of organic debris that is washed into the sea and there undergoes decomposition is also greatly increased.

There seems to me, however, to be another factor to be considered. Max Weber (1902, p. 5) has drawn attention to the fact that the region of the Malay Archipelago really consists of two very distinct areas, an eastern and a western, which show very different characteristics. If now we take the results obtained from the 'Siboga' examples of bottom deposits and examine those from each area separately, we find that there is a marked difference between the percentage of calcium carbonate present at all depths in the two series.

(*a*) In samples taken from the area comprising the Java Sea, Straits of Macassar and the Celebes Sea—that is, from the western part of the Archipelago—we get the following percentages of calcium carbonate at the depths indicated :—

Depth.	No. of samples.	Percentage of Calcium Carbonate.
0–500 metres	3	21·8
500–1,000 ,,	6	14·2
1,000–1,500 ,,	4	12·1
1,500–2,000 ,,	1	3·7
2,000–2,500 ,,	3	6·2
3,500–4,000 ,,	1	0·0

(*b*) On the other hand in the areas comprising the eastern part of the Archipelago, including the Timor Sea, we find that the percentage of calcium carbonate present in the samples is much greater at the same depths :—

Depth.	No. of samples.	Percentage of Calcium Carbonate.
0–500 metres	9	24·8
500–1,000 ,,	8	43·1
1,000–1,500 ,,	3	50·8
1,500–2,000 ,,	8	29·8
2,000–2,500 ,,	6	34·2
2,500–3,000 ,,	9	24·8
3,000–3,500 ,,	3	32·5
3,500–4,000 ,,	3	11·1
> 4,000 ,,	11	2·7

On the whole the calcium carbonate content of the bottom deposits from the western area of the Malay Archipelago agrees fairly closely with what Alcock found to be present in deposits from the Andaman sea and the north and west areas of the Bay of Bengal, while the deposits from the eastern portion of the Archipelago are intermediate between these and the deposits laid down in the open ocean. It appears then that, provided other conditions are similar, it makes little or no difference whether the deposit is laid down within the confines of a land-locked basin or along the coasts of an open sea like the Bay of Bengal.

In the accompanying Text-figure 4 I have plotted the percentages of calcium carbonate present in the deposits from different depths in (*a*) the north and west sides of the Bay of Bengal, as shown by Alcock's analyses, (*b*) the East Indian Archipelago as given by Böggild, (*c*) the south and east areas of the Bay of Bengal and (*d*) the samples of Globigerina ooze, as given by Murray and Renard from the "Challenger" results.

The first point to which I would draw attention is the similarity of all four curves. In each series, as we pass from shallower waters to those of greater depths, we find that there is at first a decrease in the percentage of calcium carbonate present; this is succeeded by a rise in the percentage, and is again followed by a second fall. In the complete series of examples obtained by the "Valdivia" and examined by Murray and Philippi (1908, p. 145) it is interesting to note that this initial fall in the carbonate content of deposits from depths of comparatively small amount appears to be absent, and for the purpose of comparison I give the figures for the two series :

Calcium Carbonate Content.

Depth in metres.	"Challenger" series.	"Valdivia" series.	Depth in metres.
0– 900	86·04.	40·0	0–1,000
900–1,800	66·86.	44·0	1,000–2,000
1,800–2,700	70·87.	63·0	2,000–3,000
2,700–3,600	69·55·	65·0	3,000–4,000
3,600–4,500	46·73.	51·5	4,000–5,000
4,500–5,400	17·36.	46·2	> 5,000
5,400–6,300	0·88.		
6,300–7,200	0·00.		

carbon dioxide set free in the above reaction will still further assist in dissolving the carbonate content of the deposit.

The rise in the percentage of calcium carbonate in the deposits from depths of 1,500 to 3,000 metres, which is found in all three series of analyses (*vide* Text-fig. 4) is undoubtedly due to more than one factor. In the first place, as one proceeds into deeper water away from land the quantity of non-calcareous terrigenous deposit will decrease and thus, other things being equal, the calcareous content will tend to rise. Simultaneously the pelagic plant life will show a marked diminution in the quantity present. Gran (*vide* Murray and Hjort. 1912, p. 378) lays down "that it is also a general rule that plankton (Phytoplankton) is far more plentiful along the coasts than in the open sea," and in consequence the silicate content of a bottom deposit will tend to diminish in quantity and this will also tend towards an increase in the carbonate content. Another factor that will have a profound bearing on the carbonate content of a deposit is the depth at which the maximum concentration of chalk-secreting planktonic organisms occurs. The work of the "Michael Sars" (*vide* Murray and Hjort, 1912, pp. 722, 723) has shown that in the North Atlantic the maximum concentration of the planktonic fauna occurs at a depth of 500 to 1,000 metres. It was at this depth that the greatest quantities and the greater number of species were obtained. If the same holds good for the carbonate-secreting forms of planktonic life, it seems obvious that we might expect to find an increase in the percentage of calcium carbonate in bottom deposits occurring at or near this level as we pass from shallower to deeper water. A study of the three curves given above in Fig. 4, shows that in both the "Challenger" series of deposits and in Alcock's series from the Bay of Bengal a distinct increase in the carbonate percentage sets in at a depth of about 1,500 metres, and continues to a depth of 3,000 metres, and, with the exception of the initial fall in shallower water, the same is shown in the "Valdivia" results. This concentration of the animal population at these depths is directly correlated with an increase in the specific gravity and viscosity of the water and is mainly dependent on the temperature of the water. In this respect the conditions existing in the Bay of Bengal and in open oceanic waters are very similar, but a peculiarity of enclosed seas, such as the Andaman Sea and the waters of the East Indian Archipelago, is the uniformity of the temperature at all depths below that of the deepest channel connecting the basin with the open ocean. In the East Indian Archipelago there are several different basins, the temperature in the bottom waters of which depends on the depth of the channels leading into them. Molengraaf (1921, p. 97) has given a complete list of the various temperatures in the troughs and basins of the Archipelago, and the lowest of these is 3·1°C, whereas the observations of the "Investigator" and other ships have shewn that across the mouth of the Bay of Bengal the bottom temperature at a depth of 4,000 metres is as low as 1·1°C. In consequence any increase in the density of the water of these basins below the level of the deepest entrance channel will be small and there will be no rapid rise in viscosity. As a result there will tend to be but little concentration of animal life such as is met with in open oceanic waters. We should, therefore, not expect to get any marked rise in the calcium carbonate content in the bottom samples, and a study of the curve of carbonate percentage in deposits from the Malay Archipelago shows that none is met with till we reach a depth of about 2,200 metres and then only of small amount in comparison with the other two series of samples.

We, finally, come to the question of the disappearance of the calcium carbonate from deposits obtained at great depths, and here a comparision of the curves given above in Fig. 4 of the deposits from the Bay of Bengal and the Malay Archipelago shows that in all three series of samples this disappearance begins at approximately the same depth, namely at about 3,000 metres (=1,640 fathoms). Murray and Hjort (1912, p. 174.) state that the increase in the rate of solution of the carbonate content of a bottom deposit sets in at a depth of 4,572 metres (=2,500 fathoms); but, judging from the results given above, this estimate appears to be too deep, and the results obtained by the "Challenger" and "Valdivia" indicate that it commences at about 2,700 to 2,800 metres in the open ocean. Böggild (1916, p. 11) from the results of his analyses of the "Siboga" bottom-deposits has come to the conclusion that in the deep waters of land-locked basins the solution of calcium carbonate proceeds much more rapidly than in waters of similar depth in the open ocean. He points out that in the "Siboga" samples from depths of 4,000 metres and over the percentage of calcium carbonate is on the average only 2·1, whereas in the case of the "Challenger" deposits from the open ocean it is at this depth as high as 46·7. (*Vide* Molengraff, 1922, p. 347). In order to account for this supposed increase in the solution of calcium carbonate in land-locked basins, Böggild has suggested that either carbon dioxide gas is able to diffuse through the water-layers from above, or that it is derived from submarine volcanoes. As regards this latter suggestion Molengraaf points out that the occurrence of submarine volcanoes along a synclinal depression between anticlinal ridges is highly unlikely and he suggests instead that there is a circulation of the water in these basins, the water rising in the middle and flowing laterally outwards, to sink again along the marginal slopes.

If, now, we compare the curves of carbonate percentage, given above in figure 4, we see that in the two series of observations the curves follow an almost exactly parallel course. In the deposits from the Malay Archipelago the percentage of calcium carbonate falls from 26 at 3,000 metres to 5 at 4,000 metres—a fall of 21; whereas in the case of the "Challenger" series it falls between the same depths from 71 to 48 or a drop of 23. At first sight this looks as if the rate of solution in the two areas—i.e. in an enclosed basin and in the open ocean—was approximately equal, but a little consideration shows that from every 100 grammes of a deposit containing only 26% of calcium carbonate it is necessary to dissolve 22·1 grammes of carbonate to give a percentage of 5·0; whereas from 100 grammes of a deposit containing 71% of calcium carbonate it is necessary to dissolve 44·2 grammes of carbonate in order to reduce the percentage to 48·0. It follows, therefore, that *the actual rate of solution of calcium carbonate is appreciably greater in the open sea than it is in the enclosed waters of a land-locked basin.*

In the series of analyses of bottom-samples from the Bay of Bengal carried out by Alcock, the percentage of calcium carbonate throughout the whole area exhibits the same general features as those of the other areas which we have been considering. In the table below I have given the average percentage at the different depths.

Depth.	No. of Samples.	Percentage of Calcium Carbonate.
0– 500 metres	2	12·0
500–1,000 "	1	17·3
1,000–1,500 "	3	28·3
1,500–2,000 "	3	3·0
2,000–2,500 "	9	6·5
2,500–3,000 "	7	32·5
3,000–3,500 "	20	38·1
3,500–4,000 "	6	23·9

In this series I have no records of any samples from depths greater than 4,000 metres, but it is clear that the percentage of calcium carbonate has already begun to decrease and, assuming that, as in the case of the series from the Malay Archipelago and those obtained by the "Challenger," this decrease is continued at a uniform rate, we should find that the percentage falls from about 38·1 at 3,000 metres depth to about 17 at 4,000 metres depth, or a drop of 21, which again agrees fairly closely with what we have found in the other areas that we have been considering. In this case, however, the actual amount of calcium carbonate that must be dissolved out of 100 grammes of the deposit in order to cause this fall in the percentage is 25·4 grammes. I have already referred to the different types of bottom-deposit found in the two areas of the Bay of Bengal and it is interesting to compare the rate of disappearance of calcium carbonate in those samples from the South and East area, where the deposit much more nearly approximates to that of the open ocean, with the result obtained above from a consideration of the area as a whole. In the nature of things these samples are all from comparatively great depths and in the table below I give the averages obtained by Alcock, and by Murray and Phillippi from their analysis of the 'Valdivia' samples.

Depth.	No. of Samples.	Percentage of Calcium Carbonate.
0– 500	2	37·0
500–1,000	0	——
1,000–1,500	1	75·0
1,500–2,000	0	——
2,000–2,500	0	——
2,500–3,000	4	49·1
3,000–3,500	17	43·7
3,500–4,000	4	33·25

In this area therefore the percentage of carbonate falls from 49·1 at on average depth of 2,750 metres to 33·25 at on average depth of 3,750 metres—a fall of 15·85: that is to say, that during the time taken for the sediment to sink through a distance of 1,000 metres, assuming that the rate of solution is approximately constant when once the mud or ooze has been deposited, 23·75 grammes of calcium carbonate have been dissolved out of every original 100 grammes. This agrees closely with the figure arrived at from the consideration of the deposits from the whole area of the Bay.

The actual amounts of calcium carbonate dissolved from 100 grammes of deposit in each of these series of observations during the time that the sediment is sinking through a depth of 1,000 metres (approximately from 3,000 to 4,000 metres depth) is as follows:

Open ocean ("Challenger" results)	44·2
Average of all areas ("Valdivia" results)	27·8
Bay of Bengal (Alcock's results)	25·4
Malay Archipelago ("Siboga" results)	22·1

The actual rate of solution at depths below 3,000 metres is therefore highest in the open ocean, that in the Bay of Bengal being intermediate between it and the rate found in an enclosed basin, such as the East Indian Archipelago.

If the rate of solution of calcium carbonate in these depths in the great oceans, is, as has hitherto been believed, largely dependent on the supply of oxygen and carbon dioxide by slow-moving bottom currents that gradually transfer masses of water from the superficial layers of the Polar regions to the bottom layers in the neighbourhood of the equator, then the above results are in exact accordance with what one would expect. In the case of the two great oceans, Atlantic and Pacific, that stretch from Pole to Pole, the two currents coming respectively from the Arctic and Antarctic regions will meet each other in or near the equator. In the case of the Indian Ocean we have, however, only the Antarctic flow (Chart VI). Carpenter (1887, p. 231) when discussing the temperature of the deeper waters of the Bay of Bengal remarks that "the submarine inflow which must come from the southward to make up for the great evaporation of the bay, is therefore, probably uniform in temperature and widely spread," and a study of the bottom temperatures in this area, recorded by the "Investigator" and other vessels, indicates that this Antarctic current extends to approximately Lat. 10°N. The supply of oxygen and carbon dioxide in the deeper layers of the Bay will thus be to a certain degree maintained, though, as the strength of the current gradually diminishes and the direction of the flow becomes directed towards the surface, the supply of these gases will not be as great as in the depths of the open ocean, though it will be greater than in waters from equal depths in a land-locked basin, where no such bottom current can penetrate and where the supply of carbon dioxide must depend either on diffusion from above or on convection currents, as has been suggested by Molengraaff (1922, p. 349). On the other hand the higher temperatures that exist in the deep waters of these land locked seas will tend to cause a higher rate of solution of calcium carbonate than in the much colder waters of the open ocean, but at the same time they will tend to expedite the rate at which particles will sink owing to the lesser increase in viscosity, and thus diminish the amount of solution taking place.

It appears, therefore, from the above considerations that the character of the bottom-deposits in these land-locked areas depends far less upon any increase in the rate of solution of calcium carbonate from the deposit itself, than on the chemical and physical conditions of the upper water-levels and the consequent altered character of the plankton throughout areas where the influx of numerous rivers and streams causes a contamination of the normal oceanic water.

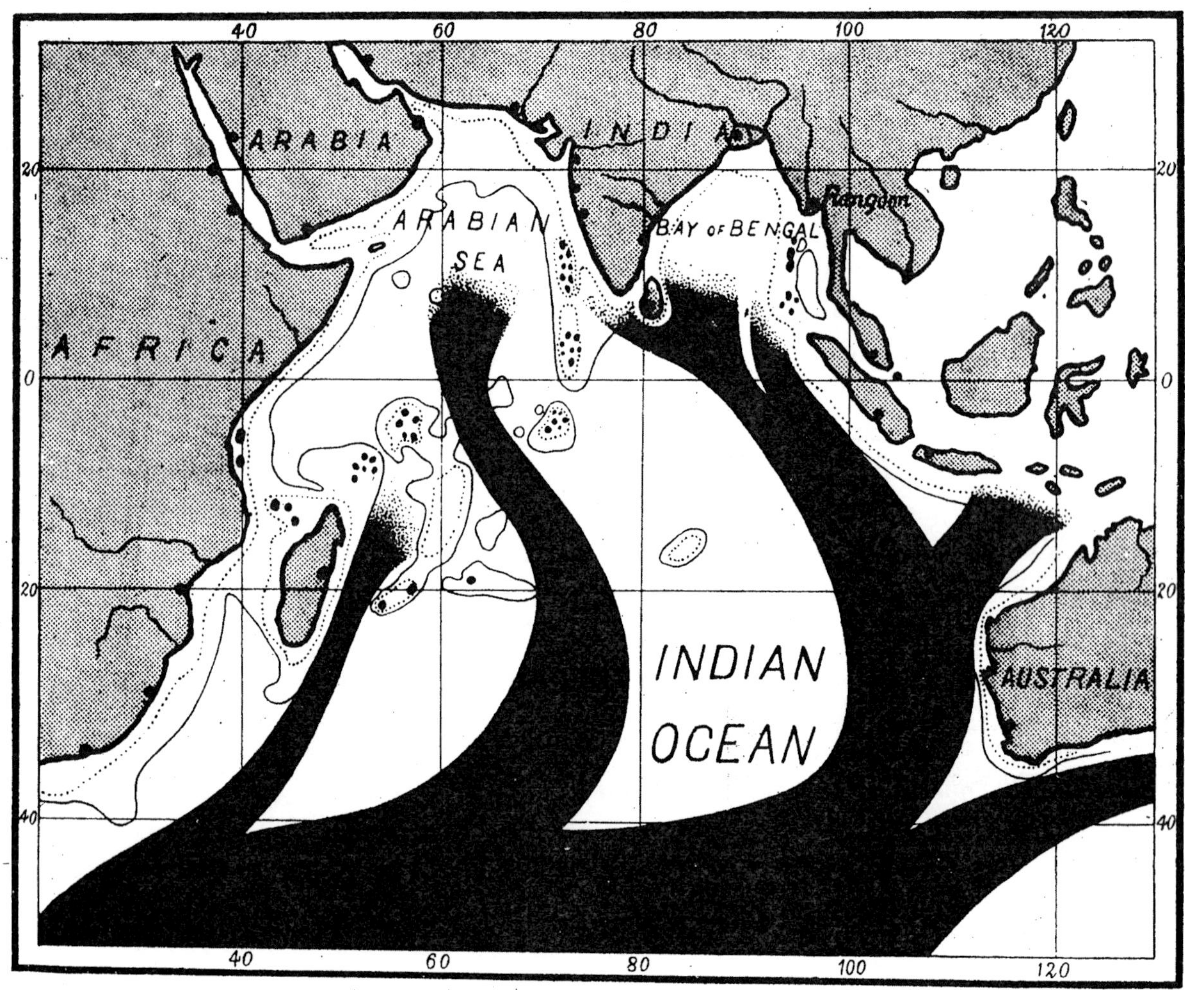

CHART VI.

Chart of the Indian Ocean showing the probable distribution of the bottom-drift of cold Antarctic water.

In conclusion I must tender my thanks to the authorities of the India Office, London, for allowing me to take a tracing of Horsburgh's original chart of the Bay of Bengal, and to Colonel A. Alcock, F.R.S., for so kindly placing his records of analyses at my disposal.

BIBLIOGRAPHY.

1. Alcock, A., 1898 .. .. "A Summary of the Deep-Sea Zoological work of the Royal Indian Marine Survey Ship *Investigator* from 1884 to 1897." *Sci. Mem. Med. Officers Army India.* Part xi. Calcutta.

2. Böggild, O. B., 1916 .. .. "Meeresgrundproben der Siboga Expedition." *Siboga Expeditie.* Monograph LXV. Leiden.

3. Buchanan, J. Y., 1887 .. .. "On the land slopes separating Continents and Ocean Basins." *The Scottish Geographical Magazine,* Vol. III, p. 217. Edinburgh.

4. Buchanan, J. Y., 1891 .. .. "On the composition of some Deep Sea Deposits from the Mediterranean."
Proc. Royal Soc., Edinburgh, Vol. XVIII, p. 131. Edinburgh.

5. Carpenter. A., 1885 .. .. "The Swatch of No. Ground"
Journal Asiatic Soc., Bengal. Vol. LIV. (New series) Pt. II. No. 2. p. 125. Calcutta.

6. Carpenter, A., 1887 .. .. "The mean Temperature of the Deep Waters of the Bay of Bengal." Natural History Notes from H.M's Indian Marine Survey Steamer *Investigator,* No. 8.
Journ. Asiatic Soc., Bengal. Vol. LVI, Pt. II, p. 230. Calcutta

7. Fergusson, I., 1863 .. .. "On recent changes in the Delta of the Ganges."
Quart. Journal Geol. Soc., London, Vol. XIX, p. 321. London.

8. Krummel, O., 1907 .. .. "Handbuch der Ozeanographie," Vol. I. Stuttgart.

9. Molengraaff, G. A. F., 1921 .. "Modern Deep Sea Research in the East Indian Archipelago.'
The Geographical Journal, Vol. LVII, No. 2, p. 95. London.

10. Molengraaff, G. A. F., 1922 .. "De Zeeën van Nederlandsch Oost.-Indie." (Geologie.)
Uitgegeven door het Koninklijk Nederlandsch Aardrijkskundig Genootschap. Leiden.

11. Murray, J., 1898 .. .. "On the Annual Range of Temperature in the Surface Waters of the Ocean and its Relation to other Oceanographical Phenomena."
The Geographical Journal, Vol. XII. No. 2. London.

12. Murray, J., 1910 .. .. "On the Depth and Marine Deposits of the Indian Ocean"
Trans. Linn. Soc., London, 2nd Series, Zoology. Vol. XIII, p. 355. London.

13. Murray, J. and Hjort, J., 1912 .. "The depths of the Ocean." London.

14. Murray, J. and Irvine, 1890 .. "On Coral Reefs and other Carbonate of lime formations in Modern Seas."
Proc. Royal Soc., Edinburgh, Vol. XVII, p. 79. Edinburgh.

15. " " " 1891 .. "On Silica and Siliceous remains of organisms in Modern Seas."
Proc. Royal Soc., Edinburgh, Vol. XVIII, p. 229. Edinburgh.

16. Murray, J. and Philippi, E., 1908 .. "Die Grundproben der Deutschen Tiefsee Expedition."
Wissenschaftliche Ergebnisse der Deutschen Tiefsee Expedition auf dem Dampfer "Valdivia" 1898-1899. Vol. X. Jena.

17. Murray, J., and Renard, A. F., 1891 "Report of Deep Sea Deposits."
Scientific results of the voyage of H.M.S. "Challenger," 1873–1876. Paris, London.

18. Richard, J., 1907 .. .. "Oceanographie." Paris.

19. Thorpe, G. F., 1905 .. .. "The submarine Cañons of the Ganges and Indus."
The Geographical Journal, Vol. XXVI, p. 568. London.

20. Tipper, G. H., 1911 .. .. "The Geology of the Andaman Islands, with references to the Nicobars."
Mem. Geol. Survey, India, Vol. XXXV, pt. 4. Calcutta.

21 Weber, M., 1902 .. .. "Introduction et description de l'expedition."
Siboga Expeditie. Leiden.

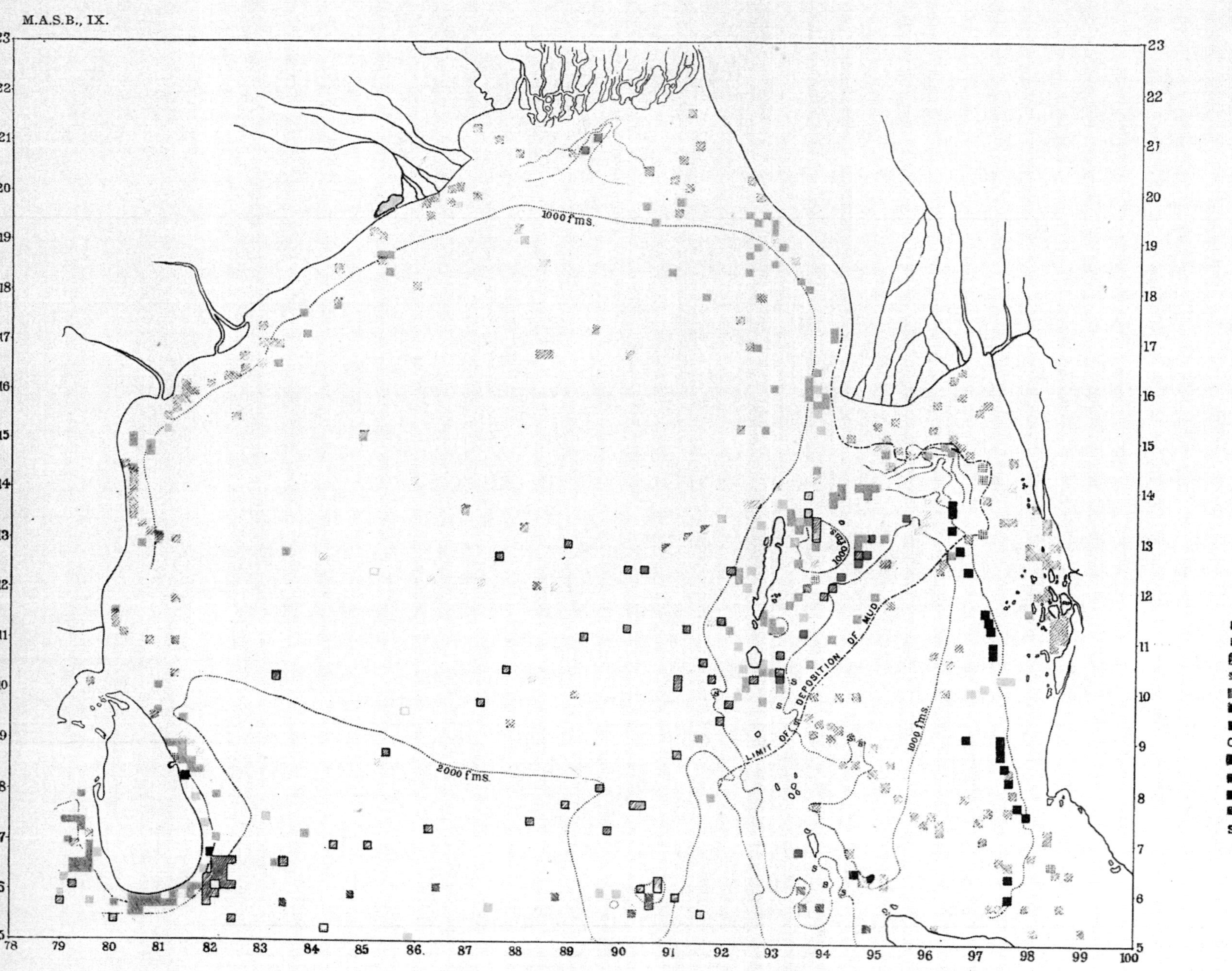

Chart IV.

Showing the distribution of bottom deposits in the Bay of Bengal and Andaman Sea. Each area is a 10′ square; in most areas only a single sample has been taken, but some include two or even three.

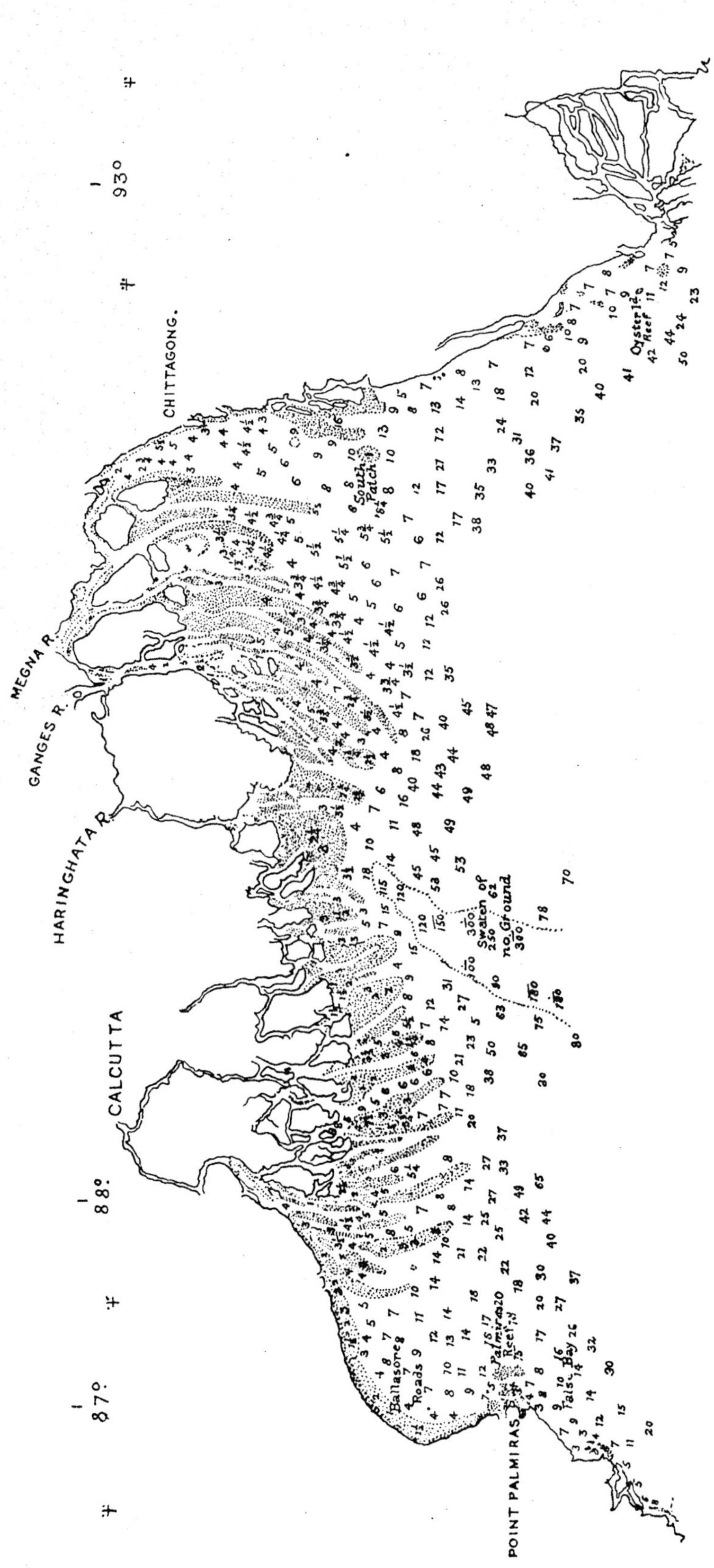

CHART V.

The delta of the Ganges and Brahmaputra Rivers, showing the trend of the mud-banks. A tracing taken from Horsburgh's original chart of the Bay of Bengal.

MARITIME METEOROLOGY IN INDIAN SEAS.

By R. B. Seymour Sewell, M.A., F.A.S.B., F.L.S., F.Z.S., Major. I.M.S.,
Director, Zoological Survey of India.

III.—MARITIME METEOROLOGY IN INDIAN SEAS.

At a very early stage in my investigations regarding the density and salinity of the surface waters in Indian seas it became clear to me that I should have to extend the scope of my studies considerably beyond the ordinary limits of hydrography. In the introduction at the commencement of this series of papers (*vide supra*, p. 1) I remarked that, "in any one year from month to month changes are taking place in the sea-water in accordance with the rhythm of the seasons, while from day to day and hour to hour variations are occurring in accordance, apparently, with influences that have their inception quite outside the waters themselves." The surface water of the ocean is continually undergoing changes in its physical and chemical conditions owing to the combined action of a number of agencies: for instance, rainfall, not only over the sea but also over the land, directly or indirectly causes a dilution of the upper layers and a consequent decrease in salinity; while evaporation will have the exactly opposite effect. Again the rate at which evaporation takes place is influenced by several factors, among which may be cited:—

(1) Barometric Pressure.
(2) Humidity of the atmosphere.
(3) Temperature.
(4) Wind.
and (5) The salinity of the sea-water itself.

Consequently, every variation, whether seasonal or diurnal, in any of these factors will tend to produce corresponding changes in the specific gravity and salinity of the surface water.

The primary function of the Surgeon-Naturalist is to investigate the fauna of the region in which the "Investigator" is working and, in consequence, my observations on the sea-water are not as complete as one might wish, but at least they indicate some of the changes that are from time to time taking place in different areas. It soon became evident that, in order to understand and explain the various changes that I observed, it was essential that my investigations should be extended so as to include the atmosphere, and, before commencing an account of my observations regarding the changes in temperature and salinity of the surface waters in Indian regions, it is necessary first to consider the meteorological conditions present in these areas. I was fortunately able to enlist the willing co-operation of the Assistant-Surgeons, who have each season been appointed to the "Investigator" to take charge of the health of the crew, and they, in succession, very kindly undertook to carry on all or most of the meteorological observations for me; and I should like to place on record my very sincere thanks to all who assisted me in this branch of research.

The R.I.M.S. "Investigator" is not specially fitted out for Meteorological work and I had, therefore, to do the best I could and trust, that, though isolated observations might show a moderate degree of error, the results obtained were of a sufficient degree of accuracy to justify one in placing some reliance, at any rate, on the *average* results.

The Thermometers—both maximum-minimum and wet and dry bulb—were enclosed in a white-painted, latticed wooden box, of the usual pattern, which was attached to the chart-room on the bridge: since double canvas awnings were kept spread throughout the whole survey-season, the box was thus completely sheltered from the direct sun's rays except for short periods during the very early morning and late evening. It has been shown (*vide* the Marine Observer, 1924, pp. 19 and 147) that simultaneous records taken by means of fixed and portable thermometers on board steam-ships exhibit a considerable difference, and that the fixed thermometer usually registers higher than the portable one. It has been argued that this is due to the influence on the fixed thermometer of heat from the engine room, the funnel, etc., and this is to some extent borne out by the fact that when the wind is astern the portable thermometer gives a lower reading than the fixed thermometer during the day, but it entirely fails to account for the fact that the fixed thermometer may register a *lower* temperature than the portable one during the night, as is shown to occur. Again, with the wind ahead the portable thermometer may register higher than the fixed one. From the accounts given by Captain Campos (1924 (*a*) and (*b*)) of the experiments on the cable ship "Colonia," there seems to be no very definite rule regarding the variation observed. In the first voyage, when the wind was ahead from any direction between two points on either bow the portable screen actually read 0·2° *more* than the fixed screen, whereas in the second voyage with the wind in the same quarter it read 0·3° *less*. Again, in the first voyage with the wind from any direction between 2 points forward or aft of the starboard-beam the portable thermometer read 0·3° *more* and in the second voyage 0·2° *less* than the fixed thermometer.

It must be remembered that a portable screen placed where the wind is strongest and on the windward side of a ship is likely to be more affected by spin-drift. No matter how efficient a latticed case may be the very fine particles of moisture thrown up by the waves and swept along by the wind will penetrate into the case and settling on the thermometers will tend to cause the dry bulb to act as a wet bulb, and therefore to register less than the true air-temperature. Without further experiments one is not, I think, justified in regarding records taken by a fixed thermometer as entirely unreliable; and, moreover, such observations have this advantage, that they can be compared with records taken in previous years by other ships, which would not be the case with observations taken by means of thermometers enclosed in a portable screen.

The amount of water-vapour present and the relative humidity of the atmosphere have been calculated from the readings of the wet and dry bulb.

The Barometer, of the usual ships pattern, is, on the "Investigator," suspended inside the chart-room on the bridge, and the index-error of the instrument was obtained by comparison with the standard barometer in the observatory at Colombo, Ceylon.

The force of the wind and its direction were recorded by one of the ships staff, and, as is customary, the strength is given in Beaufort's notation: during the first two years records were made at four-hourly intervals; but since 1922 observations were taken every two hours throughout the day and night.

On each successive voyage to and from the Survey Ground in 1921–25 the records of the sea and air-temperature and the height of the barometer have been taken at four-hourly intervals; during the day time the records were usually taken by me personally, and at midnight and 4 a.m. the observations were continued by the officer of the watch. During one season, 1922-23, extra observations were taken at 10 a.m. and 10 p.m. so as to coincide as nearly as possible with the maxima of the barometric pressure and thus provide a contrast to the readings taken at 4 a.m. and 4 p.m., when the barometric pressure is at or near its lowest phase. On the Survey Ground it was found to be impossible to arrange for both day and night observations except on special occasions. Records of the sea and air temperature were, therefore, taken at short intervals, namely two-hourly, during the day from 6 p.m. to 8 p.m. and during seasons 1923–25 a record of the wet bulb thermometer was added to those previously taken, so as to give one an idea of the changes taking place in the relative humidity of the atmosphere.

It must be pointed out that, since the ship is continually changing her position except when on the survey ground, no two observations over the open sea have been taken in the same place. The whole series of records lies within a belt extending from the Equator to 15° N. Latitude and between 70° and 95° E. Longitude. In the Bay of Bengal the various positions at which observations were taken are all included in a belt between Ceylon on the west and the Andaman Islands on the East; my own observations in this area were taken only in the months of October, January, March, and April; fortunately the "Valdivia" crossed the same belt during the month of February, thus furnishing data in an additional month, and Dallas (1886) has given an account of the meterorology of an area lying a little further to the south in a 4° square, lying between 2° and 6° North and 86° and 90° East, that is very useful for the purpose of comparison.

My observations in the Andaman Sea only cover the north-west portion of the basin, and especially that part of it lying along the route from the central group of the Nicobar Islands to Port Blair. At periodic intervals during those survey-seasons, in which the "Investigator" was engaged in surveying in and around the central group of islands, she had to return to Port Blair for recoaling, etc., at intervals of approximately three weeks, and on most occasions a series of observations were carried out during the voyages to and from port. These cover the months from October to February. Although in the present series of papers I have confined my attention in the main to conditions existing in the Bay of Bengal and Andaman Sea, I have, in the present paper, for the purpose of comparison included the results obtained in the Laccadive Sea to the west of the Indian Peninsula. My observations in this latter region are somewhat scattered and include those taken on the run up or down the west coast between Colombo and Bombay as well as those on several voyages westward towards the Maldive Islands.

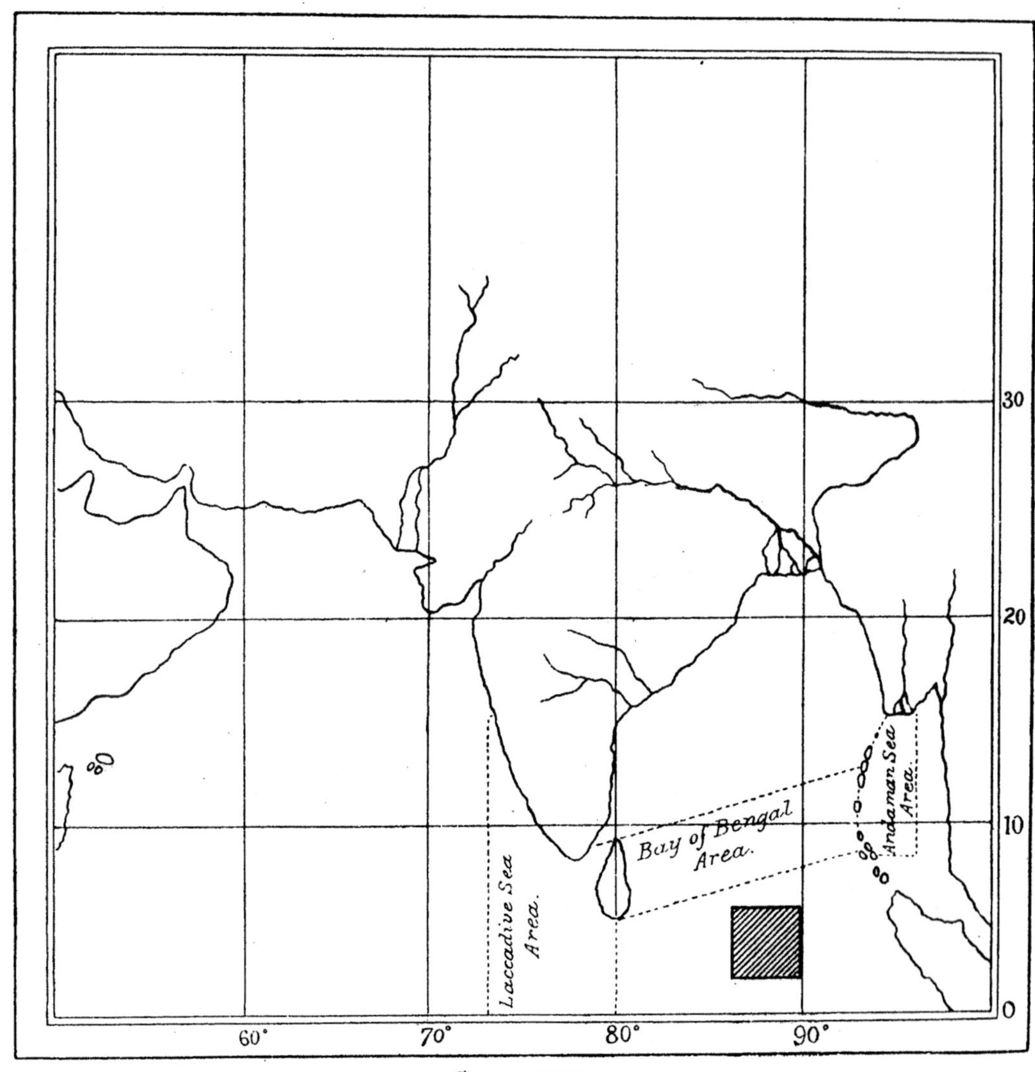

CHART VII.

Sketch-map showing the areas in which observations have been taken.

The series, however, in this area covers the whole period of the survey-season from October to May. A few observations have been taken off the south coast of Ceylon, which, strictly speaking, belongs neither to the Laccadive Sea nor the Bay of Bengal; as they are, however, too few to form a reliable group by themselves I have included those taken to the east of Longitude 80° in the Bay of Bengal series, and those to the west in the Laccadive Sea results. Dallas (1894) has published a detailed account of the general meteorological conditions throughout the year in the Arabian Sea and Persian Gulf. He has divided this area into eight squares each of 10°, and has treated each square separately. The area of the Laccadive Sea covered by my observations does not coincide with either of his squares but roughly overlaps squares 29 (Lat. 0°-10° N. Long. 70°-80°E.) and 65 (Lat. 10°-20°N. Long. 70°-80°E) in about equal proportions.

I have indicated the areas covered by the "Investigator" in the accompanying sketch map (Chart VII) and have also shown the position of the area, investigated by Dallas, to the south of the Bay of Bengal.

Seasonal variation in the air-temperature over the open waters of Indian Seas.

Since 1921 a careful record has been kept of the air-temperature at regular intervals throughout the day during each successive survey-season. These seasons, unfortunately, cover only the period of the year from the beginning of October to the end of April or early May; in consequence, I have been unable to trace with full completeness the annual changes that take place in the air-temperature over Indian waters. Fortunately the data published by Dallas (1886) for the area, mentioned above, to the south of the Bay of Bengal, and by the same author (1894) for the whole of the Arabian Sea, as well as that contained in the Annual Reports of the Marine Biologist to the Government of Ceylon, dealing with his investigations in the Gulf of Manaar, enable one to fill in the hiatus that would otherwise remain.

In the following Table 1, I have given the average temperature at four-hourly intervals throughout the day for each month of the survey-season and in each of the three divisions of the Indian waters that I have investigated. Eliot (1902, p. 61, Tables XXV and XXVI) discussing the results obtained at a number of meteorological stations in India, remarks that "the mean monthly temperatures in India have at most of these stations only one minimum and maximum in the course of the year. At a few of the stations there is a feeble secondary minimum during the rainy season, the minimum usually in August and the corresponding maximum in September or October." This second oscillation during the rainy season is clearly seen at Trivandrum, Bombay, and Rangoon, and to a slight extent at Aden, Chittagong, and Cuttack. It seems to be absent at Madras, but this is probably due to local causes. This double oscillation may, I think, be considered as characteristic of coastal areas, and one would naturally expect to find that a similar double rise and fall can be traced over the open water around the Indian Coasts. Dallas (1886, p. 52) has given the average air-temperatures recorded in each month throughout the year in the square area lying to the south of the Bay of Bengal, between Lat. 2°-6°N. and Long. 86°-90°E. He has, as is usual amongst meteorologists, given the temperatures in °F.; I have, therefore, converted these into °C, so that they can be compared with the observations taken by the "Investigator" and other vessels engaged in oceanographic work. He remarks (loc. cit., p. 47) "in tropical seas, the main meteorological changes of any one part are, as a rule, very fairly representative of a considerable area, so that it is hoped that the discussion of the small position, now selected, may afford a trustworthy basis for estimating the changes that affect that part of the Indian Ocean which stretches from the Equator northward to Ceylon and the Nicobars." The figures given by Dallas may, I think, be taken as correctly representing the conditions that prevail on the average in this area, but a comparison with his subsequent account of the meteorology of the Arabian Sea, and with the results obtained by me on the "Investigator" in different parts of Indian waters reveals certain differences that are worthy of study.

Month.	Time of Day.						Average Tempera-ture.	Average Daily Range.	
	4 A.M.	8 A.M.	12 noon.	4 P.M.	8 P.M.	12 mid-night.			
1. *Laccadive Sea.*	°C	°C	°C	°C	°C	°C	°C	°C	
October	26·0	25·9	27·1	**28·0**	26·8	26·6	26·73	2·1	
November ..	27·4	26·6	27·5	**27·7**	27·7	27·2	27·35	1·1	
December ..	27·3	27·5	**28·1**	27·9	27·8	27·2	27·63	0·9	
January.. ..	26·4	27·1	27·2	**28·3**	27·1	26·9	27·17	1·9	
February ..	26·1	26·7	**27·8**	27·4	26·6	26·5	26·85	1·7	
March	27·6	28·5	**28·9**	28·4	28·1	27·9	28·23	1·3	
April	28·5	29·1	29·6	**29·7**	29·3	29·0	29·20	1·2	
May	27·7	28·6	28·2	**28·6**	28·3	28·2	28·27	0·9	
2. *Bay of Bengal.*									
October	26·6	27·1	27·8	**27·9**	27·4	27·2	27·33	1·3	
November ..	..	..	..	..	..	..	..	..	
December ..	..	..	..	..	..	..	..	..	
January.. ..	26·4	27·3	**27·6**	27·3	26·9	26·7	27·03	1·2	
February ..	26·5	26·8	**27·4**	27·3	26·6	26·6	26·85	0·9	"Valdivia."
March	27·4	27·6	**28·8**	28·5	28·1	27·7	28·02	1·4	
April	28·1	28·7	29·7	**29·9**	29·3	28·9	29·10	1·8	
3. *Andaman Sea.*									
October	26·6	27·5	28·0	**28·3**	27·0	26·4	27·30	1·9	
November ..	26·1	**27·0**	26·4	26·6	26·5	25·6	26·37	1·4	
December ..	26·8	26·9	**28·2**	27·0	26·9	27·2	27·17	1·4	
January.. ..	26·1	27·0	27·5	**28·3**	26·6	26·5	27·00	2·2	
February ..	26·4	27·4	**28·0**	27·6	26·8	26·7	27·15	1·6	

Table 1; giving the mean temperature of the air at different times of the day in each month of the Survey Season and in different areas.

The mean monthly temperatures throughout the whole area of the Arabian Sea exhibit a clear double oscillation during the year, but, as Dallas points out (1894, p. 11), the figures for June and August have had to be calculated owing to records for these months in certain parts of the Arabian Sea being unattainable. They are, therefore, not entirely reliable, and he has in consequence divided the area into two zones, a larger southern one extending from the Equator to Lat. 20°N, in which he has a complete series throughout the year, and a smaller northern one lying between Lat. 20°–30°N. The northern zone, in which the proximity of land must exercise a profound influence, does not now concern us. In the southern zone the average monthly readings, according to Dallas, "rise steadily to a well-defined maximum in May and descend steadily during the remainder of the year, though there is a slight irregularity in October and November introduced by the cessation of the rains." The use of the term 'irregularity' is, I think, unfortunate. Dallas' figures, which I have converted into degrees Centigrade, are given below in Table 2, and, to serve as a check to these observations, I have abstracted from the Ceylon Biological Reports for the years 1920–1922 the data given for all positions in the Gulf of Manaar over deep water, so as to exclude as far as possible land influence; while not giving a complete series these data enable one to plot a curve that covers the whole year.

Locality.	January.	February.	March.	April.	May.	June.	July.	August.	September.	October.	November.	December.
	°C	°C	°C	°C	°C	°C	°C	°C	°C	°C	°C	°C
Arabian Sea, 0–20°N Dallas).	*26·05*	26·39	27·22	28·50	**28·94**	27·78	27·00	26·78	*26·67*	**27·17**	26·89	26·39
Gulf of Manaar	*26·0*	..	28·2	**29·0**	..	..	25·0	*25·9*	..	27·1	**27·2**	..
Laccadive Sea (Investigator)	27·17	*26·85*	28·23	**29·20**	28·27	..	..	..	..	26·73	27·35	**27·63**

Table 2; showing the monthly mean air-temperature in different areas of Indian waters lying to the west of the Indian Peninsula.

If we compare my results for the Laccadive Sea with these other two series of observations, it is abundantly clear that, far from there being an 'irregularity' in the average monthly temperature during the latter part of the year, there is a very definite oscillation, the temperature rising from August or September to a second maximum and then falling to a minimum in January or February. The actual month in which the second maximum is reached may vary in different localities or in different years, but the rise appears to be invariably present. There is then, clearly, a double oscillation in the air temperature throughout the open waters to the west of India during the course of the year.

When we turn to the regions lying to the east of the Indian Peninsula we find that conditions at the close of the year are very different. In Table 3, given below, I have given Dallas' observations for the area to the south of the Bay of Bengal, and my own observations for the Andaman Sea. Unfortunately my records for the Bay of Bengal are incomplete in just those months that are necessary to complete the comparison.

Locality.	January.	February.	March.	April.	May.	June.	July.	August.	September.	October.	November.	December.
	°C	°C	°C	°C	°C	°C	°C	°C	°C	°C	°C	°C
2° – 6°N. 86° – 90°E (Dallas)	*27·11*	27·56	28·00	**28·67**	28·44	28·00	*27·67*	**28·00**	27·39	27·33	*27·06*	**27·28**
Andaman Sea (Investigator)	*27·00*	27·15	..	..	..	..	..	..	..	27·30	*26·37*	**27·17**

Table 3; showing the monthly mean air-temperature in two areas on the east side of the Indian Peninsula.

In both these series of observations we find that the temperature follows the same course. During the early part of the year the air-temperature appears to rise

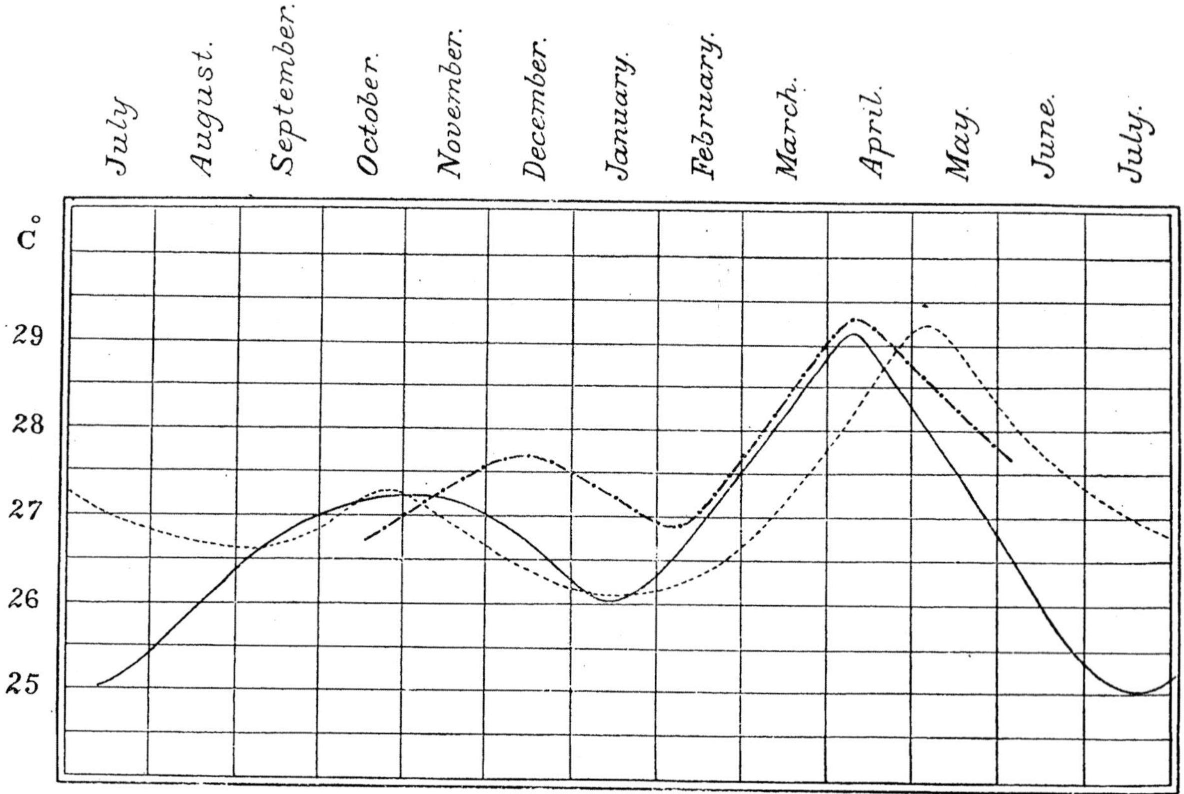

TEXT-FIG. 5.—Showing the annual oscillation of the monthly mean air-temperature on the W. side of India.

———— Gulf of Manaar.
.... Arabian Sea between 0°—20° N.
—·—·— Laccadive Sea ("Investigator").

steadily from January to April, and then falls to July. So far then, conditions in this area agree with those to the west of India; but from now on to the close of the year there is a marked difference. In this eastern area the temperature begins to rise again, as it does on the west, but we here find a second fall from August to November, followed by a rise, of only 0·22°C in the southern area of the Bay but as much as 0·8°C in the Andaman Sea, in December and a further fall to the minimum in January.

It appears then that we can divide the Indian seas into two areas, lying respectively to the west and east of the Indian Peninsula, in which conditions during the winter months are entirely different. In the first area, to the west, which includes the Arabian Sea, Laccadive Sea, and Gulf of Manaar the air temperature

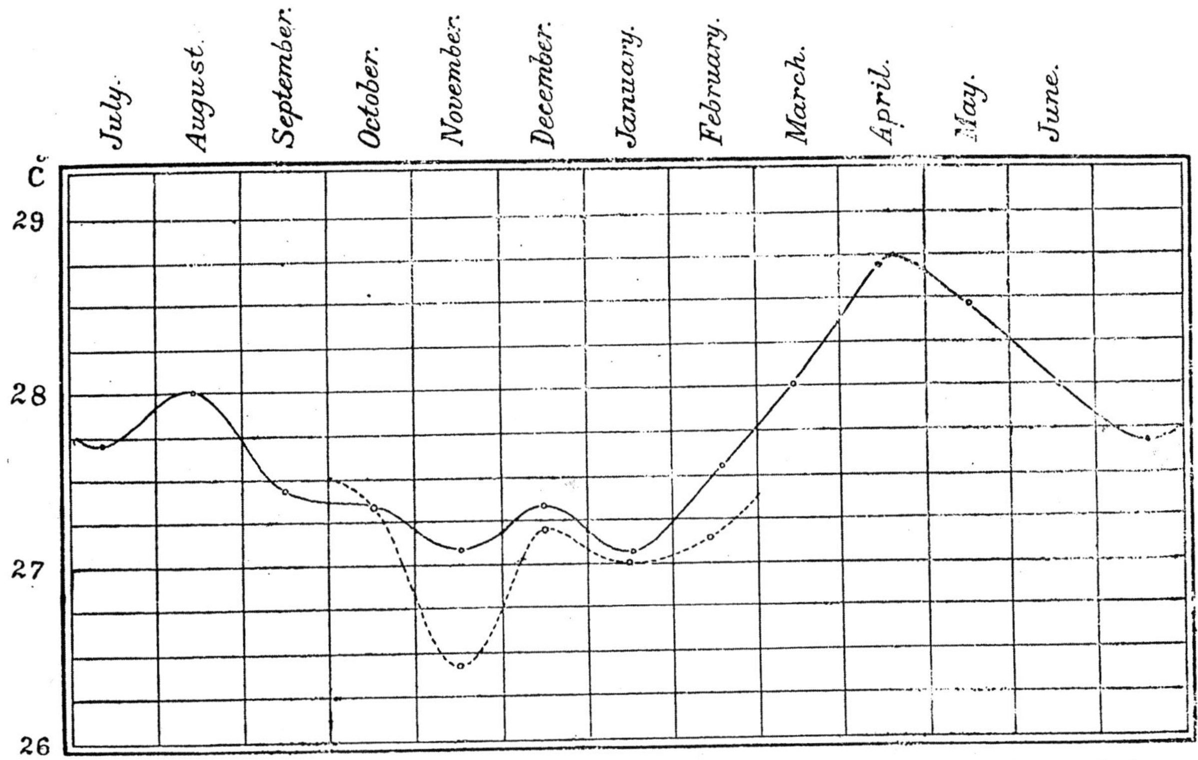

TEXT-FIG. 6.—Showing the seasonal variation in the monthly mean air-temperature in the Andaman Sea and an area to the S. and E. of the Bay of Bengal.
———— Area to S. and E. of Bay of Bengal (Dallas).
.... Andaman Sea area.
N.B.—The *vertical* scale is double that of the preceding figure.

exhibits a double oscillation during the year, as is shown by the curves in Text-fig. 5, in which I have plotted the three series of observations given above, and the manner in which the curves agree needs no comment.

This double oscillation of the air-temperature must, I think, be directly correlated (1) with the sun's declination which reaches its southerly maximum at the end of December, and (2) with the occurrence of the S.W. Monsoon in June and July.

On the east side, however, some additional factor superposes on this double oscillation of temperature an additional phase that exhibits an additional fall, having its minimum in November. In Text-fig. 6 I have plotted the series of observations, given above in Table 3, and here again the resemblance between the two curves is obvious.

The differences that are seen to exist between these two areas and those previously considered are, I think, correlated with the N.E. Monsoon, the effects of which appear to be felt to a much greater extent on the east side of the Peninsula than on the west. If this be so, the monsoon influence should be widely felt over the region to the east of India and we ought to get further evidence of its effect in other localities; and, moreover, the degree to which it is felt, should vary from year to year in accordance with the strength or otherwise of the monsoon.

Seasonal variation in the air-temperature of the coastal region of the Andaman Sea.

Throughout the period occupied by the survey of Nankauri Harbour, in the central group of the Nicobar Islands, during survey seasons 1921-1922 and 1922-1923, and again in March, 1925, a careful record of both sea- and air-temperatures was taken at intervals during the day, whenever the "Investigator" was on the survey ground. Unfortunately, it was found to be impossible to arrange satisfactorily for observations to be taken at night, and in consequence my records only cover the period of the day from 6 a.m. to 8 p.m., or in March, 1925, to 10 p.m. During the first period from November, 1921, to March, 1922, observations were taken at 6, 8, 10, and 12 in the forenoon and 3, 6, and 8 in the afternoon; during the second period, from October, 1922, to January, 1923, observations were taken at two-hourly intervals throughout the day from 6 a.m. to 8 p.m.; and the results obtained are given *in extenso* in Appendix I. In order to enable me to compare and contrast my results with the conditions existing in other parts of the Andaman Sea, a register was kept of the air-temperatures recorded *on the same days* at Victoria Point in South Burma on the east side of the Andaman Basin, and at Port Blair in the Andamans on the west side but some 240 miles further to the north of Nankauri Harbour.

In Table 4 I have given the mean temperature of the air in different months at each station.

Month.	1921-22. Victoria Point, S. Burma.	1921-22. Port Blair, Andamans.	1921-22. Nankauri Harbour, Nicobars.	1922-23. Victoria Point, S. Burma.	1922-23. Port Blair, Andamans.	1922-23. Nankauri Harbour, Nicobars.
	°C	°C	°C	°C	°C	°C
October	..	..	..	27·4	27·8	26·7
November	26·8	26·9	27·2	25·7	26·9	26·6
December	27·6	27·4	27·0	26·2	26·6	26·3
January	27·3	26·4	26·4	26·6	26·1	26·3
February	27·8	26·4	26·5	..	..	..
March	27·0	26·6	27·7	..	..	..

Table 4; showing the mean monthly air-temperature at different stations around the Andaman Sea basin in 1921-22 and 1922-23.

At Nankauri Harbour in both series of observations the air-temperature shows a steady fall during the winter months followed by a rise in the spring. The lowest average temperature occurs in January in 1922 and in December and January in the following winter season. At both the other stations, however, conditions are very different in the two seasons. At Victoria Point the temperature appears to oscillate from month to month in 1921-1922, whereas in 1922-1923 it falls from October to November and then rises steadily till January. Similarly, in 1921-1922 at Port Blair the temperature is found to rise from November to December, falls in January and rises again to March, whereas in 1922-1923 it falls steadily throughout the whole period of observation.

I have already called attention to the fact that during the winter months from October to February over the open waters on the western side of the Andaman Sea and in the area to the south and east of the Bay of Bengal, that was investigated by Dallas, the average air-temperature is found to vary in a somewhat peculiar manner, which I attribute to the effect of the N.E. Monsoon. If, now, we compare the results obtained at Victoria Point and Port Blair in 1921-1922 with these two other areas we see at once that they show exactly the same phenomenon, and in Text-fig. 7 I have plotted all four series of observations.

A comparison of these curves indicates that the cause of the lowering of air-temperature in November is most effective at Port Blair and on the west side of the Andaman Sea between Port Blair and Nankauri Harbour, and least effective in the area to the south and east of the Bay of Bengal, while Nankauri Harbour remains, apparently, entirely unaffected; furthermore, as this effect was found only in the season 1921-1922 and not in 1922-1923, it is clear that the cause is a variable one. The Indian Weather Review for 1921 remarks "In the months October to December the retreating monsoon gave rain fairly frequently in Burma and the Peninsula" and the corresponding report for 1922 states that "the retreating monsoon of the previous year, prevailed in the South of the Bay till the middle of February. The main features of the weather of the period October to December (1922) were an early retreat of the monsoon from north-east India and its vigorous activity in the south and centre of the Bay," while in January, 1923, "a revival of the monsoon occurred in the south of the Bay and gave rise to a shallow depression to the east of Ceylon on the 9th. With the disappearance of the depression on the 14th the monsoon weakened and withdrew from the Peninsula on the 15th."

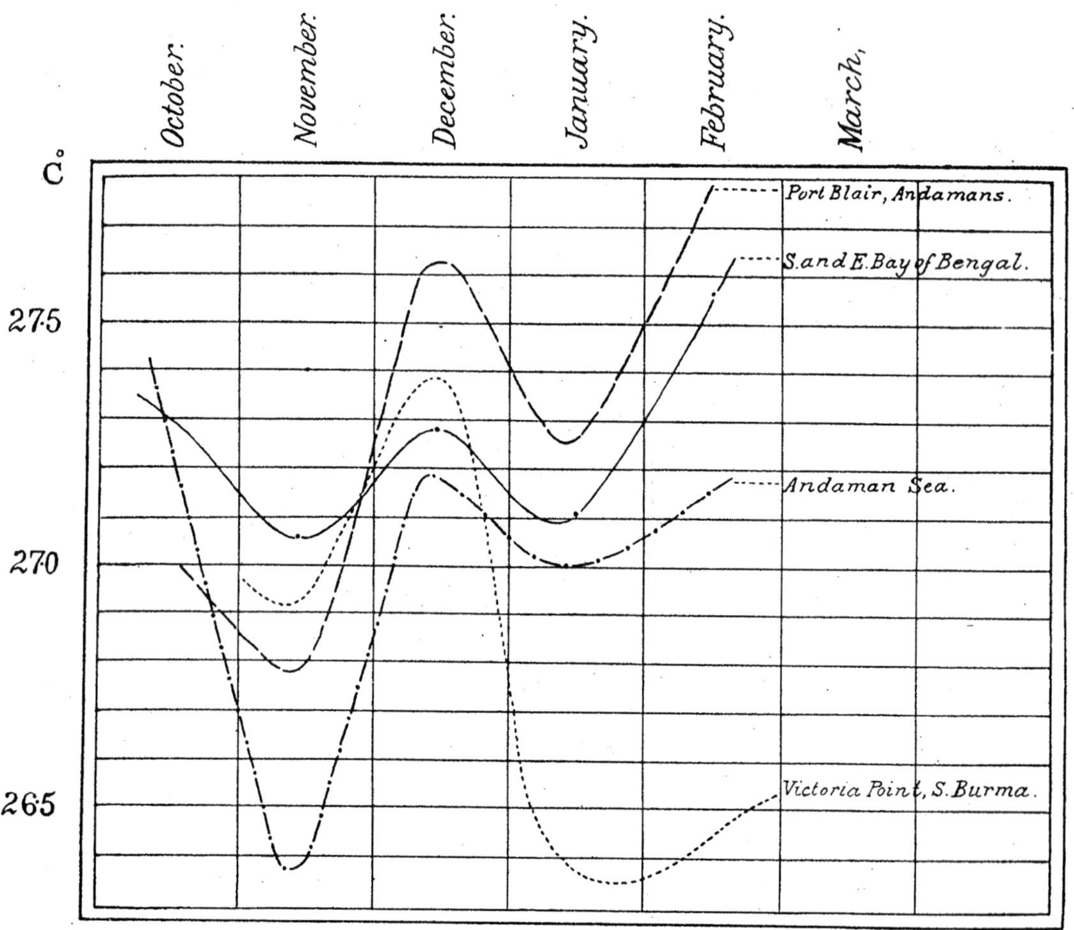

TEXT-FIG. 7.—Showing oscillations of air-temperature in the Andaman Sea and the South and East area of the Bay of Bengal in 1921-22.

There was a second temporary incursion of monsoon winds into the south of the Bay on the 22nd. The monsoon was, therefore, over considerably earlier in 1922-23 than in 1921-22 and a comparison of the rainfall in 1921 and 1922 shows that during the months of November and December this was far less in the former year. I have already (*vide supra* p. 62) put forward the view that the additional fall and subsequent rise in the mean air-temperature in the months of November and December in the regions lying to the east of the Indian Peninsula are attributable to the influence of the N.E. monsoon, and it is probable that the weaker monsoon in

1922 is sufficient to account for the difference in the air-temperatures around the Andaman Sea basin in the survey-season 1922-23, and the absence on this occasion of the usual subsidiary fluctuation.

The annual range of air-temperature over Indian waters is extremely small. As Eliot (1902, p. 65) has pointed out there are several methods of calculation by which the so-called annual range of temperature can be obtained, each of which will give different results. In any one year the range of temperature will probably fall much nearer to the figure given by the difference between the recorded highest and the lowest temperatures than to the difference between the lowest and highest mean monthly temperatures. Thus in 1923 the survey seasons cover the months of January to April and October to December, that is to say both the coldest and hottest periods of the year and the extreme temperatures recorded in each month are as follows:—

1923 Month.	Maximum Temp.	Minimum Temp.	Extreme monthly Range.
	°C	°C	°C
January	28·89	25·67	3·22
February	29·38	24·94	4·44
March	29·40	26·28	3·12
April	30·50	27·22	3·28
October	29·72	27·22	2·50
November	29·44	25·00	4·44
December	30·28	23·61	6·67

The increased range of air-temperature in December of this year is due to a cyclone, during which the minimum temperature of 23·61°C was recorded. If we include this figure we get an annual range of air-temperature from 23·61°C to 30·50°C or 6·89°C, and the next lowest temperature, that of 24·94 recorded under normal conditions during the month of February, gives an extreme range of 5·56°C. If, on the other hand, we take the annual range of temperature to be the difference between the lowest and highest values of the mean monthly temperature we get a very much lower figure. Dallas gives the following values for the annual range of temperature deduced in this latter manner for various areas in the Indian seas;

	°C.
Port Blair, Andamans	2·39
Nankauri Harbour	1·89
Whole area of Arabian Sea and Persian Gulf	1·68
Laccadive Sea	1·50
Arabian Sea between 0° and 10°N.	1·46
Central area of Laccadive Sea away from land	1·44

Eliot (1902, p. 65) for the coastal towns around the Indian Peninsula gave the following data:—

Aden	6·72
Madras	6·44
Bombay	6·33
Rangoon	5·72
Trivandrum	2·59

The results of my own observations on the air-temperature of Nankauri Harbour give a result considerably higher than the figure, given by Dallas, of 1·89°C. The variation of the mean monthly temperature in 1922, which is the only year in which I have observations concerning both the hottest and coldest months, namely January and March, of the survey-season, exhibits a range of 5·03°C, namely from 24·89°C the mean minimum temperature in January, to 29·92°C, the mean maximum temperature in March. If, however, we take the highest and lowest recorded temperatures in this year we find that the extreme annual range is one at least as great as 7·67°C, ranging from 30·56°C in November to 22·89°C in February; in March the maximum temperature recorded was 30·39°C, but it is probable that both these high temperatures would be exceeded in the month of April.

It is clear, therefore, that the annual range of air-temperature diminishes as we get away from the coastal areas into the open water of the ocean, and that in oceanic islands such as the Andamans and Nicobars the general conditions may tend to approximate either to those of the open sea or to those of a coastal region.

Daily variation of air-temperature over the open waters of Indian seas.

A study of the four-hourly observations, given above in Table I, of the temperature of the air over the open water in Indian seas in different months reveals several interesting and important facts. The first of these facts to which I will call attention is that the rise and fall in twenty-four hours is, as a rule, extremely small. Buchan (1889, p. 7), from the observations taken on board H.M.S. "Challenger," gives the average daily variation of the temperature of the air over the North Atlantic Ocean in about Lat. 30°N ; Long. 42°W as 1·8°C, while in the neighbourhood of the Equator the average range was 1·4°C in the Atlantic Ocean and only 1·2°C in the Pacific, or an average of 1·3°C throughout both areas. Throughout the whole area that I have investigated the difference between the average temperature at 4 a.m. and 4 p.m. is only 1·2°C. This figure, however, is probably somewhat lower than the total daily range, since, as Buchan showed, the lowest temperature of the air over the open sea occurs from 4–6 a.m., but the highest is found at or near 2 p.m. From the "Challenger" observations it can be seen that the average temperature at 4 p.m. is 0·1°C lower than that at 2 p.m.; applying this correction we arrive at an average daily range of temperature for the open waters of Indian seas of 1·3°C, which agrees exactly with the average for both Atlantic and Pacific Oceans and falls exactly between those given for each ocean respectively. If, however, we take the mean of the monthly averages, as observed on the R.I.M.S. "Investigator," in all three areas in Indian waters, we arrive at a slightly higher figure, namely 1·4°C; while Dallas, whose observations covered the whole period of the year, found that the average daily range of air-temperature in the south and east part of the Bay of Bengal was 1·5°C, and over the Arabian Sea between 0° and 20°N latitude it was 1·47°C.

The next point to be considered is the manner in which the daily range of temperature varies from month to month. Dallas (1886) has shown clearly that the average range may vary from 2·03°C in April to 1·0°C in May, and the question arises, are these variations due merely to an insufficient number of observations or do they exhibit any tendency to a periodic or seasonal oscillation? Dallas himself called attention to the very low daily range in the month of May and he attributed this to the occurrence during that month of increased rainfall and a consequent rapid lowering of the air temperature. A study of the average daily range of air-temperature in each month throughout Indian waters (*vide* Table 5) appears to me to indicate that we have an undoubted seasonal oscillation, though this may not coincide in different areas. In the Laccadive Sea area and in the Andaman Sea the daily range is high in October, decreases in November and December, is once again high in January and falls steadily to May: on the other hand in the Bay of Bengal area the daily range is lowest in February and rises steadily to April. Dallas (1894, p. 13) has compared the monthly average of the diurnal range of temperature with the mean temperature recorded during the same month in the six ten-degree squares in the Arabian Sea that lie south of Lat. 20°N., and he remarks "these figures appear to indicate that the smallest daily range is recorded on the days on which the mean temperature approaches most closely to the normal temperature, and that departures from the normal in either direction are accompanied with an increase in the amount of the daily range."

	January.	February.	March.	April.	May.	June.	July.	August.	September.	October.	November.	December.
	°C	°C	°C	°C	°C	°C	°C	°C	°C	°C	°C	°C
Arabian Sea .. .. 0° – 20°N (Dallas).	2·83	3·15	3·23	3·47	2·92	1·95	2·25	2·68	2·37	2·47	2·33	2·82
Laccadive Sea (*Investigator*) ..	1·9	1·7	1·3	1·2	0·9	..	..	..	..	2·1	1·1	0·9
Area south of Bay of Bengal (Dallas).	1·54	1·83	2·01	2·03	1·00	1·11	1·17	1·52	1·23	1·63	1·23	1·65
Bay of Bengal (*Investigator*) ..	1·2	0·9	1·4	1·8	..	..	..	..	..	1·3	..	..
Andaman Sea (*Investigator*) ..	2·2	1·6	..	..	..	..	..	..	..	1·9	1·4	1·4
Average	1·93	1·83	1·98	2·12	1·61	1·53	1·71	2·10	1·80	1·88	1·51	1·69
Computed values [1] ..	1·82	1·90	**1·95**	1·90	1·75	*1·67*	1·76	**1·88**	1·87	1·77	*1·69*	1·73

Table 5; showing the mean daily variation of air-temperature in different months and in different regions of Indian waters.

On the other hand a study of the mean diurnal range in all the series of observations at my disposal, including those given by Dallas as well as the "Investigator" observations, seems to me to indicate clearly that the variations observed can not be explained *in toto* on the above supposed relationship between the 'normal' temperature and the range of temperature. The variation in the

[1] These values have been computed in the usual way by the formula $\frac{a+4b+6c+4d+e}{16}$.

daily range of temperature shows an unmistakeable tendency towards a double oscillation during the year that is very similar to but does not quite coincide with the double oscillation in the monthly mean air-temperature that, as we have seen, occurs during the course of the year on the west side of India. In Table 5 I have given the mean daily variation of the air-temperature in each month in all series of observations, and have given both the average and the computed values from all the different series. The computed values clearly show a double oscillation during the year having its maxima in March and August. The process of computation, unless the rise and fall on each side of the maximum is at a uniform rate, will tend to throw the apex of the curve towards the side of the slower change. In the case of a slow rise of temperature and a rapid fall during the course of a year the apex of the computed curve will be earlier than the actual maximum, and, in the present case, computation has altered the oscillatory curve slightly by putting the first maximum in March, whereas most of the series of observations and the average show the maximum to occur in April. It seems fairly clear that the maximum daily range of temperature tends to occur at that period of the year when the sun is vertically overhead and when the sky is completely free from cloud; and, further, that the minimum ranges occur when the sun is at or near the extremes of its northerly or southerly declination, and when the sky is cloudy and there is a considerable rainfall.

Daily variation of air-temperature over the coastal region of the Andaman Sea.

As I have mentioned above (p. 62) during the survey of Nankauri Harbour a record was kept of the variations in the air-temperature in that region and also of the temperatures recorded at Victoria Point, Burma, and Port Blair, Andamans, on the same days. In the following Table 6 I have given the average daily range of air-temperature at all three stations at different months during the two seasons.

Month.	Survey Season. 1921-22. Nankauri Harbour, Nicobars.	Port Blair, Andaman Is.	Victoria Point, S. Burma.	Survey Season. 1922-23. Nankauri Harbour, Nicobars.	Port Blair, Andaman Is.	Victoria Point, S. Burma.
October	..	..	..	2·9	5·0	5·7
November	3·6	5·0	6·6	2·8	5·0	4·9
December	2·6	5·0	9·1	2·3	4·6	5·6
January	2·9	5·7	8·0	1·9	6·6	8·3
February	3·6	6·6	7·8	..	..	..
March	4·4	8·5	7·4	..	..	..

Table 6; the average diurnal range of air-temperature in each month at different stations round the Andaman Sea.

It is clear that the range of temperature varies considerably in the three stations; throughout the whole period the range is very much greater at Victoria Point than in Nankauri Harbour; and Port Blair exhibits a range that is intermediate between that of the other two stations with the exception of the months of March, 1922, and November, 1922; in this latter month it was slightly the highest, exceeding the range at Victoria Point by a fraction of a degree.

This difference in the daily range of air-temperature in these three areas is, undoubtedly, due to the varying proportions of land and sea. The proportion of the land is greatest at Victoria Point and least at Nankauri Harbour. In this latter locality we have a group of small islands separated from each other by channels and from other land by wide stretches of open water, and we should therefore expect to find that, as far as the daily variation of air-temperature is concerned, conditions approximate fairly closely to those found to exist over the open sea. Murray (1894, p. 114) has pointed out that during the cruise of the "Challenger" the mean daily range of the air-temperature over the open waters of the North-Atlantic Ocean, as determined from the observations taken on board during 126 days from March to August, 1873, and in April and May, 1876, in a mean position of Lat. 30°N.; Long. 42°W., was 1·78°C; but during 76 days when the "Challenger" was near land the mean daily range was 2·43°C. The period during which I was able to take observations in Nankauri Harbour covers 169 days, and the average range of temperature at the different stations during this time was as follows:—

	°C
Victoria Point, S. Burma	7·0
Port Blair, Andaman Is.	5·8
Nankauri Harbour, Nicobars	3·0
Open waters of Andaman Sea	1·7

I have already remarked that the range of air-temperature over the open waters of Indian seas appears to possess a seasonal variation, and the same is indicated by a study of the monthly variations over the coastal areas. In Text-fig. 8 I have plotted the mean daily variation in Port Blair and Nankauri Harbour and it is clear that in each year the fluctuation of the range follows the same general course in both areas.

A very similar result is obtained if we consider, not the mean daily range of temperature, but the extreme range recorded in each month. The actual temperatures recorded on the "Investigator" during the survey of Nankauri Harbour and the neighbouring waters are as follows:—

Year.	Month.	Air-temperature. Maximum. °C	Air-temperature. Minimum. °C	Extreme range of Temperature. °C
1921	November	32·56	23·33	9·23
	December	28·89	23·39	5·50
1922	January	28·33	23·33	5·00
	February	29·17	22·89	6·28
	March	30·39	25·11	5·28
..	..	..	..	..
	October	30·00	23·95	6·05
	November	30·56	23·17	7·39
	December	28·11	23·95	4·16
1923	January	27·39	24·61	2·78
..	..	..	..	..
1925	March	29·40	24·00	5·40

It is clear that the extreme range of temperature in any one month decrease, from a maximum in November to a minimum in January and then again increases a second maximum being reached in February or March.

We have already seen (*vide supra*, p. 57) that over the open sea in Indian waters the daily range of air-temperature exhibits a double oscillation during the year having one maximum in March or April and a second in August or September. Over land areas the daily range of temperature exhibits only a single oscillation during the year. In the following Table 7 I have given the average daily range in each month of the year for the provinces of Bombay and Bengal, for the Bay Islands and Nankauri Harbour and for the open sea. Throughout the whole series there is a steady rise in the range of variation at the commencement of the year, culminating usually in February over the land but in the Bay islands and over the open sea in March. Over the land areas the range then steadily falls to its minimum in August; but over the open sea this is reached in June. The Bay Islands, as is shown by the data given in the Indian Weather Reports, show an intermediate condition, as the minimum there occurs in July. From August to the end of the year over land areas the range of air-temperature again steadily increases, but over the open sea, the range increases from June to August and September and this is followed by a fall in November, and we can trace the same change in the range of air-temperature in the Bay Islands, though in this locality the maximum and minimum occur as a rule a month later. Eliot (1902, Table XXXIIII, p. 77) has given the mean diurnal range of air-temperature in the different months of the year at a number of stations in and around India and his figures demonstrate clearly that even in the coastal regions the diurnal range exhibits only a single oscillation having a maximum in January or

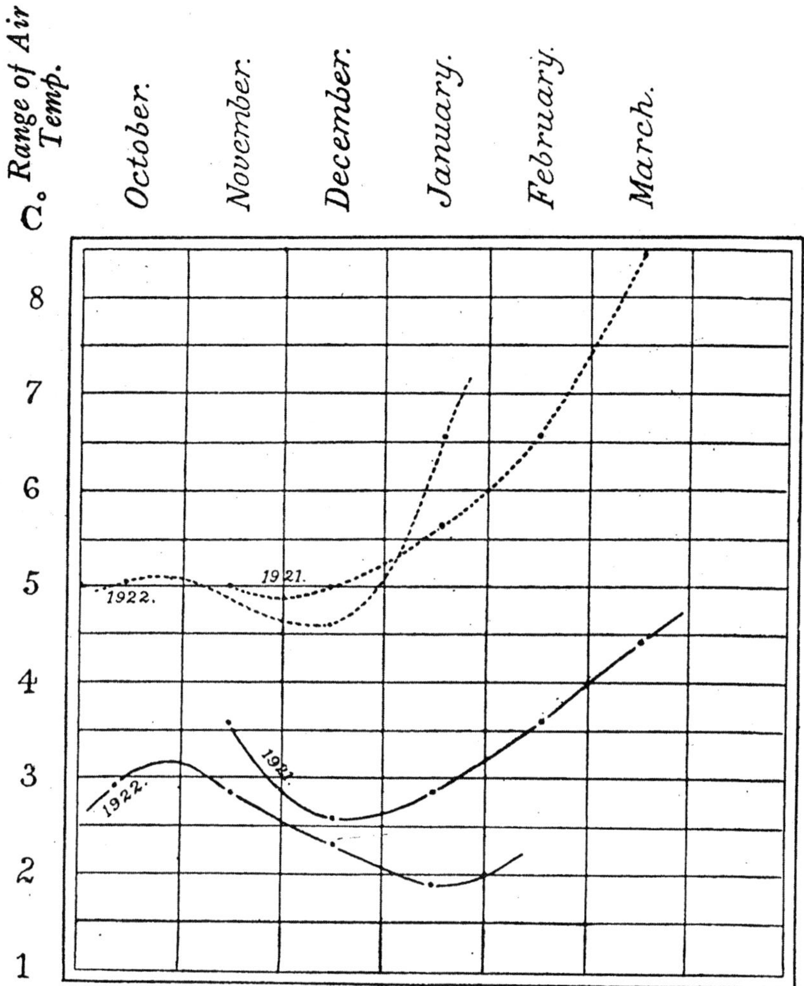

TEXT-FIG. 8.—Showing the mean range of air-temperature during the winter months in Port Blair and Nankauri Harbour.

———— Range of air-temperature in Nankauri Harbour

.... ,, ,, ,, ,, ,, Port Blair.

February, usually the latter month, and a minimum usually in July but ranging from June to August. The sole exception that he gives to this rule is Madras, where the diurnal range is highest in February (10·3°C), falls somewhat to April (8·7°C), rises again to June (10·0°C) and then falls steadily to a minimum in November (7·1°C).

Locality.	January.	February.	March.	April.	May.	June.	July.	August.	September.	October.	November.	December.
Bombay	14·4	**14·7**	14·3	12·8	11·8	8·7	6·4	*6·3*	7·8	10·3	13·7	14·2
Bengal	12·7	**13·2**	12·7	11·3	9·3	6·9	5·7	*5·4*	5·8	7·8	10·5	12·4
Bay Islands	6·0	7·1	**7·9**	7·5	5·7	4·5	*4·4*	4·4	4·6	**5·4**	5·3	*5·2*
Nankauri Harbour, 1880-1885	*4·8*	5·4	5·7	**6·1**	6·0	5·8	*5·8*	5·8	**6·3**	6·2	5·6	5·7
Open Waters of Indian Seas[1]	1·82	1·90	**1·95**	1·90	1·75	*1·67*	1·76	**1·88**	1·87	1·77	*1·69*	1·73

Table 7; giving the average daily range of air-temperature in each month in different areas.

It seems clear, then, that the double oscillation in the range of air-temperature throughout the year is a characteristic of the open seas in Indian waters, as opposed to the single oscillation that is characteristic of the land areas, and in the case of oceanic islands, such as the Andamans and Nicobars, it is the marine element that preponderates and hence in these areas also a double oscillation is usually found to occur.

If, however, we study the mean daily range of the air-temperature during the winter months at Nankauri Harbour, as is shown by the figures in Table 7, which have been calculated from the data given in "The Report of the Meteorology of India" for the last five years, namely 1880–1885, during which regular observations were taken at this station, one finds that there is a cessation of the usual fall of daily range and even a slight tendency towards an actual increase in the month of December. A comparison with the curve of mean monthly temperature in the regions round the Andaman Sea basin (*vide* Text-fig. 7, p. 64) shows that in both cases there is the same tendency towards a third oscillation at about the end of the year. In both series there is a tendency towards a fall from October to November, a rise, though only very slight in the case of the mean range of temperature, in December, and a further fall in January, followed by a rise in February. I have already pointed out that this oscillation of the mean monthly temperature is, in all probability, caused by the onset of the N.E. Monsoon in November, and it would appear that the same factor may also influence the daily range of Temperature.

Variation in the Time of Occurrence of Maximum Temperature during the Day and Subsidiary Oscillations.

We now come to the consideration of the time of day at which the air-temperature attains its maximum. It is generally stated that this occurs at or near 2 p.m. and this is certainly the case when one is considering the average variation in temperature observed over a long period of time. The results given by Buchan (1889, p. 7) of the analysis of the temperatures in the North Atlantic Ocean on

[1] These results have been computed from the actual average of all observations by the application of the formula $\frac{a+4b+6c+4d+e}{16}$

126 days, and those given by Dallas (1894, p. 13) of the observations with which he was dealing, in the Arabia Sea, agree closely as regards the time of occurrence of maximum temperature. The diurnal march of temperature in the two series of observations is as follows:—

Time of day.	2 A.M.	4	6	8	10	Noon	2 P.M.	4	6	8	10	Mid-night.
Variation from the Mean	°c	°c	°c	°c	°c	°c	°c	°c	°c	°c	°c	°c
(1) Atlantic Ocean (Bucan)	−0·63	−0·77	−0·78	−0·12	+0·43	+0·81	+1·00	+0·87	+0·41	−0·17	−0·44	−0·57
(2) Arabian Sea .. (Dallas)	−0·61	−0·71	−0·51	−0·08	+0·34	+0·67	+0·85	+0·71	+0·27	−0·15	−0·33	−0·43

In both series the time of occurrence of maximum temperature falls at 2 p.m.; but when one considers that the observations were taken in different months one at once sees that there is a very considerable degree of difference in the two series, which suggests that the march of temperature during the day is by no means such a fixed and regular occurrence as is indicated by the average two-hourly values.

Although Dallas (1884) in his first paper, dealing with the meteorological conditions in a small area to the south of the Bay of Bengal, makes no mention of the phenomenon, the table that he gives of the probable diurnal march of the air-temperature in each month, as deduced from the four-hourly observations, clearly indicates that the maximum air-temperature is reached at different times of the day in different months. Blanford (*vide* Dallas, 1886, p. 53, footnote) appears to doubt that any such variation occurs, for he remarks, "the great variability of the diurnal oscillation in different months and more especially the shifting of the epoch of maximum temperature, tend to throw doubt on the validity of the figures, as representing exactly the averages." My own observations, as well as those given by Dallas (1894) in his subsequent paper on the meteorological conditions over the Arabian Sea area, appear to confirm Dallas' original figures in as much as all the series indicate that the air-temperature over maritime regions does attain its maximum at different times of the day in different months of the year. Eliot (1902, p. 76) has also noted that at most of the meteorological stations in India (1) the epoch of the maximum is, as a rule, earliest in the rains (i.e. June to September) and (2) the epoch of the maximum over the whole of India is, as a rule, latest in the dry-season months and chiefly in January, February, and March. He attributes the difference at the different seasons of the year to an effect of the distribution of cloud.

In cases where temperatures are only recorded at four-hourly intervals it is, of course, impossible to do more than calculate, with more or less exactitude, the time of such an occurrence; but it is probable that where the observations cover a sufficiently large number of days the calculated time approaches very nearly to the actual. Without, however, undertaking any elaborate calculations one can form a rough estimate of the time of occurrence of maximum temperature by taking the average time of the highest reading in the four-hourly series in each month, and where the same temperature is recorded at two successive readings; taking the mean time between them. In the following table I give the time of occurrence of maximum

Area.	January.	February.	March.	April.	May.	June.	July.	August.	September.	October.	November.	December.
Arabian Sea (Dallas 1894) ..	1·75	2·13	2·19	2·56	**2·62**	1·87	*1·79*	2·34	2·48	**2·50**	2·19	*1·69*
Bay of Bengal, S. (Dallas 1886)	2·19	1·62	*1·44*	2·12	2·87	3·25	**3·31**	2·94	2·44	*2·25*	*2·25*	**2·31**

temperature in hours p.m. in each month as computed from the average readings in Dallas two series of observations.

At first sight these two series appear to be very different. In the Arabian Sea

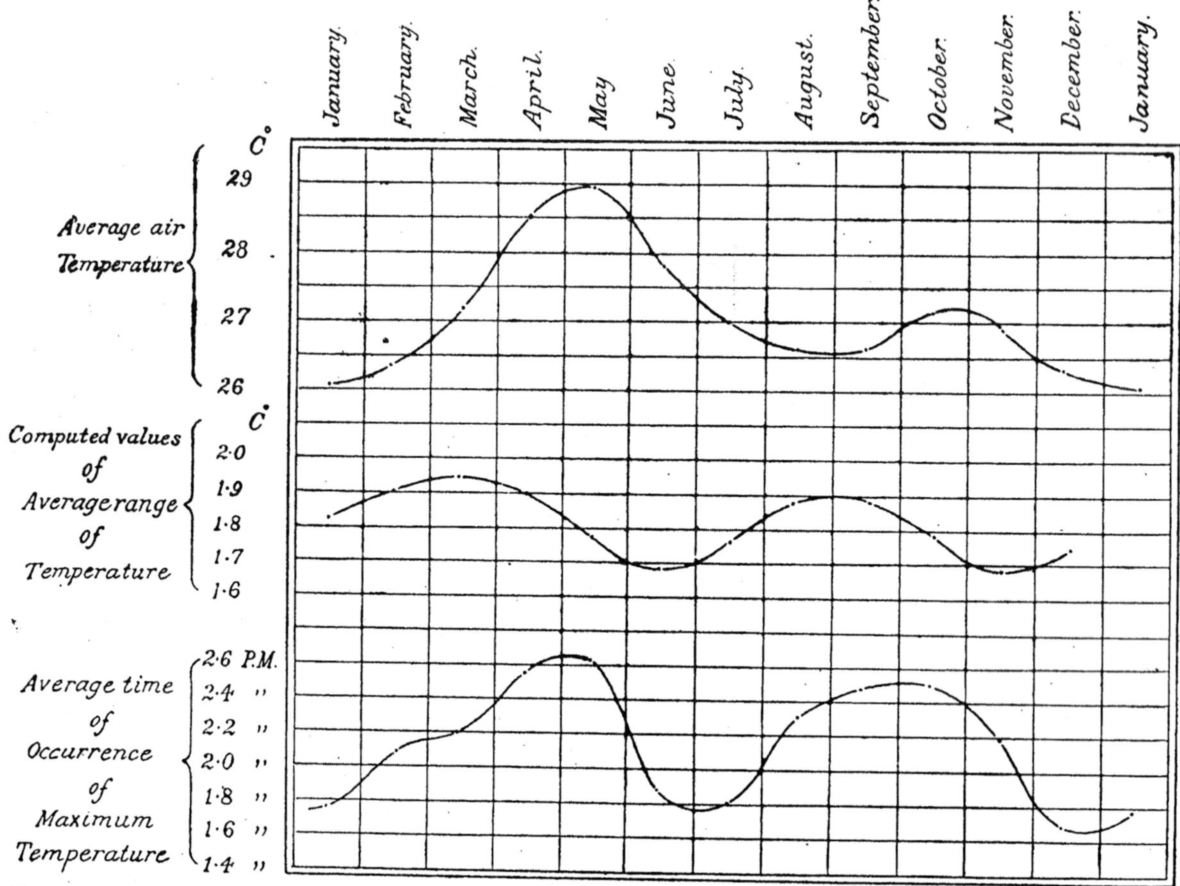

TEXT-FIG 9.—Showing the oscillation of mean air-temperature, average range of Temperature and the time of occurrence of maximum Temperature in the Arabian Sea area (Dallas).

series, the results of which I have plotted out in Text-fig. 9, the time of occurrence of maximum temperature is early in December and January, occurring between 1.40 and 1.45 p.m.; as the year grows older, the time of this occurrence gets steadily later, till in May it is at 2·61 hours p.m., i.e. at twenty-three minutes to three. In

June and July it again occurs earlier in the day and finally in October it has once more been delayed and occurs at 2.30 p.m. The variation in time, therefore, of the maximum temperature shows a clear double oscillation that agrees fairly closely with the double oscillation that, as we have already seen, is exhibited by both the average temperature and the daily range.

In the area to the south of the Bay of Bengal, however, the oscillation appears to be delayed somewhat, and thus the earliest time of maximum temperature occurs in March; the latest time is in July, when it is as late as 3.20 p.m. From then onwards it again gets earlier in each month till October or November and then is again somewhat delayed in December.

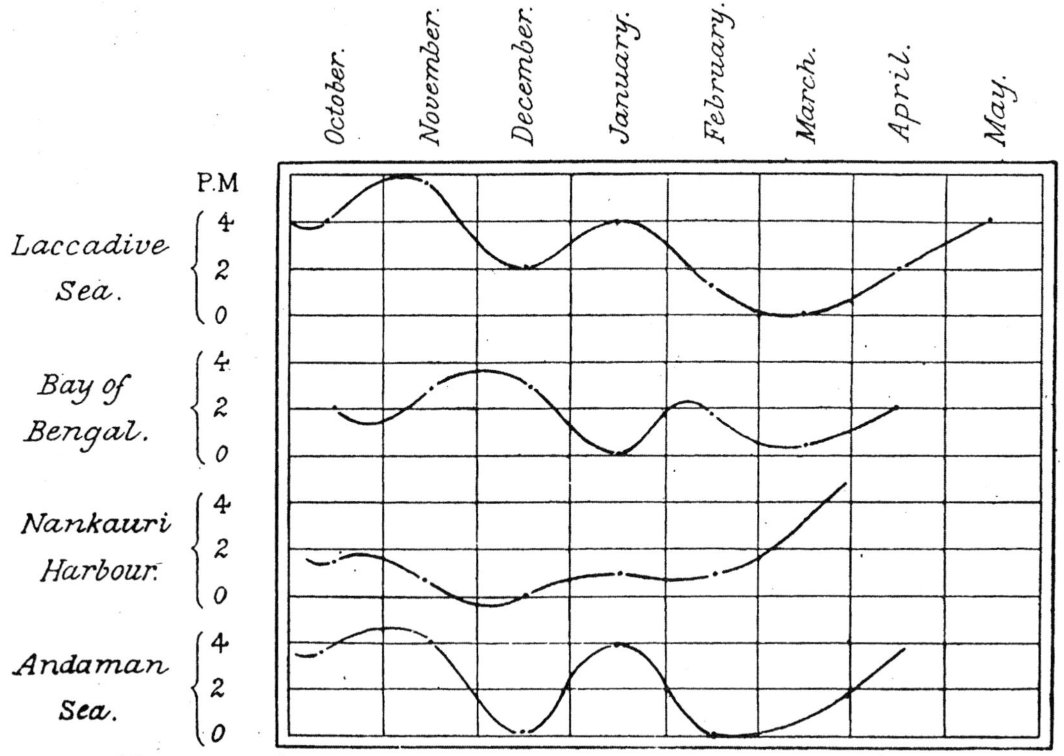

TEXT-FIG. 10.—The average time of occurrence of maximum air-temperature in different month as observed on the "Investigator."

My own observations on the "Investigator" also clearly indicate that the time of occurrence of maximum temperature exhibits very great fluctuations in different months. At first sight one is inclined to regard these fluctuations are being mere irregularities and yet in each series of observations, taken in different areas, we find that these fluctuations follow exactly the same oscillatory course.

In Text-fig. 10 I have plotted out the results obtained by me on board the "Investigator" in each area of the Indian seas and it is clear that the time of occurrence of maximum temperature varies from month to month and that the variation is exactly the same in the Laccadive Sea, the Andaman Sea and in Nankauri Harbour, whereas, as Dallas' results indicated, the change of time is delayed, by approximately a month to six weeks, in the Bay of Bengal area.

The agreement between the oscillation from season to season in the time of occurrence of maximum temperature is sufficiently striking, and a comparison of the curves given in Text-fig. 10 with those of the oscillations in the average monthly temperatures in these regions (*vide* Text-fig. 7) shows a marked degree of similarity, a rise in the average monthly temperature corresponding to a retardation of the time of maximum temperature. These minor oscillations are superposed on a single major oscillation that extends over the whole period from October to May and is clearly shown in the computed monthly averages derived from all my observations. In the major oscillation the time of maximum temperature tends to be late, namely, 2.49 p.m. in October. It then becomes progressively earlier up to February, when it occurs at 1.31 p.m., and in 1923 in the southern portion of the Laccadive Sea, between 0° and Lat. 6°N., it occurred on the average even before 12 noon, the average maximum temperature being at 10 a.m. From February onward the epoch again steadily tends to be later in the day, occurring in May at 2.49 p.m. If, now, Eliot is correct and the early occurrence of the epoch of maximum temperature is due to the presence of cloud, one would expect to find that periods of early occurrence will coincide with periods of increased rainfall, but, at any rate as regards the Andaman Sea area, the exact reverse appears to be the case. In Text-fig. 11 I have plotted the time of occurrence of maximum temperature, in each month of the survey season, in the Andaman Sea area and the actual rainfall recorded at Port Blair in the same period, and it is clear that an increase in rainfall coincides with *late* occurrence of the epoch of maximum temperature. The major oscillation is, I suggest, correlated with the movement of the sun to and from its most southerly declination, and the epoch of maximum temperature is delayed when the sun is directly overhead, but occurs earlier in the day as the declination becomes further and further removed from the vertical. On this major oscillation we have superposed a secondary oscillation that is probably caused by the rainfall of the N.E. Monsoon period, an increase in the rainfall causing a retardation of the epoch of maximum temperature in November and again to a lesser extent in January.

In addition to the primary rise and fall of the diurnal oscillation of air-temperature over the ocean, there appear in certain areas and at certain seasons to be subsidiary fluctuations. Dallas (1894, p. 15) has called attention to a remarkable triple oscillation in the temperature of the air during the day in an area lying off the coast of India to the west of Bombay. He shows that in this region as soon as one gets away from the land, the air temperature tends to exhibit three maxima and three corresponding minima during the course of twenty-four hours, and, as a rule, this oscillation is best seen as one gets further away from the land. The times of occurrence of these maxima are approximately 12 midnight, 8 a.m. and 4 p.m. and of the corresponding minima at 3.30 a.m., 11-11.30 a.m. and 9-9.30 p.m. This triple oscillation in the region of coastal water Dallas (1894, p. 16) attributes to the influence of the land- and sea-breezes on the normal single daily rise and fall. "At about this hour (9 a.m.) the land is sufficiently heated to occasion a flow of air from the sea to the land. At about this hour the breeze begins to blow, and the rise of temperature which had previously been in progress over the sea disappears, and is indeed changed into a temporary fall which lasts for three hours. After noon the sun's power counteracts the cooling due to the air movement, and at 4 p.m. the maximum temperature of the day is recorded. After that hour the temperature falls as the sun goes down, and at 6 p.m. the temperature over the land falls very slightly below the temperature over the neighbouring sea. This difference gradually increases as the evening progresses, and at about 9 p.m. the breezes from the sea probably altogether cease. Then follows a short period during which there is no perceptible movement. The amount of vapour in the air increases, radiation is interrupted and the temperature temporarily rises, so that a slight secondary maximum is developed. After midnight the temperature falls and the absolute minimum is reached at 4 a.m." Dallas's explanation may possibly account for the

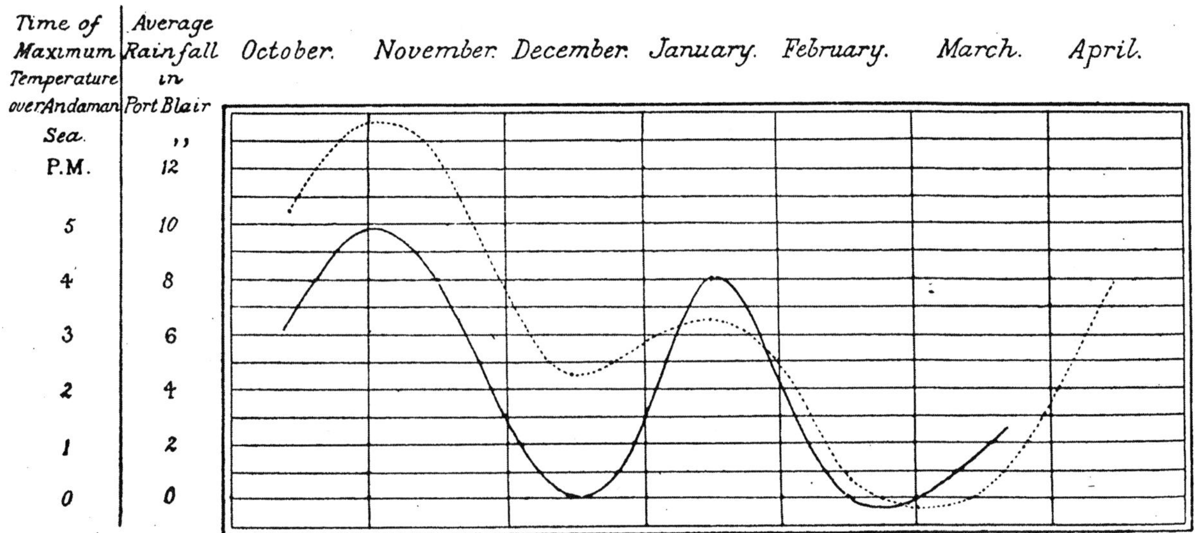

TEXT-FIG. 11.—Showing the relationship between the time of occurrence of maximum temperature and the rainfall.

.... Rainfall at Port Blair.

———— Time of occurrence of maximum temperature.

changes in temperature of the air over inshore waters where the alternating land and sea breezes can exert their influence, though it is difficult to see why the temperature of the air should rise after 9 p.m. unless it is caused by radiation from the sea, which during the absence of wind is able to make itself felt. If this be so, it will only occur if the surface temperature of the sea is higher than that of the supernatant atmosphere. Dallas (1894, p. 66 *et. seq.*) has himself shown that very similar secondary oscillations of the air-temperature during the day may be detected in areas quite removed from land influence; thus in the belt across the whole width of the Arabian sea between the Equator and Lat. 10°N. there is in the month of January, in addition to the primary rise and fall, a secondary maximum in the air temperature at midnight. This secondary oscillation appears to be absent in February and March but reappears again in April, increases in amplitude in May and decreases in June; finally it again reappears in September but only to a slight extent.

In order to detect such minor fluctuations in the air-temperature over the open sea at different times of the day, it is necessary to take observations at short intervals, and unfortunately the number of such at my disposal is not great. I give them, for what they are worth, in Table 8.

Month.	No. of days.	A.M.				P.M.			
		4	8	10	12	4	8	10	12
		°C	°C	°C	°C	°C	°C	°C	°C
October	6	*26·81*	27·59	27·85	28·19	**28·64**	28·04	27·50	27·36
November	5	**27·92**	26·63	*26·62*	27·54	27·66	**27·71**	27·18	*27·15*
December	7	27·32	27·31	**27·78**	*27·39*	**28·05**	27·85	27·43	*27·22*
January..	9	27·22	27·32	**27·42**	*27·22*	27·28	**27·39**	27·36	*27·15*
February [1]	..	*25·90*	26·61	27·51	**27·57**	27·36	26·62	26·66	26·62
March	12	*27·75*	28·20	28·19	**28·83**	28·74	28·36	28·03	27·92
April	13	28·33	*27·11*	29·01	29·48	**29·50**	29·36	29·07	28·83
May	4	*27·66*	28·24	**28·80**	28·49	**28·89**	28·53	28 15	**28·84**

Table 8; showing the oscillation during the day of the air-temperature over open sea in different months as observed on the "Investigator."

In the above series one sees that, as Dallas pointed out, the average air temperature in certain months clearly exhibits a double diurnal variation, even in areas far removed from land and, therefore, presumably free from the alternating influence of land- and sea-breezes. This double oscillation is not seen in October; it appears in November, and through occurring at different times of the day, persists through December and January; it has almost disappeared in February though a faint indication of a second maximum can be detected at 10. p.m. It is absent in March and April. But in the month of May we find a triple oscillation, the times of maximal temperature being midnight, 10 a.m. and 4 p.m., which agree closely with the times of maxima in the triple oscillation found by Dallas (*vide supra*) off Bombay. Beyond this I have no records.

From the above results one is, I think, justified in assuming that this oscillation of temperature is probably of universal occurrence during certain seasons of the year throughout Indian waters. In the coastal waters to the west of Bombay the changes during the day certainly appear to depend on variations in the wind force, and a study of the strength of the wind force at different times of the day (*vide infra* pp. 84 and 87) over the open waters of Indian seas, as well as in coastal areas, shows that there is a very clearly defined tendency for the force to rise and fall twice during the day. Such a double oscillation will of itself tend to produce corresponding oscillations, as Dallas points out, in the otherwise single rise and fall of the air-temperature; but in addition to this direct action of the wind, these variations in the strength of the

[1] The data for the month of February is taken from the results obtained by the "*Valdivia*" in the Bay of Bengal and is included so as to complete the series.

wind may, as I shall show in a subsequent paper, result in the appearance at the surface of sea-water that originally lay at some depth below, and in consequence we may get an alteration of the sea-temperature that will also tend to produce by convection and radiation a change in the temperature of the supernatant atmosphere.

One can summarize the conclusions, which can be deduced from the consideration of the data that I have given, as follows:—

(1) The monthly average temperature of the air exhibits a double oscillation during the course of the year that is obviously seasonal; the earlier fall is the result of the S.W. Monsoon and the later one is the true winter fall. The N.E. Monsoon in the region to the east of India superposes on the winter fall a secondary minor oscillation that occurs in November;

(2) The time of occurrence of maximum temperature during the diurnal temperature changes varies at different periods of the year and, like the monthly average temperature, exhibits a double oscillation, tending to be early in the day when the sun is overhead and late when the sun is at or near its northern or southern declination; moreover, here also the N.E. Monsoon superposes on the primary oscillation a secondary one that occurs in November and December;

(3) The range of temperature during the day also varies from month to month, and appears primarily to depend on the sun's position, being greatest when the sun is overhead and least at those seasons of the year when the sun is at the extremes of its northern or southern declination; and

(4) Variation in the force of the wind, whether over the open sea or in coastal areas, may produce a double or triple oscillation in the air-temperature during the course of twenty-four hours.

DIURNAL OSCILLATIONS IN THE WIND-FORCE.

Throughout Indian waters the alternation of the monsoons produces a double oscillation in the wind-force during the course of a year, but a careful study of the winds shows further that there is a quite distinct, though somewhat slight, double variation in the wind strength during the course of the day. Along the littoral region and in the coastal waters surrounding any land area of large size the alternate heating of the land during the day and cooling during the night will set up corresponding land- and sea-breezes that, in the absence of any other factor, will produce a daily double oscillation in the force of the wind. Such an oscillation is, however, not likely to be very well marked at such a place as Port Blair and still less at Nankauri Harbour, where the area of land in comparison with that of the sea is very small; and, furthermore, during the months of December and January, when the N.E. Monsoon rains cause a cooling down of the land surface, any alternating land and sea-breezes will tend to be very considerably reduced or even be entirely abolished. Nevertheless, we find that in these regions there is in every month of the survey-season, and probably throughout the whole year, a quite clear double rise and fall in the wind velocity in the twenty-four hours; and, moreover, this double oscillation can be detected over the open sea in areas far removed from land.

Chambers (1874), in a paper dealing with the diurnal variations in the winds at Bombay, shows that the observed winds can be analysed into two different sets of components. There is, firstly, an alternating land and sea-breeze and secondly, superposed on the first set, there is a double diurnal variation of the wind which attains "its maximum east position at about the same hour at which the barometer reaches its maximum and attains its maximum west position at about the same hours at which the barometer reaches its minimum and passes through zero at the times when the barometer passes through its mean positions." Chambers attributes this diurnal variation to the heating up of the atmosphere by the sun's rays and to the consequent creation of air currents; and he even puts forward the view that these changes may be the cause of the rise and fall of barometric pressure. Blanford (1876, p. 1), from observations of the winds at Calcutta, confirmed Chamber's interesting discovery. He, however, arrived at an entirely different conclusion as regards this double variation. He remarks (*loc. cit.*, p. 13), "the present results show that the double diurnal oscillation is an element of even greater relative importance in the Calcutta than in the Bombay wind system, but its relations to the barometric tides differ from those described by Mr. Chambers in many important respects; and so far from regarding it as a possible *cause* of the tides, a view which I am unable to reconcile with mechanical laws, it seems to me to be more probably a common effect of the same cause that produces the barometric tides and to depend, not on the mere existence of those oscillations of pressure but rather on their differences over land and sea. Consequently the diurnal variation of the wind is probably not a phenomenon of universal occurrence, or at all events of uniform type, but one depending on local conditions and varying with them." Chambers (1876, p. 402) showed later that this double oscillation of the wind force could be clearly traced in the observations taken in Bermuda, and again four years later (1880, p. 265) he demonstrated its presence in the winds at Karachi. In this latter paper Chambers still maintains that the effect is produced by the heating of the atmosphere by the sun's rays. Regarding the effect of this on the strength and direction of the wind, he remarks "if the north and east components of these wind variations be viewed separately, we should find that the curve, showing the variation with time of the east component would be ......similar to the curve of diurnal variation of barometric pressure in having maximum values about 10 a.m. and p.m. and minimum values about 4 a.m. and p.m. On the other hand, the variation of the north component is opposite in character in the two hemispheres, having maximum values in the north hemisphere about 7 a.m. and p.m. and minimum values about 1 a.m. and p.m.; while in the southern hemisphere the maximum values of the north component will occur at 1 a.m. and p.m. and the minimum values at about 7 a.m. and p.m." The result of the opposing variations in the two hemispheres of the north component will be to cause a complete abolition of any change at the Equator, and hence in the equatorial belt lying between 0° and 15°N., in which all my observations were taken, any diurnal variation in the strength of the north component of the wind is largely eliminated, while the east component remains unaffected. We should, therefore, expect to find a double daily variation in the strength of the wind having its maxima at or about 10 a.m. and p.m. Chambers (1880, p. 286) shows that both at Karachi and Calcutta this double oscillation, caused by the variation in strength of the east component, is clearly discernible in the wind-force and, moreover, is twice as great at the former station as at the latter, this difference being possibly due to their respective positions relatively to the sea. Calculating from the observations taken at both stations, he gives a table for the hourly variation from the mean in the strength of the wind, as follows:—

A.M.												P.M.											
1	2	3	4	5	6	7	8	9	10	11	12	1	2	3	4	5	6	7	8	9	10	11	12
−0·44	−0·57	−0·68	−0·67	−0·54	−0·23	−0·04	−0·26	+0·30	+0·44	**+0·46**	+0·32	+0·22	+0·16	+0·02	−0·14	−0·29	−0·22	−0·01	+0·27	+0·56	**+0·57**	+0·30	−0·15

At Trivandrum (*vide* Eliot, 1902, p. 204) this double oscillation is seen even more clearly than at Bombay, the wind velocity, irrespective of direction, showing the following maxima and minima;—

	Time.	Wind velocity in miles per hour.
1st Maximum	6 a.m.	10·5
1st Minimum	10 a.m.	5·6
2nd Maximum	2 p.m.	13·8
2nd Minimum	10 p.m.	2·8

At Madras, however, this double oscillation is apparently not a constant feature. Jones (1908, Table I, p. 78) has given the average velocity of the wind at this station, irrespective of direction, at all hours of the day in each month and he shows that from December to May inclusive there is only a single oscillation, having its minimum at 4–5 a.m. and its maximum at 1–2 p.m. In June and July we get evidence of a double oscillation with maxima at or near midday and midnight and in August there is a tendency to a triple oscillation. From September to November we again get a double oscillation with maxima at 3 a.m. and about 12 noon. At Port Blair, too, this double oscillation appears as a rule to be absent. Eliot (1915, Table 6, p. 189) gives a table of the mean movement of the air, irrespective of direction, at different times of the day in each month of the year at this station, the results being based on observations extending from 1894 to 1904, and his figures show no sign of any double oscillation, the single maximum, on the year's average, occuring at noon to 1 p.m. and the single minimum between 2 and 5 a.m. In a series of observations, however, that I had taken for me at four-hourly intervals from March 24 to 31, 1925, the average results obtained were as follows:—

	A.M.			P.M.		
	4	8	12	4	8	12
Strength of wind in miles per hour	*2·11*	**2·86**	2·59	*2·50*	**4·14**	3·43

so that here, too, a double oscillation may occasionally be present.

It seems clear that, at any rate at certain seasons of the year, this double oscillation of the wind-force can be detected over the greater part of, if not universally throughout, the coastal regions of Indian waters, and the question arises whether such a variation is of universal occurrence throughout the tropical zone, as it should be if it is in any way connected with the rise and fall of barometric pressure, or is it a purely local phenomenon, as suggested by Blanford, due to differences of conditions over sea and land ?

During the voyage of H.M.S. "Challenger" a two-hourly record of the wind force was kept on 1,202 days, of which 650 were on the open sea and 552 near land. Buchan (1889, p. 25) summing up the results of these observations, remarks, "with respect to the open sea, it is evident from the mean curve for the five oceans that the diurnal variation is very small, there being apparently two indistinctly marked maxima about midday and midnight respectively." He gives the mean velocity in Beauforts scale for every two hours of the day, as follows, and I have appended the variation from the mean velocity :—

Time of day. A.M.	Wind velocity.	Variation from mean.	Time of day. P.M.	Wind velocity.	Variation from mean.
2	2·98	+0·09	2	2·92	+0·03
4	2·90	+0·01	4	2·87	−0·02
6	2·85	−0·04	6	2·87	−0·02
8	2·83	−0·06	8	2·85	−0·04
10	2·87	−0·02	10	2·92	+0·03
12	2·92	+0·03	12	2·87	−0·02

Buchan concludes from the above results that "it seems probable that the line representing the true diurnal variation in the velocity of the wind is practically a uniform straight line, with the single exception of a small rise about midday, not quite amounting to a mile per hour." The range of variation in the wind-force certainly appears to be small, amounting only to 0.15, but it must be remembered that the scale of measurement is that known as Beaufort's, which is still in use on board ships at sea. The mean velocity of the wind in the whole series of observations is 2.89 and a variation of 0.15 is, therefore, a variation of 5% which certainly seems to me to be too great to warrant one regarding it as negligible.

A similar series of observations was carried out on board ship during the Norwegian North-Atlantic Expedition in 1876-78, and the results obtained are embodied in a paper by Mohn (1883). In this case the record of the velocity of the wind was taken by means of a "Robinson's anemometre" and the results are, therefore, more accurate than in the "Challenger" series. The prevailing conditions appear to have differed considerably in the three years. In 1876 there is evidence of a double daily variation in the wind, having a range of 0.85 metres per second, with maxima at 10 a.m. and 5.30 p.m. and minima at 1.30 a.m. and p.m. In 1877 the results obtained give indisputable evidence of the presence of a double daily variation in the wind-force; in this year two series of observations were taken, (*a*) in the warm current flowing along the coast of Norway and (*b*) in the cold Polar current, but the results obtained in each area are remarkably similar. The range of variation in both series combined is 1.12 metres per second and the maxima occur at or about 2 p.m. and 12 midnight, with the minima at 7.30 a.m. and 8 p.m. Finally, in 1878 a similar result was again obtained, but in this year the variation of wind force was extremely small, being only 0.26 metres per second.

In more recent years the expedition of the "Valdivia" has provided a further series of observations, and during her cruise she twice passed through the tropical zone, in which, as I have already mentioned, this diurnal variation of the wind should be most marked.

	A.M.						P.M.						Range of Variation.	%	No. of days
	2	4	6	8	10	12	2	4	6	8	10	12			
August ..	**3·17**	**3·17**	2·83	2·67	2·67	2·83	*2·33*	2·67	**3·00**	**3·00**	*2·83*	3·00	0·84	26·5	6
September ..	3·16	**3·32**	3·08	*2·87*	2·89	3·00	3·00	**3·05**	3·03	3·03	*2·95*	3·05	0·45	13·6	19
October ..	3·78	**3·89**	3·44	3·22	*3·11*	*3·11*	3·78	3·67	3·83	**4·17**	3·78	*3·78*	1·06	25·4	9
January ..	4·00	4·25	**4·36**	*4·21*	4·25	**4·38**	4·00	4·13	4·13	4·00	*4·00*	4·06	0·38	8·7	8
February ..	**3·13**	2·91	2·88	*2·54*	*2·54*	2·63	2·69	2 74	2·91	**3·00**	*2·88*	3·06	0·59	18·8	24
March ..	**2·80**	**2·80**	2·75	2·55	*2·10*	2·30	**2·40**	2·30	2·20	2·30	2·55	2·75	0·70	25·0	10
Average ..	3·34	**3·39**	3·22	3·01	*2·93*	3·04	3·03	30·9	3·18	**3·25**	*3·18*	3·28	0·46	13·6	76

Table 9; showing the daily oscillation of the wind-force (in Beaufort's Scale), as observed on the "Valdivia," in tropical areas.

In the above table I have given the average wind-force, as recorded by the "Valdivia" (*vide* Schott, 1902). The records for the first three months were taken in the tropical region of the Atlantic Ocean and the last three months in the tropical zone of the Indian Ocean. Here again one finds evidence of a double diurnal variation in the wind velocity; the two maxima occur at or about 4 a.m. and 8 p.m. and the two minima at 10 a.m. and 10 p.m. So far as it is possible to judge from this series, there appears to be a tendency for (1) the time of the maximum wind-force to occur at different times of the day in different months, and particularly so as regards the afternoon maximum; and (2) for the variation in velocity to differ in different months.

At the commencement of my studies of the maritime meteorology of Indian waters, observations on the wind-force were only taken at four-hourly intervals during the day when the "Investigator" was at sea. The results obtained are given below, and, as in the "Valdivia" series, the two maxima fall at 4 a.m. and 8 p.m.; the times of occurrence of the minima are not clearly shown owing to no observations having been taken at 10 a.m. or 10 p.m.

Month.	No. of days.	A.M. 4	A.M. 8	A.M. 12	P.M. 4	P.M. 8	P.M. 12	Average velocity of wind.	Range of Varia-tion.	%
October	21	**1·97**	1·45	1·59	**1·93**	1·83	1·59	1·73	0·52	30·1
November	8	**2·31**	1·44	1·68	**2·19**	2·00	2·13	1·96	0·75	38·3
December	8	**3·25**	2·94	2·92	2·82	2·88	2·94	2·98	0·43	14·4
January	24	1·48	**1·62**	1·27	1·60	**1·70**	1·44	1·52	0·43	28·3
February	9	1·78	1·89	**2·16**	1·99	**2·39**	1·78	2·00	0·61	30·5
March	16	1·53	1·56	1·53	**2·06**	1·91	1·62	1·70	0·53	31·2
April	28	2·12	1·82	**2·09**	1·75	2·18	**2·25**	2·04	0·50	24·5
May	27	3·35	3·01	3·22	**3·70**	3·65	3·44	3·40	0·69	20·3
Average	141	**2·22**	1·97	2·06	2·26	**2·32**	2·15			

Table 10; showing the daily oscillation of the wind-force (in Beaufort's Scale) as observed on the "Investigator" in 1921-22 and 1922-23.

Although the time of day at which the maxima are reached varies somewhat in different months, it appears that there is in certain months a distinct tendency for the wind-force to exhibit a double daily oscillation. In other months, as in December and May, this double oscillation seems to disappear and be replaced by a single oscillation, that in December has its maximum at 4 a.m. but in May at 4 p.m. In this connection it is interesting to note that it is in these months that the wind-force attained its maximum, and that, whereas December is one of the coldest months, May is one of the hottest. The series of observation is, however, not very satisfactory for with only six observations in the twenty-four hours it is impossible to trace any of the minor oscillations. During the survey seasons 1923-24 and 1924-25 a two-hourly record has been kept on board the "Investigator" of the force of the wind and the results of these observations are given in the table below :—

Month.	No. of days.	A.M. 2	A.M. 4	A.M. 6	A.M. 8	A.M. 10	A.M. 12	P.M. 2	P.M. 4	P.M. 6	P.M. 8	P.M. 10	P.M. 12	Average force of wind.	Range of variation.	%
October ..	25	**1·96**	1·86	1·74	*1·36*	1·56	1·93	**1·96**	1·90	1·90	*1·74*	1·87	1·82	1·80	0·60	33·3
November ..	47	1·48	1·49	1·36	*1·34*	1·53	1·65	**1·65**	1·64	1·51	1·40	*1·37*	**1·56**	1·50	0·31	20·7
December ..	56	2·12	2·10	2·03	*2·00*	2·18	2·22	**2·46**	2·35	1·99	*1·94*	2·09	**2·19**	2·14	0·52	24·3
January ..	31	1·83	1·80	*1·67*	1·84	2·29	**2·34**	1·97	2·29	*1·65*	1·86	1·89	**1·93**	1·95	0·67	34·4
February ..	43	1·63	1·57	*1·41*	1·53	1·71	1·79	1·89	**2·03**	1·74	*1·58*	1·62	**1·63**	1·68	0·62	36·9
March ..	56	**1·20**	1·08	1·11	*0·92*	1·21	1·31	**1·47**	1·33	1·21	*1·03*	1·12	1·15	1·18	0·55	46·6
April ..	45	1·51	1·53	*1·37*	1·42	1·59	**1·77**	1·60	1·56	1·57	*1·42*	1·57	**1·60**	1·54	0·40	26·0
Average ..	303	1·68	1·63	1·53	*1·49*	1·72	1·86	1·86	**1·87**	1·65	*1·57*	1·65	**1·70**	..	..	..

Table 11 : showing the daily oscillation of the wind force (in Beaufort's scale) as observed on the "Investigator" in 1923-24 and 1924-25.

Here again one finds a double oscillation in the wind-force during the day. In the average of this series the actual maxima occur at 4 p.m. and midnight, though there is for all practical purposes no change between 12 noon and 4 p.m., and the minima occur at 8 a.m. and 8 p.m. The difference in time of the rise and fall of the wind-force between this series and those already given may be due to differences between individual years or to different localities. It seems clear from the above independent observations that there is throughout the open waters of the ocean, at any rate in the tropical zone, a double diurnal variation in the wind-force, the times of maxima and minima differing possibly in different areas and at different times of the year, and that this oscillation is not, as Blanford suggested, dependent on purely local conditions but is a wide-spread and universal phenomenon.

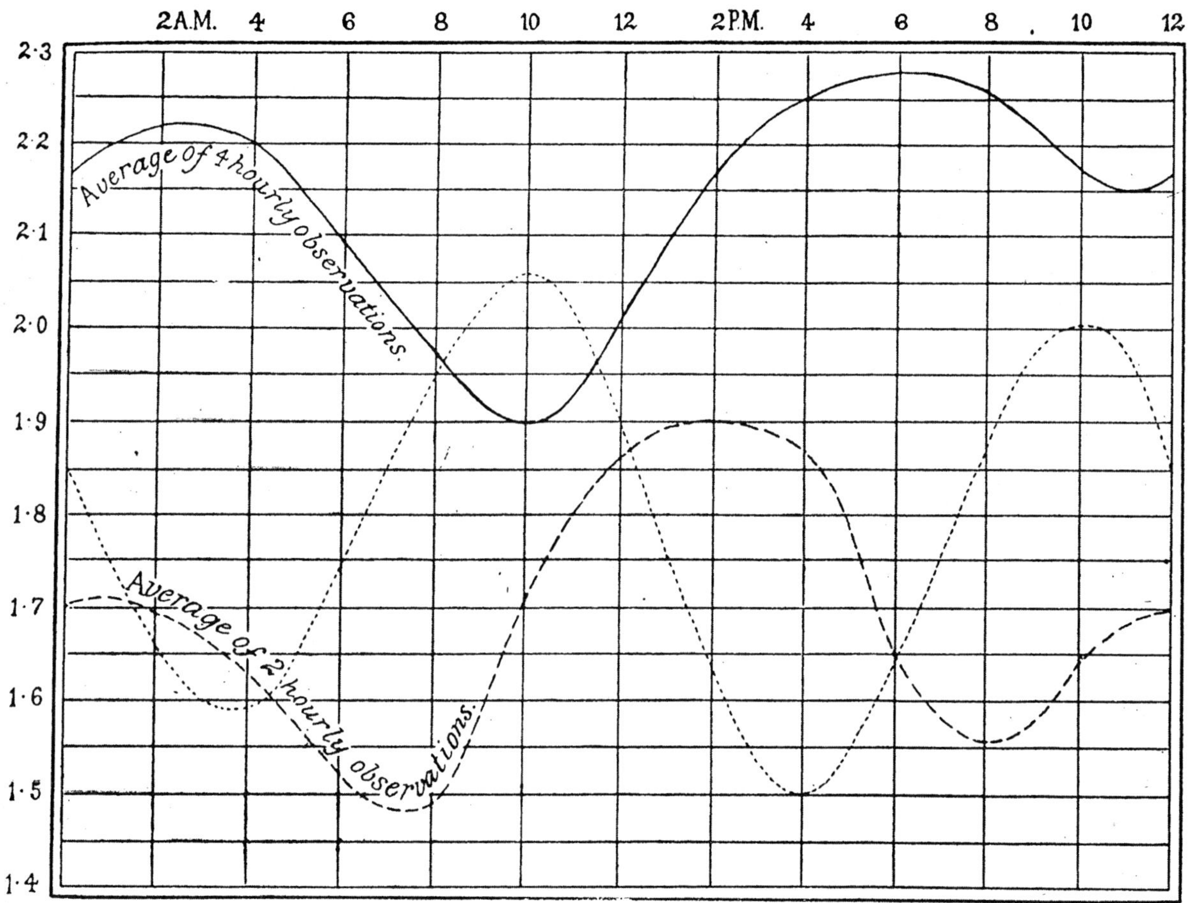

TEXT-FIG. 12—Variation in strength of wind-force as compared with the rise and fall of the barometer.

In Text-fig. 12 I have plotted the average variation of the wind-force in both the 2-hourly and 4-hourly series of observation and for the purpose of comparison I have given the average rise and fall of the barometric pressure.

The times of occurrence of the maxima and minima of the wind-force during the day, certainly seem to indicate that the double oscillation is in some way connected with the rise and fall of barometric pressure. To what extent the variation is due to these barometric changes or whether they are both the result of some other phenomenon I must leave to meteorologists to decide, but it seems not unlikely that in an area, such as the Indian seas, in which the wind blows steadily, for weeks or even months at a time, from a quarter that has an easterly or westerly element, namely from the north-east or south-west, depending on the season of the year, a rise of barometric pressure, that passes as a wave from east to west across the ocean, must tend to cause a diminution in the strength of the easterly or westerly component and a consequent fall in the total wind-force.

The bulk of my observations have been taken while the "Investigator" was at work on the survey-grounds, in areas where the amount of land is extremely small, consisting only of a few small islands situated in mid-ocean, and, therefore, in areas in which one might reasonably expect to find that purely oceanic conditions prevailed. During certain intervals, such as when returning to port for coaling etc., I was able to take observations while away from all land influence, and as these two series exhibit certain differences I have thought it advisable to consider them separately. The results obtained are given below in Table 12. During the months of October and November in both series there is a well-marked double oscillation of the wind-force during the day and, on the whole, the times of occurrence of the maxima and minima agree fairly well. In December over the survey-ground we find a double oscillation, but over the open sea this is clearly replaced by a triple one. The mid-day and mid-night maxima agree in the two areas, occurring at 2 p.m. and at 11 p.m. to 12 mid-night respectively, but over the open sea there occurs a third maximum at 6 a.m. In January we again get evidence over the open sea of a double oscillation with maxima at 12 noon and 12 mid-night, but on the survey-ground there occurs only a single oscillation, of the type that is characteristic of land areas, having a maximum at 2 p.m. and a minimum at 6 a.m. In February we find a double oscillation both on the survey-ground and over the open sea; but in March we get a double oscillation over the survey-ground and only a single oscillation, with a maximum at 2 p.m. and a minimum at 2 a.m., over the open ocean; and, finally, in April we get a triple oscillation in both series.

The tendency towards a triple oscillation in the month of December over the open sea and in the month of April over both open sea and survey-ground and in no other months of the survey season is difficult to explain. When considering the results of the 4-hourly series of observations (*vide supra* p. 84) I pointed out that in the months of December and May, i.e., in the coldest and hottest months of the year, the wind-force appears to follow a single oscillation; a comparison, however, of the figures for December in the 4-hourly series and those of the 2-hourly series over the open sea, indicate that the single oscillation as indicated by the former may be spurious; for the average wind-force at 4, 8, and 12 a.m. and p.m. in the 2-hourly series would also give us a curve leaving only a single maximum as follows:—

4 A.M.	8	12	4 P.M.	8	12
3·58	3·58	3·50	3·41	3·25	3·67

This would appear to indicate a single oscillation having a maximum at 12 midnight, which does not differ markedly from the 4-hourly series, in which there is a single oscillation with the maximum at 4 a.m. It is only by taking observations every 2 hours that we see that there is in reality a triple oscillation and it is more than probable that this triple oscillation is the normal condition during these two months, viz. December and April.

		A.M.						P.M.						Average Strength of wind.	Daily range.	Percentage variation.
		2	4	6	8	10	12	2	4	6	8	10	12			
On Survey-Ground.	No. of days.															
October ..	8	**0·88**	0·75	0·75	*0·63*	1·12	1·25	**1·31**	1·19	1·00	0·75	*0·75*	0·75	*0·93*	0·68	**73·1**
November ..	42	1·23	1·26	1·20	*1·16*	1·35	1·41	**1·45**	1·44	1·32	1·19	*1·12*	**1·33**	1·29	0·33	*25·6*
December ..	41	1·99	1·92	*1·79*	1·84	2·05	2·07	**2·37**	2·21	1·90	*1·78*	1·98	**2·03**	1·99	0·59	29·7
January ..	23	1·84	1·85	*1·67*	1·87	2·22	2·35	**2·39**	2·35	2·05	1·95	1·96	1·93	**2·04**	0·72	35·3
February ..	32	1·44	1·48	*1·27*	1·53	1·61	1·69	1·78	**1·96**	1·73	1·50	*1·47*	**1·53**	1·58	0·69	43·7
March ..	43	**1·17**	1·01	1·05	*0·81*	1·21	1·18	**1·33**	1·20	1·12	*0·90*	1·04	1·10	1·09	0·52	**47·6**
April ..	24	**1·17**	1·13	*1·04*	1·16	**1·34**	1·29	*1·15*	**1·33**	1·15	*0·83*	0·96	1·04	1·13	0·51	45·1
At Sea.																
October ..	11	3·09	2·95	2·77	*2·09*	*2·09*	2·77	**2·86**	2·77	2·95	2·86	**3·18**	3·14	2·79	1·09	39·1
November ..	2	**3·00**	2·75	*2·25*	2·50	3·25	**3·50**	2·75	2·75	2·25	*2·00*	2·75	2·75	2·71	1·50	55·4
December ..	6	*3·50*	3·58	**3·83**	3·58	*3·50*	*3·50*	**3·67**	3·41	3·41	*3·25*	**3·67**	**3·67**	**3·55**	0·58	*16·3*
January ..	8	2·37	2·37	*2·31*	2·44	2·94	**3·06**	2·75	2·62	2·69	*2·56*	2·87	**3·06**	2·67	0·75	28·1
February ..	8	2·25	2·13	*2·13*	2·13	2·50	**2·63**	2·25	*2·19*	2·50	2·63	**2·88**	2·50	2·39	0·75	31·4
March ..	10	*1·15*	1·20	1·20	1·30	1·20	1·65	**2·15**	1·90	1·55	1·45	1·45	1·35	*1·46*	1·00	**68·5**
April ..	19	*1·84*	**2·10**	1·81	*1·79*	2·02	**2·39**	2·23	*1·92*	2·11	2·21	**2·40**	2·37	2·10	0·60	28·6

Table 12; showing the variation in wind-force at different times of the day, on the survey-ground and when at sea, as observed on the "Investigator."

Dallas (1894, p. 16), as I have already mentioned above, attributes the triple oscillation in the air temperature over the coastal waters to the west of Bombay to corresponding variations in the wind-force caused by the alternating land and sea breezes, and it is, therefore, of some interest to enquire whether a corresponding oscillation in the air-temperature has been recorded on board the "Investigator" during those months in which a triple oscillation of the wind-force has been noted. As I have already mentioned, over the open waters of the Laccadive Sea and Bay of Bengal, and on the survey-ground, where one would have expected to find that oceanic conditions largely prevailed, there is in several months a double oscillation, but in only two months, as a reference to Table 12 shows, have I experienced a triple oscillation in the wind-force. The first was over the open waters of the Laccadive Sea in the month of December, 1923. During this period of seven days the wind shows an unmistakable triple oscillation having its maxima as follows:—

1st minimum at	2 a.m.
Primary maximum at	6 a.m.
2nd minimum at	11 a.m.
2nd maximum at	2 p.m.
Primary minimum at	8 p.m.
3rd maximum at	11 p.m.

The air-temperature exhibits an oscillation that clearly coincides with the wind-variation and in the accompanying Text-figure 13, I have plotted both series of observations. A rise of air-temperature coincides with a fall of wind-force and *vice versa*, and it seems clear that the oscillation in temperature in this instance is undoubtedly due to the rise and fall in the strength of the wind, a decrease in the strength of the wind being accompanied by a corresponding rise in the air-temperature, except during the early hours of the afternoon when the effect is somewhat masked by the heating and cooling due to the sun.

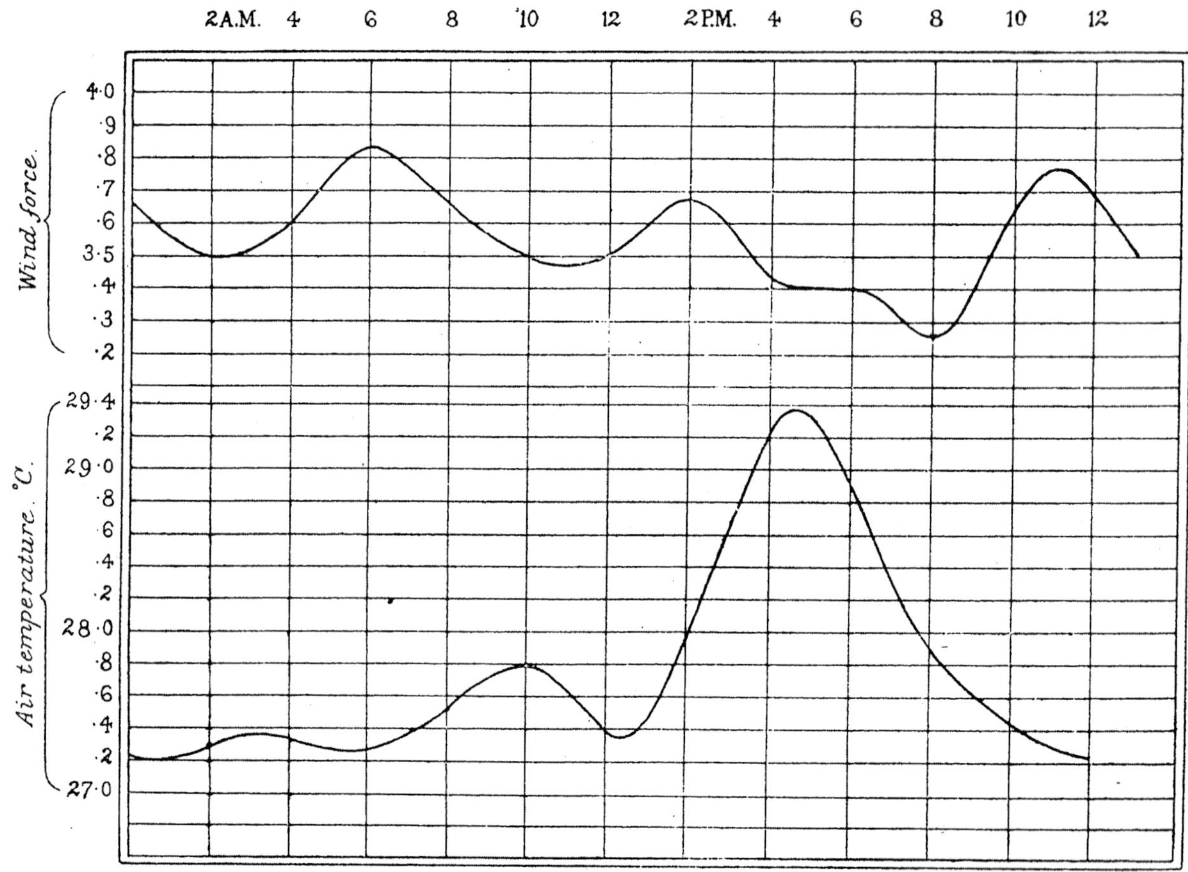

TEXT-FIG. 13.—Showing a triple variation in both wind-force and air-temperature over the Laccadive Sea in December, 1923.

The second period during which the wind again exhibits a triple oscillation is during the month of April, and again it was found over the open sea. The time of day at which the oscillations occur are as follows:—

1st minimum	2 a.m.
1st maximum	4 a.m.
Primary minimum	7 a.m.
2nd maximum	12 noon.
3rd minimum	4 p.m.
Primary maximum	10-30 p.m.

Here again a study of the air-temperatures recorded during the corresponding month in 1924, when out at sea and away from land influence, reveals a very distinct tendency for the temperature to rise as the wind falls and this is clearly seen at night, but during the day the heating effect of the sun's rays appears to counteract the cooling effect of an increased wind-force.

It is clear then that the rise and fall of the wind-strength exercises a marked influence on the temperature of the air; but the mere fact of the appearance, disappearance and reappearance of the phenomenon of a double oscillation in the air-temperature at definite periods of the year indicates that Dallas's explanation will

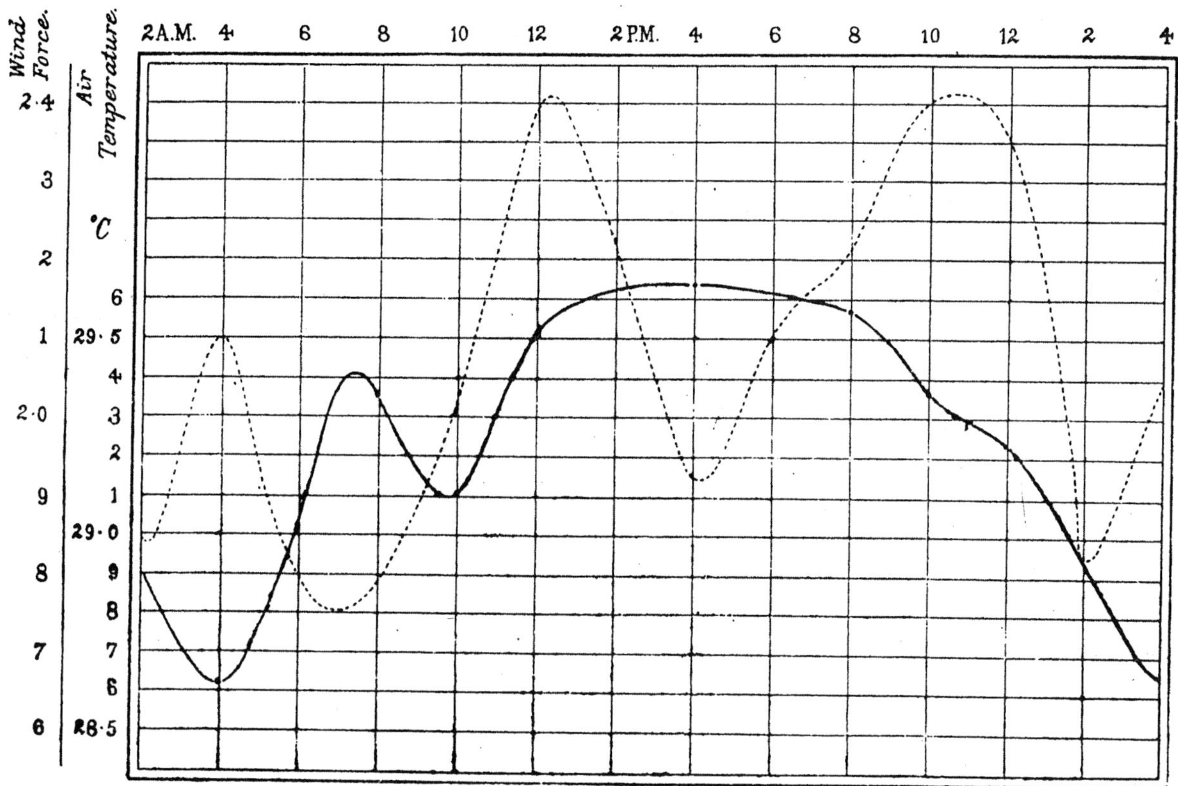

TEXT-FIG. 14.—Showing a triple variation in wind-force and a double oscillation in the air-temperature over the open Sea in April, 1922.
...... = wind-force, ———— = air-temperature.

not account for all the facts, and the periods of the year at which this double oscillation in the air temperature makes its appearance, namely from November to January and again in March to May, further indicate that the oscillation in wind-force is not the only factor concerned in these variations. If the wind was the main agent, one would expect to find that these oscillations made their appearance during those months, in which the variation in the wind-force was the greatest. I give below the average strength of wind, the mean range and the percentage variation during each month of the survey season, and it is clear that the appearance of the double oscillation in the air-temperature tends to occur at those seasons of the year when the percentage variation in the wind-force is lowest,

	Oct.	Nov.	Dec.	Jan.	Feb.	March.	April.	May.
Average strength of wind (B.S.).	1·77	2·73	2·56	2·54	2·17	1·79	1·79	3·40
Average Range of wind-force.	0·56	0·53	0·48	0·49	0·61	0·59	0·45	0·69
Percentage variation.	31·7	29·5	19·3	23·8	28·7	34·3	25·3	20·3

Variations in the amount of water vapour present and the relative humidity of the atmosphere.

Up to the present time it has been the custom of meteorological officers on board ship to obtain a record of the amount of water-vapour present in the atmosphere by means of dry- and wet-bulb thermometers; but, as I have already pointed out, a series of observations recently carried out on board steamships has tended to cast some doubt on the accuracy of these records, and in consequence it is possible that the calculated values of the amount of water-vapour present are incorrect. Whether the differences which have been noted between the readings given by fixed and portable thermometers, and which seem to occur in both wet- and dry-bulb thermometers to judge by the results obtained by Capt. V. Campos, O.B.E. on board the Cable Ship "Colonia" (*vide* "Marine Observer," 1924, p. 146), are due entirely to inaccuracy of the fixed thermometer, is by no means certain; but be that as it may, records taken by means of fixed thermometers are at any rate comparable with similar records obtained in past years by other ships employed in scientific research in various parts of the world, such as the "Challenger," "Valdivia," etc.

A study of the data given in the Indian Meteorological Memoirs shows that at inland stations the aqueous vapour pressure and the amount of water-vapour present in the atmosphere exhibit a double oscillation during the day. Eliot (1902, pp. 114-159) has very fully discussed the diurnal oscillation in the amount of water vapour in the atmosphere in various inland stations in India and in the coastal region, and he has shown that in most stations the oscillation is a double one; the amount of water-vapour increases in some stations to a considerable amount, in others only slightly, to a maximum at or about 8 a.m., it then falls till 2–4 p.m., rises again to a second maximum at about 8-10 p.m. and finally falls to an absolute minimum at 4 a.m. In the coastal areas, such as Bombay, Karachi and Trivandrum on the west, Cuttack, Calcutta, Chittagong and Rangoon on the east of Peninsular India and in Burma, this double oscillation appears to be due to the alternating effects of land- and sea-breezes. Although the extent of the morning and evening rise of vapour pressure may vary in different coastal centres, it is clearly seen in all during the cold-weather or hot-weather seasons, which cover the period of the year from October to May; but during the rainy season, from June to September, it is often obliterated and the oscillations become exceedingly irregular, owing presumably to the interference of the Monsoon winds with the alternating land and sea-breezes and to modification of the normal diffusion and convection.

It is extremely interesting to compare these results from inland and coastal stations with observations taken at sea, away from land influence. During the voyage of the "Challenger" regular observations on the humidity of the atmosphere were carried out by means of wet- and dry-bulb thermometers and the results are briefly summarised by Buchan (1889, pp. 8–10) in his report on the atmospheric conditions. The figures given by Buchan (*loc. cit.*, p. 10) of the variation, at two hourly intervals, from the mean of the twenty-four hours, appear to me to indicate a triple rather than a double oscillation during the day: his actual figures are as follows:—

Time of day.	Elastic force of water vapour.	Time of day.	Elastic force of water vapour.
2 A.M.	−0·003	2 P.M.	+0·007
4 ,,	−0·009	4 ,,	+0·015
6 ,,	−0·010	6 ,,	0·000
8 ,,	−0·003	8 ,,	−0·004
10 ,,	+0·014	10 ,,	−0·005
Noon	+0·010	Midnight	−0·007

If now we plot these figures, we obtain a curve that has three maxima at 2 a.m., 10 a.m., and 4 p.m. respectively, the height of each progressively increasing. A comparison of this curve with the curves of average air-temperature obtained during the months of December, 1923, and April, 1924, reveals a very close agreement, the periods of maximum air-temperature corresponding to periods of maximum humidity.

On board the "Investigator" during 1923-24 the weight of water-vapour in the atmosphere exhibited a very clear oscillation. In an area extending over the whole breadth of the Laccadive Sea during the cold weather period from November to January inclusive, and in the hot season of March and April in both the Laccadive Sea and Bay of Bengal, the average weight of water-vapour in the atmosphere at different times of the day is as follows:

November—January.	A.M.				P.M.			
Time of day.	4	8	10	12	4	8	10	12
Weight of water-vapour in atmosphere*	9·335	9·557	**9·634**	9·427	9·580	**9·668**	9·577	9·526
March—April.	A.M.				P.M.			
Time of day.	4	8	10	12	4	8	10	12
Weight of water-vapour in atmosphere*	10·308	**10·399**	10·283	**10·487**	10·401	10·343	**10·453**	10·418

* In both series the weight is given in grains per cubic foot.

It will be noticed that there is a considerable difference in the amount of water vapour present in the two seasons. This is possibly due in part to a slight difference of locality, for the November–January results were all obtained in the Laccadive Sea, while those for March-April include one series of observations taken across the Bay of Bengal. It conforms, however, to the results obtained by Dallas (1894), who shows that in the southern area of the Arabian Sea the vapour tension in the atmosphere decreases from October and attains its lowest level in January, and then rapidly increases in March and April.

During the cold weather we have a clear double diurnal oscillation having its maxima at 10 a.m. and 8 p.m. respectively, whereas during the hot months there appears to be a triple oscillation with maxima at 8 a.m., 12 noon, and 10 p.m. In this latter case the additional maximum appears to be at 8 a.m. The maxima at 12 noon and 10 p.m., probably correspond to those seen to occur at 10 a.m. and 8 p.m. during the winter months, and, if so, they have been delayed by about two hours. This delay is almost certainly correlated with the shifting of the epoch of maximum air-temperature, which, as I have already pointed out, occurs much earlier in the day when the sun is at or near its extreme southern declination, as it is in December, than when it is directly overhead as in March or April. It is, however, possible that in the winter months also there occurs a third maximum, but as this would then be at 6 a.m. or thereabouts it is not shown in the actual observations since no records were taken on the "Investigator" at this hour.

In order to detect minor oscillations in the amount of water-vapour present in the atmosphere it is necessary to take observations at frequent intervals throughout the whole day. I only possess such records at two-hourly intervals for a period of five days in October in the region of the Maldives and three days in February in the region of the central group of the Nicobars; and these are not sufficient to determine with any degree of exactitude the diurnal oscillations in the amount of water vapour present at different times of the day, but as they show certain interesting features I give the data below. In both instances, owing to the distance from continental land, one would have expected the conditions present to have resembled those over the open sea. A comparison, however, with the data given by Dallas (1894) indicates that this is not so, and the conditions partake more of the nature of those over coastal waters.

	A.M.						P.M.					
	2	4	6	8	10	12	2	4	6	8	10	12
October	9·283	*8·793*	9·536	9·450	**9·755**	9·349	9·326	9·309	*9·202*	9·246	**9·321**	9·255
February	*9·945*	**10·157**	10·072	*9·353*	9·858	9·907	10·087	10·000	**10·185**	*10·025*	10·171	**10·193**

Table 13; showing the weight of water-vapour, in grains per cubic foot, present in the atmosphere at different seasons of the year, around oceanic islands.

In the month of October, the "Investigator" results show a clear double oscillation having two maxima at 10 a.m. and 10 p.m. and minima at 4 a.m. and 6 p.m., so that the rise and fall of the amount of water-vapour agrees closely with the rise and fall of barometric pressure. Dallas, in comparing the results obtained in the region to the west of Bombay, finds that in the region furthest removed from land (square 80) the vapour tension has a single maximum and minimum, but nearer inshore (in square 81) there is a double oscillation, though the times of occurrence of the maxima are different from those found by me. In February my results indicate that the amount of water-vapour has a distinct tendency to a triple oscillation with maxima at 4 a.m. and at 6 and 12 p.m., and this agrees clearly with the results given by Dallas for square 80 in the previous month of January, in which he shows a triple oscillation of the vapour tension having maxima at 8 a.m. and 4 and 11 p.m. So that it would appear that round these islands the conditions approximate to those of coastal waters, rather than to those of the open ocean.

VARIATION IN THE RELATIVE HUMIDITY OF THE ATMOSPHERE OVER INDIAN SEAS.

During a period of eighty-four days, at a distance from land in the North Atlantic Ocean, the relative humidity, as observed on board the "Challenger," exhibited a single daily oscillation, having a minimum at 2 p.m. and a maximum from midnight to 4 a.m. The actual observed relative humidity at different times of the day was as follows:—

Time.	Relative humidity of air.	Time.	Relative humidity of air.
2 a.m.	82	2 p.m.	77
4 "	82	4 "	78
6 "	81	6 "	79
8 "	80	8 "	80
10 "	79	10 "	81
Noon	78	Midnight	82

A corresponding single oscillation occurs in the elastic force of water-vapour in the atmosphere, the minimum in this case occurring at 4 a.m., when the air-temperature is lowest, and the maximum at 2 p.m., when the air-temperature is highest. Buchan (*loc. cit.*, p. 10) remarks "it is only on the open sea, at a distance from land, where this typical curve of the diurnal humidity occurs with its single minimum and maximum. Over land the humidity daily curve shows two well-marked minima and maxima—the two minima occurring in the early morning and in the afternoon; and the more inland the situation and the stronger the sun, the more strongly marked is the afternoon minimum. Now the hygrometric observations made near land show a daily humidity curve intermediate between these two."

Dallas (1894, p. 32, *et. seq.*) has drawn attention to the changes that take place in the humidity of the atmosphere in the different zones of the Arabian Sea. As he points out, the changes at different periods of the year are comparatively slight; regarding the southerly zone from Lat. 0° to 20°N., he remarks that "the minimum humidity is reported in April and the maximum in August, the difference being 9 per cent." In this area the humidity is shown to fall slowly from August till January; it remains steady in February and March, falling again slightly in April.

During the survey seasons 1923-24 and again for a short period in 1925, a number of observations were carried out on the "Investigator" in order to determine not only what changes occur in the humidity of the atmosphere at different seasons, but also at different times of the day. The area, in which these observations were made, includes the southern part of the Laccadive Sea, the southern area of the Bay of Bengal and the survey ground in the Nicobars; possibly some of the differences noted may be due to a difference of locality, but, nevertheless, the results obtained are of considerable interest, and I have given them *in extenso* in an appendix to this paper.

In Table 14 I have given the average humidity in each month of the survey-season, and, for the purpose of comparison, the results for the whole year in the Arabian Sea between 0° and 20°N. Lat., as given by Dallas, and those for Port Blair, Andamans, as shown in the Indian Weather Review for 1922.

Locality.	January.	February.	March.	April.	May.	June.	July.	August.	September.	October.	November.	December.
Arabian Sea (0°-10°N.) [Dallas]	79	79	79	78	81	83	86	87	85	83	81	80
Port Blair	81	84	84	83	86	88	87	88	90	89	85	81
Indian waters ("Investigator")	83·8	84·6	80·1	80·7	..	..	..	..	..	76·5	75·1	82·2

Table 14; showing the relative humidity of the atmosphere in each month.

The three series of data are, strictly speaking, not comparable with each other, since the Port Blair series is, I believe, based on a single observation at 8 a.m. each day, Dallas's series in the Arabian Sea is founded on observations taken at four-hourly intervals, while the "Investigator" series is taken from observations at 4, 8, 10 and 12 night and morning while steaming, or at two-hourly intervals from 6 a.m. to 8 p.m. while at anchor. Nevertheless, the comparison of these records is of considerable interest. The data for the Arabian Sea give no indication whatever of any rise in humidity during the period of the N.E. Monsoon, from December to February, so far as the actual observations are concerned; Dallas (1894, p. 64), however, remarks that "the computed figures show a very slight secondary oscillation in the spring months with its maximum in February and its minimum in January." As I have already shown when dealing with the air-temperature in different months on the east and west side of the Peninsula, the effect of the N.E. Monsoon is felt to a much greater extent on the east side, and the "Investigator" series shows, in strict accord with this regional difference, a very marked increase in humidity from November to December, that continues, though to a less extent, to February and is then succeeded by a fall. This rise is undoubtedly due to the N.E. Monsoon, and it is interesting to note that it occurred in 1923-24 much earlier than is normally the case at Port Blair, at which station the humidity is usually low throughout December and January and rises in February, falling again slightly in April.

It is well known that the humidity of the atmosphere exhibits a considerable range of variation during the course of the day. During midday the temperature of the air steadily rises and simultaneously the humidity falls, whereas at night the humidity rises as the air-temperature falls. Dallas (1894) gives the changes in humidity in various areas of the Arabian Sea at four-hourly intervals in each month, and almost invariably his results show a single diurnal oscillation, the humidity attaining its maximum between midnight and 4 a.m. and its minimum between 12 noon and 4 p.m. His results, therefore, agree in the main with the observations taken on board the "Challenger" in the open ocean. In some few instances, however, his figures give a distinct indication of a double oscillation, thus in the area lying between Lat. 10°–20°N.; Long. 50°–60°E., in the month of January there are two maxima at 4 a.m. and 8 p.m., and two minima at 12 noon and 12 midnight respectively, the noon minimum being the lowest. A study of the "Investigator" records, taken while on the open sea, seems to me to indicate clearly that, at any rate at certain seasons of the year, the normal variation in the humidity of the atmosphere follows, not a single, but a double oscillation, of which the mid-day fall is the greatest. I have only a small series of observations that cover the whole period of the day at two-hourly intervals: these were taken in the regions of the Maldives and Nicobars during the months of October and February and in all cover a period of eight days. The averages of all these observations are given below in Table 15 and, for comparison, I give the simultaneous record of the air-temperature and the wind-force.

Time of day.	A.M.						P.M.					
	2	4	6	8	10	12	2	4	6	8	10	12
Humidity ..	82·1	*80·8*	**83·3**	80·2	81·0	*77·4*	*77·4*	79·0	79·2	80·8	81·1	**82·5**
Air Temperature °C.	27·3	27·4	*27·2*	27·4	27·8	27·9	**28·2**	27·9	27·9	*27·5*	**27·9**	27·3
Wind force (Beaufort Scale).	**1·80**	1·72	1·58	*1·45*	1·64	1·86	1·[illegible]	**1·97**	1·82	*1·66*	1·75	1·73

Table 15; showing a double oscillation of humidity, air-temperature, and wind-force during the day.

I have already pointed out that at certain seasons of the year we find that a double oscillation occurs during the day in both the air-temperature and the wind-force, and both are clearly seen in the above series of observations. Alterations in air-temperature and wind-force both have an effect upon the humidity of the atmosphere, a rise tending to cause a fall of humidity and *vice versa*, and the oscillations in humidity clearly are the result of changes in the other two conditions. The maximum humidity occurs at 6 a.m., when the air-temperature is at its lowest and the wind-force is also comparatively small. The lowest humidity occurs between mid-day and 2 p.m. and at this latter time the air-temperature has reached its maximum and the wind-force is high. From 8 to 10 p.m. we appear to get a certain degree of antagonism between the air-temperature and the wind-force: both have been falling since 4 p.m. and, in consequence, the humidity has risen, but at 8 p.m. the wind-force has reached its minimum and this has allowed the air-temperature to rise again slightly, so that at 10 p.m. it is 0.4°C above what it was at 8 p.m. This rise in the air-temperature has checked the rise in humidity, so that the second maximum is not reached till

12 midnight, by which time the temperature of the air has again fallen; at the same time this maximum is not as high as it might be because the wind-force has again increased, and a still further increase in the wind-force at 2 a.m. causes a fall in the humidity which is continued to 4 a.m., when a lowering of wind-force again occurs and this appears to permit a slight rise in the air-temperature. Two hours later, at 6 a.m., the air-temperature has now reached its minimum and the progressive fall in the wind-force now permits the humidity to reach its maximum. It is interesting to compare these results with observations taken on board the "Valdivia" (*vide* Schott, 1902) during her passage across the Bay of Bengal and Laccadive Sea in the month of February.

Time of day.	A.M.						P.M.					
	2	4	6	8	10	12	2	4	6	8	10	12
Humidity ..	82·4	82·9	**83·3**	78·8	74·5	74·1	*70·5*	74·5	77·3	78·5	80·0	80·5
Air Temperature °C.	**26·41**	26·35	*26·24*	26·69	27·39	**27·72**	27·61	27·49	26·96	26·69	26·56	*26·36*
Wind force (Beaufort Scale).	3·13	**3·25**	3·13	2·88	*2·50*	2·75	2·75	2·75	**2·81**	2·69	*2·44*	2·50

Table 16; showing the average air-temperature, wind-force and relative humidity during 8 days in the Bay of Bengal and Laccadive Sea in February (from the "Valdivia" observations).

Here we find that the relative humidity of the atmosphere exhibits a single oscillation only, having its maximum at 6 a.m., corresponding exactly to the period of lowest temperature, and its minimum at 2 p.m., shortly after the epoch of maximum temperature, which appears to have occurred somewhere about 1 p.m. The wind-force shows a clear double oscillation during the day with maxima at 4 a.m. and 6 p.m., *i.e.* two hours later than in the "Investigator" record. Here again we can detect a slight antagonism between wind-force and air-temperature. The wind-force is high at 6 p.m. and the air-temperature is steadily falling; but at 10 p.m. the wind-force is at its lowest and it remains low till after 12 p.m. Between 12 midnight and 2 a.m. there is a very slight rise in the air-temperature, which is, however, followed by a further fall at 4 a.m., when the wind-force has again risen: but this rise in the air-temperature at 2 a.m. is too slight, amounting to only 0·05°C, to produce any effect on the relative humidity.

Since this secondary fall of the humidity, that, as we have seen from the "Investigator" records, occurs in the early hours of the morning, is but slight in extent in comparison with the main fall at mid-day and, further, only lasts for a short period of time, observations taken only at four-hourly intervals may entirely fail to demonstrate its presence. In the following Table 17 I have given the average of all observations taken while the "Investigator" was at sea; in most months the double oscillation is quite clearly indicated, and it is also clearly seen in the average of the whole series. One is, I think, justified in concluding that this double oscillation is the normal condition and that, in addition to the primary fall in the humidity of the atmosphere that occurs at mid-day, a secondary and smaller fall occurs in the early hours of morning.

Month and year.	A.M.				P.M.				No. of days.
	4	8	10	12	4	8	10	12	
October, 1923 ..	..	**79·45**	78·33	75·66	*73·23*	**76·66**	75·89	..	6
November, 1923 ..	*74·04*	**79·55**	79·34	*73·91*	75·06	75·17	**77·74**	77·04	5
December, 1923 ..	*81·97*	**83·01**	81·91	*81·05*	81·29	81·67	82·49	**83·13**	7
January, 1924 ..	*80·24*	83·78	84·45	84·91	85·41	**85·79**	83·95	84·66	9
March, 1924 ..	82·53	82·05	*79·89*	81·90	81·62	81·65	83·51	**84·29**	12
April, 1924 ..	**86·12**	*81·65*	**82·89**	81·56	81·38	*80·68*	82·78	82·17	8
April, 1925 ..	81·29	80·48	..	78·76	*77·02*	80·04	..	**81·38**	10
Average ..	*81·03*	**81·42**	81·12	79·68	*79·29*	80·24	81·06	**82·11**	..

Table 17; showing the average relative humidity of the atmosphere at intervals during the day in different months over the open sea.

In the following Table 18 I have given the average degree of humidity of the atmosphere at different times of the day in each month, during which observations were taken, while on the survey ground in or near oceanic islands in mid-ocean, such as the Nicobars.

Time of day.	A.M.				P.M.			
	6	8	10	12	2	4	6	8
October ..	**79·86**	79·79	79·14	76·21	*76·00*	76·89	**78·68**	78·04
November ..	**77·93**	76·08	74·21	*71·98*	73·10	74·19	75·17	**76·12**
December ..	**82·54**	80·48	*79·13*	**80·49**	79·95	*78·78*	**81·99**	81·66
March ..	**88·00**	83·57	80·43	*79·24*	**80·37**	*80·15*	82·19	**85·12**
April ..	**83·15**	80 62	79·63	79·22	*77·68*	**78·27**	*78·13*	**80·44**

Table 18; showing the average humidity of the atmosphere at different times of the day in different months round oceanic islands.

It will be noticed that the humidity in this series of observations, presents a first Maximum at 6 a.m. In the first two months there is between 6 a.m. and 8 p.m. a single fall and rise, which corresponds to the similar fall and rise noted over the open sea. In each month the humidity falls steadily from a first maximum at 6 a.m. to a first minimum, which is reached at different times in different months: thus in October it is late, occurring at 2 p.m., in the following two months it is reached at progressively earlier times in the day, occurring at 10 a.m. in December in which month there is clear evidence of a triple oscillation, a 2nd maximum occurring at 12 noon and being succeeded by a 2nd minimum at 4 p.m. after which the humidity again rises. I have unfortunately, no data for January and February, but in March and April we still find evidence of a triple oscillation though the epoch of first minimum and second maximum occur at progressively later times in the day. As most of my observations were taken around oceanic islands in the Laccadive Sea, one would expect that the midday epoch of minimum humidity would show a correlation with the time of epoch of maximum temperature and a reference to the data for the Arabian Sea given in the Table on page 74 shows that this is undoubtedly the case. The occurrence of a triple oscillation in December and again in March and April, is of interest, and is probably due to the interaction of the air temperature and the wind force.

A very good example of the great effect that wind may produce in the degree of

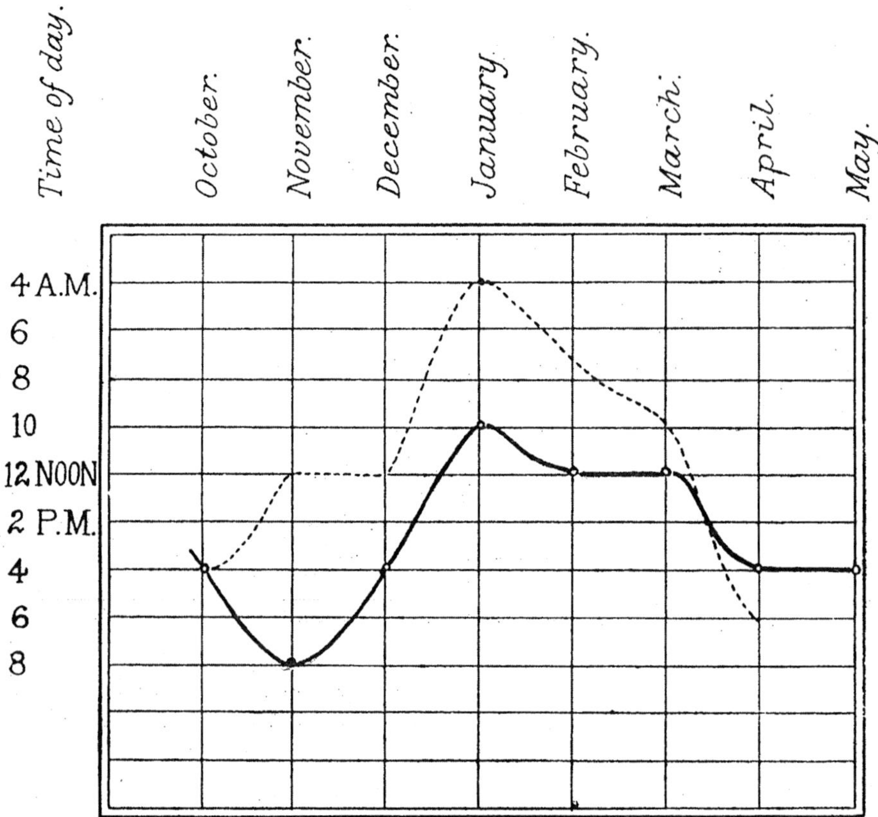

TEXT-FIG. 15.—Showing the time of maximum temperature and minimum humidity in different months over the open Sea.
——— Time of Epoch of Maximum Temperature of atmosphere.
- - - - - - - - " " " " Minimum humidity " "

humidity of the atmosphere is seen in the "Investigator" record of April 14th, 1924. The "Investigator" was then in the neighbourhood of the Maldives and some 400 miles west of Colombo. Throughout the day until 1 p.m. there was a flat calm and the humidity was high, falling slowing from 82·9 at 6 a.m. to 75·6 at noon. After 1 p.m. a breeze of force 1 (Beaufort scale) sprang up and blew from N.W. to W.S.W. till 6 p.m. At 2 o'clock the humidity had fallen to 67·8 and at 6 p.m. to 66·2. The breeze then died away and the humidity rose to 73·2 at 8 p.m. But the main factor concerned in the daily variation of humidity is undoubtedly the air-temperature. As we have already seen (*vide supra*, p. 72) there seems to be clear evidence that the epoch of maximum air-temperature occurs at different times of the day in different months; and we should also expect to find that, in conformity with this variation, the epoch of minimum humidity also exhibits a corresponding alteration in the time of its occurrence. In the above Text-fig. 15 I have plotted the average times of occurrence in each month of the epochs of maximum temperature and minimum humidity and it is clear that there is a general agreement between the two. In both cases the epoch is late in October; in January, on the contrary, it is early; and from then on the epoch becomes progressively later.

Daily oscillation in the amount of rainfall.

Blanford (1879) has called attention to the fact that in Calcutta the frequency with which rain falls exhibits a regular oscillation during the twenty-four hours of the day. Taking the average of the whole year, he finds that "the hour at which rain is least frequent is shortly before midnight, and that at which it is most so, from 2 to 3 p.m....... For about three hours after midnight the frequency of rainfall increases rapidly, but after 3 a.m. more slowly, till about sunrise; after which there is a slight falling off to a secondary minimum at 9 a.m....... After 9 a.m. the frequency increases rapidly to the absolute maximum between 2 and 3 p.m. From this maximum it declines, without interruption, to the minimum before midnight." The oscillation, therefore, of the rainfall has a distinct tendency towards the double type such as exists in the rise and fall of barometric pressure and which also seems to be present in the wind-force. In a later paper Blanford (1886) embodied his original series of observations and included others in which he investigated not merely the frequency of rainfall but also the actual amount precipitated in different hours of the day. In both papers he shows that the oscillation observed differs at different seasons of the year: during the rainy season the oscillation follows that found in the whole-year's average, but during the periods of the cold season (October to February) and the hot season (March–May) conditions are somewhat different. In the *cold season* "the absolute minimum of frequency falls about noon; from which time it rises to a maximum between 6 and 9 p.m., and then falls to a secondary minimum between midnight and 2 a.m. From 2 to 6 a.m. rain is slightly more frequent; and after that hour it declines gradually to the minimum at noon...... In the *hot season* there is but one well-defined maximum and one less determinate minimum, the former very decidedly between 6 and about 8 p.m., the latter somewhat indefinitely between sunrise and 11 a.m."

Unfortunately I have no observations on the rainfall at sea, but it is clear that, if a similar periodicity is of universal occurrence, it will have a profound effect upon the average curves of both air- and sea-temperature and on the salinity of the surface water and will produce corresponding oscillations in the diurnal changes.

Hill (1881, p. 345) has pointed out that the records available at Allahabad exhibit a very similar diurnal oscillation to that found at Calcutta, and at Lucknow the rainfall shows two maxima at about 10 a.m. and 10-30 p.m. respectively. Buchan (1889, p. 30) has shown from the "Challenger" observations that the variation in the frequency of occurrence of rainfall differs according as the ship was near land or well out in the open sea. On the open ocean there is only one maximum at 2 a.m. and one minimum at 4 p.m.; while near land, however, the diurnal oscillation exhibits a double rise and fall, "the results showing two maxima and two minima, the secondary maximum occurring from 10 a.m. to 2 p.m., the two maximum periods being the times of maximum and minimum temperature, and the two minima the early morning and early evening respectively." In the vicinity of land, however, conditions appear to be somewhat different. A study of Buchan's own data reveals a

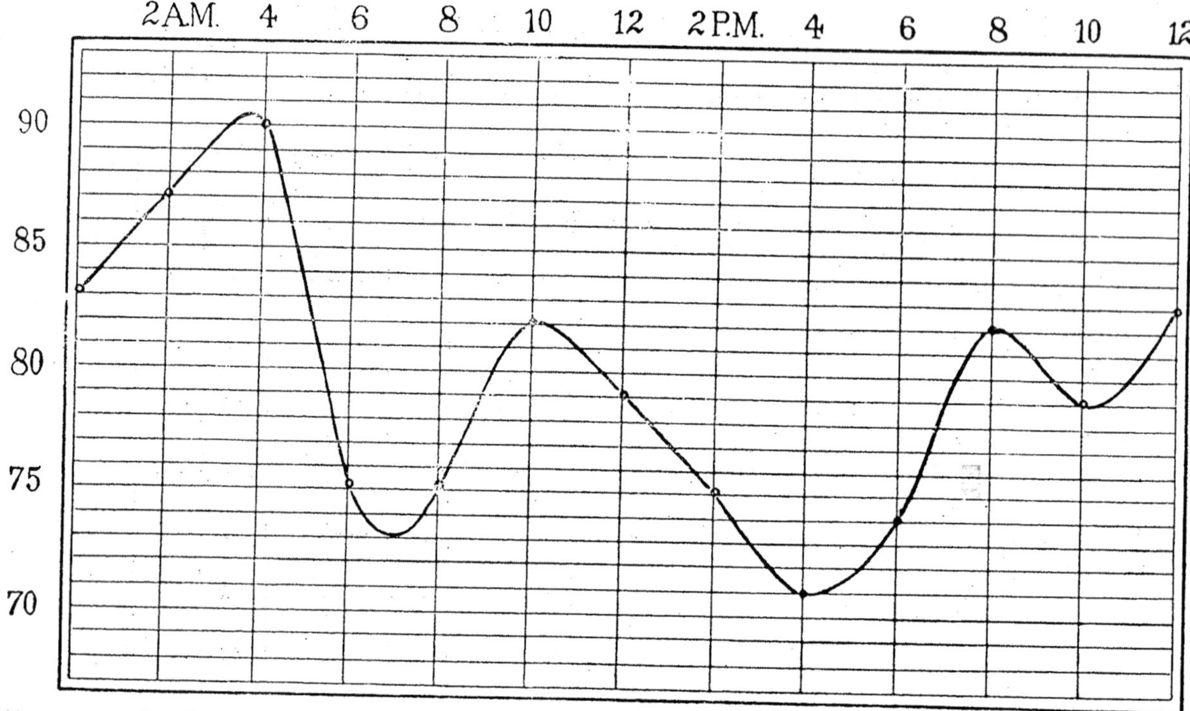

TEXT-FIG. 16.—Showing the oscillations in the frequency of occurrence of rainfall at different times of the day, based on the "Challenger" observations taken when near land.

tendency towards a triple oscillation, as a glance at the accompanying Text-fig. 16 shows; and the time of these maxima of rainfall appears to bear a certain degree of relationship to the rise and fall of vapour-pressure in the atmosphere (*vide supra*. p. 91); in the early morning, the maximum vapour-pressure is at 2 a.m. and the maximum rainfall at 4 a.m.; at 10 a.m. both maxima coincide, while in the evening the maximum vapour-pressure is at 4 p.m. and the rise of rainfall occurs at 8 p.m.

The double diurnal variation in the humidity of the atmosphere, and still more the similar variation at certain seasons of the year in the amount of rainfall at different times of the day will have a very important effect on the salinity of the surface water of the ocean, for at the time of day when humidity is highest the amount of evaporation will tend to be diminished. The increased force of the wind, which occurs at the same time as and is to a large extent the cause of low humidity, will further tend to increase the rate of evaporation and so to raise the surface salinity, whereas an increased humidity and a fall of wind together with, at certain seasons of the year, actual rainfall, will cause a lowering of the surface salinity. We should thus *a priori* expect to be able in a series of observations to find that the surface salinity exhibits a diurnal oscillation. That we do find clear evidence of a double diurnal oscillation in the surface salinity, I shall demonstrate in a subsequent paper, and I shall then take the opportunity of discussing its causation in detail.

BIBLIOGRAPHY.

1. Blanford, H. F., 1876 .. .. "On the winds of Calcutta, an analysis of ten years hourly observations of the wind vane and four years anemograms" *Indian Meteorological Memoirs*, Vol. I, No. 1. Calcutta.

2. Blanford, H. F., 1879 .. .. "On the diurnal variation of Rainfall Frequency at Calcutta." *Journ. Asiatic Soc. Bengal*, Vol. XLVIII, No. iv. p. 41. Calcutta.

3. Blanford, H. F., 1886 .. .. "On the Diurnal variation of the Rainfall at Calcutta." *Indian Meteorological Memoirs*, Vol. IV, No. 2, p. 39. Calcutta.

4. Buchan, A. 1889 .. .. "Report on atmospheric circulation based on the observations made on board *H.M.S. Challenger* during the years, 1873–1876, and other Meteorological observations." *Report Sci. Results of H.M.S. "Challenger"*, Physics and Chemistry, Vol. II. London.

5. Campos, Capt. V., 1924 .. "Comparison of Temperatures in portable and fixed screens." *The Marine Observer*, Vol. I, No. 2, p. 19. London.

6. Campos, Capt. V., 1924 .. "Comparison of Temperatures in portable and fixed screens." *The Marine Observer*, Vol. I, No. 11, p. 146. London.

7. Chambers, F., 1874 .. .. "The diurnal variations of the wind and barometric pressure at Bombay." *Phil. Trans. Royal Society, London*, Vol. 163, pt. 1. London.

8. Chambers, F., 1876 .. .. "The diurnal variations of the wind and barometric pressure." *Proc. Royal Society, London*, Vol. XXV. London.

9. Chambers, F., 1880 .. .. "The winds of Kurrachee." *Indian Meteorological Memoirs*, Vol. I, No. 9. Calcutta.

10. Dallas, W. L., 1886 .. .. "The Meteorological Features of the southern part of the Bay of Bengal." *Indian Meteorological Memoirs*, Vol. IV, Pt. I, No. 3. Calcutta.

11. Dallas, W. L., 1894 .. .. "Investigation into the mean temperature, humidity and vapour tension conditions of the Arabian Sea and Persian Gulf." *Indian Meteorological Memoirs*, Vol. VI, Pt. I, No. 2. Calcutta.

12. Eliot, J., 1899 .. .. "The discussion of the hourly meteorological observations recorded at Trivandrum during the years 1853–64."
Indian Meteorological Memoirs, Vol. X, No. 1. Calcutta.

13. Eliot, J., 1902 .. .. Discussion of the results of the hourly observations recorded at 29 stations in India given in Vols. V, IX and X of the "Indian Meteorological Memoirs."
Indian Meteorological Memoirs, Vol. XII, No. 2. Calcutta.

14. Hill, S. A., 1881 .. .. "Some results of the Meteorological observations taken at Allahabad during the years 1870–79."
Indian Meteorological Memoirs, Vol. I, No. 10. Calcutta.

15. Indian Weather Review, 1921, (1924) Calcutta.

16. Indian Weather Review, 1922, (1924) Calcutta.

17. Jones, R. Le., 1908 .. .. "A discussion of some of the anemographic observations recorded at Madras."
Memoirs of the Indian Meteorological Department, Vol. XX, No. 5. Calcutta.

18. Krummel, O., 1907 .. .. Handbuch der Ozeanographie, Vol. I. Stuttgart.

19. Malpas, A. H., 1922 .. .. 'Marine Biology.' Report of the Government Marine Biologist for 1921.
Ceylon Administration Report, for 1921.

20. Mohn, H., 1883 .. .. "Meteorology."
The Norwegian North Atlantic Expedition, Christiana.

21. Murray, Sir J. 1898 .. . "On the annual Range of Temperature in the surface waters of the Ocean and its relation to other Oceanographical Problems."
The Geographical Journal, Vol. XII, No. 2. p. 113. London.

22. Pearson, J., 1923 .. .. 'Marine Biology.' Report of the Government Marine Biologist for 1922.
Ceylon Administration Report for 1922.

23. Schott, G., 1902 .. .. "Oceanographic and maritime Meteorologie im auftrage des reichs marine amtas."
Wissenschaftliche Ergebnisse de Deutsch Tiefsee Expedition auf den dampfer "Valdivia", 1898-1899. Vol. I. Java.

APPENDIX I.

Results of observations on the temperature of the air and of the surface-water in Indian Seas.

November, 1921.	Victoria Point, Burma.			Port Blair, Andamans.			Nankauri Harbour, Nicobars.			
	Direction of Wind.	Maximum Temp. °C	Minimum Temp. °C	Direction of Wind.	Maximum Temp. °C	Minimum Temp. °C	Direction of Wind.	Maximum Temp.	Minimum Temp.	Weather.
1	..	29·33	24·67	..	30·44	24·22	N W.	..	..	
2	N.E.	26·44	23·94	E.S.E.	30·33	26·22	N.W. to E.	..	..	
3	N.E.	26·61	23·89	..	31·22	24·78	E.	..	..	Heavy rain.
4	S.E.	27·72	23·56	N.N.E.	30·11	24·89	E.	..	..	Heavy rain.
5	..	..	23·50	E.S.E.	29·89	26·33	S.E.	..	..	Heavy rain.
6	E.N.E.	31·94	23·89	..	..	..	S.E.	..	..	Occasional storms.
7	N.E.	30·83	25·00	S.S.W.	27·78	23·56	W.	..	..	Fine.
8	N.E.	31·94	24·22	S.S.W.	28·56	24·00	W.	..	..	Occasional storms.
9	..	31·11	23·44	N.W.	28·67	25·11	Calm	..	..	Cloudy.
10	N.E.	25·67	23·33	N.N.W.	29·44	23·44	Calm	32·56	24·05	Fine.
11	E.N.E.	29·33	22·50	N.N.E.	29·78	25·44	Calm	29·06	23·33	Fine.
12	N.N.E.	29·56	22·78	N.N.W.	30·00	23·56				
17	..	..	..	..	..	..	Calm	..	..	Fine.
18	..	..	..	..	..	..	Calm	29·22	24·61	Fine.
19	..	..	..	..	..	..	N.E.	..	24·50	Light airs, Fine.
20	..	..	..	..	..	..	N.E. to E.N.E.	28·06	25·39	Cloudy.
21	N.E.	30·22	22·50	N.N.W.	29·78	21·89	E.	28·33	25·39	Fine, Rain later.
22	N.E.	30·83	23·89	E.N.E.	29·78	24·33	S.E.	..	24·72	Fine, slight rain,
23	E.S.E.	31·66	23·39	E.N.E.	29·89	26·22	N. to N.E.	..	25·83	Fine.
24	..	31·66	23·89	E.N.E.	30·22	26·22	N.E. to E.	29·17	25·44	Fine.
25	N.E.	30·28	23·89	N.N.E.	30·67	26·56	N.E. to E.	28·72	26·89	Fine.
26	..	31·05	23·33	N.N.E.	30·78	26·22	N.E.	29·06	27·06	Light showers.
27	..	30·83	22·50	E.N.E.	30·56	25·89	N.E.	..	..	..
28	E.N.E.	30·28	23·33	N.N.E.	30·33	25·67	N.E.	..	26·17	..
29	E.N.E.	31·00	23·28	N.N.E.	30·00	22·78	N.E.	..	26·22	..
30	E.N.E.	30·95	22·89	E.N.E.	30·22	25·56	N.E.	..	26·22	..
Average for the month	..	30·08	23·47	..	29·90	24·94	..	29·00	25·41	
Average daily range		6·61°			4·96°			3·59°		

November, 1921.	6-30 A.M.		8 A.M.		10 A.M.		12 Noon.		3 P.M.		6 P.M.		8 P.M.	
	Sea.	Air.	Sea.	Air.	Sea.	Air.	Sea.	Air.	Sea.	Air.	Sea.	Air.	Sea.	Air.
1	..	..	..	..	..	..	..	..	..	..	..	..	..	..
2	..	..	..	..	..	..	..	..	..	..	..	..	..	..
3	*28·1	24·17	†28·0	25·56	‡27·9	25·27	28·2	26·94	28·6	27·78	28·3	27·22	28·2	27·22
4	27·9	25·56	..	..	28·0	25·27	28·3	25·83	28·1	26·39	27·9	26·11	27·9	26·11
5	27·8	25·44	27·9	25·83	..	..	28·1	27·50	28·0	27·50	..	..	27·8	26·22
6	27·9	26·66	28·1	27·78	..	..	28·3	27·94	§28·4	27·22	28·2	26·94	27·8	26·94
7	28·0	25·83	..	..	..	..	28·3	28·22	..	..	..	..	27·7	27·78
8	27·8	..	27·8	..	..	..	28·1	27·78	27·8	27·22	27·7	26·22	22·6	26·17
9	27·4	..	27·4	..	..	..	27·6	..	27·6	26·94	27·6	26·11	27·6	25·39
10	26·8	26·39	..	..	..	..	27·9	..	28·0	28·61	27·7	27·78	27·6	26·56
11	26·8	23·61	27·0	26·11	27·2	27·67	27·4	26·39	..	..	..	..	..	..
12														
	6-0 A.M.													
17	27·7	24·28	27·9	27·22	28·0	28·94	28·2	29·11	28·2	28·06	27·8	26·94	27·6	25·83
18	27·6	24·72	27·9	26·17	28·1	28·11	27·6	28·33	27·6	27·92	..	..	27·7	25·95
19	27·7	24·72	27·8	27·67	27·6	27·22	27·9	27·61	27·8	27·44	27·8	26·94	27·8	26·44
20	27·9	25·56	27·9	26·28	..	..	..	..	..	..	27·9	25·56	27·7	25·83
21	27·8	25·61	27·9	28·06	..	..	28·1	28·06	28·2	26·28	27·8	25·28	27·8	24·72
22	27·7	24·72	28·1	27·56	27·8	26·22	28·2	28·06	28·6	29·56	27·9	26·78	27·8	26·44
23	27·9	26·11	28·0	27·50	28·1	27·89	28·3	29·06	28·6	28·89	28·1	27·56	28·1	27·22
24	27·8	26·78	..	..	27·9	28·56	28·3	28·89	28·3	29·06	27·9	27·94	27·9	27·78
25	27·9	27·22	..	..	27·8	28·11	27·8	28·56	28·1	28·39	28·0	27·89	27·9	27·78
26	27·9	26·94	27·9	27·78	27·7	28·22	27·9	28·89	28·1	28·56	..	..	27·7	27·44
27	..	..	..	..	..	..	..	..	..	..	..	..	27·9	26·39
28	27·6	26·39	27·7	27·11	27·6	27·89	27·7	27·94	28·1	26·39	28·0	27·39	27·8	27·22
29	27·9	26·66	27·8	27·22	27·6	28·22	28·1	28·39	28·3	28·33	27·8	27·50	27·6	27·00
30	27·7	26·56	..	..	27·6	27·78	27·8	28·06	27·7	27·89	27·8	27·50	27·7	27·22
Average	27·85	25·70	27·80	26·99	27·80	27·11	28·04	28·08	28·03	27·89	27·88	26·88	27·79	26·56
Difference	2·10		0·81		0·69		—0·04		0·14		1·00		1·23	

* Taken at 5-0 A.M. † Taken at 7-0 A.M. ‡ Taken at 9-0 A.M. § Taken at 4-0 P.M.

January, 1922.	Victoria Point, Burma.			Port Blair, Andamans.			Nankauri Harbour, Nicobars.			
	Direction of Wind.	Maximum Temp.	Minimum Temp.	Direction of Wind.	Maximum Temp.	Minimum Temp.	Direction of Wind.	Maximum Temp.	Minimum Temp.	Weather.
		°C	°C		°C	°C		°C	°C	
2	N.E.	31·66	23·33	Calm	29·22	22·00		28·06	25·61	
3	Calm	31·22	23·83	N.N.E.	29·56	23·67		28·17	25·72	
4	Calm	..	22·78	N.N.E.	29·83	25·33		27·61	24·00	Rain storms.
5	Calm	30·33	24·56	N.N.E.	29·78	25·22		26·94	23·78	Storms.
6	N.	32·00	24·17	E.S.E.	29·00	26·11		26·94	23·44	Storms.
7	Calm	31·66	23·61	N.N.E.	31·33	25·72		27·61	23·33	
8	Calm	32·17	21·94	N.N.E.	30·00	25·33		27·78	..	
9	Calm	31·72	21·67	N.N.E.	30·11	24·56		..	24·56	
10	Calm	32·17	23·61	N.N.E.	29·89	25·11		29·17	..	
11	E.N.E.	33·89	22·67	..	..	..		..	25·95	
12	..	..	..	N.N.W.	30·00	22·78		28·06	25·28	
13	N.E.	30·00	23·33	N.N.E.	29·67	23·67		..	25·72	
14	Calm	30·61	23·05	N.N.E.	29·44	23·11		28·17	25·61	
15	Calm	30·28	23·28	N.N.E.	29·44	22·33		28·06	25·67	
16	E.	30·50	23·61	N.N.E.	29·44	22·67		..	24·56	
17	E.	30·50	23·05	N.N.W.	29·56	20·67		28·06	25·28	
18	N.E.	31·05	24·17	N.N.E.	28·67	21·56		28·33	25·11	
19	Calm	30·00	23·39	N.N.E.	28·11	23·11		26·61	..	Rain storms.
20	..	..	..	..	..	..		26·94	..	Rain.
21	Calm	29·44	22·17	N.N.E.	26·56	22·33		..	24·22	
22	Calm	33·39	23·33	E.S.E.	26·22	21·17		28·33	25·22	
23	Calm	31·44	22·83	N.N.E.	28·78	24·22		..	24·39	
24	N.E.	31·56	23·44	N.N.W.	29·44	23·67		..	26·39	
Average for month.		31·28	23·30		29·24	23·52		27·80	24·89	
Average daily range		7·98			5·72			2·91		

January, 1922.	6 A.M.		8 A.M.		10 A.M.		12 Noon.		3 P.M.		6 P.M.		8 P.M.	
	Sea.	Air.	Sea.	Air.	Sea.	Air.	Sea.	Air.	Sea.	Air.	Sea.	Air.	Sea.	Air.
2	27·6	25·67	27·7	26·78	27·9	27·50	28·0	27·78	27·8	27·61	27·7	26·72	27·7	26·44
3	27·6	25·83	27·6	26·89	..	..	27·8	27·89	27·8	27·83	27·7	26·94	27·7	26·39
4	27·6	25·83	..	..	27·8	26·39	27·7	25·56	27·7	26·94	27·6	24·17	27·5	27·44
5	27·4	26·39	27·4	26·66	27·6	26·94	27·6	25·22	27·5	25·72	27·4	26·78	27·3	26·89
6	27·1	25·44	27·4	26·50	27·6	27·11	27·6	26·00	27·1	26·66	27·2	26·39	27·1	26·66
7	26·8	23·56	27·0	26·00	26·9	27·33	27·1	27·61	27·6	27·50	27·4	27·00	27·8	26·94
8	27·4	26·56	27·4	26·33	..	..	27·5	27·72	27·6	27·61	..	..	27·4	26·83
9	26·9	26·05	27·3	26·61	27·3	27·44	27·4	27·72	27·5	28·11	..	..	27·0	26·78
10	26·9	24·17	27·2	25·67	..	..	..	..	..	..	27·4	26·89	26·9	26·66
11	26·9	25·83	27·3	26·39	27·3	27·50	27·3	27·39	27·6	27·61	..	..	27·0	26·72
12	27·1	26·11	27·4	26·50	27·5	27·17	27·6	27·56	27·7	27·72	27·5	27·22	27·1	26·78
13	27·1	26·22	27·3	26·66	27·1	27·33	..	..	..	..	27·4	27·50	27·0	26·66
14	27·0	26·17	..	..	27·3	27·61	27·6	28·00	27·6	28·06	..	..	27·4	26·89
15	27·0	25·78	27·3	26·66	..	..	..	..	..	..	..	..	..	..
16	27·3	25·83	..	..	27·6	27·67	27·7	28·06	..	..	..	..	..	..
17	27·3	..	27·1	26·28	27·2	27·22	..	..	..	..	27·6	26·61	27·3	26·11
18	27·1	25·56	27·2	26·39	..	..	27·4	27·61	..	..	27·5	26·39	27·4	25·89
19	27·0	25·67	27·5	26·22	..	..	27·2	26·33	..	..	..	..	27·0	26·11
20	27·1	26·39	27·1	26·83	27·3	26·94	27·4	26·66	27·3	25·72	26·8	24·33	26·7	24·44
21	26·7	24·50	27·4	25·61	27·2	26·39	27·5	26·28	..	..	..	..	27·0	26·66
22	27·4	25·28	..	..	..	..	..	..	..	..	..	..	27·4	25·56
23	27·4	24·44	27·1	27·22	..	..	..	..	27·9	29·94	..	..	27·6	27·22
24	27·3	25·83	27·7	..	27·8	28·83	27·9	28·33	..	..	..	..	27·6	27·06
Average	27·17	25·54	27·35	26·45	27·43	27·29	27·55	27·16	27·59	27·46	27·52	26·36	27·31	26·53
Difference	1·63		0·90		0·14		0·39		0·13		1·16 *		0·78	

* **Here again the increased difference is probable attributable to rain storms in the late afternoon, as in the previous month.**

February, 1922.	Victoria Point, Burma.			Port Blair, Andamans.			Nankauri Harbour, Nicobars.			
	Direction of Wind.	Maximum Temp. °C	Minimum Temp. °C	Direction of Wind.	Maximum Temp. °C	Minimum Temp. °C	Direction of Wind.	Maximum Temp. °C	Minimum Temp. °C	Weather.
2	N.E.	31·22	23·89	N.N.E.	29·56	23·11	N.E.	28·89	25·56	Rain storms.
3	Calm	31·89	23·89	N.N.E.	29·00	23·78	N.E.	28·89	25·56	Rain at night.
4	N.E.	31·78	23·05	E.N.E.	29·33	23·00	N.E.	..	24·22	Rain at night.
5	N.	29·89	22·72	N.N.E.	29·78	24·22	N.E.	27·61	24·72	..
6	E.	30·28	22·83	N.N.E.	29·67	25·22	N.E.	27·72	23·33	Rain.
7	Calm	32·78	25·00	N.N.E.	30·22	22·89	N.E.	27·22	23·78	Rain.
8	Calm	33·22	24·17	E.N.E.	28·00	23·00	N.E. to E.S.E.	27·33	23·22	Rain.
9	E.N.E.	33·78	23·33	N.N.E.	28·56	23·44	S.E. to E.	28·06	..	Rain.
10	E.N.E.	31·61	24·56	N.N.E.	30·22	24·00	S.E.	..	..	..
11	N.E.	30·89	23·39	N.N.E.	30·33	24·33	S.E.	..	25·44	..
12	N.E.	31·61	24·39	N.N.E.	30·78	24·56	S.E. to E.N.E.	..	25·28	..
13	N.E.	35·00	23·33	N.N.W.	30·22	23·44	N.E.	..	26·56	Storms.
14	N.E.	31·72	23·28	N.N.W.	29·78	22·89	N.E.	28·61	..	..
15	Calm	31·56	24·50	N.N.W.	31·56	22·89	N.E.	28·33	24·39	..
16	N.E.	31·56	24·56	N.N.W.	29·89	21·89	N.E.	..	24·28	..
17	N.E.	28·28	23·83	N.W.	29·06	22·00	N.E.	..	24·44	..
18	Calm	31·39	23·61	N.W.	31·11	21·89	N.E.	28·72	24·72	..
19	E.N.E.	30·11	23·22	N.W.	29·89	22·00	N.E. to N.W.	29·00	24·22	..
20	N.E.	31·05	23·28	N.N.W.	29·56	22·22	N.E.	..	22·89	..
21	N.E.	31·61	24·94	N.E.	29·78	23·00	S.E. to N.E.	..	26·11	..
22	N.E.	32·78	25·00	N.W.	30·00	23·33	N.E.	28·89	25·83	..
23	N.E.	32·33	24·00	N.W.	31·11	22·67	N.E.	29·17	..	..
24	Calm	31·39	24·72	N.N.W.	31·44	21·67	N.W.	..	..	..
Average for month		31·66	23·89		29·75	23·11		28·34	24·70	
Average daily range		7·77			6·64			3·64		

February, 1922.	6 A.M.		8 A.M.		10 A.M.		12 Noon.		3 P.M.		6 P.M.		8 P.M.	
	Sea.	Air.	Sea.	Air.	Sea.	Air.	Sea.	Air.	Sea.	Air.	Sea.	Air.	Sea.	Air.
2	27·4	26·11	27·3	27·00	27·6	28·06	27·8	28·67	..	..	..	..	27·6	25·78
3	27·4	26·11	27·7	26·67	27·6	28·06	..	..	27·9	28·33	27·6	27·22	27·6	26·50
4	27·6	26·28	27·6	26·78	..	..	27·8	28·06	27·9	28·44	..	..	27·7	26·67
5	..	..	27·7	24·89	..	..	27·8	26·39	..	..	27·6	26·56	27·6	26·39
6	27·6	26·56	27·7	25·61	27·8	27·72	27·7	26·17	27·7	26·89	27·6	26·67	27·6	25·00
7	27·2	25·06	27·4	26·50	27·4	26·94	27·5	24·78	27·6	24·67	27·5	25·28	27·3	24·56
8	26·6	24·56	26·9	26·94	..	..	27·4	27·22	..	..	27·6	25·28	27·3	24·94
9	26·5	24·50	27·5	26·05	27·7	26·94	27·8	26·94	..	..	27·8	27·50	27·6	26·78
10	27·5	25·67	27·6	27·22	27·8	28·11	27·9	28·50	28·1	28·61	27·7	27·50	27·6	27·11
11	27·5	26·39	27·6	26·67	..	..	27·9	28·50	28·6	28·33	..	..	28·0	27·22
12	..	..	27·8	26·67	..	..	28·3	28·72	..	..	..	..	..	..
13	27·6	25·67	27·6	26·67	..	..	..	..	..	..	..	..	..	..
14	..	..	27·7	27·33	27·8	28·11	28·0	28·17	28·1	28·11	28·0	27·56	27·7	27·33
15	27·6	26·50	27·7	27·33	..	..	27·7	28·06	..	..	..	..	27·8	26·89
16	27·7	24·39	27·8	26·94	27·8	28·61	27·6	28·56	..	..	..	..	27·8	26·00
17	27·7	24·72	27·8	26·28	28·0	27·22	..	..	..	..	28·4	26·94	28·3	26·11
18	..	..	28·2	27·44	28·1	27·06	28·4	27·78	28·4	28·67	..	..	28·1	26·33
19	..	..	28·3	26·50	..	..	..	..	28·4	28·94	..	..	28·2	26·00
20	28·1	24·44	28·3	..	28·2	26·89	28·3	27·39	28·2	28·83	28·4	27·22	27·8	26·05
21	27·6	22·89	28·1	25·39	28·4	26·56	28·4	27·44	..	..	28·2	28·44	28·2	26·39
22	28·0	26·22	..	..	28·5	28·06	..	..	28·4	28·61	28·5	28·67	28·4	27·50
23	28·0	26·05	28·3	26·67	28·5	28·17	28·5	28·78	28·4	28·89	28·5	..	28·4	27·59
24	28·2	25·72	28·2	26·39	..	..	..	..	28·6	28·56	..	..	28·6	26·44
Average	27·54	25·44	27·76	26·57	27·94	27·61	27·93	27·65	28·33	28·14	27·95	27·07	27·87	26·39
Difference	2·10		1·19		0·33		0·28		0·19		0·88		1·48	

March, 1922.	Victoria Point, Burma. Direction of Wind.	Maximum Temp. °C	Minimum Temp. °C	Port Blair, Andamans. Direction of Wind.	Maximum Temp. °C	Minimum Temp. °C	Expedition Harbour, Nicobars. Direction of Wind.	Maximum Temp. °C	Minimum Temp. °C	Weather.
2	Calm	31·61	21·61	N.N.W.	30·78	24·11	N.E. to E.	..	25·56	
3	Calm	30·83	25·50	N.N.W.	31·33	23·00	E.N.E. to W.	..	25·56	
4	Calm	30·44	23·33	N.N.W.	31·33	21·33	..	..	25·95	
5	Calm	30·00	23·33	N.N.W.	30·11	21·11	..	..	25·33	
6	..	..	..	..	..	..	..	..	25·11	
Average		30·72	23·33		30·89	22·39		29·92*	25·50	
Average daily range		7·39			8·50			4·42		

* This figure is taken from the highest average of the daily series of observations.

October, 1922.	Victoria Point, Burma. Direction of Wind.	Maximum Temp. °C	Minimum Temp. °C	Port Blair, Andamans. Direction of Wind.	Maximum Temp. °C	Minimum Temp. °C	Nankauri Harbour, Nicobars. Direction of Wind.	Maximum Temp. °C	Minimum Temp. °C	
19	..	..	..	..	..	..				Occasional rain.
20	S.E.	29·67	22·83	Calm	29·00	24·22				
21	..	..	..	..	..	..	..	..	..	..
22	Calm	29·39	24·89	Calm	31·44	24·78				
23	N.E.	29·72	24·44	N.N.E.	31·44	24·78				Slight rain.
24	N.E.	26·66	24·56	N.N.E.	30·22	25·11				
25	..	..	..	..	..	..	..	..	..	..
26	Calm	31·66	24·89	E.	29·78	26·66				
27	N.E.	30·83	24·89	E.	30·44	26·89				
28	N.E.	32·50	24·72	N.N.E.	30·78	24·78				Slight rain.
29	N.E.	31·56	25·06	N.N.E.	30·11	25·56				Occasional rain.
30	N.E.	28·33	23·89	E.S.E.	30·11	25·11				
31	Calm	31·95	25·28	N.N.E.	29·11	25·00				
Average		30·23	24·55		30·24	25·29		25·23	28·14*	
Average daily range		5·68			4·95			2·91		

* These figures are taken from the two-hourly readings and, therefore, the minimum temperature is probably too high, so that the difference would also be somewhat greater.

March, 1922.	6 A.M.		8 A.M.		10 A.M.		12 Noon.		3 P.M.		6 P.M.		8 P.M.	
	Sea.	Air.	Sea.	Air.	Sea.	Air.	Sea.	Air.	Sea.	Air.	Sea.	Air.	Sea.	Air.
2	28·4	25·56	28·4	..	..	..	28·6	29·72	28·7	29·44	28·7	29·83	28·7	27·39
3	28·3	25·78	28·6	27·06	28·3	27·78	29·1	30·00	29·6	30·39	..	..	28·8	27·28
4	28·1	26·05	28·5	28·17	28·0	28·50	28·7	29·33	..	..	..	..	28·6	27·78
5	..	..	28·7	26·94	..	..	..	..	..	..	..	..	..	..
6	28·3	25·11	28·6	26·39	..	..	..	..	..	..	..	..	..	..
Average	28·28	25·63	28·56	27·13	28·15	28·14	28·80	29·69	29·15	29·92	28·70	29·83	28·70	27·48
Difference	2·65		1·43		0·01		—0·89		—0·77		—1·13		1·22	

October, 1922.	6 P.M.		8 A.M.		10 A.M.		12 Noon.		2 P.M.		4 P.M.		6 P.M.		8 P.M.	
	Sea.	Air.	Sea.	Air.	Sea.	Air.	Sea.	Air.	Sea.	Air.	Sea.	Air.	Sea.	Air.	Sea.	Air.
19	27·89	25·67	28·06	26·61	27·83	26·61	27·78	27·78	27·78	26·61	..	..	27·78	26·66	27·72	26·11
20	27·67	25·28	27·78	27·64	..	..	28·56	28·56	..	..	28·11	27·50	27·83	27·22	27·83	27·22
21	27·67	24·72	27·07	26·67	..	..	27·78	26·67	27·72	25·72	..	..	27·78	26·11	27·72	26·11
22	27·50	26·11	27·83	26·94	..	..	28·06	26·94	..	..	..	..	27·89	27·72	28·06	26·67
23	27·61	25·56	27·89	27·78	27·78	28·61	28·11	28·61	28·06	28·33	27·89	28·33	27·83	27·50	27·67	25·83
24	27·34	25·56	27·39	27·39	27·61	27·78	27·78	27·89	..	..	27·83	27·94	..	..	27·72	25·72
25	27·50	25·67	27·01	27·50	27·78	28·33	27·83	30·00	28·22	29·39	27·83	27·78	27·78	26·67	27·89	25·33
26	27·78	24·94	27·78	25·56	28·06	27·28	27·89	27·83	27·78	27·83	27·83	26·72	27·83	26·17	28·06	26·17
27	27·72	25·61	27·83	27·00	28·06	27·22	28·06	27·78	28·33	28·33	28·33	27·28	28·33	26·44	27·72	25·61
28	27·83	25·56	28·00	25·61	28·28	26·44	28·28	26·72	28·89	28·94	..	..	27·78	27·22	28·06	26·67
29	27·72	25·33	27·78	25·61	28·06	25·89	28·28	27·28	28·39	26·67	28·39	27·83	28·22	26·17	27·89	26·17
30	27·78	24·50	27·72	23·95	28·06	26·72	28·33	27·83	28·06	26·44	28·28	27·28	28·33	26·17	28·22	4·782
31	27·22	23·95	27·78	24·22	27·22	24·22	27·72	24·50	27·78	26·17	27·89	26·44	27·78	25·06	..	..
verage	27·63	25·72	27·78	26·42	27·87	26·91	28·04	27·57	28·10	27·44	28·04	27·45	27·93	26·59	27·88	26·03
ifference	2·36		1·36		0·96		0 47		0·66		0·59		1·34		1·85	

November, 1922.	Victoria Point, Burma.			Port Blair, Andamans.			Nankauri Harbour, Nicobars.			
	Direction of Wind.	Maximum Temp.	Minimum Temp.	Direction of Wind.	Maximum Temp.	Minimum Temp.	Direction of Wind.	Maximum Temp.	Minimum Temp.	Weather.
		°C	°C		°C	°C		°C	°C	
1	Calm	31·22	23·89	E.	30·78	25·00				
2	Calm	29·17	24·00	S.S.E.	30·11	24·44				
3	N.E.	25·67	23·33	N.N.E.	31·11	24·22				
4	E.	26·39	23·28	N.N.W.	29·44	23·33				Rain.
5	N.E.	27·22	23·89	N.N.E.	27·22	23·89				
6	N.E.	29·56	23·83	S.S.E.	29·78	25·22				
7	Calm	28·33	24·39	E.N.E.	30·78	26·22				Rain.
8	Calm	28·06	22·67	N.N.W.	30·22	25·00				
9	Calm	29·33	23·67	W.	29·22	25·56				
15	Calm	27·00	23·05	..	..	..		..	..	Rain.
16	Calm	27·83	23 28	N.W.	30·22	23·89				
17	..	..	..	..	..	..		..	..	..
18	E.	29·89	23·61	W.N.W.	26·89	23·33				
19	Calm	30·00	23·61	S.S.W.	26·89	23·33				
20	N.	31·00	24·17	S.S.E.	28·33	23·44				
21	E.S.E.	29·94	22·78	Calm	29·89	24·00				
22	N.E.	24·50	22·22	E.N.E.	29·22	25·00				
23	E.N.E.	25·83	23·05	E.N.E.	29·78	24·78				
24	N.N.E.	27·67	23·39	E.N.E.	29·44	24·11				
25	..	..	..	..	..	..	..	..	..	..
26	Calm	29·44	23·83	E.	29·44	24·11				
27	E.	27·72	22·50	E.	30·00	23·00				
28	N.N.E.	26·22	20·56	..	..	..	..	..	..	..
29	..	..	..	..	..	..	..	..	..	..
30	N.N.E.	26·39	21·78	E.	29·11	25·33				
Average daily range.		28·11	23·22		29·39	24·36		27·96*	25·15	
Difference.		4·89			5·03			2·81		

* These figures are taken from the two-hourly readings.

November, 1922.	6 A.M.		8 A.M.		10 A.M.		12 Noon.		2 P.M.		4 P.M.		6 P.M.		8 P.M.	
	Sea.	Air.	Sea.	Air.	Sea.	Air.	Sea.	Air.	Sea.	Air.	Sea.	Air.	Sea.	Air.	Sea.	Air.
1	27·33	25·06	27·72	26·17	27·89	27·83	27·89	27·83	28·33	27·72	28·22	27·28	28·06	26·61	27·83	25·61
2	27·50	25·61	27·78	27·78	..	..	28·06	28·68	28·22	28·67	28·06	26·72	27·89	26·61	28·06	26·44
3	27·50	24·50	28·06	25·61	27·50	26·17	..	..	27·78	26·17	27·72	25·61	27·17	24·78	27·50	25·17
4	27·50	25·06	27·67	25·56	27·67	25·06	..	..	27·56	25·56	27·50	25·39	27·50	25·06	27·50	25·06
5	27·17	25·06	..	..	27·22	25·06	..	..	27·22	24·78	27·50	24·50	27·33	23·95	27·06	23·95
6	27·00	23·95	27·33	25·06	27·78	28·11	28·00	28·94	28·06	29·44	27·89	26·83	28·28	25·61	27·72	25·33
7	27·17	25·61	27·72	27·83	27·89	27·56	28·83	28·11	28·11	28·67	28·44	27·56	27·89	26·61	27:33	25·61
8	27·33	25·56	27·89	27·83	27·33	25·61	27·78	28·11	28·11	25·72	28·06	26·17	27·72	25·56	27·50	23·17
9	27·33	24·50	27·56	24·50	..	..	..	..	..	..	..	..	..	..	..	..
15	27·33	25·56	27·33	26·72	27·83	28·94	29·22	30·06	27·83	25·39	27·56	23·95	27·22	25·06	26·94	24·50
16	27·22	23·95	..	..	27·33	24·22	27·33	24;50	27·44	24·78	27·33	24·50	27·33	25·33	27·33	25·61
17	26·94	23·95	27·22	25·06	27·22	25·89	..	..	27·50	26·72	27·50	27·00	27·33	25·61	27·22	25·33
18	27·22	25·06	27·50	26·44	27·72	27·83	27·78	28·11	27·67	27·00	27·72	27·78	27·50	26·72	27·67	26·44
19	27·33	25·33	27·33	26·72	27·50	26 72	27·50	27·28	27·72	27·83	27·89	27·56	27·72	27·00	27·72	25·95
20	27·17	25·56	27·50	25·89	27·72	30·00	28·28	30·33	28·33	30·56	28·06	29·61	27·50	27·83	28·11	26·72
21	27·50	25·61	27·89	27·72	27·67	27·83	27·78	28·94	28·44	28·94	28·22	29·50	27·67	27·83	27·67	27·00
22	27·33	25·83	27·67	26·72	27·78	27·28	28·06	27·72	28·33	27·56	27·89	26·72	..	..	27·50	25·61
23	27·22	25·61	27·72	26·72	28·00	29·50	28·00	28·39	28·33	28·94	28·06	27·56	..	..	28·06	25·61
24	27·50	25·06	27·78	25·61	27·78	26·72	27·72	26·17	27·72	26·72	27·72	26·17	27·50	25·61	27·50	25·06
25	27·33	25·33	27·72	26·72	28·06	25·06	27·67	26·44	27·67	26·44	27·22	25·89	..	..	27·22	25·33
26	27·50	25·06	27·67	26·72	27·89	27·56	27·78	27·56	27·89	27·83	..	..	27·50	26·72	27·72	26·44
27	27·50	25·72	27·33	26·17	27·67	27·00	..	..	27·67	26·67	27·67	25·89	27·72	25·89	27·72	25·89
28	27·22	25·06	27·56	26·44	27·67	26·72	..	..	27·72	27·00	27·72	26·44	27·50	26·05	27·22	25·61
29	27·50	25·33	27·72	26·72	27·78	27·28	27·89	27·83	28·00	28·39	27·78	27·28	27·22	26·67	27·22	25·89
30	27·67	25·72	27·72	26·72	27·72	27·28	27·72	27·78	27·78	27·83	27·67	27·28	..	..	27·22	26·17
Average.	27·33	25·15	27·63	26·41	27·68	27·01	27·96	27·93	27·90	27·29	27·80	26·67	27·58	26·06	27·52	25·56
Difference.	2·18		1·22		0·67		0·03		0·61		1·14		1·52		1·96	

December, 1922.	Victoria Point, Burma. Direction of Wind.	Maximum Temp. °C	Minimum Temp. °C	Port Blair, Andamans. Direction of Wind.	Maximum Temp. °C	Minimum Temp. °C	Nankauri Harbour, Nicobars. Direction of Wind.	Maximum Temp. °C	Minimum Temp. °C	Weather.
1	N.E.	27·67	22·89	N.N.E.	29·22	24·22		..	..	
2	Calm	29·56	22·89	N.N.E.	29·44	23·89		27·83	25·56	
3	S.E.	29·56	23·89	N.N.E.	29·67	25·56		27·50	25·33	
4	N.N.E.	29·56	24·00	E.	29·67	26·11		..	25·56	
5	..	..	..	..	..	..		27·78	25·50	Rain.
6	..	..	..	E.	30·22	26·22		27·17	23·95	
7	N.N.E.	29·67	23·33	E.S.E.	27·89	24·56		26·17	..	
8	E.	28·44	23·89	S.S.E.	27·33	24·00		27·28	25·56	
14	..	..	..	..	..	..		..	25·61	
15	Calm	29·50	23·33	N.N.E.	29·44	24·44		27·00	25·06	Heavy rain.
16	Calm	28·28	23·56	N.N.E.	29·11	23·67		27·78	25·56	
17	Calm	26·67	23·33	N.N.E.	29·11	23·11		27·78	25·61	Rain.
18	..	..	..	..	..	..		26·44	24·50	
19	N.E.	31·61	23·89	N.N.E.	28·56	23·89		27·83	24·78	
20	Calm	28·44	23·33	N.N.E.	28·22	22·89		27·89	24·50	
21	..	..	..	..	..	..		27·78	25·33	Fine rain.
22	Calm	29·22	22·67	N.N.E.	28·33	23·89		28·11	25·06	
Average		29·01	23·42		28·94	24·34		27·45	25·15	
Average daily range.		5·59			4·60			2·30		

December, 1922.	6 A.M.		8 A.M.		10 A.M.		12 Noon.		2 P.M.		4 P.M.		6 P.M.		8 P.M.	
	Sea.	Air.	Sea.	Air.	Sea.	Air.	Sea.	Air.	Sea.	Air.	Sea.	Air.	Sea.	Air.	Sea.	Air.
1	27·22	25·56	27·50	25·72	27·50	27·56	27·67	27·83	27·67	27·56	27·50	26·72	27·50	26·17	27·22	26·72
2	27·50	25·33	27·22	25·61	27·50	26·72	27·67	27·28	27·78	27·28	27·67	27·28	27·50	26·72	27·22	26·72
3	27·17	25·33	27·22	26·44	27·33	27·28	..	..	..	..	..	..	..	..	..	..
4	27·17	25·56	27·22	26·44	27·50	27·56	27·67	27·17	27·78	27·00	27·72	27·00	27·67	27·00	27·50	26·17
5	27·17	25·50	27·22	25·89	27·17	26·61	..	..	27·11	23·11	27·17	25·78	27·11	25·61	26·67	23·95
6	26·67	23·95	26·78	24·78	26·94	26·17	26·78	25·61	26·89	25·61	26·94	25·06	26·78	24·50	26·67	24·44
7	..	..	26·67	26·44	27·11	27·28	26·94	27·28	26·94	27·28	26·78	25·89	26·78	25·56	26·94	25·61
8	26·67	25·61	26·72	26·72	..	..	..	...	..	..	..	..	..	..	..	..
14	27·50	25·61	27·50	25·89	27·50	27·00	..	..	27·56	24·94	27·33	24·78	27·33	25·06	27·33	25·06
15	27·17	25·06	27·22	23·67	27·67	26·72	28·06	26·67	28·06	27·78	27·72	27·28	27·33	26·17	27·67	25·06
16	27·50	25·56	27·00	26·05	27·67	27·78	27·89	26·83	..	..	..	..	..	..	27·67	26·61
17	27·50	25·61	27·67	26·61	..	..	27·78	26·44	27·67	25·06	27·33	24·50	27·22	24·50	27·33	25·06
18	27·22	25·33	27·50	26·44	27·50	26·44	27·50	27·83	27·50	25·61	27·50	26·17	27·67	25·61	26·78	24·22
19	26·61	24·78	26·61	25·56	27·50	27·00	27·50	27·00	27·67	27·00	27·89	27·72	27·56	26·44	27·50	26·17
20	27·22	24·50	27·33	25·61	27·50	26·72	27·89	27·72	27·78	27·78	27·50	27·56	27·50	25·89	27·50	25·61
21	27·22	25·33	27·33	25·33	27·50	26·44	27·72	27·28	27·89	28·11	27·50	26·17	27·50	26·11	27·33	25·61
22	27·50	25·06	27·67	27·28	27·67	27·83	27·83	28·11	27·83	28·39	27·78	28·00	..	..	..	..
verage	27·19	25·23	27·20	25·91	27·44	27·01	27·61	27·16	27·58	26·54	27·45	26·42	27·34	25·79	27·24	25·50
ifference	1·96		1·29		0·43		0·45		1·04		1·03		1·55		1·74	

January, 1923.	Victoria Point, Burma.			Port Blair, Andamans.			Nankauri Harbour, Nicobars.			
	Direction of Wind.	Maximum Temp.	Minimum Temp.	Direction of Wind.	Maximum Temp.	Minimum Temp.	Direction of Wind.	Maximum Temp.	Minimum Temp.	Weather.
		°C	°C		°C	°C		°C	°C	
14	N.E.	30·50	22·22	E.N.E.	29·78	24·89				
15	N.E.	30·00	21·94	N.N.E.	29·78	22·78				
16	N.E.	30·11	22·22	N.N.W.	29·11	21·89				
17	Calm	32·22	23·44	N.N.W.	29·00	21·89				
18	..	..	..	..	..	..				
Average daily range		30·71	22·45		29·42	22·86		27·26	25·41	
Difference		8·26			6·56			1·85		

January, 1923.	6 A.M.		8 A.M.		10 A.M.		12 Noon.		2 P.M.		4 P.M.		6 P.M.		8 P.M.	
	Sea.	Air.	Sea.	Air.	Sea.	Air.	Sea.	Air.	Sea.	Air.	Sea.	Air.	Sea.	Air.	Sea.	Air.
14	..	..	..	..	..	..	27·39	27·71	27·61	27·56	27·39	27·28	27·33	26·61	27·28	26·61
15	27·22	25·61	27·22	26·33	27·22	27·11	27·33	27·28	27·33	27·28	27·50	..	27·33	27·00	27·22	26·44
16	27·22	25·61	27·11	26·28	27·33	27·28	27·39	27·39	27·39	27·39	27·22	27·28	27·22	26·94	27·22	26·17
17	27·22	25·83	27·22	26·28	27·50	27·50	27·50	27·39	27·50	27·39	..	..	..	..	..	..
18	27·22	24·61	27·17	25·22	27·33	26·22	27·39	26·44	27·44	26·39	27·39	26·22	27·28	25·83	27·22	25·33
Average	27·22	25·41	27·18	26·03	27·34	27·03	27·40	27·24	27·45	27·20	27·37	26·93	27·29	26·59	27·33	26·14
Difference	1·81		1·15		0·31		0·16		0·25		0·44		0·70		1·09	

March, 1925.	6 A.M.		8 A.M.		10 A.M.		12 Noon.		2 P.M.		4 P.M.		6 P.M.		8 P.M.		10 P.M.	
	Sea.	Air.	Sea.	Air.	Sea.	Air.	Sea.	Air.	Sea.	Air.	Sea.	Air.	Sea.	Air.	Sea.	Air.	Sea.	Air.
4	27·0	27·3	27·2	27·4	27·2	27·9	27·0	25·7	27·6	26·7	27·2	26·8	27·2	26·9	27·0	26·3	27·2	26·0
5	27·0	24·9	27·2	26·1	27·4	27·3	28·0	27·9	28·4	28·0	27·6	28·2	27·6	28·2	28·0	27·9	28·0	27·9
6	27·4	25·1	27·2	25·1	27·2	25·7	27·2	26·3	27·4	26·7	27·4	26·9	27·4	26·9	27·0	26·6	27·4	26·1
7	27·0	25·6	27·8	27·4	27·8	27·8	28·0	28·0	28·0	26·7	27·2	25·2	27·6	25·2	27·4	25·9	27·2	25·4
8	27·6	26·4	27·6	26·2	27·6	27·1	27·6	27·2	27·6	28·0	27·6	27·0	27·6	27·0	27·6	27·2	28·0	27·2
9	27·4	26·1	27·4	26·4	26·8	24·4	27·4	26·1	27·4	27·0	27·6	27·1	27·4	26·8	27·0	25·6	27·0	25·0
10	27·0	26·6	27·0	27·3	27·4	28·3	27·6	28·6	27·6	28·7	27·8	28·4	27·6	28·1	27·4	28·0	27·4	27·9
11	26·8	25·3	27·4	27·6	27·6	28·9	27·8	28·6	27·6	28·4	27·4	28·7	27·4	28·1	27·4	27·9	27.2	27·8
12	27·6	27·3	27·6	28·3	27·8	28·9	28·0	29·1	28·0	29·2	27·8	28·7	27·4	27·9	27·0	27·3	27·0	26·9
13	26·8	26·3	27·2	28·0	27·2	27·6	27·6	25·1	27·2	24·4	27·0	24·8	26·8	24·3	26·8	24·1	26·8	24·8
14	26·6	24·4	27·6	27·7	28·0	28·6	28·0	28·8	27·8	28·4	27·6	28·9	27·6	27·2	27·6	26·0	27·2	25·2
15	26·8	24·0	27·6	26·4	27·6	28·3	28·6	28·9	28·0	29·2	28·6	29·0	28·0	28·6	27·6	26·4	27·0	26·0
16	26·8	24·6	27·6	26·9	27·6	27·8	28·0	28·3	28·0	28·7	28·2	28·7	28·0	28·3	27·8	26·7	27·8	26·1
17	27·2	25·1	27·6	26·9	28·0	29·0	28·0	28·9	28·2	29·1	28·2	28·7	28·2	28·1	27·8	26·7	27·0	26·1
18	27·2	25·0	28·0	26·9	28·2	28·7	28·0	29·0	28·0	28·8	28·2	28·7	28·0	27·0	27·6	26·0	27·6	25·9
19	27·6	24·9	27·6	27·0	28·4	28·4	28·6	29·0	28·8	28·7	28·8	28·7	28·6	27·4	28·0	26·4	28·4	26·8
27	26·8	24·8	27·0	26·3	27·0	27·7	27·0	26·8	27·0	25·4	26·8	26·0	27·0	26·1	26·6	25·7	26·6	25·7
28	26·8	25·3	27·0	26·8	28·0	28·1	28·2	28·7	28·6	29·4	28·0	29·3	28·0	28·2	27·8	27·8	27·8	27·3
29	27·4	26·4	27·8	27·4	28·4	28·7	28·6	28·9	29·4	29·0	28·6	29·2	28·4	29·2	28·2	28·6	28·4	28·0
30	28·0	27·4	28·4	28·0	28·4	28·7	28·6	28·7	28·2	27·6	28·2	27·8	28·2	28·7	28·2	27·7	28·0	26·8
31	27·4	25·9	28·0	26·9	28·2	28·9	28·2	28·4	29·0	28·9	29·0	29·4	28·2	29·2	28·2	27·0	27·8	26·8
Average	27·15	25·65	27·51	27·00	27·70	27·99	27·93	27·95	27·99	27·95	27·85	27·91	27·72	27·50	27·52	26·75	27·47	26·41
Difference	+1·50		+0·51		−0·29		−0·02		+0·04		−0·06		+0·22		+0·77		+1.06	

APPENDIX II.

Results of observations on the relative humidity of the atmosphere by means of wet- and dry-bulb thermometers.

October, 1923.

Date.	A.M. 2	4	6	8	10	12	P.M. 2	4	6	8	10	12
9	..	..	..	..	..	..	..	74·02	..	78·09	..	..
10	..	..	..	76·80	81·04	79·72	..	71·39	..	78·20	77·84	..
11	..	..	..	82·44	79·74	73·32	..	70·58	..	76·21	76·83	..
12	..	..	..	84·40	75·05	75·71	..	80·58	..	80·08	76·51	..
13	..	..	..	..	..	..	..	..	..	..	..	..
14	..	..	..	..	..	..	..	72·39	..	72·15	73·20	..
15	..	..	..	77·86	79·28	72·89	..	70·42	..	74·24	75·06	..
16	..	..	..	75·93	76·54	..	..	..	..	..	..	..
17	..	..	..	..	..	..	..	..	..	..	..	..
18	..	..	..	..	..	..	..	..	..	..	..	..
19	..	..	..	92·06	..	71·66	71·00	78·71	80·25	77·45	..	..
20	..	..	85·28	83·15	80·36	79·06	75·26	81·15	81·08	78·53	77·86	82·01
21	82·01	74·06	77·86	81·65	96·46	83·31	79·62	77·57	72·91	77·64	..	..
22	..	..	79·30	76·05	77·38	72·15	68·75	72·13	71·11	74·05	..	..
23	..	..	79·77	80·96	75·92	73·36	70·94	77·25	75·95	77·90	..	..
24	..	..	83·75	84·58	80·51	80·93	77·64	77·57	80·93	80·52	..	..
25	..	..	82·59	79·05	79·55	77·57	77·64	83·14	74·55	79·62	70·43	77·70
26	73·78	70·04	82·01	78·21	77·70	79·84	75·85	74·01	79·91	79·19	..	..
27	..	..	81·88	82·80	85·73	81·58	82·51	77·64	92·34	86·25	..	..
28	..	..	79·02	79·62	80·77	67·37	88·31	84·88	87·36	89·54	..	..
29	..	..	73·52	75·01	76·38	73·78	70·55	72·01	72·01	73·78	..	..
30	..	..	74·43	75·78	73·13	73·91	73·91	66·59	75·78	71·29	73·91	73·67
31	79·55	75·49	78·86	68·37	75·77	..	..	..	..	..	..	..
Average	78·45	*73·20*	**79·86**	79·77	79·49	76·06	76·00	*75·84*	**78·68**	78·04	75·21	77·79

November, 1923.

Date.	A.M.						P.M.					
	2	4	6	8	10	12	2	4	6	8	10	12
1	..	..	74·94	73·71	67·85	67·85	66·74	67·61	67·21	67·85	..	..
2	..	..	75·71	75·59	73·19	69·13	67·85	70·26	72·01	73·78	..	..
3	..	..	76·29	74·38	71·56	68·02	68·48	66·74	..	66·42	..	..
4	..	..	72·81	73·71	..	67·51	..	71·02	..	75·25	81·59	81·23
5	..	75·65	..	89·58	84·81	80·32	..	80·44	..	79·55	77·45	77·31
6	..	66·86	..	..	..	..	..	..	..	..	..	..
7	..	..	..	..	..	..	..	..	..	..	..	..
8	..	..	..	..	..	..	..	..	..	..	..	..
9	..	..	..	..	..	..	..	73·86	..	73·86	70·05	71·90
10	..	71·90	..	77·45	79·39	..	..	73·66	..	72·01	81·88	77·70
11	..	81·74	..	71·61	73·83	73·91	..	76·30	..	..	..	..
12	..	..	77·64	81·09	77·86	74·01	76·82	74·71	73·93	79·77	..	..
13	..	..	81·73	79·62	83·31	82·01	77·99	76·67	81·82	83·44	..	..
14	..	..	75·64	72·07	70·29	65·17	62·21	63·38	69·61	68·43	..	..
15	..	..	79·77	75·69	70·98	68·14	77·83	75·94	77·86	82·01	..	..
16	..	..	77·69	81·71	82·61	79·06	72·81	72·38	68·48	69·09	..	..
17	..	..	79·02	67·25	65·96	66·67	66·71	66·74	72·07	72·07	..	..
18	..	..	79·69	76·43	74·94	63·04	72·65	84·97	87·15	80·37	..	..
19	..	..	77·64	70·12	71·47	65·84	76·57	74·02	73·94	77·79	..	..
20	..	..	70·11	70·16	69·63	71·69	65·76	65·00	68·01	72·78	..	..
21	..	..	76·43	69·49	68·55	68·68	68·68	68·75	68·61	70·17	..	..
22	..	..	74·50	72·07	68·68	67·62	70·22	67·07	68·55	70·94	..	..
23	..	..	81·65	77·22	72·92	75·05	73·26	75·93	75·78	77·70	..	..
24	..	..	66·67	76·92	76·84	72·69	72·06	73·91	73·80	77·70	..	..
25	..	..	85·66	80·82	80·45	79·62	82·65	86·18	79·55	79·69	..	..
26	..	..	79·63	73·91	70·43	70·29	79·78	66·59	84·94	80·44	..	..
27	..	..	76·86	73·71	71·20	71·45	70·22	71·66	70·94	73·83	..	..
28	..	..	80·45	77·11	74·12	77·11	72·54	76·21	75·93	78·18	..	..
29	..	..	83·75	79·69	73·91	73·83	76·59	86·93	82·71	85·18	..	..
30	..	..	88·05	86·85	91·18	87·55	89·74	90·45	85·77	87·12	..	..
Average	..	*74·04*	**77·93**	76·08	74·64	*72·25*	73·10	73·98	75·17	75·98	**77·74**	77·04

December, 1923.

Date.	A.M.						P.M.					
	2	4	6	8	10	12	2	4	6	8	10	12
1	..	..	79·17	74·30	75·64	72·65	70·29	75·07	73·18	75·78	..	..
2	..	..	80·17	77·70	..	78·94	..	75·35	..	85·31	79·24	81·60
3	..	81·73	..	77·79	84·57	77·86	..	83·27	..	81·99	89·79	89·79
4	..	81·87	..	83·70	..	..	..	..	..	..	..	..
5	..	..	..	..	..	..	..	..	..	..	..	..
6	..	..	..	..	..	..	..	..	..	..	..	..
7	..	..	..	..	..	..	..	82·08	..	81·98	86·26	77·86
8	..	73·78	..	74·01	70·41	70·10	..	72·42	..	76·01	77·86	77·86
9	..	..	..	..	77·99	..	..	..	..	..	..	..
10	..	..	77·64	77·86	..	85·03	81·82	75·79	81·15	77·79	..	..
11	..	..	86·01	77·57	75·79	76·57	75·86	75·93	79·99	77·70	..	..
12	..	..	94·63	85·31	91·02	90·24	86·71	87·09	92·29	80·44	..	..
13	..	..	82·57	79·55	..	89·06	86·01	79·69	79·63	79·63	..	..
14	..	..	81·59	81·59	73·92	83·45	83·14	85·77	81·19	85·93	..	..
15	..	..	82·57	78·00	..	76·05	80·93	82·96	85·15	84·92	..	..
16	..	..	84·11	86·99	79·15	79·22	74·77	75·85	79·85	81·11	..	..
17	..	..	81·11	80·46	76·91	77·34	74·20	75·36	77·86	79·16	..	..
18	..	..	79·84	82·91	79.95	78·40	74·80	76·01	92·29	92·57	..	..
19	..	..	83·81	88·95	84·12	80·20	79·77	77·70	81·88	79·77	..	..
20	..	..	78·44	76·30	..	92·58	86·43	85·92	85·97	83·81	..	..
21	..	..	81·08	81·14	77·89	..	76·34	77·56	77·80	79·84	..	..
22	..	..	79·77	79·77	79·81	78·58	77·36	76·43	78·32	76·57	..	..
23	..	..	77·69	78·48	75·36	74·40	75·86	76·57	77·02	77·75	..	..
24	..	..	84·50	75·72	72·64	72·75	69·54	71·00	72·01	..	..	..
25	..	..	80·60	87·09	85·97	84·01	87·30	79·72	90·51	..	..	..
26	..	..	92·13	85·89	84·56	82·22	90·39	80·07	85·18	94·09	..	..
27	..	..	77·62	77·81	74·77	75·11	86·60	84·11	93·78	89·02	..	..
28	..	..	88·22	76·43	78·52	81·36	81·47	77·01	75·79	75·06	..	..
29	..	..	..	..	..	83·16	..	88·24	..	82·42	81·88	81·88
30	..	81·88	..	88·29	97·51	94·60	91·40	84·61	..	82·29	79·91	89·79
31	..	94·70	..	86·05	79·08	81·64	..	83·06	..	..	..	..
Average	..	82·79	82·54	80·75	*79·84*	**80·62**	80·52	*79·28*	81·99	81·71	82·49	**83·10**

January, 1924.

Date.	A.M. 2	4	6	8	10	12	P.M. 2	4	6	8	10	12
3	..	..	..	..	..	..	..	..	..	86·09	81·74	85·91
4	..	77·45	..	88·84	89·15	94·68	..	85·25	..	87·72	86·09	84·75
5	..	77·70	..	84·04	81·94	81·98	..	89·05	..	86·11	86·09	86·11
6	..	86·11	..	84·55	85·36	82·76	...	81·93	..	83·22	81·88	81·88
7	..	79·69	..	77·70	81·35	80·21	..	..	..	..	..	..
Average	..	80·24	..	83·78	84·45	84·91	..	85·41	..	**85·79**	83·95	84·66

February, 1924.

Date.	A.M. 2	4	6	8	10	12	P.M. 2	4	6	8	10	12
20	..	..	..	..	81·93	88·00	83·66	84·05	84·04	88·31	89·65	88·27
21	86·09	86·09	89·79	85·26	81·88	81·09	..	..	..	..	..	..
22	..	..	..	..	..	..	..	..	..	..	..	..
23	..	..	..	..	..	..	..	..	..	..	..	..
24	..	..	88·22	81·74	79·77	77·83	78·67	79·90	82·01	83·95	83·89	83·81
25	81·74	83·81	86·09	..	..	..	..	..	..	..	..	..
26	..	..	..	..	..	..	..	..	..	..	..	..
27	..	..	..	..	79·99	74·06	79·49	81·19	82·84	91·12	90·55	90·55
28	89·06	95·21	91·35	79·77	79·97	..	..	..	..	..	..	..
Avei.ige	*85·63*	88·37	**88·86**	82·26	80·69	*80·14*	80·61	81·71	82·97	87·79	**88·03**	87·54

March, 1924.

Date.	A.M.						P.M.					
	2	4	6	8	10	12	2	4	6	8	10	12
1	..	..	..	..	..	..	..	..	..	..	..	..
2	..	..	..	..	..	..	..	..	..	..	..	..
3	..	..	..	..	..	..	..	..	..	..	..	..
4	..	..	..	..	..	..	..	81·32	..	81·93	81·88	81·88
5	..	81·74	..	83·95	81·93	81·35	..	84·03	..	84·76	86·09	86·09
6	..	83·89	..	83·95	80·58	81·23	..	84·11	..	85·59	81·88	81·88
7	..	81·88	..	86·99	83·89	75·90	..	85·75	..	82·03	81·99	83·88
8	..	83·95	..	..	..	..	..	..	..	..	..	..
9	..	..	..	..	..	..	..	..	..	..	..	..
10	..	..	..	..	..	..	..	69·13	..	75·29	79·69	81·74
11	..	77·56	..	78·85	81·32	77·70	..	72·49	..	72·62	81·88	83·95
12	..	73·66	..	68·35	66·66	..	..	..	..	..	..	..
13	..	..	75·64	70·29	66·92	70·35	73·32	74·00	74·00	75·78	..	..
14	..	..	82·07	74·05	74·60	74·28	76·01	76·07	76·01	75·93	..	..
15	..	..	81·11	77·36	76·07	72·45	77·36	73·61	76·84	77·99	..	..
16	..	..	81·88	76·07	72·41	72·90	75·79	76·64	75·62	77·27	..	..
17	..	..	81·93	74·91	74·33	71·25	75·65	77·36	74·90	73·72	..	..
18	..	..	79·91	74·33	74·33	76·19	74·33	76·19	80·06	80·55	..	..
19	..	..	83·32	74·33	73·82	72·40	76·19	76·19	77·23	85·67	..	..
20	..	..	84·11	82·09	82·11	82·15	79·57	82·16	81·39	84·19	..	..
21	..	..	84·69	81·39	78·85	76·19	80·63	74·33	78·63	82·91	..	..
22	..	..	83·11	80·82	81·39	80·14	80·89	80·14	80·11	81·38	..	..
23	..	..	83·65	83·65	84·97	82·13	80·93	80·59	83·39	80·06	..	..
24	..	..	86·26	79·58	78·85	77·39	74·99	77·36	80·99	82·15	..	..
25	..	..	82·03	80·06	75·75	80·14	76·20	71·76	76·20	81·39	..	..
26	..	..	75·52	72·29	75·76	74·28	75·11	74·31	72·48	80·11	..	..
27	..	..	82·08	83·65	80·63	82·91	83·43	84·20	84·21	81·38	..	..
28	..	..	86·28	82·15	83·43	83·43	80·18	82·17	82·17	83·43	..	..
29	..	..	..	83·29	..	85·42	..	84·17	..	85·47	88·41	84·22
30	..	86·35	..	84·70	84·96	89·80	..	92·02	..	85·49	86·29	90·64
31	..	91·18	..	86·29	..	..	..	..	..	..	..	..
Average		82·53	82·10	79·30	*77·89*	**78·18**	*77·54*	**78·75**	*73·39*	80·71	83·51	**84·29**

April, 1924.

Date.	A.M. 2	4	6	8	10	12	P.M. 2	4	6	8	10	12
1	..	..	..	..	..	..	..	..	..	..	..	..
2	..	..	..	..	..	..	..	88·58	..	84·71	86·29	86·29
3	..	86·81	..	83·65	83·01	84·62	..	77·36	..	79·61	80·02	78·18
4	..	82·03	..	82·56	80·58	..	..	..	..	..	..	..
5	..	..	82·03	80·01	79·30	76·98	80·11	78·18	74·33	78·15	..	..
6	..	..	81·53	82·62	81·40	80·99	76·89	76·31	76·65	82·15	..	..
7	..	..	81·28	80·85	80·11	75·51	73·26	74·27	74·33	76·71	..	..
8	..	..	79·30	79·61	78·15	79·68	78·19	76·20	78·92	78·10	..	..
9	..	..	84·71	80·11	80·14	80·18	80·18	79·40	81·66	80·85	..	..
10	..	..	83·32	80·56	78·10	77·40	78·18	80·18	80·14	82·16	..	..
11	..	..	85·49	84·71	80·64	77·46	78·17	78·22	78·94	80·90	..	..
12	..	..	85·48	80·56	85·49	76·65	73·72	74·54	75·73	78·92	..	..
13	..	..	84·70	79·36	77·15	78·14	75·49	75·49	76·92	77·39	..	..
14	..	..	82·86	82·66	80·09	75·60	67·78	67·76	66·19	73·16	..	..
15	..	..	80·83	77·44	76·20	79·40	78·63	78·66	80·09	80·14	..	..
16	..	..	84·97	78·10	76·70	75·49	78·08	78·14	81·43	76·71	..	..
17	..	..	80·11	77·46	77·52	74·99	74·27	74·27	80·09	89·27	..	..
18	..	..	85·49	82·92	83·68	81·38	81·27	87·72	84·19	81·38	..	..
19	..	..	79·58	80·11	78·98	95·21	93·68	97·54	86·99	92·89	..	..
20	..	..	86·30	81·36	85·90	84·25	80·21	81·66	83·68	84·19	..	..
21	..	..	85·50	82·13	73·92	77·44	72·48	72·09	67·93	74·45	..	..
22	..	..	..	81·64	80·15	78·18	..	73·93	..	78·81	78·18	80·11
23	..	86·29	..	80·88	96·01	86·71	..	84·17	..	80·06	82·08	84·13
24	..	92·48	..	80·82	82·09	77·45	..	..	..	..	..	..
25	..	..	..	..	..	..	..	..	..	83·42	90·71	82·11
26	..	90·86	..	87·08	84·71	80·66	..	82·19	..	78·59	80·09	82·21
27	..	82·21	..	81·55	82·78	84·07	..	82·79	..	79·54	82·07	82·17
28	..	82·15	..	75·14	73·76	79·26	..	80·66	..	..	..	..
29	..	..	..	..	..	..	..	..	..	..	..	..
30	..	..	..	..	..	..	..	..	..	..	..	..
Average	..	**86·12**	83·15	80·95	80·66	79·90	*77·68*	79·18	78·13	80·59	**82·78**	82·17

March, 1925.

Date.	A.M. 2	4	6	8	10	12	P.M. 2	4	6	8	10	12
1	..	..	..	..	..	..	..	..	..	..	..	..
2	..	..	..	..	..	..	..	..	..	..	..	..
3	..	..	..	..	..	..	..	..	..	..	..	..
4	..	..	85·5	84·7	86·8	85·8	87·9	86·2	90·5	85·9	90·8	..
5	..	..	91·6	91·8	83·4	80·3	81·1	82·8	79·3	84·3	84·3	..
6	..	..	95·7	89·6	87·7	85·9	88·8	85·2	86·9	90·8	92·7	..
7	..	..	94·8	83·4	85·4	80·3	91·9	92·6	94·7	93·9	94·7	..
8	..	..	89·9	91·9	88·9	86·1	82·7	86·9	86·0	85·5	86·1	..
9	..	..	93·9	94·0	91·4	89·8	86·9	86·9	89·7	93·9	94·7	..
10	..	..	94·5	87·7	78·6	76·4	79·5	81·2	81·9	85·4	86·2	..
11	..	..	94·9	91·5	78·0	83·5	85·4	85·5	87·9	89·8	92·3	..
12	..	..	96·1	86·2	84·7	82·1	78·7	74·7	84·6	85·3	87·7	..
13	..	..	89·8	81·9	85·3	90·5	95·6	94·7	93·7	94·6	90·4	..
14	..	..	94·6	87·0	77·1	74·8	80·4	78·0	90·6	87·7	92·6	..
15	..	..	97·7	91·9	82·3	78·8	77·3	79·7	85·5	94·9	93·9	..
16	..	..	95·6	93·4	84·6	80·4	78·0	80·5	82·0	90·5	92·4	..
17	..	..	95·6	86·8	81·3	74·1	74·8	79·5	82·7	93·4	93·3	..
18	..	..	95·6	95·2	78·8	75·5	74·7	72·6	85·2	89·8	90·5	..
19	..	..	95·6	84·4	81·2	75·5	77·9	82·0	86·9	90·7	91·4	..
20	..	..	..	..	..	..	..	..	..	..	..	..
21	..	..	..	..	..	..	..	..	..	..	..	..
22	..	..	..	..	..	..	..	..	..	..	..	..
23	..	..	..	..	..	..	..	..	..	..	..	..
24	..	..	..	..	..	..	..	..	..	..	..	..
25	..	..	..	..	..	..	..	..	..	..	..	..
26	..	..	..	..	..	..	..	..	..	..	..	..
27	..	..	94·7	89·8	83·4	88·8	93·8	89·8	89·8	93·9	91·8	..
28	..	..	92·6	87·1	78·7	79·5	75·6	76·6	83·5	89·8	89·7	..
29	..	..	94·9	93·4	81·3	80·5	80·5	79·6	81·3	89·8	91·5	..
30	..	..	90·6	87·0	82·0	83·6	85·3	87·0	81·2	88·9	93·4	..
31	..	..	94·8	94·3	83·7	84·5	80·5	80·6	82·0	90·6	87·7	..
Average	..	..	**93·9**	89·2	83·6	*81·7*	83·2	83·0	86·0	90·0	**90·9**	..

April, 1925.

Date.	A.M. 2	4	6	8	10	12	P.M. 2	4	6	8	10	12
17	..	..	..	75·7	..	69·8	..	67·8	..	69·7	..	70·4
18	..	70·3	..	71·6	..	69·5	..	66·0	..	76·9	..	78·1
19	..	78·0	..	77·6	..	75·5	..	74·3	..	79·4	..	80·1
20	..	82·0	..	82·7	..	..	..	..	..	..	..	..
21	..	..	..	..	..	..	..	..	..	..	..	..
22	..	..	..	..	..	84·2	..	82·9	..	87·9	..	92·9
23	..	86·2	..	85·2	..	79·6	..	82·7	..	87·1	..	86·3
24	..	85·9	..	92·3	..	79·9	..	80·7	..	..	..	..
25	..	..	..	..	..	..	..	..	..	..	..	..
26	..	..	..	84·1	..	79·3	..	83·7	..	79·6	..	80·1
27	..	86·2	..	84·8	..	92·9	..	75·8	..	76·2	..	82·2
28	..	82·1	..	..	..	76·8	..	76·2	..	80·7	..	82·2
29	..	82·9	..	77·4	..	80·1	..	80·1	..	82·9	..	80·1
30	..	78·0	..	73·4	..	..	..	..	..	..	..	..
Average	..	81·29	..	80·48	..	78·76	..	77·02	..	80·04	..	**81·38**

THE TEMPERATURE AND SALINITY OF THE COASTAL WATER OF THE ANDAMAN SEA.

By R. B. Seymour Sewell, M.A., F.A.S.B., F.L.S., F.Z.S., Lt.-Col., I.M.S.,
Director, Zoological Survey of India.

density of distilled water at 4°C. as 1·00000. The accuracy of the instrument, therefore, appears to be equal to that of total immersion hydrometers, such as those used by Nansen in his work on the density of the waters of the Norwegian Sea, in which the calculated error in the specific gravity (σ_0) is 0·005. Care must, of course, be taken to ensure that the hydrometer is perfectly clean and free from grease, and this was done by careful cleaning in soap and water and ether. As a rule, owing to pressure of work, only one determination was made, but occasionally two observations were taken consecutively on the same water-sample and the mean of the two readings was accepted. The samples of surface-water were taken by means of a canvas bucket, which was lowered over the side and, as far as was possible, was so manipulated that the water was taken from the upper six inches; at the same time a reading of the temperature of the water was taken by immersing a thermometer in the water as soon as the bucket was on deck. Care was, of course, taken to see that the sample was obtained from well forward, so as to avoid the possibility of any contamination from the discharge of pipes, etc. In order to avoid, as far as possible, any change of temperature in the sample during each experiment, all water-samples were stored in my laboratory on the R.I.M.S. "Investigator" in spring-stoppered bottles for at least 24 hours prior to examination, so that they might take up the laboratory temperature. From the result obtained by the hydrometer I have calculated the density of the water at the mean temperature during its examination as compared with that of distilled water at 4°C. and have taken the corresponding values of the salinity and specific gravity, etc., from Knudsen's Tables.

In 1922, at the suggestion of Mr. D. J. Matthews, of the Hydrographic Department, Admiralty, I commenced to use the titration method, as recommended by the Conseil Permanent pour l'Exploration de la Mer. The titration apparatus and the standard sea-water were supplied by the Laboratoire Hydrographique of Copenhagen. Care was taken to ensure that the temperature of all water samples and of the standard sea-water was the same—a difference of not more than 0·1°C. being aimed at. To ensure this all water-samples and the standard water were stored in glass, spring-stoppered bottles, supplied by the Laboratoire Hydrographique. The time for which the samples were stored varied according to circumstances from 24 hours to one week, but, as care was taken to see that all rubber washers and springs were in good order, any change in the salinity of the sample due to evaporation must have been so slight as to be negligible. Matthews (1926, p. 171) in a recent paper on the physical oceanography of the Indian Ocean, states that as a result of certain experiments carried out by him he has come to the conclusion that the composition of the sea-water of the Indian region is so nearly identical with that of the water of the North Atlantic that data derived from the examination of the latter can be applied equally well and with equal reliability to the water of the former area. In a comparison of the density of six samples determined by a pycnometer and that calculated by Kundsen's Tables from the " halogen " content, he found that the difference was not greater than 0·015 per thousand, and from this he concludes that "the water of the Indian Ocean is of the same composition as that of the North Atlantic Ocean as far as the larger constituents, on which the salinity depends, are in question." Be this true or not, his experience based on samples collected along the main steamship routes or in the neighbourhood of the Seychelles and Mauritius, that is to say, on what we may term 'oceanic water' is very different from mine in respect of water samples collected for the most part in areas within a comparatively small distance from land, either continental, such as round India itself, or in the vicinity of oceanic islands, such as the Nicobars. In order to test the results obtained by the two methods, *i.e.* by the hydrometer and by titration, I carried out a number of observations on the same water-samples and the results obtained are given on p. 136 in Table 19. I found that in nearly every instance the 'Buchanan' hydrometer gave a slightly higher value of the specific gravity (σ_0) of the sample than that obtained by calculation from the results of the titration method. In this respect my results agree fairly well with those obtained by the International Central Laboratory in Christiania in certain samples of water from the depths of the Norwegian Sea that were examined by them by both titration and by total immersion hydrometers (*vide* Helland-Hansen and Nansen, 1909, p. 334 *et seq.*). In all I examined, in this series of observations, 34 samples by both methods; of these samples, 24 are from Nankauri Harbour in the Nicobar Islands and the remaining 10 are from the open waters between Nankauri Harbour and Port Blair in the Andamans. The average difference between the results obtained by the two methods throughout the whole series is 0·1355, the hydrometer reading being the higher of the two in 28 cases. In six samples the titration method gave results higher than those given by the 'Buchanan' hydrometer and it is interesting to note that all these samples were from the waters of Nankauri Harbour.

In the case of the samples from Nankauri Harbour, the difference between the results obtained by the two methods ranges from +0·431 to −0·873, a range of 1·304, and the average variation is −0·135; whereas in the samples from the open sea the difference varies only from −0·349 to −0·037, a range of 0·312, and the average variation is −0·146. The average variation in the two areas is, therefore, very similar and it would seem probable that the greater range found in Nankauri Harbour is due to drainage from the land causing an alteration of the saline constituents of the sea-water in the vicinity.

Date.	Time at which sample was taken.	Position.		Specific Gravity (σ_0) by		Difference. T—H.
		Lat. N.	Long. E.	Buchanan Hydrometer.	Titration.	
Jan. 2	12 noon ..			26·646	26·580	−0·066
,, 3	12-30 p.m. ..			26·623	26·490	−0·133
,, 4	1-0 p.m. ..			25·836	25·870	+0·034
,, 5	2-0 p.m. ..			25·869	26·300	+0·431
,, 6	2-35 p.m. ..			26·289	26·270	−0·019
,, 7	3-0 p.m. ..			26·181	26·230	+0·049
,, 8	3-0 p.m. ..			26·464	26·380	−0·080
,, 9	6-45 p.m. ..			26·473	26·420	−0·053
,, 10	8 p.m. ..	"Investigator" Station 614, Nankauri Harbour.		26·541	26·480	−0·061
,, 11	8 p.m. ..			26·611	26·510	−0·100
,, 12	9 p.m. ..			26·736	26·330	−0·406
,, 14	10 a.m. ..			26·700	26·430	−0·270
,, 18	....			26·856	26·720	−0·136
,, 19	3 p.m. ..			26·692	26·270	−0·422
,, 20	3-45 p.m. ..			26·577	26·260	−0·317
,, 23	6-25 a.m. ..			26·449	26·040	−0·409
,, 24	8 a.m. ..			26·416	26·558	+0·134
,, 25	8 a.m. ..			26·545	26·430	−0·115
,, 25	12 noon ..	8°. 32′. 30″	93°. 18′. 00″	26·779	26·430	−0·349
,, 25	12 midnight ..	10 . 21 . 00	93 . 04 . 00	26·148	25·990	−0·158
,, 26	4 a.m. ..	10 . 58 . 00	93 . 06 . 00	25·741	25·610	−0·131
,, 26	8 a.m. ..	11 . 32 . 45	92 . 47 . 45	26·462	26·390	−0·072
,, 31	12 noon ..	11 . 31 . 45	93 . 48 . 15	25·691	25·62	−0·071
,, 31	8 p.m. ..	10 . 20 . 00	93 . 01 . 00	26·227	26·19	−0·037
,, 31	12 midnight ..	9 . 43 . 00	93 . 08 . 00	26·167	26·08	−0·087
Feb. 1	4 a.m. ..	9 . 05 . 00	93 . 16 . 00	26·589	26·39	−0·199
,, 1	8 a.m. ..	8 . 34 . 45	93 . 12 . 30	26·734	26·52	−0·214
,, 1	12 noon ..	....	....	26·805	26·66	−0·145
,, 2	12 noon ..			26·889	26·72	−0·169
,, 3	12-45 p.m. ..			26·770	26·78	+0·010
,, 4	1-45 p.m. ..	"Investigator" Station 614, Nankauri Harbour.		26·784	26·94	+0·156
,, 6	3-45 p.m. ..			26·787	26·78	−0·007
,, 7	6 p.m. ..			26·232	25·81	−0·422
,, 8	6 a.m. ..			26·193	25·32	−0·873

Table 19; Results obtained from examination of water samples by means of a 'Buchanan' hydrometer and by silver-nitrate titration.

There can be little doubt that in these waters there is a greatly increased range in the divergence between the results obtained by the hydrometer and by the titration method, and it is of some interest to attempt to discover to what this may be due. It is, I am convinced, due to changes in the chemical composition of the sea-water itself.

At the present time the method of calculating the salinity of the sea-water from the "halogen" content of the sample is based on the assumption that the chemical composition of the water is invariably uniform. While this assumption may, within limits, be justifiable in the case of samples taken from the surface well out in the open ocean, it by no means follows that it is equally justifiable in the case of either samples taken at some depth below the surface or from the surface itself in in-shore waters. It must be borne in mind that the chemical composition of the sea-water is liable to vary: Atkins and others have shown that at certain seasons of the year the prolific growth and consequent metabolic activity of the diatoms and other organisms of the micro-plankton cause an appreciable variation in the percentage quantitites of phosphates and silicates in the upper levels of the in-shore waters in the North Atlantic Ocean and its offshoots, and there is no reason to doubt that similar periodic changes occur in the surface-waters of the tropical belt. It has also been shown that in the case of samples taken at different depths in the Atlantic Ocean (*vide* Atkins and Harvey, 1925) there is a steady decrease in the percentage of phosphates and a corresponding increase in the silicates as one proceeds from the surface to a depth of 3,000 metres, and this also is to be attributed to the activity of the diatom flora in the surface layer, since these minute plants are unable to thrive at depths greater than 100 fathoms and in consequence can affect only the more superficial layers; and there is every reason for supposing that a similar change in the chemical composition occurs also in the waters of the Indian Ocean as one proceeds from the surface downwards. In the case of in-shore waters and especially of those in a land-locked area, such as Nankauri Harbour, the diatom flora is likely to be extremely abundant. Sir John Murray (1898, pp. 136, 137,) has pointed out that it is in the in-shore waters that diatoms attain their greatest concentration in the tropics and I can corroborate this from my own experience on the coast of southern Burma and in the Mergui Archipelago, where in the majority of instances the plankton obtained is of the phyto-plankton type: in such regions the chemical effect of diatom growth will be proportionately great and will result in an alteration of the amount of phosphates and silicates present. We have, however, at present no data regarding the seasonal changes in the plankton in Indian seas, and, in consequence, no attempt can be made to correlate these with the differences observed in the results obtained by means of the 'Buchanan' hydrometer and by the titration method on the same water sample, to which I have referred above. Moreover, in an area of heavy rainfall, such as is experienced in the Nicobars, there is certain to be a considerable amount of drainage from the land. The Nicobars, or, at any rate, those islands that surround Nankauri Harbour are largely composed of magnesium clay (*vide* Tipper, 1911, and p. 4 *supra*) and in consequence drainage from the land will certainly carry into the sea an appreciable quantity of magnesium salts that are bound to affect the percentage of the chemical composition of the sea-water and will thus affect the relationship between the halogen content and the specific gravity.

A study of the results given in Table 19 and a comparison of the various differences between the specific gravity (σ_0) as given by the 'Buchanan' hydrometer and by the titration method reveals that out of six cases in which the titration gives the higher reading five occur on days when the water-sample was taken between the hours of 12 noon and 2 P.M.

If now we take the average difference, as recorded at different times of the day, we get the following results:—

Time of day.				Titration result—Hydrometer result.	
				Actual average.	Computed average.
Between 12 midnight	and		1 A.M.	−0·122	−0·235
,,	4 A.M.	,,	5 A.M.	−0·165	−0·286
,,	6 A.M.	,,	7 A.M.	−0·641	−0·339
,,	8 A.M.	,,	9 A.M.	−0·133	−0·296
,,	10 A.M.	,,	11 A.M.	−0·270*	−0·102
,,	12 noon	,,	1 P.M.	−0·132	−0·093
,,	1 P.M.	,,	2 P.M.	+0·095	+0·010
,,	2 P.M.	,,	3 P.M.	+0·136	**+0·013**
,,	3 P.M.	,,	4 P.M.	−0·155	−0·084
,,	6 P.M.	,,	7 P.M.	−0·237	−0·166
,,	8 P.M.	,,	9 P.M.	−0·099	−0·215
,,	9 P.M.	,,	10 P.M.	−0·406*	−0·235

Table 20; giving the actual average and the computed average at different times of the day of the observed difference between the specific gravity (σ_0) as observed by the hydrometer and calculated from the halogen content.

* In both these cases only a single observation was obtainable.

These figures certainly suggest that there is during the course of the day a distinct oscillation in the difference between the specific gravity as shown by the hydrometer and that calculated from the halogen content of the water sample. Since the hydrometer result is a direct reading of the specific gravity, it follows that this oscillation must be due to a variation in the percentage quantity of the halogens in the sea-water, these salts possessing a different proportion to the total salts at different times of the day; this may be brought about either by periodic variation in the amount of the halogens themselves or by a variation in the amount of certain other salts. The salts which suggest themselves as being liable to exhibit variation are the phosphates, carbonates and silicates. At times of increased activity of the plankton these salts will be built up by anabolism into the tissues of the smaller animals and plants and there will thus be caused a diminution in the quantity contained in the sea-water, and hence the halogen percentage will ber aised. On the other hand, owing to increased metabolism and consequent increased excretion, such salts as the ammonia derivatives, nitrates, etc., will be increased and the halogen percentage tend to be lowered. While these salts, namely the nitrates, ammonia, phosphates, silicates, etc., are ignored when calculating the salinity from the chlorine content they will nevertheless exercise an influence on the hydrometer and thus cause it to register a result differing from the calculated specific gravity. These differences, as shown by the computed averages given above, show an oscillation in which the difference between the titration and hydrometer results (T.—H.) shows a maximum excess at 2 p.m. and a maximum deficiency at 6–7 a.m.: in other words, the actual percentage of the halogens in the sea-water is greatest at 2 p.m. and is least at 6–7 a.m. Now, it is extremely interesting to note that these times correspond to the period of maximum sunlight and maximum temperature, and to a period of comparative darkness and minimum temperature respectively, and it appears probable that the change in the composition of the water during the day in these in-shore and coastal waters is directly correlated with the activity of the phytoplankton and other vegetable organisms of the coast, and that we have here evidence of a change, during each 24 hours, that is exactly comparable to the changes taking place in temperate seas at different seasons of the year. In a region, such as Nankauri Harbour, where there are large and extensive coral reefs the activities of the phytoplankton will be reinforced by the activities of the microscopic plants of the reefs, and especially of the Zooxanthellæ that exist in millions in the tissues of the corals and Alcyonacea, as well as of the larger algæ and plants growing on the reefs and mud-flats. The combined effect of the metabolic activity of the fauna and flora thus would appear to be, at any rate, one of the factors that cause an actual rise in the halogen percentage, that reaches its maximal epoch simultaneously with the period of maximum activity of the phytoplankton. That these changes in the character of the surface-water are not, however, limited to such enclosed coastal areas as Nankauri Harbour is shown by the fact that an exactly similar oscillation was found to occur on January 25-26, and January 31–February 1, when we were steaming between Nankauri Harbour and Port Blair.

In order to test this difference between the titration results and those obtained by means of the 'Buchanan' hydrometer, a further series of double observations was taken on water-samples collected 4-hourly from the surface while steaming down the east coast of Ceylon on the 22nd and 23rd of April, 1925. The results obtained by the two methods in this instance are given below in Table 21 and are shown graphically in Text-fig. 17.

Time.		Specific gravity by Titration.	Specific gravity by 'Buchanan' hydrometer.	Difference.
		T.	H.	T.—H.
22-4-25	12 noon	27·76	27·83	−0·07
	4 p.m.	27·20	27·49	−0·29
	8 p.m.	27·49	27·84	−0·35
	12 midnt.	27·52	27·70	−0·18
23-4-25	4 a.m.	27·51	26·80	+0·71
	8 a.m.	27·67	27·53	+0·14

Table 21; the specific gravity of water-samples collected at 4-hourly intervals on April 22-23, 1925, and examined by (*a*) Titration, and (*b*) 'Buchanan' hydrometer.

Here again we find a perfectly definite single oscillation in the difference between the results given by the two methods, but on this occasion it is clear that from 10 a.m. to about 1 a.m. the specific gravity as given by the hydrometer is higher than that given by the titration method, while from 1 a.m. to 10 a.m. the titration result is the higher of the two. From 4 a.m. to 8 p.m., that is to say approximately during the hours of sunshine, during which period the temperature of the water is raised and the chlorophyll-bearing organisms in the plankton are in full activity, the percentage quantity of the halogens falls steadily, rising again during the night hours from 8 p.m. to 4 a.m. It would appear from the above data that in these coastal waters there is at this time of the year a daily chemical change in the composition of the surface water, the proportion of halogens in the water being raised during the night hours, reaching its maximum at or near 4 a.m. and becoming steadily lower throughout the day, attaining its minimum at or near 8 p.m. Comparing this with the actual specific gravity as recorded by the 'Buchanan' hydrometer we see that the two curves clearly tend to alternate but that there is a small drop in the specific gravity between

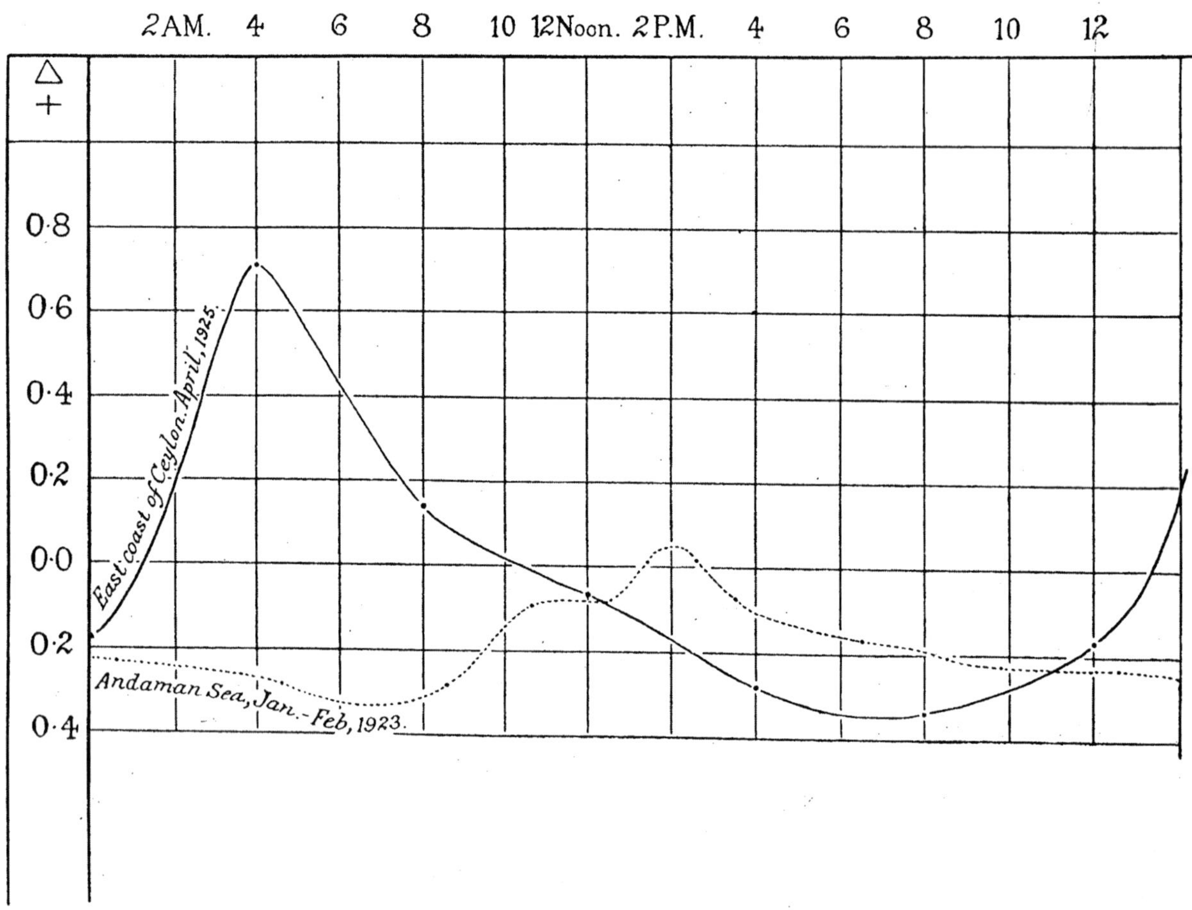

TEXT-FIG. 17.—Showing the difference between the specific gravity (σ_0) of the surface-water as observed by a 'Buchanan' hydrometer and calculated from the halogen content, in the Andaman Sea in January-February, 1923 (......) and off the east coast of Ceylon in April, 1925 (———).

12 noon and 8 p.m. that is not accompanied by any appreciable alteration in the halogen content.

A comparison of these results with those already obtained from a survey of my observations taken in Nankauri Harbour and on the west side of the Andaman Sea, reveals that we have here a complete reversal of the former changes. In the first series, taken in January and February, during the rainy season, the halogen percentage is lowest at 4–9 a.m. and is highest from 1–3 p.m. rising steadily during the hours in which the sea-temperature is raised and there is bright sunlight, whereas off the east coast of Ceylon in April, during the dry season, the halogen content is highest at 4 a.m. and lowest at 8 p.m., decreasing steadily during the hours of daylight and rising during the night. It is, I think, probable that we have here evidence of a seasonal change in the conditions of the surface water. I have in a previous paper (*vide* Sewell, 1926, pp. 117 *et seq.*) called attention to the seasonal periodicity in the occurrence of the salps in the uppermost strata of the sea and it is probable that a similar seasonal occurrence exists in all the classes of the zooplankton. In January and February owing to the dilution of the surface water the bulk of the zooplankton will probably be concentrated at some little depth below the surface, whereas in April it will be in maximum concentration on the surface itself and especially so during the day-time, and this may account for the difference noted in the surface-water in April, since in this month with an increased zooplankton there will be an increase in the quantity of ammonia salts, nitrates, etc., owing to increased metabolism and consequent excretion of the zooplankton, that will completely mask the effect of the increased activity of the phytoplankton.

My investigations regarding the temperature of the sea-water may be said to have commenced in 1921. Prior to that year a certain number of observations had been made, but for the most part these were taken only once a day and in consequence give no indication of the changes that take place during the twenty-four hours. In 1921, however, I commenced to take regular observations on the temperatures of the air and the sea at intervals throughout the day and in this work I received very great assistance from the Assistant-Surgeons, who were during the following survey-seasons attached to the R.I.M.S. "Investigator" for medical duties in connection with the crew, and to whom I tender my sincere thanks.

As a reference to the ordinary maps, and to the chart (*vide* Plate 5) already published in this series shows, the coastal regions on the two sides of the Andaman Sea differ widely from each other. On the east side the Andaman Sea basin is bounded by the long stretch of land that extends roughly in a southerly direction to the southernmost point of the Malay Peninsula. Although this strip of land is comparatively narrow, especially in the middle of its length, where it separates the Andaman Sea from the Gulf of Siam by a distance of only 40 odd miles, it contains a series of mountain ranges that run roughly north and south and that act as a large catchment area for the very considerable rainfall that occurs in this region during the period of the south-west monsoon. Flowing into the northern end of the Andaman Sea are the large and important rivers, the Irrawaddi, the Sitang and the Salween; and although the area of the Burma Coast that was under survey during the years 1910-14 lay too far to the south, especially during the latter part of this period, to be directly affected by the influx of river-water from these sources, there are at intervals along the coast quite large rivers, such as Yé River, Tavoy River and the Tenasserim River, all of which have a very marked influence on the salinity of the sea-water in the regions round their mouths; and, in addition to these, numerous other smaller streams and tributaries pour fresh-water into the sea either along the coast of the mainland itself or from the various islands that lie dotted over the in-shore waters of the Andaman basin. Off the portion of the coast extending between the mouth of Hinze Basin to a point a little south of the mouth of the Tavoy River, the number of islands are comparatively few and are arranged in groups, known as the North, Middle and South Moscos Islands, that lie for the most part in a straight line. The southern portion of the area is, however, thickly and irregularly dotted over with numerous islands of all sizes, some small, others quite large, which together constitute the Mergui Archipelago. On the west side the Andaman Sea basin is marked off from the Bay of Bengal by a scattered chain of islands, separated by wide and deep channels into groups constituting primarily the Andamans and the Nicobars, which are separated from each other by Ten-Degree Channel, and these two divisions are again subdivided into smaller groups by channels of lesser dimensions.

Throughout the period 1921–1925 the "Investigator" was engaged in surveying Nankauri Harbour in the central group of the Nicobar Islands and the surrounding area. Nankauri Harbour, as I have already pointed out (*vide supra*, p. 11) is a channel that runs approximately in an east to west direction between Nankauri and Camorta islands, the two entrances, in varying degree, being protected by other islands, on the east by the island of Trinkat and on the west by Kachal. The harbour itself consists of a somewhat narrow channel, the depth of which varies from some 25 fathoms at the eastern end to 35 fathoms in the western portion, and on either hand wide bays open out, in which there is considerably less depth of water. Immediately to the north of Nankauri Harbour, and at one point separated from it by a narrow neck of land only a few yards across, lies a second large inlet, Expedition Harbour, which possesses only one outlet on the western side; and further north still near the northern end of Camorta is a third, smaller and shallower inlet, known as Dring Harbour, that also opens to the west. Separating Nankauri and Camorta islands from those on the east and west, there is, on the east, Beresford Channel, a narrow and comparatively shallow channel largely obstructed by extensive coral banks and reefs, and on the west, Revello Channel, a wide channel in which the depth of water is over 100 fathoms. Nankauri Harbour, owing to its possessing an entrance at each end, communicates on the east with the waters of the Andaman Sea, and on the west, through Revello Channel, with the open waters of the Bay of Bengal.

Advantage was taken of the opportunity provided by the survey to make a more or less systematic investigation regarding the temperature and salinity of the surface water of the harbour and of the neighbouring part of the Andaman Sea, and the results obtained are given in the following pages.

Seasonal variation in the Temperature of the Surface-water.

I have already pointed out (*vide supra*, p. 57) that there is a very clear double oscillation in the average air-temperature over Indian Seas and that during the course of the year we have two maxima and two minima. One would naturally

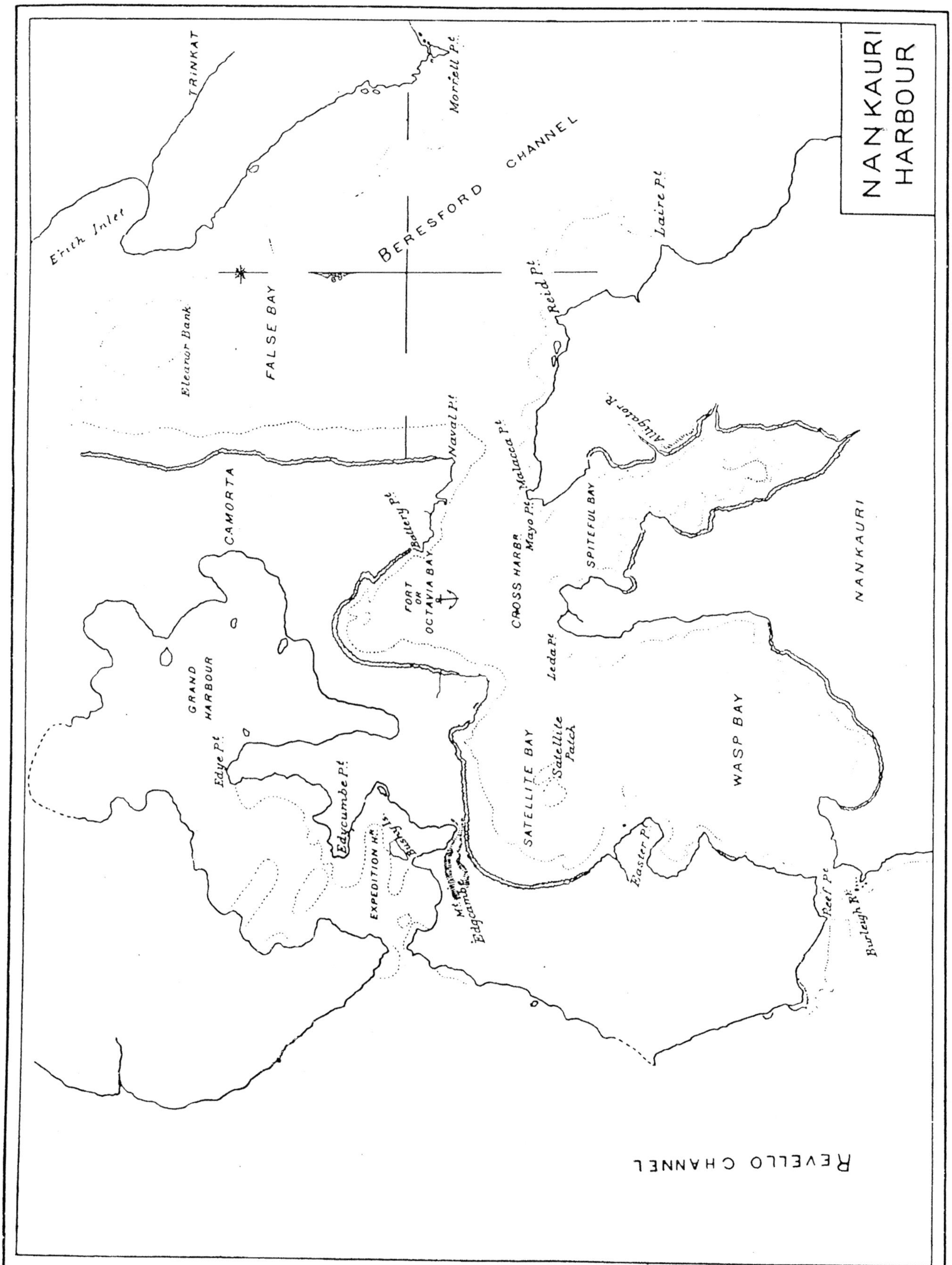

NANKAURI HARBOUR

TRINKAT
Morrell Pt.
Erith Inlet
BERESFORD CHANNEL
Laire Pt.
Reid Pt.
Eleanor Bank
FALSE BAY
Alligator R.
Naval Pt.
Malacca Pt.
CAMORTA
Battery Pt.
Mayo Pt.
SPITEFUL BAY
CROSS HARBR.
FORT OR OCTAVIA BAY
NANKAURI
GRAND HARBOUR
Leda Pt.
Satellite Patch
SATELLITE BAY
WASP BAY
Edye Pt.
Edgcumbe Pt.
EXPEDITION HR.
Bushy Is.
Mt. Edgcumbe
Easter Pt.
Reef Pt.
Burleigh Rk.
REVELLO CHANNEL

Chart VII. Sketch chart of Nankauri Harbour, Nicobar Islands.

expect to find a similar double oscillation in the surface-temperature of the sea, and, so far as my observations go, this would appear to be the case. In the following Table 22, I have given the average temperature observed in Nankauri Harbour in each month during the course of the survey; the actual observations have already been published in Appendix I (*vide* pp. 104–119.)

Survey Season.	1921-22.	1922-23.	Average.
	°C.	°C.	°C.
October	..	27·91	27·91
November	27·88	27·68	27·78
December	26·67	27·38	27·03
January	27·42	27·31	27·37
February	27·90	..	27·90
March	28·62 [1]	27·63 [2]	28·13

Table 22; giving the average monthly temperature of the surface-water in Nankauri Harbour.

[1] Based on observations during five days only in Expedition Harbour.

[2] Based on observations taken in Revello Channel to the west of Nankauri Harbour.

In such a harbour as this, with wide, shallow flats on either side, it is probable that the surface-temperature will vary somewhat according to the position, but as, throughout both seasons 1921-22 and 1922-23, the "Investigator" was always anchored in the same spot, near the mouth of Octavia Bay on the north side of the harbour, such local variations may, I think, be ignored. In 1921-22 the mean temperature fell from 27·88°C. in November to 26·67°C in December and then rose steadily to 28·62°C. in March. In 1922-23 the minimum temperature appears to have occurred a month later, namely in January: during this season the mean temperature fell from 27·91°C. in October to 27·31°C. in January and then rose again to 27·63°C in March. The delay in the minimum temperature in 1922-23 is in all probability due to the same cause that gave rise to the differences in the air temperature in these seasons, as I have already noted (*vide* p. 62 *et seq.*) namely, a revival of the N.E. Monsoon in the south part of the Bay of Bengal in January, 1923. It is clear then that during the period of the year occupied by the survey the average temperature of the surface-water shows a steady fall from October to December-January, the actual month in which the lowest average is reached varying from year to year, and then rises again to March; as Krummel (1907, p. 411) points out, in the region of the Indian Ocean "Im Mai wirkt die kulminierende Sonne, mit unbehinderter Strahlung und gefördert durch Windstillen, auf die Meeresoberfläche ein, so dass der grösste Teil des Arabischen Meers und Bengalischen Golfs über 29° bis 29·8°.........Der Ausbruch des Südwestmonsuns erniedrigt die Temperaturen des Arabischen Meers um 2° bis 4° .... im Bengalischen Golf und in der Chinasee um 1° bis 2° ...." There is then in these waters a yearly double oscillation in the surface temperature.

DAILY VARIATION IN THE TEMPERATURE OF THE SURFACE-WATER OF NANKAURI HARBOUR.

The amount of the daily variation in the temperature of the surface-water of the sea depends very largely on local conditions and is much greater in temperate regions than in the tropics. Krummel (1907, p. 382), quoting from the work of Al. v. Humboldt, gives the range of temperature in tropical seas between mid-day and mid-night as about 0·76°C, the extreme limits of variation lying between 0·2° and 1·2°. These figures, however, do not represent the maximum variation that may occur, since it has been found that the lowest temperature of the surface-water occurs not at mid-night, but from 4–6 a.m. and the highest temperature is reached 2 to 3 hours after mid-day. Buchan (1889, pt. V, p. 5), in his report on the investigations carried out by H.M.S. "Challenger," gives the range of variation in the open waters of the N. Atlantic Ocean, in a mean position of Lat. 30°N.; Long. 42°W., as 0·44°C (=0·8°F.) while near the Equator in both Atlantic and Pacific Oceans the average range was found to be 0·40°C. Krummel (1907, p. 383) has also given the average variation of the surface temperature based on three series of observations in

(1) the North Pacific Ocean, in the months of June-July;
(2) the South Pacific Ocean, in the months of October-November; and
(3) the North Atlantic Ocean, in the months of February-March.

The three series show ranges of 0·50°C, 0·36°C and 0·55°C respectively; but it is impossible to say how far the differences noted in the three series are due to local causes and how far to seasonal variations in the range of temperature. I have already pointed out (*vide supra*, pt. III, p. 67) that the daily range of the air-temperature over Nankauri Harbour shows a very clear seasonal variation, and one would certainly expect to find a similar variation in the range of temperature of the surface water of the harbour.

In the following Table 23, I have given month by month the variation above and below the mean monthly temperature at different times of the day:—

1921-22.	6 A.M.	8	10	12	3 P.M.	6	8	Mean Temp. °C	Average range of Temp. °C
November ..	−0·03	−0·08	−0·08	**+0·16**	+0·15	0·0	*−0·09*	27·88	0·25
December ..	−0·06	−0·05	+0·10	+0·11	**+0·12**	*−0·07*	*−0·07*	26·67	0·19
January ..	*−0·25*	−0·07	+0·01	+0·13	**+0·17**	+0·10	−0·11	27·42	0·42
February ..	*−0·36*	−0·14	+0·04	+0·03	**+0·43**	+0·05	−0·03	27·90	0·79
March [1] ..	*−0·34*	−0·06	(−0·47) [2]	+0·18	**+0·53**	+0·08	+0·08	28·62	0·87

[1] This series of observations was taken in Expedition Harbour, to the north of Nankauri Harbour.
[2] This figure is probably unreliable.

1922-23.	6 A.M.	8	10	12	2 P.M.	4	6	8	Mean Temp. °C	Average range of Temp. °C
October ..	*−0·22*	−0·13	−0·04	+0·13	**+0·19**	+0·13	+0·02	−0·03	27·91	0·41
November..	*+0·35*	−0·05	0·00	**+0·28**	+0·22	+0·12	−0·10	−0·16	27·68	0·63
December ..	*−0·17*	−0·16	+ 0·08	**+0·25**	+0·22	+0·09	−0·02	−0·12	27·38	0·32
January ..	−0·09	*−0·13*	+0·03	+0·09	**+0·14**	+0·06	−0·02	−0·08	17·31	0·27
February ..	..	..	..	..	..	..	..	..	..	..
March [1] ..	*−0·48*	−0·12	+0·07	+0·30	**+0·36**	+0·22	+0·09	−0·11	27·63	0·84

Table 23; giving the variation of temperature above and below the mean temperature of the surface-water of Nankauri Harbour.

[1] This series of observations was taken in Revello Channel to the west of Nankauri Harbour.

In a land-locked basin such as Nankauri Harbour one would expect to find that the daily range of temperature of the surface was considerably greater than over the open ocean owing to the rapid cooling of the air over the land, though, in the case of a harbour such as this in an oceanic group of islands, this effect would be much less than in one situated on the coast of a large land area; as a matter of fact, it is found that the average range throughout the whole series of observations is 0·50°C, which agrees exactly with the figure given by Krummel for the North Pacific Ocean and is but 0·1°C more than the figure given by Buchan for the tropical areas of the Pacific and Atlantic Oceans.

A study of Table 23, however, reveals clearly that there is a very distinct variation from month to month in the amount of the daily range of surface-temperature; the range increases from October to November; it then shows a decrease in December and January, the extent and length of the period of which varies from year to year, almost certainly in accordance with the variations in the strength and persistence, or otherwise, of the North-East Monsoon; and from February to March it again steadily increases. The average range for October-November is 0·43°C which is only 0·07° above the figure given by Krummel for the South Pacific Ocean (*viz.* 0·36°) and that for the months of February-March is 0·83°C. It seems probable, then, that the differences between the three series of observations in the three great oceans, as given by Krummel, are largely, if not entirely, due to seasonal rather than to local conditions. The least variation in the surface-temperature is found in these latitudes, (*i.e.* of Nankauri Harbour) to occur at that period of the year when the sun's declination has reached its southerly limit and when the North-East Monsoon is blowing and the sky is, in consequence, overcast; and the greatest variation occurs when the sun is overhead and the sky is usually free from cloud.

In Text-fig. 18, I have plotted out the average results obtained by me in my observations and a comparison of the Text-figure with the figures given in Table 23

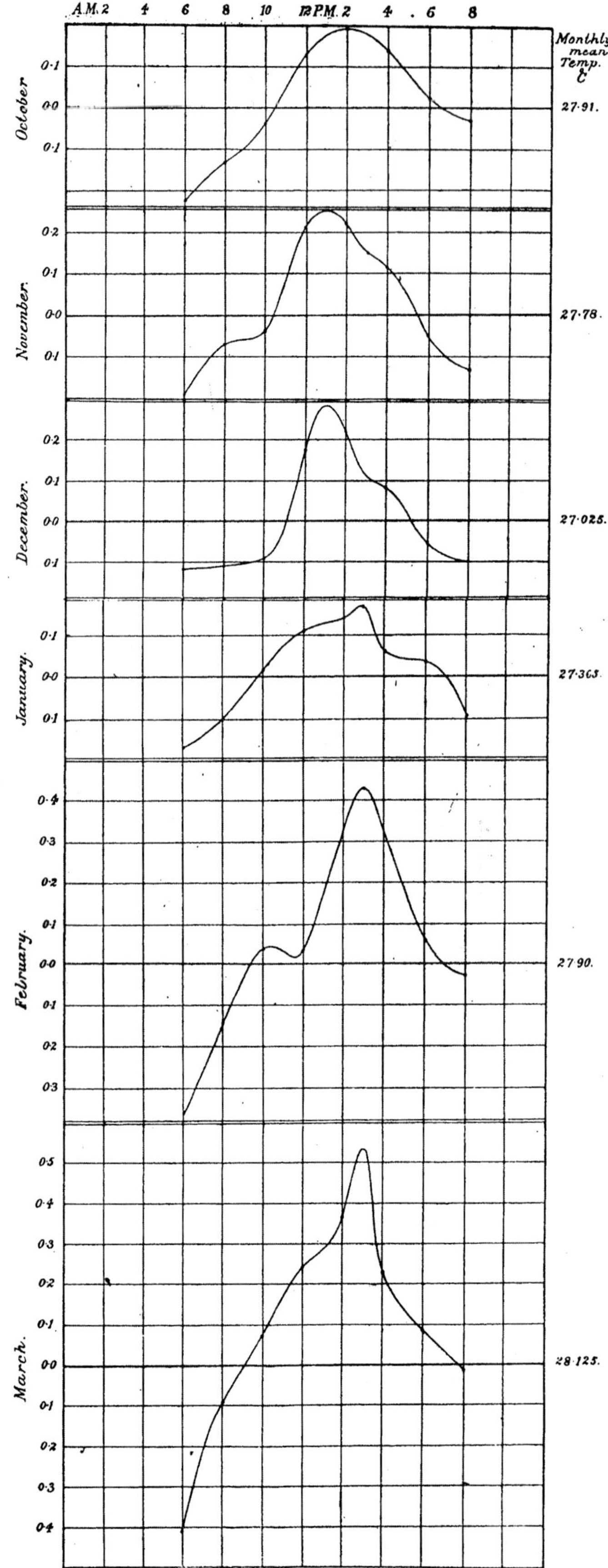

TEXT-FIG. 18.—Showing the variation above and below the mean monthly temperature of the mean temperature of the surface-water in Nankauri Harbour and the neighbourhood at different times of the day in different months of the survey-season.

shows clearly not only this seasonal alteration in the daily range of the temperature of the surface water but also that there is from month to month a change in the time of day at which the maximum temperature is reached. I have already pointed out (*vide supra*, p. 72 *et seq.*) that the maximum temperature of the air is reached at different times of the day, being earlier at certain seasons than at others, and one would naturally expect to find that the same rule holds good for the surface temperature. In the following Table 24 I have given the data for each month :—

	October.	November.	December.	January.	February.	March.
Mean Temperature of surface water.	27·91°	27·88°	27·05°	27 37°	27·90°	28·14°
Average daily range of temperature.	0·41°	0·44°	0·38°	0·34°	0·79°	0·85°
Time of occurrence of maximum temperature.	2·0 P.M.	1·0 P.M.	1·0 P.M.	3·0 P.M.	3·0 P.M.	3·0 P.M.

Table 24; showing the mean temperature, the daily range and the time of occurrence of maximum temperature of the surface-water of Nankauri Harbour and the neighbourhood.

As in the case of the air-temperature, it is clear that there is a distinct correlation between the three series; the mean temperature of the surface water and the amount of the daily range of temperature rise and fall together; the time of occurrence of the maximum daily temperature tends to be earlier in November and December, that is to say during the period of the North-East Monsoon, when the sky is overcast and the sun is approaching its most southerly declination, and it is at this season of the year that the mean temperature of the surface water is at the lowest, and during the early months of the year, January–March, the epoch becomes progressively later, being as late as 3.30 p.m. in March, when the sun is almost overhead, and the mean temperature is approaching its highest point.

I have already called attention (*vide supra*, p. 76) to the fact that although the air-temperature over the open sea normally presents a single oscillation during the course of the day in certain areas, such as off the west coast of India, a double or even a triple oscillation can be detected. In the case of the air-temperature in Nankauri Harbour and the neighbourhood, the average temperature in each month, except January, 1922, presents a single rise and fall as is clearly shown in the figures given in Table 25 :—

TIME OF DAY.

1921-22.		A.M.				P.M.		
		6	8	10	12	3	6	8
		°C	°C	°C	°C	°C	°C	°C
November	..	25·70	26·99	27·11	**28·08**	27·89	26·88	26·56
December	..	26·66	26·87	27·72	**27·78**	27·68	26·48	26·68
January	..	25·54	26·45	**27·29**	27·16	**27·46**	26·36	26·53
February	..	25·54	26·57	27 61	27·65	28·14	27·07	26·39
March	..	25·63	27·13	28·14	29·69	29·92	29·83	27·48
Average	..	25·79	26·80	27·56	28·07	**28·22**	27·32	26·73

1922-23.		A.M.				P.M.			
		6	8	10	12	2	4	6	8
		°C	°C	°C	°C	°C	°C	°C	°C
October	..	25·27	26·42	26·91	**27·57**	27·44	27·45	26·59	26·03
November	..	25·15	26·41	27·01	**27·93**	27·29	26·67	26·06	25·56
December	..	25·23	25·91	27·01	**27·16**	26·54	26·42	25·79	25·50
January	..	25·41	26·03	27·03	**27·24**	27·20	26·93	26·59	26·14
Average	..	25·27	26·19	26·99	**27·48**	27·12	26·87	26·26	25·56

1925.		A.M.					P.M.			
		6	8	10	12	2	4	6	8	10
		°C	°C	°C	°C	°C	°C	°C	°C	°C
March	..	25·65	27·00	**27·99**	27·95	27·95	27·91	27·50	26·55	26·41

Table 25; giving the average air-temperature at different times of the day in different months at Nankauri Harbour and the neighbourhood.

If, however, we consider the records for each day separately we find that on an appreciable number of days we have clear evidence of a double diurnal oscillation and on one day, January 6, 1922, a triple one. I have extracted a record of all these days from my observations, as follows:—

		A.M.				P.M.		
		6	8	10	12	3	6	8
		°C	°C	°C	°C	°C	°C	°C
3rd November, 1921	..	*24·17*	**25·56**	*25·27*	26·94	**27·78**	27·22	27·22
19th ,, ,,	..	*24·72*	**27·67**	*27·22*	**27·61**	27·44	26·94	26·44
21st ,, ,,	..	*24·72*	**27·56**	*26·22*	28·06	**29·56**	26·78	26·44
21st December, ,,	..	26·56	27·11	**27·50**	*26·39*	**27·44**	*24·39*	—
5th January, 1922	..	*26·39*	26·66	**26·94**	*25·22*	25·72	26·78	**26·89**
6th ,, ,,	..	*25·44*	26·50	**27·11**	*26·00*	**26·66**	*26·39*	**26·67**
11th ,, ,,	..	*25·83*	26·39	**27·50**	*27·39*	**27·61**	..	26·72
6th February, ,,	..	**26·56**	*25·61*	**27·72**	*26·17*	**26·89**	26·67	*25·00*
7th ,, ,,	..	25·06	26·50	**26·94**	24·78	*24·67*	**25·28**	*24·56*
18th ,, ,,	..	..	**27·44**	*27·06*	27·78	**28·67**	..	*26·33*

	A.M.				P.M.			
	6	8	10	12	2	4	6	8
29th October, ,, ..	*25·33*	25·61	25·89	**27·28**	*26·67*	**27·83**	26·17	26·17
30th ,, ,, ..	24·50	*23·95*	26·72	**27·83**	*26·44*	**27·28**	26·17	24·78
4th November, 1922 ..	25·06	**25·56**	*25·06*	..	**25·56**	25·39	25·06	*25·06*
7th ,, ,, ..	*25·61*	**27·83**	*27·56*	28·11	**28·67**	27·56	26·61	25·61
8th ,, ,, ..	25·56	**27·83**	25·61	**28·11**	*25·72*	**26·17**	25·56	*23·17*
25th ,, ,, ..	*25·33*	**26·72**	*25·06*	**26·44**	**26·44**	25·89	..	*25·33*
15th December, ,, ..	25·06	*23·67*	**26·72**	*26·67*	**27·78**	27·28	26·17	25·06
18th .. ,, ..	25·33	26·44	26·44	**27·83**	*25·61*	**26·17**	25·61	*24·22*

	A.M.				P.M.				
	6	8	10	12	2	4	6	8	10
4th March, 1925 ..	27·3	27·4	**27·9**	*25·7*	26·7	26·8	**26·9**	26·3	*26·0*
7th ,, ,, ..	25·6	27·4	27·8	**28·0**	26·7	*25·2*	*25·2*	**25·9**	*25·4*
8th ,, ,, ..	26·4	*26·2*	27·1	27·2	**28·0**	*27·0*	*27·0*	**27·2**	**27·2**
9th ,, ,, ..	26·1	**26·4**	*24·4*	26·1	27·0	**27·1**	26·8	25·6	*25·0*
11th ,, ,, ..	*25·3*	27·6	**28·9**	28·6	*28·4*	**28·7**	28·1	27·9	27·8
13th ,, ,, ..	26·3	**28·0**	27·6	25·1	*24·4*	**24·8**	24·3	*24·1*	24·8
14th ,, ,, ..	*24·4*	27·7	28·6	**28·8**	*28·4*	**28·9**	27·2	26·0	25·2
17th ,, ,, ..	*25·1*	26·9	**29·0**	*28·9*	**29·1**	28·7	28·1	26·7	26·1
27th ,, ,, ..	*24·8*	26·3	**27·7**	26·8	*25·4*	26·0	**26·1**	25·7	25·7
30th ,, ,, ..	27·4	28·0	**28·7**	**28·7**	27·6	27·8	**28·7**	27·7	*26·8*
31st ,, ,, ..	*25·9*	26·9	**28·9**	28·4	28·9	**29·4**	*29·2*	27·0	26·8

It must be borne in mind that the direct rays of the sun, falling in the early hours of the morning on the covered box in which the thermometers were housed, might have tended to cause an artificial rise in the temperature, but if this had been the cause of the first rise in the recorded temperature one would have expected to find it always at about the same time of day, whereas a study of the data given above shows that there is a distinct tendency for the first maximum to be recorded at different times of the day in different months, viz :—

In October at 12 noon.
,, November ,, 8 A.M.
,, December ,, 10 A.M.
,, January ,, 10 A.M.
,, March ,, 10 A.M.

The second rise usually occurs somewhere between 2 and 4 P.M ; though on the average there is a distinct tendency for the time of this epoch to vary from month to month. Taking the average of all observations it occurs as follows :—

In October at 4 P.M.
,, November ,, 2.10 P.M.
,, December ,, 3.0 P.M.
,, January ,, 4.40 P.M.
,, February ,, 4.0 P.M.
,, March ,, 5.5 P.M.

One must bear in mind that a double oscillation of this type in the temperature of the sea-water may be caused by rainfall. I have previously called attention to the fact that there is a clear tendency for rainfall to occur at two different periods of the day (*vide supra*, p. 99 *et seq.*) and in considering the average temperatures such a double rainfall is liable to cause a double oscillation in the sea-temperature. But the frequency of occurrence of this double diurnal oscillation in the month of March, when rainfall is almost completely absent in these regions, shows clearly that this is not the only, and probably not the most important, factor in the production of such temperature changes; it is probably due to a large extent to the diurnal oscillations in the force of the wind which may act either by actual cooling of the surface-water, or by blowing away the upper layer of warmer water and so causing an upwelling of colder water from below.

The daily range of the temperature of the surface-water, in an area such as Nankauri Harbour, is bound to vary very considerably from day to day, as well as from month to month, in as much as it is dependent on (1) the amount of heat absorbed from the sun's rays, which will be greater on a fine day than on a cloudy one, (2) on the amount of cooling of the surface which will vary with the air temperature, wind velocity, amount of rainfall, etc., and (3) on the amount of heat given off at night by radiation to the atmosphere. In the case of a harbour, such as this, there seems to be another factor concerned, namely the amount of interchange that is going on between the water of the harbour and the open water outside. The open water of the Bay of Bengal has a uniformly lower surface-temperature than that of Nankauri Harbour, and the degree of interchange will depend very largely on the tides, being greatest at spring tides and least at neaps. In the following Table 26, I have given the average observed daily-range of the temperature of the surface-water of Nankauri Harbour on different days of the lunar phase and have also given the computed average, according to the formula $\frac{1a+4b+6c+4d+1e}{16}$. It is clear that the daily range of temperature exhibits four maxima and four minima in each lunar month, the maxima corresponding closely to spring and neap tides, which occur usually from one to two days *after* the lunar phase. It seems then that, if there is a large amount of interchange, as there is during spring tides, there is an increased daily range of temperature owing to the outer open water being colder than the harbour water; whereas, during a diminished interchange as at neap tides, the effect of cooling at night, consequent on the more rapid cooling of the land, is increased and in consequence the daily range of temperature of the harbour water is again increased.

Days of Lunar phase.	1	2	3	4	5	6	7	8	9	10	11	12	13	14
No. of Observations ..	4	4	5	5	7	6	5	6	7	6	7	6	5	4
Averaged daily variation °C..	0·575	0·663	0·643	0·654	0·478	0.400	0.400	2·460	0·463	0·395	0·564	0·583	0·482	0·650
Computed average °C ..	0·5505	0·620	**0.63**	60·592	0·508	0·438	*0·424*	0·438	0·466	0·545	**0·628**	0·627	9·572	0·541

Days of Lunar phase.	15	16	17	18	10	20	21	22	23	24	25	26	27	28
No. of Observations ..	4	3	4	5	2	2	3	2	3	2	5	5	2	4
Average daily variation °C..	0·460	0·403	0·777	0·444	0·200	0·300	0·330	0·600	0·600	0·577	0.400	0·543	0·503	0·415
Computed average °C ..	*0·514*	0·525	**0·544**	0·455	0·330	*·0311*	0·399	0·512	**0·565**	0·545	0·527	0·533	*0·517*	0·507

Table 26; showing the average daily range of the temperature of the surface-water of Nankauri Harbour on different days of the lunar phase.

Relationship between the Temperature of the Sea and Air.

Throughout the tropical belt it is usually found that the surface-temperature of the sea-water has, at almost all times of the day, a higher temperature than that of the supernatant air; Koppen (1890, pp. 445–454) has given figures of the average difference throughout the whole 24 hours in different areas, and according to him in the whole breadth of the Equatorial zone between 20°N. and 20°S. the sea-temperature exceeds that of the air-temperature by 0·22°C. while in the tropical portion of the Indian Ocean, between 15°N. and 15°S. it is stated to be 0·20°C. Krummel (1907, p. 386), quoting from a work by Hann (1891), to which I have been unable to refer, has given a table for the difference between the sea- and air-temperatures at different times of the day, the data, from which the table has been calculated, being the observations taken on H.M.S. "Challenger" during a period of 126 days in the North Atlantic Ocean. The table shows that in that region the surface temperature is usually higher than that of the air and that the difference between the two series is greatest at 1 a.m. and is least at 1 p.m.; it also shows that the sea-temperature attains its maximum an hour later than the air-temperature and that in the middle of the day the surface-temperature may even fall slightly below the temperature of the air.

I have already (*vide* Appendix I, pp. 104–119) published the observations taken on board the R.I.M.S. "Investigator" during the survey-seasons 1921-22 and 1922-23 and again in 1925; during the first season observations were taken during periods covering 89 days, and in the second season on 71 days, and the results obtained are given below in Table 27. I have given the results of the two series of observations separately for two reasons; firstly because the times during the day, at which the observations were taken, differ slightly, thus in 1921-22 only one observation, at 3 p.m., was taken in the afternoon, whereas in 1922-23 observations were taken at 2 and 4 p.m.; and, secondly, because, as we have already noted regarding the average temperature and the daily range of the sea-water, local conditions differed somewhat during the two periods.

	A.M.				P.M.		
Survey Season 1921-22	6	8	10	12	3	6	8
	°C	°C	°C	°C	°C	°C	°C
Sea Temperature ..	27·69	27·82	27·80	28·02	**28·18**	27·93	27·85
Air Temperature ..	25·79	26·80	27·57	28·07	**28·22**	27·33	26·73
Difference ..	1·90	1·02	0·23	−0·05	−0·04	0·60	1·12

	A.M.				P.M.			
Survey Season 1922-23	6	8	10	12	2	4	6	8
	°C	°C	°C	°C	°C	°C	°C	°C
Sea Temperature ..	27·30	27·46	27·61	27·79	**27·80**	27·70	27·57	27·48
Air Temperature ..	25·34	26·35	27·19	**27·57**	27·28	27·07	26·51	26·00
Difference ..	1·96	1·11	0·42	0·22	0·52	0·63	1·06	1·48

Table 27; showing the relationship between the sea and air temperatures in Nankauri Harbour and the neighbourhood.

From the above table it will be noticed that in the first series of observations the sea and air both reach their maximum at approximately the same time of day, viz. 3 p.m., and that during the middle of the day, from 12 noon to 3 p.m. the air-temperature is slightly higher than that of the sea; whereas in the second series the maximum sea-temperature is reached at about 1 p.m. or a little later and that of the air at or near 12 noon, while throughout the day the air-temperature remains slightly less than the sea-temperature. The first series, therefore, agrees, with Hann's results as regards the relationship of the two series to each other at mid-day, the sea-temperature being then slightly the lower; and the second series agrees as regards the time of occurrence of the maximum temperatures, that of the air preceding that of the sea by about an hour.

By interpolation in the first series of observations, namely those taken during the survey season 1921-22, one is able to arrive at an approximate figure for the hours of 2 and 4 p.m. and I have, from the data thus obtained, calculated the average difference between the sea- and air-temperatures at intervals of two hours throughout the day, from 6 a.m. to 8 p.m., in each month throughout the period of the year that is covered by the survey-season, namely, from October to March. The results are given in Table 28.

	A.M.				P.M.			
	6	8	10	12	2	4	6	8
	°C	°C	°C	°C	°C	°C	°C	°C
October ..	2·36	1·36	0·96	0·47	0·66	0·59	1·34	1·85
November ..	2·14	1·01	0·68	0·00	0·28	0·83	1·26	1·59
December ..	1·46	1·02	0·19	0·23	0·55	0·74	1·33	1·34
January ..	1·72	1·02	0·23	0·28	0·24	0·46	0·93	0·94
February ..	2·10	1·19	0·33	0·28	0·10	0·36	0·88	1·42
March[1] ..	2·08	0·97	−0·14	−0·46	−0·44	−0·49	−0·46	1·00
Average ..	1·98	1·10	0·38	0·13	0·23	0·42	0·88	1·36

Table 28; giving the average difference in each month and throughout the whole period of observation between the sea- and air-temperatures in Nankauri Harbour and the neighbourhood.

[1] This series is derived from the combination of the observations taken in Expedition Harbour in 1922 and in Revello Channel in 1925.

For the purpose of convenience I have reproduced below that portion of the table given by Krummel (1907, p. 386) that corresponds in the time of day to my series of observations, and a comparison of the averages given in Table 28 with this series shows that although in my observations the average temperature of the sea surface is always greater than that of the air, even in the middle of the day, the two series follow a somewhat similar course. The difference, however, between the two temperatures in the early morning and evening is nearly 2½ times as great in Nankauri Harbour as it is in the Atlantic Ocean, and this I attribute to the effect of the proximity of land, though in such a locality as Nankauri this effect would be comparatively small.

	A.M.				P.M.				
	5	7	9	11	1	3	5	7	9
	°C	°C	°C	°C	°C	°C	°C	°C	°C
Sea Temperature ..	19·8	19·8	20·0	20·1	20·1	20·2	20·1	20·0	19·9
Air Temperature ..	19·0	19·2	19·6	20·2	20·6	20·6	20·3	19·7	19·3
Difference ..	0·8	0·6	0·4	−0·1	−0·5	−0·4	−0·2	0·3	0·6

Table 29; relationship between sea- and air-temperature in the N. Atlantic Ocean based on the "Challenger" observations (from Krummel).

A study of the figures for each month shows that there is a quite appreciable seasonal variation in the relationship. The differences in the mean temperatures of the surface-water and of the air in Nankauri Harbour in each month of the survey-season are as follows:—

	°C		°C
October ..	1·20	January ..	0·74
November ..	1·01	February ..	0·92
December ..	1·00	March (Expedition Harbour) ..	0·36

Average for whole period 0·87°C.

The difference in each month thus exhibits a very considerable range of variation: it is greatest in October, and is least in March, when it approximates to Koppen's figure, being only 0·36°C; but the figure for March is not strictly comparable to those of other months owing to the difference in the local conditions between Nankauri and Expedition Harbours and Revello Channel. The average difference for the whole period is very much greater than that given by Koppen (*vide supra*, p. 151) for the whole breadth of the Indian Ocean between 15°N. and 15°S., namely 0·22°C, but this also I attribute to the land-locked condition of the harbour.

It is also interesting to note that the variation, month by month, in the average difference between the air- and sea-temperatures is, in the main, due to the great increase in the difference between the two series of temperatures in the early morning, namely, at 6 a.m. In October this difference is as great as 2·36°C: it falls steadily to December, in which month it is only 1·46°C, and it then rises again till March, when it is again high, being 2·08°C. A further study of my results indicates that even in two consecutive years the temperature relationship may vary very considerably. I give below the average difference at different times of the day between the sea- and air-temperatures in the two survey-seasons, 1921-22 and 1922-23:—

	A.M.				P.M.			
Survey Season	6	8	10	12	2	4	6	8
1921-22	1·87	1·02	0·23	−0·05	−0·20	0·20	0·61	1·13
1922-23	2·08	1·26	0·59	0·28	0·64	0·80	1·28	1·66

In both years the range of difference is approximately the same, namely, from 1·87 to −0·20=2·07 in 1921-22, and from 2·08 to 0·28=1·80 in 1922-23.

SURVEY SEASON, 1921-22.

	A.M.				P.M.			
Month.	6	8	10	12	2	4	6	8
November	.. **2·10**	0·81	0·69	*0·04*	*−0·04*[1]	0·52[1]	1·00	**1·23**
December	.. **0·95**	0·75	*−0·04*	0·00	0·05	0·61	**1·12**	0·92
January	.. **1·63**	0·90	*0·14*	**0·39**	*0·26*	0·64	**1·16**	0·78
February	.. **2·10**	1·19	0·33	0·28	*0·23*	0·53	0·88	**1·48**
March	.. **2·65**	1·43	0·01	*−0·89*	**−0·81**[2]	−0·95	*−1·13*	**1·22**

SURVEY SEASON, 1922-23.

	A.M.				P.M.			
Month.	6	8	10	12	2	4	6	8
October	.. **2·36**	1·36	0·96	*0·47*	**0·66**	*0·59*	1·34	**0·85**
November	.. **2·18**	1·22	0·67	*0·03*	0·61	1·14	1·52	**1·96**
December	.. **1·96**	1·29	*0·43*	0·45	**0·04**	*1·03*	1·55	**1·74**
January	.. **1·81**	1·15	0·31	*0·16*	0·25	0·44	0·70	**1·09**

SURVEY SEASON, 1924-25.

	A.M.				P.M.				
Month	6	8	10	12	2	4	6	8	10
March	.. **1·50**	0·51	*−0·29*	−0·02	**0·04**	*−0·06*	0·22	0·77	**1·06**

Table 30; showing the difference between the sea- and air-temperatures at different times of the day in different months.

[1] The figures for 2 and 4 p.m. in this series are calculated from the data at 12 noon, 3 p.m., and 6 p.m.

[2] The difference at 3 p.m. was only −0·77.

A study of the average figures would lead one to assume that the difference between the two sets of temperatures follows a single uniform curve throughout the day, but if we consider the results obtained in each month separately (*vide* Table 30 we find that although in certain months the curve of relationship is a uniform rise and fall, in others there is a very distinct tendency towards a double fluctuation.

In 1921-22 this double oscillation in the temperature difference is clearly seen in the month of January, 1922, when the least difference, 0·14, is reached at 10 a.m.; this is succeeded by an increase to 0·39 at 12 noon and at 3 p.m. the difference has again fallen to 0·13, after which it rises steadily till 8 P.M. A similar double oscillation is seen in the month of March, 1922, in which the least difference is reached at 12 noon, after which the figure rises again to —0·77 at 3 p.m. and then falls to —0·95 at 4 p.m. rising again to 1·22 at 8 p.m. In 1922-23 in the month of October there is a slight secondary oscillation, the least difference being reached at 12 noon, after which the figure increases slightly at 2 p.m., falls again at 4 p.m. and then again steadily rises till 8 p.m. In November of this season the difference followed a single curve but in December we again get a double oscillation, the least difference being reached at 10 a.m. and this is followed by a rise till 2 p.m., after which there is an extremely small fall amounting to only 0·01 at 4 p.m. and finally the difference steadily increases till 8 p.m. In 1925 in the month of March this double oscillation is extremely well marked. This series of observations was taken not in Nankauri Harbour but in Revello Channel to the west; the minimum difference is reached in this series at 10 a.m. and is suceeded by a rise to 0·04 at 2 p.m. after which there is a second fall to —0·06 at 4 p.m. and a final rise to 1·06 at 10 p.m., This double oscillation in the temperature relationship can thus be detected in the months of October, December and January and again in March, that is to say in those months in which, as we have already seen (*vide supra*, p. 77), a double oscillation can be detected in the air-temperature over the waters of the Indian seas; and the well-marked character of the phenomenon in the month of March also agrees with the frequency with which, as I have shown above (p. 149), we can detect a double oscillation in the temperature of the surface water in this area.

PERIODIC VARIATION OF THE SALINITY AND DENSITY OF THE SURFACE WATER IN THE COASTAL REGIONS OF THE ANDAMAN SEA.

During the first two years of my tenure of the appointment of Surgeon-Naturalist the only apparatus on board the R.I.M.S. " Investigator " for determining the density and salinity of the sea-water was a set of four hydrometers of the ordinary pattern, graduated in degrees of salinity at a temperature of 16°C. In later years, from 1913 on, other and more delicate methods of examination were introduced; during the survey season 1913-14, and again in 1921-22 and the early part of season 1922-23 I used a 'Buchanan' hydrometer, but since January, 1923, I have employed the titration method of estimating the halogen content and of calculating from it the total salinity of the sea-water by Knudsen's Tables.

Helland-Hansen and Nansen (1909, p. 98) have emphasised that in an examination of the salinity and density of surface-waters care must be taken to ensure that all samples are taken from the same depth, so as to avoid as far as possible any slight differences due to surface evaporation in the uppermost few inches. All surface-samples were taken on the R.I.M.S. "Investigator" by means of a canvas bucket and rope, and when the ship was at anchor the bucket was lowered so as to collect the water from a depth of only 6 inches before being hauled in. This was also done as far as possible when the vessel was steaming, but it was difficult to ensure accuracy under such conditions. All samples were taken from the foreward part of the ship so as to eliminate any chance of the water being contaminated by the outflow of condenser-water, etc.

My earlier observations, made in 1910-11, while not being of a sufficient degree of accuracy to satisfy the requirements of modern hydrographic research, are yet of interest, for, although they may not demonstrate minor changes, they clearly reveal the presence in the salinity and specific gravity of the water of major oscillations that are of a character identical with those found in the neighbouring area of the Burma coast in 1913-14, when I was using a 'Buchanan' hydrometer. In both cases the period of oscillation is identical and, since the earlier observations tend to confirm the results of the later season, I have included both series in the following account.

During my first season as Surgeon-Naturalist, in 1910-11, the R.I.M.S. "Investigator" was engaged in a survey of the southern Burma Coast, the area under survey extending roughly from Yé River in the north to the mouth of Tavoy River in the south. A number of observations of the temperature of the surface water were made during this season at 7-30 a.m. each morning, and, simultaneously, readings of the density of the water were taken by means of an ordinary hydrometer. The hydrometers used were not capable of giving one a reading reliable to nearer than one-half of a degree and, in consequence, small fluctuations of density may have passed unnoticed; from these hydrometer readings I have, by means of Knudsen's tables, calculated the corresponding salinities and the results obtained are given in Appendix III.

My readings, with the exception of two intervals, each of eight days duration, during which the R.I.M.S. "Investigator" returned from the survey-ground to Rangoon for the purpose of recoaling, etc., extend from February 3rd to April 7th, 1911. Throughout the whole period the ship was continually shifting her position by a few miles in accordance with the exigencies of the survey, but all readings were taken in an area lying within Lat. 13° 33′ to 14° 13′ N. and Long. 97° 43′ to 98° 08′ 30″ E. A reference to the map (*vide* Text-fig. 19) shows that no rivers of any importance open to the sea along this stretch of the Burma coast, but that immediately to the south of the area lies the entrance to Tavoy River. It is, therefore, only at the southern end of this region that there was a likelihood of the influx of river-water producing any marked diminution in the salinity of the sea. Except on the two occasions on which she returned to Rangoon, as noted above, the R.I.M.S. "Investigator" put in from each Saturday mid-day to Monday morning to Byikhwaaw Bay at the extreme end of Tavoy peninsula; one might, therefore, have expected to find that on these days the salinity would show a diminution owing to the proximity of the bay to the mouth of Tavoy River, but these weekly visits appear to have no relationship to the periodic oscillation that was found to occur in the salinity of the surface-water.

The results obtained have been plotted in Text-figure 20 and this clearly shows the manner in which the salinity rises and falls. For the purpose of comparison I have given the lunar phases throughout the period. It is clearly seen that the oscillation in salinity occupied a time-period of 18½ days; it has no relation whatever to the lunar phases and is, therefore, apparently not of tidal origin. Helland-Hansen and Nansen (1909, p. 99, *et seq.*) have, however, called attention to similar periodic changes in the salinity of the surface-water of the Norwegian Sea. They found that in certain areas the variation in salinity corresponded in time with changes in the moon's phases, though not always in the same direction; thus in an area extending, approximately, from Lat. 67°-68° N. and Long. 10°-12° W. the maximum salinity "corresponds with the culmination of the moon," whereas across the southern part of the Norwegian Sea the culmination of the moon corresponded with the minimum salinity. During the years 1903–1908, Reichard (1913) had also obtained evidence of a periodic oscillation in the salinity of the surface-water in the neighbourhood of Heligoland, and this too seems to be dependent on tidal changes.

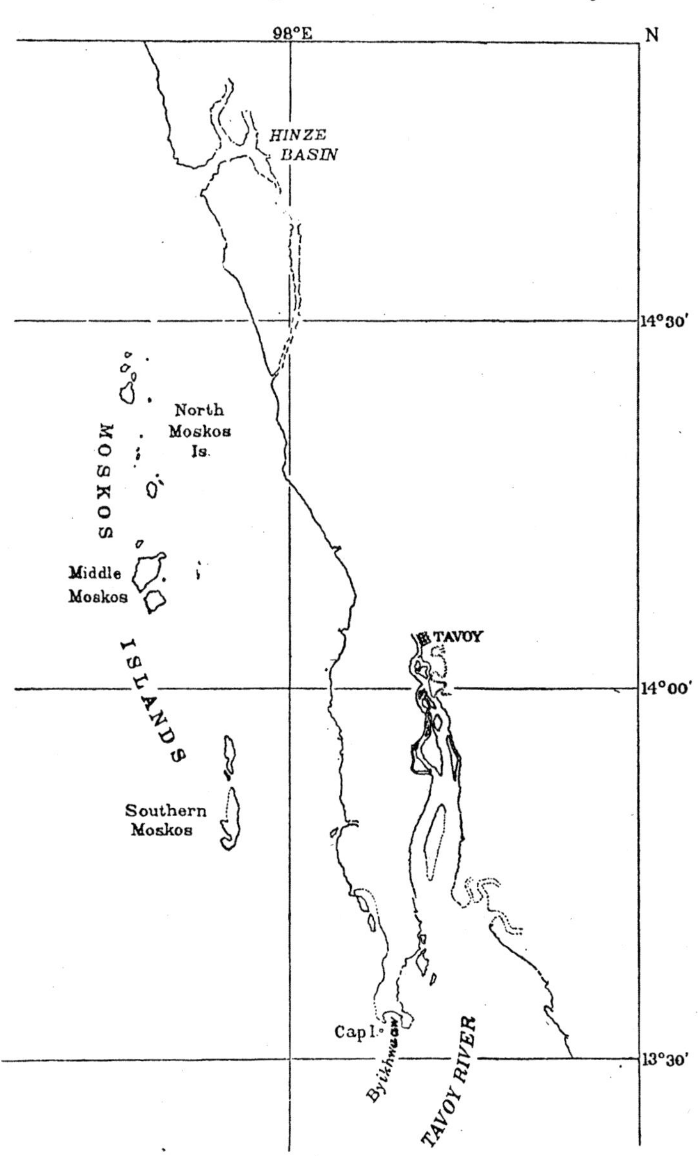

TEXT-FIG 19.—An outline map of the region of the coast of Southern Burma under survey in 1910-11.

A.—Variation in Salinity due to Tides.

In a region where numerous streams and rivers are continually pouring fresh-water into the sea one would naturally expect to find that the inflowing tide causes a considerable alteration in the salinity of the coastal waters by bringing in denser, more saline sea-water and by, at the same time, producing a temporary reversal of the river-current in estuarine regions. One would thus expect a rise in salinity with the inflowing tide and a fall corresponding to the ebb. The results obtained by me from the examination of a series of water-samples taken at the surface at different states of the tide in 1911-12 and again in 1913-14 show clearly that the tide produces a very marked effect, but the direction in which the effect is produced varies enormously in different localities. In a region such as the coast of southern Burma, where the coast line is a complicated one and the whole continental shelf is thickly dotted over with numerous islands, the tidal currents in inshore waters set in very different directions in different localities and masses of water may be moved parallel to the coast line. I have already shown (Sewell, 1913, pp. 361–364) that in Byikhwaaw Bay, which lies at the end of Tavoy Point close to the mouth of Tavoy River, both flood- and ebb-tides produce at first a fall of salinity in the sea-water, followed later by a rise, owing to the tidal flow causing the movement, during the early part of the flood, of a mass of water of low salinity from the mouth of the river northward past the bay, and, *vice versa*, during the ebb this area of low salinity is again carried to the southward. On the other hand at the mouth of Tavoy River and in the area to the south of it around Tavoy Island, the flood causes a fall in salinity and the ebb a rise, again owing to the movement of a mass of water of low salinity parallel to the coast line.

Still further to the south, in the region lying south of Lat. 12° N. where we were surveying in 1913-14, I found that the flood-tide caused a marked rise in salinity and the ebb a corresponding fall. Three sets of observations were taken in the anchorage in Mergui Harbour (Lat. 11° 29′ 57″ N.; Long. 98° 34′ 38″ E.) during the course of the season and the results obtained are shown in the accompanying Text-figs. 21–23.

The results obtained on October 12th, 1913, and on April 8th, 1914, agree fairly closely; in the former the specific gravity (σ_0) varied from 19·20 at low water to 24·89 about half-an-hour after high water, corresponding to a change of salinity from 23·90, to 30·99, a rise and fall of 7·09, or 22·88%; on April 8th, 1914 the specific gravity (σ_0) varied from 23·02 at low water to 25·94 at high water; this corresponds to a rise of salinity from 28·66 to 32·29, or 11·24%. The difference between the mean salinity in the two months, namely 22·05 in October and 24·48 in April, is almost certainly seasonal and dependent on rainfall and the rate of evaporation. On the 19th of October, 1913, however, there was a much greater change; the salinity varied with the tide from 3·53 to 23·40, a rise and fall of 19·07 or 84·91%. This great variation is all the more remarkable in that it occurred near neap tides, when the tidal effect would be at its lowest. The

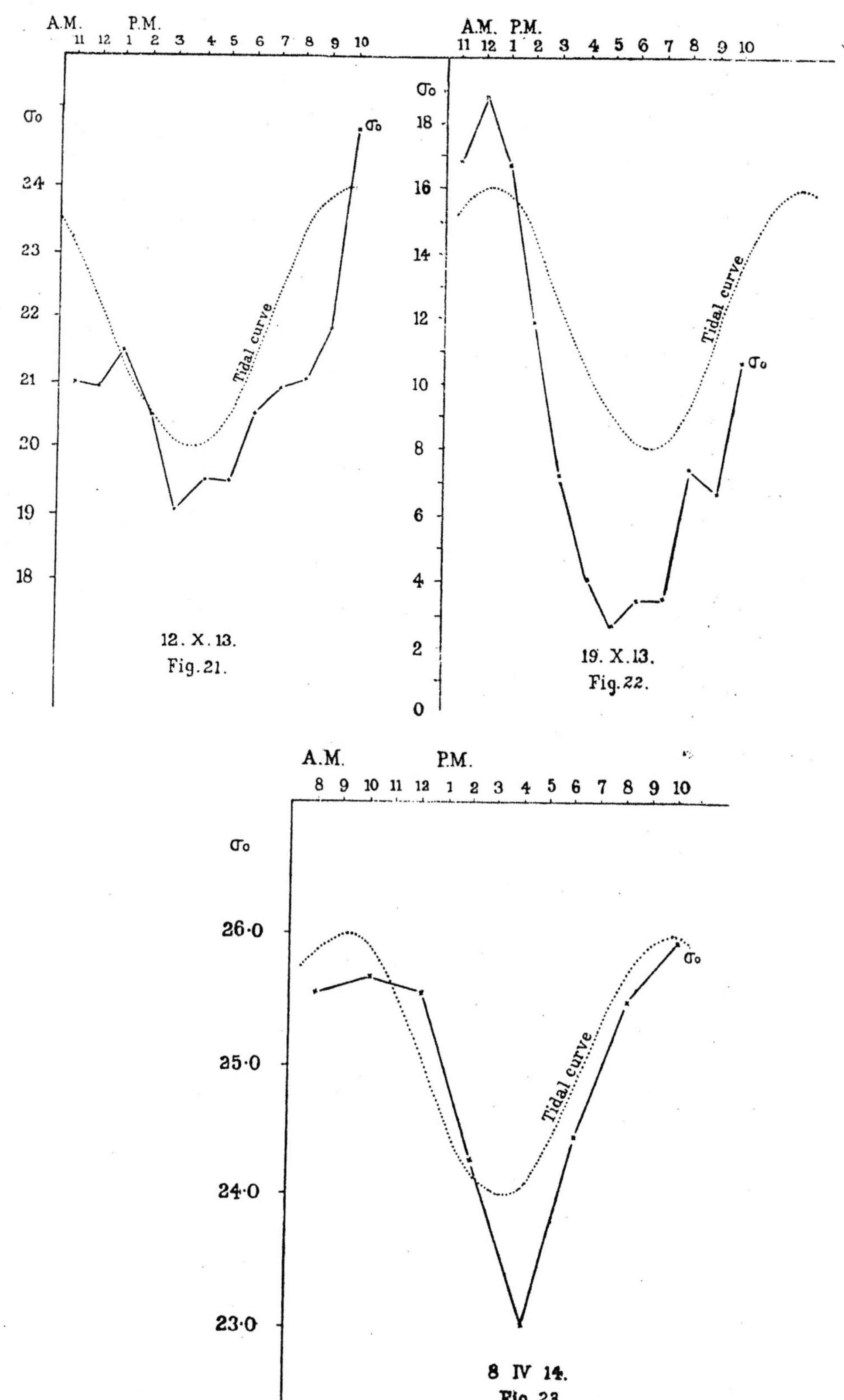

Text figs. 21-23. Changes in specific-gravity (σo) of surface water during ebb and flood tides in Mergui Harbour.

low salinity shown during the ebb indicates that there was a large increase in the amount of fresh-water pouring into the sea, and during the period from October 11th to 18th a considerable quantity of rain had fallen. In addition to demonstrating the effect of the tidal-wave on the salinity these series also show to what

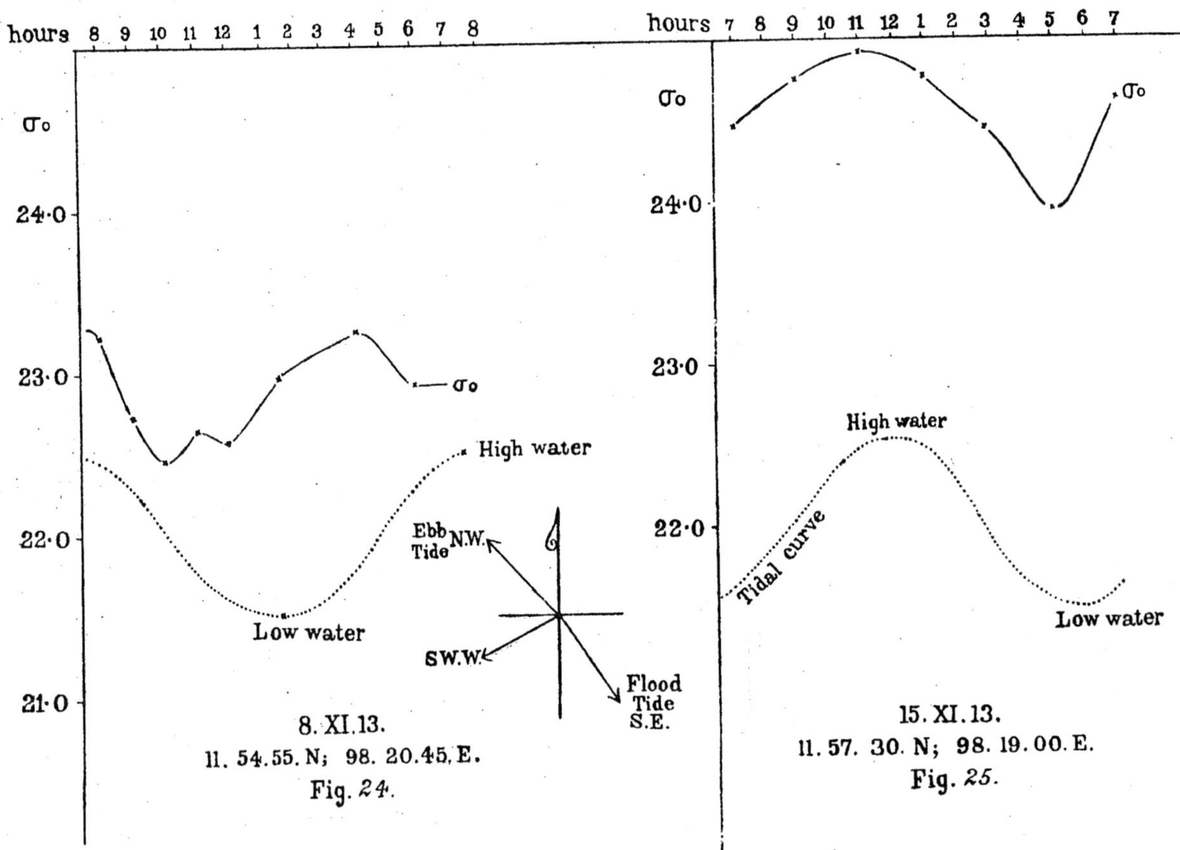

Text figs. 24-25. Changes in specific-gravity (σ_0) of Surface water with Ebb and Flood tides. Burma coast, 1913.

great and sudden changes in salinity estuarine-living animals have to be capable of adaptation.

At two other stations in this area very similar results were obtained, the flood tide causing a rise in salinity and a fall occurring on the ebb, though the range of variation was considerably less. At a position lying in Lat. 11° 57′ 30″ N.; Long. 98° 19′ 00″ E. (*vide* Text-fig. 25) the rise and fall of salinity correspond almost exactly with the rise and fall of the tide and the range of variation is 1·18, namely from 29·82 to 31·00, the specific gravity (σ_0) varying from 23·95 to 24·81; while a little further south in Lat. 11° 54′ 55″ N.; Long. 98° 20′ 45″ E. (*vide* Text-fig. 24) the rise and fall of salinity was from 27·99 to 29·06, a variation of 1·07, the specific gravity ranging from 22·45 to 23·25.

In the case of this last station (*vide* Text-fig. 24) the rise and fall of the salinity of the surface water does not exactly correspond to the rise and fall of the tide, as is the case in the other stations. The rise of salinity, as shown by the rise in specific gravity, begins at about the middle of the ebb and the fall at about the middle of the flood-tide, and it may be that we have here the result of a movement of water parallel with the coast, similar to that which I found in the neighbourhood of the mouth of Tavoy River; the directions of the flood and ebb tidal currents certainly seem to indicate that this is so. In this station the flood-tide sets towards the south-east and brings with it water of increased density. At half tide the direction of the current changes and now runs towards the south-west, and this will bring with it coastal water of a somewhat lowered specific gravity; while the ebb-tide sets towards the north-west and in consequence the specific gravity of the water again falls at first, owing to the admixture of coastal water of low specific gravity.

It appears then that on the whole in the area under investigation the rise and fall of the tide and the rise and fall of salinity (S) and specific gravity (σ_0) of the surface water follow each other fairly closely.

B.—Seasonal variation.

In order to trace the seasonal changes in the salinity of the surface-water and also to detect, if possible, any periodic variation, samples of water were taken daily in 1913-14 throughout the periods that the R.I.M.S. "Investigator" was on the survey-ground, and were carefully examined by means of a 'Buchanan' hydrometer. So as to diminish, as far as possible, any variation that might be due to tidal changes, all samples were taken at or near the time of high-water. The results have been plotted out in Text-Fig. 26.

At first sight it would appear that the specific gravity, and consequently the salinity, of the surface water is liable to a very considerable variation of a somewhat irregular character, for the curve exhibits large fluctuations even from day to day. These alterations of salinity are, however, attributable, in the main at any rate, to changes in the ship's position; in a coastal region such as this, where several large rivers flow out into the sea, the areas round their mouths must of necessity have a low salinity owing to admixture with fresh-water. Thus as one approaches the mouth of the Tavoy River, or further to the south the Tenasserim River, one may experience a marked fall in salinity in the course of a few miles. Where, however, samples were taken in one place for any length of time, as, for instance, during the period from February 25th to March 3rd, 1914, inclusive, these subsidiary variations are absent or, at any rate, very considerably reduced.

The first change, to which I would call attention, is the seasonal one. Throughout the whole period of my investigations in this area, extending from November 14th, 1913, to April 4th, 1914, the specific gravity, and equally the salinity, shows a steady rise. This I associate entirely with the seasonal character of the rain-fall. The average monthly rainfall, as given in the Meteorological Reports for India, in this area is as follows:—

MONTH.	LOWER BURMA.	BAY ISLANDS.
August ..	.. 27·99 inches ..	.. 11·13 inches.
September ..	.. 17·43 ,, ..	.. 11·70 ,,

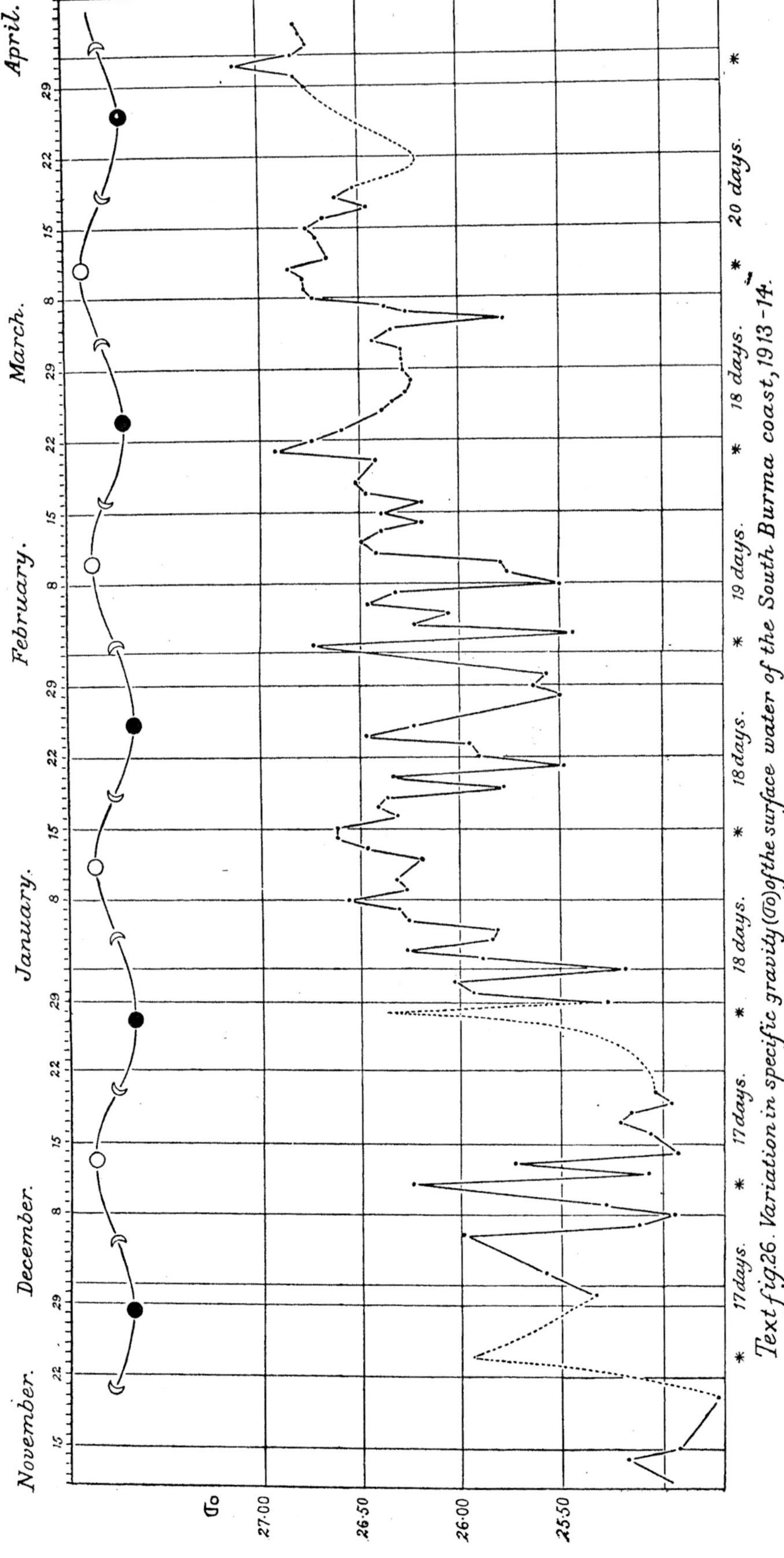

Text fig. 26. Variation in specific gravity (σ_0) of the surface water of the South Burma coast, 1913-14.

MONTH.	LOWER BURMA.	BAY ISLANDS.
October	8·01 inches.	7·08 inches.
November	6·31 „	5·62 „
December	0·48 „	4·10 „
January	0·16 „	0·91 „
February	0·25 „	0·54 „
March	0·58 „	0·63 „
April	1·66 „	1·40 „

It is clear that during the months of August and September there is a heavy rainfall throughout the whole area; during the next two months this shows a marked decrease and from December on there is, as a rule, but little rain. In consequence, during the earlier months, in addition to the direct lowering of the salinity of the surface-water by rain and by a consequent cessation of evaporation, large quantities of fresh-water are brought down by the rivers and this still further dilutes the surface-water. We thus get during the months prior to the commencement of the survey-season a marked lowering of the salinity, but this process appears to have reached its maximum by the month of November and, on the cessation of the rains, evaporation one again sets in and the salinity of the sea steadily rises. A similar variation in the salinity will result as a consequence of the rains of the south-west Monsoon, so that we probably have a double oscillation during the course of the year.

C.—Periodic variation.

In addition to this seasonal variation, the specific gravity and salinity of the surface-water on the Burma Coast show a regular periodic oscillation, the average duration of each wave from apex to apex throughout the whole series of observations being 18·13 days. During the earlier part of the survey-season 1913-14, that is in November and December, 1913, this oscillation in the specific gravity is not so clearly indicated as it is during the months of February to April, 1914, and my observations seem to indicate that as the season advances the period of oscillation tends gradually to increase in length. In the chart given above (*vide* Text-fig. 26) I have indicated the apex of each oscillation by an asterisk (*) and for the purpose of reference I have also given the various phases of the moon. At the commencement of the period of observation, in the latter part of November and in December, 1913, the time occupied by each complete wave from maximum to maximum of the specific gravity (σ_0) is almost exactly 17 days, whereas at the close of the period in April, 1914, the length of each oscillation is on the average about 19 days. I have already mentioned (*vide supra*, p. 157) that during the survey-season, 1910-11, I obtained evidence in the area lying to the north of an exactly similar oscillation in the specific-gravity having a time period of 18·5 days.

It seems to me that we have here a phenomenon that corresponds very closely to the periodic variation in the salinity of the waters of the Gulmar Fjord, that was discovered and described by Pettersson (1909): but it is, I think, obvious that in this region the periodic variation in the specific gravity of the surface-water bears no relationship whatever to the phases of the moon and, therefore, can in no way be connected with the increase and decrease of the tidal effect at springs and neaps; Pettersson attributed the periodic oscillation observed by him in the main to the influence of the lunar phases: "These variations," he remarks, "take the form of great submarine waves which enter the Fjord at intervals corresponding with the variations in the declination and the distance of the moon," though he subsequently points out that "the attraction of the sun and moon upon the waters of the ocean do not seem, however, to be the sole agent of the phenomenon in question." His observations, taken down to a depth of 50 meters, showed that he was undoubtedly dealing with an oscillation of the deep waters of the basin at the head of which the Gulmar Fjord is situated. The work of Watson (1904) and Wedderburn (1904–1909), during the progress of the Scottish Lake Survey, has demonstrated the presence of similar oscillations in the lower levels of the water of deep lakes, in which a layer of warm water is superposed on a deeper cold stratum, and indeed such oscillations seem to be of almost universal occurrence in all large areas of inland waters. Halbfass (1911) has brought forward evidence to show that there is a temperature seiche in the Madü See having a period of 25 hours, which corresponds to the length of the long axis of the lake, and Hayford (1922) has demonstrated that a similar phenomenon can be traced in the great lakes of North America. Wedderburn (1911, p. 55) himself anticipated that such oscillations would be found to occur in the large ocean basins and both Wedderburn (1909 (c)) and O. Krummel (1911, p. 192) claimed that the observations of Pettersson on the Gulmar Fjord were of this nature. Pettersson, however, in a later paper (1912) showed that the oscillation in the Gulmar Fjord appears to be definitely connected with the lunar phase and that this seems to be the causative agent in producing the observed rise of salinity, not only in the short-period variation, namely that of 14 odd days, but in other long-period phases in which a regular rise and fall of the salinity of both upper and lower levels can be detected. It would, therefore, seem that Pettersson is right as regards the changes in the Gulmar Fjord, but the absence of any correlation between the oscillations of salinity on the Burma Coast and the lunar phases renders it clear that this phenomenon cannot be attributed to the same cause. Honda, Terada, Yoshida and Isitani (1908) have demonstrated the occurrence of periodic seiches in small bays round the Japanese Coast and similar occurrences have been recorded in many similar bays in other parts of the world, for an account of which I would refer readers to Krummel's work (1911, pp. 168-195). Sir George Darwin (1910, p. 415 *et seq.*) and James Murray (1911, p. 134) have suggested that there is evidence of a seiche, of approximately 3 days' duration, in the Ross Sea, and the work of Nansen (1902) on the deep-sea temperatures taken during the expedition of the "Fram" to the North Pole, and that of Helland-Hansen and Nansen (1909) on the density of the water at different levels in the Norwegian Sea has produced a certain amount of evidence to show that periodic seiches do actually occur in the larger ocean basins, and there is a certain amount of evidence in favour of the periodic changes in the salinity of the surface-water off the coast of Burma being due to this cause. If this be so, we should expect to find that the period of oscillation exhibits a clear relationship to the size and depth of the basin of the Andaman Sea. Wedderburn (1908, p. 424) has given the following formula for calculating the period of such a seiche in a basin where the density of the water is clearly divided into two layers or strata, an upper layer of water of low density floating on a deeper layer of more dense water; a condition such as this is, as is well known, present throughout the whole region of the Indian Seas.

$$t = \frac{2L}{\sqrt{\dfrac{G \times (\sigma_2 - \sigma_1)}{\dfrac{\sigma_1}{p} + \dfrac{\sigma_2}{P}}}}$$

L is the length of the basin in meters, and σ_1 and σ_2 and p and P are respectively the densities and depths of the upper and lower layers of water. If, therefore, the periodic oscillation of the specific gravity of the surface-water off the Burma Coast is due to an internal seiche in the deeper waters of the Andaman Sea basin, the observed time of oscillation should agree closely with the time-period calculated from the above formula. The Andaman Sea has approximately a total length of 1,234 sea miles or 2,252,600 metres and the maximum depth is about 2,200 fathoms or 4,015 metres. Owing to the influx of river-water into the basin the specific gravity of the water of the upper levels is considerably reduced and the lowering of the density is still further enhanced by the comparatively high temperature of the surface-water. It would seem probable, therefore, that conditions are favourable for the production of an internal seiche. Unfortunately I have no direct data regarding either the density (σ_t) of the water in the depths of the basin or of the relative depths of the upper and lower strata. I have, therefore, had to fall back on observations made in neighbouring localities. For the first of these data I have taken the density of the deep water as 1·0278, this figure being derived from various observations made by the S.S. "Valdivia" in the Indian Ocean between 0°58′ and 2°0′ N. and 96° and 99°43′ E. (*vide* Schott. 1902, Table 21): and, for the second, the depth of 125 metres is based on the serial temperature and salinity observations of the "Investigator" in the Bay of Bengal down to a depth of 500 fathoms, which show that in the eastern part of the Bay the "discontinuity layer" lies at a depth of about 65 to 70 fathoms (*vide* also Schott, 1902, p. 180).

I have already pointed out that the density of the surface-water varies throughout the period of my observations and this will of itself cause a difference in the length of the period of oscillation, if it be due to a seiche. I have, therefore, taken the average density for each month and have calculated by Wedderburn's formula the corresponding period of oscillation, and in Table 31 below I give the results so obtained and the actual observed period of oscillation in the specific gravity of the surface-water. It seems more than probable that the actual depth of the upper layer of less saline water varies somewhat from season to season and, as I shall hope to show in a subsequent paper, shows a distinct periodic oscillation as regards its lower limit, which reaches one maximum of depth about March to April. In the following calculations I have not taken this increased depth (p) into account; it would, however, tend to lengthen the calculated period of oscillation and thus cause the figure for March to approach more nearly to the observed period.

Month.	Average density (σ_t) of surface water.	Calculated period of oscillation.	Observed period of oscillation of specific gravity (σ_t)
November, 1913. ..	1·018955	16·3 days	17·0 days.
December, 1913. ..	1·020154	18·0 days	17·5 days.
January, 1914 ..	1·020661	18·1 days	18·0 days.
February, 1914 ..	1·020697	18·1 days	18·5 days.
March, 1914 ..	1·020840	18·4 days	19·0 days.

Table 31; showing the monthly variation in the observed and calculated period of oscillation in the specific gravity of the surface-water on the Burma Coast.

So close in the agreement between the observed and calculated period of oscillation, that it is, I think, sufficient ground for believing that we have here evidence of a seiche in the deep water of the Andaman Sea basin, and the question arises what is the cause of this oscillation. Wedderburn (1909 (c), p. 596) remarks that "in lakes the density seiche is started by winds which accumulate the surface waters towards the lee end of the lake. When the wind moderates or ceases, the oscillation begins and continues until equilibrium is established, or until another storm springs up which starts a new seiche." It is not improbable that the causative agent in the formation of this seiche in the Andaman Sea basin is the wind of the North-East Monsoon during November and December. This must tend to cause an alteration of the depth of the surface stratum at the two ends of the basin, and since the seiche would not reach its maximum till the wind moderated we should not expect to find the best evidence of such an oscillation until the close of the monsoon period in January, which agrees with what I have pointed out above, namely, that the oscillation becomes most marked in February and the succeeding months. Such a to-and-fro swing of the water of the deeper level will at these intervals of time, namely 18 days approximately, bring more saline water nearer to the surface; as a result of admixture with the less saline water of the upper stratum, brought about in all probability by wave action, which will also be more effective during and just after the period of the monsoon than during the calm periods before and afterwards, there will ensue periodically a rise in the salinity of the surface-water.

Seasonal and Periodic Variations in the Salinity of the Surface Water of Nankauri Harbour and the Neighbourhood.

During the survey season of 1921-22 and the early part of 1922-23 the R.I.M.S. "Investigator" was engaged in the survey of Nankauri Harbour in the central group of the Nicobars. I was thus able to carry out observations on the conditions of the surface-water on the west side of the Andaman Sea basin. During the whole of this period my observations on the salinity were made by means of a 'Buchanan' hydrometer and from the results obtained I have calculated, by Knudsen's Tables, the actual salinity and density of the water.

It is of interest to see how far the conditions on the west side of the Andaman sea differ from those on the east and to what extent these conditions may vary from year to year.

A. Seasonal variation.

In Table 32 I have given the average temperature, salinity (S) and density (σ_t) of the surface-water in each month of the two survey-seasons; my actual observations are given *in extenso* in Appendix VI.

Month.	Survey season 1921-22.			Survey season 1922-23.		
	Salinity.	Temperature.	Density.	Salinity.	Temperature.	Density.
		oC			oC	
October	33·58	28·7	21·13	33·78	28·1	21·48
November	33·46	28·4	21·15	33·77	27·5	21·64
December	33·14	28·1	20·99	33·87	27·3	21·81
January	32·95	27·8	20·94	..	..	..
February	33·21	28·5	21·02	..	..	..

TABLE 32; giving the Salinity, Temperature and Density in each month of the surface-water of Nankauri Harbour, Nicobars.

N.B.—*The average Temperature, as given, refers only to the water-samples taken for examination.*

A comparison of the monthly averages in the two years shows that during the period 1921-22 both the salinity and the temperature of the surface-water show a steady decrease from October to January; in the month of January this phase has reached its maximum and in the following month, February, the salinity and temperature of the water commence to rise. In the following season, 1922-23, although the temperature of the surface-water falls steadily from October to December, the salinity exhibits an increase. As a result of these changes it is found that the density of the water in 1921-22, in spite of the fall of salinity, increases slightly from October to November, but then falls steadily till January, after which it again begins to rise: in 1922-23, however, the fall of temperature is combined with a rise of salinity and there is, therefore, a continuous rise in the density from October to December.

It is clear that the conditions present in these two seasons were very different; the fall in the temperature is part of the double seasonal change that takes place throughout the whole of the Indian Seas, for as I have already noted (*vide supra*, pp. 57–62), the temperature of these waters exhibits two maxima, in

May and again in September-October, with two minima, corresponding to the two monsoons, in July and in January-February. The difference in the salinity of the surface-water in the two seasons appears to be correlated with the rainfall in the Bay of Bengal and over the Bay Islands and Upper Burma. A glance at the map shows that three large rivers, the Irrawaddi, the Salween and the Sitang, drain the whole of upper Burma and the country to the north and flow out into the northern end of the Andaman Sea basin: any excess of rainfall, therefore, over the region of upper Burma, that is drained by the Irrawaddi and Salween Rivers will, in the course of time, find its way into the Andaman Sea, the time taken depending on the distance away from the river mouth. I have in the following table given the amount of rainfall over both areas, during the months of July to December inclusive, in 1921 and 1922.

Area.	1921. July–December.		1922. July–December.		Excess of 1921 over 1922.
	Total Rainfall inches.	Departure from Normal.	Total Rainfall inches.	Departure from Normal.	
Bay Islands	56·25	+3·92	54·05	+2·60	+2·20
Upper Burma	34·22	+4·59	33·42	+3·64	+0·80

During both periods the amount of rainfall was above the normal but to a less extent in 1922 than in 1921. As a result of this increased rainfall in 1921 and especially of the increase which fell over the Bay Islands, namely 2·20 inches, there must have been a marked lowering of the salinity locally and this must have been still further lowered by the increased outflow of fresh-water from the Irrawaddi and Salween Rivers. The dilution of the surface-water over the northern end of the Andaman Sea will thus have been considerably greater in 1921 than in 1922; and during the months of the North-East Monsoon this dilute surface-water will have been swept by the wind in a south-westerly direction and, in consequence, have caused an abnormal lowering of both the salinity and density of the surface-water in and around the region of the central group of the Nicobars.

B. *Variation in Salinity due to Tides.*

According to "The Bay of Bengal Pilot" the flood-tide sets into Nankauri Harbour through the western entrance and the ebb flows out through the eastern channel: but during the period of our survey of the harbour from October, 1921, to February, 1922, and again from October to December, 1922, the flood-tide entered the harbour through both east and west entrances, while the ebb-tide flowed through the harbour and out through the western entrance only. There was thus a more or less constant current flowing into the harbour through the eastern entrance, the strength of the flow varying with the state of the tide. It is probable that the direction of flooding of the tide is influenced by the prevailing monsoon and, therefore, varies at different times of the year; the account given by "The Bay of Bengal Pilot" is probably correct for the period of the South-West Monsoon but during the North-East Monsoon the conditions are changed to what we found during our visits.

Owing to the continuous flow of water through the eastern entrance one would not expect to find any very marked tidal changes in the salinity of the surface-water of the harbour, and especially of that region of the harbour where the usual anchorage is situated, since this is comparatively close to the eastern entrance. On the one hand, however, owing to the presence of large mud-flats, that are covered at high-water, and over which evaporation must be considerable, one would expect that a rise in salinity of the harbour water would result, especially when high-water occurred at or near noon; on the other hand, and particularly during the wet season in December and January, the drainage of rain-water from the land into the harbour must tend to cause a lowering of salinity in the water of the harbour.

In order to determine what changes, if any, occurred during the rise and fall of the tide, a series of samples was taken at a position well out in Cross Harbour, between Octavia and Spiteful Bays, at two-hourly intervals between 6 a.m. on the 24th and 6 a.m. on the 25th February, 1924, when we were at anchor in the harbour. The samples were titrated with standard silver nitrate solution and the results are given in Table 33:

Date 1924.	Time.	Temp. of Sample.	Air Temp.	Relative Humidity of Air.	Barometric Pressure (Corrected).	Cl.	S.	σ_0	σ_t	
		°C	°C		Inches of Hg.					
Feb. 24th	6 A.M.	28·0	26·67	88·22	29·83	18·54	33·49	26·91	21·28	Low water 6-15 A.M.
"	8 A.M.	27·5	27·22	81·74	29·87	18·53	33·48	26·90	21·44	
"	10 A.M.	28·0	27·78	79·77	29·90	18·55	33·51	26·93	21·30	
"	12 noon.	28·2	28·22	77·83	29·88	18·53	33·48	26·90	21·21	High water 11.55 A.M.
"	2 P.M.	28·2	28·33	78·67	29·84	18·50	33·42	26·86	21·17	
"	4 P.M.	28·0	28·06	79·90	29·83	18·55	33·51	26·93	21·40	
"	6 P.M.	27·6	27·78	82·01	29·84	18·59	33·58	26·99	21·49	Low water 6-05 P.M.
"	8 P.M.	28·0	27·50	83·95	29·89	18·54	33·49	26·91	21·28	
"	10 P.M.	..	27·22	83·89	29·93	18·54	33·49	26·91	..	
"	12 mid-night.	..	..	..	..	..	..	..	..	High water 12-15 A.M.
Feb. 25th	2 A.M.	..	27·22	81·74	29·87	18·54	33·49	26·91	..	
"	4 A.M.	..	26·94	83·81	29·87	18·45	33·33	26·78	..	
"	6 A.M.	28·0	27·22	86·09	29·88	18·43	33·30	26·75	21·13	Low water 6-35 A.M.

Table 33; observations on the salinity of the surface-water of Nankauri Harbour at different states of the tide.

Note.—It was found impossible to arrange for a reading of the temperature during the hours from 8 p.m. to 4 a.m. I am therefore unable to give the density of the surface-water during this interval.

The results obtained have been plotted out in Text-fig. 27 and, for the purpose of reference, I have also given the rise and fall of the tide, as recorded by the Tidal Observation Station on shore and only about one-quarter of a mile away from the ship: I have also given the variations in the humidity of the atmosphere, as calculated from wet- and dry-bulb observations, and the rise and fall of the barometric pressure as recorded on the ship during the same period. From a comparison of the four curves, it appears that there is a distinct tendency for the salinity of the harbour-water to rise and fall with the tide, the influx of the open-sea water on the flood causing a rise in salinity, that is followed by a fall on the ebb, this latter being

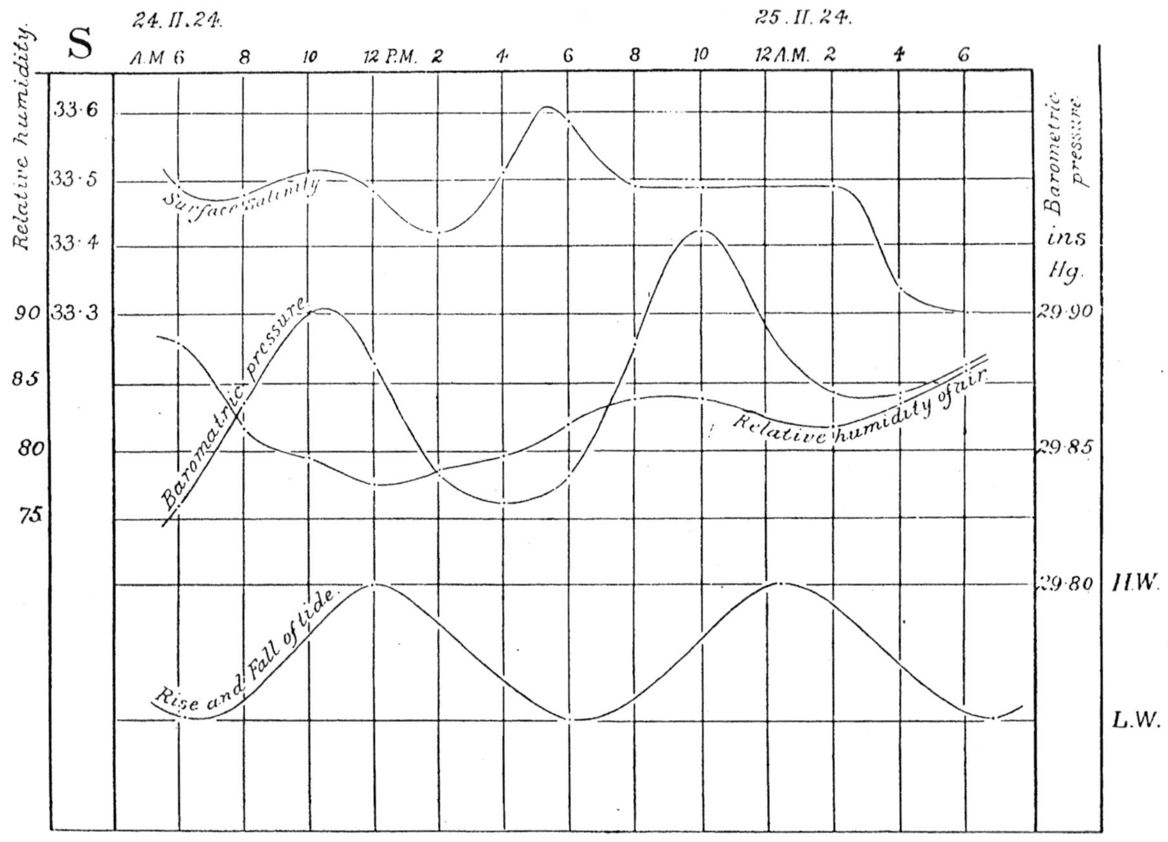

TEXT-FIG. 27; showing the variation of salinity of the surface-water, the rise and fall of the tide and the variation in the humidity of the atmosphere in Nankauri Harbour, Nicobars, on February 24-25, 1924.

due doutless to drainage from the land. The correlation of the two curves is, however, interrupted during the early part of the afternoon, when the temperature of the water is highest and the humidity of the atmosphere is low, and there is a marked rise in salinity between 2 and 5-30 p.m. followed, after a moderate fall, by a period of stability due, undoubtedly, to evaporation of the water in the wide shallow bays on either side of the harbour and the discharge of this condensed water into the main channel on the ebb-tide. A second point, however, to which I would call attention is the manner in which the rise and fall of the surface salinity agrees with the rise and fall of barometric pressure. The agreement between the two curves is closest in the morning from 7 a.m. to 2 p.m.; between 2 and 10 p.m. the parallelism is interrupted owing to the effect of evaporation, but from then on the two curves again tend to coincide; the change in salinity, however, being about 2 hours later than the recorded barometric change.

On February 13th, 1922, observations were made on the salinity and temperature of the water of the harbour at different depths during both the flood and the ebb tides. The samples were taken at intervals of five fathoms between the surface and the bottom, the total depth of water being 23 fathoms. The times, at which the serial observations were taken, were (a) at 8-30 a.m. or 3½ hours after low water, and (b) at 11-30 a.m. or 1½ hours after high water. The water samples were taken by means of an 'Ekman' reversing water-bottle and the density was estimated by a 'Buchanan' hydrometer, the results being converted into salinities by means of Knudsen's Tables. The results obtained are given below in Table 34 and I have charted them in Text-fig. 28.

Date.	Time.	State of Tide.	Depth.	Temp. in situ	Hydrometer reading		Cl.	S.	σ_0	σ_t
					$\frac{S^t}{S^4}$	Temp.				
				°C		°C				
Feb. 13, 1922.	8-30 a.m.	Flood tide 3½ hours after low water.	Surface	28·50	1·020856	29·1	18·49	33·40	26·842	21·055
			5 fms.	27·60	1·020599	29·7	18·45	33·33	26·783	21·294
			10 ,,	27·75	1·021155	30·0	18·94	34·22	27·491	21·905
			15 ,,	27·92	1·020451	30·2	18·47	33·37	26·808	21·214
			20 ,,	27·22	1·020433	30·3	18·48	33·39	26·825	21·454
	11-30 a.m.	Ebb tide 1½ hours after high water.	Surface	28·30	1·020592	29·8	18·48	33·39	26·830	21·241
			5 fms.	27·89	1·020152	31·0	18·45	33·33	26·784	21·202
			10 ,,	27·59	1·020151	31·1	18·475	33·38	26·822	21·333
			15 ,,	27·67	1·020089	31·1	18·43	33·30	26·754	21·238
			20 ,,	27·89	1·020397	31·2	18·66	33·71	27·086	21·482

Table 34; Results of serial observations on the Salinity and Temperature of the water of Nankauri Harbour.

On the flood tide the salinity shows but little change between the surface and 5 fathoms depth; below this, however, there is a marked increase that reaches its maximum at 10 fathoms; below this again the salinity falls to about 17 fathoms and, finally, there is a slight rise in the salinity of the bottom-water. On the ebb tide, however, there is but little change in the salinity of the water between the surface and 15 fathoms and this is followed by a rise from the 15-fathom level to the bottom. The temperature of the water during the flood falls steadily from the surface to a depth of 5 fathoms, it then equally steadily increases to 15 fathoms and from this falls again rapidly to the bottom; on the ebb tide, however, the temperature falls steadily from the surface to a depth of 12 fathoms and then rise

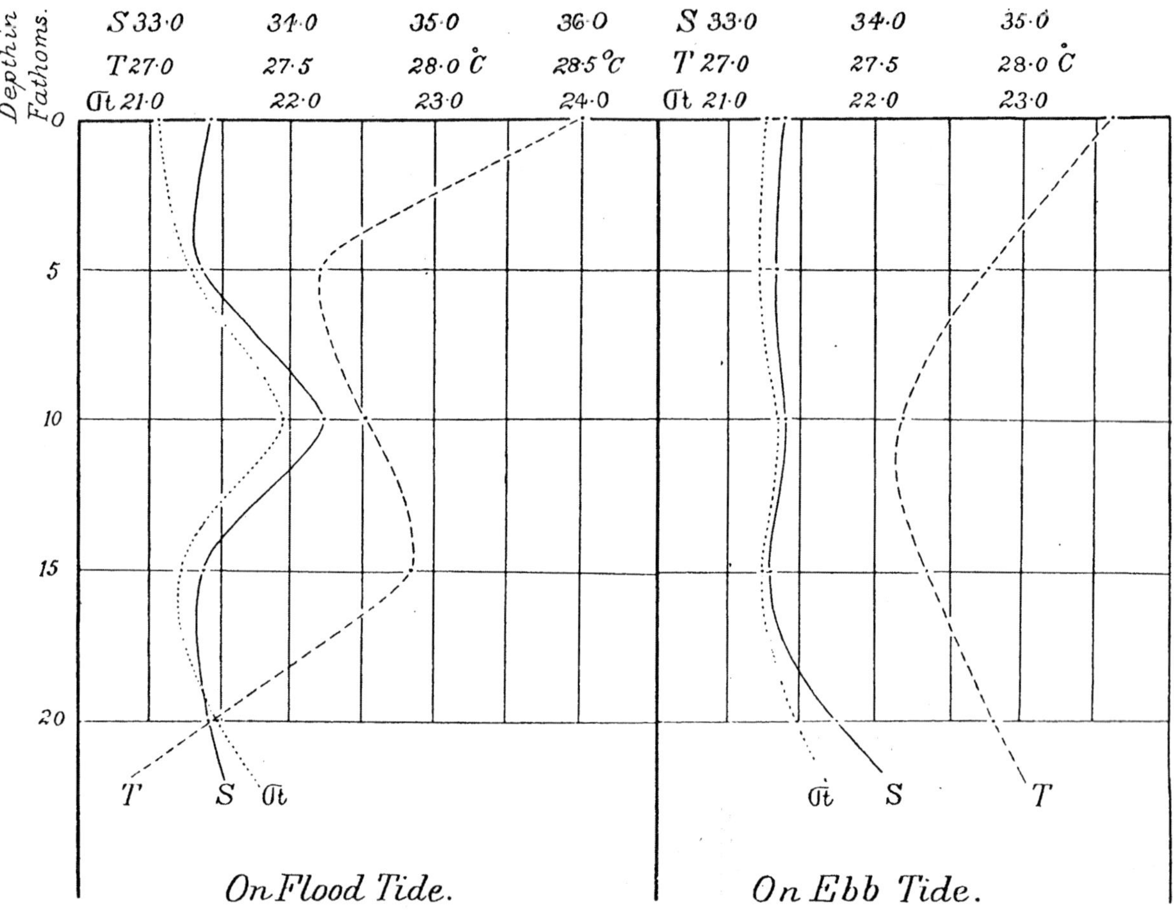

TEXT-FIG. 28; showing the variation at different states of the tide in the salinity and temperature of the water of Nankauri Harbour.

again as one approaches the bottom. A comparison of these two series of changes seems to indicate that during the flood tide there is a well-marked inflow into the harbour at or about the 10-fathom level of sea-water that has a higher salinity and a higher temperature than that found at the levels above and below, and that during the period of slack-water some at any rate of this water sinks to the bottom and hence causes the rise in salinity and temperature that is observed in the bottom-water on the ebb tide.

On the 27th and 28th of February, 1924, a series of observations was made at two-hourly intervals on the surface-water off the entrance to Dring Harbour near the north end of Camorta Island in the Nicobars; before the series was complete the R.I.M.S. "Investigator," owing to the exigencies of the survey, moved to Expedition Harbour. The results obtained by the titration method are given below in Table 35 and I have also plotted them in Text-fig. 29, and for the purpose of reference I have also given the rise and fall of the tide, the relative humidity of the atmosphere, as calculated from wet- and dry-bulb readings, and the height of the barometer recorded on board.

Date 1924.	Time.	Temp. of Sample.	Air Temperature.	Relative Humidity At Atmosphere	Barometric Pressure (corrected.)	Cl.	S.	σ_0	σ_t	
		°C	°C		Inches of Hg.					
Feb. 27. ..	10 A.M.	27·0	28·33	79·90	29·94	18·56	33·53	26·94	21·63	Low water 7-40 A.M.
	12 Noon	27·0	28·89	74·06	29·93	18·54	33·49	26·91	21·60	
	2 P.M.	28·2	28·78	79·49	29·87	18·50	33·42	26·86	21·17	High water 2·05 P.M.
	4 P.M.	27·0	28·33	81·19	29·83	18·59	33·58	26·99	21·68	
	6 P.M.	27·4	28·78	82·84	29·87	18·57	33·55	26·96	21·52	} Rain storm about 6 P.M.
	8 P.M.	27·0	26·89	91·12	29·91	18·55	33·51	26·93	21·62	
	10 P.M.	..	27·22	90·55	29·94	18·59	33·58	26·99	..	Low water 8-30 P.M.
	12 midnigt.	..	27·22	90·55	29·93	18·59	33·58	26·99	..	
,, 28 ..	2 A.M.	..	26·67	89·06	29·88	18·55	33·51	26·93	..	High water 2-30 A.M.
	4 A.M.	..	26·67	95·21	29·86	18·56	33·53	26·94	..	
	6 A.M.	27·0	26·67	91·35	29·87	18·56	33·53	26·94	21·63	Low water 8-30 (approx.) A.M.
	8 A.M.	27·9	27·78	79·77	29·92	18·59	33·58	26·99	21·39	} Both these samples were taken at the anchorage in Expedition Harbour.
	10 A.M.	28·0	28·61	79·97	29·94	18·59	33·58	26·99	21·36	

Table 35; observations on the salinity of the surface-water at anchorage opposite entrance to Dring Harbour and in Expedition Harbour.

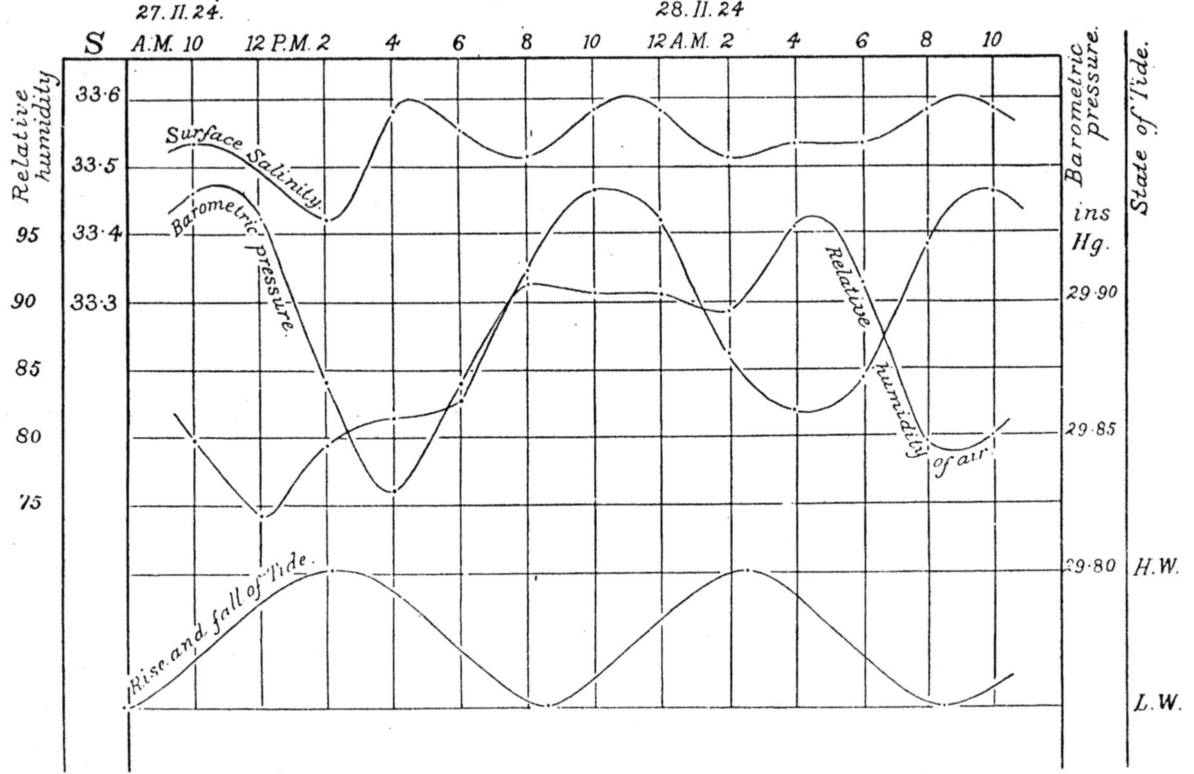

TEXT-FIG. 29; showing the variation in salinity and temperature of the surface-water off the entrance to Dring Harbour, Camorta Island, Nicobars, February 27 and 28, 1924.

A comparison of these curves indicates that in the main the daily variation in the salinity of the surface-water in this area differs considerably from the changes that we have already seen to take place in Nankauri Harbour. Corresponding with the rise of the tide there appears in this area to be a fall in the salinity; this is best seen between the hours of 10 a.m. and 4 p.m. on the 27th and again between 10 a.m. and 10 p m. on the 28th; between 4-30 p.m. and 11 p.m. on the 27th there occurs a well-marked fall and subsequent rise in the salinity that was probably due to the occurrence of a rain storm, and this masks the apex of the rise and fall in the salinity that should have occurred with the fall and subsequent rise in the tide during this period. It is probable that in this locality there occurs during the rise of the tide and subsequent slack water a very considerable degree of evaporation of the surface water, both in the bay itself, which is comparatively shallow, and in the very extensive mangrove swamps at the head of the bay, resulting in a general rise of salinity, and during the ebb this condensed water is carried out of the harbour and so raises the salinity of the water off the entrance. It is interesting to note, however, that in this series of observations there is again a quite well-marked relationship, exactly similar to that noted above in Nankauri Harbour, between the curve of salinity and that of barometric pressure: at 10 a.m. both salinity and barometric pressure are high and both fall together till 2 p.m., after which the salinity rises, as a result possibly of increased evaporation during the early part of the afternoon; but between 4-30 and 8 p.m. this rise is interrupted by the rain storm already mentioned. From 8 p.m. on the 27th to 10 a.m. on the 28th the salinity and barometric pressure again present the same rise and fall.

On February 20th and 21st, 1924, a third series of observations were carried out in Revello Channel in the neighbourhood of Perseus Reef at the extreme north end of the Camorta Island. The results of these observations are given in Table 36, and I have plotted them in Text-fig. 30.

Date 1924.	Time.	Temp. of Sample.	Air Temperature.	Relative Humidity of Atmosphere.	Barometer Pressure (corrected).	Cl.	S.	σ_0.	σ_t.	
		°C	°C		Inches of Hg.					
Feb. 20	10 a.m.	28·0	28·06	81·93	29·90	18·47	33·37	26·81	21·19	High water, 9.35 a.m.
,,	12 noon	28·0	28·33	88·00	29·86	18·42	33·28	26·74	21·12	
,,	2 p.m.	27·8	28·61	83·66	29·81	18·45	33·33	26·78	21·23	
,,	4 p.m.	28·2	28·06	84·05	29·79	18·43	33·30	26·75	21·05	Low water, 3.35 p.m.
,,	6 p.m.	28·0	28·06	84·05	29·80	18·49	33·40	26·84	21·22	
,,	8 p.m.	28·5	27·22	88·31	29·85	18·46	33·35	26·80	21·02	
,,	10 p.m.	28·0	26·67	89·65	29·86	18·54	33·49	26·91	21·28	High water, 10.05 p.m.
,,	12 mid-night.	..	26·94	88·27	29·85	18·46	33·35	26·80	..	

Date 1924.	Name.	Temp. of Sample.	Air Temperature.	Relative Humidity of Atmosphere.	Barometer Pressure (corrected).	Cl.	S.	σ_0.	σ_t.	
Feb. 21	2 a.m.	..	27·22	86·09	29·85	18·46	33·35	26·80	..	
,,	4 a.m.	..	27·22	86·09	29·85	18·46	33·35	26·80	..	Low water, 4.25 a.m.
,,	6 a.m.	27·8	26·11	89·79	29·80	18·44	33·31	26·77	21·22	
,,	8 a.m.	27·8	27·22	85·26	29·85	18·43	33·30	26·75	21·23	
,,	10 a.m.	27·8	27·78	81·88	29·86	18·45	33·33	26·78	21·26	High water, 10.15 a.m.
,,	12 noon	28·2	27·78	81·09	29·83	18·45	33·33	26·78	21·10	

Table 36; observations on the specific gravity of the surface-water off Perseus Reef, Camorta Island, Nicobars.

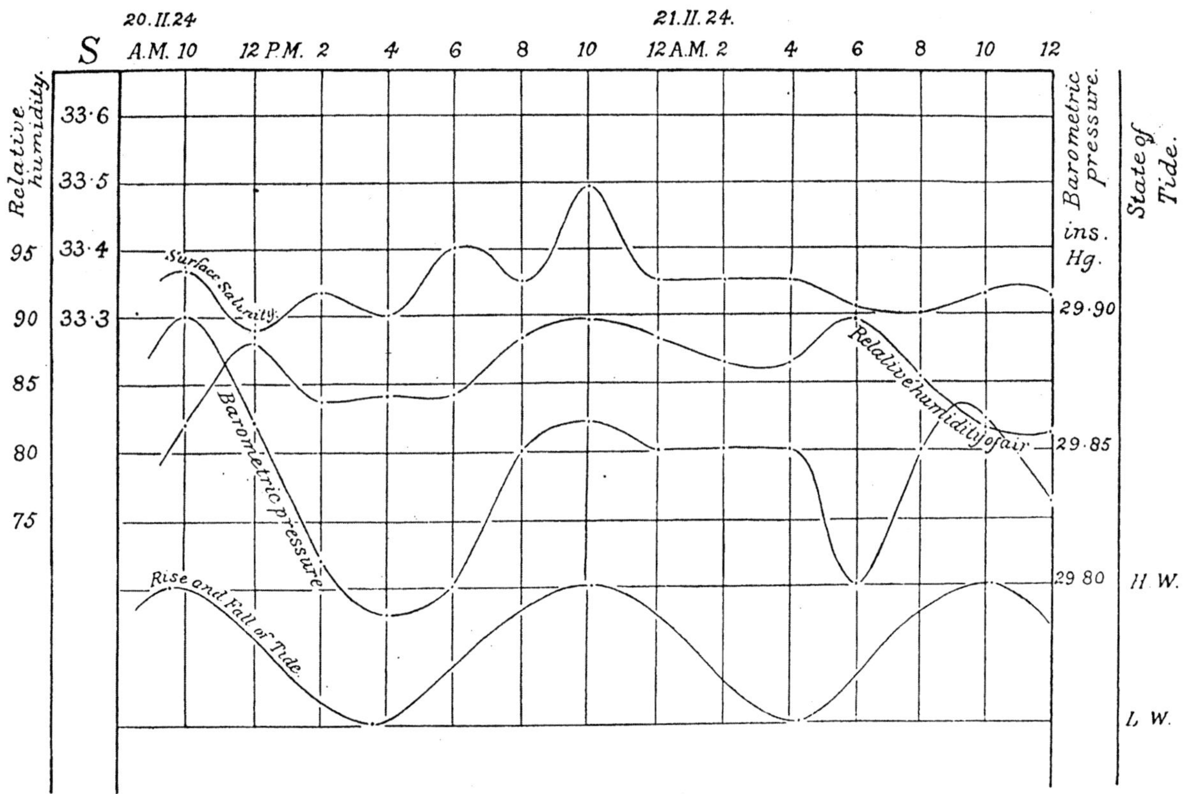

TEXT-FIG. 30.—Showing the variation in salinity and temperature of the surface-water off Perseus Reef, north end of Camorta Island, Nicobars, February 20-21, 1924.

Ignoring, for the moment, minor fluctuations in the salinity of the surface-water, it is clear that the general salinity exhibits during the period of observation two minima and one maximum. The first fall occurs between 10 a.m. and 12 noon and is succeeded by an irregular rise, which reaches its maximum at 10 p.m.; this is followed by a second fall till 7-30 a.m. on the following day, after which the salinity again rises till 11 a.m. and then once more commences to fall. A comparison of the curves of salinity and barometric pressure reveals that there is here, as in the case of the other areas, off Dring Harbour and in Nankauri Harbour, a very close agreement between the two, the only difference occurring between 12 noon and 4 p.m. on the 20th during which time the barometer exhibits the normal fall, whereas the average level of the salinity has commenced to rise, but this rise can perfectly well be accounted for by increased evaporation during the hottest part of the day. There is, however, an almost equally close agreement between the rise and fall of the salinity and the rise and fall of the tide. It seems possible, therefore, that in this area the rise and fall of salinity may be due to (1) changes brought about by the flood and ebb of the tide. This, as I have pointed out above, appears to be the case in Nankauri Harbour. The results of my investigations in this region show that in the month of February the salinity of the surface water exhibits a distinct rise as one passes northwards towards the Andamans, and, furthermore, that the surface-water of the Bay of Bengal to the west possesses, at this period of the year, a higher salinity than the water of the Andaman Sea to the east. In the region round the northern end of Revello Channel, the tidal flow sets to the south on the flood and to the north on the ebb and thus the flood-tide would bring in water of higher salinity and lower temperature; or (2) the changes observed may, on the other hand, be due to an actual change in the surface-water that is brought about by some other agency that synchronises with and may be dependent on changes in the barometric pressure itself, and the fact that this relationship can be traced in all three series indicates that this is a wide-spread phenomenon. I have already (*vide supra*, p. 79 *et seq.*) shown that there is a very distinct variation in the strength of the wind, correlated with the rise and fall of barometric pressure, and in a subsequent paper I shall show that this is further correlated with changes in the salinity of the surface-water of the open sea in and around the coasts of India.

Before leaving this series of observations we must turn our attention for a moment to the minor fluctuations that occur in the salinity of the surface-water off Perseus Reef (*vide* Text-fig. 30). These fluctuations are clearly seen between 10 a.m. and 10 p.m. on the 20th February, and, moreover, a comparison of the salinity- and temperature-records shows that a rise in salinity is accompanied by a fall of temperature. The intervals between the successive secondary maxima in salinity is exactly four hours. Taking Wedderburn's formula as our basis of calculation and applying the correction necessary for an open bay (*vide* Krummel, 1907, p. 163) we can estimate the approximate length of the time-period of such a seiche. The length of Revello Channel is approximately 8 sea-miles or 14·8 kilometres and the breadth of the southern entrance almost exactly half this. The greatest depth of the channel at the southern end, where a sounding of 140 fathoms has been recorded. The density of the surface-water at the time of my observations was approximately 1·022 and I have taken the depth of the surface-layer as being 60 fathoms ; for the density of the deeper stratum I have taken the figure 1·02592 ; this is based on "Investigator" records of the deep water of the Bay of Bengal (*vide infra*, p. 182). On this data the calculated time-period of a seiche in Revello Channel is 4·04 hours, which agrees so closely with the observed oscillation of 4 hours that it seems probable that we have here evidence of a local seiche.

C. *Periodic Oscillations in the Salinity.*

During survey-seasons 1921-22 and 1922-23 the R.I.M.S. "Investigator," was usually anchored in Nankauri Harbour at the anchorage near the mouth of Octavia Bay (*vide* Chart VII, p. 142). The period, during which observations were taken, extends continuously from October, 1921, to February, 1922, and again from October to December, 1922, except when, at intervals, it became necessary for the ship to return to our base at Port Blair in the Andamans for the purpose of coaling, etc. Conditions, therefore, for the investigation regarding the occurrence of any periodic oscillation in the salinity of the sea-water were extremely favourable.

As I have already pointed out (*vide supra*, pp. 160, 170.) my researches in previous years, as well as those conducted during my stay in this area, show that in the coastal waters of Indian seas the rise and fall of the tide may of itself produce an alteration in the salinity of the water ; and, in order to avoid, as far as possible, any alteration in the salinity of the surface-water due to these influences, which might mask other periodic oscillations, samples were taken for examination as nearly as possible at the time of high-water each day during our stay in the harbour ; the estimation of the time of high-water was rendered easy and comparatively accurate since the tidal data, obtained by the tide-party on shore during the previous day, were available each morning. The water-samples were carefully examined by means of a 'Buchanan' hydrometer and the salinity was calculated by means of Knudsen's Tables. The results of my examinations are given *in extenso* in Appendix VI, and the salinities obtained each day have been plotted in Text-figures 31 and 32. A study of the data and of the two charts reveals that in this region there is clear evidence of a periodic oscillation in the salinity of the surface-water, and, further, that this evidence is much more clearly shown during the first period October, 1921–February, 1922, than during the second period October–December, 1922. In both text-figures I have given for comparison the different phases of the moon throughout the periods of observation, and it is clear that in this case, as in the case of the similar oscillations observed in the salinity of the surface-water off the coast of southern Burma, there is no relationship between the two. At the commencement of my observations in October, 1921, the maxima of salinity correspond clearly with the times of spring tides, but towards the end of the same series they occur at or near the times of neap tides. During the second series of observations from October to December, 1922, the maximal salinities occur, to commence with, at the moon's first quarter, or near neap tides, whereas the last maximum, on December 16th, occurs only two days prior to new moon. It seems clear, therefore, that the phenomenon in no way

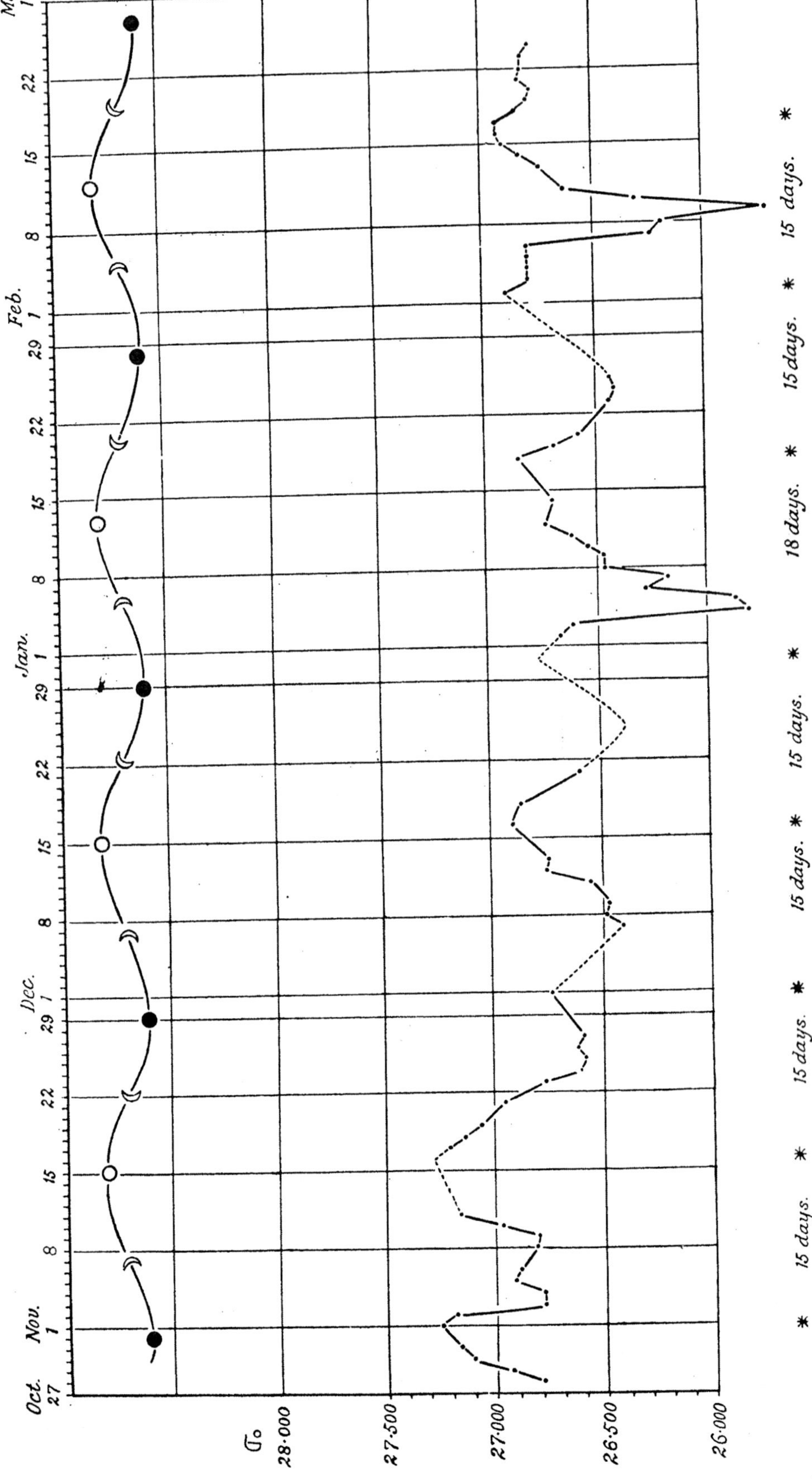

Text fig. 31. Values of σ_0 of Sea water from day to day in Nankauri Harbour Sta. 614. 1921-22.

depends on either lunar or tidal influences, and we must look elsewhere for the causation of the oscillation in salinity.

I have already pointed out (*vide supra*, p. 165) that the similar oscillations in the salinity of the water on the coast of Burma appear to be due to the presence in the deep waters of the Andaman Sea basin of an internal seiche and that the period of the to-and-fro swing in that area, occupying from 17 to 19 days, agrees closely with the period calculated by means of Wedderburn's formula from the size of the basin and the density of the water. In the case of the water of Nankauri Harbour the average observed time of oscillation in the salinity throughout both periods is

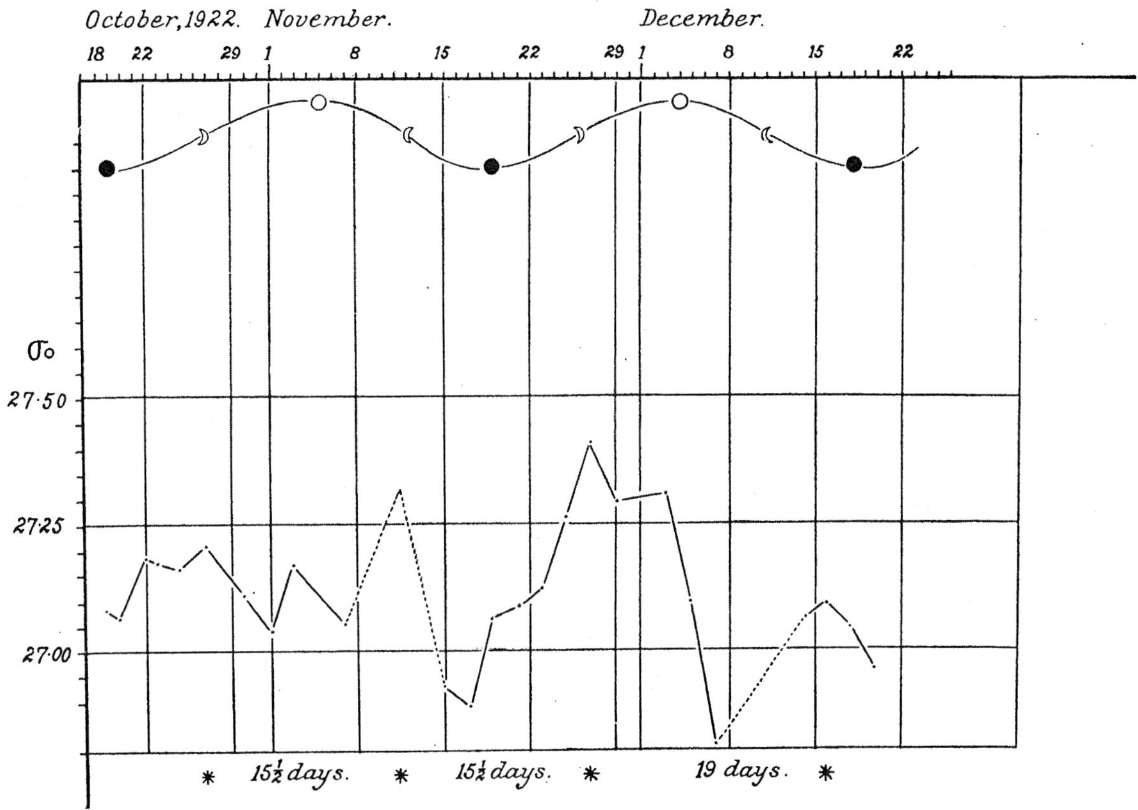

Text fig. 32. Values of σo of Sea-water from day to day in Nankauri Harbour, Sta. 614, 1922.

15·4 days, though it is interesting to note that on one occasion, namely from December 31st, 1921, to January 18th, 1922, the time of oscillation appears to have been 18 days, and therefore, on this occasion agrees closely with the period found on the opposite side of the Andaman Sea basin. Since the average period of oscillation in the Nicobars is only 15·4 days, it is clear that, if this be also due to a seiche, as seems probable, one must look outside the Andaman Sea basin for the source of its origin. In the region of Nankauri Harbour the tidal-wave (*vide* Krummel, 1911, p. 315) sets from the south-west to the north-east, and the tidal currents around the central group of islands appear to run in the main in a similar direction; at the southern end of Nankauri Island the flood-tide sets to the north, part passes up through Revello Channel on the west, and the remainder sweeps up the east side of Nankauri Island into Nankauri Harbour and up Beresford Channel on the east side of Camorta Island. At the north end of the group, the flood tide sweeps in from the west and then runs southward through Revello Channel and along the edge of Perseus Reef at the north-west end of Camorta Island. As a result of this, the water that enters Nankauri Harbour on the flood tide has come largely, if not entirely, from the Bay of Bengal lying to the westward; and it appears probable that the origin of any oscillation in the salinity of the water of the harbour is to be sought in a seiche in the deep-waters of the Bay, using the term in the restricted sense to include only the area between the coast of India and Ceylon and the Andaman-Nicobar ridge.

In October, 1921, and again in the same month in 1922, the R.I.M.S. "Investigator" carried out a series of serial observations on the temperature and salinity of the waters at different depths in the Bay, the salinity being calculated by the titration method; the results, as regards the density of the water of different levels, are given below in Table 37 and I have plotted the average of all these observations in Text-fig. 33.

Lat. N.	.. 6°.51'.00"	6°.38'.30"	8°.11'.42"	8°.51'.30"	10°.10'.18"	10°.21'.00"	
Long E.	.. 83°.22'.30"	83°.34'.30"	86°.29'.42"	86°.52'.00"	89°.55'.12"	90°.17'.30"	Average
Date.	.. 19.X.21	10.X.22	20.X.21	11.X.22	21.X.21	12.X.22	
Surface	.. 21·404	22·328	21·013	22·322	20·719	20·737	**21·420**
50 fathoms	.. 22·851	23·440	...	23·782	23·429	23·880	**23·476**
100 ,,	.. 26·008	25·766	...	25·934	...	25·966	**25·918**
150 ,,	.. ...	...	...	...	26·699	...	(20·699)
200 ,,	.. 26·873	26·880	...	26·645	...	...	**26·799**
300 ,,	.. ...	...	...	...	27·106	...	(27·106)
400 ,,	.. 26·900	...	...	...	...	...	(26·900)
500 ,,	.. ...	...	27·415	27·281	...	...	**27·348**

Table 37; showing the density of the water at different depths in the Bay of Bengal, from observations on board the R.I.M.S. "Investigator" in October, 1921 and 1922.

The S.S. "Valdivia," during her cruise, carried out serial observations at different depths in a position opposite the mouth of the Bay of Bengal in Lat. 7°43' N.; Long. 88°45' E. (*vide* Schott, 1902, Temperaturkurven Tafel 22) and the results gave a density at the surface of 22·6, at 100 metres 23·88, at 200 metres 26·16 and at 3,692 metres 27·89. The results of these two series are, therefore, in comparatively close agreement. It is clear that the lower limits of the layer of low-density lies in the Bay of Bengal, at a depth of approximately 120 fathoms (or 220 metres) while the maximum depth of water at the mouth of the Bay is 2,200 fathoms (or 4,020 metres). For the purpose of calculating the period of oscillation in this area by Wedderburn's formula I have taken the density of the upper stratum (σ_1) as 1·0230 and of the lower stratum (σ_2) as 1·0275; p and P being respectively 205 metres and 3,815 metres.

The total length of the Bay of Bengal is approximately 960 nautical miles (=1,777,935·4 metres). Applying Wedderburn's formula to the above data, this gives a calculated time period for a uni-nodal seiche in the length of the Bay of 28·4 days. If, however, the seiche be a bi-nodal one, this figure must be multiplied by a factor 0·55, if the area in which the seiche occurs possesses a bottom contour such as that of the Bay of Bengal (*vide* Krummel, 1911, p. 165): this gives us a calculated period for a bi-nodal seiche in the Bay of 15·62 days, which agrees so closely with the observed period of oscillation in the surface salinity, namely 15·43 days, that one is, I think, justified in regarding this oscillation as evidence of such a seiche. There seems to be little doubt that in both the Andaman Sea and the Bay of Bengal there is, at any rate at certain seasons of the year, a seiche in the more saline, deeper stratum, the time period of which is respectively in the neighbourhood of 18 days

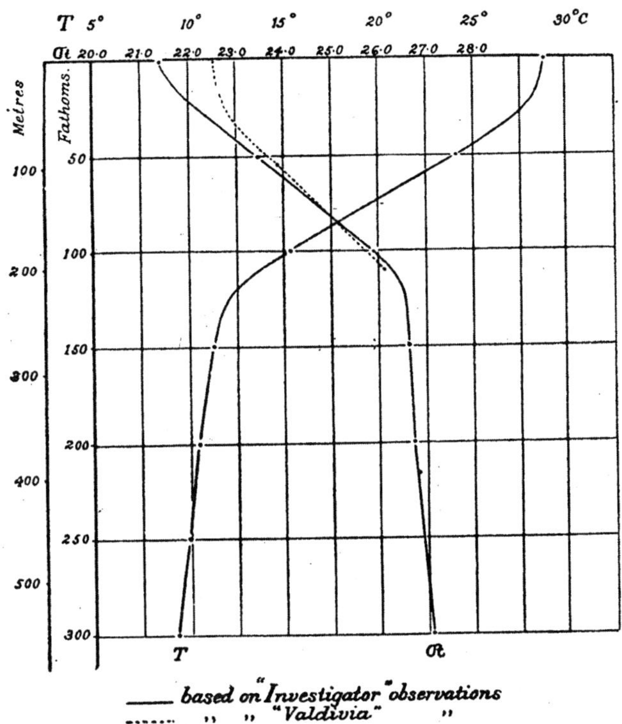

TEXT-FIG. 33.—The temperature and density (σ_t) of the sea-water at different depths on the east side of the Bay of Bengal in October.

and 15·5 days, that of the Andaman Sea Basin being uni-nodal, whereas in the Bay of Bengal it is bi-nodal.

It is interesting to note that the extent of the variation in the salinity of the surface-water in Nankauri Harbour steadily increases from October to February, as follows:—

Month.	*Range of Salinity.*
October	0·46
November	0·51
December	0·50
January	0·74
February	1·21

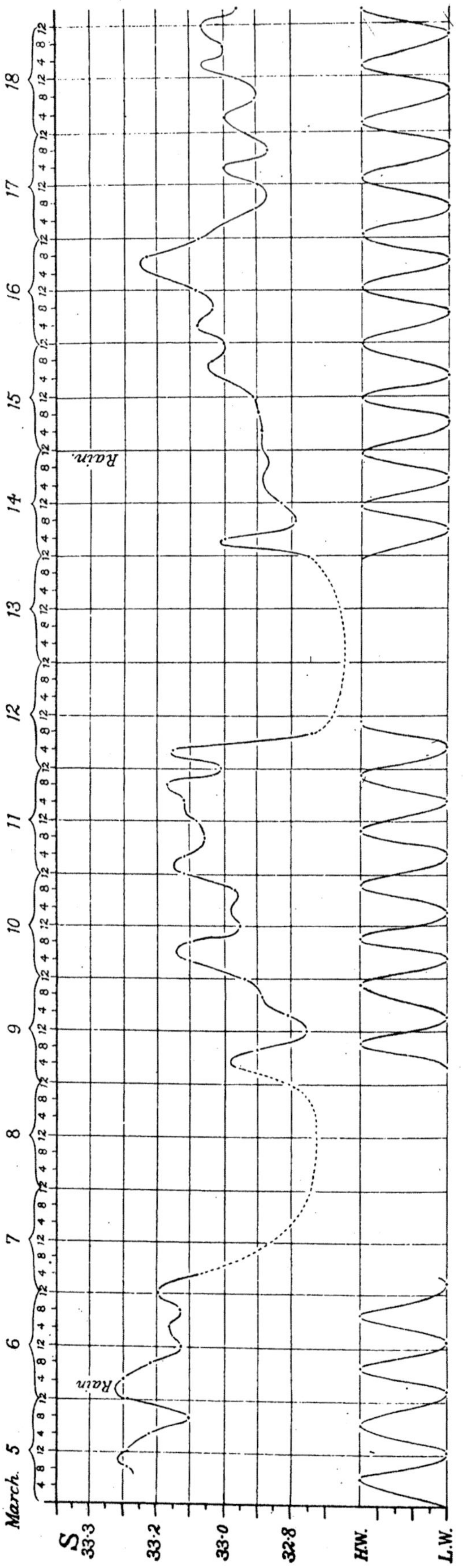

Text fig.34. 4-hourly variation in the Salinity (S) of the surface water in Revello Channel, Nicobars in March, 1925.

This increase in the range of the salinity may be accounted for, either by (1) an increased deficiency in the salinity of the surface-water; the effect of a periodic upwelling of deeper, more saline water and its admixture with the water of the surface-layer will clearly be small if the difference between the salinity of the two strata is decreased, and I have already shown that the density of the surface-layer fell steadily in 1921 from October to January; or by (2) an increase in the amplitude of the oscillation of the deep stratum. If, as I have suggested above, the strong winds of the north-east monsoon are the cause of the seiche, one would expect to find that the maximum oscillation of the deep stratum and, therefore, the maximum range of salinity of the surface-water would occur, as it does, at the close of the north-east monsoon in February. I have already mentioned (*vide supra*, p. 179) that evidence of this periodic oscillation in the salinity of the surface-water was obtained during both survey seasons, viz. October, 1921, to February, 1922, and again in October, to December, 1922; but in the latter period the amplitude is much smaller, dependant on the smaller difference between the salinities of the two strata of water. In October–December, 1922, the average salinity of the surface-water was appreciably higher than it had been during the previous season, October, 1921, to February, 1922, the average of both periods being 33·27 and 33·81 respectively. The difference in the two years is undoubtedly correlated with the rainfall (*vide supra*, p. 170) and it seems probable that the evidence of this seiche, as exhibited by the oscillation of the surface salinity, will be slight in any year in which the rainfall is below the normal and in which, in consequence, there is not the usual lowering of the salinity of the surface-water.

During the month of March, 1925, the "Investigator" was engaged in surveying the approach to Nankauri Harbour on the west side, and throughout the month she was, as a rule, anchored in Revello Channel. During the period March 4th to 19th she remained at an anchorage off Hoinipot village, just to the south of East Bay, on the east side of Kachal Island, except on Saturdays and Sundays (the 7th, 8th and 12th, 13th), when she returned to her old anchorage in Nankauri Harbour. Throughout this period samples of the surface-water in Revello Channel were taken at intervals of four hours: these were carefully titrated and the results are given *in extenso* in Appendix VII and I have plotted the salinity in Text-figure 34. For the purpose of comparison I have also given in the text-figure the data regarding the rise and fall of the tide.

As is clearly seen in Text-fig. 34 the salinity exhibits a very clearly marked major oscillation, the maxima, of which three occurred during the period of observation, occurring as follows:—

1st Maximum	on March	6th	at	2 a.m.
2nd ,,	,, ,,	11th	,,	8 p.m.
3rd ,,	,, ,,	16th	,,	5 p.m.

In addition to these major oscillations there are superposed a number of minor oscillations and a comparison of these with the tidal data given below appears to indicate that the two may be correlated, the rise of the tide being, as I have previously shown to be the case in the area off the mouth of Dring Harbour further to the north, associated with a fall of salinity; on the other hand there appears to be a very clear indication that this rise and fall of salinity is associated with the double diurnal oscillation of the barometric pressure. If we take the average of the salinity at different times of the day throughout the period of observation, we get the following results:—

Time of day ..	4 a.m.	8 a.m.	12 noon.	4 p.m.	8 p.m.	12 midnight.
Salinity ..	33·07	32·99	32·96	33·07	33·03	33·02.

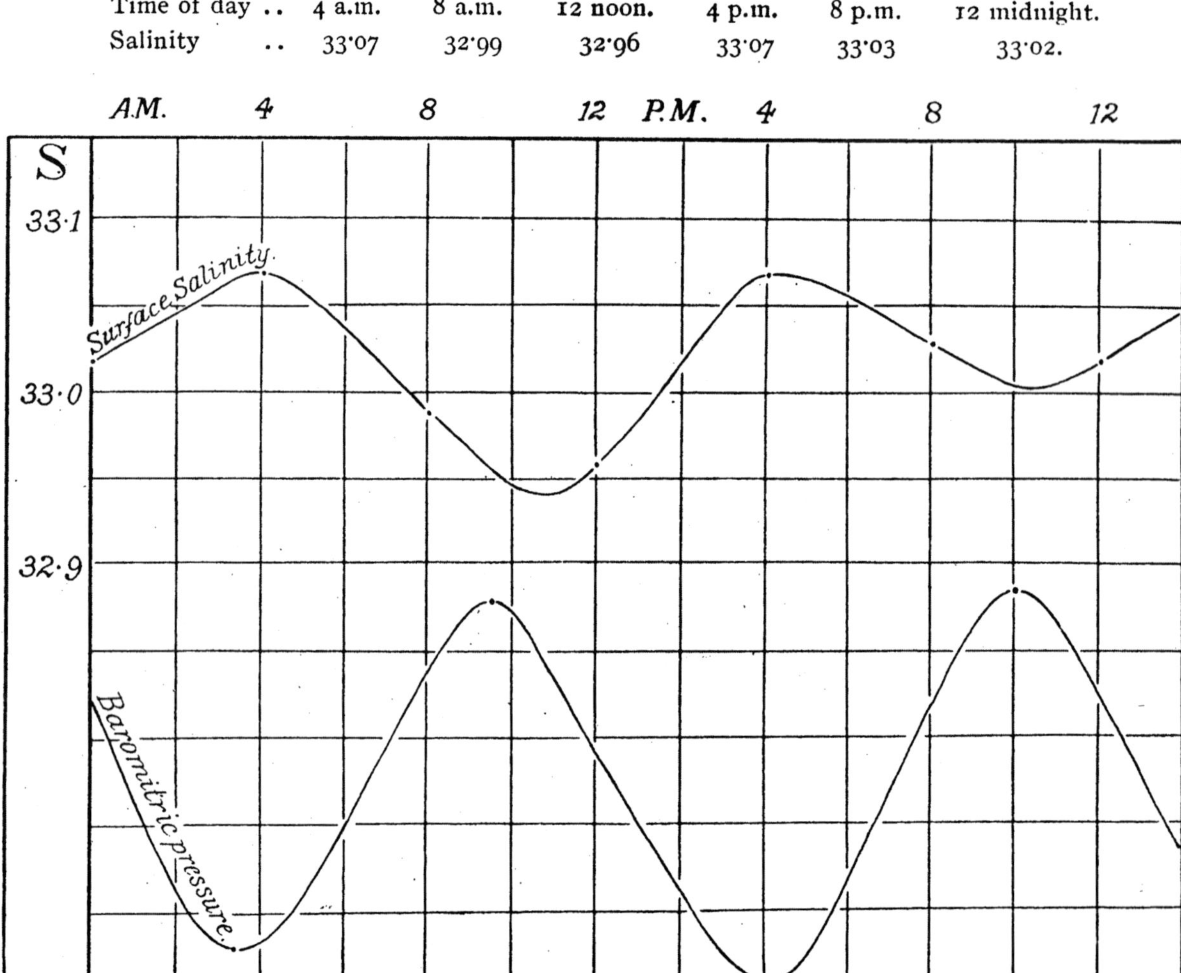

TEXT-FIG. 35.—Showing the average daily rise and fall of salinity of the surface-water and the rise and fall of barometric pressure in Revello Channel, Nicobars, in March, 1925.

In Text-figure 35 I have plotted these results and, for the purpose of comparison, I have given the average rise and fall of the barometer throughout the day. A comparison of the two curves shows beyond doubt that the rise and fall of barometric pressure and salinity synchronise with each other, the minimum barometric pressure corresponding to the epoch of maximum salinity and *vice versa*. I have already (*vide supra*, pp. 176, 178) referred to the manner in which the rise and fall of salinity and the rise and fall of barometric pressure appear to synchronise in the region to the north, around Perseus Reef and off Dring Harbour; but in these instances the epoch of maximum salinity coincided with the maximum barometric pressure. This latter condition was found to be present in the month of February whereas the present form of oscillation was noted in March. In a subsequent paper I hope to deal with this phenomenon at some length and will here content myself with stating that there appears to be a very definite correlation between the rise and fall of the barometer and the salinity of the surface-water, a correlation that is reversed at certain seasons of the year.

The major oscillation is, I think, attributable to a seiche in the Bay of Bengal. The width of the Bay of Bengal at its mouth between the east coast of Ceylon and the west of the Andaman-Nicobar ridge, is approximately 740 nautical miles (=1,370,480 metres), and the depth is 4,020 metres. During the whole period of my observations the average density of the surface-water was found to be 1·0230. For the purpose of calculation, by Wedderburn's formula, of the theoretical period of oscillation occupied by a seiche across the mouth of the Bay and in default of any further observations on this point, I have taken the depth and the density of the bottom layer of water as 3,815 metres and 1·0275 density, this being the figure derived from the data (as noted above, p. 182) obtained by the "Valdivia" and used by me in calculating the time period of a seiche in the long axis of the Bay. This gives us a calculated period for a uni-nodal seiche of 10·95 days, and for a bi-nodal seiche of $10{\cdot}95 \times 0{\cdot}55 = 6{\cdot}022$ days. We have already seen that the evidence of the rise and fall of salinity in Nankauri Harbour points to the existence of a bi-nodal seiche in the long axis of the Bay and the comparatively close approximation of the calculated period 6·022 days and the observed period of oscillation of salinity of 5·33 days in Revello Channel seems to me to indicate that there is also a bi-nodal seiche across the Bay of Bengal, the presence of which is revealed by the periodic rise and fall in the salinity of the surface-water.

Bibliography.

Note.—I have been unable to refer to those publications marked * .—R.B.S.S.

1. Atkins, W. R. G., and Harvey, H. W., 1925 .. .. "The variation with depth of certain salts, utilised in plant growth in the sea."
Nature, Vol. CXVI, p. 784. London.

2. Buchan, A., 1889 .. .. "Report on atmospheric circulation based on the observations made on board H.M.S. "Challenger" during the years 1873-76, and other meteorological observations."
Report, Sci. Results of H.M.S "Challenger," Physics and Chemistry, Vol. II. London.

3. Darwin, Sir G. .. .. "Tidal observations of the British Antarctic Expedition, 1907."
Proc. Roy. Soc. London: Series A, Vol. LXXXIV, p. 403. London.

4. Halbfass, 1911 .. .. "Gibt es im Madü-See Temperatur-Seiches"?
Internat. Rev. der gesamten Hydrobiologie und Hydrographie, Vol. III, Hydrographisches Supplement, Erste Serie, pp. 1-40. Leipzig.

5. *Hann, 1891 .. .. Lehrbuch der Meteorologie, p. 61.

6. Hayford, 1922 .. .. "Effects of winds and Barometric pressures on the Great Lakes."
Carnegie Institute of Washington, Bull. 317. Washington.

7. Helland-Handen, B. and Nansen, F., 1909 .. .. "The Norwegian Sea; its physical oceanography based upon the Norwegian Researches, 1900–1904."
Rep. Norwegian Fishery and Marine Investigations, Vol. II, Pt. 1, No. 2. Christiania.

8. Honda, K., Terada, T., Yoshida, Y. and Isitani, D., 1908 .. "Secondary undulations of Oceanic Tides." *Journ. Coll. Sci. Imp. Univ. Tokyo*, Vol. XXIV. Tokyo.

9. *Koppen, 1890 .. .. Ann. de Hydrographie, pp. 445–454.

10. Krummel, O., 1907 .. .. Handbuch der Ozeanographie, Vol. I. Stuttgart.

11. Krummel, O., 1911 .. .. Handbuch der Ozeanographie, Vol. II. Stuttgart.

12. Matthews, D. J., 1926 .. "Physical Oceanography"; the Percy Sladen Trust Expedition to the Indian Ocean in 1905.
Trans. Linn. Soc. London, Vol. XIX, p. 169. London.

13. Murray, James, 1911 .. .. "The observation of Tides and Seiches in Frozen Seas."
"*Internat. Rev. der gesamten Hydrobiologie und Hydrographie*, Vol. IV. Leipzig.

14. Murray, Sir J., 1912 .. .. "On the Annual Range of Temperature in the Surface Waters of the Ocean and its relation to other Oceanographical Phenomena."
The Geographical Journal, Vol. XII, No. 2, p. 113. London.

15. Nansen, F., 1902 .. .. "The Oceanography of the North Polar Basin."
The Norwegian North Polar Expedition, 1893–1896. Scientific Results, Vol. III. Christiania.

16. Pettersson, O., 1909 .. .. "Gezeitenaehnliche Bewegungen des Tiefenwassers."
Conseil Permanent Int. pour l'exploration de la Mer. *Publications de Circonstance*, No. 47. Copenhagen.

17. Pettersson, O., 1912 .. .. "The Connection between Hydrographical and Meteorological Phenomena."
Quar. Journ. Roy. Met. Soc., Vol. XXXVIII, No. 163, p. 173. London.

18. Reichard, A. C., 1913 .. .. "Hydrographische Beobachtungen bei Helgoland in den Jahren 1893–1908."
Wissenschaftliche Meeresuntersuchungen herausgegeben von der Komm. zur Wiss. untersuchungen der deutschen Meere in Kiel und der Biol. anstalt auf Helgoland (New Series), Vol. X.

19. Schott, G., 1895 .. .. "Die Jahrliche Temperatur schwankung der Ozean wassers."
Petermann's Mitteilungen, Vol. XLI, p. 153. Gotha.

20. Schott, G., 1902 .. .. "Oceanographie und Maritime Meteorologie."
Wissenschaftliche Ergebnisse der Deutschen Tiefsee Expedition, Vol. I. Jena.

21. Sewell, R. B. S., 1913 .. .. "Notes on the Biological work of the R.I.M.S. 'Investigator' during Survey-seasons, 1910-11 and 1911-12."
Journ. Proc. Asiatic Soc. Bengal (New Series), Vol. IX, Nos. 8 and 9, p. 329. Calcutta.

22. Sewell, R. B. S., 1926 .. .. "The Salps of Indian Seas."
Rec. Ind. Mus., Vol. XXVII, Pt. 2, p. 65. Calcutta.

23. Tipper, G. H., 1911 .. .. "The Geological Formation of the Andamans, with references to the Nicobars."
Rec. Geol. Survey India, Vol. XXXV, Pt. 4. Calcutta.

24. Watson, P. R., 1904 .. .. "Movements of the waters of Loch Ness, as indicated by Temperature Observations."
Geographical Journal, Vol. XXIV, p. 430. London.

25. Wedderburn, P. M., 1904 .. "Seiches observed in Loch Ness."
Geographical Journal, Vol. XXIV, p. 441. London.

26. " " 1907 .. "An Experimental Investigation of the Temperature changes occurring in Fresh-water Lochs."
Proc. Roy. Soc. Edinburgh, Vol. XXVIII (1907-08), p. 2. Edinburgh.

27. " " 1908 .. "On the Temperature of the Fresh-water Lochs of Scotland."
Trans. Roy. Soc. Edinburgh, Vol. XLV, p. 407. Edinburgh.

28. " " 1909 .. (*a*) "Temperature Observations in Loch Garry (Inverness-shire). With notes on Currents and Seiches."
Proc. Roy. Soc. Edinburgh, Vol. XXIX, p. 98. Edinburgh.

29. " " " .. (*b*) "Dr. O. Pettersson's observations on Deep-water Oscillations."
Proc. Roy. Soc. Edinburgh, Vol. XXIX, p. 602. Edinburgh.

30. " " " .. (*c*) "Temperature Oscillations in Lakes and the Ocean."
Scottish Geographical Magazine, Vol. XXV, p. 591. Edinburgh.

31. " " 1911 .. "Some analogies between Lakes and Oceans."
Internat. Rev. der gesamten Hydrobiologie and Hydrographie, Vol. IV, p. 55. Leipzig.

APPENDIX III.

Observations on the Surface-Water, at 7.30 a.m. daily, off the coast of southern Burma during Survey Season, 1910-11.

Month.	Date.	Position.		Hydrometer reading.	°C	S	Cl.	σ_0
		Lat. N.	Long. E.					
1911 February	3	14°. 07′. 00″	87°. 48′. 00″	1022·5	27·8	32·94	18·23	26·46
,,	4	14 . 10 . 00	97 . 46 . 00	1022	27·2	32·03	17·73	25·74
,,	5	13 . 33 . 00	98 . 08 . 30	1022	26·7	31·84	17·62	25·58
,,	6	13 . 33 . 00	98 . 08 . 00	1022	26·7	31·84	17·62	25·58
,,	7	13 . 45 . 00	97 . 55 . 00	1022	26·7	31·84	17·62	25·58
,,	8	13 . 55 . 00	97 . 55 . 30	1022	26·9	31·92	17·67	25·65
,,	9	14 . 13 . 00	97 . 56 . 00	1022·5	26·7	32·51	17·99	26·12
,,	10	14 . 10 . 30	97 . 57 . 30	1023	26·1	32·94	18·23	26·46
,,	11	14 . 00 . 00	98 . 00 . 00	1023	26·7	33·18	18·36	26·66
,,	12	13 . 33 . 00	98 . 08 . 30	1023	26·7	33·18	18·36	26·66
,,	13	13 . 33 . 00	98 . 08 . 30	1023	26·9	33·26	18·41	26·72
,,	14	13 . 43 . 00	98 . 00 . 00	1023	26·7	33·18	18·36	26·66
,,	15	13 . 52 . 30	97 . 59 . 30	1023	27·2	33·37	18·47	26·81
,,	16	13 . 45 . 00	97 . 55 . 00	1023	27·2	33·37	18·47	26·81
,,	17	13 . 57 . 30	97 . 55 . 00	1022·5	26·7	32·51	17·99	26·12
,,	18	13 . 48 . 00	97 . 57 . 00	1023	27·2	33·37	18·47	26·81
,,	19	13 . 33 . 00	98 . 8 . 30	1023	27·2	33·37	18·47	26·81
,,	20	13 . 33 . 00	98 . 8 . 30	1023	26·9	33·26	18·41	26·72
,,	21	13 . 47 . 00	98 . 00 . 00	1023	26·7	33·18	18·36	26·66
,,	22	13 . 52 . 00	97 . 59 . 30	1023	26·9	33·26	18·41	26·72
March	2	14 . 04 . 30	97 . 58 . 00	1023·5	26·7	33·85	18·73	27·19
,,	3	14 . 05 . 00	98 00 . 00	1023	26·7	33·57	18·58	26·97
,,	4	13 . 58 . 30	98 . 01 . 00	1023	26·7	33·18	18·36	26·66
,,	5	13 . 33 . 00	98 . 8 . 30	1023	26·9	33·26	18·41	26·72
,,	6	13 . 33 . 00	98 . 8 . 30	1023	26·7	33·18	18·36	26·66
,,	7	13 . 58 . 00	98 . 01 . 15	1023·5	26·7	33·85	18·73	27·19
,,	8	13 . 46 . 30	98 . 02 . 30	1023	27·2	33·37	18·47	26·81
,,	9	13 . 45 . 30	98 . 02 . 00	1023	26·7	33·18	18·36	26·66
,,	10	13 . 42 . 00	98 . 02 . 30	1023	26·7	33·18	18·36	26·66
,,	11	13 . 39 . 30	98 . 03 . 00	1023	26·9	33·26	18·41	26·72
,,	12	13 . 33 . 00	98 . 08 . 30	1023	27·2	33·37	18·47	26·81
,,	13	13 . 33 . 00	98 . 08 . 30	1023	26·7	33·18	18·36	26·66
,,	14	13 . 34 . 30	98 . 03 . 30	1022	27·2	32·04	17·73	25·74
,,	15	13 . 36 . 00	98 . 03 . 00	1022	27·2	32·04	17·73	25·74
,,	16	13 . 38 . 15	98 . 02 . 30	1023	26·7	33·18	18·36	26·66
,,	17	13 . 38 . 45	98 . 07 . 15	1023	27·2	33·37	18·47	26·81
,,	18	13 . 39 . 45	98 . 04 . 30	1023	27·2	33·37	18·47	26·81
,,	19	13 . 33 . 00	98 . 08 . 30	1023	27·2	33·37	18·47	26·81
,,	20	13 . 33 . 00	98 . 08 . 30	1023	27·8	33·61	18·60	27·01
,,	21	13 . 39 . 00	97 . 56 . 00	1023	27·8	33·61	18·60	27·01
,,	29	13 . 55 . 00	97 . 51 . 00	1023	27·8	33·61	18·60	27·01
,,	30	13 . 54 . 00	97 . 52 . 00	1023	27·8	33·61	18·60	27·01
,,	31	13 . 50 . 00	97 . 52 . 00	1022	27·8	32·28	17·86	25·93
April	1	13 . 45 . 00	97 . 43 . 00	1022	27·8	32·28	17·86	25·93
,,	2	13 . 33 . 00	98 . 08 . 30	1022	26·7	31·84	17·62	25·58
,,	3	13 . 33 . 00	98 . 08 . 30	1022	27·2	32·04	17·73	25·74
,,	4	13 . 51 . 00	97 . 51 . 45	1022	27·8	32·28	17·86	25·93
,,	5	14 . 01 . 15	97 . 49 . 15	1022	27·2	32·04	17·73	25·74
,,	6	14 . 02 . 15	97 . 45 . 00	1022	27·8	32·28	17·86	25·93
,,	7	13 . 56 . 45	98 . 00 . 00	1022	27·8	32·28	17·86	25·93

APPENDIX IV.

Observations of the density of the surface-water at different states of the Tide, by means of a 'Buchanan' Hydrometer, in the Mergui Archipelago, 1913-14.

Date 1913.	Time.	Position.		Hydrometer reading.		$\frac{S^t}{S^4}$.	σ_0.	Cl.	S.	State of Tide.
		Lat. N.	Long. E.	Sp. grav.	o·C					
Oct. 12	11 a.m.			1·01962	27·10	1·01606	20·99	14·46	26·13	
	12 noon			1·01951	27·00	1·01598	20·87	14·38	25·99	
	1 p.m.			1·02005	27·30	1·01646	21·50	14·81	26·76	
	2 ,,			1·01917	27·35	1·01553	20·51	14·13	25·53	
	3 ,,			1·01792	27·87	1·01415	19·20	13·22	23 90	Low water 3.14 p.m.
	4 ,,	11°.29′.57″ [Anchorage Harbour.]	98°.34′.38″ in Mergui	1·01824	27·70	1·01452	19·54	13·46	24·33	
	5 ,,			1·01817	27·75	1·01440	19·414	13·37	24·16	
	6 ,,			1·01915	27·75	1·01541	20·51	14·13	25·53	
	7 ,,			1·01954	27·80	1·01575	20 90	14·40	26·02	
	8 ,,			1·01976	27·90	1·01597	21·17	14·58	26·35	
	9 ,,			1·02037	27·95	1·01657	21·83	15·04	27·18	High water, 9.28 p.m.
	10 ,,			1·02323	28·00	1·01940	24·89	17·15	30·99	
Oct. 19	11 a.m.			1·01567	29·35	1 01147	16·81	11·58	20·93	
	12 noon			1·01751	29 40	1·01329	18·80	12·95	23·40	High water, 12.11 p.m.
	1 p.m.			1·01559	29·40	1·01138	16·73	11·52	20·82	
	2 ,,			1·01112	29·45	1·00691	11·89	8·19	14·81	
	3 ,,	98°.34′.38″	11°.29′.57″	1·00685	29·50	1·00264	7·29	5·03	9·11	
	4 ,,			1·00386	29·50	0·99967	4·06	2·81	5 11	
	5 ,,	[Anchorage Harbour.]	in Mergui	1·00268	29·50	0·99849	2·78	1·94	3·53	
	6 ,,			1·00332	29·90	0·99901	3·48	2·42	4·40	
	7 ,,			1·00333	29·95	0·99900	3·48	2·42	4·40	Low water, 6.39 p.m.,
	8 ,,			1·00705	29·95	1·00271	7·52	5·18	9·39	
	9 ,,			1·00563	30·30	1·00119	6·70	4·62	8·37	
	10 ,,			1·010035	30·15	1·00562	10·76	7·41	13·41	
April 8	8 a.m.			1·023761	30·65	1·019132	25·554	17·60	31·80	High water, 9.11 a.m.
,,	10 a.m.	12 . 29 . 57	98 . 34 . 38	1·023895	30·20	1·019403	25·679	17·69	31 96	
,,	12 noon			1·023811	29·65	1·019489	25·571	17·62	31·83	
,,	2 p.m	[Anchorage Harbour.]	in Mergui	1·022587	29·60	1·018285	24·257	16·71	30·19	
,,	4 p.m.			1·021430	29·60	1·017133	23·022	15·86	28·66	Low water, 3.30 p.m.
,,	6 p.m.			1·022741	29·60	1·018439	24·423	16·82	30·39	
,,	8 p.m.			1·023769	29·60	1·019462	25·524	17·58	31·76	
,,	10 p.m			1·024156	29·65	1·019832	25·939	17·87	32·29	
Nov. 8	8 p.m.			1·02173	29 80	1·017381	23·35	16·09	29·06	High water, 7.19 p.m.
	9 p.m.			1·02166	28·55	1·017677	23·23	16·01	28·92	
	10 p.m.			1·02121	27·85	1·017433	22·723	15·65	28·28	
	11 p.m.			1·02099	27·60	1·017286	22·479	15·49	27·99	
	12 mid-night.	11°.54′.55″	98°.20′.45″	1·02105	28·35	1·017128	22·655	15·61	28 20	
,, 9	1-30 a.m.			1·02055	28·75	1·016914	22·577	15·55	28·11	Low water, 1.21 a.m.
	4 a.m.			1·02140	28·80	1·017344	22·957	15·82	28·59	
	6 a.m.			1·02167	28·60	1 017673	23·242	16·01	28 93	
	7 a.m.			1·02135	28·95	1·017249	22·907	15·78	28·51	High water, 7.36 a.m.
,, 15	7 p.m.			1·02284	27·95	1·019028	24·472	16·86	30·46	
	9 p.m.			1·02311	27·85	1·019326	22·758	17·06	30·82	
	11 p.m.			1·02325	28·00	1·019422	24·912	17·16	31·00	High water, 11.48 p.m.
,, 16	1 a.m.	11°.57′.30″	98°.19′.00″	1·02314	28·10	1·019284	24·799	17·08	30·87	
	3 a.m.			1·02283	28·25	1·018931	24·472	16·86	30·46	
	5 a.m.			1·02235	28·35	1·018423	23·961	16·51	29·82	Low water, 5.53 a.m.
	7 a.m.			1·02295	28·95	1·018843	24·622	16·96	30·64	

APPENDIX V.

Daily observations of the temperature and salinity of the surface water, at or near High Tide, in the Mergui Archipelago, 1913-14.

Date. 1913.	Position.		Hydrometer reading.		$\frac{S^t.}{S^4.}$	σ_0	Cl.	S.	Time of observation.
	Lat. N.	Long. E.	Sp. grav.	°C					
Oct. 15	12°. 51′. 30″	98°. 30′. 45″	1·02134	28·35	1·01742	22·88	15·76	28·48	11.0 a.m.
,, 16	12 . 40 . 00	98 . 27 . 00	1·021855	27·20	1·01826	23·39	16·11	29·11	11.15 a.m.
,, 18	12 . 29 . 57	98 . 34 . 38	1·00816	30·75	1·00357	8·74	6·02	10·90	6.30 a.m.
,, 22	12 . 52 . 20	98 . 12 . 10	1·02021	29·65	1·01590	21·705	14·95	27·01	1.20 a.m.
,, 23	12 . 51 . 00	98 . 10 . 45	1·02026	28·55	1·01628	21·723	14·96	27·03	2.45 a.m.
,, 24	12 . 45 . 50	98 . 17 . 54	1·02139	28·75	1·01735	22·947	16·63	30·05	5.15 a.m.
,, 25	12 . 29 . 57	98 . 34 . 38	1·01223	28·35	1·00834	13·085	9·01	16·29	6.50 a.m.
Nov. 1	12 . 40 . 00	98 . 26 . 30	1·02324	28·50	1·019256	24·91	17·16	31·00	12.20 a.m.
,, 3	12 . 23 . 15	98 . 02 . 00	1·02316	28·30	1·01925	24·83	17·10	30·90	..
,, 4	11 . 58 . 00	98 . 05 . 30	1·02311	28·50	1·01914	24·785	17·07	30·84	1.15 a.m.
,, 5	} 12 . 00 . 10	98 . 20 . 30 {	1·02191	29·95	1·01750	23·229	16·83	30·41	2.15 a.m.
,, 6			1·02175	29·45	1·01697	22·79	15·70	28·37	2.55 a.m.
,, 7			1·02173	29·80	1·017381	23·35	16·08	29·05	5.40 a.m.
,, 8	} 11 . 54 . 55	98 . 20 . 45 {	1·02166	28·55	1·017677	23·23	16·00	28·91	8.0 p.m.
,, 9			1·02137	28·80	1·017314	22·925	15·79	28·53	7.0 a.m.
,, 10	11 . 58 . 15	98 . 21 . 10	1·02234	27·70	1·018603	23·930	16·48	29·78	8.40 p.m.
,, 12			1·02284	27·90	1·019043	24·471	16·86	30·46	10.0 p.m.
,, 14	} 11 . 57 . 30	98 . 19 . 00 {	1·02349	27·90	1·019691	25·167	17·34	31·33	11.0 p.m.
,, 15			1·02325	28·00	1·019422	24·912	17·16	31·00	11.0 p.m.
,, 20	11 . 54 . 00	98 . 18 . 45	1·02309	28·10	1·019224	24·736	17·04	30·79	2.45 a.m.
,, 30	} 11 . 52 . 10	98 . 18 . 40 {	1·02362	28·70	1·019585	25·340	17·46	31·55	11.25 p.m.
Dec. 2			1·02382	29·70	1·019482	25·582	17·62	31·83	1.10 p.m.
,, 6			1·02325	27·50	1·020567	25·98	17·90	32·34	5.7 p.m.
,, 7	} 11 . 57 . 50	98 . 22 . 05 {	1·02347	27·25	1·019857	25·122	17·30	31·26	6.20 p.m.
,, 8			1·02329	27·65	1·019563	24·943	17·18	31·04	7.19 p.m.
,, 9			1·02358	28·80	1·019515	25·286	17·42	31·47	8.6 p.m.
,, 11			1·02339	30·55	1·019791	26·228	18·07	32·65	9.30 p.m.
,, 12			1·02328	29·10	1·019127	25·067	17·29	31·24	10.10 p.m.
,, 13			1·02391	31·35	1·019054	25·732	17·73	32·03	10.55 p.m.
,, 14	} 11 . 54 . 45	98 . 24 . 55 {	1·02323	29·05	1·019092	24·928	17·17	31·02	11.40 p.m.
,, 16			1·02335	29·10	1·019197	25·059	17·26	31·18	12.15 p.m.
,, 17			1·02346	29·95	1·019047	25·204	17·36	31·36	12.50 p.m.
,, 18			1·02346	28·85	1·019381	25·168	17·34	31·33	1.30 p.m.
,, 19	11 . 53 . 45	98 . 25 . 00	1·02331	27·50	1·019626	24·960	17·19	31·06	2.20 p.m.
,, 20	11 . 54 . 45	98 . 23 . 45	1·02343	27·35	1·019789	25·084	17·23	31·13	3.30 p.m.
,, 29			1·02361	27·25	1·019997	25·273	17·41	31·46	12.0 p.m.
,, 30	} 11 . 48 . 52	98 . 23 . 23 {	1·02430	25·50	1·021167	25·944	17·87	32·29	12.25 p.m.
,, 31			1·02449	23·00	1·021995	26·040	17·94	32·41	1.0 p.m.
1914.									
Jan. 1	11 . 53 . 04	98 . 18 . 17	1·02361	23·50	1·020994	25·192	17·36	31·36	1.30 p.m.
* ,, 3			1·02458	25·90	1·021338	26·256	18·09	32·68	2.35 p.m.
* ,, 4	} 11 . 43 . 30	98 . 20 . 00 {	1·02412	26·20	1·020798	25·849	17·81	32·18	3.10 p.m.
* ,, 5			1·024105	27·35	1·020463	25·807	17·78	32·12	4.15 p.m.
* ,, 6	11 . 37 . 50	98 . 19 . 30	1·02458	25·90	1·021338	26·256	18·09	32·68	5.40 p.m.
* ,, 7	11 . 34 . 00	98 . 21 . 03	1·02466	25·00	1·021659	26·306	18·12	32·74	5.40 p.m.
* ,, 8	11 39 . 54	98 . 19 . 27	1·024905	24·90	1·021929	26·536	18·28	33·03	8.10 p.m.
* ,, 9			1·02463	24·90	1·021655	26·273	18·10	32·70	9.0 p.m.
* ,, 10			1·02460	27·00	1·021054	26·321	18·13	32·75	9.45 p.m.
* ,, 11	} 11 . 43 . 30	98 . 20 . 00 {	1·022705	27·45	1·019038	24·312	16·75	30·26	10.30 p.m.
* ,, 12			1·02441	28·70	1·020372	24·180	18·03	32·57	11.10 p.m.
* ,, 13	} 11 . 28 . 48	98 . 20 . 36 {	1·02468	28·55	1·020685	26·463	18·23	32·94	11.50 p.m.
* ,, 14			1·02481	29·00	1·020681	26·618	18·34	33·13	12.15 a.m.
* ,, 15	11 . 30 . 00	98 . 20 . 06	1·02484	27·90	1·021036	26·611	18·33	33·12	12.50 a.m.
* ,, 16			1·024525	28·90	1·020427	26·310	18·12	32·74	1.30 a.m.
* ,, 17			1·024705	26·45	1·021312	26·412	18·20	32·88	2.10 a.m.
* ,, 18	} 11 . 37 . 30	98 . 19 . 30 {	1·02459	28·35	1·020655	26·360	18·16	32·81	2.50 a.m.
* ,, 19			1·02404	28·35	1·020107	25·771	17·75	32·07	3.50 a.m.

Date.	Position.		Hydrometer reading.		$\frac{S^t}{S_4}$.	σ_0	Cl.	S.	Time of observation.	Surface temperature. °C.	Density of water *in situ*.
	Lat. N.	Long. E.	Sp. grav.	°C.							
*Jan. 20	11°. 31'. 48"	98°. 21'. 24"	1·02453	29·30	1·020011	26·329	18·14	32·77	5.10 a.m.	..	..
* ,, 21	11 . 31 . 25	98 . 24 . 06	1·02377	28·50	1·019794	25·493	17·56	31·73	6.40 a.m.	..	..
* ,, 22	11 . 30 . 10	98 . 24 . 57	1·02413	29·65	1·019806	25·911	17·85	37·25	8.0 p.m.	..	..
* ,, 23			1·02416	29·90	1·019760	25·952	17·88	32·30	9.10 p.m.	..	..
* ,, 24	11 . 30 . 30	98 . 25 . 00	1·02469	28·45	1·020725	26·471	18·24	32·95	10.0 p.m.	..	..
* ,, 25	11 . 43 . 28	98 . 19 . 41	1·02443	29·80	1·020059	26·239	18·08	32·66	10.40 p.m.	..	..
* ,, 28			1·02379	28·35	1·019858	25·504	17·57	31·74	12.15 p.m.	..	..
,, 29	12 . 29 . 57	98 . 34 . 38	1·02391	28·80	1·019844	25·647	17·67	31·92	12.40 p.m.	..	..
,, 30			1·023835	28·50	1·019859	25·558	17·61	31·82	1.5 p.m.	..	..
Feb. 2	11 . 13 . 50	98 . 11 . 40	1·024905	29·40	1·020655	26·734	18·42	33·27	2.10 p.m.	..	..
,, 3	11 . 38 . 34	98 . 34 . 27	1·023701	29·50	1·019425	25·448	17·53	31·67	2.40 p.m.	..	..
,, 4			1·02445	29·10	1·020292	26·236	18·08	32·66	3.30 p.m.	..	..
,, 5	11 . 26 . 05	98 . 35 . 00	1·02429	28·75	1·020238	26·053	17·95	32·42	5.30 p.m.	..	..
,, 6	11 . 32 . 20	98 . 35 . 00	1·02458	29·90	1·020178	26·457	18·23	32·94	7.25 p.m.	..	..
,, 7	11 . 16 . 00	98 . 35 . 00	1·02450	30·05	1·020052	26·322	18·13	32·75	8.40 p.m.	..	..
,, 8	11 . 35 . 00	98 . 34 . 15	1·02379	28·55	1·019799	25·511	17·57	31·75	9.40 p.m.	..	..
,, 9			1·02402	29·00	1·019894	25·762	17·75	32·06	10.15 p.m.	..	..
,, 10	11 . 25 . 10	98 . 35 . 00	1·02404	29·15	1·019869	25·799	17·77	32·11	11.00 p.m.	..	..
,, 11			1·02457	30·35	1·020029	26·429	18·21	32·90	11.35 p.m.	..	..
,, 12			1·02473	27·75	1·020969	26·487	18·25	32·97	12.0 noon.	..	..
,, 13			1·02458	29·45	1·020316	26·389	18·18	32·84	0.30 p.m.	..	..
,, 14	11 . 23 . 33	98 . 33 . 45	1·02444	28·15	1·020564	26·192	18·04	32·59	1.10 p.m.	..	..
,, 15			1·02459	29·15	1·020417	26·389	18·18	32·84	1.45 p.m.	..	..
,, 16			1·02445	27·7	1·020705	26·187	18·04	32·59	2.20 p.m.	..	..
,, 17			1·02465	29·70	1·020309	26·472	18·24	32·95	3.10 p.m.	..	..
,, 18			1·024803	29·40	1·020452	26·517	18·27	33·01	4.10 p.m.	..	..
,, 20	11 . 16 . 15	98 . 36 . 30	1·024614	29·20	1·020425	26·415	18·20	32·88	8.5 p.m.	..	..
,, 21			1·024589	30·15	1·020109	26·923	18·55	33·50	9.15 p.m.	..	..
,, 22			1·024954	28·25	1·021048	26·747	18·43	33·30	9.50 p.m.	..	..
,, 23			1·024804	28·50	1·020824	26·594	18·32	33·10	10.30 p.m.	..	..
,, 25			1·024588	29·50	1·020308	26·398	18·19	32·86	11.10 a.m.	29·4°	20·24
,, 26			1·024548	29·00	1·020420	26·337	18·14	32·78	11.50 a.m.	29·0	20·42
,, 27			1·024439	30·25	1·019929	26·264	18·09	32·68	12.10 p.m.	30·4	19·88
,, 28	Paway Island,	near shore.	1·024399	30·65	1·019764	26·235	18·08	32·66	12.40 p.m.	30·8	19·71
Mar. 1			1·024453	30·05	1·020005	26·272	18·10	32·70	1.5 p.m.	30·1	19·99
,, 2			1·024448	30·40	1·019891	26·279	18·10	32·71	1.20 p.m.	30·5	19·86
,, 3			1·024431	31·30	1·019589	26·290	18·11	32·72	1.35 p.m.	31·5	19·52
,, 4	11 . 25 . 28	98 . 27 . 21	1·024598	30·20	1·020103	26·433	18·21	32·90	1.30 a.m.	..	..
,, 5			1·024501	30·65	1·019865	26·344	18·15	32·78	2.0 a.m.	..	..
,, 6			1·023980	29·95	1·019596	25·794	17·77	32·10	2.40 a.m.	..	..
,, 7	11 . 31 . 15	98 . 35 . 15	1·024434	30·50	1·019846	26·268	18·10	32·70	5.20 a.m.	..	..
,, 7			1·024598	28·40	1·020648	26·370	18·17	32·82	8.10 p.m.	..	..
,, 8	11 . 12 . 50	98 . 35 . 30	1·025014	26·10	1·021715	26·726	18·41	32·27	8.20 p.m.	..	..
,, 9			1·024986	28·5	1·021005	26·789	18·45	33·34	9.15 p.m.	..	..
,, 10	11 . 09 . 05	98 . 35 . 07	1·024954	29·60	1·020642	26·793	18·46	33·34	10.0 p.m.	..	..
,, 11	11 . 15 . 30	98 . 40 . 00	1·025074	27·60	1·021355	26·851	18·50	33·41	10.35 p.m.	..	..
,, 12	11 . 23 . 40	98 . 33 . 35	1·024855	28·95	1·020740	26·663	18·37	33·18	11.15 p.m.	..	..
,, 14			1·024946	27·70	1·021199	26·717	18·41	33·26	12.10 a.m.	..	..
,, 15	11 . 34 . 35	98 . 35 . 30	1·024950	28·65	1·020925	26·756	18·43	33·30	12.30 a.m.	..	..
,, 16	11 . 27 . 45	98 . 36 . 15	1·024861	29·40	1·020611	26·688	18·38	33·21	1.0 a.m.	..	..
,, 17			1·024792	29·65	1·020466	26·463	18·23	32·94	1.30 a.m.	..	..
,, 18	11 . 27 . 40	98 . 38 . 45	1·024812	28·45	1·020847	26·601	18·33	33·11	2.0 a.m.	..	..
,, 19	11 . 27 . 30	98 . 32 . 45	1·024661	31·49	1·019786	26·540	18·28	33·03	2.50 a.m.	..	..
,, 29	11 . 17 . 20	98 . 29 . 40	1·024894	29·70	1·020583	26·766	18·44	33·31	11.50 p.m.	..	..
,, 30			1·024972	29·80	1·020599	26·819	18·47	33·38	12.10 a.m.	..	..

Date.	Position:		Hydrometer reading.		$\frac{S^t}{S^4}$.	σ_0	Cl.	S.	Time of observation.	Surface temperature. °C.	Density of water *in situ*.
	Lat. N.	Long. E.	Sp. grav.	°C.							
Mar. 31	11°. 19'. 58"	98°. 29'. 52"	1·025336	31·15	1·020425	26·135	18·69	33·77	12.30 a.m.	..	..
April 1	11 . 24 . 10	98 . 27 . 50	1·025026	28·90	1·020926	26·845	18·49	33·40	12.45 a.m.	..	..
,, 2	11 . 25 . 10	98 . 27 . 51	1·024924	29·50	1·020643	26·758	18·43	33·30	1.10 a.m.	..	..
,, 3	11 . 32 . 00	98 . 39 . 40	1·024978	29·45	1·020711	26·801	18·46	33·35	1.30 a.m.		..
,, 4			1·024991	28·85	1·020906	26·806	18·47	33·36	2.20 a.m.	..	..

APPENDIX VI.

Daily Observations on the Temperature and Salinity
of the Surface-water of Nankauri Harbour
at or near High Tide.
1921-1922.

Date 1922	Time.	Surface Temp.	Air Temp.	Hydrometer reading. °C.	Hydrometer reading. Sp. grav.	σ_0	Cl.	S.	σ_t
Oct. 27 ..	7-30 p.m. ..	28·2°	..	31·7	1·019921	26·792	18·46	33·34	21·107
,, 28 ..	8-11 p.m. ..	29·2	..	31·6	1·020084	26·938	18·56	3 ·53	20·912
,, 29 ..	8-40 p.m. ..	28·2	..	31·7	1·020206	27·108	18·67	33·74	21·401
,, 30 ..	8-11 p.m. ..	29·2	..	30·2	1·020775	27·156	18·71	33·80	21·111
Nov. 1 ..	10- 0 a.m. ..	28·6	..	30·6	1·020720	27·246	18·77	33·91	21·394
,, 2 ..	11-20 a.m. ..	29·1	..	30·3	1·020771	27·189	18·73	33·84	21·179
,, 3 ..	12 noon. ..	28·6	..	29·8	1·020558	26·775	18·45	33·32	20·960
,, 4 ..	12 noon. ..	28·7	..	29·6	1·020634	26·785	18·48	33·39	21·143
,, 5 ..	12-35 p.m. ..	28·4	..	29·8	1·020692	26·919	18·54	33·50	21·155
,, 6 ..	1- 3 p.m. ..	28·6	..	29·6	1·020734	26·892	18·53	33·47	21·069
,, 8 ..	3- 0 p.m. ..	28·8	..	29·1	1·020824	26·808	18·47	33·37	20·925
,, 9 ..	5-30 p.m. ..	20·0	..	29·1	1·020821	26·805	18·47	33·36	22·185
,, 10 ..	6-30 p.m. ..	27·9	..	28·0	1·021337	26·969	18·58	33·57	21·374
,, 17 ..	10- 0 a.m. ..	28·4	28·94	30·6	1·020682	27·205	18·74	33·86	21·425
,, 18 ..	11- 0 a.m. ..	28·1	28·06	30·5	1·020656	27·140	18·70	33·78	21·463
,, 19 ..	12 noon. ..	..	..	30·3	1·020651	27·060	18·64	33·68	..
,, 21 ..	12-45 p.m. ..	..	..	29·85	1·020718	26·955	18·57	33·55	..
,, 23 ..	5- 0 p.m. ..	..	26·66	28·5	1·020947	26·712	18·40	33·24	..
,, 24 ..	6- 0 p.m. ..	28·3	27·94	28·75	1·020748	26·601	18·33	31·11	20·896
,, 25 ..	6- 0 p.m. ..	..	..	29·0	1·020635	26·569	18·30	33·07	..
,, 26 ..	8- 0 p.m. ..	28·1	28·06	29·3	1·020592	26·661	18·37	33·18	21·018
,, 27 ..	8- 0 p.m. ..	28·3	26·39	29·3	1·020550	26·586	18·32	33·10	20·882
Dec. 1 ..	10-45 a.m. ..	..	..	28·9	1·020826	26·738	18·42	33·28	..
,, 7 ..	3- 0 p.m. ..	..	..	30·9	1·019837	26·406	18·19	32·87	..
,, 8 ..	4- 0 p.m. ..	28·2	..	27·9	1·020923	26·489	18·25	32·97	20·824
,, 9 ..	5-30 p.m. ..	..	..	28·0	1·020868	26·465	18·24	32·95	..
,, 11 ..	7- 0 p.m. ..	28·0	26·66	28·1	1·020913	26·549	1··29	33·04	20·946
,, 12 ..	7-45 p.m. ..	28·0	26·28	28·0	1·021065	26·748	18·43	33·30	21·134
,, 13 ..	8- 0 a.m. ..	27·9	26·39	28·05	1·021104	26·736	18·42	33·28	21·115
,, 16 ..	8-15 a.m. ..	..	..	29·9	1·020634	26·893	18·53	33·47	..
,, 18 ..	....	..	..	29·4	1·020777	26·866	18·51	33·44	..
,, 21 ..	....	28·2	..	29·9	1·020359	26·598	18·32	33·11	20·926
Date 1922									
Jan. 2 ..	12 noon. ..	..	..	29·05	1·020690	26·646	18·36	33·17	..
,, 3 ..	....	28·2	..	29·05	1·020669	26·623	18·34	33·13	20·949
,, 4 ..	1- 0 p.m. ..	..	..	29·05	1·019937	25·836	17·80	32·16	..
,, 5 ..	2- 0 p.m. ..	28·0	..	29·05	1·019967	25·869	17·82	32·20	20·313
,, 6 ..	2-35 p.m. ..	27·7	..	30·0	1·020038	26·289	18·11	32·72	20·801
,, 7 ..	3- 0 p.m. ..	28·0	..	20·75	1·020022	26·181	18·04	32·58	20·504
,, 8 ..	3- 0 p.m. ..	..	..	30·0	1·020201	26·464	18·23	32·94	..
,, 9 ..	6-45 p.m. ..	27·4	..	30·0	1·020209	26·473	18·24	32·95	21·07.
,, 10 ..	8- 0 p.m. ..	27·3	..	30·05	1·020255	26·541	18·28	33·03	21·164
,, 11 ..	8- 0 p.m. ..	..	..	30·2	1·020268	26·611	18 33	33·12	..
,, 12 ..	9- 0 p.m. ..	..	..	28·35	1·021005	26·736	18·42	33·28	..
,, 14 ..	10- 0 a.m. ..	27 7	..	28·1	1 021054	26·700	18·39	33 22	21·184
,, 18 ..	....	27·8	..	28·1	1·021199	26·856	18·50	33·42	21 298
,, 19 ..	3- 0 p.m. ..	28·0	..	28·3	1·020981	26 692	18·39	33·22	21·078
,, 20 ..	3-45 p.m. ..	..	..	28·5	1·020808	26·577	18·31	33·08	..
,, 23 ..	6-25 a.m. ..	27·8	..	29·5	1·020355	26·449	18·22	32·92	20·918
,, 24 ..	8- 0 a.m. ..	..	..	29·45	1·020325	26·416	18·20	32·88	..
,, 25 ..	8- 0 a.m. ..	28·0	..	29·4	1·020478	26·545	18·29	33·04	20·922
Feb. 2 ..	12-0 noon. ..	..	..	28·0	1·021263	26·889	18·52	33·47	..
,, 3 ..	12-45 p.m. ..	..	..	28·0	1·021152	26·770	18·44	33·31	..
,, 4 ..	1-45 p.m. ..	..	..	28·05	1·021148	26·784	18·45	33·33	..

Date 1922.	Time.	Surface Temp.	Air Temp.	Hydrometer reading. °C.	Hydrometer reading. Sp. grav.	σ_0	Cl.	S.	σ_t
Feb. 6	3-45 p.m.	..	..	28·10	1·021135	26·787	18·45	33·33	..
„ 7	6- 0 p.m.	..	..	27·85	1·020699	26·232	18·07	32·65	..
„ 8	6- 0 a.m.	..	..	27·85	1·019741	26·193	18·04	32·59	..
„ 9	8- 0 a.m.	..	..	31·15	1·019115	25·723	17·72	32·01	..
„ 10	8-30 a.m.	..	..	30·1	1·020028	26·315	18·13	32·75	..
„ 11	9-30 a.m.	..	..	30·1	1·020319	26·628	18·34	33·14	..
„ 13	10-35 a.m.	28·6	27·94	31·2	1·020039	26·738	18·42	33·28	20·925
„ 14	11- 0 a.m.	..	..	30·85	1·020231	26·822	18·48	33·38	..
„ 15	12-0 noon.	..	..	30·70	1·020875	26·899	18·53	33·48	..
„ 16	12-30 p.m.	..	..	27·6	1·021419	26·919	18·54	33·50	..
„ 17	1- 0 p.m.	28·2	27·83	27·65	1·021409	26·925	18·55	33·51	21·231
„ 18	2- 0 p.m.	..	..	27·9	1·021253	26·844	18·49	33·40	..
„ 19	3- 0 p.m.	..	..	28·15	1·021115	26·784	18·45	33·33	..
„ 20	4- 0 p.m.	..	..	28·65	1·020939	26·771	18·44	33·31	..
„ 21	6- 0 p.m.	..	..	28·90	1·020909	26·827	18·48	33·39	..
„ 22	7-15 p.m.	..	..	28·85	1·020911	26·811	18·47	33·37	..
„ 23	8- 0 a.m.	28·7	26·66	28·90	1·020883	26·799	18·46	33·35	20·949
„ 24	8-45 a.m.	28·6	27·17	28·75	1·020911	26·776	18·45	33·33	20·961
Oct. 19	8- 0 a.m.	28·06	26·61	29·11	1·025237	26·079	18·65	33·70	21·420
„ 20	8-55 a.m.	27·83	..	29·06	1·025226	26·067	18·65	33·69	21·483
„ 22	10-30 a.m.	27·94	..	29·17	1·025331	27·183	18·72	33·82	21·556
„ 23	11- 0 a.m.	..	..	29·44	1·025289	27·171	18·72	33·81	..
„ 25	12-0 noon.	27·83	..	28·17	1·025349	27·164	18·71	33·80	21·565
„ 27	2-20 p.m.	28·33	..	28·00	2·023393	26·206	18·74	33·86	21·449
„ 30	6- 0 p.m.	28·33	..	27·94	1·025264	26·114	18·68	33·74	21·364
Nov. 1	8- 0 p.m.	27·83	..	27·94	1·025243	27·040	18·63	33·66	21·458
„ 3	9- 0 p.m.	27·72	..	28·00	1·025358	27·169	18·71	33·81	21·614
„ 7	11- 5 a.m.	28·56	..	27·11	1·025283	27·054	18·64	33·67	21·232
„ 15	6-45 p.m.	27·22	..	29·80	1·025072	26·928	18·55	33·51	21·550
„ 17	8- 0 a.m.	27·22	..	29·80	1·025013	26·866	18·51	33·44	21·493
„ 19	9-30 a.m.	27·50	..	29·78	1·025200	27·061	18·64	33·68	21·585
„ 21	10-25 a.m.	27·50	..	29·83	1·025213	26·081	18·66	33·70	21·603
„ 23	11-40 a.m.	28·00	..	29·11	1·025277	27·123	18·68	33·75	21·405
„ 25	1-50 p.m.	27·67	..	29·22	1·025407	27·266	18·78	33·94	21·720
„ 27	4-30 p.m.	27·22	..	29·00	1·025543	27·403	18·88	34·10	21·994
„ 29	6-30 p.m.	27·22	..	29·11	1·025436	27·299	18·80	33·97	21·897
Dec. 3	8-40 p.m.	27·22	..	28·61	1·025453	27·302	18·81	33·97	21·899
„ 5	9-50 a.m.	27·17	..	28·40	1·025278	27·099	18·67	33·73	21·725
„ 7	11-10 a.m.	27·17	..	28·70	1·024990	26·803	18·46	33·35	21·450
„ 14	6-30 p.m.	27·33	..	28·86	1·025211	27·051	18·64	33·67	21·629
„ 16	7-40 a.m.	27·00	..	28·83	1·025255	27·086	18·66	33·71	21·768
„ 18	9-20 a.m.	27·50	..	28·83	1·025211	27·040	18·63	33·66	21·564
„ 20	10-30 a.m.	27·56	..	28·86	1·025125	26·953	18·57	33·54	21·475

APPENDIX VII.

Observations on the Temperature and Salinity of the Surface-water at 4-hourly intervals in Revello Channel, Nicobars, in March, 1925.

Date.	Time.	Surface Temp.	Air Temp.	Cl.	S.	σ_0	σ_t	State of Tide.
1925. March.		°C.	°C.					
,, 5	8 a.m.	27·2	26·1	18·42	33·28	26·74	21·38	..
	12 noon	28·0	27·9	18·42	33·28	26·74	21·12	L.W. 12.10
	5 p.m.	27·6	28·2	18·39	33·22	26·70	21·41	..
	8 p.m.	28·0	27·9	18·32	33·10	26·59	20·98	H.W. 6.50
	12 midnight.	..	..	18·42	33·28	26·74	..	..
,, 6	4 a.m.	..	..	18·43	33·30	26·75	..	L.W. 2.05
	8 a.m.	27·2	25·1	18·38	33·21	26·68	21·32	H.W. 7.30
	12 noon	27·2	26·3	18·33	33·12	26·61	21·26	..
	4 p.m.	27·4	26·9	18·35	33·15	26·64	21·22	L.W. 1.35
	8 p.m.	27·0	26·6	18·33	33·12	26·61	21·32	..
	12 midnight.	..	..	18·37	33·19	26·67		H.W. 7.30
,, 7	4 a.m.	..	..	18·32	33·10	26·59	..	L.W. 2.25
,, 9	8 a.m.	27·4	..	18·21	32·90	26·43	21·03	H.W. 9.10
	12 noon	27·4	..	18·13	32·75	26·32	20·93	..
	4 p.m.	27·6	..	18·16	32·81	26·36	20·90	L.W. 3.25
	8 p.m.	27·0	..	18·20	32·88	26·41	21·14	H.W. 9.25
	12 midnight.	..	..	18·23	32·94	26·46	..	..
,, 10	4 a.m.	..	..	18·32	33·10	26·59	..	L.W. 4.10
	4 a.m.	27·0	27·3	18·32	33·10	26·59	21·31	H.W. 9.30
	12 noon	27·6	28·6	18·24	32·95	26·48	21·01	L.W. 3.45
	8 p.m.	27·4	28·0	18·25	32·97	26·49	21·08	H.W. 9.50
	12 midnight.	..	..	18·33	33·12	26·61	..	..
,, 11	4 a.m.	..	..	18·33	33·12	26·61	..	L.W. 4.45
	8 a.m.	27·4	27·6	18·30	33·06	26·56	21·15	H.W. 10.05
	12 noon	27·8	28·6	18·31	33·08	26·58	21·04	..
	4 p.m.	27·4	28·7	18·33	33·12	26·61	21·20	L.W. 4.15
	8 p.m.	27·4	27·9	18·36	33·17	26·65	21·23	H.W. 10.15
	12 midnight.	..	..	18·27	33·01	26·52	..	..
,, 12	4 a.m.	..	..	18·35	33·15	26·64	..	L.W. 5.15
	8 a.m.	27·6	28·3	18·12	32·74	26·30	20·84	..
	12 noon	28·0	29·1	18·09	32·68	26·26	20·68	H.W. 10.30
,, 13	12 midnight.	..	..	18·13	32·75	26·32	..	H.W. 11.10
,, 14	4 a.m.	..	..	18·27	33·01	26·52	..	L.W. 6.05
	8 a.m.	27·6	27·7	18·15	32·79	26·35	20·89	
	12 noon	28·0	28·8	18·17	32·83	26·38	20·79	H.W. 11.25
	8 p.m.	27·6	26·0	18·19	32·86	26·40	20·94	L.W. 5.55
	12 midnight.	..		18·20	32·88	26·42	..	H.W. 11.40
,, 15	4 a.m.	..	..	18·20	32·88	26·42	..	L.W. 6.25
	8 a.m.	27·6	26·4	18·21	32·90	26·43	20·96	
	12 noon	28·6	28·9	18·22	32·92	26·45	20·66	H.W. 12 noon.
	8 p.m.	27·6	26·4	18·29	33·04	26·55	21·07	L.W. 6.40
	12 midnight.	..	..	18·27	33·01	26·52	..	H.W. 12.10
,, 16	4 a.m.	..	..	18·31	33·08	26·58	..	L.W. 6.40
	8 a.m.	27·6	26·9	18·29	33·04	26·55	21·07	
	12 noon	28·0	28·3	18·31	33·08	26·58	20·97	H.W. 12.45

Date.	Time.	Surface Temp.	Air Temp.	Cl.	S.	σ_0	σ_t	State of Tide.
1925. March.		°C.	°C.					
,, 16	8 p.m.	27·8	26·7	18·40	33·24	26·71	21·16	L.W. 7.20
	12 mid-night.	..	..	18·31	33·08	26·58	..	H.W. 12.55
,, 17	4 a.m.	..	..	18·29	33 04	26·55	..	
	8 a.m.	27·6	26·9	18·21	32·90	26·43	20·96	L.W. 7.25
	12 noon	28·0	28·9	18·21	32·90	26·43	20·84	H.W. 1.30
	4 p.m.	28·2	28·7	18·29	33·04	26·55	20·88	
	8 p.m.	27·8	26·7	18·20	32·88	26·42	20·89	L.W. 8.25
	12 mid-night.	..	..	18·24	32·95	26·48	..	H.W. 2·05
,, 18	4 a.m.	..	..	18·27	33·01	26·52	..	
	8 a.m.	28·0	26·9	18·22	32·92	26·45	20·85	L.W. 8·35
	12 noon	28·0	29·0	18·25	32·97	26·49	20·89	H.W. 2·40
	8 p.m.	27·6	26·0	18·27	33·01	26·52	21·05	L.W. 9.50
	12 mid-night.	..	..	18·30	33·06	26·56	..	
,, 19	4 a.m.	..	..	18·24	32·95	26·48	..	

V. TEMPERATURE AND SALINITY OF THE SURFACE-WATERS OF THE BAY OF BENGAL AND ANDAMAN SEA, WITH REFERENCES TO THE LACCADIVE SEA.

In the previous paper, No. 4 of this series, I have discussed the changes that have been observed in the temperature and salinity of the coastal waters of the Andaman Sea, and I now propose to deal in a similar manner with the various observations, taken by my predecessors on the "Investigator" and those that I have been able to make myself, on the open waters of the Indian seas away from the influence of land. In many features the conditions existing in these two different areas, namely the open water and the coastal region, have been found to present a noticeable contrast, and the results obtained indicate that in open waters the surface of the ocean is constantly exposed to outside influences that result in the production of a number of extremely interesting phenomena. I do not pretend that the conclusions, at which I have arrived, are in any way final. It was only when the "Investigator" was steaming from port to port or from her base to the survey-ground, wherever from year to year that might be, that I was able to take observations, and there is a limit to the amount that a single scientific officer can do in the way of research when at sea. I am fully aware that more observations are required before one can safely assume that many of the features, which appear to be present in these waters, have been proved; but I have thought it advisable to put on record all that I have been able to achieve, in order that my results may serve as a guide to those who in the future may be able to carry on the work, either on board the "Investigator" herself or in other ships that may in the future be engaged in oceanographic research in Indian waters.

Seasonal Variation in the Temperature of the surface-water in Indian Seas.

The suface-water throughout the Bay of Bengal and the Andaman Sea possesses a low salinity owing to the influx of large volumes of fresh-water from the great rivers of India and Burma, and this influx must also to some extent affect the temperature of the surface-water. We have already considered the alterations that occur in the general meteorological conditions over Indian seas and from analogy we should expect to find a similar range of variation in the temperature of the surface-water, not only at different times of the day but also at different seasons of the year; and in practice it is found that the temperature of the surface-water exhibits a small though regular annual variation in accordance with the changing seasons.

As I have already pointed out (*vide supra*, p. 57) the air-temperature over the open waters of the Indian seas exhibits a clear double oscillation in the course of the year, there being two maxima, one in April and the other in September-October, corresponding to the two dry hot seasons, and two minima during the periods of the south-west and north-east monsoons respectively. One would naturally expect to find that the temperature of the surface-water follows the same or at least a similar course and also exhibits a double oscillation, and all the available evidence shows clearly that this is actually the case. As Krummel (1907, p. 411) has pointed out, the temperature of the surface-water in the northern part of the Indian Ocean increases to a maximum in the month of May, when the temperature over the greater part of the Arabian Sea and the Bay of Bengal lies between 29.0° and 29.8°C, while in the Andaman Sea and in a small area off the east coast of India it may reach to 30°C or over. At the commencement of the south-west monsoon there is a distinct fall in the temperature of the surface-water, which Krummel gives as 2° to 4°C, in the Arabian Sea and 1° to 2°C in the Bay of Bengal. During the subsequent part of the year there is a second oscillation in the surface-temperature: at the close of the south-west monsoon the temperature again rises, and it finally falls for a second time with the onset of the north-east monsoon and the incidence of the cold season, when the sun is at its maximum southerly declination. Broun (1906, p. 428) during a period of five years carried out observations on the temperature of the sea-water in his evaporating tanks at Trivandrum in the Madras Presidency (Lat. 8° 29′ N.; Long. 76° 56′ E.) and his results show clearly that there is a distinct tendency for the temperature to exhibit a double oscillation during the course of the year; his results are given below in Table 37 and it is clear that the temperature in this area is lowest in the month of January; it then rises till April, after which it falls, as a result of the onset of the south-west monsoon, till July and then rises again till November, after which month it again falls as a result of the onset of the cold weather and the commencement of the north-east monsoon. Broun's observations, however, were carried out in artificial evaporating tanks and his results are not, therefore, conclusive as regards the conditions that exist over the open sea.

January.	February.	March.	April.	May.	June.	July.	August.	September.	October.	November.	December.	
°C.	°C.	°C.	°C.	°C.	°C.	°C.	°C.	°C.	°C.	°C.	°C.	
27·3	27·4	29·3	29·4	29·0	27·5	26·7	27·0	27·3	27·24	27·8	27·6	Broun
26·06	..	28·75	29·03	..	25·61	..	25·70	..	..	27·5	..	Pearson

Table 37; showing the annual march of the temperature at Trivandrum (Broun) and in the Gulf of Mannar (Pearson).

An exactly similar double oscillation during the course of the year is, however, clearly evident in the temperature observations made by Dr. J. Pearson (*vide* Marine

Biological Reports of the Ceylon Government, 1921 and 1922) in the Gulf of Mannar and I have shown both his and Broun's series graphically in Text-figure 36. In computing the average for the Gulf of Mannar I have made use of only those observations that were taken by Pearson in deep water of over 100 fathoms; in this series at the commencement of the year the temperature is low, being 26·06°C in January; it then steadily rises to 29·03°C in April, but at the commencement of the south-west monsoon the temperature falls till it is as low as 25·61° in June, after which it again rises to 27·50°C in November and finally falls again till January. The annual range in the monthly mean temperature is thus 3·42°C; in the shallower waters near the shore the range is slightly greater, being 3·9°C, the monthly mean temperatures being somewhat higher in the hot months and lower during the winter and the monsoon seasons.

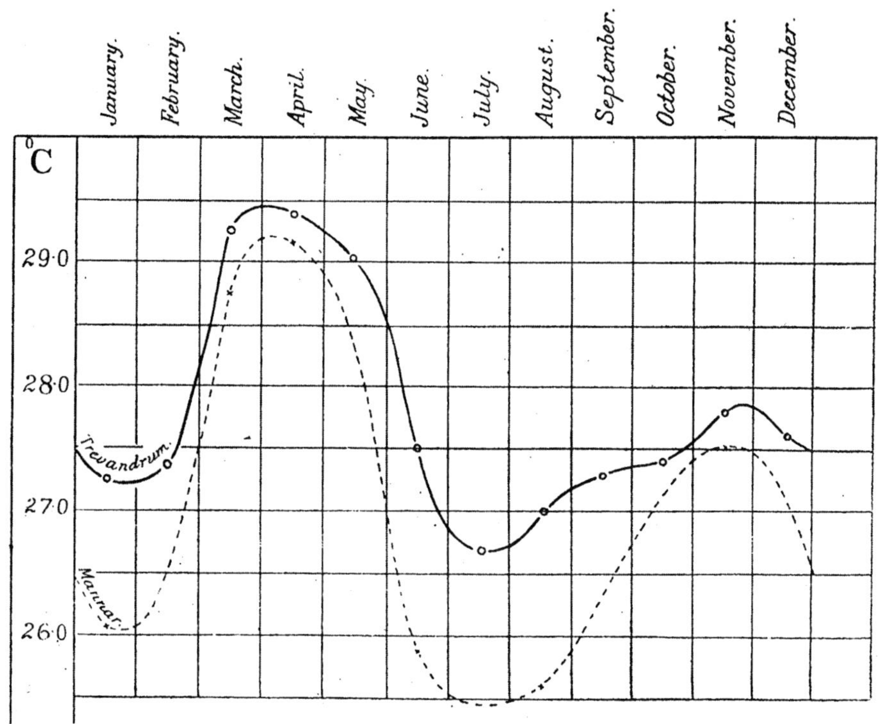

Text-fig. 36. Showing the annual variation in the surface temperature at Trevandrum and in the Gulf of Mannar.

The observations that have been taken on board the "Investigator" fall within that part of the year that constitutes the survey-season, namely from October to May, and in the following tables 38A and 38B I have given month by month the data derived from all observations that had been taken prior to the commencement of my own work as Surgeon-Naturalist in 1910; the period for which I have data extends from 1884 to 1909 inclusive, thus covering a period of 26 years. Throughout this period observations of the temperature of the surface-water were taken at irregular intervals and as a rule only when a deep-sea trawl was being carried out and in consequence these observations cannot be regarded as strictly comparable to my own results, which are based on observations taken at 4-hourly intervals throughout the day. I have, therefore, kept the two series entirely separate. The actual number of observations taken during this early period of work in the "Investigator" are 263 in the Bay of Bengal and 299 in the Andaman Sea and I have given the results of these two sets of observations separately.

Month.	October.	November.	December.	January.	February.	March.	April.	May.	
	°C.	°C.	°C.	°C.	°C	°C.	°C.	°C.	
Average temperature ..	27·40	27·36	26·09	(23·71)	27·51	27·71	28 86	27·50	
Highest recorded temperature.	30·00	30·00	28·17	26·00	27·22	30·60	30·56	..	
Lowest recorded temperature.	25·50	25·00	21·72	(23·05)	21·72	26·11	26·94	..	
Maximum range ..	4·50	5·00	6·45	..	5·50	4·49	3·62	..	
No. of observations ..	21	27	47	11	18	27	108	4	Total .. 263

Table 38A; the results of observations on the surface-temperature taken by the "Investigator" in the Bay of Bengal prior to 1910.

Month.	October.	November.	December.	January.	February.	March.	April.	
	°C.	°C.	°C.	°C.	°C.	°C.	°C.	
Average temperature ..	26·67	27·31	26·95	27·24	26·68	27·39	28·30	
Highest recorded temperature ..	28·33	28·90	27·78	29·44	28·06	30·00	30·28	
Lowest recorded temperature ..	22·78	24·44	25·00	25·00	25·00	26·10	25·56	
Maximum range	5·55	4·46	2·78	4·44	3·06	3·90	4·72	
No. of observations	13	47	37	57	69	40	25	Total 299

Table 38B; the results of observations on the surface-temperature taken by the "Investigator" in the Andaman Sea prior to 1910.

It will be seen from the above data that the results obtained in these two areas exhibit *inter se* considerable differences. In the first series, namely from the Bay of Bengal, the average temperature is found to fall steadily from 27·4°C in the month of October to 23·7°C in January; it then rises equally steadily to 28·9°C in April and falls slightly in the month of May, but the number of observations in this latter month are too few to be of any great value for the purpose of comparison with the results obtained in other months. Again, the extreme range of temperature, *i.e.*, between the highest and the lowest recorded temperatures, in each month rises from 4·5°C in October to 6·45°C in December; the data for January are too meagre to form a basis of comparison, but the range falls steadily from 5·5°C in February to 3·62°C in April. In the second series, from the Andaman Sea, the average surface-temperature exhibits a series of fluctuations; it rises from 26·7°C in October to 27·31°C in November falls to 26·95°C in December, rises to 27·24°C in January, falls a second time to 26·68°C in February and finally rises steadily to 28·3°C in April. (It is probable that the average temperature given above for the month of October is too low for comparatively few observations were available for this month.)

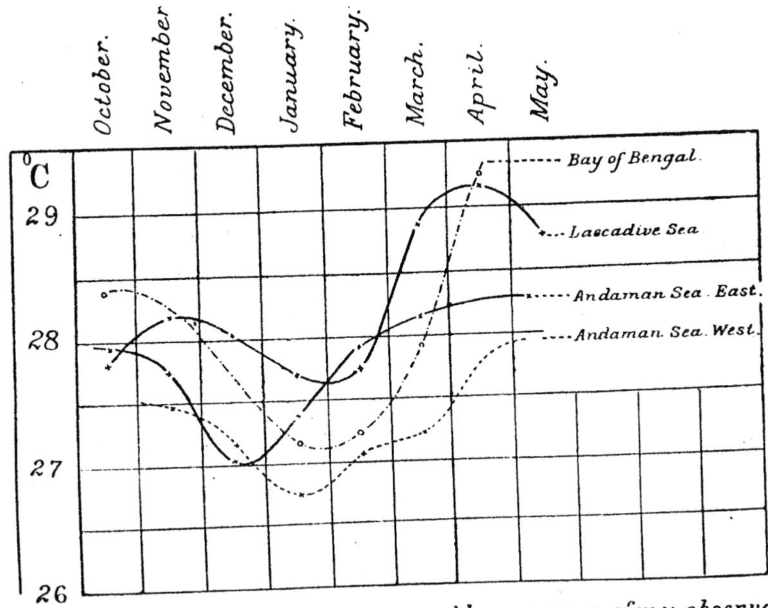

Text-fig. 37. Showing the mean monthly averages of my observations on the temperature of the surface water in different regions of the Indian Seas.

A comparison of the oscillations of the average monthly temperature during the winter months over the Andaman Sea in both the surface-water and the supernatant air reveals a close similarity in the changes that have been observed (*vide* Text-figures 7 [p. 64] and 37). This double oscillation in the air-temperature I have already attributed to the influence of the north-east monsoon and it seems probable that the oscillation in the surface-temperature is to be attributed to the same agency; as in the case of the air-temperature, the oscillation is only to be detected over the region of the Andaman Sea and possibly over the south-east part of the Bay of Bengal, and is absent from the greater part of the Bay area.

The mean of the average monthly temperatures throughout the period from October to May is 27·02°C for the Bay of Bengal and 27·22°C for the Andaman Sea; or for the whole width of this region 27·12°C. Krummel gives the mean average temperature for the whole width of the Indian Ocean, lying between Lat. 10° and 20° N. as 27·33°C., which agrees very closely with the above figures, especially in view of the fact that for the most part these observations on the "Investigator" were taken during the winter months and, therefore, tend to be lower than the figure given by Krummel, that is based on observations taken throughout the whole year.

The observations taken by me on the "Investigator" during the survey-seasons of 1910 to 1914 and again from 1921 to 1925 cover only that part of the year from October to May. The average monthly temperatures recorded in each of these months in different parts of the Indian seas are given in Table 39 below, and I have shown them graphically in Text-fig. 37.

Month.		Andaman Sea.	Bay of Bengal.	Laccadive Sea.
		°C.	°C.	°C.
October ..	..	28·30	28·44	27·79
November	..	28·02	..	28·17
December	..	27·61	..	28·00
January ..	..	27·70	27·15	27·28
February	..	28·04	27·25	27·73 ("Valdivia")
March ..	..	..	27·98	28·89
April ..	..	29·00	29·25	29·19
May ..	..	..	..	28·74

Table 39; the average temperature of the surface-water in different regions and in different months.

In all cases there is a fall of temperature during the closing months of the year; in the Laccadive Sea this appears to set in in November, while in the Andaman Sea it commences from October. The lowest temperature is reached in January and from then on it rises again to April, after which it appears to fall again. These results, therefore, confirm the conclusion arrived at from a scrutiny of the data given by Broun and Pearson, to which I have referred above, namely that there is in these waters a double seasonal oscillation in the surface-temperature of the ocean exactly similar to the double oscillation in the temperature of the air.

From year to year, however, conditions prevailing over Indian seas exhibit a very considerable range of variation and especially so during those months in which the temperature depends on the strength or otherwise of the monsoons. In 1921 and 1922 a careful record of the sea-temperature was taken, at four-hourly intervals, while crossing the Bay of Bengal from Ceylon to the Andamans. In both years the passage was made in the month of October, in 1921 from October 18th to 22nd and in 1922 from October 9th to 13th. The results obtained in these two years are as follows:—

Year	Average temperature.	Average maximum.	Average minimum.	Range of temperature.
	°C.	°C.	°C.	°C.
1921	28·79	29·33	27·66	1·67
1922	28·04	28·07	27·55	0·52

It is obvious that very different conditions existed in the surface-waters in these two years, and this difference is in all probability correlated with the difference in the strength of the south-west monsoon. I have discussed this matter more fully later (*vide infra* p. 271) when dealing with the differences noted in the salinity of the surface-water in these years; suffice it to say here that in 1921 the south-west monsoon presented a slight defect and in consequence the degree of cooling of the surface during the monsoon and the rate of the rise of temperature, after the monsoon had blown itself out, must have been correspondingly affected, the cooling being diminished and the subsequent heating expedited, thus giving a high mean temperature and high average maximum: in 1922 on the other hand the south-west monsoon was in excess thus causing an increased lowering of the surface-temperature and a delay in its subsequent rise, and so causing a low average temperature and a low average maximum.

The Annual Range of Temperature of the Surface-Water.

Sir John Murray (1898) has published a chart showing the annual range of temperature in all the great oceans and he remarks (*loc. cit.* p. 123) "in the Indian Ocean a belt of small range (less than 10° Fahr.) stretches across from the east coast of Africa (south of Cape Guardafui and north of Madagascar) to the shores of the Malay Peninsula and Sumatra (and passing through the Arafura Sea into the Pacific), lying mostly to the north of Lat. 10°S., and filling up the greater part of the Arabian Sea and of the Bay of Bengal. The observations within this Indian Ocean tropical area show a range of 72° to 87° Fahr." Murray's chart, however, is based on the extreme temperatures that have been recorded in the two months of February and August and it, therefore, does not give us a true picture of the seasonal changes that are taking place in these Indian waters, since, owing to local conditions, there is in this area not a single but a double oscillation in the temperature, and, though February may be the coldest month, August is certainly not the hottest.

Schott (1895, p. 153 and Pl. X) has compiled a chart that shows the annual variation between the highest and lowest mean monthly temperatures and according to this chart throughout the whole belt between Ceylon on the west and the Andaman Islands on the east, that is to say along the route that the "Investigator" traversed across the Bay of Bengal, the range of temperature is approximately 2·0°C.

The true mean annual temperature, the mean range of temperature and the mean monthly temperature can only be arrived at from a number of observations spread over many years but it is interesting to note that the results of the observations on the "Investigator" agree closely with the figures given by Schott. If we exclude from the data given above in Table 38 that for the month of January, we get for the range of variation in the average monthly temperatures a figure that agrees closely with Schott's, namely 1·63°C for the range in the Andaman Sea and 2·77°C for the Bay of Bengal, or an average for the whole area of 2·2°C. On the other hand the maximum annual range of temperature, as given by the earlier "Investigator" observations, is from 21·72°C to 30·60°C in the Bay of Bengal—a range of 8·88°C (or 16·0°F) and from 22·78°C to 30·28°C in the Andaman Sea—a range of 7·5°C (or 13·5°F). These results, therefore, agree closely with the data given by Sir John Murray. In both instances, however, it is interesting to note that the annual range is appreciably higher in the Bay of Bengal than in the Andaman Sea, and this raises a question as to whether there is, in reality, any clear difference between these two areas, such as the above figures indicate. A study of the average monthly temperatures in each of these divisions of Indian waters, as calculated from the four-hourly records taken by me on the "Investigator" since 1914, shows that the annual range of the average monthly temperature of the surface-water in each area is as follows:—

Andaman Sea	from	27·61°C	to	29·00°C = 1·39°C
Bay of Bengal	,,	27·15°C	,,	29·25°C = 2·10°C
Laccadive Sea	,,	27·28°C	,,	29·16°C = 1·88°C

Here again we find that the annual range of the temperature of the surface-water, as deduced from the average monthly temperatures is considerably smaller in the Andaman Sea than in the Bay of Bengal. Finally, if we turn to the data given in the charts of surface-temperature, specific gravity, etc., of the Bay of Bengal and Andaman Sea (published by the Meteorological Department of the Government of India) we find that in a belt extending from Ceylon on the west to the coast of Burma on the east, between Lats. 6° and 8°N. there is clear evidence of a distinct fall in the annual range of temperature as deduced from the average surface-temperature of the four three-monthly periods, December–February, March–May, June–August and September–November. Dividing the whole belt into areas of 4° of Longitude we get the following annual variation in the surface water:—

Between Long.	80°—84°E.	from	81°F	to	85°F = 4°F	(or 2·22°C)
,, ,,	84°—88°E.	,,	81°F	,,	87°F = 6°F	(or 3·33°C)
,, ,,	88°—92°E.	,,	82°F	,,	87°F = 5°F	(or 2·78°C)
,, ,,	92°—96°E.	,,	81°F	,,	85°F = 4°F	(or 2·22°C)
,, ,,	96°—100°E.	,,	82°F	,,	85°F = 3°F	(or 1·67°C)

It thus seems clear that the annual range of temperature, whether we base our estimate on the extremes of temperature recorded during a series of years or on the average temperature recorded in each month or period of three months, is very considerably less in the Andaman Sea area than in the Bay of Bengal and, moreover, that even in the latter area there is a steady increase in the annual range as we pass westwards.

Daily Variation in the Temperature of the Surface-Water of the Open Ocean in Indian Waters.

On every occasion, from the year 1914, that I have been on board the "Investigator" it has been my custom to keep a four-hourly record of the temperature of the surface-water whenever the ship was steaming from port to port, and at the same time a sample of the water was collected and examined in order to determine its salinity. A study of these temperature records reveals several interesting features and shows that the daily variation of the surface-temperature differs considerably at different times of the year. In Table 41 I have given the average surface-temperatures at different times of the day, as deduced from my observations, in all the three great divisions of the Indian seas and for the purpose of reference and to help out the series I have included in the data for the Bay of Bengal the results obtained by the S.S. "Valdivia" during her passage across this region in the month of February, 1899.

Month.		Time of day.						Range of temperature.
		4 A.M.	8 A.M.	12 Noon.	4 P.M.	8 P.M.	12 P.M.	
1. *Andaman Sea.*		°C	°C	°C	°C	°C	°C	°C
October	..	27·72	27·97	28·82	28·72	28·67	27·89	1·46
November	..	27·52	28·00	28·39	28·78	28·07	27·35	1·43
December	..	27·47	27·61	27·78	27·83	27·67	27·32	0·51
January	..	27·25	27·95	28·03	28·00	27·70	27·25	0·78
February	..	27·50	28·00	29·00	28·05	28·00	27·70	1·50
*April	..	27·50	29·02	29·73	30·83	28·89	28·05	3·33
2. *Bay of Bengal.*								
October	..	27·60	28·48	28·93	28·58	28·50	28·53	1·33
January	..	26·66	27·13	27·40	27·34	27·34	27·01	0·86
†February	..	26·93	26·93	27·70	27·57	27·43	26·93	0·77
March	..	26·66	27·87	28·90	28·60	28·10	27·75	2·24
April	..	28·28	29·11	30·12	30·26	29·42	28·35	1·98
3. *Laccadive Sea.*								
October	..	27·25	27·78	28·26	28·33	27·89	27·25	1·08
November	..	27·09	27·98	28·75	28·78	28·43	27.99	1·69
December	..	27·51	28·01	28·13	28·40	28·29	27·67	0·89
February	..	27·23	27·81	28·08	28·15	27·71	27·37	0·93
March	..	28·10	28·86	29·41	29·53	29·08	28·37	1·43
April	..	28·31	29·23	29·67	29·69	29·43	28·61	1·38
May	..	27·94	28·45	29·37	29·40	28·79	28·44	1·46

Table 40; giving the average-temperature at different times of the day and the average range of temperature during the day of the surface-water in the different parts of the Indian seas.

* This series refers to the passage in 1914 across the north-west part of the Andaman Sea from Rangoon to Duncan Passage.

† This series was taken by the S.S. "Valdivia" (*vide* Schott, 1902).

The data given above are more complete in certain regions than in others, but a comparison of the results in all three areas shows that in every case the lowest temperature is reached at about 4 a.m., while the highest temperature falls somewhere between 12 noon and 4 p.m. Krummel (1907, p. 383) has given a table, which

I reproduce below, showing the average variation, above and below the mean temperature of the surface-water, at different times of the day in the open ocean:—

2 a.m.	4 a.m.	6 a.m.	8 a.m.	10 a.m.	12 noon.	2 p.m.	4 p.m.	6 p.m.	8 p.m.	10 p.m.	12 p.m.
-0·16	-0·19	-0·19	-0·12	0·06	0·19	0·29	0·25	0·17	0·04	-0·10	-0·13

According to this table the surface-temperature reaches its highest point at about 2 p.m. and its lowest at from 4 to 6 a.m. In Table 41 I have given the average variation above and below the mean monthly temperature of the temperature of the surface-water at different times of the day in each month of the survey-season:—

Month.	A.M.			P.M.			Range of Tempera-ture.
	4	8	12	4	8	12	
	°C	°C	°C	°C	°C	°C	°C
October ..	-0·69	-0·13	*0·49*	0·37	0·18	-0·29	1·18
November ..	-0·79	-0·11	0·48	*0·69*	0·16	-0·43	1·48
December ..	-0·32	0·01	0·15	0·31	0·18	-0·31	0·63
January ..	-0·47	0·12	*0·29*	0·25	0·10	-0·30	0·76
February ..	-0·45	-0·09	*0·59*	0·25	0·04	-0·34	1·04
March ..	-0·77	-0·07	*0·72*	0·63	0·16	-0·38	1·49
April ..	-1·11	-0·33	0·69	*1·11*	0·10	-0·81	2·22
May ..	-0·77	-0·29	0·63	*0·66*	0·05	-0·30	1·43

Table 41; giving the oscillation above and below the mean monthly temperature of the surface-temperature in different months of the year.

A comparison of the data for each month with the Table given by Krummel shows clearly that it is only in the month of December that conditions in these waters agree with the average variation for the open ocean and that throughout the whole period there is clear evidence of a seasonal variation from the mean both as regards the daily range of temperature and the time of day at which the maximum temperature is reached.

Taking the daily range of temperature first, the Table clearly shows that there is a very considerable variation from month to month. Commencing with a range of 1·18°C in October, this shows a slight increase to 1·48°C in November; in December the range falls to a minimum and now only averages 0·63°C; from this point it steadily increases till it reaches a maximum, so far as the period of the year for which I have data is concerned, of 2·22°C. in April, after which it appears to again fall, the average for the month of May being only 1·43°C; as, however, the number of observations taken during this latter month is considerably less than in the other months it is not safe to place too much reliance on the average. I have previously called attention to the fact that the mean daily variation, or the range, of the air-temperature over Indian seas exhibits a periodic oscillation during the course of the year that has two maxima in March and August and two minima in June and November (*vide supra*, p. 67), and a study of the surface-temperatures clearly indicates that a similar oscillation can be traced in the daily range during, at any rate, the winter months and the following hot, dry season, namely from October to May. In Text-figure 38 I have shown graphically the average range of temperature of the surface-water in each month of the survey-season in (*a*) Nankauri Harbour in the Nicobars and (*b*) over the open waters of the Indian seas. In both series of observations the variation in the daily range follows the same course and it is interesting to note that the range of temperature is actually greater over the open sea than in the enclosed waters of Nankauri Harbour. In both cases the range of variation increases from October to November; it then falls rapidly in December; steadily rises again till April, in which month it appears to attain its maximum; and probably falls again in May. It is unfortunate that I have no records for the remaining months of the year but it appears to be probable that the daily range of surface-temperature follows a similar course to that of the air-temperature and exhibits during the whole year a double oscillation, the two maxima being in April and October, that is to say approximately a month later than the maxima in the range of variation in the air-temperature and corresponding to the maxima in the double oscillation in the average temperature (*vide supra*, p. 57). Similarly I have already (*vide supra*, p. 66) shown that in the case of the air-temperature we can trace a distinct correlation between the mean monthly temperature and the maximum daily range of temperature; as I then remarked "the variation in the daily range of temperature shows an unmistakable tendency towards a double oscillation during the year that is very similar to but does not quite coincide with the double oscillation in the monthly mean air-temperature." In Text-figure 39 I have shown graphically the average temperature and the maximum range of temperature of the surface-water of both the

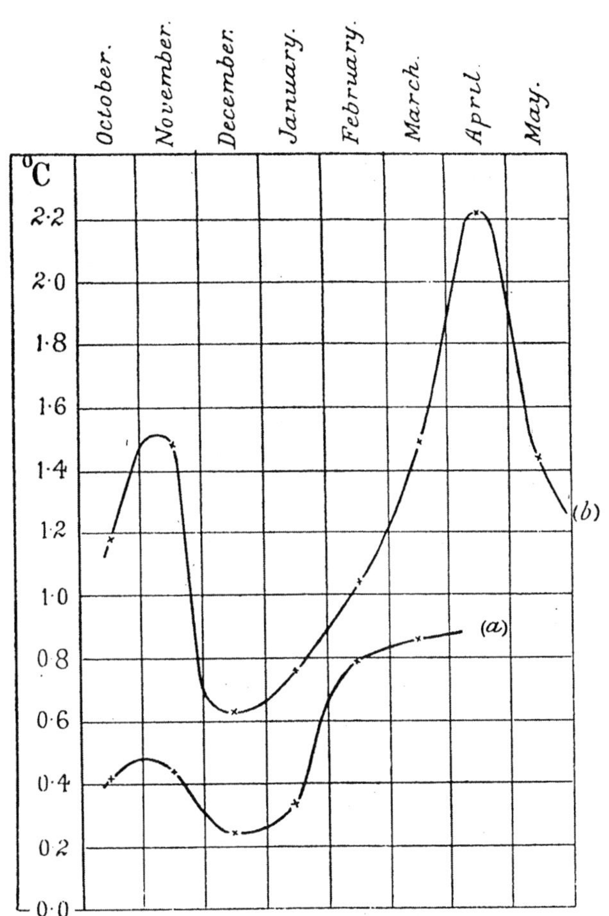

Text fig. 38. *Showing the seasonal variation in the range of surface temparature in*
(a) Nankauri Harbour, Nicobars
(b) Open water of Indian Seas

Bay of Bengal and the Andaman Sea, as shown in the earlier series of observations and it is of considerable interest to compare this result with the results obtained by the study of the air-temperature over the same two regions. In the case of the surface-water in these two areas, namely the Bay of Bengal and the Andaman Sea, the correlation between the average temperature and the maximum range are the exact opposite of each other; in the Bay of Bengal the maximum range appears to be greatest when the average temperature is lowest, whereas in the Andaman Sea the two sets of figures vary simultaneously and in the same direction and, if we admit, as

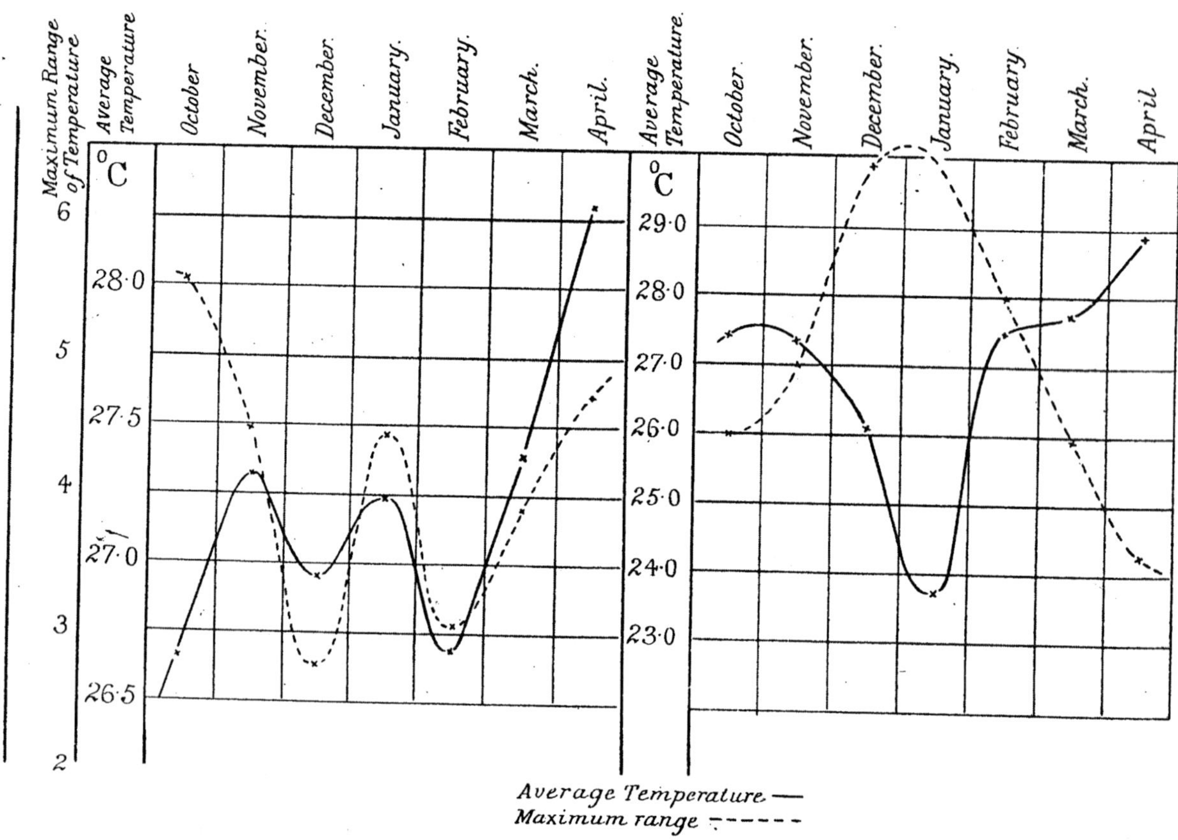

Surface Temperature of the Andaman Sea. *Surface Temperature of the Bay of Bengal.*

Text-fig. 39. Showing the monthly average Temperature of the Surface water of the Andaman Sea and Bay of Bengal, during the Survey season period.

I have pointed out above (*vide supra*, p. 211) that the average temperature is too low in the month of October, the parallel nature of the two series of variations becomes complete, a high average temperature being associated with a high maximum range. At first sight then it would appear that the relationships between the daily range and the average monthly temperature are different in these two regions, but implicit reliance cannot be placed on these earlier observations, owing to the irregular time of day at which the records were taken. In my own observations, taken since 1914, comparison of the average monthly temperature and the average daily range reveals clearly that the two oscillate together, a high average temperature being associated with a large range, and this is particularly well seen in the data for the Andaman Sea and the Laccadive Sea (*vide* Text-fig. 40): unfortunately I have no records of the surface conditions in the Bay of Bengal in the months of November

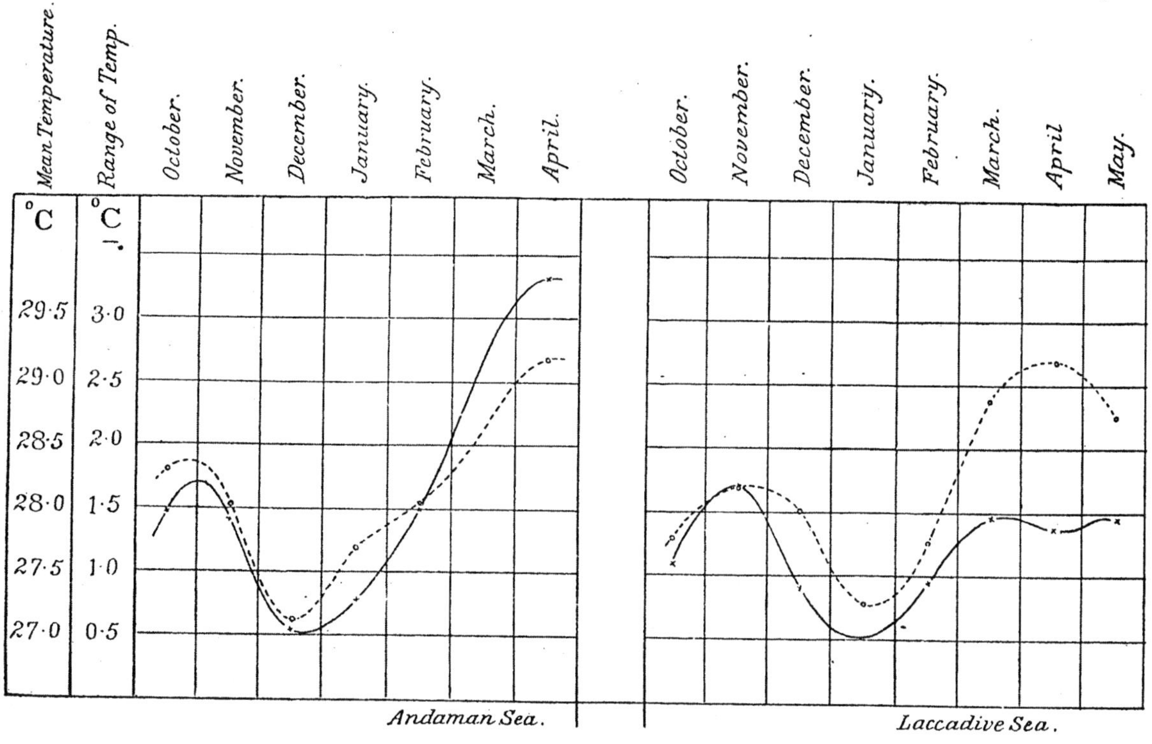

Text-fig. 40. Showing the variation in Indian Seas of the average monthly temperature and the average daily range of temperature in each month of the survey season.

and December. The evidence at our disposal then clearly indicates that the conditions existing at the sea-surface during the course of the year undergo the same or at least a similar variation to those of the air.

Time of Occurrence of the Epoch of Maximum Temperature.

The second interesting feature brought out by the study of the data given above lies in the difference, in different months of the season, in the time of day at which the maximum temperature is reached. If we consider the average results for all three regions in which investigations have been carried out (*vide* Table 41, p. 216), it is clear that the maximum temperature is reached at about 12 noon in the month of October; the time of the epoch then becomes later, occurring about 4 p.m., in November and December; it is then again earlier, at about 12 noon from January to March; and finally gets later again in April and May. In the following Table 42 I have given the time of occurrence of the epoch of maximum temperature in the surface-water in different months and in all three regions of the Indian seas separately:—

	October.	November.	December.	January.	February.	March.	April.	May.
Andaman Sea	13	15	16	12	12	..	16	..
Nankauri Harbour	14	13	13	15	15	15	..	..
Bay of Bengal	12	..	..	13	13	12	14	..
Laccadive Sea	14	14	17	..	15	14	14	14
Average time of occurrence	13·3	14·0	15·3	13·3	13·7	13·7	14·7	

Table 42; giving the average time of occurrence of the epoch of maximum temperature of the surface-water in different regions of Indian waters in each month of the survey-season.

Note.—In the above table the hours are given as reading from 0 to 24 throughout the day, and in the average time the decimals must be multiplied by 6 to convert the decimal into minutes.

I have previously called attention (*vide supra*, p. 72) to the fact that the epoch of maximum temperature of the air tends to occur at different times of the day in

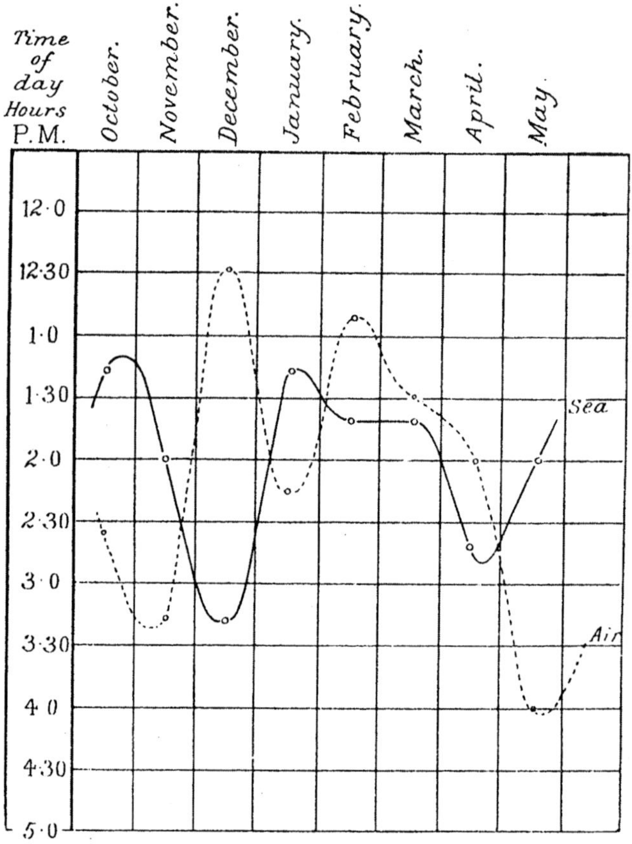

Text-fig. 41. Showing the average time of occurrence of the epoch of maximum temperature of the Sea and Air in different months.

different months of the year and I showed that this oscillation in the time of occurrence could be detected in all parts of the Indian region and that though the changes in time were not always simultaneous in the different regions, they always tended to follow the same general type of oscillation that was clearly a seasonal one. The same phenomenon appears to be present in the surface-temperature also and here again we can trace a distinct agreement between the variations in the average temperature, the range of temperature and the time of occurrence of the epoch of maximum temperature.

In Text-figure 41 I have given the average time of day in each month of the occurrence of maximum temperature of the surface-water and for comparison the time of the epoch of maximum air-temperature. This latter figure has been arrived at by taking the average time of occurrence in all parts of the Indian seas, the results of which are as follows:—

Month.	October.	November.	December.	January.	February.	March.	April.	May.
Time of epoch of maximum air-temperature in hours. p.m.	2·37	3·18	12·30	2·15	12·53	1·30	2·0	4·0

A comparison of the two curves shows that they both follow a remarkably similar course, but that the changes in the time of occurrence of maximum air-temperature follow exactly a month later than the corresponding changes in the sea-temperature.

Relationship Between the Temperature of the Sea Surface and that of the Air.

I have already (*vide supra*, p. 151) considered at some length the relationship between the temperature of the sea and that of the air at different times of the year in Nankauri Harbour and the neighbourhood, and we have seen that on the whole the relationship follows a course that agrees in its general character with the relationship found by Buchan (1889, p. 7) to be present in the North Atlantic Ocean. In individual months, however, we found that there was a considerable range of variation, the difference between the two temperatures being greatest in the month of October and least in March, and that it was in the latter month only that the sea-temperature fell during the middle of the day to a lower level than the temperature of the air. It is of considerable interest to examine the relationship of the sea- and air-temperatures over the open waters of Indian seas and to see to what extent they conform to or differ from the results obtained in inshore waters. It was, however, only during the last two years of my work as Surgeon-Naturalist that the necessity became apparent of making a detailed study of the meteorological conditions that were present as well as a study of the sea-water itself and, therefore, in this respect my observations are by no means as complete as I could wish.

The results obtained in Nankauri Harbour gave the following average relationship between the sea and air temperatures;

In the case of the N. Atlantic series of observations, it has been shown that the sea-temperature, taking the average of all observations, falls in the middle of the day to a fraction of a degree *below* the air-temperature. In the case of my observations, however, the results are very different. In Indian seas the various occasions on which the average surface-temperature was found to be lower than the air-temperature are given below in tabular form :—

Month and year.	Locality.	Sea temperature minus air temperature.	Time of day.
October, 1922	Laccadive Sea	−0·31	4 p.m.
	Do.	−0·26	12 Midnight.
November, 1923	Do.	−0·33	4 a.m.
January, 1924	Bay of Bengal	−0·59	4 a.m.
	Do.	−0·19	8 a.m.
	Do.	−0·23	10 a.m.
	Do.	−0·21	10 p.m.
March, 1924	Do.	−0·76	4 a.m.
	Do.	−0·03	10 p.m.
April, 1923	Laccadive Sea	−0·03	12 noon.
April, 1924	Do.	−0·28	10 p.m.
	Do.	−0·27	12 Midnight.
	Do.	−0·11	4 p.m.
	Do.	−0·03	8 p.m.
	Do.	−0·85	12 Midnight.
	Do.	−1·14	4 a.m.
April, 1925	Bay of Bengal	−0·80	12 Midnight.
	Laccadive Sea	−0·94	12 Midnight.
May, 1923	Do.	−0·16	8 a.m.
	Do.	−0·07	10 p.m.

A study of these data clearly reveals that out of a total of 20 observations, the sea-temperature is in 15 instances lower than the air-temperature between the hours of 10 p.m. and 10 a.m. and in only 5 cases does this occur between the hours of 10 a.m. and 10 p.m. Moreover, the average amount below the air-temperature to which the sea-temperature falls is greatest at 12 midnight and 4 a.m. and is least from 12 noon to 8 p.m. It seems clear that, whatever may be the condition in the N. Atlantic, in Indian seas the tendency is for the sea-temperature, when it becomes lower than the air-temperature, to do so at or about midnight, and since the temperature of the air normally falls more rapidly than that of the sea it would further seem probable that at this time of the day there is some phenomenon occurring which causes an abnormal lowering of the surface-temperature of the sea. In the following pages I have carefully compared the temperatures of the sea-surface and the air at different times of the day in each of the months of the survey-season, namely October to May inclusive. In this way I have attempted to determine the manner and direction in which their relationship over the open waters of Indian seas at different times of the year and in different regions varies from what has hitherto been considered to be the normal. As I shall show later, the alteration in the relationship of the two temperatures appears to depend ultimately on variation in the strength of the wind and I have therefore included all available data dealing with these changes. Wherever it has been possible, I have given the average strength of the wind, as observed on the R.I.M.S. "Investigator," during the same period and at the same time as that in which my observations were made on the temperature of the air and of the surface-water, but in a few instances this has not been possible and I have then given the average deduced from all observations taken in the same month but regardless of the area in which the observations were made.

October.

For the month of October I have five series of observations covering all three main regions of the Indian seas. The average results of all five series are given below:—

		A.M.				P.M.			
		4	8	10	12	4	8	10	12
		°C	°C	°C	°C	°C	°C	°C	°C
Laccadive Sea.									
October, 1922 (3½ days)	Sea	26·22	26·66	....	27·33	27·70	27·18	....	26·27
	Air	25·97	25·90	....	27·09	28·01	26·80	....	26·53
	Diff.	0·25	0 76	....	0·24	-0·31	0·38	....	-0·26
October, 1923 (5 days)	Sea	27·68	28 29	28·52	28·71	28·59	28·28	28·15	27·82
	Air	26·85	27·78	28·32	28·61	28·47	27·95	27·50	27·41
	Diff.	0·83	0·51	0·20	0·10	0·12	0·33	0·65	0·41
	Wind-force, B.S.	*2·17*	*0·91*	....	*1·00*	*1·75*	*1·50*	...	*1·33*
Bay of Bengal.									
October, 1921 (3½ days)	Sea	26·88	28·93	....	29·90	29·33	29·40	....	28 00
	Air	26·11	25·55	....	28·06	28·52	27·96	....	27·33
	Diff.	0·77	3·38	....	1·84	0·81	1·44	....	0·67
October, 1922 (3½ days)	Sea	27·56	28·08	....	27·95	28·37* (27·83)	28·24* (27 75)	....	28·06
	Air	27·06	27·20	....	27·41	26·93	26·85	....	26·85
	Diff.	0·50	0 88	....	0·54	1·44 (0·90)	1·39 (1·00)	....	1·21
Andaman Sea.									
October, 1921-22 (2 days)	Sea	27·36	27·97	....	28·82	28·72	28·66	....	27·89
	Air	26·61	27·52	....	27·99	27·80	27·02	....	26 41
	Diff.	0·75	0 45	....	0·83	0·92	1·64	....	1·48
	Wind force, B.S.	*1·92*	*1·69*		*1·90*	*2·01*	*1·97*		*1·70*
October, 1921	Wind-force	*2·28*	*2·14*	....	*2·14*	*2·28*	*2·07*	....	*1·78* 7 days.
October, 1922	Wind-force	*1·56*	*1·25*	....	*1·56*	*1·75*	*1·87*	....	*1·62* 8 days.

Table 43; Giving the average of observations on the surface-temperature, air-temperature and wind-force at different times of the day in various parts of the Indian seas in the month of *October*.

Note.—In this and succeeding tables the strength of the wind is given in Beaufort's notation.

* I have omitted from my calculation one reading at each of these times, as it seems clear that we were in a cold current (vide Appendix VIII, p. 347). The figures beneath in brackets give the average of *all* observations.

Unfortunately I have only two series of observations in which I possess data for the strength of the wind taken simultaneously with the observations on the temperature of the sea and air; for the other three series I have had to rely on the average strength of the wind observed in all the three areas combined. As I have remarked above, the observations on the strength of the wind are those recorded by the ship's officers and are, as is usual in such cases, given in the Beaufort Scale. In Text-fig. 44 I have shown graphically the results obtained in the Bay of Bengal in October, 1921, and in the Laccadive Sea in October, 1922. Between these two series of observations there is a marked degree of similarity, the curve of the temperature-difference exhibit-

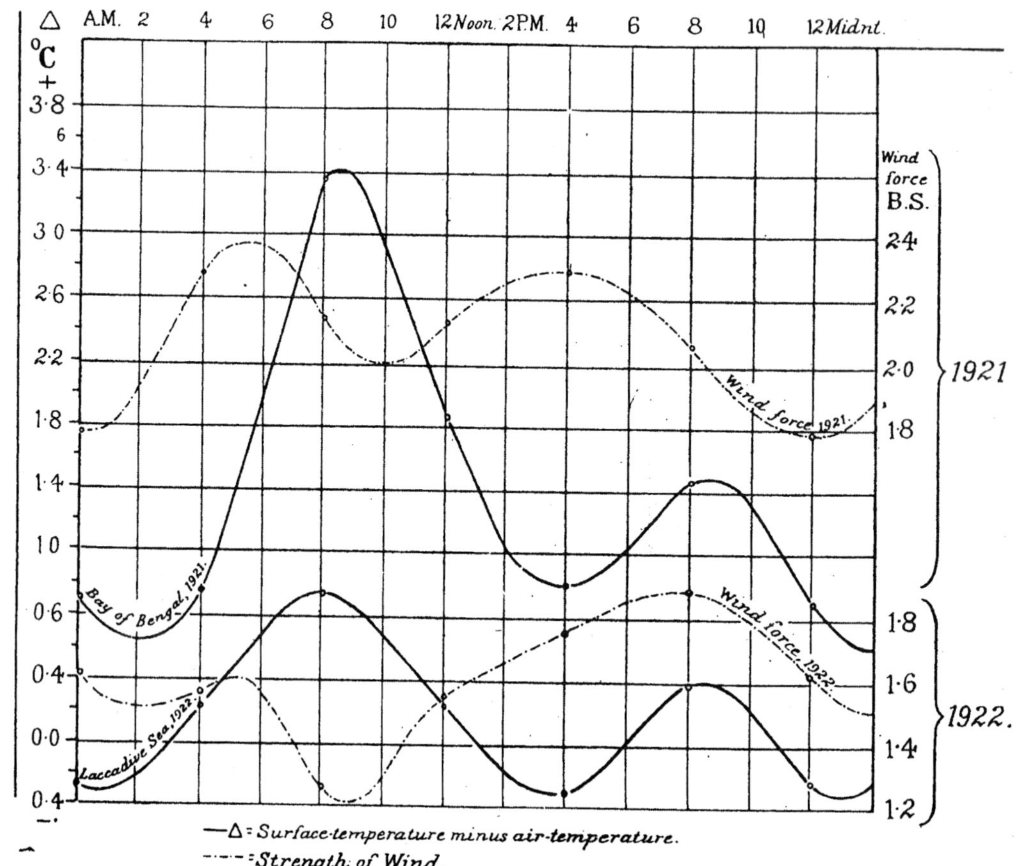

Text-fig 44 Showing the relationship between the temperature of the surface and of the air during the day in the Bay of Bengal. October, 1921, and the Laccadive Sea, 1922, as compared with the strength of the Wind.

ing in each case a clear double oscillation. A comparison of the times of occurrence of maximum and minimum differences between the temperatures of the sea and air (Δ = Sea-temperature − air-temperature) shows clearly that in both cases these occur at almost identical times of the day; but the first maximum in the Bay of Bengal in 1921, occurring at 8-30 a.m., is very much greater than the corresponding maximum in the Laccadive Sea in 1922. In both cases, however, the morning oscillation of the temperature-difference is greater than the evening one. If now we compare the temperature-difference with the average wind-force in these two areas, we see that, while this latter follows the same general course in both years and shows clearly the double oscillation to which I have referred above (*vide supra* p. 79), in 1921 there is but little difference between the morning and evening oscillation in the strength of the wind, whereas in 1922 the morning oscillation in the wind-force is very much less than the evening one, and it seems clear that the lesser variation in the temperature-difference in 1922 is associated with a diminished oscillation in the wind-force. Again, in both series it is interesting to note that the variation in the temperature-difference, while corresponding with the oscillation in the wind-force, exhibits a distinct "lag", the maximum temperature-difference occurring about 3 hours later than the maximum wind-force in the morning and from 2-5 hours later in the evening.

The data for the Laccadive Sea area in October, 1923 (*vide* Text-figure 45) shows a similar double oscillation in the temperature-difference, in that the morning maximum is considerably greater than the evening maximum; the actual times of occurrence of

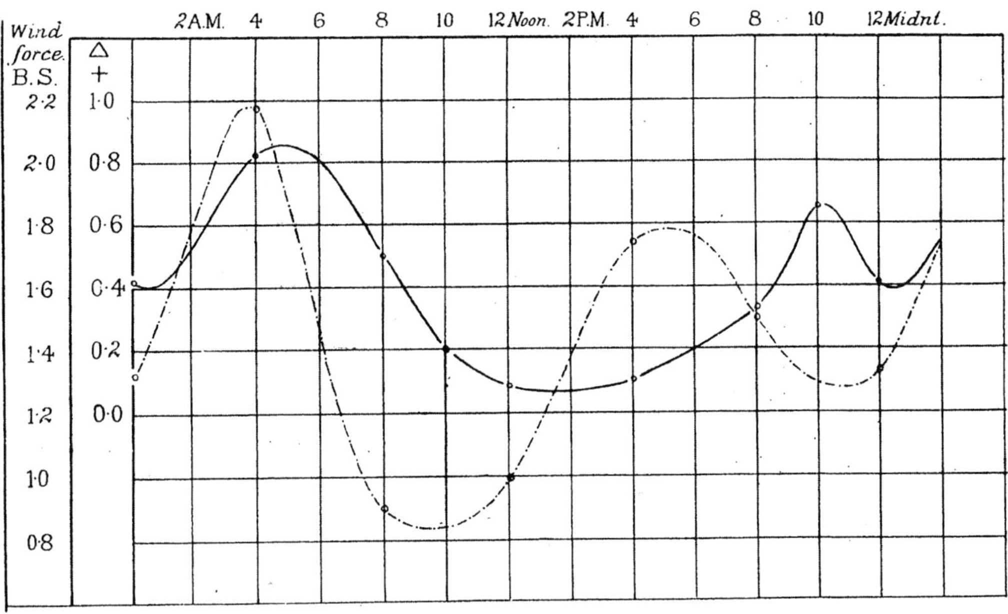

Text-fig. 45. Showing the relationship between the surface-temperature and the air-temperature during the day in the Laccadive Sea in October, 1923 as compared with the wind-force.

these maxima are, however, somewhat different from those in the preceding series and are now found at 5 a.m. and 10 p.m. instead of at 8 a.m. and 8-30 p.m.; furthermore, the morning oscillation in the wind-force is much greater than the evening one and this corresponds to the maximum difference in the temperature relationship. We again find a very distinct "lag" in the changes in the temperature-difference behind the variation in the wind-force, amounting to about 1 hour in the morning and 5 hours in the evening.

The last two series of observations in this month to be considered are those taken in the Bay of Bengal in 1922 and in the Andaman Sea in 1921-22 (*vide* Text-fig. 46). In this latter series I have combined the observations taken in two consecutive years as the observations in each year cover only a single day. While differing somewhat from the results that we have hitherto considered, these two series again show the double diurnal oscillation in the temperature-difference, following on the variation in the wind-force, but in both cases, the evening maximum is considerably greater than the morning one, so much so that this latter appears in this instance as a comparatively slight irregularity in an otherwise uniform curve. Here again there appears to be a distinct correlation between the range of the temperature-difference and the variation in the strength of the wind, the major oscillation in the wind-force being associated with the major variation in the temperature-difference. An interesting feature of the Andaman Sea data is the great increase in the " lag" of the morning maximum of the temperature-difference, which occurs at 11-30 a.m., behind the corresponding oscillation of the wind-force at 4 a.m.; in this case the " lag " is as great as 7½ hours and during this time the strength of the wind has fallen to a minimum at 8 and has again begun to rise.

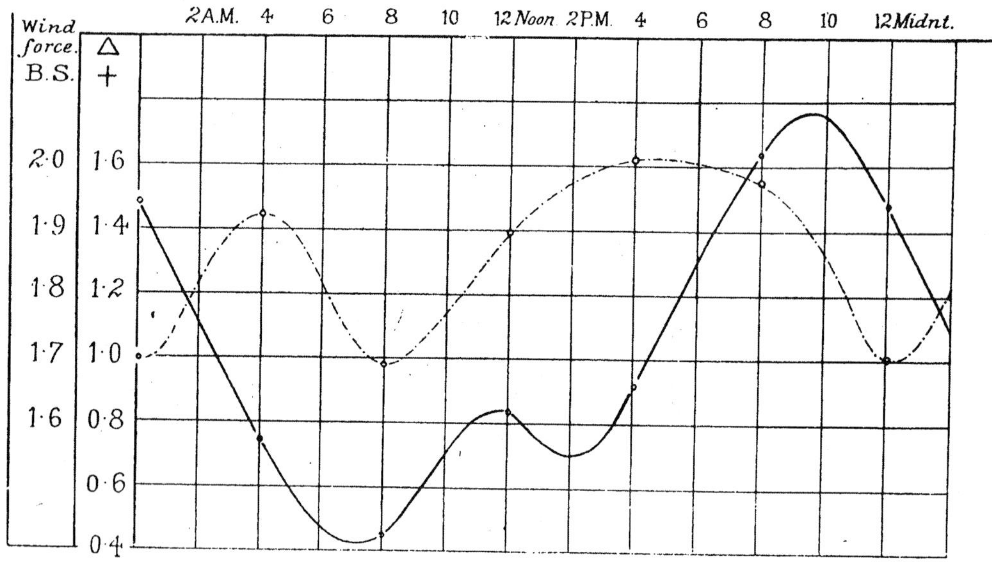

Text-fig. 46 *Showing the relationship between the surface-temperature and the air-temperature during the day in the Andaman Sea, in October 1921-22, as compared with the wind-force.*

Such a condition of affairs appears to indicate that the effect of the variation in the strength of the wind on the temperature-difference is not, at any rate entirely, a direct one: I shall, however, discuss this matter at some length later in this paper, but it may be as well to point out here that, if the variation in the temperature-difference is due to movements of water-masses induced by oscillations in the wind-force, each increase in the strength of the wind will set up laterally-moving surface currents that, owing to the inertia of the water-masses, may not commence for some time after the change in the wind-force has become noticeable but will tend to continue for some time after the wind-force has again dropped.

November.—For the month of November I possess only two series of observations, taken, respectively, in the Laccadive Sea and in the Andaman Sea. The average results of these two series are given below in Table 44.

		A.M.				P.M.			
		4	8	10	12	4	8	10	12
Laccadive Sea.		°C	°C	°C	°C	°C	°C	°C	°C
November, 1923, (4 days.)	Sea ..	27·08	27·98	28·38	28·75	28·78	28·43	28·12	22·99
	Air ..	27·41	26·63	26·62	27·54	27·66	27·71	27·22	27·15
	Diff. ..	−0·33	1·35	1·76	1·21	1·12	0·72	0·90	0·84
	Wind-force B.S.	*2·1*	*2·3*	*1·7*	*1·9*	*2·0*	*1·9*	*2·2*	*2·0*
Andaman Sea.									
November, 1921-22,	Sea ..	27·55	28·00	..	27·85	28·53	28·03	..	27·43
	Air ..	26·11	27·03	..	26·41	26·59	26·50	..	25·58
(3 days.)	Diff. ..	1·44	0·97	..	1·44	1·94	1·53	..	1·85
	Wind-force B.S.	*2·75*	*1·12*	..	*1·50*	*2·37*	*2·12*	..	*2·25*

Table 44; giving the average of observations on the surface-temperature, air-temperature and wind-force at different times of the day in various parts of the Indian seas in the month of *November.*

A comparison of the two series shows clearly that there is a very considerable degree of difference between them. In the Andaman Sea series, which is taken from

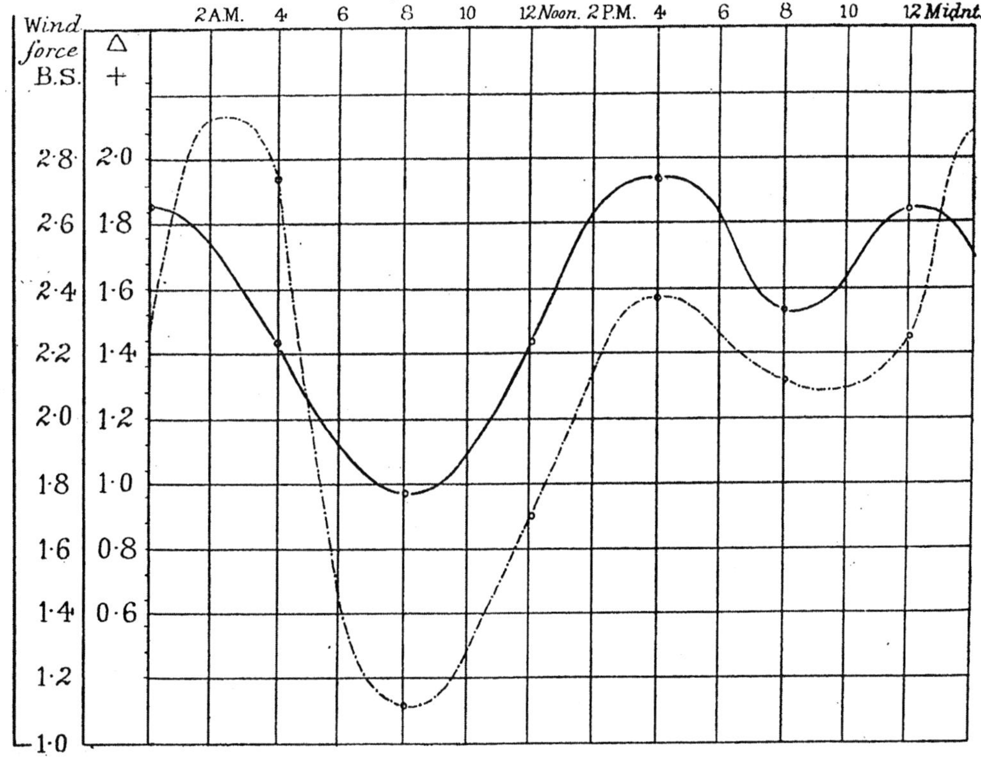

Text-fig. 47. Showing the relationship between the surface-temperature and the air-temperature during the day in the Andaman Sea, in November, 1921-22. as compared with the wind-force.

the combined results obtained in two successive years, 1921 and 1922, we find that the character of the rise and fall of the temperature-difference differs considerably

otherwise be a single large oscillation having a maximum at about 4 a.m. and a minimum at 1 p.m. A comparison of the observed oscillations in the temperature-difference and in the wind-force shows, however, that there is a quite distinct tendency for a rise in the wind-force to be accompanied by a rise in the temperature-difference, comparable to the simultaneous rise and fall that we have observed in various areas in preceding months, and in this instance there again appears to be a clear correlation between the extent of the variation in the wind-force and of the accompanying change in the temperature-difference, the major oscillation of the wind-force between the hours of 10 p.m. and 10 a.m. being accompanied by the major variation in the temperature-difference. In the Laccadive Sea region in this month in 1923 the wind-force exhibited a quadruple oscillation in the twenty-four hours

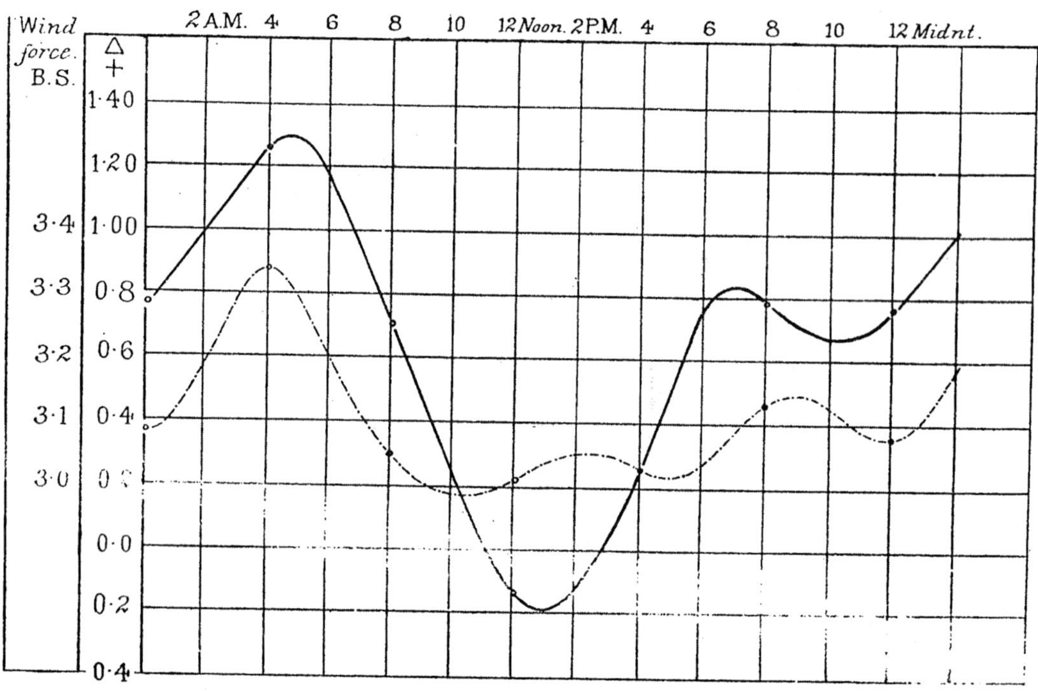

Text-fig. 49. Showing the relationship between the surface-temperature and air-temperature during the day in the Andaman Sea in December, 1921-22, as compared with the wind-force.

during the period in which my observations were conducted, and a comparison of this variation with the rise and fall of the temperature-difference, as recorded simultaneously, reveals that there is a marked agreement between the two (*vide* Text-fig. 50). In this series the rise and fall of the wind-force is based on observations taken at two-hourly intervals throughout the day, and this reveals a small subsidiary oscillation at 6 p.m. which is unaccompanied by a similar variation in the course of the temperature-difference, no observations on the temperatures having been taken at this hour of the day. I have already (*vide supra* p. 87) called attention to the presence of a triple oscillation in the wind-force in Indian seas in this month of the year and it is interesting to note that a similar triple oscillation in the wind-force, apparently accompanied by a corresponding triple variation in the temperature-difference, was also found to occur in the same region, namely the Laccadive Sea, during the preceding month of November (*vide* p. 230 and Text-fig. 48).

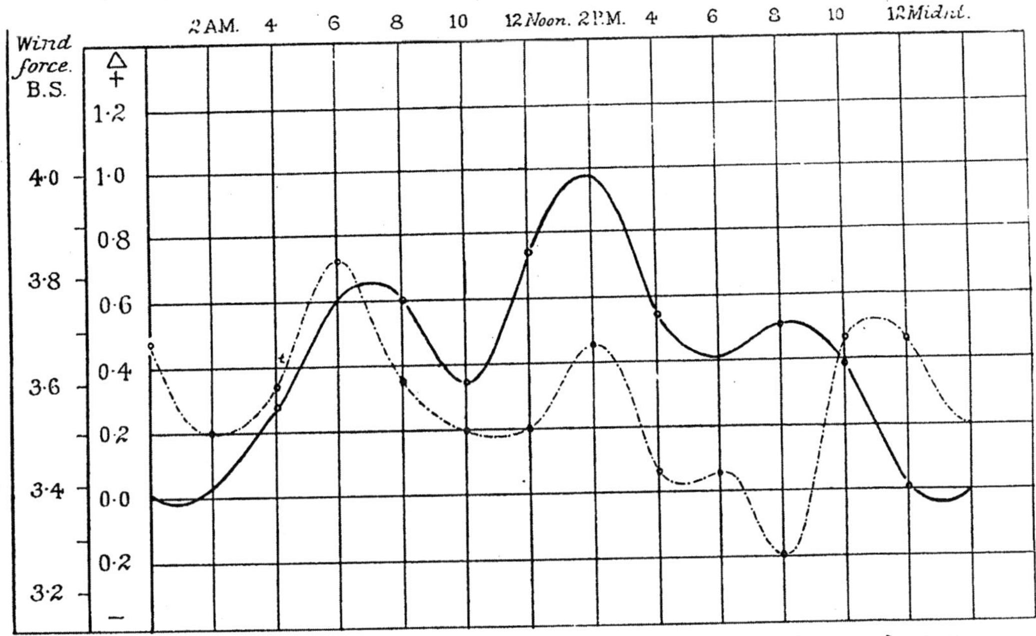

Text-fig. 50. Showing the relationship between the surface-temperature and air-temperature in the Laccadive Sea in December, 1923, as compared with the wind-force.

January. During the month of January I possess three sets of records taken respectively in different years in the Andaman Sea and the Bay of Bengal. The averages of these records are given below in Table 46 :—

		A.M.				P.M.			
		4	8	10	12	4	8	10	12
		°C	°C	°C	°C	°C	°C	°C	°C
Bay of Bengal.									
January, 1923, (4 days).	Sea ..	26·67	27·13	..	27·30	27·25	27·25	..	26·81
	Air ..	25·72	26·76	..	27·13	26·90	26·39	..	26·07
	Diff. ..	0·95	0·37	..	0·17	0·35	0·86	..	0·74
	Wind-force B. S.	*2·21*	*2·50*	..	*2·21*	*2·00*	*2·28*	..	*2·14*
January, 1924, (4 days).	Sea ..	26·63	27·13	27·19	27·50	27·43	27·48	27·15	27·18
	Air ..	27·22	27·32	27·42	27·22	27·28	27·39	27·36	27·15
	Diff. ..	0·59	0·19	0·23	0·28	0·15	0·09	0·21	0·03
	Wind-force B. S.	*2·70*	*2·70*	*2·70*	*3·20*	*2·80*	*2·90*	*3·10*	*3·30*
Andaman Sea.									
January, 1922, (3 days).	Sea ..	27·33	27·93	..	28·03	28·00	27·77	..	27·17
	Air ..	26·67	27·20	..	27·37	27·33	26·28	..	26·67
	Diff. ..	0·66	0·73	..	0·66	0·67	1·49	..	0·50
	Wind-force B. S.	*2·50*	*3·00*	..	*2·50*	*2·50*	*3·50*	..	*2·50*

Table 46; showing the average of observations on the surface-temperature, air-temperature and wind-force at different times of the day in various parts of the Indian seas in the month of *January*.

In the case of the first two series, namely those taken in the Andaman Sea 1922 (Text-fig. 51) and in the Bay of Bengal in 1923, (Text-fig. 52) the type

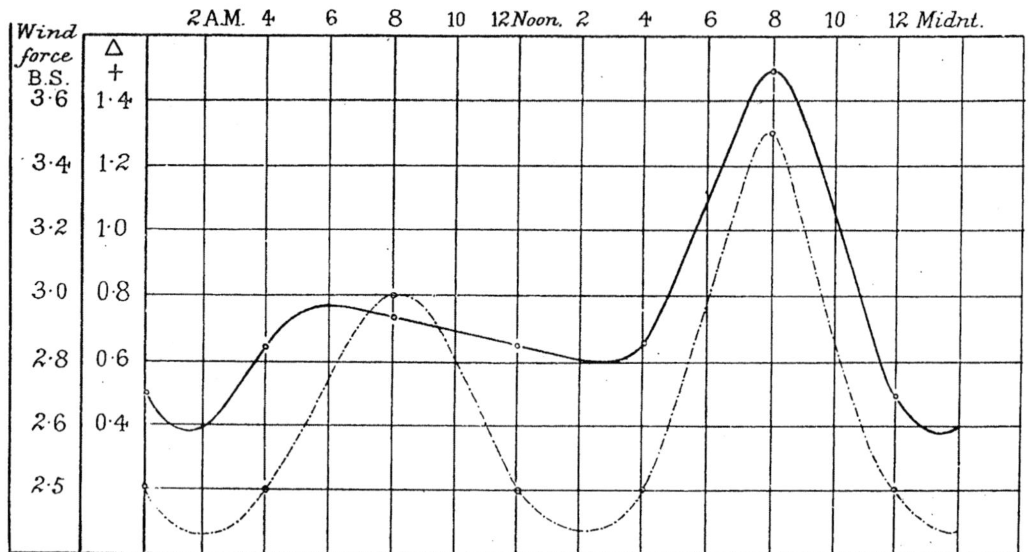

Text-fig. 51. Showing the relationship between the surface-temperature and the air-temperature in the Andaman Sea in January, 1922 as compared with the wind-force.

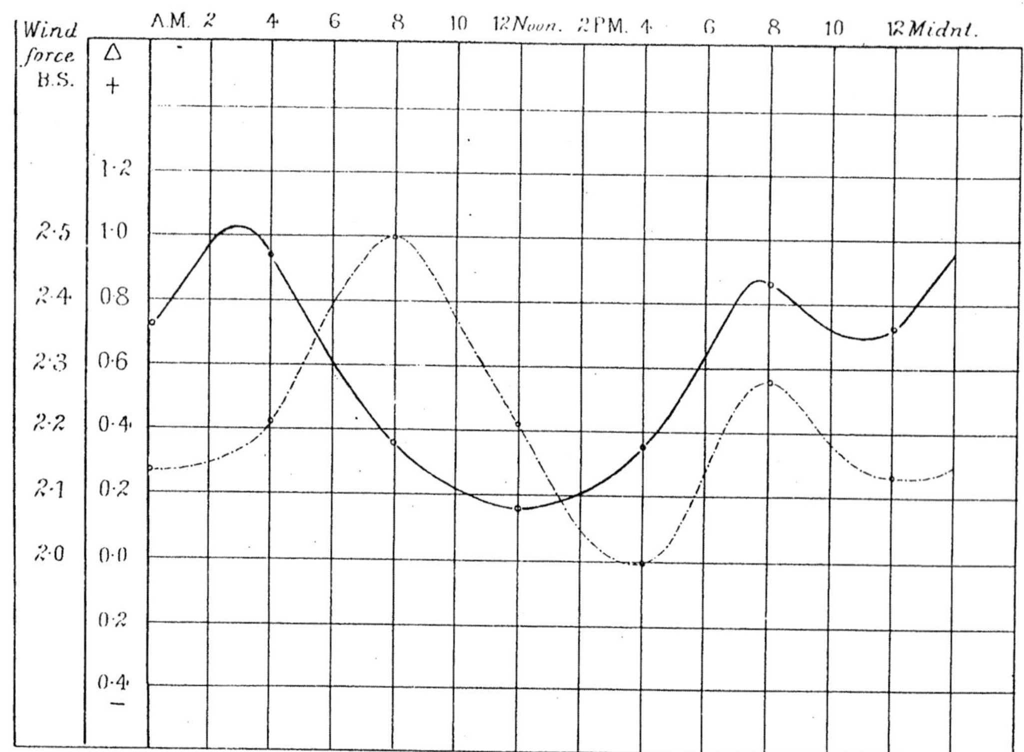

Text-fig. 52. Showing the relationship between the surface-temperature and the air-temperature in the Bay of Bengal in January, 1923, as compared with the wind-force.

oscillation is clearly a double diurnal one, similar to that which we have so commonly met with; and in both cases there is a fairly close agreement between the rise and fall of the wind-force and the oscillation of the temperature-difference. In both series the changes in the wind-force follow the same general trend, there being two maxima at 8 a.m. and p.m. respectively; but in the case of the record for the Andaman Sea in 1922, the primary maximum occurs at 8 p.m., whereas in the case of the Bay of Bengal in 1923 it is the morning rise, occurring at 8 a.m., that is the primary maximum. The oscillations of wind-force and temperature-difference during the evening rise, between 4 p.m. and 12 midnight, occur simultaneously in both series, but in both cases in the morning oscillation, between 12 midnight and 4 p.m., there is a well marked "lag," amounting to about 4 hours, in the change in the wind-force behind the variation in the temperature-difference. Both series of observations show clearly that the maximum variation in the temperature-difference

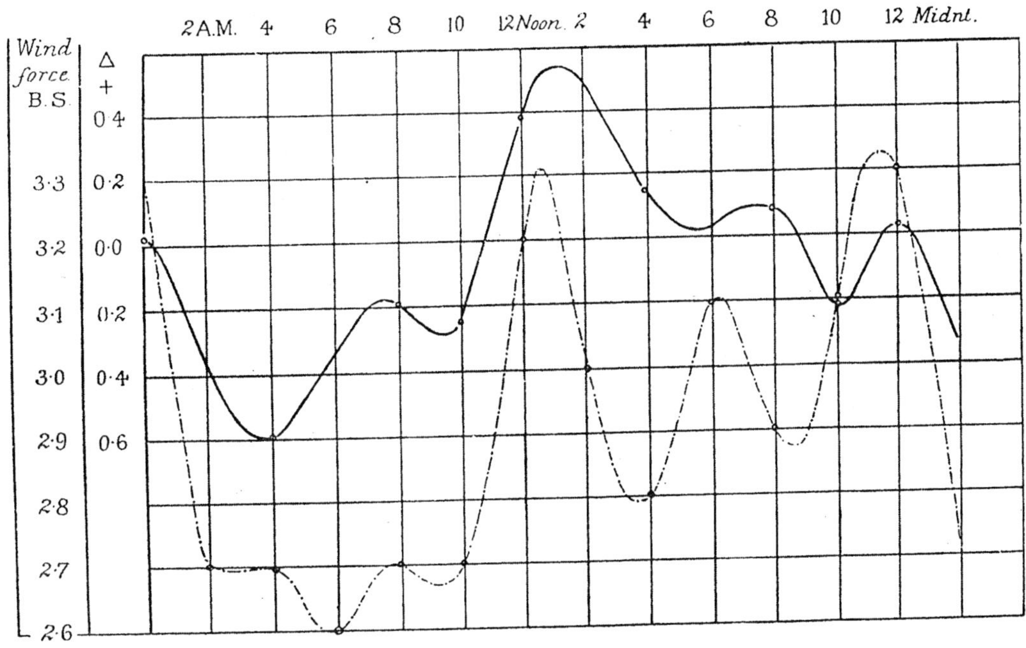

Text-fig. 53. Showing the relationship between the surface-temperature and the air-temperature in the Bay of Bengal in January, 1924 as compared with the wind-force.

coincides with the maximum oscillation in the strength of the wind, though this occurs in the evening in the Andaman Sea in 1922 and in the morning in the Bay of Bengal in 1923. In the series of observations taken in the Bay of Bengal in January 1924, the double oscillation of the temperature-difference and the wind-force is replaced by a quadruple oscillation. The variation of the wind-force, as given in Text-fig. 53, is taken from the two-hourly observations taken on the same days as those on which the temperature readings were taken; and there can be no two opinions regarding the very clear manner in which the two series of oscillations coincide both as regards time and the extent of the observed fluctuations. Throughout the whole twenty-four hours the changes in the temperature-difference exhibit a slight "lag" of $\frac{1}{2}$ to 2 hours behind the corresponding variation in the wind-force.

February. Here again I have only two series of observations, taken respectively in the Andaman Sea in 1922 and in the Laccadive Sea in 1923. The average results of these two series are given below in Table 47 :—

		A.M. 4	A.M. 8	A.M. 10	A.M. 12	P.M. 4	P.M. 8	P.M. 10	P.M. 12
Laccadive Sea.		°C	°C	°C	°C	°C	°C	°C	°C
February,* 1923, (7 days.)	Sea ..	27·23	27·22	..	27·83	27·96	27·43	..	27·43
	Air ..	26·30	26·83	..	27·30	27·69	26·72	..	26·67
	Diff. ..	0·93	0·89	..	0·53	0·27	0·71	..	0·76
	Wind-force B.S.	*1·92*	*1·70*	..	*2·21*	*2·17*	*2·67* (*2·10*)	..	*1·83*
Andaman Sea.									
February, 1922, (2 days.)	Sea ..	27·50	28·00	..	29·00	28·60	28·10	..	27·70
	Air ..	26·11	27·33	..	28·00	27·56	26·83	..	26·67
	Diff. ..	1·39	0·67	..	1·00	1·04	1·27	..	1·03
	Wind-force B.S.	*1·25*	*2·00*	..	*2·00*	*1·50*	*1·25*	..	*1·00*

* This includes the last two days in January, 1923.

Table 47; giving the average of observations on the surface-temperature, air-temperature and wind-force at different times of the day in various parts of the Indian seas in the month of *February*.

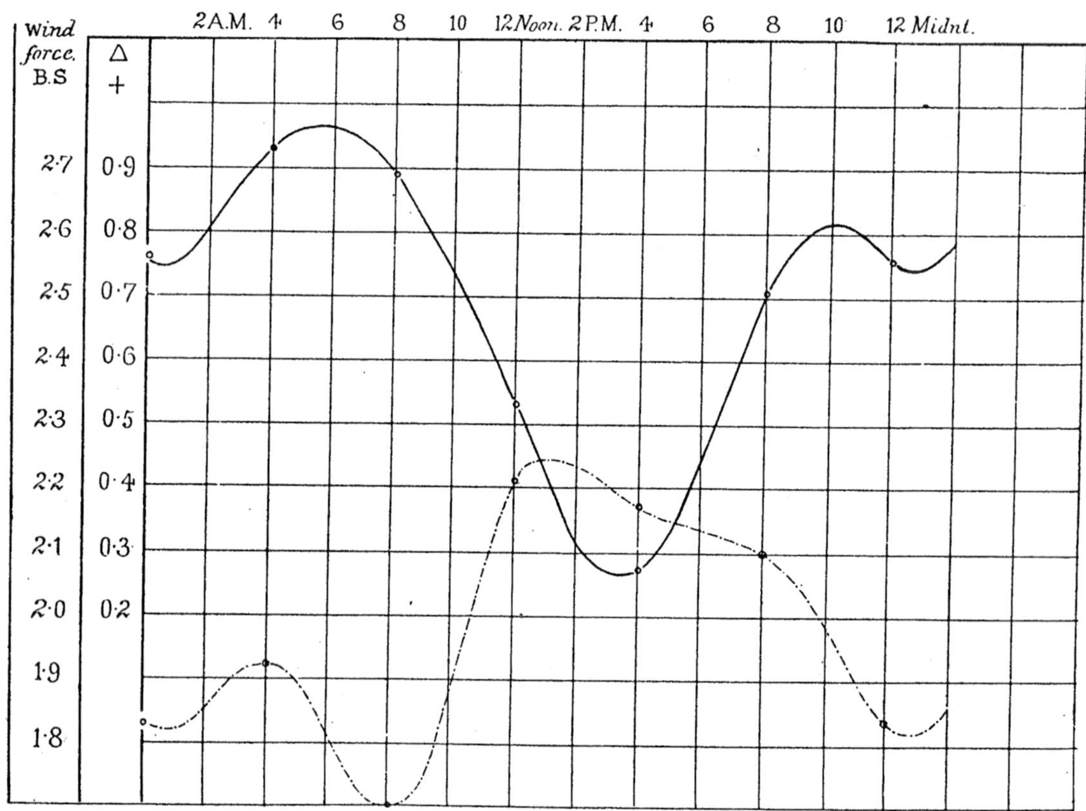

Text-fig. 54. Showing the relationship between the surface-temperature and the air-temperature in the Laccadive Sea in February, 1923, as compared with the wind-force.

In the case of the Laccadive Sea series (*vide* Text-fig. 54) the relationship of the surface-temperature and air-temperature again appears to tend toward a single oscillation and closely resembles the condition found to be present in the Andaman Sea in December, 1921-22 (*vide* Text-fig. 49). The temperature-difference reaches a primary maximum at, approximately, 6 a.m.: it then falls steadily till 3 p.m., after which it rises to a secondary maximum at 10 p.m., and falls slightly again till about 1 a.m. The wind-force on the other hand exhibits an unequal double oscillation during the twenty-four hours, but reaches its primary maximum at 1 p.m. and then appears to fall more or less steadily till 1 a.m. The average strenth of the wind at 8 p.m. as given in Table 47 above, namely 2·67, is probably too high; this figure is the average of all the observations taken at this hour on six days, but on one day the wind-force at 8 p.m. was logged as 5·5 B.S.; at 4 p.m., on the same day the

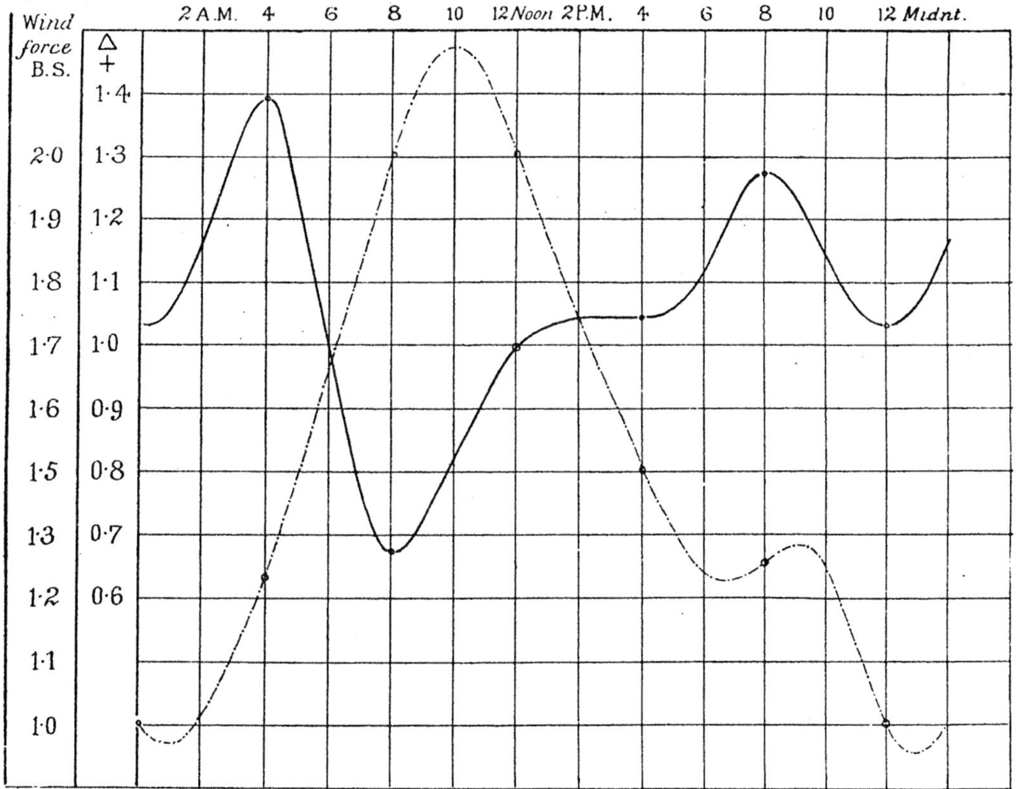

Text-fig. 55. Showing the relationship of the surface-temperature and air-temperature in the Andaman Sea in February, 1922 as compared with the wind-force.

force was 3 and at 12 midnight it was only 2·5. On all other days the force at 8 p.m. ranged from 2 to 2·5. Excluding this high reading and taking the average of the other observations the figure arrived at is 2·10. On comparing the temperature-difference and the wind-force there seems to be a distinct tendency for the two to alternate with each other: the primary maximum of the temperature-difference at 6 a.m. corresponds to the minimum wind-force at 8 a.m. and the minimum temperature-difference at 3 p.m. clearly corresponds with the maximum wind-force at 1 p.m.

The observations taken in the Andaman Sea in 1922 (*vide* Text-fig. 55) also appear to me to indicate that there is a similar relationship between the temperature-

and the major variation in the temperature-difference between 12 noon and 12 midnight. In the Laccadive Sea area in the following year, 1924, both factors exhibit a triple oscillation, of which the middle variation is by far the best marked: and on this occasion the manner in which a rise in the wind-force is accompanied by a rise in the temperature-difference is again very clearly seen (*vide* Text-fig. 57). I have previously called attention to the fact that a triple oscillation in the wind-force during the twenty-four hours of the day can be detected in Indian waters in the month of March (*vide supra*, p.87) and in the data from the Bay of Bengal area taken during this month in 1924 (Text-fig.58) this triple oscillation can be detected not only in the wind-force but also in the temperature-difference, though on the whole this latter tends to follow a single major oscillation having its maximum at 10 a.m. and its

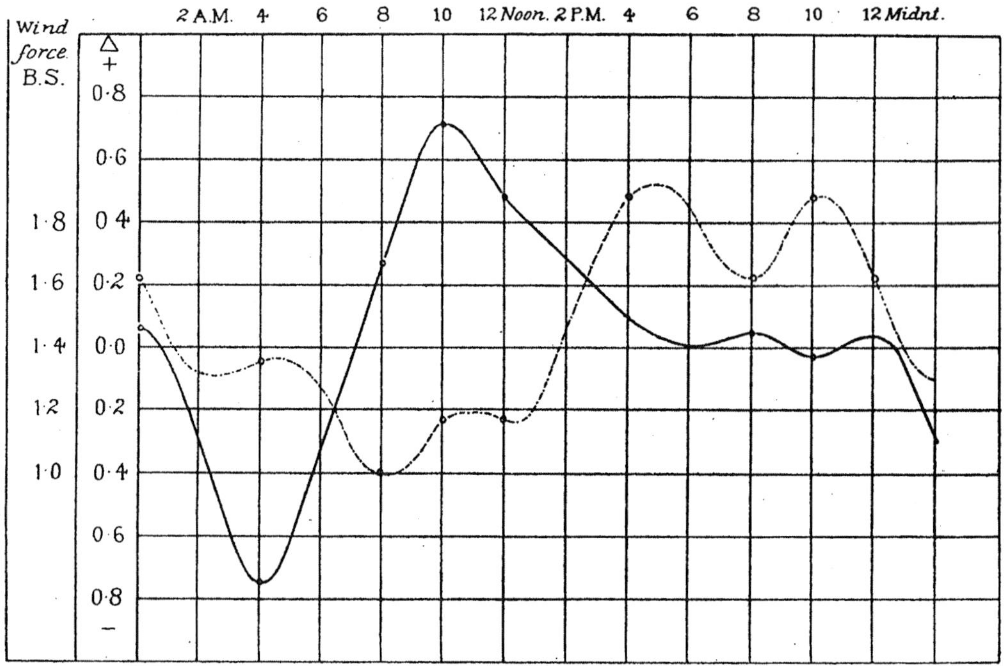

Text-fig. 58. Showing the relationship of the surface-temperature and the air-temperature in the Bay of Bengal, in March, 1924, as compared with the wind-force.

minimum at 4 a.m., while the wind-force exhibits a tendency to a quadruple oscillation. A comparison of the oscillations in the wind-force and the temperature-difference appears to indicate that the two vary in opposite directions. The minimum wind-force at 8 a.m. is followed after a "lag" of 2 hours by the maximum temperature-difference and from this point till 5 p.m. the wind-force steadily rises and the temperature-difference falls; between 5 p.m. and midnight, both series of data exhibit a small double oscillation; and finally at 4 a.m. the minimum temperature-difference is accompanied by a rise, though only a small one, in the wind-force.

April. For this month I possess four series of observations, three of which were taken in the Laccadive Sea and the fourth in the Bay of Bengal. The average results of these different series are given below in Table 49:—

		A.M.				P.M.			
		4	8	10	12	4	8	10	12
Laccadive Sea.		°C	°C	°C	°C	°C	°C	°C	°C
April, 1923, (3 days.)	Sea	28·53	29·20	29·15	29·37	29·33	29·05	28·33	28·33
	Air ..	27·69	28·48	28·60	29·40	29·21	28·11	28·06	27·87
	Diff. ..	0·84	0·72	0·55	–0·03	0·12	0·94	0·27	0·46
	Wind-force B.S.	*4·67*	*3·67*	*4·00*	*4·33*	*4·50*	*5·00*	*5·25*	*4·33*
April, 1924, (7 days.)	Sea ..	28·86	29·64	29·77	30·20	30·26	29·86	28·86	28·86
	Air ..	28·75	29·37	29·55	29·77	29·64	29·57	29·37	29·37
	Diff. ..	0·11	0·27	0·22	0·43	0·62	0·29	–0·51	–0·51
	Wind-force B.S.	*2·00*	*1·53*	*1·75*	*1·71*	*2·29*	*1·87*	*1·81*	*2·08*
April, 1925, (3 days.)	Sea ..	27·75	29·42	..	29·65	29·72	29·62	..	28·87
	Air ..	28·89	29·16	..	29·14	29·83	29·65	..	29·72
	Diff. ..	–1·14	0·26	..	0·51	–0·11	–0·03	..	–0·85
	Wind-force B.S.	*3·00*	*3·00*	..	*3·62*	*3·37*	*3·50*	..	*4·25*
Bay of Bengal.									
April, 1925, (5½ days.)	Sea ..	28·42	29·32	..	30·03	30·02	29·58	..	28·00
	Air ..	28·11	28·73	..	29·65	29·91	29·32	..	28·94
	Diff. ..	0·31	0·59	..	0·38	0·11	0·26	..	–0·94
	Wind-force B.S.	*2·00*	*1·67*	..	*2·42*	*1·83*	*2·10*	..	*2·50*

Table 49; giving the average of observations on the surface-temperature, air-temperature and wind-force at different times of the day in various parts of the Indian seas in the month of *April*.

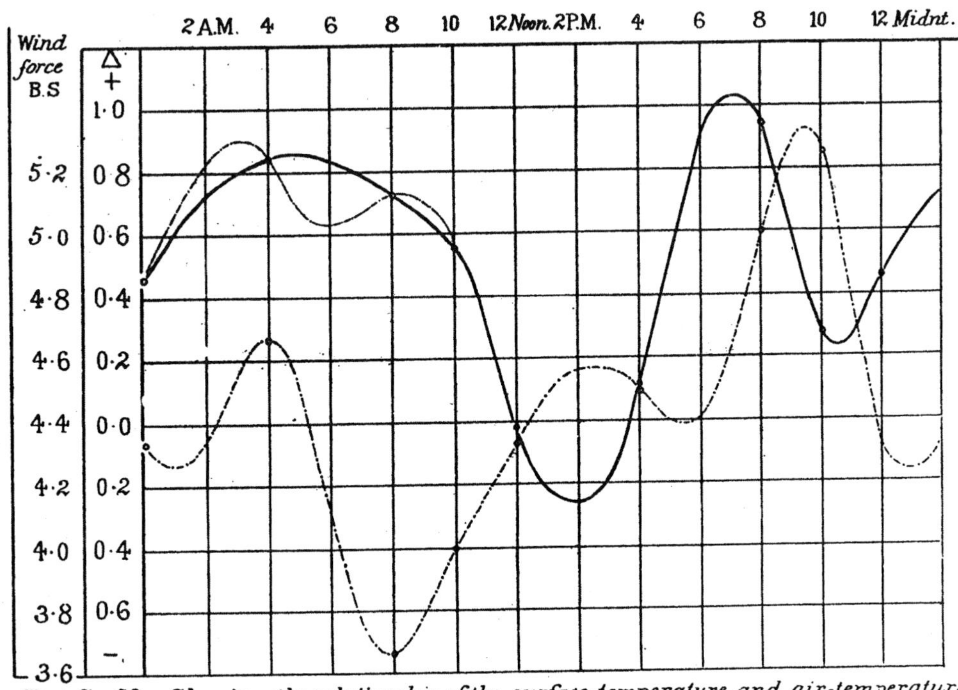

Text-fig. 59 Showing the relationship of the surface-temperature and air-temperature in the Laccadive Sea in April, 1923, as compared with the wind-force

Of these four series of observations, those taken in the Laccadive Sea in 1923 and 1924 (*vide* Text-figs. 59 and 60) at first sight appear to be atypical, in that the variation during the day in the wind-force exhibits a triple oscillation whereas the variation in the temperature-difference is a double one. In the first series taken in 1923, this apparent double oscillation of the temperature-difference is almost certainly due to the fact that observations which are taken at 4-hourly intervals may be too widely spaced to give a record of slight and temporary changes. In Text-fig. 59 I have shown by a continuous line the oscillation of the temperature-difference as given by the actual readings and by a dotted line I have indicated the probable oscillation. A comparison of this latter with the variation of the wind-force suggests that both exhibit a triple oscillation and, furthermore, it seems more than probable that the two sets of oscillations are related to each other and vary in opposite

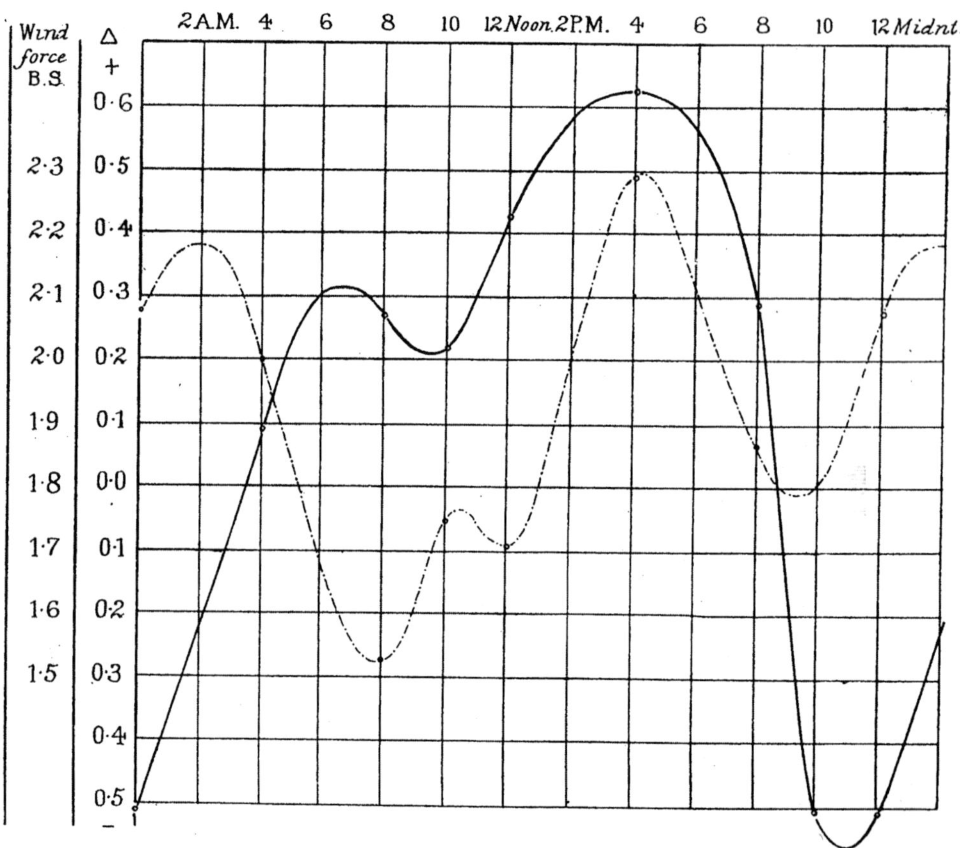

Text-fig. 60. Showing the relationship of the surface-temperature and air-temperature in the Laccadive Sea, in April, 1924, as compared with the wind-force.

directions, a fall in the wind-force being associated, after a "lag" of some 1-2 hours, with a rise in the temperature-difference. In the 1924 series, we again seem to get an indication of the same type of relationship, though in this case it is not quite so clear. The wind-force and the temperature-difference appear to oscillate in opposite directions, a rise in the wind-force being accompanied by a fall in the temperature-difference; and this is clearly indicated during the early part of the day from 2 a.m. to 12 noon, but between noon and 9 p.m. we get a rise and fall in the wind-force that is unaccompanied by any corresponding change in the temperature-difference. In the Laccadive Sea in the month of April, 1925 (Text-fig. 61) the conditions present and the

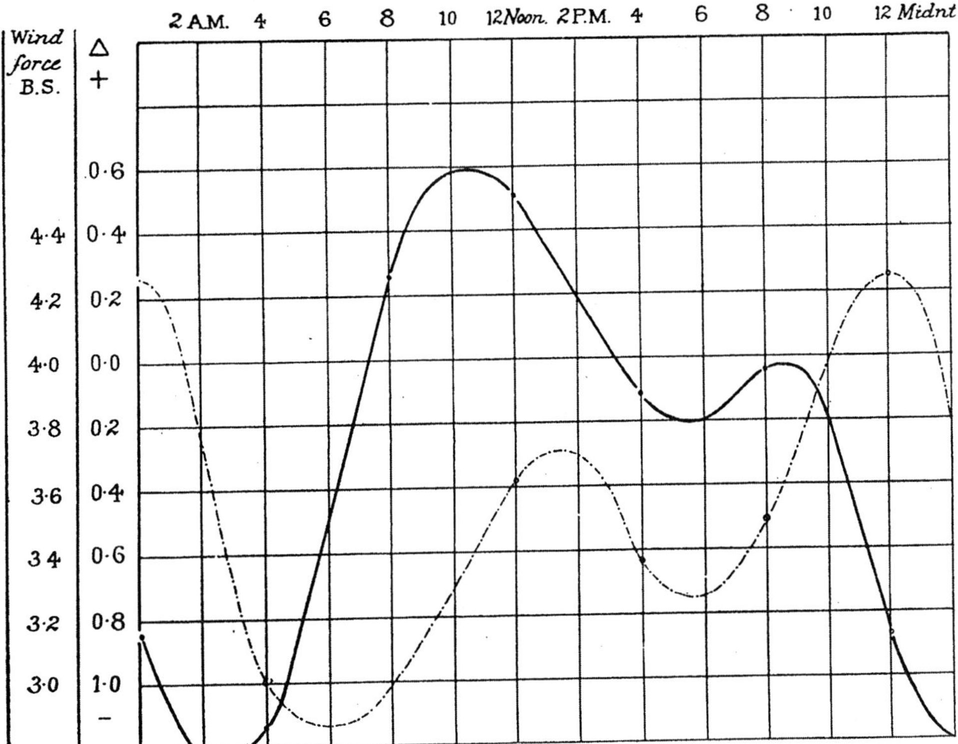

Text-fig. 61. Showing the relationship of the surface-temperature and the air-temperature in the Laccadive Sea in April, 1925, as compared with the wind-force.

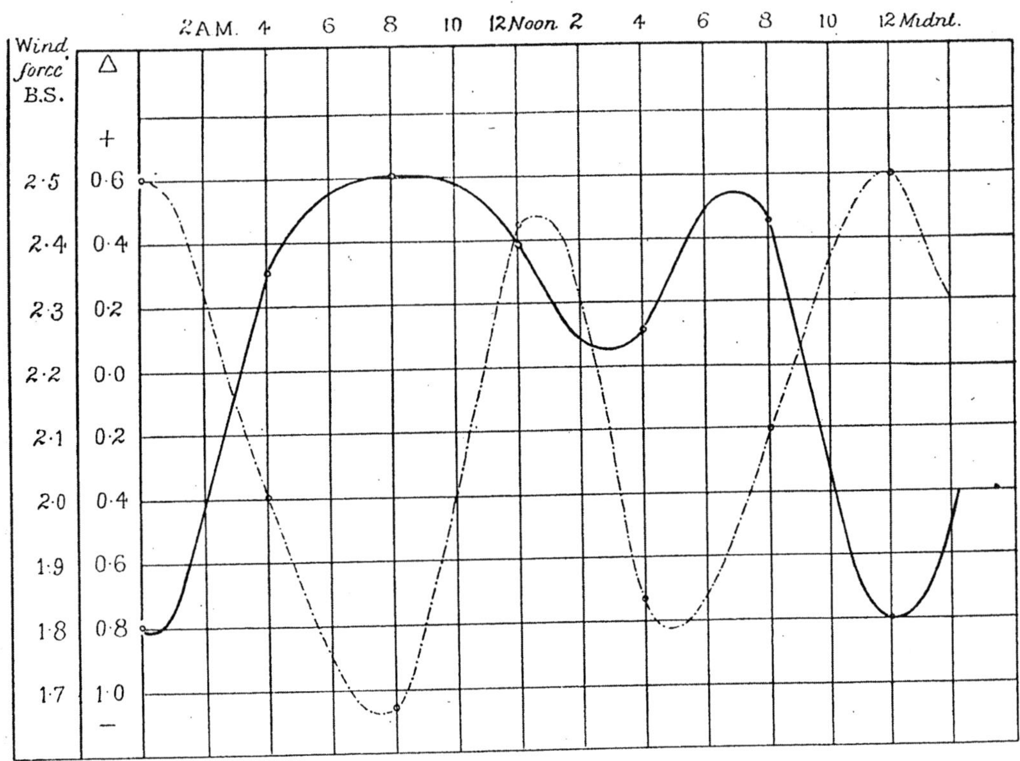

Text-fig. 62. Showing the relationship of the surface-temperature and air-temperature in the Bay of Bengal in April, 1925, as compared with the wind-force.

oscillations of the temperature-difference and the wind-force appear on the whole to exhibit the same relationship, namely, a fall in the wind-force being accompanied by a rise in the temperature-difference and *vice versa.* In this series both changes exhibit a double oscillation, and there is a considerable "lag" amounting to 4 to 6 hours in the changes in the temperature-difference behind the variation in the wind-force; but it is clear that in both cases the primary oscillation occurs between midnight and noon, so that the maximum variation in the wind-force is followed by the maximum variation in the temperature-difference.

In the results of the observations taken in the Bay of Bengal in 1925 (*vide* Text-fig. 62) we have the clearest indication that the wind-force and temperature-difference are correlated and that in this area in the month of April the oscillation of the two occur in opposite directions: and furthermore in this series the "lag" of the change in temperature-difference behind that of the wind-force is but small.

May. In this month I possess only a single set of observations, which are given below in Table 50:—

		A.M.				P.M.			
		4	8	10	12	4	8	10	12
Laccadive Sea.		°C	°C	°C	°C	°C	°C	°C	°C
May, 1923, (3½ days.)	Sea ..	28·03	28·09	29·30	29·38	29·40	28·57	28·08	28·33
	Air ..	27·69	28·24	28·80	28·49	28·89	28·33	28·15	28·24
	Diff. ..	0·34	0·16	0·50	0·89	0·51	0·24	0·07	0·09
	Wind-force B.S.	*4·6*	*5·0*	*4·0*	*4·7*	*4·5*	*5·0*	*5·3*	*5·0*

Table 50; giving the average of observations on the surface temperature, air temperature and wind-force at different times of the day in the Laccadive Sea in the month of *May*.

In this series there is again a very clear relationship between the temperature-difference and the wind-force; throughout the whole twenty-four hours the oscillations of the two series of observations coincide in a particularly striking manner. Invariably a rise in the wind-force is accompanied by a fall in the temperature-difference (*vide* Text-fig. 63), and the extent of the variations agree remarkably closely.

A survey of all these various results in the different months of the survey-season, from October to May inclusive, appears to me to indicate that there is throughout this whole period distinct evidence of correlation between the wind-force and the temperature-difference, and that the relationship differs at different periods of the year. During the early part of the season, from October to January or sometimes even later, a rise in the wind-force is accompanied by a rise in the temperature-difference, but during subsequent months this relationship becomes changed and in April and May we get, as a rule, a rise in the wind-force accompanied by a fall in the temperature-difference. The actual time of year at which the change from the first to the second type takes place probably varies from year to year, and may even vary in different localities; thus it could be detected in the month of February in 1922 in both the Andaman Sea and Laccadive Sea regions, but it appears to become fully established only at a later date, usually in the month of April. It also seems probable that the degree of variation in the temperature-difference is related to the extent of variation in the strength of the wind. As a rule this oscillation, of both temperature-difference and wind-force, is a double one during the twenty-four hours, but in certain months this may, as we have seen, be replaced by a triple or even a quadruple variation. That this connection between the temperature-difference and the wind-force is not a simple one is indicated by the facts (1) that there is during the year a complete reversal of the relationship, and (2) that whereas in some cases there is a "lag" in the changes of the temperature-difference behind the corresponding variation in the strength of the wind, in other instances the temperature-difference appears to anticipate the change in the wind-force. Although the variation in the strength of the wind is a simple one, involving changes in only one factor, a change in the temperature-

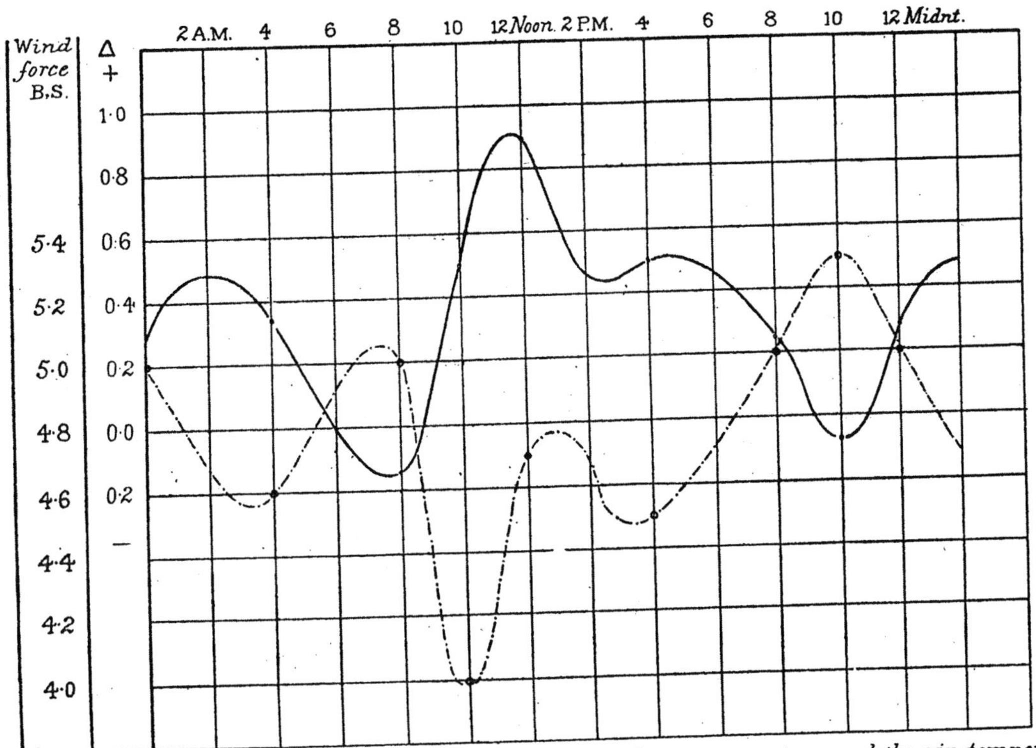

Text-fig. 63. *Showing the relationship of the surface-temperature and the air-temperature in the Laccadive Sea in May, 1923, as compared with the wind-force.*

difference between the surface of the sea and the supernatant atmosphere may be brought about by alterations in the temperature of either medium. As we have already seen, in Indian waters the surface-temperature is normally higher than the air temperature at all times of the day, and a fall in the temperature-difference may be due to either (1) a fall in the temperature of the surface-water, (2) a rise in the air temperature or (3) to unequal changes in both sea and air; and it is in consequence a matter of extreme difficulty to arrive at any definite conclusion regarding the observed changes. There seems little doubt that in the altered relationship of the strength of the wind and the difference in the temperature of the air and of the sea-surface we are dealing with the results of more than one phenomenon. The first factor that occurs to one is the cooling effect of the wind on the surface of the ocean. Eliot (1898. p. 174) has shown that during certain parts of the year there is at Trivandrum a double diurnal oscillation in the rate of evaporation of salt-water and it is more than probable that the same holds good over the open water of Indian seas. The times of maximum and minimum epochs of the rate of evaporation in the open differ in certain months of the year, the average times, as given by Eliot (*loc. cit.* Table ccx.), being as follows:—

MONTH.	Minimum.	Maximum.	Minimum.	Maximum.
October-November	8-0 a.m.	2-0 p.m.	10-0 p.m.	11-30 p.m.
December–February	9-0 „	3-0 „	10-0 „	12-0 Midnt.
March–May	8-0 „	3-0 „	10-0 „	11-0 p.m.

Eliot has also shown that an increased rate of evaporation causes a marked lowering of the temperature of the sea-water (*loc. cit.*, p. 164) and if this occurs, as I think one may assume, over the open waters in a manner similar to that found to be present at Trivandrum, it would account for a lowering of the temperature of the sea-water and would thus cause a diminution in the temperature-difference between the sea and the air at or near the times given above, namely at 2-0 to 3-0 p.m. and again at 11-0 p.m. daily in each month of the Survey season, *viz.*, during the months of October to May.

A study of the times of day at which the minimum temperature-differences exist certainly indicates that there is a distinct tendency for them to occur at two periods of the day more frequently than at other times. I give below the actual number of instances in which in the course of my observations the minimal difference has been found to occur in the course of the day at the times stated and, as my observations are but few, I have divided the day into intervals of four hours:

	A.M.			P.M.		
Time of day.	12-4	4-8	8-12	12-4	4-8	8-12
No. of instances of occurrence of minimum temperature-difference.	12	5	6	10	6	9

If now we plot these figures in the form of a curve, it is clear that the minimum temperature-difference tends to occur most frequently at 1 a.m. and 2 p.m., that is, at times which agree fairly closely with those given by Eliot, and from this it seems clear that one of the factors involved in the double diurnal oscillation of the temperature-difference is the rate of evaporation that is going on from the surface of the sea, increased evaporation causing a lowered temperature-difference (*vide* Text fig. 64).

The rate of evaporation from the sea surface depends on several factors, namely:—

(1) The temperature of the air and sea,
(2) The humidity of the atmosphere,
(3) The amount of barometric pressure, and
(4) The velocity of the wind

and it is probable that of these the velocity of the wind is the most potent factor. In the coastal region, as I have already pointed out (*vide supra*, p. 79 *et seq*), there is in most areas a quite distinct double diurnal oscillation in the strength of the wind due to the alternating influence of the land and sea breezes, and over the open sea we also find evidence of a similar double oscillation that appears to be dependent on the rise and fall of the barometric pressure. It is not surprising, therefore, that one can detect a double oscillation in the relationship of the temperature of the sea and air. If, however, this were the only factor involved and the oscillation were due solely to increased evaporation, we should expect to find that the type of relationship between the strength of the wind and the temperature-difference would be uniform throughout the year, but we have already seen that this is by no means the case. In the following Table 51 I have given the average temperature-difference and the strength of the wind, as well as the average range of each, during the months of the survey season.

Month.		Average wind-force B.S.	Average temperature-difference Sea-Air. °C	Average range of wind-force.	Average diurnal range of temperature-difference. °C
October	..	1·77	0·812	0·56	1·22
November	..	2·73	1·237	0·53	1·53
December	..	2·56	0·514	0·48	1·51
January	..	2·54	0·426	0·49	0·88
February	..	2·17	0·874	0·61	0·69
March	..	1·79	0·485	0·59	1·38
April	..	1·79	0·184	0·45	1·32
May	..	3·40	0·292	0·69	1·05

Table 51; giving the average temperature-difference and wind-force, and the daily range in different months of the survey season.

There appears to be little or no correlation between the temperature-difference and the wind-force. The average wind-force exhibits a double seasonal oscillation. In October the wind-force is low; it exhibits a first maximum in November and then falls somewhat irregularly till March and April, after which it rises rapidly in May. The temperature-difference on the other hand rises rapidly from October to November and then falls sharply in December and January; it rises again to a second maximum in February and then falls again in March and April, and rises once more slightly in May. There is thus a not very clearly marked tendency for the curves of wind-force and temperature-difference to alternate with each other though, on the whole, it seems probable that the temperature-difference tends to be greatest when the wind-force is least and is least when the wind-force is greatest. The relationship between the temperature-difference and the average daily range of the wind-force is, however, much more clearly shown; and both temperature-difference and range of wind-force vary in the same direction throughout the whole Survey season. They both are high in November, fall in December and January, rise again in February, fall till April and then rise again in May. This appears to be due, at least in part, to the cooling effect of the wind on the surface of the water: with a strong wind there is increased evaporation and a consequent decrease in the surface-temperature, and in the absence of any wind the sea-surface becomes heated more than the overlying atmosphere, and, therefore, the temperature-difference tends at such times to be high.

There appears to be little evidence of any direct relationship between the average daily variation in the strength of the wind and the average daily range of the temperature-difference, such as one would expect were the changes in the temperature-difference due solely to the action of the wind-force. The range of the temperature-difference exhibits a clear double oscillation during the survey season: it rises rapidly from October to November, remains high in December and then falls rapidly till February, after which it rises again in March and April and falls in May; on the other hand the average diurnal range of the wind-force is high in October, falls rapidly in November and then steadily increases again till May.

It is, however, also possible that the changes in the temperature-difference are due less to the actual variation in the strength of the wind than to the cooling effect of rainfall during the north-east monsoon months of December and January and to occasional storms that precede the south-west monsoon in April and May. I have previously called attention to the fact that in India there seems to be clear evidence of a double diurnal oscillation in the rainfall (*vide supra*, p. 99, *et seq*). In Calcutta rain-storms are most frequent at or near 2–6 a.m. and 6–9 p.m. and least frequent at 12–2 a.m. and 12 noon. The effect of such storms is to cause a marked fall in the air-temperature and a corresponding but much less fall in the sea-temperature.

Sea-temperature.			Air-temperature.		
Before rain. °C	After rain. °C	Diff. °C	Before rain. °C	After rain. °C	Diff. °C
28·0	26·5	1·50	26·66	24·94	1·72
29·33	29·00	0·33	27·50	26·56	0·94
29·17	29·00	0·17	28·72	27·78	0·94
27·78	26·94	0·84	27·50	25·00	2·50
28·5	29·0	−0·50	28·61	26·39	2·22
29·3	29·5	−0·20	28·89	27·22	1·67
	Average	0·36			1·67

Table 52; showing the effect of rain-storms on sea- and air-temperature.

In Table 52 I have given the data of six observations regarding the effect of rain-storms on the temperature and it is clear that the air is cooled by such to a much greater extent than the sea. A rain-storm will, therefore, cause a marked increase in the temperature-difference and we should, if these storms are responsible for causing the double oscillation in the temperature-difference, expect to find at least some evidence of agreement between the usual times of occurrence of the epochs of maximum temperature-difference and the frequency of rain-storms. I give below the frequency of occurrence of the maximal temperature-difference in each 4-hourly period throughout the day.

	A.M.			P.M.		
Time of day.	12-4	4-8	8-12	12-4	4-8	8-12
No. of instances of occurrence of maximum temperature-difference.	6	12	5	6	11	10

By plotting these figures (*vide* Text-fig.64) it appears that the epoch of maximum temperature-difference tends to occur most frequently at 6 a.m. and 8 p.m., times which agree quite closely with the times at which rain-storms are most frequent.

During those months in which the daily changes in the temperature-difference and the wind-force oscillate in opposite directions, namely in April and May, it is

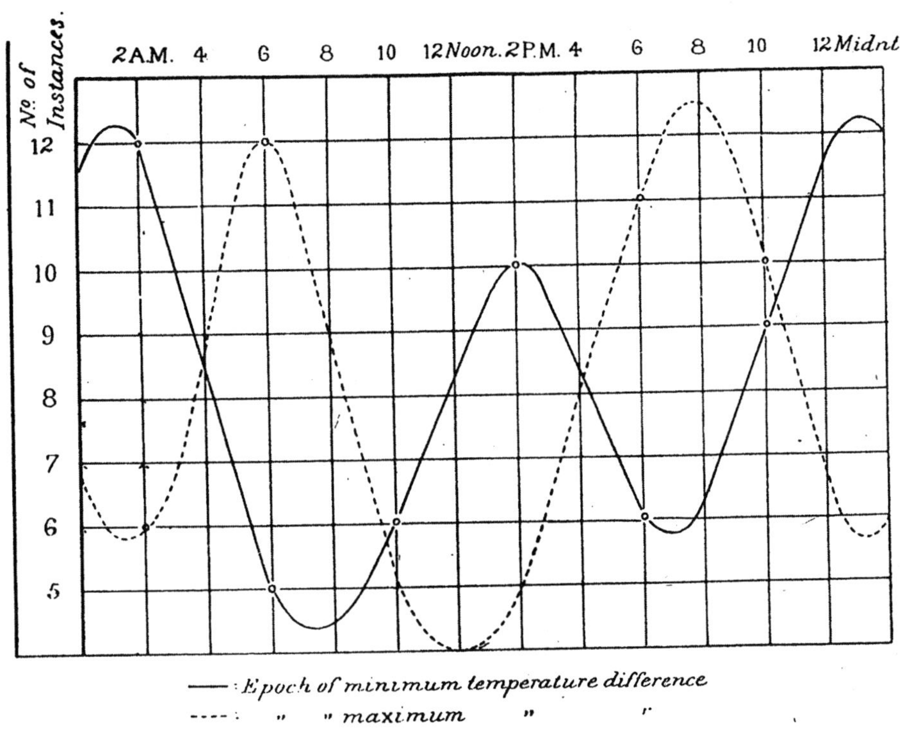

Text-fig. 64. Showing the frequency of occurrence at different times of the day of the epochs of maximum and minimum temperature-difference between the Air and the surface water.

probable that this is, at least in part, due to increased evaporation, but during the other months of the year, from October to February, the relationship is the exact opposite of this, the temperature-difference increasing with the wind-force, and it would appear that we are here faced with the effect of some other factor. I have already called attention to the fact that in Indian coastal waters, as well as in the region of the North Atlantic Ocean (*vide supra*, p. 222) the temperature-difference exhibits only a single oscillation during the course of the day, and this I believe to be the normal relationship. Whatever may be the cause of the difference between the temperature relationship of (1) the sea and air in Indian inshore waters, which as we have seen agrees with the observations made on the

"Challenger" in the North Atlantic Ocean and (2) that found to be present in the open waters of the Indian seas, there can be no doubt regarding the actuality of the difference itself. If now we subtract the "normal" temperature-difference ($\triangle$) from the observed temperature-difference ($\triangle'$) between the sea and the air over the open waters of Indian seas we find that the discrepancy between the two sets of observations invariably follows a single daily oscillation, which reaches its lowest point at or near 12 midnight to 4 a.m., and its highest point at or near 12 noon to 4 p.m. In the following Table 53 I have given the observed difference, which I term the "normal," between the sea-and air-temperatures over the inshore regions ($\triangle$) and the average difference from this normal ($\triangle' - \triangle$) in each month over the open waters:—

	A.M.			P.M.			
	4	8	12	4	8	12	Range of diff.
Normal Diff. ($\triangle$) ..	1·83	0·98	0·33	0·54	1·19	1·88	
October ($\triangle' - \triangle$) ..	−1·21	+0·22	+0·38	+0·04 (?)	−0·15	−1·18	1·59
November „ ..	−1·27	+0·18	+1·00	+1·01	−0·06	−0·53	2·28
December „ ..	−1·06	−0·32	−0·03	−0·14	−0·55	−1·49	1·46
January „ ..	−1·32	−0·43	+0·23	0·00	+0·03	−1·24	1·55
February „ ..	−0·67	+0·30	+0·94	+0·62	+0·30	−0·48	1·61
March „ ..	−1·78	−0·57	+0·21	+0·54	−0·53	−1·55	2·32
April „ ..	−1·75	−0·55	+0·01	−0·27	−0·85	−2·38	2·39
May „ ..	−1·49	−1·14	+0·56	−0·03	−0·95	−1·79	2·35
Average ($\triangle' - \triangle$) ..	−1·32	−0·29	+0·41	+0·22	−0·34	−1·33	

Table 53; showing the discrepancy in each month of the survey season between the normal temperature-difference and the observed temperature-difference over the open waters of Indian seas ($\triangle' - \triangle$).

In Text-fig. 65 I have shown graphically this discrepancy between the normal temperature-difference and the observed temperature-difference in each month, and it seems clear that there is some factor that influences the temperature-difference between the sea and the air over the open waters and causes a marked alteration of this difference during the night hours. It is not always an easy matter when considering the relationship of two series of data to decide whether any change in the normal relationship is due to abnormalities in the one series or the other, or even to changes in both; but in the present case it appears almost certain that any variation from the normal must be due to some factor that is influencing the sea-temperature. If this be admitted, then it appears that there is in these open waters of the Indian seas some agent that causes a lowering of the temperature of the surface-water below that to which it would normally fall by reason of the usual night cooling, and it seems probable that we have here evidence of the effect of the "change over" that takes place between the surface-water and the water of a deeper stratum as a result of the evaporation and consequence increase in salinity of the surface-water during the day. The vertical currents set up in the upper levels of the sea, as a result of the increased salinity due to evaporation during the day, and the rise in specific-gravity caused by cooling during the night, will result in the appearance on the surface during the night or early hours of the morning of water that has come from some depth below the surface and therefore may possess a somewhat lower temperature than the actual surface water, thus producing the lowering of the temperature-difference between the sea and the air. If this be the cause of the variation from the normal that we have

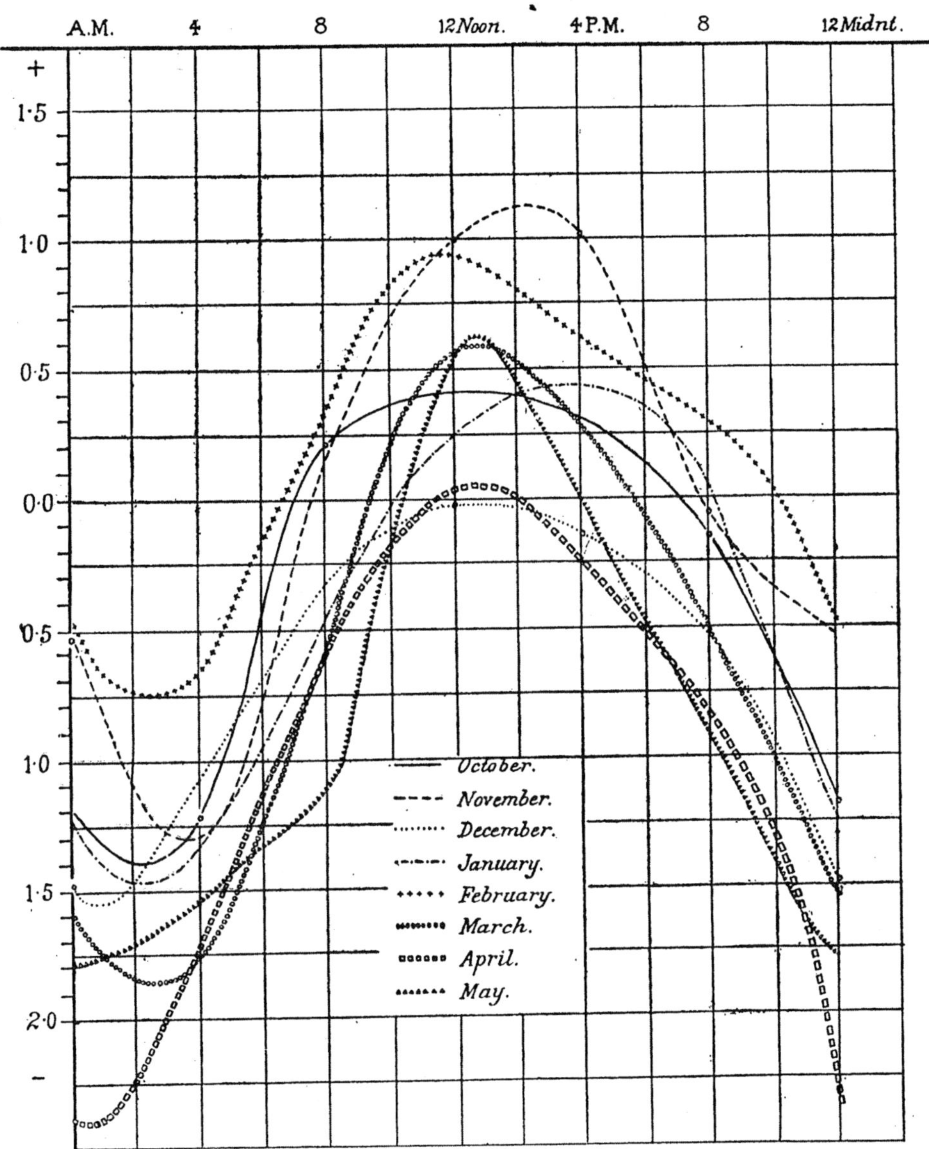

Text-fig. 65. Showing the average difference in each month of the observed temperature difference over the open sea in Indian waters from the normal temperature difference. (Δ-Δ′).

been considering, we should expect to find that this variation exhibited a seasonal change in accordance with the conditions present from month to month in the water of the surface layer; in those months in which the weather is fine and the surface temperature is high the amount of evaporation will be great and the strength of the convection current also will be proportionately high and, furthermore, in those months in which the range of temperature of the sea and air is large the increase in the specific-gravity of the surface-water caused by great cooling will be increased and consequently the extent of the convection current will also be increased. We have already seen (*vide supra*, p. 207) that the average temperature of the sea and air rises from October to November, then falls to January or February, rises again till April and finally falls in May; we have also seen that the average range of the surface-temperature increases from October to November, falls rapidly to December, rises again till April and finally falls in May. In Text-fig. 66 I have shown the average range of the surface-temperature in each month of the survey-season and the extent of the

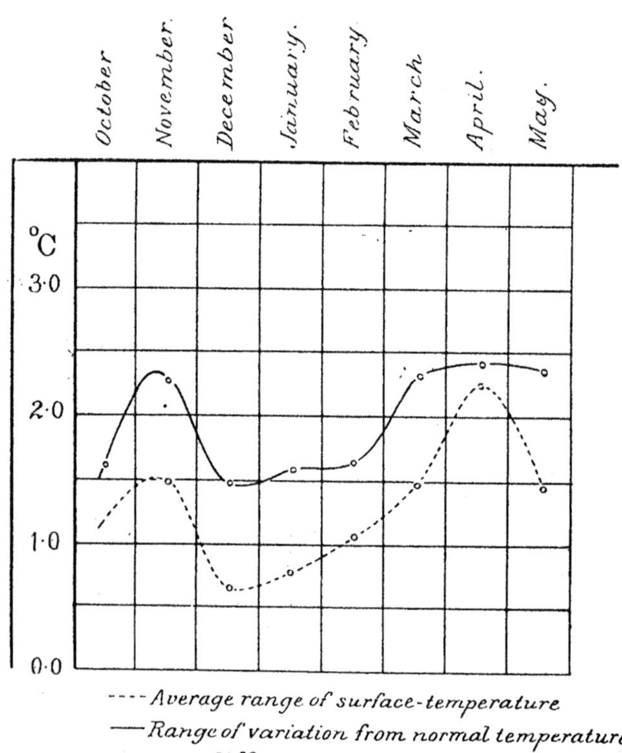

Text-fig. 66. Showing the average daily range of surface-temperature and the range of variation from the normal temperature difference. (Sea - Air) in each month.

variation from the normal of the observed temperature-difference between the sea and the air and the manner in which these two vary synchronously and in the same direction leaves little room for doubt that the one is connected with the other. The double effect then of (1) the daily fall and rise in the temperature-difference between the sea and the air under the influence of the sun's heating in the day and the subsequent night cooling and (2) the cooling of the surface water by the upwelling of deeper water from below under the influence of convection currents, is sufficient of itself to produce a double oscillation in the temperature-difference between the sea-surface and the supernatant air during the 24 hours.

If, now we take the results obtained in each month and compare these with the average of the whole series of calculations, we arrive at a figure which we may take to be a variation from the normal effect of the convection current on the daily heating and cooling of the sea and air by means of the sun's action. In Table 54 I have given the average discrepancy during the whole survey-season between the difference of sea- and air-temperatures as found in inshore waters and as observed over the open ocean (△′−△) and the variation in each month from this average. I have already pointed out that this average difference between the data derived from observations in the two areas *i.e.*, the coastal waters and the open sea, is probably due to the effect of the upwelling of water from below under the influence of convection currents; and the variation in each month from this average figure will be an indication of some other influence that is affecting this relationship between the increased density of the surface-water and the consequent convection current.

	A.M.			P.M.		
	4	8	12	4	8	12
Average diff. of normal and observed temperature-difference in open waters (△′−△)	−1·32	−0·29	+0·41	+0·22	−0·34	−1·33
October	+0·11	+0·51	−0·03	−0·18	+0·19	+0·15
November	+0·05	+0·47	+0·59	+0·79	+0·28	+0·83
December	+0·26	−0·03	−0·44	−0·36	−0·21	−0·16
January	0·00	−0·14	−0·18	−0·22	+0·37	+0·09
February	+0·65	+0·59	+0·53	+0·40	+0·64	+0·75
March	−0·46	−0·28	−0·20	+0·32	−0·19	−0·22
April	−0·43	−0·26	−0·40	−0·49	−0·51	−1·05
May	−0·17	−0·85	+0·15	−0·25	−0·61	−0·46

Table 54; showing the average discrepancy between the normal and the observed temperature-difference and the monthly variation from this average.

The average wind-force, given in the Beaufort Scale, as deduced from observations taken simultaneously with the temperature observations at four-hourly intervals in the different months is as follows:—

	A.M.			P.M.		
	4	8	12	4	8	12
October	1·98	1·50	1·65	1·95	1·85	1·61
November	2·31	1·43	1·68	2·18	1·99	2·12
December	3·45	3·31	3·25	3·21	3·19	3·37
January	2·47	2·73	2·64	2·43	2·89	2·68
February	1·99	1·89	2·35	2·28	2·73	1·97
March	1·32	1·67	1·79	2·09	1·86	1·61
April	2·73	2·28	2·76	2·66	2·87	3·05
May	4·60	5·00	4·70	4·50	5·00	5·00

It appears more than likely that the factor that is producing this variation from the normal temperature relationship is the wind-force. If this be so, then we should be able to trace a distinct relationship between the strength of the wind and the degree of variation from the normal; in Text-fig. 67 I have shown graphically the variation from the normal and the average wind-force at four-hourly intervals throughout the day in each month of the survey season.

A comparison of these with each other clearly reveals the fact that there is a very close relationship between these two phenomena, though the actual relationship varies with the season of the year. During the early part of the survey-season in the month of October the variation from the average clearly varies in the opposite direction to the wind-force, a rise in the strength of the wind being accompanied by a fall in the variation from the normal and a very similar result is seen in the month of November, though in this case there is a marked advance in the change of the temperature curve in front of that of the wind-force and especially so during the afternoon change. In the following month of December the relationship of the two factors has completely altered and a rise in the wind-force is now accompanied by a rise in the temperature variation, both exhibiting a single oscillation only in the twenty-four hours; during the next two months there is on the whole the same tendency, namely for the wind-force and the temperature-variation to coincide, though the agreement is not so clear as in the month of December; but in the month of March we again get a clear correspondence of the two, a rise in the wind-force being clearly accompanied by a rise in the temperature-variation. In April we again get a reversal of this relationship and a return to the condition that was present in October and there is now a clear tendency for the wind-force and the temperature-variation to oscillate in opposite directions, and this condition reaches its clearest agreement in the month of May, in which the relationship of the two factors is once more exactly as it was in the month of October. There is, I think, little doubt that we have here evidence of a seasonal change; in both October and November the weather is as a rule fine, and the temperature of the sea and air is high; in the months of December to March we get the effect of the north-east monsoon and the temperature of the sea and air is considerably reduced, while the high winds and the comparatively irregular manner in which they blow will naturally tend to obscure any relationship between them and the sea-temperature. In the months of April and May we again get a return to the warm weather conditions, similar to the condition that was present in the month of October.

As Helland Hansen and Nansen (1909, p. 105 p. 98 *et seq.* and footnote) have pointed out when summing up their observations on the mid-water strata of the Norwegian Sea, an increase in the lateral movement of the surface strata of the sea will tend to stretch these strata and so to cause a diminution of their thickness; if the affected stratum be the suface-layer, any thinning of this, owing to increased movement by reason of an increased wind-force, will cause the deeper stratum to approach nearer the surface, and by the effect of the waves, the height of which will simultaneously be increased, this deeper stratum may become actually mixed with the surface-water, and so affect both the temperature and salinity of the surface and modify the normal relationship between the sea- and air-temperature. If, now, this variation in the normal

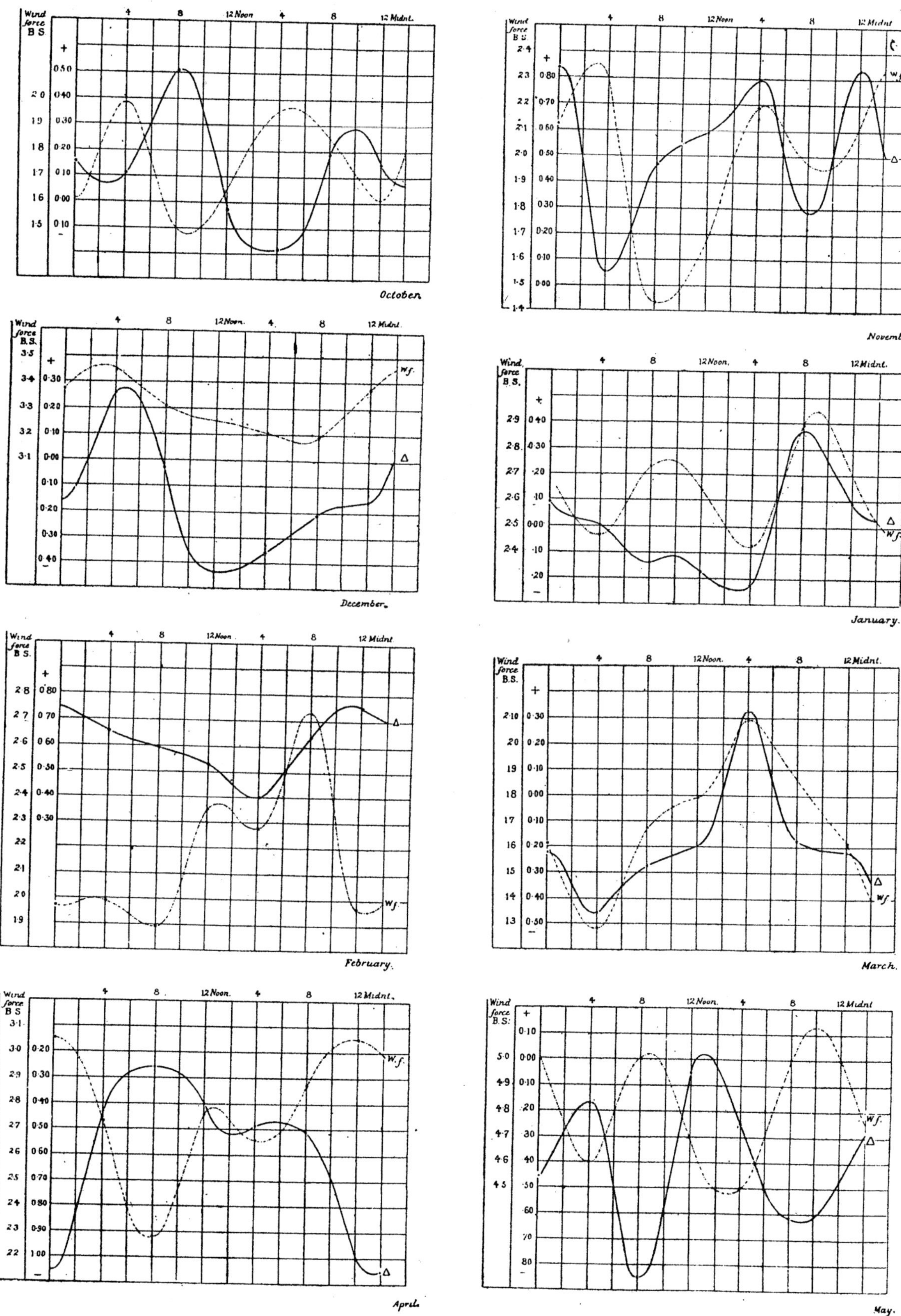

Text-fig. 67. Showing the Variation between the normal effect of convection currents and the average observed effect in different months.

relationship of the temperatures of the sea and air is correlated with and actually due to an upwelling of water from below, caused by the surface-stratum being blown away at periodic intervals, we should find that the resulting effect on the temperature difference will depend on whether the deeper stratum of water possesses a lower or a higher temperature than the surface-water. If the temperature of the lower stratum is, as it usually would be, lower than the temperature of the surface-water, any upwelling will tend to cause a lowering of the surface-temperature and so will cause a diminution in the difference between the sea- and the air-temperatures; whereas, on the other hand, if the lower stratum possesses a higher temperature, upwelling will tend to cause an increase in the temperature-difference.

Unfortunately the evidence at my disposal regarding the temperature of the water in the various upper levels of the Indian Ocean at different times of the year is somewhat scanty. It has been shown that in European waters during the summer months the surface-temperature is higher than that of the water at a depth of 10 to 30 fathoms, whereas during the winter months the surface-layers are colder than those immediately beneath (*vide* Krummel, 1907, p. 444 and 473) and one would naturally expect that a similar change might occur in Indian waters with the changing seasons. In India there are two periods of hot weather, occurring just before and just after the south-west monsoon, and correspondingly, two cold seasons during the periods of the south-west monsoon in July and August and in the winter months of January and December respectively; of these the latter is the colder. We should, therefore, expect that in October and possibly in November and again in April and May the upper layers of the water in the ocean would be warmer than those at a depth of a few metres, and that in the cold periods the exact opposite might be the case. Krummel (1907, p.474) gives a table, showing the manner in which the warmer waters of the Gulf of Aden can be traced flowing northwards over the surface of the Red Sea. This table also indicates that at certain seasons of the year the relative temperatures of the water at the surface and at 10 metres depth become reversed. The data that he gives are as follows:—

Station.		18	314	33	321	339	85
Position ..	N.	27° 24·5′	18° 3′	23° 21′	15° 51·5′	12° 41·3′	22° 4′
	E.	34° 2′	40° 14·7′	37° 37′	41° 43′	43° 15·9′	38° 0′
Date.		26.X.95	26.X.97	1.XI.95	29.X.97	2.XII.97	6.XII.95
Metres.		°C	°C	°C	°C	°C	°C
0		27·3	30·5	28·6	29·3	26·2	28·1
10		27·1	30·4	28·2	29·4	26·8	28·3
20		27·0	30·2	28·1	29·2	26·6	28·3

From the above it is clear that in October, 1895, the surface-temperature is higher than that at a depth of ten metres, whereas by the month of December this condition has been reversed and the temperature at a depth of ten metres is higher than that at the surface. That a similar state of affairs is at least possible during the hot season, in the month of April and the early part of May, in the southern part of the Bay of Bengal is indicated by the data in the Table below, which I have extracted from the List of Oceanic Depths (1910), published by the Admiralty, London.

		1	2	3	4	5	6	7	8
Position	N.	6° 26·0′	5° 59·5′	6° 08·4′	5° 55·6′	6° 01·3′	5° 49·5′	5° 39·6′	5° 48·8′
	E.	81° 51·2′	81° 42·9′	81° 39·2′	81° 35·6′	81° 32·0′	80° 54·8′	80° 46·7′	80° 35·8′
Date.		28.IV.1909	10.V.1909	10.V.1909	11.V.1909	11.V.1909	17.V.1909	18.V.1909	19.V.1909
Fathoms.		°C	°C	°C	°C	°C	°C	°C	°C
0		29·11	27·50	27·28	27·83	27·83	27·28	27·56	27·44
5		29·44	28·94	27·00	27·83	27·28	..	..	..
10		29·33	28·11	26·89	27·67	27·33	27·72	27·44	27·00
15		29·00	27·83	26·72	27·83	26·56	..	..	..
25		27·39	26·89	25·39	27·11	26·22	26·22	26·72	26·17

In this series Nos. 1, 2 and 6 reveal a higher temperature at a depth of 5 to 10 fathoms than at the surface, while No. 4 shows a fall in temperature from the 5 fathom level to the depth of 10 fathoms followed by a rise at 15 fathoms and Nos. 3, 7 and 8 indicate a steady fall throughout the series from the surface to 25 fathoms. It appears probable that at this time of the year the conditions in this area are due to a surface-current, flowing down the east coast of India from the Bay of Bengal and round the south side of Ceylon, that has a lower temperature than the water immediately subjacent to it.

It must be borne in mind that owing to the changes in temperature which the surface-water undergoes during the course of the day, these diurnal changes may mask any seasonal change, since in the middle of the day, even in the months of November to January, the surface-water may become slightly warmer than the underlying stratum. In order, therefore, to arrive at a definite conclusion regarding the seasonal changes it is necessary to take the average of a number of observations at different times of the day and night; and here again we are handicapped by the fact that night observations are much fewer than day ones and in consequence the average of all observations is likely to give too high a reading for the surface layer.

Thanks to Dr. J. Pearson, the Director of the Colombo Museum, and Mr. A. H. Malpas, Marine Biologist to the Government of Ceylon, I am able to give the results of a series of investigations regarding the temperature of the water at different levels and at different seasons of the year in the Gulf of Mannar. Unfortunately, they have given in only a few cases data regarding the levels between the surface and a depth of 50 metres; their observations are published *in extenso* in the Reports on Marine Biology in the Ceylon Administration Reports for the years concerned.

A study of the temperatures recorded in this area in the month of October, 1913, shows that the average surface-temperature is higher than that found at a depth of some 20 metres, though in three cases it is exactly the same. The actual data are given below in Table 55:

Date.		20.X.13	20.X.13	21.X.13	21.X.13	22.X.13	22.X.13	23.X.13
Position	Lat. N.	8°49′	8°21′	8°18′	8°46′	7°46′	7°57′	7°30′
	Long. E	79°15′	79°08′	78°42′	78°44′	78°37′	79°14′	79°16
		°C	°C	°C	°C	°C	°C	°C
Surface		..	27·7	27·7	26.6	26·6	27·2	27 2
18 Metres		26·6	27·7	27·2	26·6	26·6	26·6	26·6
92 ,,		24·4	23·3	23·8	23·3	23·3	23·8	23·3
183 ,,		21·1	20·5	20·5	20 5	20·5	21·6	..
Time of day ..		6-50 A.M.	3-15 P.M.	10-0 P.M.	1-0 A.M.	1-0 A.M.	2-50 A.M.	2-50 A.M.

Table 55; showing the temperature at different depths in the Gulf of Mannar in October, 1913.

During the hottest part of the day, namely about 2 p.m., we should expect to find that the surface-water is considerably warmer than that below, but this series certainly appears to indicate that in the early hours of the morning there is but little difference in the temperature of the water at the surface and at a depth of 18 metres. The average for the whole series is 27·17°C for the surface and 26·84°C for the 18-metre depth level. In the data for later years Dr. Pearson has confined his observations to the levels of the surface, 50, 100, 200 and 300 metres. A study of the temperatures at these levels in the different months of the year reveals several interesting features. In Table 56 I have reproduced the results obtained at eight stations in the month of October, 1920.

Date.		16.X.20	16.X.20	16.X.20	16.X.20	17.X.20	17.X.20	?	?
Position	Lat. N. ..	8°01′	8°09′	8°49′	8°46′	8°18	7°46′	7°30′	7°57
	Long. E...	79°08′	79°39′30″	79°58′	78°44′	78°42′	78°13′	79°14′	79°14′
		°C	°C	°C	°C	°C	°C	°C	°C
Surface	..	26·15*	26·55	27·15	26·75	26·95	26·15	26·25	27·05
50 Metres	..	26·50	26·48	25·61	23·56	25·11	23·21	24·41	24·17
100 ,,	..	23·10	22·86	22·67	21·26	24·37	19·30	21·41	22·07
200 ,,	..	15·36	16·04	16·35	14·55	16·00	15·15	14·97	15·44
300 ,,	..	12·12	..	14·94	11·28	12·58	12·19	11·92	..
Time of day	From	1-37	6-18	4-10	10 0	9-37	10-22	5-5	8-15
	To	3-20	7-15	5-25	11-9	10-40	11-37	6-0	10-10
		A.M.	A.M.	P.M.	P.M.	A.M.	P.M.	A.M.	P.M.

Table 56; showing the temperature at different depths in the Gulf of Mannar in October, 1920.

With the single exception of the first series,* in which the surface-temperature is slightly lower than that at 50 metres, due almost certainly to the fact that the observation was taken in the early hours of the morning when the temperature of the surface-water is at its lowest, the surface-temperature is throughout considerably higher than that at the lower levels. The results obtained in the same area in the month of November, 1921, are given in the Table 57:

Date.		17.XI.21		18.XI.21				19.XI.21	21.XI.21	27.XI.21		
Position	Lat. N.	7°46′	8°31′	8°08′	8°46′	8°21′	8°49′	8°49′	8°40′	7°13′	7°33′30″	8°09
	Long. E.	78°37′	79°42′30″	78°08′	78°20′	78°44′	79°08′	79°15′	70°45′	79°43′	79°43′	79°39′30″
		°C	°C	°C	°C	°C	°C	°C	°C	°C	°C	°C
Surface		28·08	27·65*	27·40*	27·60*	27·55*	27·72	28·15*	27·60*	28·05	27·66*	27·62
50 Metres		28·03	27·70	27·67	27·79	27·64	27·58	28·24	27·62	27·88	27·75	27·84
100 ,,		24·10	24·01	25·30	24·17	26·72	25·00	27·53	25·35	26·80	26·21	26·23
200 ,,		16·62	14·48	14·27	15·17	14·48	14·28	15·01	15·45	15·11	14·15	14·37
300 ,,		11·74	11·71	11·57	12·09	11·85	12·08	12·42	12·71	11·78	11·76	11·65
Time of day		0-55 P.M.	5-30 P.M.	6-25 A.M.	2-10 P.M.	5-0 P.M.	9-30 P.M.	3-15 P.M.	6-45 P.M.	9-55 A.M.	4-55 P.M.	Mid-night.

Table 57; showing the temperatures at different depths in the Gulf of Mannar in November, 1921.

In this month it is seen that in the very great majority of cases* the surface-temperature is lower than the temperature recorded at a depth of 50 metres; this is the case in no less than eight out of the eleven serial observations. That this is not due merely to the heating-up of the water during the day and its more rapid cooling on the surface at night is shown by the very different time of the day at which the observations were taken. Thus the surface-temperature was found to be lower in different series of records at 6-25 a.m. and at 12-10, 3-15, 4-55, 5-0, 5-30, 6-45 p.m. and at midnight. The average of the whole series of observations is 27·73°C at the surface and 27·79°C at a depth of 50 metres.

In the following year, 1922, however, conditions were very different. The temperatures in this year are considerably lower as regards the deeper levels than in the preceding year the average being 24·810°C as compared with 27·785°C at the 50 metre level and 20·534°C as compared with 25·584°C at the 100 metre level. The details of the observations in 1922 are given below in Table 58 and it will be observed that in consequence of this lowering of the temperature in the deeper levels the surface temperature is throughout higher than the temperature at a depth of 50 metres.

Date.	19.XI.22		20.XI.22			21.XI.22				22.XI.22	
Position { Lat. N.	7°10′	7°40′	8°00′	8°15′	8°25′	8°44′	8°40′	8°44′	8°18′	8°00′	7°48′
Position { Long. E.	79°30′	79°20′	79°30′	79°08′	79°30′	79°25′	79°00′	78°35′	78°42′	78°19′	78°51′
	°C	°C	°C	°C	°C	°C	°C	°C	°C	°C	°C
Surface	27·44	27·34	28·85	27.13	27·63	26·95	23·37	27·45	26·89	26·88	27·61
50 Metres	26·50	25·64	23·11	22·24	23·24	24·62	26·78	25·22	24·88	26·62	27·30
100 ,,	(26·25) ?	19·71	20·84	20·21	20·65	20·77	19·86	20·92	21·15	20·70	..
200 ,,	17·51	13·60	14·21	13·97	14·39	14·73	15·10	14·65	16·89	15·44	..
300 ,,	12·46	11·50	11·65	11·79	12·23	12·25	12·64	13·74	13·15	12·59	..
Time of day	3–30 P.M.	11–25 P.M.	3–45 A.M	10–5 A.M.	3–0 P.M.	2–35 P.M.	6–25 A.M.	1–10 P.M.	6–18 P.M.	0–40 A.M.	7–5 A.M.

Table 58 ; showing the temperatures at different depths in the Gulf of Mannar in November, 1921.

It seems clear then that in this month of November conditions may differ very considerably from year to year; in certain years the surface-temperature is higher than that at a depth of 50 metres, whereas in other years the reverse may be the case.

An examination of the results obtained in the month of January, 1921, shows that a very similar state of affairs existed in that month to what was found to be present in November, 1921. The results of Dr. Pearson's observations are given below in Table 59.

Date.	5. I. 21	5. I. 21	5. I. 21	5. I. 21	6. I. 21	6. I. 21	6. I. 21	7. I. 21
Position { Lat. N.	7°57′	8°21′	8°09′	8°31′	8°49′	8°46′	8°18′	7°46′
Position { Long. E.	79°14′	79°08′	79°39′30″	79°42′30″	79°15′	78°44′	78°42′	78°37
	°C	°C	°C	°C	°C	°C	°C	°C
Surface ..	25·46*	25·66	26·35*	26·06*	27·75	25·66	26·05*	25·36*
50 metres ..	26·97	25·27	26·54	26·27	25·56	25·35	26·47	25·40
100 ,, ..	23·15	23·31	20·97	22·99	23·71	22·11	23·43	22·12
200 ,, ..	14·26	14·18	14·01	16·06	17·84	15·90	17·99	15·74
300 ,, ..	12·58	12·83	11·95	..	12·67	12·03	12·53	12·52
Time of day { From ..	3-3	8-40	2-40				5-30	5-50
Time of day { To ..	4-5 A.M.	9-35 A.M.	3-45 P.M.	..	..	..	6-15 P.M.	5-45 A.M.

Table 59; showing the temperature at different depths in the Gulf of Mannar in January, 1921.

Here again one sees that in five instances (*) out of eight the surface-temperature is lower than the temperature at a depth of 50 metres and again this occurs at any time of the day, so that it does not depend on the greater cooling of the surface-water during the night. It would appear, then, that the surface-water in the Gulf of Mannar during the winter months may and, in all probability, usually does possess a lower temperature than that at a depth of 50 metres, but it would not be safe on the evidence that we at present possess to assume that this is a normal condition throughout the whole area under review. It must be remembered that during those months in which this condition has been observed in the Gulf of Mannar, the north-east monsoon wind is blowing steadily and that this causes a surface-current of water to flow from the Bay of Bengal through the Palk Straits into the northern end of the Gulf. Pearson has remarked that the effect of this current is to cause a distinct lowering of the salinity of the surface-water in that part of the Gulf that lies to the north, and it is possible that this lowering of the temperature of the surface-layer is due to the same phenomenon; but this will not account for the great difference in the actual temperatures recorded at the 50-metre level in the months of November, 1921 and 1922, and it seems more probable that this is due to a difference in the two years in the amount of heating up of the surface-water during the preceding months of September and October following on the cessation of the south-west monsoon and the gradual sinking of the more dense surface-water as the season becomes cooler, a phenomenon that, as I shall have occasion to show later, causes a clear reversal of the seasons in the deeper layers of water in the coastal area of the Indian seas.

In the months of March and April the surface-water, as one would expect, is invariably warmer than the water at a depth of 50 metres, as is shown by the data given below in Tables 60–62.

Date.		3. IV. 20	3. IV. 20	4. IV. 20	4. IV. 20	4. IV. 20	4. IV. 20	5. IV. 20
Position	Lat. N.	8°49′	8°21′	7°46′	8°46′	7°57′	8°18′	7°30′
	Long. E.	79°15′	79°08′	78°37′	78°44′	79°14′	78°42′	78°16′
		°C	°C	°C	°C	°C	°C	°C
Surface	..	29·81	28·85	29·02	29·63	30·27	29 57	28·81
50 metres	..	28·18	28·29	27·68	28·10	29·91	28·63	28·55
100 ,,	..	24·04	22·68	24·19	21·34	24·87	24·10	25·38
200 ,,	..	14·61	14·68	13 81	14·01	13·09	14·12	15·14
300 ,,	..	..	12·19	11·60	11·99	11·71	11·92	12·09
Time of day	From	6-18	11-10	3-37	9-50	5-18	10-10	2-28
	To ..	7-22	12-40	5-0	11-20	6-20	11-00	3-28
		A.M.	A.M.	A.M.	A.M.	P.M.	P.M.	A.M.

Table 60; showing the temperatures at different depths in the Gulf of Mannar in the month of April, 1920.

Date.	29. III. 21			30. III. 21				2. IV. 21			3. IV. 21
Position: Lat. N.	8°49′	8°46′	8°18′	8°21′	8°09′	7°57′	7°30′30″	7°13′	7°30′	7°46′	8°08′
Position: Long. E.	79°15′	78°44′	78°42′	79°08′	79°39′30″	79°14′	79°43′	79°45′	79°16′	78°37′	78°08′
	°C	°C	°C	°C	°C	°C	°C	°C	°C	°C	°C
Surface ..	28·37	29·12	29·35	28·65	28·83	28·45	28·95	28·95	29·90	29·52	28·98
50 metres ..	27·16	27·56	27·36	27·36	27·32	27·04	27·16	27·43	27·21	27·60	27·80
100 ,, ..	21·46	22·79	24·13	22·11	21·39	22·77	22·16	21·00	22·60	22·09	21·85
200 ,, ..	14·71	15·23	13·51	14·70	14·02	14·86	14·37	13·86	14·03	14·24	14·23
300 ,, ..	12·48	11·81	11·55	11·82	12·89	11·64	12·48	12·88	(14·63) ?	..	12·08
Time of day ..	5-0 A.M.	..	8-20 P.M.	..	6-5 A.M.	11-15 A.M.	4-27 P.M.	8-57 A.M.	2-18 P.M.	..	..

Table 61; showing the temperatures at different depths in the Gulf of Mannar in the months of March and April, 1921.

Date.	28. IV. 22			29. IV. 22				30. IV. 22			
Position: Lat. N.	8°15′	7°10′	7°40′	7°48′	8°00′	8°18′	8°40′	8°40′	8°44′	8°25′	8°00′
Position: Long. E.	79°08′	79°30′	79°20′	78°51′	78°19′	78°42′	78°35′	79°00′	79°25′	79°30′	79°30′
	°C	°C	°C	°C	°C	°C	°C	°C	°C	°C	°C
Surface ..	29·04	29·35	28·88	29·11	28·47	29·06	28·40*	28·59	29·05	28·80	29·10
50 metres ..	28·84	28·82	28·76	28·62	26·73	28·96	28·50	27·62	28·72	28·69	26·25
100 ,, ..	25·23	23·77	22·86	22·67	21·78	(26·83) ?	22·06	20·92	22·00	23·52	23·69
200 ,, ..	17·61	14·61	14·42	14·51	17·19	..	14·23	14·16	14·45	15·01	14·58
300 ,, ..	13·36	11·85	12·07	11·87	14·05	..	11·59	11·24	12·04	13·21	11·76
Time of day ..	1-14 a.m.	1-45 p.m.	6-48 p.m.	7-20 a.m.	12-10 p.m.	6-45 p.m.	10-55 p.m.	1-53 a.m.	10-11 a.m.	2-55 p.m.	8-15 p.m.

Table 62; showing the temperatures at different depths in the Gulf of Mannar in the month of April, 1922.

From the above tables it is clear that even as late as April the temperature of the surface-water may fall during the early hours of the morning to such an extent that it is but little warmer than that at 50 metres and in one case (*, 1922 at 10.55 p.m.) it was even slightly colder than the water at this depth; on the whole, however, the surface-water is decidedly warmer.

The above evidence certainly indicates that during the winter months the surface-water of Indian seas, or at any rate of certain areas, may be colder than that lying at a depth of some 20–50 metres and, furthermore, that this condition is specially likely to be present during the night and early-morning hours. An increase in the wind-force at such times may still further cool the surface by increasing the rate of evaporation that is going on, but it will simultaneously cause a thinning of the surface-layer and by an increase in the amplitude of the waves will set up an increased admixture of the surface-water and that lying at some depth below. We should, as a result of these various changes, expect to find that an increase in the wind-force may be followed by an alteration in the normal temperature relationship of the air and of the sea-surface, resulting in an increase in the temperature-difference. On the other hand, during the hot dry months of October and again in April and May an increase in the wind-force will cause a lowering of the surface-temperature and at the same time will give rise to an admixture with the surface-water of water from some depth below, that at this time of the year has a considerably lower temperature; both factors thus tend to cause a lowering of the surface-temperature and a consequent decrease of the temperature-difference between sea and air, so that in these months the wind-force and temperature-difference will tend to vary in opposite directions. A comparison of the various curves given above in Text-fig. 67 shows that the observed results agree with what I have noted above and at least indicate the possible existence of a double diurnal upwelling of water from some depth below towards, or even actually to, the surface.

The Density and Salinity of the Surface-Water of the Bay of Bengal.

On seven occasions I have been able to carry out a series of observations on the surface-waters of the mouth of the Bay of Bengal from the southern extremity of Ceylon in Longitude 80°E to Duncan Passage, between South Andaman and Little Andaman Islands in Longitude 93°E. Water-samples were taken and the temperature of the surface-water was noted every four hours: during the day these samples were taken and the observations made by me personally, while those at 12 midnight and 4 a.m. were carried out by the various officers of the watch, to whom my thanks are due. In 1914 and again in 1921 the density of the water was carefully tested by means of a 'Buchanan' Hydrometer, but in 1922 I adopted the titration method and since that year all salinities, densities and specific gravities have been calculated by means of Knudsen's tables from the halogen-content.

The various times of the year when these observations were made and the direction of the passage across the Bay were as follows:—

(*a*) 1914, April 18th to 22nd: Andamans to Ceylon.
(*b*) 1921, October 18th to 22nd: Ceylon to Andamans.
(*c*) 1922, October 9th to 13th: Ceylon to Andamans.
(*d*) 1923, January 23rd to 26th: Andamans to Ceylon.
(*e*) 1924, January 3rd to 7th: Ceylon to Andamans.
(*f*) 1924, March 4th to 8th: Andamans to Ceylon.
(*g*) 1925, April 17th to 24th: Andamans to Ceylon.

In an area, such as the Bay, that is exposed at different times of the year to such opposite influences as the south-west and north-east monsoons, and into which most of the large rivers of India pour their water, one would naturally expect to find that the conditions present in the surface-waters would show a considerable range of variation both as regards the salinity of the water and its temperature. The results obtained by me during the seven voyages I have given *in extenso* in Appendix VIII and they are shown graphically in the accompanying charts (Text-figs. 68 to 74). These results exhibit *inter se* a considerable range of variation, which I attribute largely to the different seasons of the year, in which the observations were made.

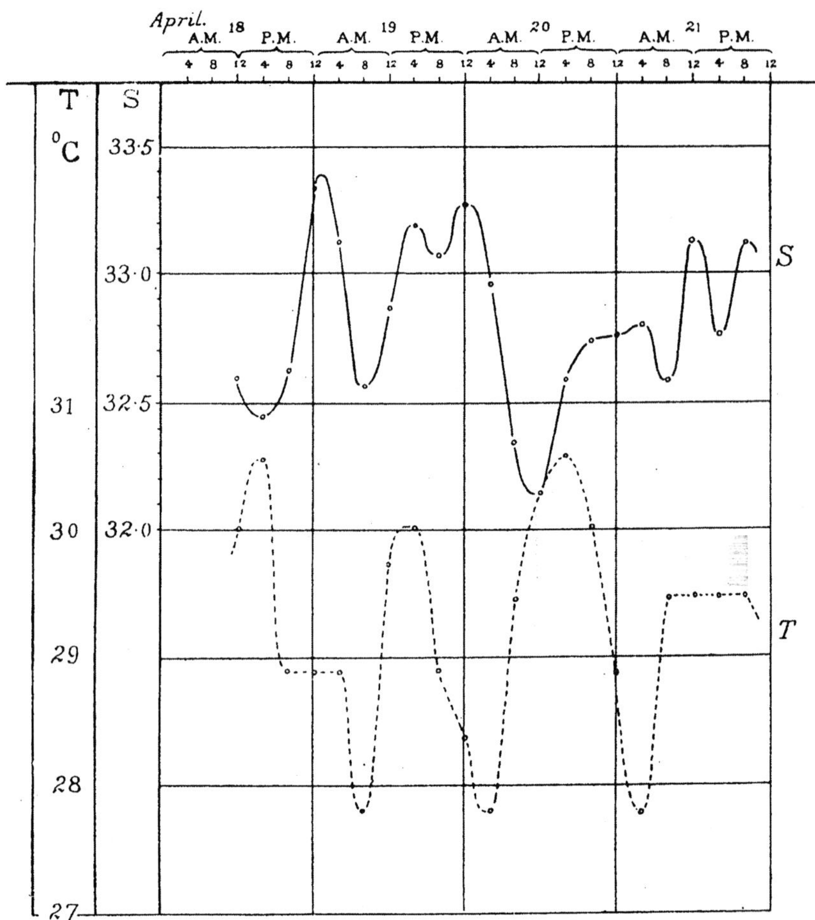

Text-fig. 68. The salinity and temperature of the surface water across the Bay of Bengal in April, 1914, from E to W.

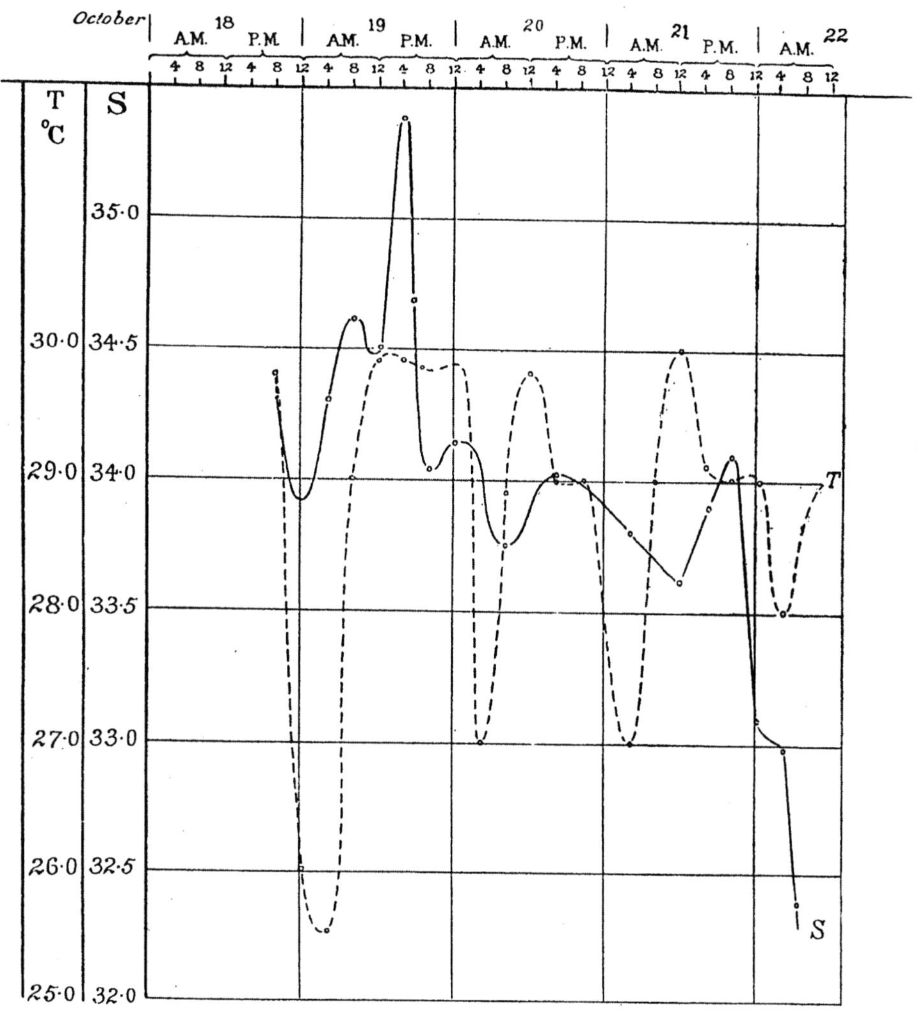

Text-fig. 69. The salinity and temperature of the surface water across the Bay of Bengal in October, 1921 from W. to E.

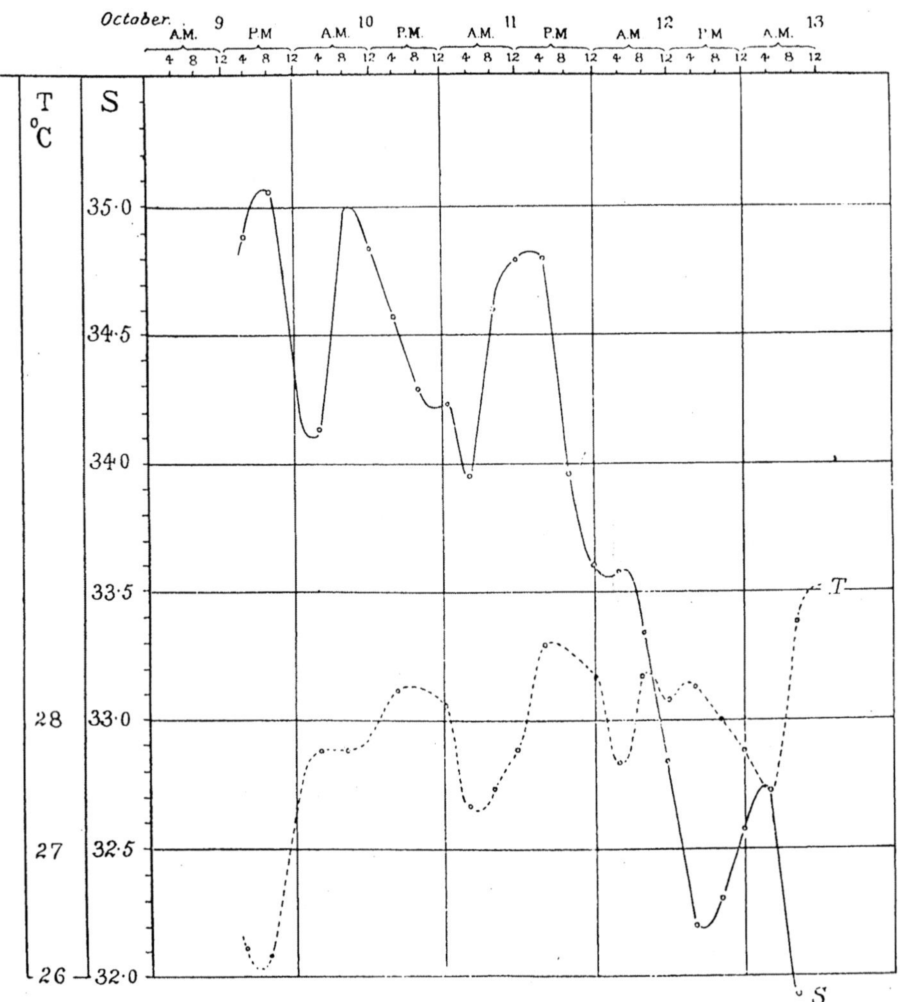

Text-fig. 70. The salinity and temperature of the surface water across the Bay of Bengal in October, 1922, from W. to E.

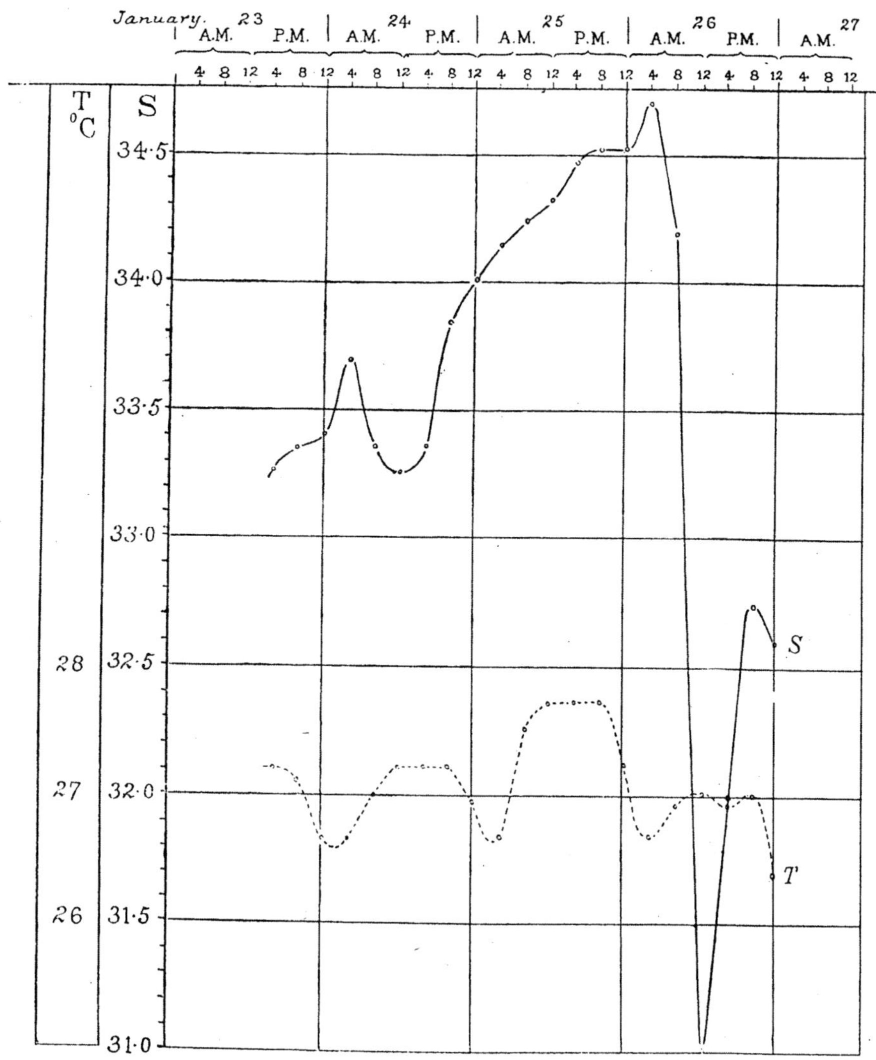

Text-fig. 71. The salinity and temperature of the surface water of the Bay of Bengal in January, 1923, from E. to W.

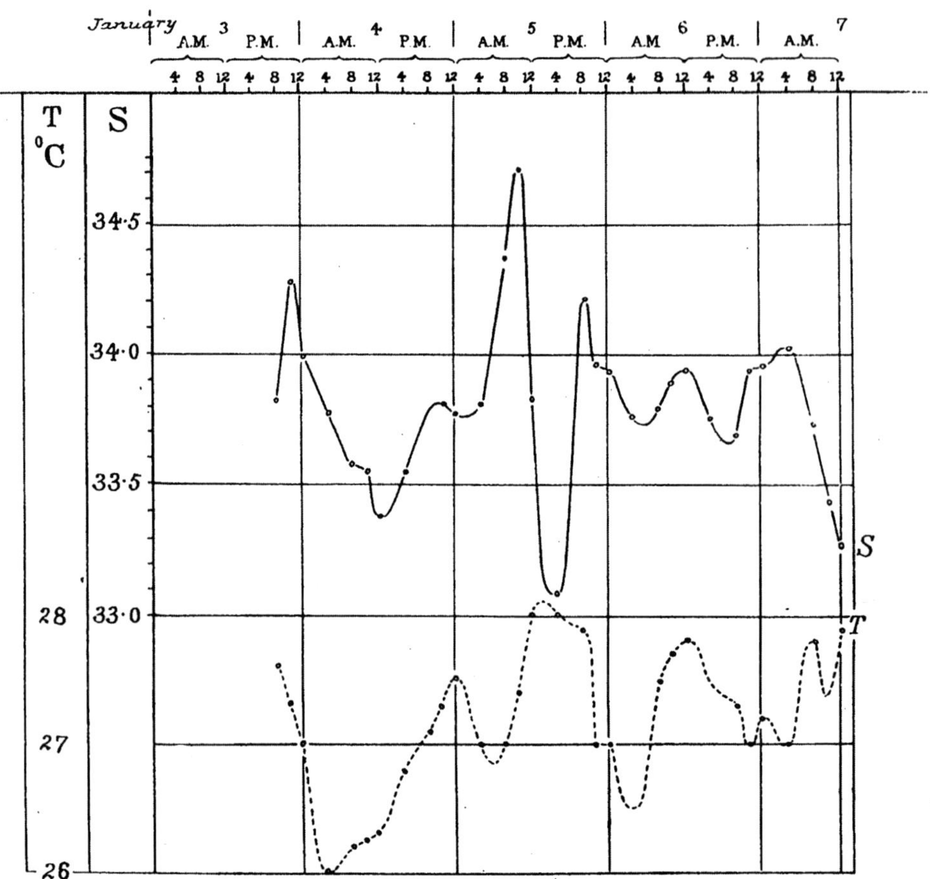

Text-fig 72 The salinity and temperature of the surface water across the Bay of Bengal in January, 1924, from W. to E

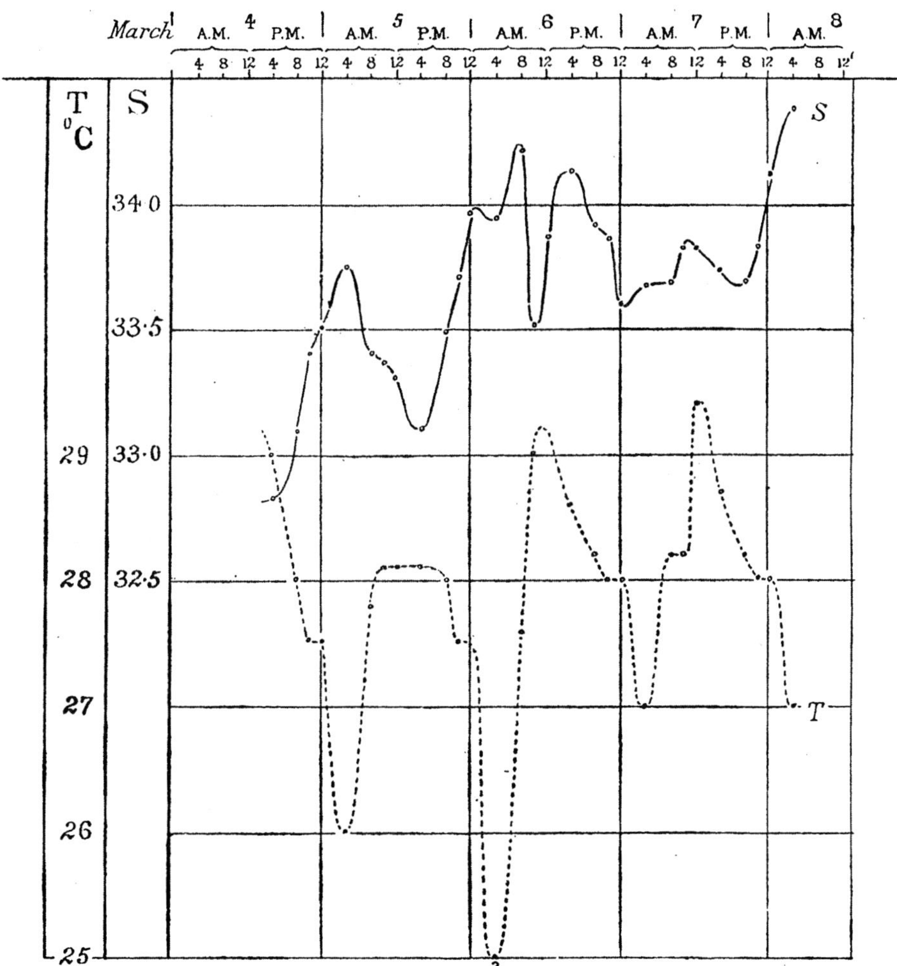

Text-fig. 73. The salinity and temperature of the surface water across the Bay of Bengal in March, 1924, from E. to W.

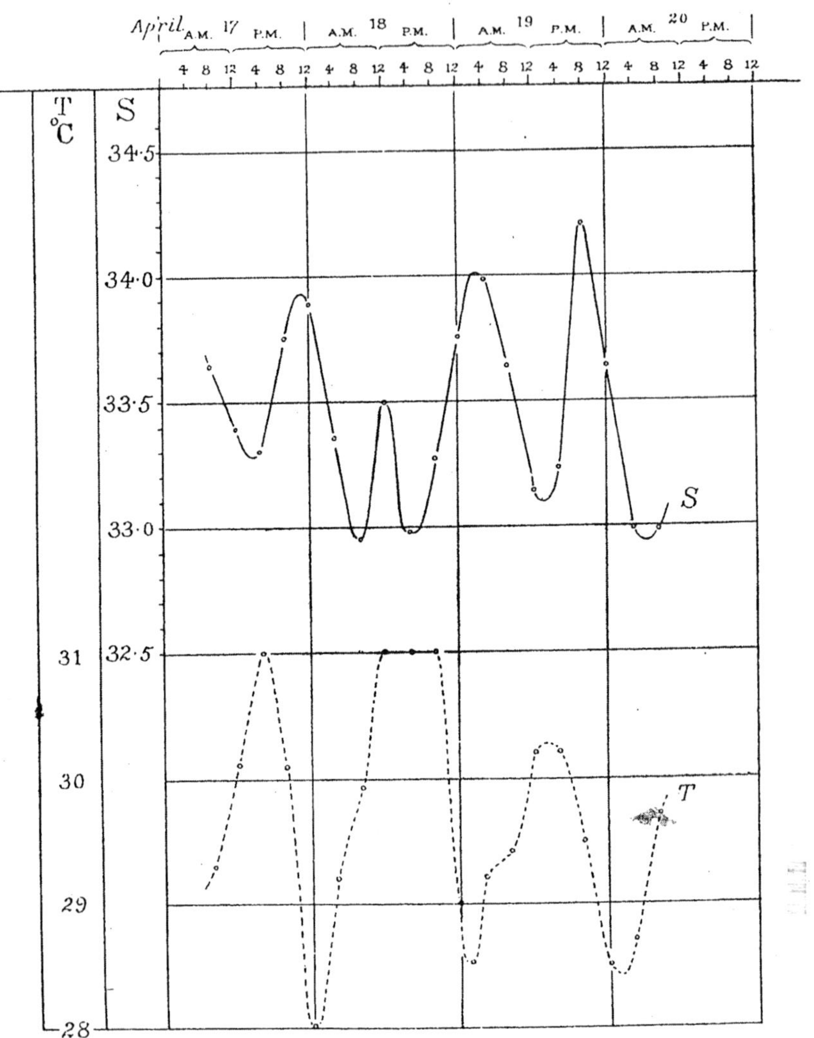

Text-fig. 74. The salinity and temperature of the surface water across the Bay of Bengal in April, 1925, from E to W.

The average specific gravity (σ_{0S}), the density *in situ* (σ_t) and the surface temperature obtained during each voyage are given in Table 63 below.

Time of year.		Specific gravity (σ_0)	Density *in situ* (σ_t)	Surface Temperature °C
October 18 to 22, 1921	..	27·265	21·383	28·69
October 9 to 13, 1922	..	26·176	21·578	27·85
January 23 to 26, 1923	..	26·970	21·649	27·07
January 3 to 7, 1924	..	27·017	21·775	27·18
March 4 to 8, 1924	..	27·024	21·411	27·93
April 18 to 22, 1914	..	26·357	20·353	29·26
April 17 to 24, 1925	..	27·175	21·097	29·28

Table 63; giving the average density, specific gravity and temperature of the surface water of the Bay of Bengal in each series of observations.

Although the number of observations is too small to enable one to draw any definite conclusions on the subject, the results obtained serve to indicate the great variation that may be found from season to season and from year to year. With the changing seasons the salinity of the surface-water increases during the later months of the year and falls in the spring; thus the average specific gravity (σ_0) in October is 26·720, in January the average has risen to 26·993, while in March, in the only year in which I was able to take observations, it was 27·024 and, finally, the average of two years in April is 26·766. The density *in situ* (σ_t) of the water also apparently increases during the cold winter months and falls again in the spring and this latter fall is much more marked than the preceding rise. The average temperature, on the other hand, falls from October to January or February and later rises again in April, thus serving to augment the effect of the change in salinity on the surface-density *in situ*. A comparison of the average salinity in the same month but in different years reveals a great range of variation, and the first two series of observation, namely those taken in the month of October in 1921 and 1922, show how greatly conditions may vary from year to year. In 1921 the average specific gravity (σ_0) of the water and the surface-temperature were both considerably higher than in 1922. The higher temperature however, more than counterbalances the higher specific gravity (σ_0) so that the average density *in situ* (σ_t) is slightly lower in 1921 than in 1922. This difference in the two years is undoubtedly correlated with the different conditions of the south-west monsoon. Owing to the geographical configuration of India the greater number of the rivers of India, including such large rivers as the Ganges and the Bramaputra, open into the Bay of Bengal either directly or through the Andaman Sea and in consequence the major portion of the rainfall over not only the Indian Peninsula but also the region lying to the north, finds its way eventually into the Bay, though this will, of necessity, take some time, depending on the length of the river. The salinity in the Bay in October will, therefore, be influenced to a large extent by the rainfall of the preceding south-west monsoon in July and August. In any one year the effect of the monsoon may vary very considerably in different parts of India, the rainfall being in excess in certain areas and below normal in others; thus the "Summary of the weather conditions in India" issued by the Meteorological Department, records that in 1921 "the aggregate rainfall of the season was in slight excess in Orissa, Chota Nagpur and Malabar, about normal in Burma, Assam, Bengal, Bihar and south-east Madras and below normal elsewhere." I have, therefore, extracted from the detailed summary the weather conditions of those parts of India, whose main rivers flow directly or indirectly into the Bay of Bengal and in which excess or deficit of rainfall is likely to produce an effect on the density of the surface-water of the Bay, and the data are given in Tables 64 and 65.

District		Rainfall during S. W. Monsoon, 1921.			Difference in average temperature from the Normal.
		Actual Inches	Normal Inches	Difference Inches	Degrees F.
Burma	..	59·2	59·8	−0·6	0 5
Assam	..	63·3	64·3	−1·0	−0·1
Bengal	..	55·4	54·7	+0·7	+0·4
Bihar and Orissa	..	45·5	42·1	+3·4	+0·4
United Provinces	..	28·5	33·4	−4·9	+0·5
Central India	..	24·8	31·4	−6 6	−0·3
Central Provinces	..	27·9	39·8	−11·9	+0·7
Hyderabad	..	14·5	26·7	−12·2	+1·9
Mysore	..	19·7	22·4	−2·7	+0·4
Madras	..	28·2	28·2	0	+0·6
Total	..	367·0	402·8	−35·8	+0·5

Table 64; giving the rainfall over India during the south-west Monsoon in 1921.

There was, therefore, in 1921 a deficiency of rainfall during the south-west monsoon, in the areas that will affect the Bay of Bengal, of 35·8 inches of rain below the normal and at the same time the average temperature, recorded over the same area, was 0·5°F above normal.

In the following year, 1922, the conditions experienced were the exact opposite of this. The "Summary of the Weather Conditions in India" of this year reports that "the total rainfall of the period (of the south-west Monsoon) was above normal by about 15″ or 44 per cent. in the United Provinces East, 15·5″ or 38 per cent. in Bihar, 12·3″ or 22 per cent. in Bengal, 10·4″ or 32 per cent. in the United Provinces West and 10·1″ or 24 per cent. in Chota Nagpur. It was in excess by 10 to 20 per cent. in Orissa, the North-West Frontier Province, Rajputana East, Central India and Malabar. About the normal amount was received in the Bay Islands, Burma, Assam, the Punjab, Kashmir, Rajputana West, Gujarat, the Central Provinces and the Konkan. Elsewhere rainfall was in deficit."

Taking again those areas that are likely to affect the Bay of Bengal we find that during this year the rainfall and the average temperature during the south-west monsoon season were as follows:—

District		Rainfall during the S. W. Monsoon, 1922.			Difference in average temperature from the Normal.
		Actual Inches	Normal Inches	Difference Inches	°F
Burma	..	62·2	60·1	+2·1	0
Assam	..	61·0	64·4	−3·4	−0·1
Bengal	..	67 5	55·2	+12·3	−0 2
Bihar and Orissa	..	54·1	42·1	+12·0	−1·1
United Provinces	..	45 9	33·5	+12·4	−1·3
Central India	..	35·1	31·0	+4·1	−0·8
Central Provinces	..	39·9	39·9	0	−0·6
Hyderabad	..	20·3	26·5	−6·2	+0·1
Mysore	..	15·7	22·4	−6·7	−0·1
Madras	..	27·9	28·3	−0·4	+0·4
Total	..	429·6	403·4	+26·2	−0·37

Table 65; giving the rainfall over India during the south-west Monsoon in 1922.

In contrast to 1921, there was thus in 1922 an excess of rainfall in the area concerned, amounting to 26·2 inches above the normal, while the average temperature was 0·37°F below normal. Between these two years, therefore, there was a difference in rainfall of 62 inches and a difference in the average temperature of 0·87°F, the rainfall being the greater and the average temperature the less in 1922. In the accompanying Text-figure 75 I have given the average specific gravity (σ_0) of the surface-water of the Bay of Bengal and the variation in the rainfall above or below normal in the area that drains into the Bay of Bengal for the two years and this shows clearly the way in which the two are related to each other.

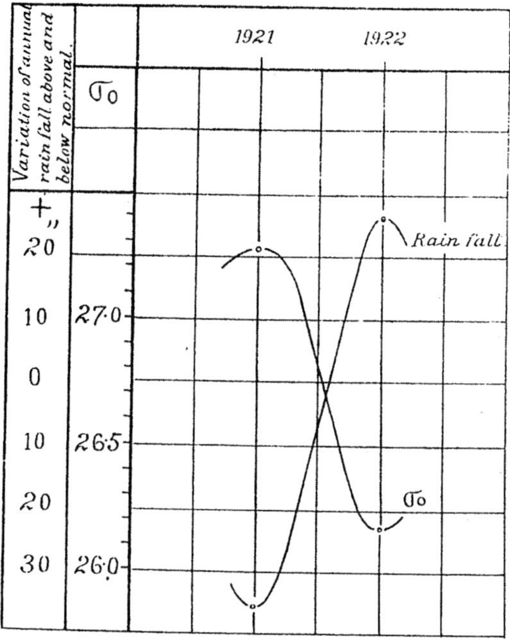

Text-fig. 75. Showing the average specific gravity of the surface water in the Bay of Bengal in the month of October and the variation from the normal of the rain fall over India during the South-West Monsoon in 1921 and 1922.

The difference in the average salinity of the Bay of Bengal in the month of January in the years 1923 and 1924 is again amply accounted for by the difference in the rainfall over Peninsular India and the country to the north and east of the Bay. In 1922 (*vide* Indian Weather Report, 1922) "the main features of the weather of the period October to December were an early retreat of the monsoon from north-east India and its vigorous activity in the south and centre of the Bay." In the following month of January there was a recrudescence of the monsoon in the south of the Bay accompanied by abundant rain in the south-east of the Peninsula. In December, 1923, according to the Indian Weather Review (1923), "the activity of the monsoon was chiefly displayed in the extreme south-east of the Peninsula; heavy rain, followed by disastrous floods, occurred in the Tinnevelly district about the middle of the month. The striking feature in northern India at this time was the occurrence of unusually early and heavy winter rains in the Punjab and the United Provinces." This was followed by a temporary revival of the north-east monsoon in the first week of January (*vide* Indian Weather Report, 1924). In Tables 66 and 67 I have given the rainfall during the preceding months of October to December in these two years, and from these it is clear that, whereas during the closing three months of 1922 the rainfall over this area was in excess of the normal by 5.56 inches and the average specific gravity of the surface-water of the Bay in the month of January was 26. 97 (S=33.57); during the corresponding period of 1923 there was a deficiency of 17.92 inches and the specific gravity in January was 27.02 (S=33.62).

District.	Rainfall during October to December, 1922. Actual. Inches.	Normal. Inches.	Difference. Inches.
Burma	10·74	9·25	+1·49
Assam	5·29	6·61	−1·32
Bengal	4·74	5·65	−0·91
Bihar and Orissa	1·95	3·49	−1·54
United Provinces	0·94	1·84	−0·90
Central India	1·23	1·36	−0·13
Central Provinces	2·42	2·50	−0·80
Hyderabad	4·39	3·76	+0·63
Mysore	11·88	8·04	+3·84
Madras	19·79	15·31	+4·84
Total	63·37	57·81	+5·56

Table 66; giving the rainfall over India during the later months of the year and the commencement of the North-East Monsoon in 1922.

District.	Rainfall during October to December, 1923. Actual. Inches.	Normal. Inches.	Difference. Inches.
Burma	8·48	9·23	−0·75
Assam	5·37	6·96	−1·59
Bengal	3·83	6·07	−2·24
Bihar and Orissa	4·20	3·90	+0·30
United Provinces	3·32	1·21	+21·1
Central India	0·71	2·56	−1·85
Central Provinces	1·16	2·77	−1·61
Hyderabad	0·47	3·92	−3·45
Mysore	2·05	8·29	−6·24
Madras	15·04	17·64	−2·60
Total	44·63	62·55	−17·92

Table 67; giving the rainfall over India during the later months of the year and the commencement of the North-East Monsoon in 1923.

I have already mentioned that in Indian seas there is a quite distinct double oscillation during the course of the day in the salinity of the surface-water and, as I shall show later, the character of this oscillation differs from season to season. At certain seasons this oscillation shows a distinct rise and fall accompanying the rise and fall of barometric pressure, while at others it varies in the opposite direction. Any conclusions regarding the distribution of the salinity of the surface-water of these seas are, therefore, liable to be fallacious if they are based on observations taken irregularly or at any one particular time of the day, since such a time may at one period of the year correspond to a rise in the daily oscillation, whereas at another period it may correspond to a fall. In order, therefore, to arrive at an approximate figure for the average salinity in different degrees of Longitude across the Bay of Bengal at different times of the year, I have taken the average of all observations in the months of October, January and April in each zone of 2-degrees of longitude. The results thus obtained are given below and I have plotted them in Text-fig. 76.

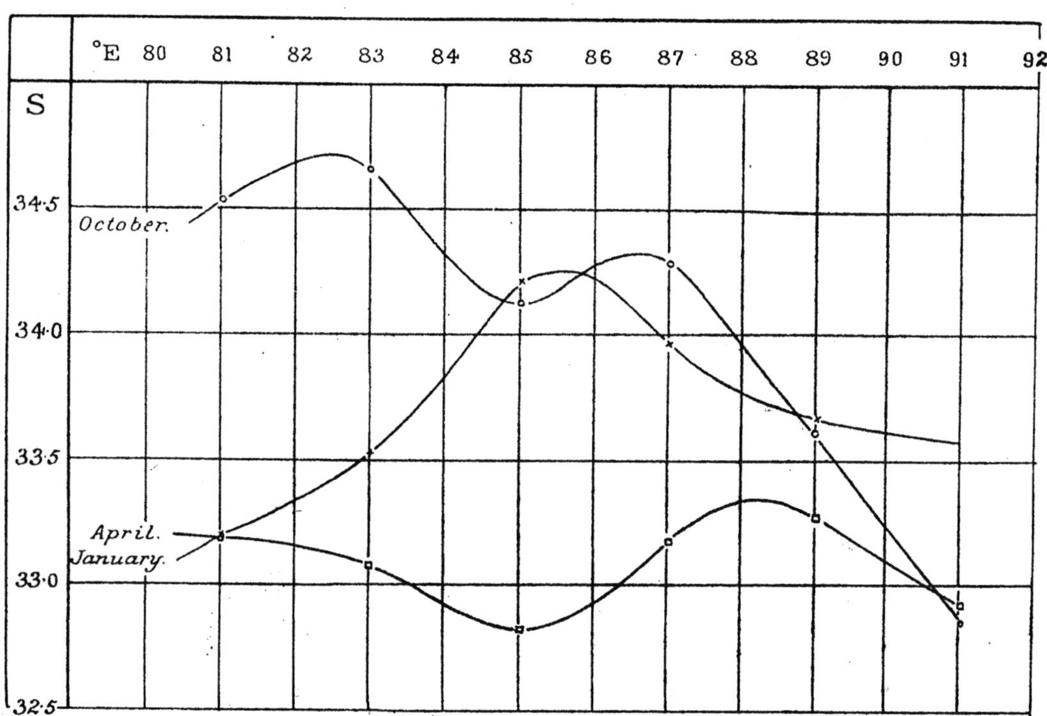

Text-fig. 76. Showing the average salinity of the surface water in different months of the year in every 2° of Longitude across the Bay of Bengal.

Month.	80°–82°	82°–84°	84°–86°	86°–88°	88°–90°	90°–92° Long. E.
October	34·54	34·67	34·13	34·29	33·61	32·85
January	33·20	33·54	34·22	33·96	33·67	33·58
April	33·19	33·08	32·84	33·18	33·26	32·93

In the month of October as we pass from west to east the salinity at first steadily rises from 34·54 to 34·67. This is almost certainly the result of a movement of water from west to east, and a reference to the Admiralty Current Chart for the Indian Ocean for October shows that there is a strong surface-current flowing during this month from west to east across the south end of Ceylon. This latter current would appear to be one of comparatively high salinity and to have a salt content of 34·5 or over. In the region of Long. 85°E there is a distinct drop in the average salinity followed in Long. 87°E by a small rise and, finally, on the east side of the Bay the salinity falls rapidly as one approaches the Andaman Islands. This latter fall, from a salinity of 34·29 in 87°E to 32·85 in 91°E, is undoubtedly due to an outflow from the Andaman Sea, but the smaller drop in salinity in the region of the middle of the Bay in 85°E is not so easy to explain; it is probably due to a movement of water of low salinity from the area lying to the north-west of the Andaman Islands into the centre of the Bay under the influence of one of the rotatory currents produced by the wind (*vide infra*, p. 284, Text fig. 79) at this time of the year.

In the month of January the salinity is low over the area to the south of Ceylon and the west side of the Bay, but rises steadily as one proceeds eastwards and reaches its maximum, 34·22, in about Long. 85°E, after which it again falls as we proceed towards the Andaman Sea. At this time of the year the north-east monsoon is in full swing and there is a strong current flowing from the region of the Andamans towards the south end of Ceylon. This current will, to commence with, be one of low salinity, owing to the outflow of water from the Andaman Sea, in which the surface-water is greatly diluted by the influx of river-water. During the passage of the current westwards across the Bay its salinity will steadily rise partly from evaporation and partly from admixture; but off the west coast of Ceylon we normally encounter at this time of the year a second strong current of water of low salinity setting from the head of the Bay of Bengal along the east coast of India and then southwards and westwards round Ceylon. In January, 1923 this reduction in the salinity of the water to the immediate east of Ceylon was well-marked, but in the following year, 1924, it was not noticeable. The passage across the Bay was made later in the month in 1923 than in 1924, and possibly the surface-current caused by the north-east monsoon winds, which sets down the east coast of India, had not become fully established in the latter year, but it is more probable that the cause of this difference in the two years is to be found in the deficient rainfall of the north-east monsoon and especially of the marked deficiency over Hyderabad, Mysore and Madras in the three months October—December 1923 to which I have already drawn attention (*vide supra*, p. 274, Tables 66 and 67).

Finally, in April we get a tendency to a reversion to the condition that we found to be present in October. The water on the east side of the Bay is still to some extent diluted by the outflow from the Andaman Sea, but the salinity steadily rises as we proceed westwards to Long. 88°E. We now find that there is once again in the middle of the Bay, from Long. 82°E to 88°E, an area of lowered salinity that appears to be due to water of low salinity coming up into the Bay from the south, and as this is driven westwards the salinity again rises.

SEASONAL VARIATION IN THE SALINITY OF THE SURFACE-WATER OF THE WEST SIDE OF THE ANDAMAN SEA AND THE BAY OF BENGAL.

I. *Andaman Sea.*

During survey-seasons 1921-22 and 1922-23 the R.I.M.S. "Investigator" made the passage from Nankauri Harbour in the Nicobar Islands to Port Blair in the Andamans, or *vice versa*, on fifteen different occasions; on thirteen out of the fifteen trips samples of the surface-water were taken every four hours and carefully examined by means of a 'Buchanan' hydrometer; the results are given *in extenso* in Appendix IX. The course steamed during these trips lies nearly north and south along

Latitude N.	17–18 X. 1922	25–26 X. 1921	9–10 XI. 1922	11–12 XI. 1921	15–16 XI. 1921	2–3 XII. 1921	5–6 XII. 1921	8–9 XII. 1922	12–13 XII. 1922	31-XII 1921 1.I. 1922	25–26 I. 1922	31-1 1-II. 1922	25–26 II. 1922
Port Blair.		24·972					25·417		25·729				
Port Blair – 11°	25·753 25·939	25·896		26·093	26·257	25·938	25·746	27·028	26·893	26·630	26·523	25·691	26·336
11° – 10°	26·052	25·850 26·002	26·274	26·181 27·498	26·439	26·281 27·128	26·212	26·912 26·898	26·916 27·013	26·819 26·831	25·741 26·148	26·414 26·227	26·530 26·586
10° – 9°	26·798 27·091	26·891	26·606	27·467	27·278 27·349	26·423	26·944 26·636	27·522	27·457	26·672	26·166 26·387	26·167 26·589	26·942 26·582
9° – 8°	27·241	27·282	27·286 27·189	27·094 27·160	27·208	26·608 26·878	26·643	27·261 27·252	26·899 26·381	26·470 26·751	26·779	26·734	26·697
Nankauri Harbour.							26·581				26·545	26·805	
Range of Variation	1·488	2·310	1·012	1·405	1·092	1·190	1·198	0·624	1·076	0·361	1·038	1·114	0·606
Average range	1·899		1·170			1·022				0·699		0·860	

Table 68; Showing the surface salinity in different latitudes during successive voyages between Port Blair and Nankauri Harbour.

the west side of the Andaman Sea and the results of the salinity observations are given in Table 68, in which they have been arranged in accordance with the position of Latitude N., from which they were obtained. On each voyage the salinity of the different samples was found to vary very considerably: in some cases this oscillation exhibits only a single maximum, whereas in others two maxima were detected. It is probable that this variation is due to more than one cause. It may be due, at least in part, to an influx through the various channels across the Andaman-Nicobar Ridge of more saline water from the Bay of Bengal into the Andaman Sea: a comparison of the average specific gravity of the surface-water of the Bay of Bengal and of the west side of the Andaman Sea in the months of October, 1921 and 1922, shows that in both cases the water of the Bay of Bengal had a higher average specific gravity than that of the Andaman Sea area.

	Bay of Bengal.	Andaman Sea.	Difference.
October 1921	27·265	26·479	0·786
October 1922	26·176	26·149	0·027

I have already pointed out (*vide supra*, p. 273 and Text-fig. 75) that the lower specific gravity of the surface-water of the Bay of Bengal in October 1922, as compared with the conditions present in the same month in 1921, is correlated with the increased rainfall of the south-west monsoon in the former year, and it will be noticed that an exactly similar variation in the specific gravity is shown in the Andaman Sea region, but that the difference between the two areas is much more marked in the year 1921 in which the rainfall was defective; this difference is clearly due to the higher salinity in the Bay of Bengal in 1921.

The average specific gravity of the surface-water on the west side of the Andaman Sea shows a very clear seasonal oscillation. I give below in tabular form the average in each month.

Month.	σ_0
October	26·31
November	26·89
December	26·92
January	26·51
February	26·49

These figures indicate that in this area the effect of the north-east monsoon does not make itself generally felt before January: during the first three months of the survey-season the specific gravity steadily rises but in January we get a fall that is continued, though to a less extent, in February.

Latitude N.	October.	November.	December.	January.	February.	Average.	Difference.
11°-12°	25·640	**26·675**	26·183	*26·107*	26·336	26·188	1·035
10°-11°	25·968	26·598	**26·772**	*26·133*	26·558	26·406	0·804
9°-10°	26·927	**27·175**	26·942	*26·327*	26·762	26·829	0·848
8°-9°	**27·261**	27·187	26·794	26·756	*26·697*	26·939	0·564
Average	26·449	26·909	26·673	26·331	26·563		
Difference	1·621	0·589	0·759	0·623	0·426		

Table 69; showing the average specific gravity (σ_0) of surface water in each degree of Latitude N. and in each month on the west side of the Andaman Sea. The Maximum and Minimum densities in each degree of Latitude are indicated by heavy type and by italics respectively.

During both survey-seasons, namely 1921-22 and 1922-23, the specific gravity of the surface-water in October and the following months was very much lower in the northern area, around the Andaman Islands, than in the southern region near the central group of the Nicobars. In Table 69 I have given the average specific gravity (σ_0) of the surface-water, calculated from all observations, in each degree of Latitude N. and this shows clearly that the amount of the lowering of the specific gravity in the northern region is most marked in the month of October. The actual difference in the specific gravities of the northern and southern regions in October, 1921 was as great as 1·848, and in the same month in 1922, was 1·395. This lowering of specific gravity and salinity in the northern area is attributable to the influx of large volumes of river-water into this region. Although the south-west monsoon is over by October, a large volume of water will still be coming down the great rivers and in consequence the waters of the more northernly region of the Andaman Sea will be considerably diluted by the outflow from the Irrawaddi, the Salween and the Sitang

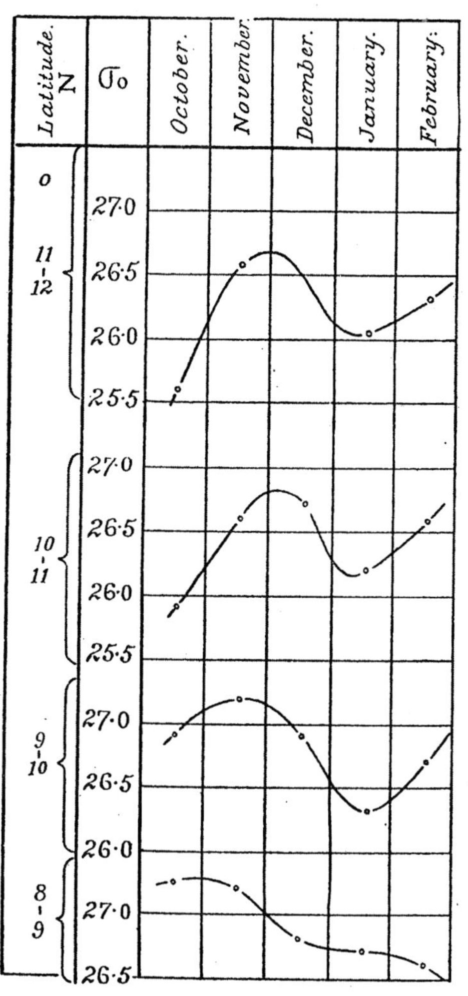

Text-fig. 77. Showing the variation in the average monthly salinity in each degree of Latitude in the western part of the Andaman Sea during survey seasons, 1921-22 and 1922-23.

rivers, though this dilution will steadily be becoming less and less at this season of the year. As we follow the seasonal changes in each degree of Latitude N. (*vide* Text fig. 77) we see that there is a very considerable difference in the seasonal rise and fall of specific gravity in the northern and southern regions. In the northern area extending between Lat. 9°N and 12°N the specific gravity rises from October to November, and this is, I think, clearly attributable primarily to the diminished outflow of river-water and, secondarily, to the increased evaporation during the warm, dry weather that intervenes between the two monsoons; and in conformity with this the rise is most marked in the northern Latitudes 11-12°N. In the month of December there is a sharp fall in the specific gravity (σ_0) of the surface-water in Lat. 11-12°N whereas in Lat. 10-11°N the specific gravity is still slightly higher and it is not until the following month of January that we find the corresponding fall, for, as one would expect from the geographical features of the Andaman Sea basin, the influence of the rainfall of the north-east monsoon and the consequent increased outflow of river-water is earliest detected and is most apparent in the northern part of the basin. Finally, in the month of February the specific gravity commences to rise again, owing, to the cessation of the north-east monsoon and the consequent increased evaporation. In Latitude 9-10°N the specific gravity of the surface-water rises slightly from October to November and then in December we get a slight fall; thus we can in this Latitude detect a fall in the specific gravity of the surface-water a whole month earlier than in the next degree of Latitude to the north. In Latitude 8-9°N there is a small but steady decrease in the specific gravity throughout the whole period of my observations. This continuous fall in Latitude 8-9°N and the fall in Latitude 9-10°N in December, a whole month before the effect of the north-east monsoon can be traced in Latitude 10-11°N, cannot be due to the outflow of river water from the great Burmese rivers and is almost certainly due to an influx into the southern end of the Andaman Sea basin, under the influence of the north-east monsoon, of a mass of water of low salinity from the Java Sea and the western Pacific Ocean through the Straits of Macassar. As the season progresses, the fall of salinity in the southern area of the Andaman Sea is first noticeable in the more southerly latitude and the mass of water of low salinity makes its way progressively further and further northwards as the season progresses.

At the end of Survey-Season 1913-14 the R.I.M.S. "Investigator" proceeded to Rangoon and then sailed for Bombay, *viâ* Duncan Passage between the South and Little Andaman Islands. I was thus able to carry out a series of observations on the specific gravity and temperature of the surface-water in the north-west area of the Andaman Sea in April, 1914. The results obtained are given in full in Appendix IX and they show very clearly that the water in the neighbourhood of the Rangoon river estuary and of the delta of the Irrawaddi river is of very low specific gravity. As one proceeds south-west the salinity rapidly rises and finally approximates to that found in the open waters of the Bay of Bengal. This rise of salinity is undoubtedly due to the inflowing current of water into the Andaman Sea through Preparis and Ten Degree Channels, an influx of open-sea water that reaches its maximum a little later in the year and, as I have already pointed out (*vide supra* p. 31), has a marked effect on the nature of the sea-bottom in the north-west area of the Andaman Sea basin.

2. *Bay of Bengal.*

Our knowledge of the seasonal distribution of the surface-salinity of the water of the Bay of Bengal is still comparatively meagre. Schott (1902) includes in his atlas a single chart showing the general distribution of the surface-salinity throughout the whole of the Atlantic and Indian Oceans, and Ringer (1922) has reproduced an outline sketch of Schott's map. Neither of these maps takes into account the very great changes that occur in the salinity itself, as well as in its distribution, during the changing seasons of the year, and in both cases the north-east portion of the Bay of Bengal is left blank, as if to indicate that our knowledge of this region is nil. Dallas in 1887 prepared a series of charts (published by the Meteorological Department of the Government of India) dealing with the conditions of mean specific gravity, temperature and surface-currents present in the Bay of Bengal and Andaman Sea during the four seasons of the year, December-February, March-May, June-August, and September-November. These four periods correspond to the periods of the south-west and north-east monsoons and the two intervening dry seasons. Unfortunately in the title of the publication these maps are referred to as "showing the specific-gravity, temperature and currents of the sea-surface;" whereas in the letter-press accompanying the maps it is stated that "lines of equal density were drawn." Matthews (*Trans. Lin. Soc.* 1926) has recently published an account of the surface-salinity of the Indian Ocean based partly on the samples collected by H.M.S. "Sealark" during the Percy Sladen Expedition to the Indian Ocean in 1905 and partly on samples collected by the various ships of the Peninsular and Oriental Steamship Navigation Company on several voyages across the Indian Ocean to India, Australia and China. He also has prepared a series of four maps, each dealing with a period of three months and has drawn in the "isohalines" for the greater part of the Indian Ocean in each period; but in all cases the eastern and northern part of the Bay of Bengal is again left a blank and it seems probable that Matthews was unacquainted with Dallas' charts; at any rate he makes no reference to them. A comparison of the figures given by Dallas with the isohalines drawn by Matthews and with the results of my own observations reveals what appears at first sight to be a marked discrepancy. My own results and those of Matthews agree very fairly well but the contours as shown by Dallas do not, so far as one can judge, coincide even approximately with either the true specific gravity (σ_0) or with the true density (σ_t). From a comparison of the three sets of data, namely those given by Matthews and Dallas and the results of my own observations, it appears that the contour lines as given in Dallas' maps do not represent the distribution of either the true specific gravity or the true density of the surface-waters, as these terms are understood to-day, but are the lines of the relative weight of the sea-water referred to a corresponding weight of distilled water *at the same temperature*—in other words they refer to $\frac{St}{St}$. I have, therefore, in the absence of any definite proof assumed that this is the case and from Dallas' figures I have calculated the true specific gravity (σ_0) and have converted this into terms of actual salinity. This has been done by taking Dallas' figures and multiplying them by the factor corresponding to the temperature given in Scheel's Table (*Zeitschrift für Instrumentenkunde,* 1897). We thus arrive at the true value of the density (σ_t) and from Knudsen's Tables can then find the correct values for the specific gravity and salinity. The results obtained now agree quite closely with my observations and with those given by

Matthews, allowing for such differences as are bound to occur in waters such as these in different years. By combining all three sets of data and basing my contours very largely on those given by Dallas I have charted the salinity of the surface-water in the

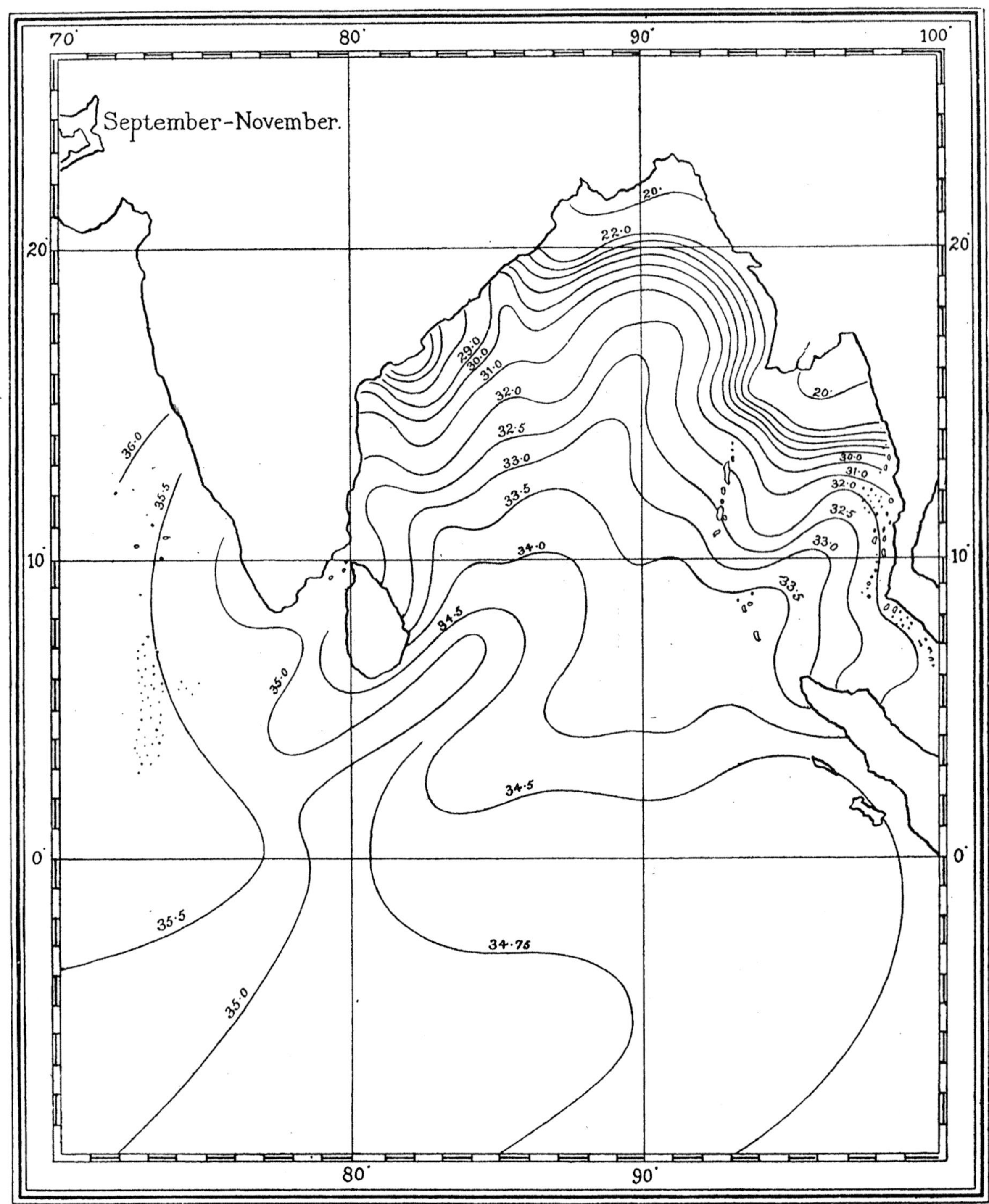

Text-fig. 78; The surface salinity of the Bay of Bengal; September-November.

Bay of Bengal for the four periods September-November, December-February, March-May and June-August and the results are given in Text-figs. 78, 80, 82 and 84.

(a) *September-November*, (Text-fig. 78).

The descriptive text published with the Meteorological Department's charts points out that there is at this period of the year, more than in any other, a great range of density (and salinity) between the areas to the north and the south of the Bay. "The amount of fresh water emptied into the Bay by the two great rivers, the Ganges and the Irrawadi, is still very large." On the west side there is also a marked fall in the salinity of the coastal-water around the mouths of Godaveri and Kistna: "a strong southerly current runs at this season along the lower Coromandal Coast and round Ceylon, and in this neighbourhood, consequently, the low density waters from about the mouth of the Godaveri are carried southwards, being recruited by those of the Cavery and other smaller rivers." As a result of this great influx of fresh-water from all the rivers, which are still full from the rainfall of the south-west monsoon, the salinity of the northern areas of both the Bay of Bengal and the Andaman Sea is remarkably low, being less than 20·0 and with the exception of a small tongue of denser water that can be traced passing in a north-easterly direction round the south end of Ceylon, nowhere east of Long. 80° or north of the Equator does the surface salinity exceed 34·75. Matthews (1926, Pl. 13) has indicated the presence of this tongue of water, of a salinity of 35·0, passing around the south of Ceylon and then roughly in a north-easterly direction for a short distance into the Bay; but the extent of this area of high salinity will, of course, differ from year to year and during the seasons in which I was investigating this area it seems to have extended considerably further to the north-east than he shows it. It is of great interest to compare the distribution of the average surface-salinity in this period, September-November, with the corresponding surface-currents in the same period. In Text-fig. 79 I have given a sketch-chart showing the main trend of the surface-currents at this time of the year and in compiling this chart I have made free use of the current charts for the Indian Ocean, published by the Admiralty, London, and also those given by Max Weber (1923) in his account of the oceanographic conditions in the Malay region. During the period covered by these three months there are very considerable changes in the direction of the currents; in September we still get the remains of the effect of the south-west monsoon, whereas by November the north-east monsoon has in most years begun to set in. We can distinguish the following main sets of currents :—

(1) A strong easterly current passes across between the south of Ceylon and the equator and at the south-east corner of the island this spreads out fan-wise; part continues across the mouth of the Bay of Bengal till near the Sumatran Coast it becomes deflected either to the north or the south; a second part gradually bends northwards to enter the Bay of Bengal.

(2) Commencing at the head of the Bay of Bengal is a current that flows towards the south-west, clearly the result of the N.E. monsoon winds; meeting the coast of India this current sweeps along it till it finally reaches the east coast of Ceylon and then bends westwards, keeping close along the coast, to reach the Gulf of Mannar.

(3) A third current arises at the north end of the Andaman Sea and also passes

towards the west and south-west, but this, soon after leaving the Andaman Sea, is altered and by the combined effect of currents (1) and (3) we get in the centre and

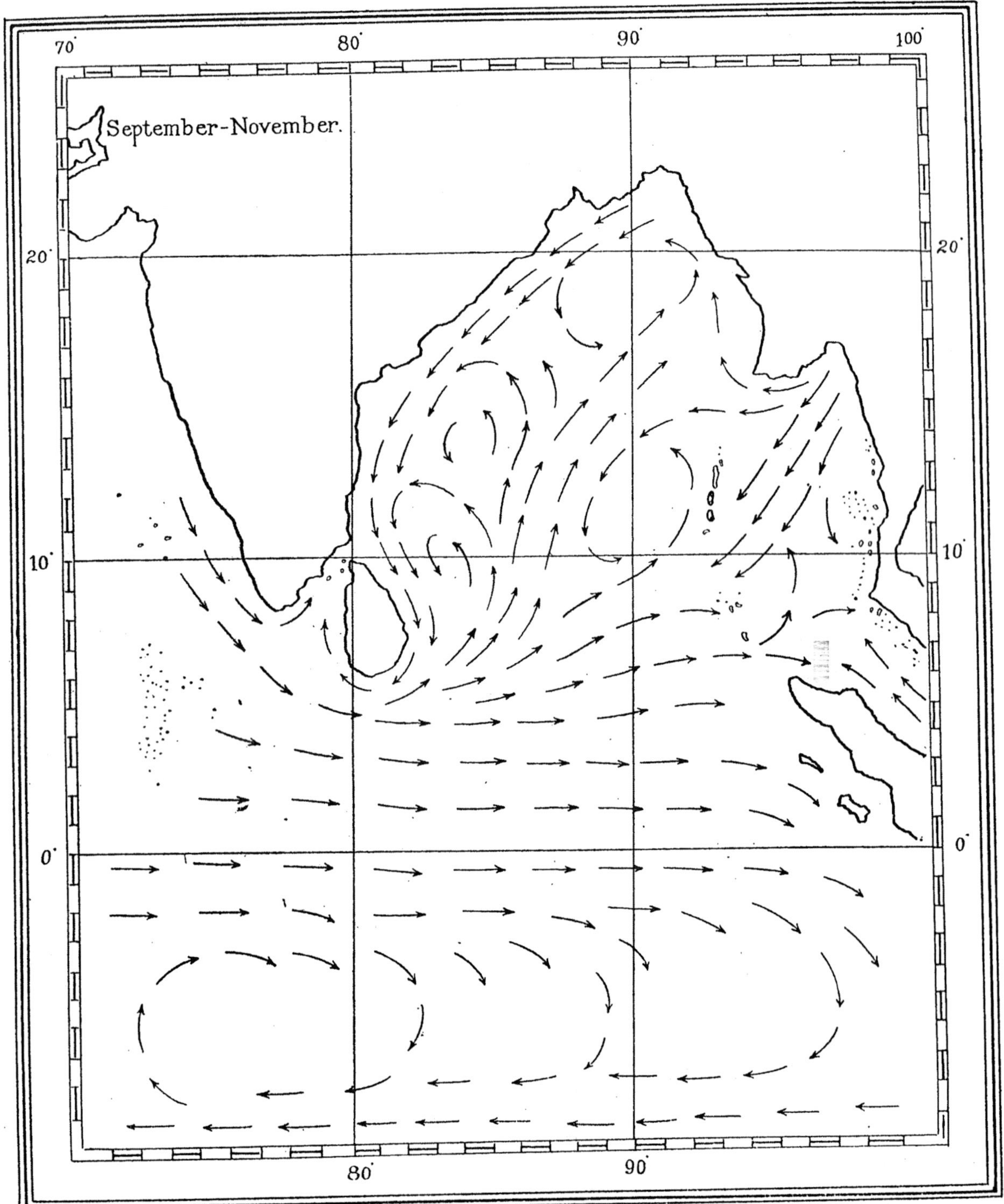

Text-fig. 79; The surface currents of the Bay of Bengal; September-November.

north-west parts of the Bay a number of rotatory currents in which the general trend of movement of the surface masses is counter-clockwise.

A comparison of Text-figs. 78 and 79 shows clearly the manner in which the trend of the isohalines and the surface currents agree.

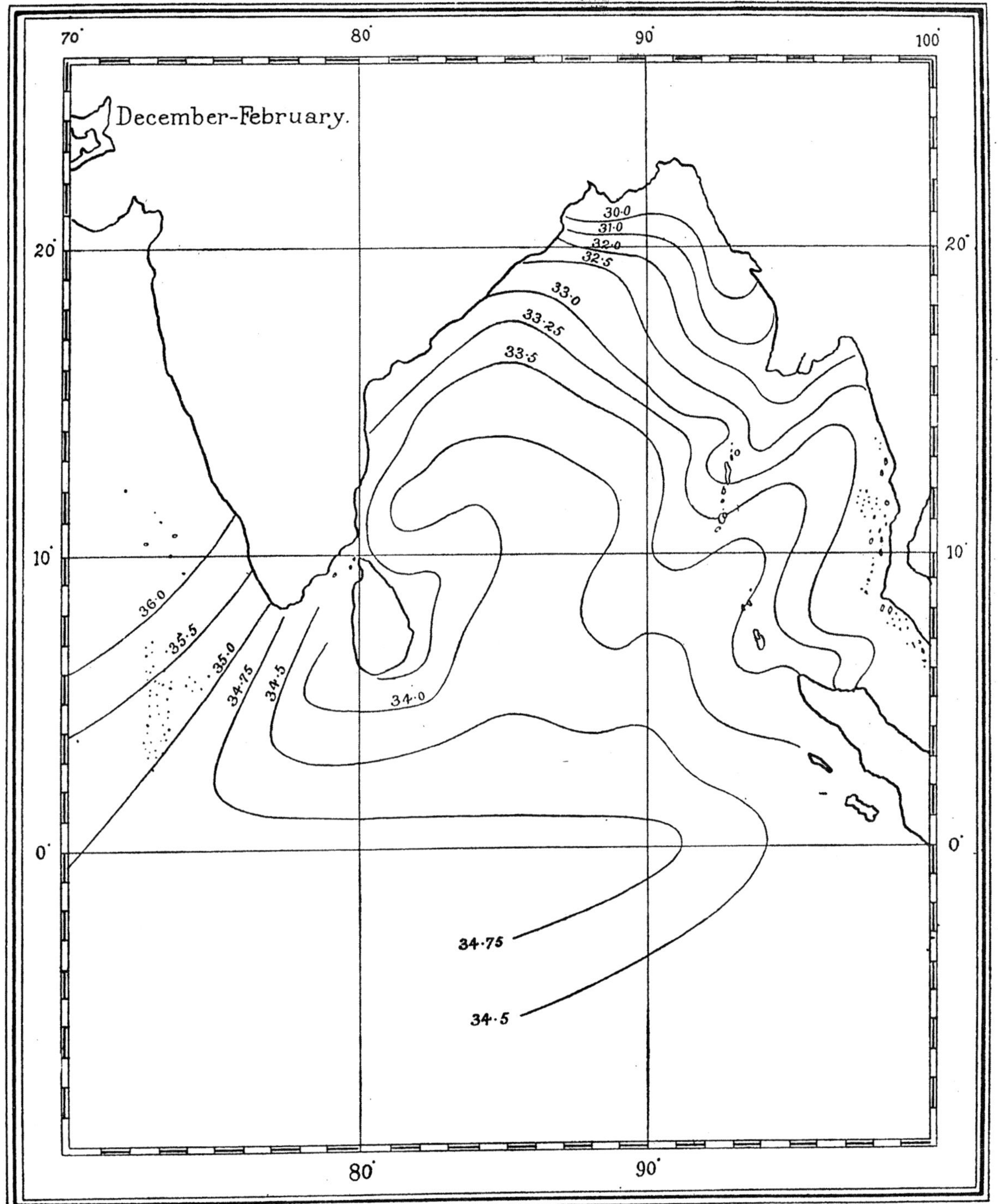

Text-fig. 80; The surface salinity of the Bay of Bengal; December to February.

December-February. (Text-fig. 80.)

During the succeeding three months, namely from December to February, a great change occurs in the direction and the strength of the surface-currents in and around India, owing to the establishment of the north-east monsoon. The time at which the monsoon sets in varies from year to year and from area to area and

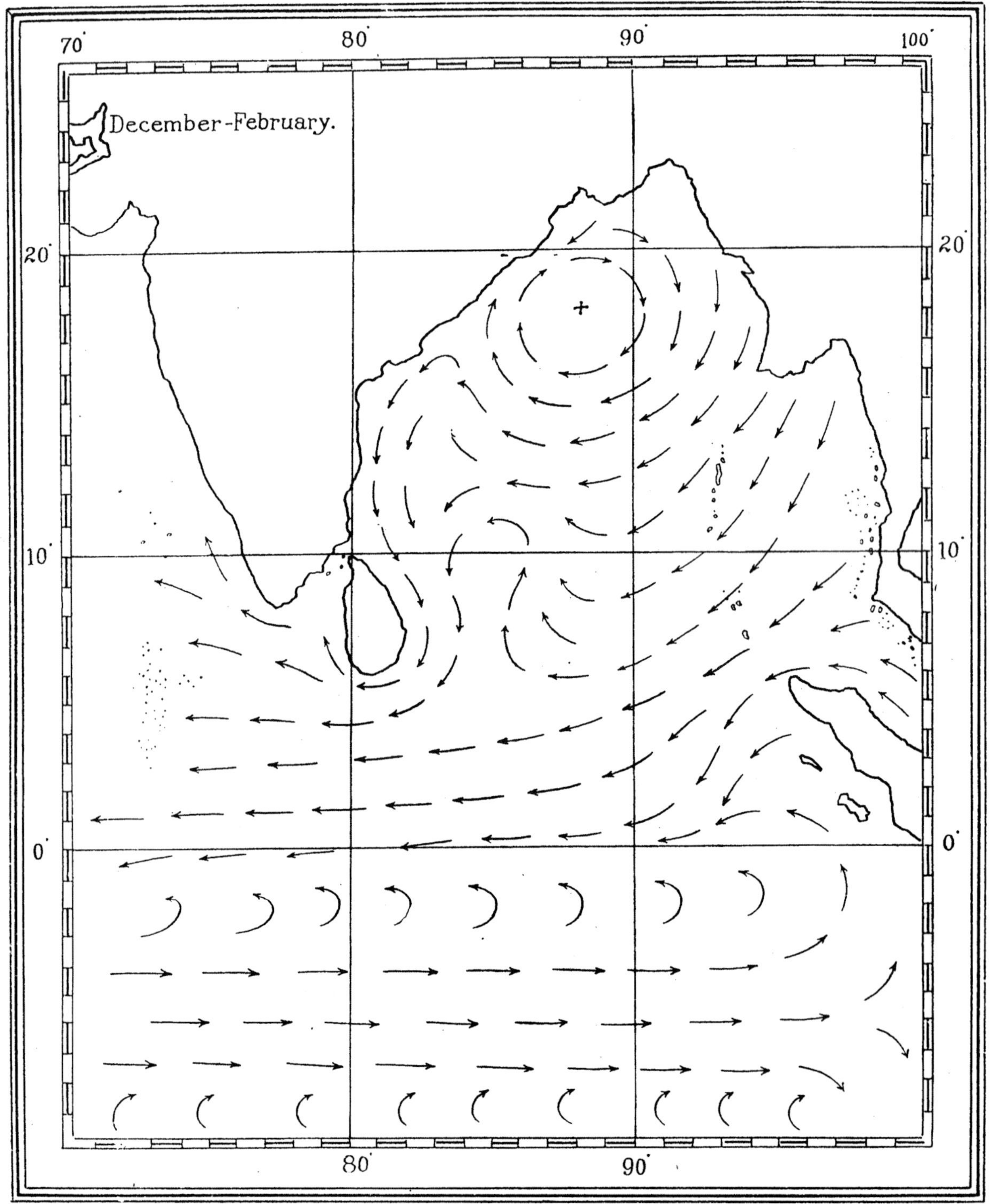

Text-fig. 81. The surface currents of the Bay of Bengal, December to February.

therefore in individual localities there may be during this period very considerable changes both as regards the surface-currents and the surface-salinity. The great influx of river-water into the Bay of Bengal and the Andaman Sea that follows the south-west monsoon, has now largely diminished and in consequence we no longer find the very marked lowering of salinity in the three areas around the mouths of the Irrawaddi, the Gangetic Delta and the mouths of the Godaveri and Kistna on the east coast of India. The greater part of the Bay of Bengal is occupied by water of a less salinity than 34, but midway between Ceylon and the Nicobars a tongue of water having a salinity of between 34 and 34. 5 passes northwards and then curves to the west towards Palk Bay and the straits between India and Ceylon. In the description that accompanies the Meteorological Department's charts it is stated that on the east side of the Bay "an arm of high density water extends through the Ten Degree Channel across the Andaman Sea to the Tenasserim Coast. To the north and west of this, the lines bend southward, as far south at Lat. 7°. This results from the outpouring of the Irrawaddi, the water of which, uniting with the low density current down the Arakan Coast, travels southward past the Andamans to the centre of the Bay in Lat. 7°." In the accompanying Text-fig. 81 I have compiled a generalised map of the surface-currents from the data given in the three monthly current charts of the Indian Ocean published by the Admiralty and those given by Max Weber, and a comparison of this chart with that published by Michaelis (1923, chart. 1) of the surface-currents in the whole of the Indian Ocean in the month of January reveals a close agreement. As he points out "Im Golf von Bengalen zeigt die Strömung einen zyklonalen Wirbel. Es ist leicht zu erkennen, dass er unter dem Antrieb des Nordost-monsuns und unter der Einwirkung der Erdrotation und Küstenkonfiguration zustande kommt. Das Wasser wird durch den Wind von der birmanischen Küste ab und uber den Meerbusen westwärts gegen die Koromandelküste getrieben und wird dort durch die Küstengestaltung gezwungen, vorwiegend nach Norden abzufliessen. Dadurch kommt es zur Wirbelbildung. Das Zentrum des Wirbels liegt in 88°O und 18°N. Beständigkeit und Stromstärke sind nun im Westteil des Wirbels grösser, wo die Bewegung durch die vorgelegene Küste eingeengt wird. Hier entwickeln sich auch in der flachen Küstenbuchtung zwischen Ceylon und der Mündung des Godawari in ungefähr 17°N Neerströme, deren Wasser zum Teil östlich um Ceylon herum in den offenen Ozean gelangt." A comparison of Michaelis' description and of the general trend of the surface-currents, as given in Fig. 81, with the isohalines in Text-fig. 80 shows how clearly the distribution of the salinity and the direction of the surface-currents are related to each other.

March–May. (Text-fig. 82.)

A study of the Admiralty current charts for these months indicates that during this season of the year very great changes are taking place in the direction of flow of the surface-currents and, therefore, a chart for the whole period of these three months can only be approximately accurate. During the earlier part of the period, in March and April, there is a quite distinct double cyclonal circulation going on in the waters at the head of the Bay (*vide* Text-fig. 83), the currents moving round clockwise about two centres situated approximately in (1) 16°N and 88°E and

(2) 15°N and 91°E respectively, *i.e.*, in very much the same position as the circular movement that we found to be present during the period December-January;

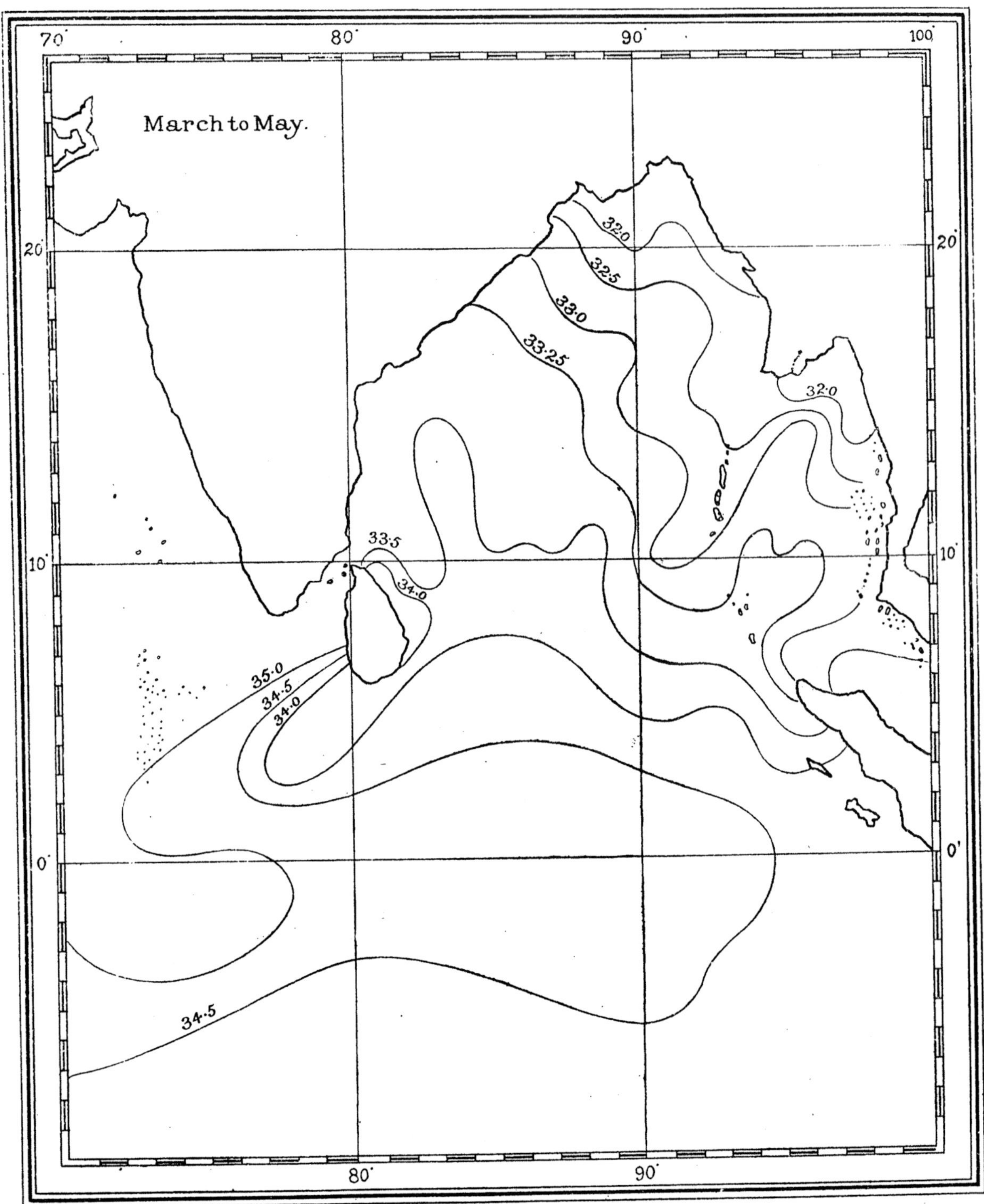

Text-fig. 82; The surface salinity of the Bay of Bengal; March to May

but this tends to disappear in May. Similarly in the earlier part of the period the currents across the mouth of the Bay sweep from east to west, but in May this is reversed and with the commencement of the south-west monsoon winds there is developed a well-marked surface-drift from west to east, which opposite the

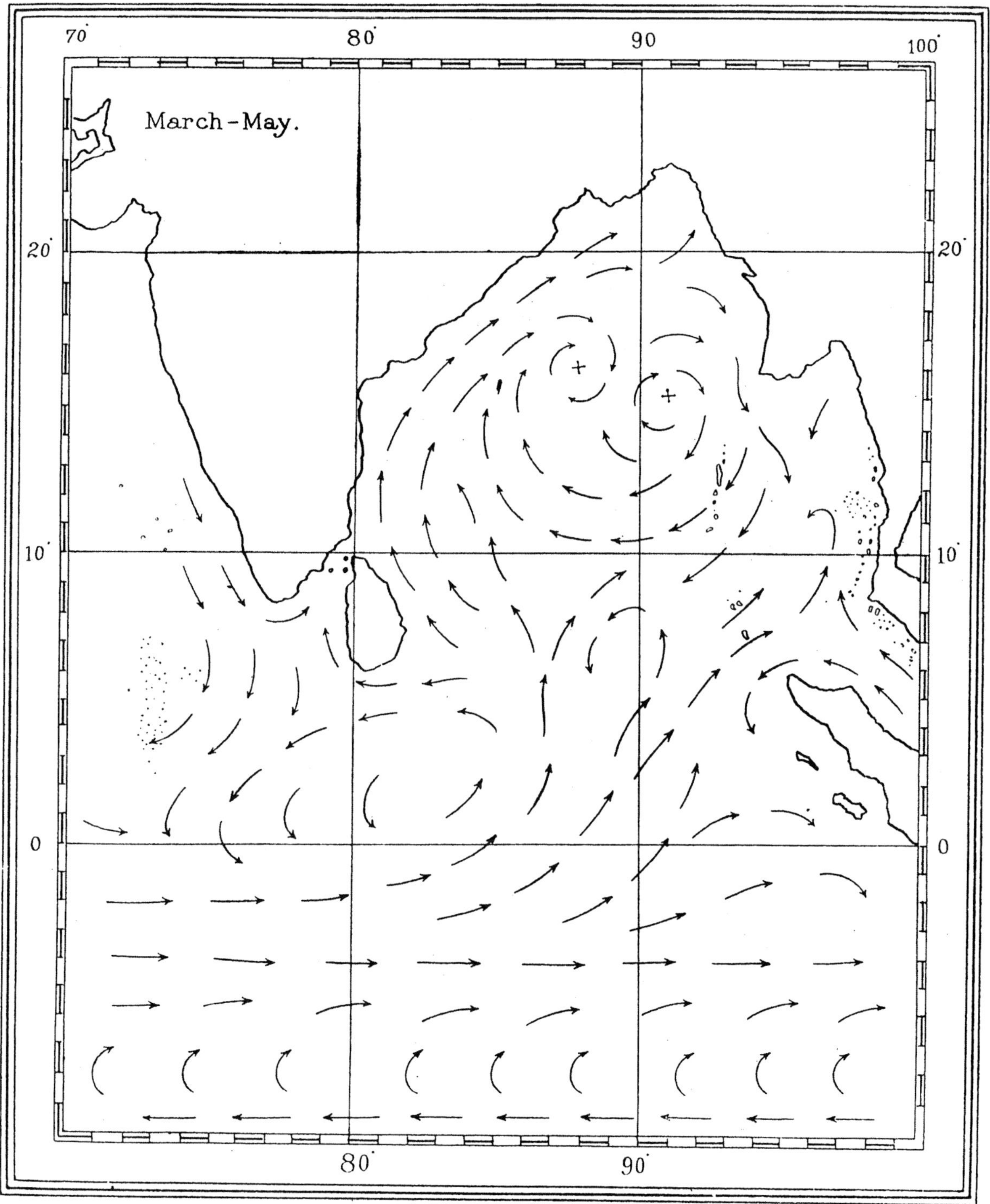

Text-fig. 83; The surface currents of the Bay of Bengal; March to May.

centre of the mouth bends northwards and runs up into the Bay. In the region of the Andamans there is a well-marked drift in a south-west direction, a continuation of the currents produced by the N.E. Monsoon; while later an easterly to N.E. drift is developed in the region of the Nicobars. The lines of equal salinity (Text-fig. 82) show clearly the effect of the circular movement of the water-masses at the head of the Bay. The whole of the Bay, with the exception of a small area lying to the east of Ceylon is occupied by water having a salinity of less than 34.0 but to the south in the region of the Equator there is a wide belt of water having a salinity of 34.5 or over and the trend of the isohalines indicates that this water is to some extent carried northwards towards the Bay: a continuation of this surface-drift can be clearly seen in the tongues of water of a salinity of 33.5 that pass north towards the coast of India or north-east towards Upper Burma. We also get a drift in the N.E. direction of water of comparatively high salinity past the Nicobars and this seems to divide the outflow of low-salinity water from the Irrawaddi into two streams one of which flows to the south-west past the Andamans while the other and, according to the introduction to the Meteorological Department's charts of surface density, the most important drift flows down the coast of south Burma. We can still trace an area of low salinity to the south-west of Ceylon but further south there is, as already mentioned, a well-marked easterly drift of water of high salinity, 34.5 and over, that passes across the mouth of the Bay nearly to Sumatra.

June–August. (Text-fig. 84.)

The last of the four periods of the year, namely, June–August, covers the commencement and a great part of the duration of the south-west monsoon. At this season of the year surveying is impossible and in consequence the "Investigator" is laid up in Bombay, undergoing refitting, etc. I have, therefore, no observations of my own during these months, but a study of the data given by Matthews and of the Meteorological Department's charts shows that the greater part of the Bay is occupied by water of a salinity lower than 34.5; the only exception to this is a belt of water of salinity between 34.5 and 35.0 that lies immediately to the east of Ceylon and extends northwards across Palk Bay to the coast of India. At the head of the Bay and at the northern end of the Andaman Sea the salinity falls rapidly owing to the increasing outflow of river-water, following on the break of the monsoon, and it is interesting to note that in the former region the lowering of the salinity is very much less than in the latter, and, apparently, can be detected round the mouth of the Ganges, but is absent opposite that portion of the delta that corresponds to the outflow of the Brahmaputra River. Across the south of Ceylon a wide belt of water of a salinity of 34.5 or over passes eastward towards Sumatra and a tongue of this can be traced passing to the north-east into the Bay, while to the west of this tongue, between it and Ceylon an outflowing tongue of water of low salinity passes southwards. According to the Meteorological Department charts there is about Lat. 2°N; Long. 88°E an oval area of water of low salinity, the presence of which seems to be correlated with an outflow of water through the Straits of Malacca and out of the Andaman Sea towards the centre of the Bay. In Text-fig. 85 I have given a generalised chart of the surface-currents for the same period. This chart is based on the data given in the monthly current

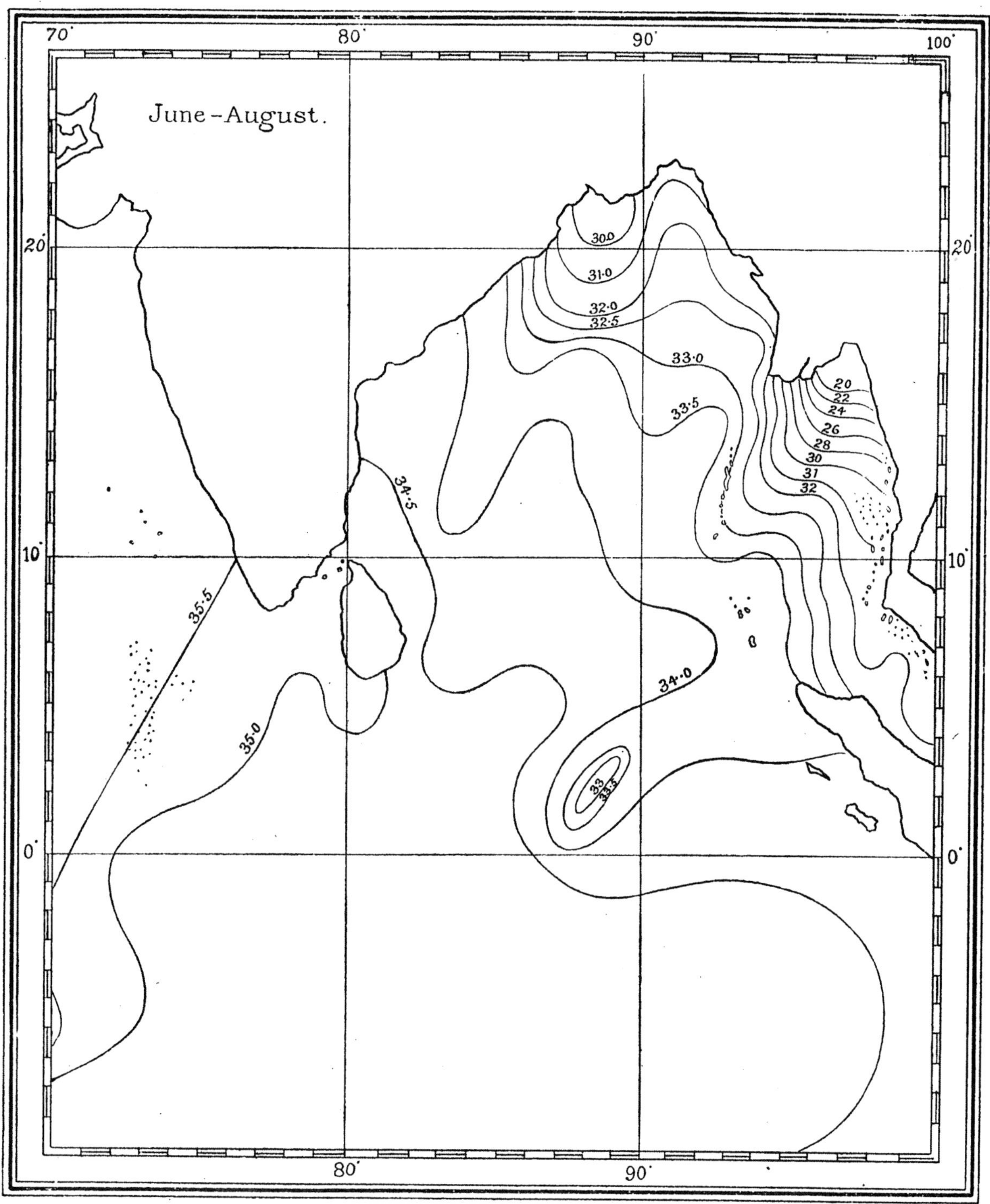

Text-fig. 84; The surface salinity of the Bay of Bengal; June to August.

charts of the Indian Ocean (Admiralty) and those given by Max Weber, and a comparison of this chart with that given by Michaelis (1923, chart 2.) for the month of July reveals a very close agreement. At this period of the year a strong surface-drift flows eastward across the southern end of Ceylon, and in the belt lying between

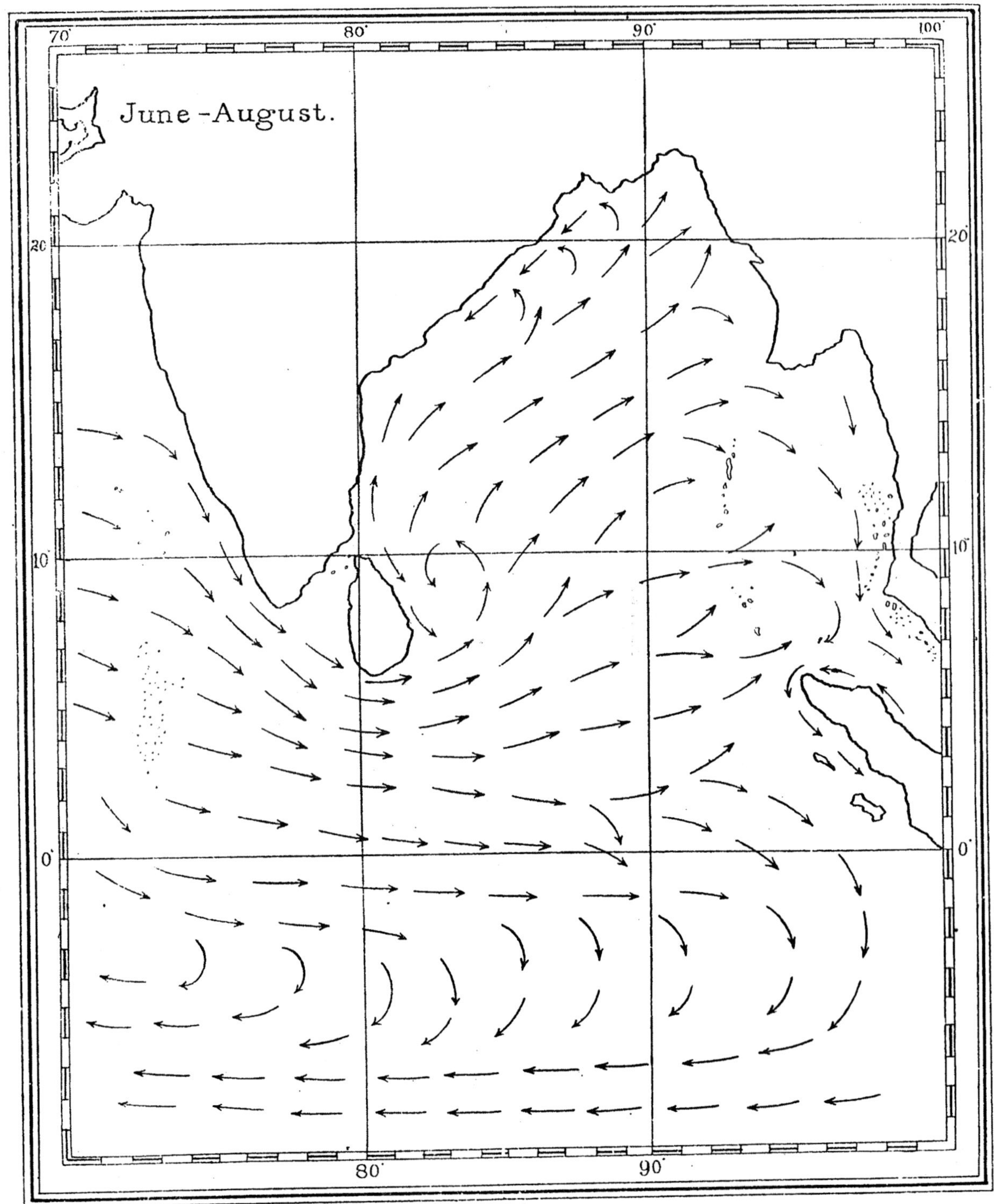

Text-fig. 85; The surface currents of the Bay of Bengal; June to August.

0° and 10°S Lat. this current bends round to join the westerly-flowing drift of the southern equatoral current. On the north the easterly drift passes in the main in a north-easterly direction across the Bay of Bengal and then east and south-east across the Andaman Sea, finally making its exit from the Indian region through the straits of Malacca, but a smaller drift, coming from this latter channel, curves round the north end of Sumatra and then passes for some distance south and possibly south-west along the Sumatran coast. In two areas, however, there is evidence of a rotational movement; the first and greater of these lies close to the east coast of the Indian Peninsula extending north along the Orissa Coast; the second cyclonal movement is situated to the east of Ceylon. The 'isohaline' contours suggest that there is a third movement of this nature lying opposite the centre of the mouth of the Bay about Lat. 4°N; it is possible that this is to be attributed to the current, already mentioned, that sweeps in a north-westerly direction through the Straits of Malacca and then along the north-east coast of Sumatra. This current finally bends to the west and on passing out of the Andaman Sea meets the easterly flowing drift and may set up a cyclonal movement, the position of which is indicated by the area of low salinity that we found in Lat. 4°N. Long. 88°E.

DAILY VARIATION IN THE SURFACE SALINITY.

Whenever the R.I.M.S. "Investigator" has been proceeding from one port to another or from her base to the survey-ground it has been my custom to take observations on the temperature and salinity of the surface-water at four-hourly intervals throughout the day, namely at 4, 8 and 12 in both forenoon and afternoon. During the day both water-samples and temperature-readings have been taken by me personally and the samples and observations at 12 midnight and 4 a.m. have been taken for me by the officer on duty. I have repeatedly noticed that the surface salinity exhibits a quite appreciable range of variation and, moreover, that in any given series of days this variation tends to exhibit a definite rise and fall at certain stated times of the day. Very similar oscillations of the surface salinity have been noted by Helland Hansen and Nansen (1909, p. 98 *et. seq.*) in the Norwegian Sea, where also they found evidence of regular oscillations in the deeper strata; but the two series did not appear to coincide. As regards the surface variations these authors suggest with some hesitation that there may be some connection between the oscillations and the lunar phases; and as regards the oscillations themselves they remark "Although it may be difficult to understand how these periodical oscillations are created, their regularity in the curves, and their coincidence with the upper and lower culminations of the moon are so marked that they can hardly be ignored as being merely accidental". It must be pointed out that my observations have, for the most part, been taken while the ship has been steaming and, therefore, with the single exception of a period of two days at the end of April 1914, when we were engaged in searching for a reported rock off the west coast of India, no two consecutive observations are in the same locality.

A study of the various charts of the surface-salinity that I have given above (*vide* Text-figs. 68–74) shows that in addition to the major seasonal oscillations which I have already dealt with, (*vide supra*, p. 277 *et. seq.*) the salinity and specific gravity exhibit a number of minor fluctuations, rising at intervals of a few hours to a maximum and then falling again. If we take a census of the number of instances on which the maximum salinity falls in each 4-hourly interval, we arrive at the following figures;

	A.M.			P.M.		
Time of day.	4	8	12	4	8	12
No. of instances of maximum salinity.	9	4	4	5	7	7
Computed average.	6·8	5·5	4·7	5·0	6·4	7·2

It is clear then that the maximum daily rise of salinity exhibits a definite tendency to occur at or near midnight and this is almost certainly due to the daily 'change over' of the waters of the upper levels under the influence of evaporation by day and cooling during the night hours. During the day evaporation raises the salinity of the surface-water, but owing to the simultaneous raising of the temperature the actual density (σ_t) of this uppermost stratum remains low and hence the more saline but actually less dense water continues on the surface till the lowering of the temperature at night renders its density greater than that of the lower stratum, and convection currents are set up resulting in the change over of the two strata. If now we consider the variations in the density of the surface-water *in situ* (σ_t) and the fluctuations in the temperature of the water, it is clear that the latter is the prime agent in the various oscillations in the density and in consequence the fluctuations exhibit the greatest range between early morning when the temperature is lowest and about 2 p.m. when it is highest. The extent of this range will be found to vary with the seasons of the year since as I have already shown (*vide supra*, p. 214 *et. seq.*) the range of temperature of the surface-water varies very considerably in different months. The average range of temperature of the surface-water in the Bay of Bengal in different months is as follows :—

Month.	Average range of temperature of surface water. °C	
October	1·33	
January	0·86	
February	0·77	(Valdivia)
March	2·24	
April	1·98	

In every case, as the temperature rises it will produce a fall in the density *in situ* of the sea-water that will reach a minimum at 2 p.m. and as the temperature falls, so the density will rise to a maximum at or shortly after midnight.

In the accompanying Text-figure 86 I have given the average of all my observations, taken at four-hourly intervals throughout the day, in the Bay of Bengal from 1914 to 1923. The results show very clearly the manner in which the surface-temperature and the density of the water *in situ* alternate with each other and I have also indicated (by the dotted line) the variation in the density that would have been produced had it been due to changes in the temperature alone, the salinity remaining constant throughout the whole period. This shows clearly that the actual density, as observed by a "Buchanan" hydrometer, is greater at 4 a.m. and 8 p.m. and considerably less at 12 noon and slightly less at 12 mid-night than the density calculated from the change of temperature alone. In the following Table 69 I have given the observed average density at different times of the day and the density calculated from the variation in the temperature if the salinity had remained the same throughout.

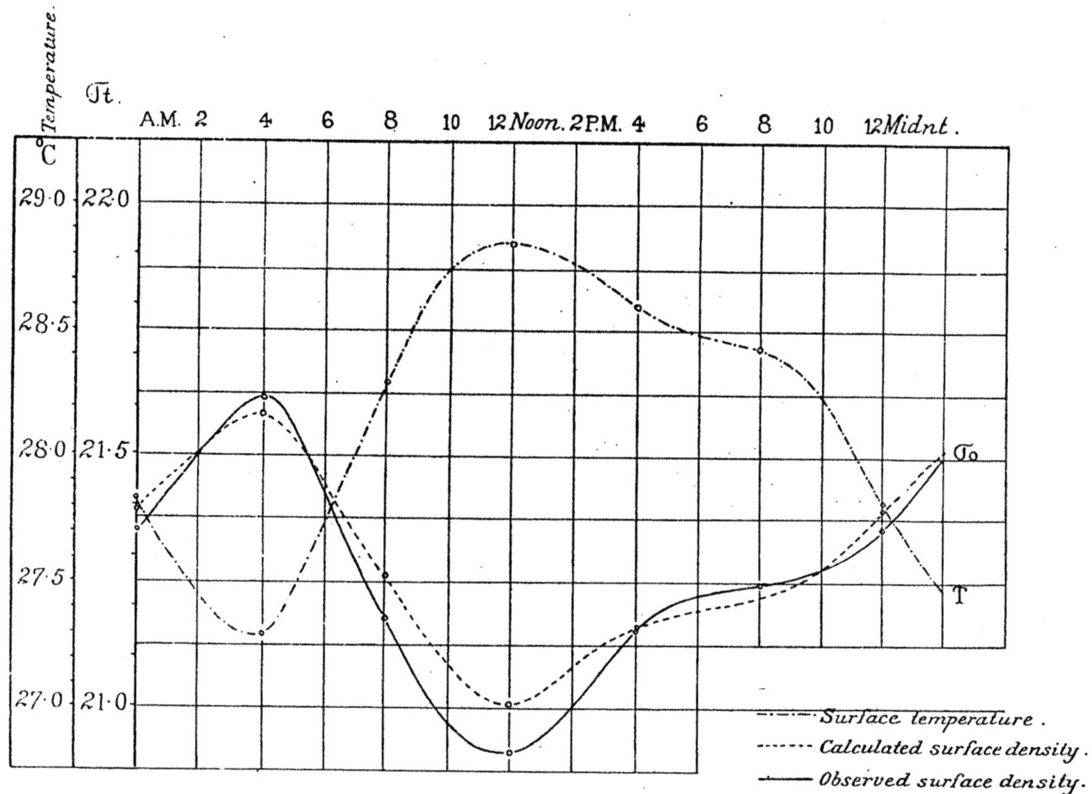

Text-fig. 86. Showing the observed surface temperature, the observed density in situ (σt) and the calculated changes in the density produced in a sample of water by changes in temperature only.

Time of day.	Average temperature.	Observed density *in situ.* I	Calculated density *in situ.* II	Difference. I-II
4 A.M.	27·297	21·615	21·585	+0·030
8 A.M.	28·292	21·172	21·262	−0·090
12 Noon.	28·837	20·910	21·083	−0·073
4 P.M.	28·592	21·153	21·163	+0·010
8 P.M.	28·420	21·239	21·220	+0·019
12 Midnt.	27·808	21·355	21·392	−0·037

Table 69; showing the actual and calculated densities in the surface-water of the Bay of Bengal at different times of the day.

It seems clear from the above that some influence other than the change of temperature is at work producing changes in the density *in situ* of the surface-water and this can only be a change in the actual salinity. If, now, we exclude all observations that have been taken in those areas to the east and west of the Bay, where, as we have already seen, water of low salinity, owing to the admixture of river-water, is met with at different times of the year and consider only those readings taken well out in the central region of the Bay, where one might expect to find a more or less uniform state of affairs, we find that the maximum salinity, as shown by the specific gravity, tends to occur at 4 a.m. and 4 p.m. respectively. I have purposely limited the observations that I am now considering to those taken and examined by means of a 'Buchanan' hydrometer, and have excluded those taken more recently and examined by the titration method, so that no error may be introduced owing to the employment of different methods. The result of all these observations in this central region of the Bay gives the following average specific gravity of the water at different times of the day :

4 a.m.	8 a.m.	12 noon	4 p.m.	8 p.m	12 midnt.
27.322	27.027	26.996	27.201	27.116	27.004

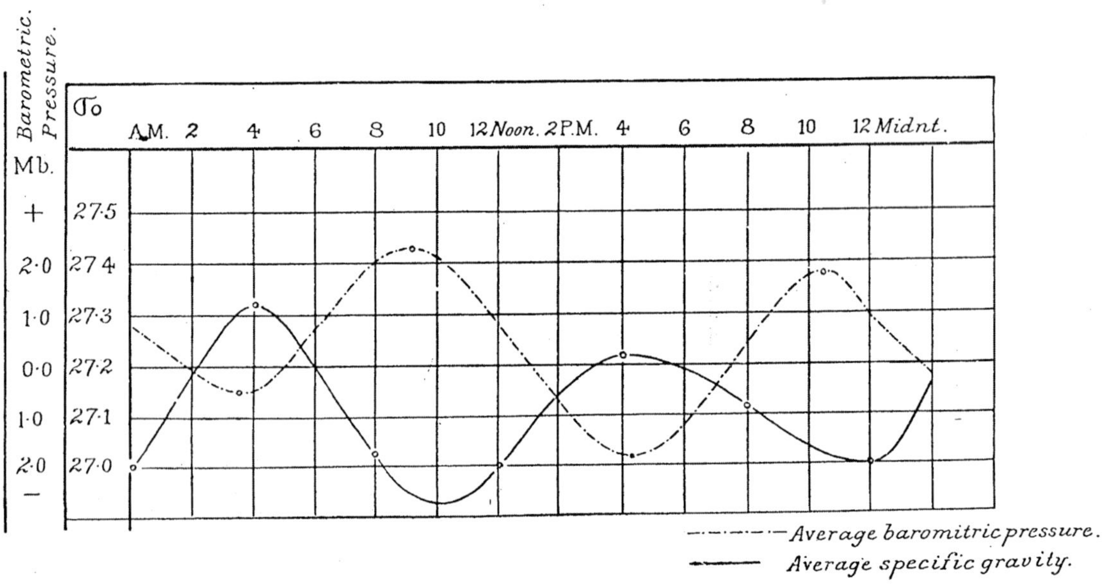

Text-fig. 87. Showing the average specific gravity (σ_0) of the surface water at 4-hourly intervals, as compared with the rise and fall of barometric pressure, in the Bay of Bengal in October.

The variation is but slight, amounting to only 0.326 and, if we were dealing with isolated observations, might be attributed to experimental error, but I have been able to trace an exactly similar rise and fall in the surface-water wherever I have been able to take a series of observations, and there seems no doubt that we have here evidence of an actual natural phenomenon.

I have already (*vide supra, p.* 185) called attention to the manner in which the salinity of the surface-water in various regions round the central group of the Nicobar Islands appears to rise and fall with the changes in the barometric pressure and the same periodicity holds good in this region of the Bay of Bengal. In Text-figure 87 I have plotted the curves of the average daily variation in the specific gravity (σ_0) of the surface-water, as given above, and for the purpose of comparison the average fluctuation of the barometer at different times of the day at Calcutta, as given in the Barometer Manual (1919). The manner in which the two curves alternate with each other is sufficiently striking to need no comment. A high barometric pressure corresponds almost exactly with a lowered specific gravity and, *vice versa*, a fall in the barometric pressure is accompanied by a rise in the specific gravity.

As I already mentioned, most of my observations were taken while the R.I.M.S. 'Investigator" was steaming from port to port and it is, of course, possible that the ship has from time to time passed through belts of water of low average salinity; but this double daily variation in the salinity is found to occur in all parts of the Indian seas and, moreover, the type of the curve of variation in different areas during the same period of the year agrees closely. It would be absurd to suppose that a passage across belts of low average salinity had always occurred at the same time of the day in areas wide apart and at different times of the year, and it seems beyond doubt that there is every day a perfectly definite tendency towards a double oscillation of the salinity of the surface-water in Indian seas. That this double diurnal oscillation is not due to local patches of water of low salinity encountered during a voyage from one place to another is clearly indicated by the result of my observations in Revello Channel in the Nicobars, where we were anchored for days at a time in March, 1925 (*vide supra*, p. 185) and those off the west coast of India in 1914. During the end of April and the first two days in May, 1914, the "Investigator" was engaged in searching for a rock that had been reported off the west coast of India in the neighbourhood of the Vangala Light. During this period we were anchored well out from the coast and though we twice changed our anchorage, each time the distance moved was only a few miles. The actual dates and positions were as follows:—

Date.	Position. Lat.N.	Long. E.
From April 29th, 8 p.m. to April 30th 4 a.m.	15°54′00″	73°30′45″
From April 30th, 4-30 a.m. to May 1st 2 p.m.	15°22′15″	73°24′50″
From May 1st 2-30 p.m. to May 2nd, 8-30. a.m.	15°51′45″	73°31′30″

The average salinity of the surface-water at different times of the day during this period was as follows:—

Time of day,	4 a.m.	8	12 noon.	4 p.m.	8	12 midnight.
Salinity	36.15	36.32	35.09	36.26	36.48	36.45

These results have been plotted in Text-figure 89 and for the purpose of comparison I have again given the average rise and fall in the barometric pressure as recorded at Calcutta, since I did not take any observations on board the ship during this period. A comparison of the double diurnal oscillation of the surface-salinity in the open waters of the Bay of Bengal in October (Text-fig. 87) and at the south end of Revello Channel in the Nicobars in March (Text-fig. 88) shows that the two curves are absolutely identical, having two maxima at 4 a.m. and 4 p.m. and two minima at 10—10-30 a.m.

and 10 p.m.—12 midnight, and in each case the major oscillation occurs between midnight and 12 noon. The rise and fall of the salinity on both these occasions alternates with the rise and fall of barometric pressure. The observations taken off the west coast of India in April-May, 1914 show an equally clear double diurnal oscillation, but in this case the salinity varies with the barometric pressure and not against it. The difference between the results obtained in these two areas on the east and west sides of India respectively might be due to different local conditions, but, as I shall show later, the evidence clearly indicates that it is seasonal.

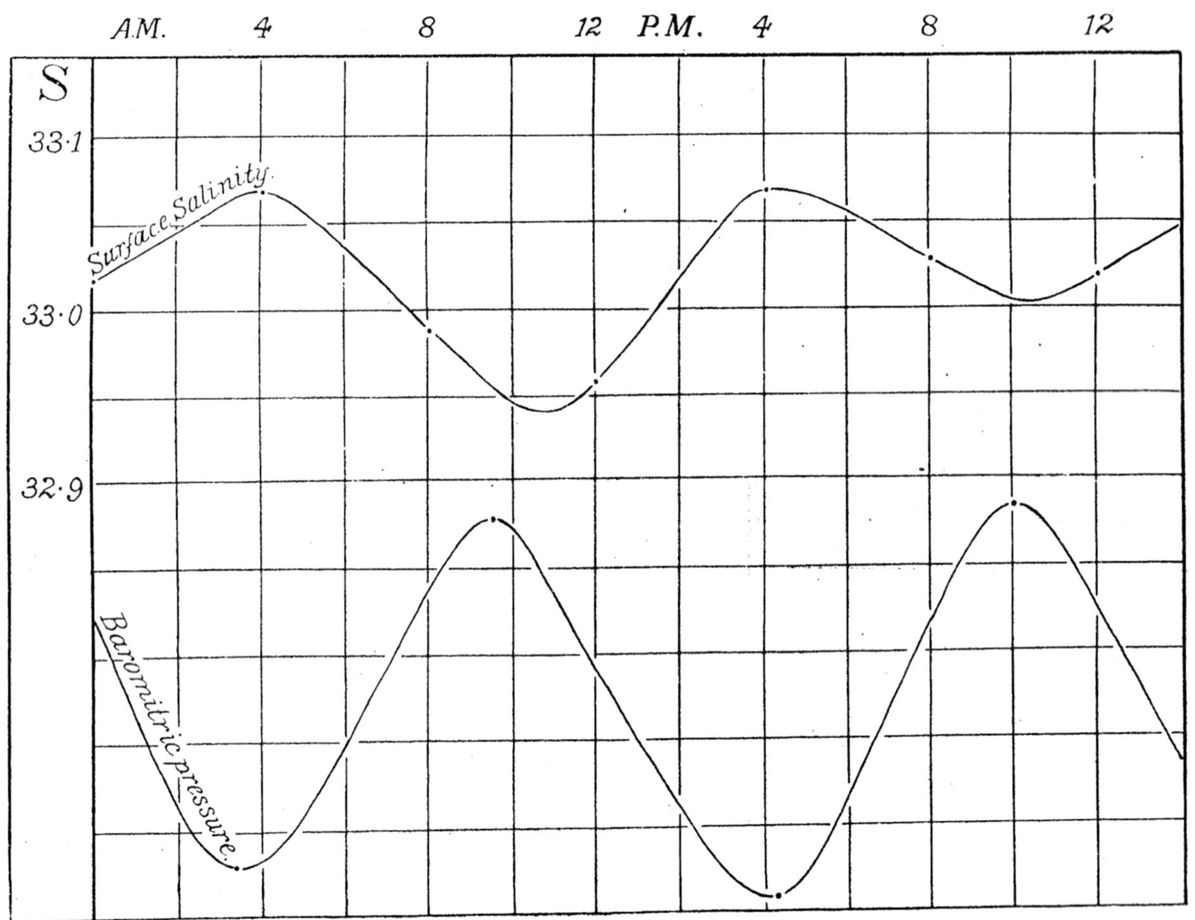

TEXT-FIG. 88.—Showing the average daily rise and fall of salinity of the surface-water and the rise and fall of barometric pressure in Revello Channel, Nicobars, in March 1925.

The surface of the sea is, of course, constantly undergoing changes that are the result of several distinct agencies, among the most important of which is evaporation. Evaporation is always tending to cause an increase in the salinity of the surface-water and the rate of evaporation will vary with changes in the temperature, the humidity of the atmosphere, the strength of the wind, etc.; on the other hand convection currents will tend to remove the condensed water from the surface and substitute less saline water from below. These factors would, of themselves, be sufficient to account for a single daily oscillation, but not for a double oscillation such as has been found to occur.

An oscillation in the salinity, such as I have described, may be due to one of two or possibly to both causes, namely (1) a lateral movement of the surface-water or (2) to an up-welling of water of a different salinity from below; and such evidence as I at present possess seems to me to point to the latter as being the more important, if not the sole, cause and, moreover, that this is in some way connected with the daily changes in barometric pressure. Whatever may be the cause of the alteration in the pressure of the atmosphere, it will affect not only the atmosphere itself but also the actual level of the ocean. A diminution in the pressure on the surface will permit of a corresponding degree of expansion of the water lying vertically below and, as a result, there will be within the area of diminished pressure a rise in the surface-level of the water that must be accompanied by a tendency for the water to flow outwards away from the affected area. This phenomenon has been carefully studied in connection with changes in the level of great lakes and especially in the case of the great American Lakes by Hayford

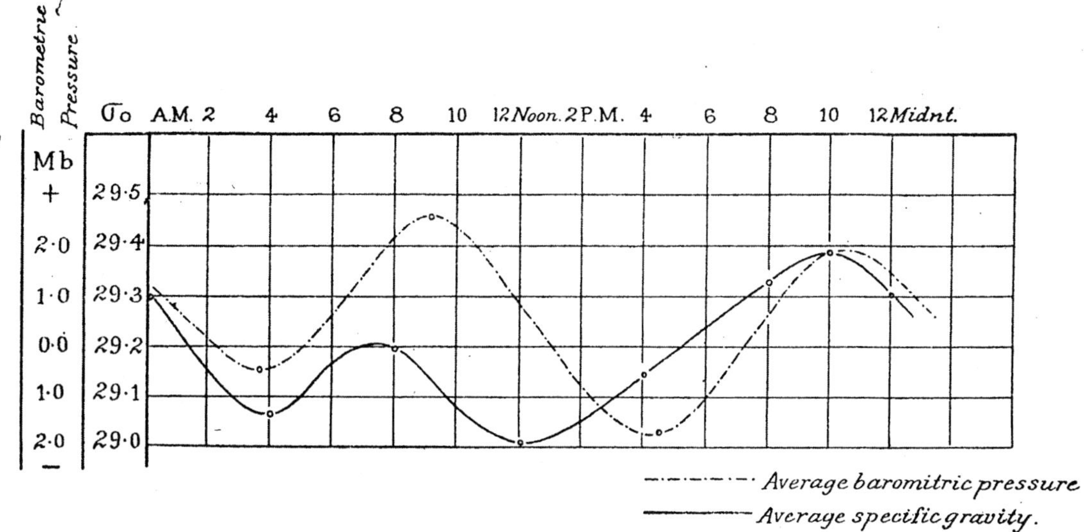

Text-fig. 89. Showing the average specific gravity (σo) of the surface water at 4-hourly intervals as compared with the rise and fall of barometric pressure, off the west coast of India,-April 29-May 2, 1914.

(1922), but exactly similar changes of level occur in all large areas of water, the surface rising and falling with the changes in barometric pressure. The actual extent of the fluctuation in level can be calculated by means of the formula given by Hayford (*loc. cit.* p. 11);

$$H_1 - H_2 = -(M_1 - M_2)\ \frac{d_m}{d_w}$$

Where $H_1 - H_2$ = the degree of elevation of the surface
$M_1 - M_2$ = the change in height of the barometer and
d_m = the density of mercury (13.6)
d_w = the density of water

Since the density of sea-water is greater than that of fresh-water the actual amount of elevation or depression of the surface-level will be somewhat less than in the case of fresh-water. Taking the density of sea-water as 1.02 and that of mercury as 13.6, the actual rise and fall of the sea-level will approximately be 0.11 feet for every rise of 0.1 inch of mercury in the barometric pressure. The normal variation in the barometric pressure is much greater in the Tropics than in the Temperate or Polar regions and in consequence the rise and fall of the sea-level will be proportionately greater. The average range of the barometric pressure in the Tropics is given in the Barometer Manual as 0.34 inches at 10 degrees N and 0.36 at the Equator; but in the neighbourhood of land this is somewhat increased and in Calcutta it is 3.2 millebars in July and 4.4 millebars in January. A careful record of the changes in the barometric pressure was kept on board the "Investigator" and without going into unnecessary detail it may be stated that the average rise and fall of the mercury-level was approximately 0.1 inch. This would be sufficient to cause a rise and fall of the sea level of 1.3 inch; and, as Krummel (1911, p. 515), quoting from the work of Alexander von Humbolt, points out, there is thereby produced an insignificant movement of water, which is forced to move from east to west every six hours. But this insignificant movement appears to be, of itself, quite insufficient to produce the changes that, as I have shown, occur in the salinity of the surface-water and it seems clear that there is some other factor at work causing a periodic upwelling of water from some depth below the surface that tends to coincide with, though it is not directly traceable to, the changes in the atmospheric pressure.

While my own observations have been confined entirely to Indian seas, a study of the data and results published by other workers in different parts of the world seems to indicate the possibility that these periodic diurnal changes in salinity are not confined to Indian waters. Pettersson (1909) has shown that in addition to the long-period fluctuations in the temperatures of the deep strata in the Skagerak, which he observed in the Gullmar fjord and which show a period of oscillation of approximately 14 days, there is also a double diurnal oscillation in the respective levels of the isohalines and, moreover, the vertical movement of the lowest, most saline water is greater than that of the intermediate, less saline stratum, a rise of the 32 ‰ water being accompanied by a marked thinning of the 26 to 30 ‰ layer. The data that he gives for the rise on July 27/28, 1908, appears to exhibit a double oscillation with maxima at 3 p.m. and 4 a.m. respectively. This double oscillation Pettersson believes to be due to the effect of tides but he gives no data regarding the ebb and flood on these days and it is not shown that the rise and fall of the tide in any way correspond to the salinity changes. It is, however, worth pointing out that the times of these maxima correspond very closely to the minima of the double diurnal oscillation in the barometric pressure, as is the case in the double diurnal variations in the salinity of the surface water in the Indian ocean. A similar rise and fall has also been indicated in the results of observations in the region round Heligoland [*vide* Reichard, A. C., 1913]. Again, H. W. Harvey (1923, p. 225) points out that in the English Channel in the neighbourhood of Plymouth "a comparison of the depth of the water layers of equal temperature at the station after a lapse of twelve hours, when the whole mass of water will have been swept back to nearly the same position by the tidal stream, is somewhat suggestive of undulatory movement, as is thought to occur in the Norwegian Sea"; but in this case, as in all observations in shallow waters where the tidal streams run strongly, one is unable to arrive at any definite conclusions, since these are liable to be invalidated by changes in the thickness of the successive strata produced by different rates of lateral movement at different levels, which may of themselves result in a relative thinning of certain strata.

The work of the "Michael Sars" (*vide* Murray and Hjort, 1912, pp. 278-280) has provided evidence that indicates the presence of undulatory movements of water at different depths in the upper waters of the North Atlantic Ocean in the region of the Faroe-Shetland Channel, and Hjort (*loc. cit.* p. 279) remarks "these observations go far to prove the presence of such undulations of the water layers,... But these variations are not compressed in one single period, as if they were due to an ordinary progressive wave or an ordinary standing wave alone. The shape of the curve points to complicated functions of the velocity as the cause of the variations, but it is possible, nay probable, that we are confronted with an interplay of several different factors." Similar vertical oscillations have been recorded, as already mentioned, in the mid-water region of the Norwegian Sea by Helland Hansen and Nansen (1909, p. 89 *et. seq.*), their detection being brought about by changes in the temperature of the water at different levels. After reviewing all their observations these authors (*loc. cit.* p. 105) sum up their results as follows:—"All we can say at present is that many of our observations indicate the probability that great, hitherto unknown, oscillatory movements may occur in the intermediate strata of the sea and also at its surface."; and in a footnote they add, "It might be more reasonable to suppose that an increased movement of the surface strata will stretch the strata, so to speak, and consequently diminish their thickness, while a decrease or stop in the movement may make the surface strata considerably thicker; and if there is much difference between the strata near the surface, a bucket may in the former case take more water from the underlying strata than in the latter." Equally if there be such a difference in the upper strata, then wave action may also cause an oscillation in the observed salinity of the surface-water by varying the degree to which admixture takes place and, as I have already mentioned, in Indian waters there seams to be clear evidence of such an oscillation. It was suggested to me by Mr. Durst, of the Air Ministry, London, that the probable cause of these observed changes in the salinity at different times of the day is to be sought in the variation in the strength of the wind. I have previously (*vide supra*, p. 79 *et. seq.*) shown that we can detect a distinct double oscillation in the strength of the wind and that this is clearly correlated with the rise and fall of the barometer, and it seems more than probable that the rise and fall in the salinity of the surface-water is brought about by an increase in the strength of the wind causing an equivalent increase in the rapidity of the surface current. Such an alteration in the current will cause a thinning of the surface-stratum and a consequent approach nearer to the surface of the water from a deeper level; simultaneously, the increase in the wind-force will cause an increase in the height of the waves and this will cause an increase in the amount of admixture that is going on between the two levels. Again, an increase in the wind-force will of itself tend to cause a rise in the salinity of the surface-water, since it will also cause an increase in the rate of evaporation. I have already called attention (*vide supra* p. 246) to Eliot's paper in which he shows that at Trivandrum in the Madras Presidency' the rate of evaporation of sea-water exhibits two maxima and two minima in the 24 hours, the maxima occurring at 3—3-30 p.m. and at 11 p.m. during the period October–February, but only a single maximum at 3-0 p.m. in March–May. Since the rate of evaporation is mainly dependent on the strength of the wind, the variation in the strength of the wind and the salinity of the surface-water should in any series of observations show a distinct and clear tendency to vary together and the two curves should run a more or less parallel course. In the following Text-fig. 90 I have plotted the strength of the wind and the oscillation

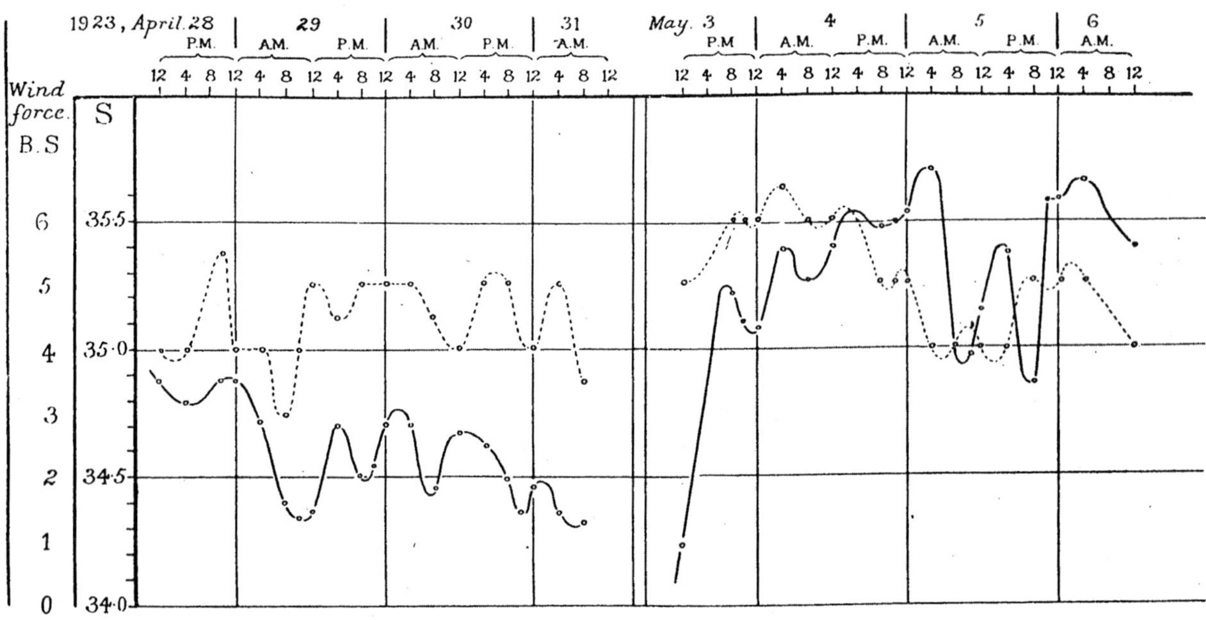

Text-fig. 90 The variation in the wind-force and in the surface salinity of the Laccadive Sea in April–May, 1923.

in the salinity of the surface-water in the Laccadive Sea off the west coast of India during the last few days in April and the first few days of May, 1923 and the manner in which the two curves follow each other is clear, the salinity however showing a 'lag' of approximately 3 to 4 hours. If these oscillations in salinity were due solely to evaporation, it is difficult to understand why this lag should be present and, moreover, as we have seen the relationship of the oscillation in surface-salinity to the changes in barometric pressure and, therefore, to the wind-force appears to be completely reversed at certain seasons of the year. It seems probable, therefore, that the changes are due to actual movements of the water masses rather than to any change in the same mass of water.

The agreement between the times of maximum wind-force and the maximum variation of the salinity, whether this be in the nature of an increase or a decrease, is by no means always exact. This may be due to more than one cause ; in the first place it may be that the number of readings of the wind-force have been too few to give the exact time of the variation of the wind ; while, since the strength of the wind was estimated by the officer of the watch, who merely had his previous experience to guide him, the estimation of the strength may sometimes have been somewhat wide of the mark. Another factor that may influence the time relationships of the two phenomena is the presence of a "lag" in the movement of the water strata. As Hayford (1922, p. 17) in his work on the effect of barometric pressure on the water level of the great American Lakes, has pointed out, "friction will tend in general to reduce the range of fluctuation of water-surface and to produce a lag of the response behind the barometric changes which produce it," while, further, inertia will also tend to produce a lag behind the change in pressure.

In the case of the changes in salinity that we have been considering, it appears that we may at times find a quite appreciable lag in the relationship of the two phenomena ; such a lag is well seen in the chart of the changes in the Bay of Bengal in January (*vide* Text-figure 101) and a similar lag, of approximately 2 hours, is seen in the results for May, 1923 (*vide* Text-fig. 117) off the west coast of India. If the oscillation in the salinity is due to the variation in the strength of the wind we should expect to find that the salinity changes follow after the changes in the strength of the wind-force, and, by reason of its inertia, the movement of the water, both in the lateral direction in case of the surface stratum that is being moved by the wind and in the vertical direction in the case of the upward-moving deeper water that is coming up to the surface to take its place, will continue for some time after the original impulse has disappeared. It is interesting to note that this lag in the variation of the salinity behind the variation in the wind-force appears to be most marked in, if it is not absolutely confined to, those cases in which the two sets of changes run in opposite directions, a rise in the wind-force being accompanied by a fall in the surface-salinity. This is, after all, what one would expect since the first effect of such an increase in the wind-force will be to cause an increased evaporation from the surface and so increase the salinity and it is only after this surface-water has been blown away and less saline water from below reaches the surface, or sufficiently near the surface to become mixed by wave action with the uppermost stratum, that the fall of salinity would become noticeable.

In the other set of changes in which the rise of the wind-force is accompanied by a rise in salinity, it is quite a common occurrence to find that the wind-force shows a lag behind the salinity variation.

If, now, we consider the average variation in the salinity of the surface-water at different times of the day and in different months, we get the results given below in Table 70, in which I have given the average oscillation of salinity in each month irrespective of the area in which the observations were taken and the average oscillation for the whole period of the survey-season from October to May inclusive:

	Time of day.						No. of days.	Average Salinity.	Range of variation.
	A.M.			P.M.					
	4	8	12	4	8	12			
October ..	34·52	34·58	34·65	35·27	34·60	34·48	22	34·68	0·79
November ..	34·92	34·67	34·65	34·73	34·54	34·85	7	34·73	0·38
December ..	34·42	34·21	34·21	34·31	34·56	34·76	10·5	34·41	0·55
January ..	33·67	33·58	33·26	33·25	33·51	33·60	11	33·48	0·42
February ..	34·16	34·34	33·90	34·20	34·26	34·12	8	34·16	0·44
March	34·39	34·42	34·30	34·25	34·34	34·35	13·5	34·34	0·17
April	34·17	34·28	34·30	34·35	34·36	34·29	31	34·29	0·19
May	35·70	35·41	35·30	35·57	35·57	35·44	4·5	35.48	0·40
Average ..	34·49	34·44	34·32	34·49	34·47	34·49	..	....	....

Table 70; the average salinity, based on all observations in Indian waters, of the surface-water at different times of the day.

It seems clear that there is a distinct tendency for the surface-salinity throughout the whole period of the survey-season from October to May to exhibit a double oscillation during the twenty-four hours. In the case of the average for the whole period the salinity reaches its maximum at about 2 a.m. and again at 4 p.m., the corresponding minima occurring at 12 noon and 9 p.m. The second minimum is, however, only very slightly marked. In some months the average salinity, deduced from all observations in different regions, exhibits only a single oscillation during the day. This type of single oscillation is found in October, December and again in February, although in certain regions, as in the case of the Bay of Bengal (*vide ante*, p. 296) in the month of October we may find a perfectly clear double oscillation. There is also a considerable difference in these months in the character of the single oscillation, for, whereas in October the average salinity is lowest at about midnight and rises steadily till 4 p.m., thereafter falling rapidly again to midnight, in the other two months the average salinity is highest at or soon after midnight, falls rapidly till approximately midday and thereafter rises again. It would appear, therefore, that the conditions present in the surface levels in December and February are the exact opposite of those of October, and that at some intermediate period there is a direct change from one phase to the other. The changes that the periodic oscillation in the surface-salinity undergo during the course of the eight months for which I possess data, are best seen by comparing the difference month by month between the 4-hourly average and the average for the month itself. The results of this comparison are given below in Table 71:

	Time of day.					
	A.M.			P.M.		
	4	8	12	4	8	12
October	−0·16	−0·10	−0·03	**+0·59**	−0·08	−*0·20*
November	**+0·19**	−0·06	−*0·08*	**0·00**	−*0·19*	+0·12
December	+0·01	−*0·20*	−*0·20*	−0·10	+0·15	**+0·35**
January..	**+0·19**	+0·10	−0·22	−*0·23*	+0·03	+0·12
February	0·00	**+0·18**	−*0·26*	+0·04	**+0·10**	−*0·04*
March	+0·05	**+0·08**	−0·04	−*0·09*	0·00	+0·01
April	−*0·12*	−0·01	+0·01	+0·06	**+0·07**	0·00
May	**+ 0·22**	−0·07	−*0·18*	**+0·09**	**+0·09**	−*0·04*

Table 71; showing the difference at 4-hourly intervals throughout the day between the average 4-hourly salinity and the average monthly salinity.

In the months of October, January and March the variation from the monthly average clearly exhibits a single oscillation during the day but *inter se* the average results in these three months exhibit a very clearly defined difference for in October the variation from the monthly average is lowest at about midnight, rises steadily till 4 p.m. and then rapidly falls again till midnight; whereas in January the oscillation exhibits a minimum variation at about 4 p.m. and a maximum at or shortly after 4 a.m., so that the two curves of oscillation at these two periods of the year are almost a looking-glass reflection of each other. In March the variation is lowest at 4 p.m. as in January, but rises steadily till about 10 a.m. when there is a sharp fall. In all the other months of the season, the variation from the monthly average shows a clear double oscillation. In November and December this double oscillation is of a very similar type; the variation is highest at about 12 midnight to 2 a.m., falls till 7 or 8 a.m., rises again somewhat till 1 to 2 p.m., falls a second time till 5 to 8 p.m. and then again rises. In February the variation is low at 2 a.m., rises till 8–10 a.m., falls again till 2–4 p.m. and rises again till 8–10 p.m., the curve of variation being the exact opposite of that in November. As already mentioned in March the variation from the average follows a single curve. In April we can detect a faint trace of this double oscillation and in May we get a pronounced double oscillation, that is once again of very much the same type as that met with in November and December, being at its highest at a little after 4 a.m., at its lowest at 9-30 a.m. and then rising slowly to a second but not so pronounced maximum at 6 p.m., after which it again falls till about 11 p.m. It seems clear that we have here evidence of a seasonal change in the variation, a period showing a daily double oscillation alternating with periods of transition, and the direction of the variation from the monthly average changing with the season of the year. Thus in October we have a transitional period with a single oscillation; in November-December we get a dry-season effect; this is succeeded by a transitional period in January, again with a single oscillation, which leads into the wet-season phase in February, corresponding to the north-east monsoon; and after the third transitional phase in

March, we reach the dry-season phase again in May. I have already (*vide supra*, p. 79 *et. seq.*) drawn attention to the fact that throughout Indian waters, whether in the coastal regions or over the open sea, we can detect in most months of the year a clear double oscillation in the strength of the wind that appears to be directly related to, if not actually dependent on, the rise and fall throughout the day of the barometric pressure. If now this alteration at certain definite times of the day in the strength of the wind gives rise to an up-welling of water from below we should find that the rise and fall of salinity at different periods of the year agrees with the conditions present in the upper levels of the sea. Throughout the whole width of the Indian Seas there are usually present two different layers in the water; as a result of dilution by river-water and by rain the surface-water possesses a lowered salinity and below this lies a mass of more saline water; but the relative salinities of these two layers and more especially of the superficial strata of the upper layer will vary enormously with the different seasons of the year. Unfortunately, I have no observations of my own regarding the changes in the salinity of the upper few fathoms of these seas, but, thanks to the kindness of Dr. J. Pearson, Director of the Colombo Museum, Ceylon, I have been able to examine the results obtained by the Ceylon Government Fishery Steamers in the Gulf of Mannar. These have all been published in the Annual Administration Reports and I have taken the opportunity of including here those that have a bearing on the subject that we are now considering.

In the month of October a series of six observations in the Gulf of Mannar gave the following results:

Date.		20. X. 13	21. X. 13	21. X. 13	22. X. 13	22. X. 13	23. X. 13
Position	Lat. N	8°21′	8°18′	8°46′	7°46′	7°57′	7°30′
	Long. E	79°08′	78°42′	78°44′	78°37′	79°14′	79°16′
Surface	..	35·55	35·36	35·25	35·41	34·07	34·70
18 Metres	..	35·55	35·39	35·16	35·49	35·21	34·97
92 "	..	35·09	34·94	35·06	35·06	34·96	35·00
183 "	..	35·08	35·08	35·08	35·12	35·08	....

Table 72; showing the salinity (S) at different depths in the Gulf of Mannar in October, 1913: where more than one result is given for the same depth, I have taken the average.

Again in the same month in 1920 a second series of observations gave the following:—

Date.		16. X. 20	16. X. 20	16. X. 20	16. X. 20	17. X. 20	17. X. 20	....
Position	Lat. N ..	8°1′	8°09′	8°49′	8°46	8°18′	7°46′	7°30
	Long. E ..	79°8′	79°39′30″	79°58′	78°44′	78′42′	78°13′	79°16′
Surface ..	..	35·19	35·35	35·46	35·34	35·34	35·43	35·30
50 Metres ..	..	35·35	35·23	35·30	35·30	35·26	35·17	35·12
100 ,, ..	..	35·21	35·30	35·26	35·26	35·12	35·30	35·07
200 ,, ..	..	35 35	35·25	35·16	35·21	34·99	35·70	35·12
300 ,, ..	..	35·35	....	35·12	35·21	35·70	35·16	35·48

Table 73; showing the salinity (S) at different depths in the Gulf of Mannar in October, 1920.

In the first series (1913) it is clear that in the majority of cases there is a distinct rise in the salinity of the upper levels as we pass from the surface down to a depth of 18 metres and this is followed by a drop in the salinity at the 100 metre level; in the second series (1920) the exact opposite is the case and the surface-water is in all cases, with the single exception of the first set, more saline than the water at lower depths, and this condition can in sets 3, 4, 5 be traced down to the 200 metre level, while in 7 it exists down to the 100 metre level and in series 2 and 6 it is present down to the 50 metre level. As Pearson has pointed out, in every case where the samples, taken in 1913, were obtained from positions south of Lat. 8°30′ N. the surface water possessed a lower salinity than the water taken from a depth of 18 metres. North of this area the conditions are changed owing to the influx through the Pamban Pass and the Palk Straits of water from the Bay of Bengal. It is clear then that at this time of the year there is a tendency in certain years for the surface-water of the open sea to have a distinctly higher salinity than the water lying at a somewhat deeper level; but that conditions may vary from year to year. This is undoubtedly the result of the gradual rise in temperature and the consequent increase in evaporation that has been steadily going on since the cessation of the south-west monsoon in September. In the month of November in the year 1921, a series of 14 observations by Pearson gave the following results:

Date.		16. XI. 21	17. XI. 21				18. XI. 21				19. XI. 21	21. XI. 21
Position	Lat. N	7°13′	7°33′33″	8°09′	8°31′	7°46′	8°08′	8°46′	8°46′	8°21′	8°49′	8°40′
	Long. E	79°45′	77°43′	79°39′30″	79°42′30″	78°37′	78°08′	78°20′	78°44′	79°08′	79°15′	70°45′
Surface	..	33·06	32·83	32·72	32·41	32·74	32·90	32·72	32·88	33·03	33·28	33·95
50 Metres	..	34·25	33·13	34·60	35·25	35·67	34·60	35·23	34·51	33 34	34·14	35·12
100 ,,	..	35·35	35·53	35·53	35·30	35·37	35·44	35·17	35·64	35·41	35·21	35·53
200 ,,	..	35·12	35·12	35·17	35·21	35·21	35·08	35·07	35·21	35·21	35 07	35·17
300 ,,	..	35·17	35·25	35·23	35·25	35·21	35·12	35·12	35·30	35·25	35·30	35·25

Date.		27. XI. 21		
Position	Lat. N.	8°09′	7°33′30″	7°13′
	Long. E.	79°39′30″	79°43′	79°45′
Surface	..	33·19	33·39	33·26
50 Metres	..	35·07	33·37	33·37
100 ,,	..	35·61	35·43	35·70
200 ,,	..	35·17	35·12	35·21
300 ,,	..	35·25	35·21	35·25

Table 74 : showing the salinity (S) at different depths in the gulf of Mannar in November, 1921.

A second series of observations in the same month in the following year gave the results below :—

Date.		19. XI. 22		20. XI. 22			21. XI. 22				22. XI. 22	
Position	Lat. N ..	7°10′	7°40′	8°0′	8°15′	8°25′	8°44′	8°40′	8°44′	8°18′	8°00′	7°48′
	Long. E.	79°30′	79°20′	79°30′	79°08′	79°30′	79°25′	79°0′	78°35′	78°42′	78°19′	78°51′
Surface ..	..	35·16	35·32†	35·16	35·48†	35·34†	35·28	35·32†	35·17†	35·48†	35·46†	33·37
50 Metres	..	35·71	35·55	35·26	35·39	35·57	35·57	35·35	35·73	35·79	35·57	34·97
100 ,,	..	35·68	35·05	35·26	35·21	35·16	35·43	35·21	35·10	35·44	35·34	35·21
200 ,,	..	35·03	35·10	34·99	35·14	35·12	35·10	35·05	35·07	35·34	35·21	35·52
300 ,,	..	35·07	35·05	35·21	35·21	35·17	35·16	35·28	35·21	35·44	35·32	35·68

Table 75 : Showing the salinity (S) at different depths in the Gulf of Mannar in November, 1922.

From a comparison of these results with those taken in October it is clear that the surface-water has undergone a great change. The surface stratum now has a considerably lower salinity than the water at a slightly deeper level; in 1921 this dilution of the surface-water could be detected in the majority of observations down to a depth of 150 metres, the salinity steadily rising as we pass from the surface to this depth and a comparison of the surface-salinity shows that in this area in 1921 the degree of dilution of the surface-water was very much greater than in 1922, in which year the effect can only be traced to a depth of 50 metres. It is, however, interesting to note that in one instance* in this latter year, this dilution of the surface-layer is absent, the salinity steadily decreasing from the surface down to 200 metres, and in no less than six instances (†) the salinity, while increasing from the surface to a depth of 50 metres, then decreases again to the 200-metre level.

This condition of the sea, in which the surface-layer consists of water of a less salinity than that at a deeper level, will remain constant till the end of the north-east monsoon and by January will have become thoroughly established throughout Indian seas. This is clearly seen in the results of the observations taken by Pearson in the month of January, 1921, which I reproduce below :

Date.		5.I.21	5.I.21	5.I.21	5.I.21	6.I.21	6.I.21	6.I.21	7.I.21	7.I.21
Position	Lat. N ..	7°57′	8°21′	8°09′	8°31′	8°49′	8°46′	8°18′	7°46′	7°30′
	Long. E ..	79°14′	79°08′	79°39′30″	79°42′30″	78°15′	78°44′	78°42′	78°37′	79°16′
Surface	..	33·68	35·08	34·00	34·54	34·40	33·10	33·87*	34·63	34·14
50 Metres	..	33·84	35·17	34·85	34·94	34·92	35·17	33·80	35·19	35·01
100 "	..	35·26	35·23	35·07	35·07	35·05	35·35	34·85	34·99	35·19
200 "	..	35·26	35·26	34·90	34·96	34·94	33·12	34·67	34·94	34·99
300 "	..	35·23	35·25	35·8	..	34·96	35·43	34·94	34·85	35·01

Table 76: showing the salinity (S) at different depths in the Gulf of Mannar in January, 1921.

A comparison of these results with those in November of the same year shows that now the dilution of the surface-water can be clearly traced down to a depth of 100 metres; in only a single instance* is it absent, and a comparison of the surface-salinity with that in the preceding month of October, 1920, (*vide* Table 73) shows how much dilution has taken place.

By the end of March and the commencement of April the north-east monsoon will have ceased and in consequence conditions will again begin to change in the surface-levels. In the following Tables 77, 78 and 79 I have given the results obtained in the Gulf of Mannar at the end of March or early in April in the three years 1920 to 1922.

Date.		3.IV.20	3.IV.20	4.IV.20	4.IV.20	4.IV.20	4.IV.20	5.IV.20
Position	Lat. N ..	8°49′	8°21′	8°46′	8°46′	8°18′	7°57′	7°30′
	Long. E ..	79°15′	79°08′	78°44′	78°20′	78°42′	79°14′	79°16′
Surface	..	34·90	34·67	34·90	34·34	34·29	33·93	34·23
50 Metres	..	34·78	34·36	34·72	34·99	34·45	34·63	34·67
100 "	..	34·79	..	35·26	35·46	35·48	35·53	35·71
200 "	..	35·35	35·34	35·23	35·35	35·07	35·14	35·32
300 "	..	..	34·27	35·35	34·90	35·32	35·14	35·10

Table 77; showing the salinity (S) at different depths in the Gulf of Mannar in April, 1920.

Date.		29.III.21	29.III.21	29.III.21	30.III.21	30.III.21	2.IV.21	2.IV.21	2.IV.21	3.IV.21
Position	Lat. N	8°49′	8°46′	8°18′	8°21′	7°58′	7°13′	7°46′	7°30′	8°08′
	Long. E	79°15′	78°44′	78°42′	79°08′	79°14′	79°45	78°37′	79°16′	78°08′
Surface	..	35·07	34·88	35·03	34·87	34·87	34·76	34·72	34·90	34·63
50 Metres	..	35·03	34·81	35·08	34·89	35·10	35·07	35·14	34·99	35·30
100 ,,	..	35·21	35·03	35·30	35·08	35·39	35·26	35·34	35·17	35·16
200 ,,	..	35·08	35·39	35·37	35·03	35·34	35·21	35·17	35·26	35·16
300 ,,	..	35·43	35·25	35·03	35·23	35·10	35·39	..	35·25	35·07

Table 78; showing the salinity (S) at different depths in the Gulf of Mannar in March-April, 1921.

Date.		28.IV.22.		30.IV.22.	28.IV.22.	30.IV.22.			29.IV.22.			
Position	Lat. N ..	7°10′	7°40′	8°0′	8°15′	8°25′	8°44′	8°40′	8°44′	8°18′	8°0′	7°48′
	Long. E ..	79°30′	79°20′	79°30′	79°08′	79°30′	79°25′	79°00′	78°35′	78°42′	78°19′	78°51′
Surface ..	..	34·51	34·72	34·56	34·65	34·51	34·43	34·51	34·60	34·83	34·72	34·67
50 Metres ..	..	35·05	34·79	35·12	34·69	34·67	34·47	34·79	34·67	34·85	34·99	34·78
100 ,, ..	..	35·14	35·23	35·19	34·96	35·14	34·83	33·08	35·01	34·90	35·08	34·90
200 ,, ..	..	34·88	35·23	35·19	35·01	35·23	35·08	35·01	35·07	35·25	35·03	35·01
300 ,, ..	..	35·23	35·34	35·26	35·08	35·25	35·16	35·14	35·16	35·07	35·05	35·19

Table 79; showing the salinity (S) at different depths in the Gulf of Mannar in April, 1922.

By this time of the year the surface-salinity has again risen considerably and it is clear that once again there is a change going on in the relationships of the upper levels of water. With the cessation of rain and the rise of temperature and, in consequence, an increase in the rate and extent of evaporation from the surface, the water of the upper levels is gradually becoming once again more saline than at a depth of some 50 metres. Already in certain cases there is but little difference in the salinity at these two levels and in the course of another month, that is to say by May, it seems clear that we shall have the same condition present once again as we found to exist in the month of October, namely the surface-water will then be more saline than the water at a depth of 50 to 100 metres.

It is, I think, clear from the above that there is during the course of the survey-season a double change in the relationship of the salinity of the surface-water and of that at a depth of some 50-100 metres, and that this alteration of relationship agrees closely with the double change in certain months of the year in the variation of the salinity of the surface-water to which I have already called attention (*vide supra*, p. 304). In the months of October and November the surface-water may be more saline than that at a lower depth and therefore in such cases upwelling, under the influence of increased wind, will cause a diminution of salinity on the surface, and the same condition in all probability occurs in the hot season of the latter part of April and May; whereas, during the wet season of the north-east monsoon in January and February conditions are reversed and the surface-water is less, and frequently very much less, saline than that below, an upwelling under these circumstances causing a rise in the surface-salinity. A rise in the strength of the wind may, therefore, be accompanied by a fall of salinity in the dry seasons from October to November and again in April and May, whereas in the other months we should expect to find it accompanied by a rise in salinity, and this agrees fairly well with what I have already shown to be the case. But since in any series of years conditions are liable to some degree of variation, depending on the time of onset and intensity of the N.E. monsoon and on the height to which the sea-temperature is raised and, therefore, on the rapidity with which the change from a less saline to a more saline surface-water is reached in the dry months of March to May, it is necessary to consider independently the results obtained in any given year and in each locality.

October. Commencing with the month of October, the records taken in different years on the "Investigator" are as follows:—

		Time of Day.							
		A.M.				P.M.			
		4	8	10	12	4	8	10	12
Laccadive Sea.									
1921 (4 days)	Salinity	35·26	34·96	....	45·24	35·01	35·29	....	35·25
1922 (4 days)	Salinity	34·87	35·13	....	35·46	35·92	35·26	....	35·12
1923 (5 days)	Salinity	35·43	35·37	35·41	35·37	35·55	35·26	35·36	35·26
	Wind force, B.S.	*2·17*	*0·91*	....	*1·00*	*1·75*	*1·50*	....	*1·33*
Bay of Bengal.									
1921 (3½ days)	Salinity	33·79	33·61	....	34·02	34·44	34·11	....	33·71
1922 (3½ days)	Salinity	33·60	33·71	....	34·15	34·11	33·90	....	33·47
Andaman Sea.									
1921-1922 (2 days)	Salinity	32·97	33·93	....	32·05	32·26	32·30	....	32·86
	Wind force, B.S.	*1·92*	*1·69*	....	*1·90*	*2·01*	*1·97*	....	*1·70*
Average (22 days)	Salinity	34·52	34·58	....	34·65	35·27	34·60	....	34·48

Unfortunately in this month I have only two series of observations, relating to the wind-force, that were taken simultaneously with the water-samples. In the other years readings were taken at the same time and on the same days but unfortunately the details were not kept and I have only the average results of all observations taken in all three regions; the average wind-force as observed at different times of the day in the years 1921 and 1922 during the month of October are as follows:—

	Time of day.					
	A.M.			P.M.		
	4	8	12	4	8	12
1921 (7 days)	2·28	2·14	2·14	2·28	2·07	1·78
1922 (8 days)	1·56	1·25	1·56	1·75	1·87	1·62

I have already pointed out that in the month of October the average variation of the surface-salinity (*vide* Table 70, p. 304) follows a single oscillation during the 24-hours and this type of oscillation is very clearly seen in the average results obtained in the Laccadive Sea in 1922 (*vide* Text-fig. 91). In this year the salinity is at its minimum, 34·87, at 4 a.m., it rises steadily to 35·92 at 4 p.m., and thereafter falls again to 4 a.m. The times of occurrence of maximum and minimum salinity agree closely with those of maximum and minimum surface-temperature, *viz.*, 26·22°C at 4 a.m. and 27·70°C at 4 p.m. (*vide supra* p. 225). The average wind-force also exhibits a single oscillation, having a minimum at 8 a.m. and a maximum at 8 p.m. and there is, I think, little doubt that the salinity change in this instance

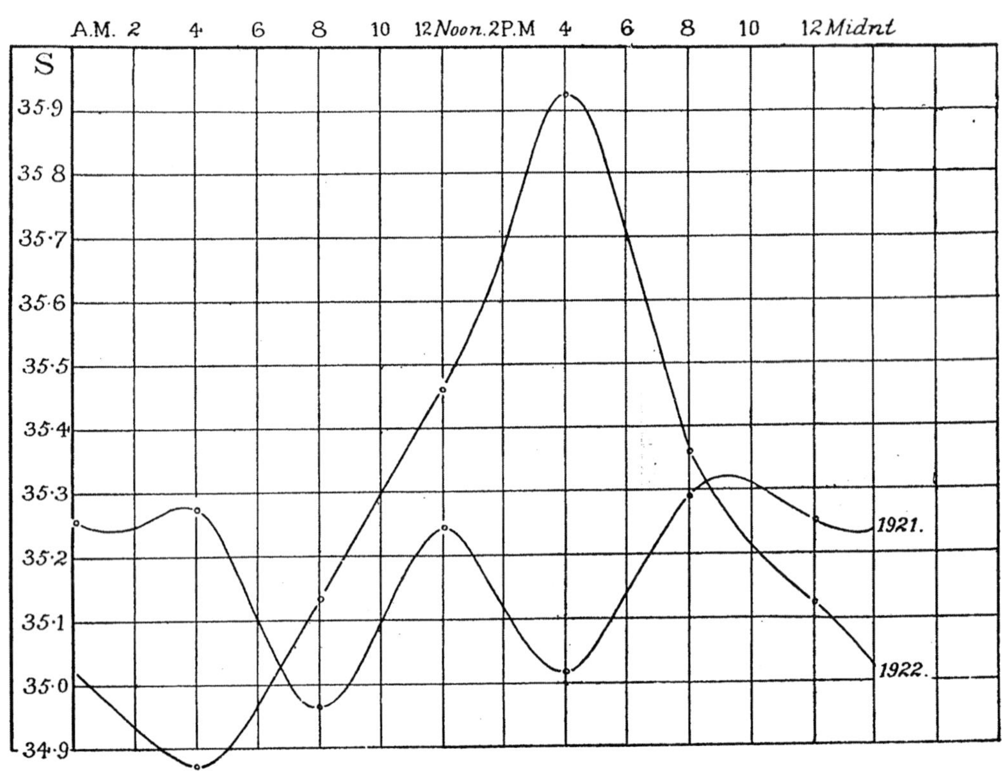

Text-fig. 91 Showing the salinity of the surface water at different times of the day in October, 1921, and 1922, in the Laccadive Sea

is due mainly, if not entirely, to evaporation. In 1921 (Text-fig. 91) the salinity exhibits a tendency towards a triple oscillation during the day and in 1923 (*vide* Text-fig. 92) there is evidence of even a quadruple oscillation. Unfortunately, as mentioned above, I do not possess separate data for the wind-force recorded in the Laccadive Sea in the month of October, 1921, and the average force for all regions shows only a double oscillation, but in the 1923 series the wind-force, recorded simultaneously with the taking of the water-samples, shows a quadruple oscillation (*vide* Text-fig. 92) exactly similar to that exhibited by the salinity and there can be no doubt that the two are related to each other, a rise in the wind-force setting up either (1) increased evaporation and so giving rise to an

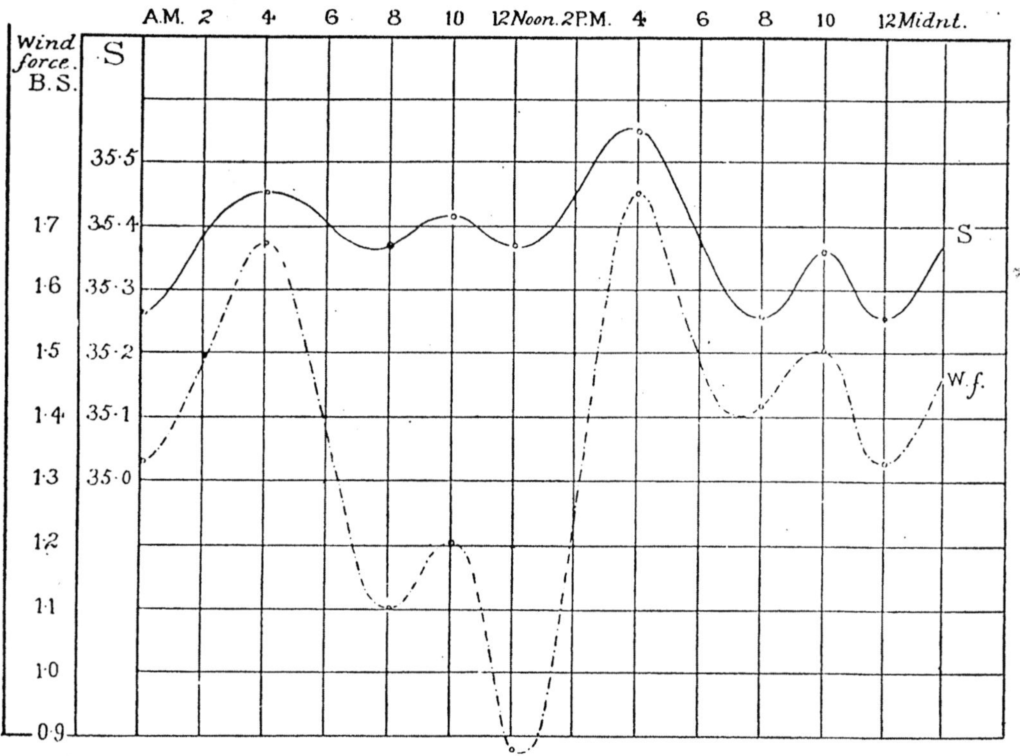

Text-fig. 92. Showing the average salinity of the surface water at different times of the day in the Laccadive Sea in the month of October, 1923, as compared with the wind-force.

increased salinity in the surface-water or (2) an upwelling of more saline water

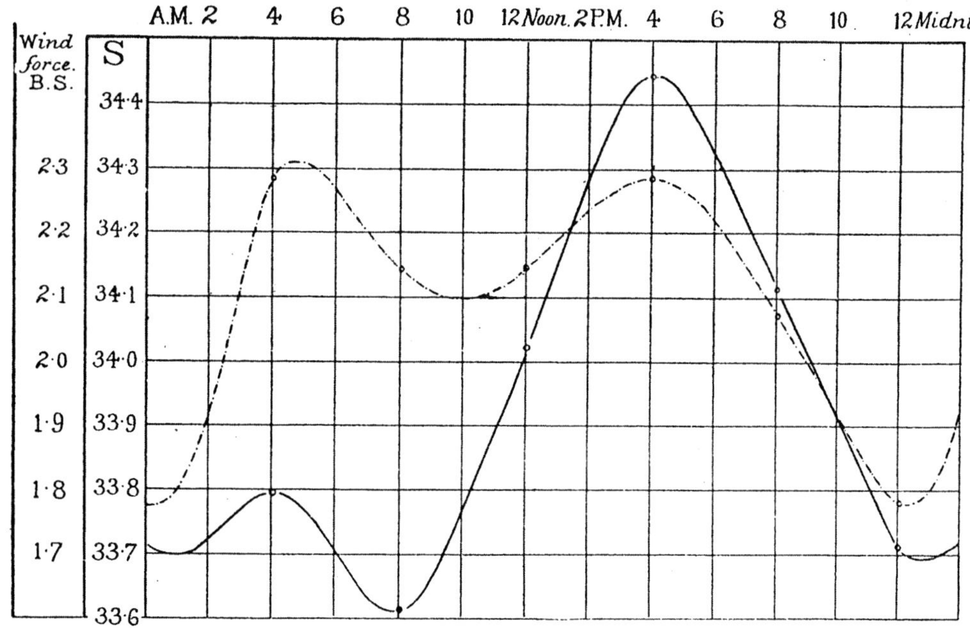

Text-fig. 93. Showing the variation in salinity of the surface water at different times of the day in the Bay of Bengal in the month of October, 1921, as compared with the average wind-force throughout Indian Seas in the same month.

N.B.—In this and following Text-figures the salinity is indicated by the continuous line and the wind-force by the interrupted line.

from below, or possibly both. In the Bay of Bengal area in 1921 (Text-fig. 93) the salinity exhibits a clear double oscillation having maxima at 4 a.m. and again at 4 p.m., and of these the second or later maximum is by far the higher and the rise and fall of the salinity shows a very clear agreement with the changes in the strength of the wind. In 1922 (Text-fig. 94) the variation exhibits a single rise and fall that is very similar to the single oscillation found in the Laccadive Sea in 1922 (*cf.* Text-fig. 91), but with a maximum at about 2 p.m. and a minimum at about 1 a.m.; a slight irregularity in the curve at 4 a.m. suggests that we have here a faint trace of the first rise in salinity that was quite well-marked in the 1921 series, but the greater part of the change seems, in this case also, to be due to increased evaporation by wind and heat. In the record from the Andaman Sea I have combined

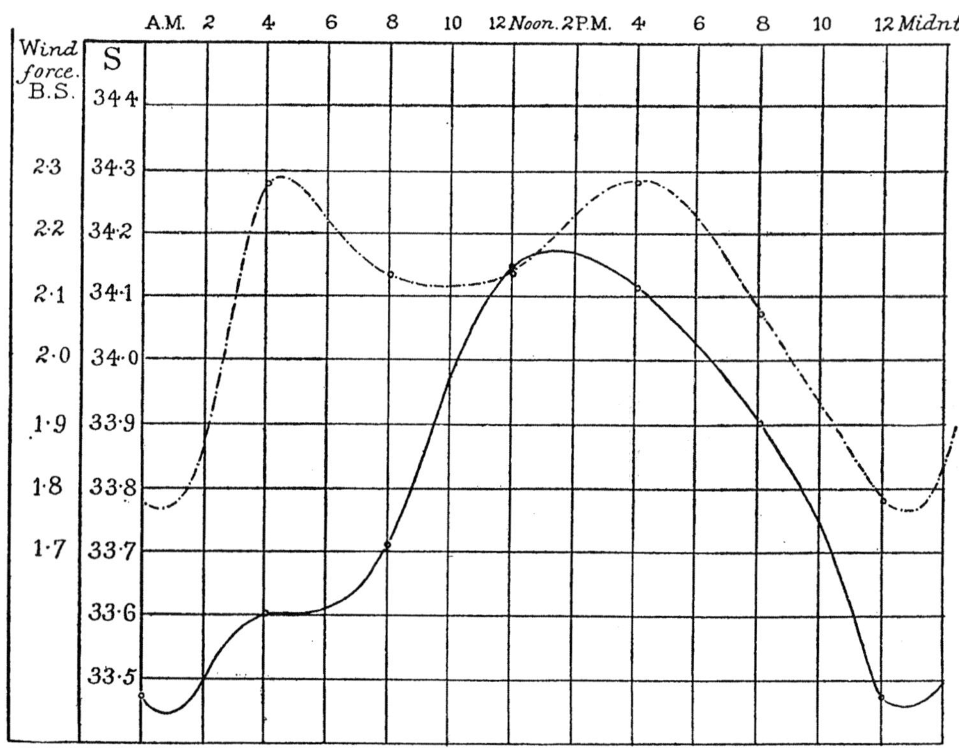

Text-fig. 94. Showing the variation in salinity of the surface water at different times of the day in the Bay of Bengal in the month of October, 1922.

the data of both years 1921 and 1922, as I was only able to take observations on a single day in each year. The results obtained are shown graphically in Text-fig. 95 and we have here clear evidence of a tendency for the salinity to vary in the opposite direction to the wind-force. The rise and fall of salinity exhibits a triple curve, in which the primary maximum occurs at 8 a.m. and two secondary maxima at 2 a.m. and 4 p.m. Of these, the first two coincide with the periods of least wind but there is apparently no change in the wind-force to correspond with the third maximum of salinity at 4 p.m. and this slight rise is probably due to increased evaporation owing to the rise in the temperature of the surface-water during the hottest part of the day. The association in this series of observations of a fall of salinity with a rise of wind-force clearly indicates that we are not dealing with the result of evaporation, since the observed changes are the exact opposite of those which increased evaporation would produce : it seems probable that we have here evidence of an upwelling of water of low salinity as a result of increased lateral movement of the surface-layer under the influence of increased wind and a consequent thinning of the superficial stratum to an extent that allows either the actual appearance on the surface of water from some depth below or at least an admixture of surface and deeper water by means of wave-action. It is almost unnecessary to point out that in any one year the rate of evaporation of the surface-water in any month will differ from that in the same month of another year and equally that under similar conditions of temperature and humidity of the

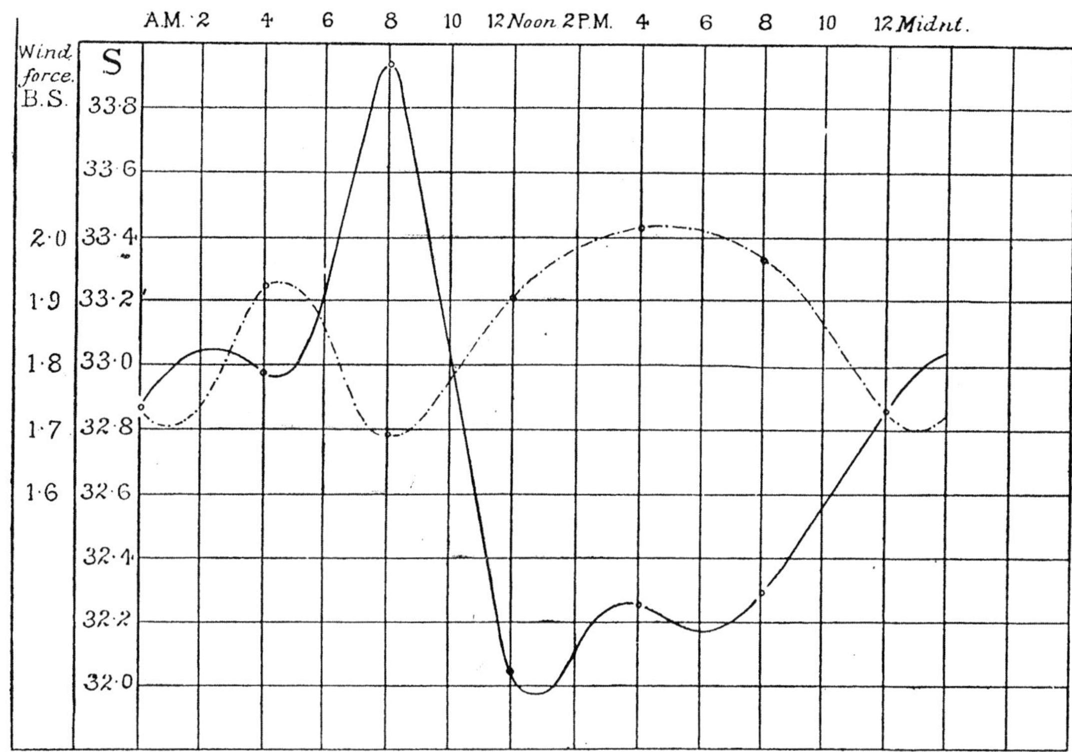

Text-fig. 95. Showing the salinity of the surface water at different times of the day in the Andaman Sea in the month of October, 1921-22, as compared with the wind-force.

atmosphere the rate of evaporation of sea-water is slower when the salinity is higher than when it is low. The change, therefore, of conditions in conformity with the changing seasons of the year in Indian waters, from that present during the wet season when, as we have seen, we have a layer of dilute surface-water floating above a more dense layer, to that found at the end of the hot dry season, when this condition is reversed and we get a layer of hot, more saline water floating on a colder and less saline stratum, will occur at different times in different years and in different parts of the Indian seas in the same year, since the salinity is least in the Andaman Sea and highest in the Laccadive Sea, the Bay of Bengal being intermediate between the two. The different behaviour, therefore, under the influence of the wind-force of the surface-water of the Bay of Bengal in October, 1921, and in the Andaman Sea in 1921-22, may well be due to the different time of occurrence of the critical period in the two areas; in the former area the surface-water appears to have still been less saline than that immediately below, whereas in the Andaman Sea, in which the salinity of the water is considerably less, the change had already taken place.

November. During this month I have only been able to obtain two series of observations, the records of which are given below:

		Time of day.							
		A.M.				P.M.			
		4	8	10	12	4	8	10	12
Laccadive Sea.									
1923	Salinity ..	35·87	35·77	35·46	35·77	35·81	35·50	35·35	35·45
(4 days)	Wind-force, B.S.	*2·1*	*2·3*	*1·7*	*1·9*	*2·0*	*1·9*	*2·2*	*2·0*
Andaman Sea.									
1921-22	Salinity ..	33·65	33·21	..	33·15	33·28	33·25	..	34·06
(3 days)	Wind-force, B.S.	*2·75*	*1·12*	..	*1·50*	*2·37*	*2·12*	..	*2·25*
Average Salinity (7 days) ..		34·92	34·67	..	34·65	34·73	34·54	..	34·85

Text-fig. 96. Showing the variation of the surface salinity at different times of the day in the Laccadive Sea in November, 1923, compared with the wind-force.

In both these two series of observations there is clear evidence that the rise and fall of salinity and the rise and fall of the strength of the wind tend to coincide with each other. We have previously seen that in the Laccadive Sea in the preceding month of October, 1923, the salinity and wind-force rose and fell together (*vide* Text-fig. 92) and the same relationship is still maintained in November (Text-fig. 96), though with this difference that, whereas in October the oscillation was of a quadruple character, with two primary maxima at 4 a.m. and 4 p.m. and two secondary maxima at 10 a.m. and 10 p.m. respectively, in November the salinity shows two primary maxima at about 5 a.m. and 2 p.m. and the wind-force exhibits a triple oscillation having maxima, corresponding to the maxima of salinity, at 7 a.m. and 2 p.m. and, in addition, there is a transient rise to a third maximum at 10 p. m., which is, however, unaccompanied by any corresponding rise in the salinity. In the Andaman Sea series (Text-fig. 97) we get a double oscillation during the day in both the wind-force and the surface-salinity and these variations agree fairly closely, though at or near midnight the change in the wind-force appears to lag somewhat behind the variation in the salinity. A comparison with the results obtained in the same region in the preceding month (*vide* Text-fig. 95) indicates that there has been a complete reversal of the relationship between the oscillation of the surface-salinity and the variation of the wind-force; and in the present month a rise in the wind-force is accompanied by a rise in salinity instead of

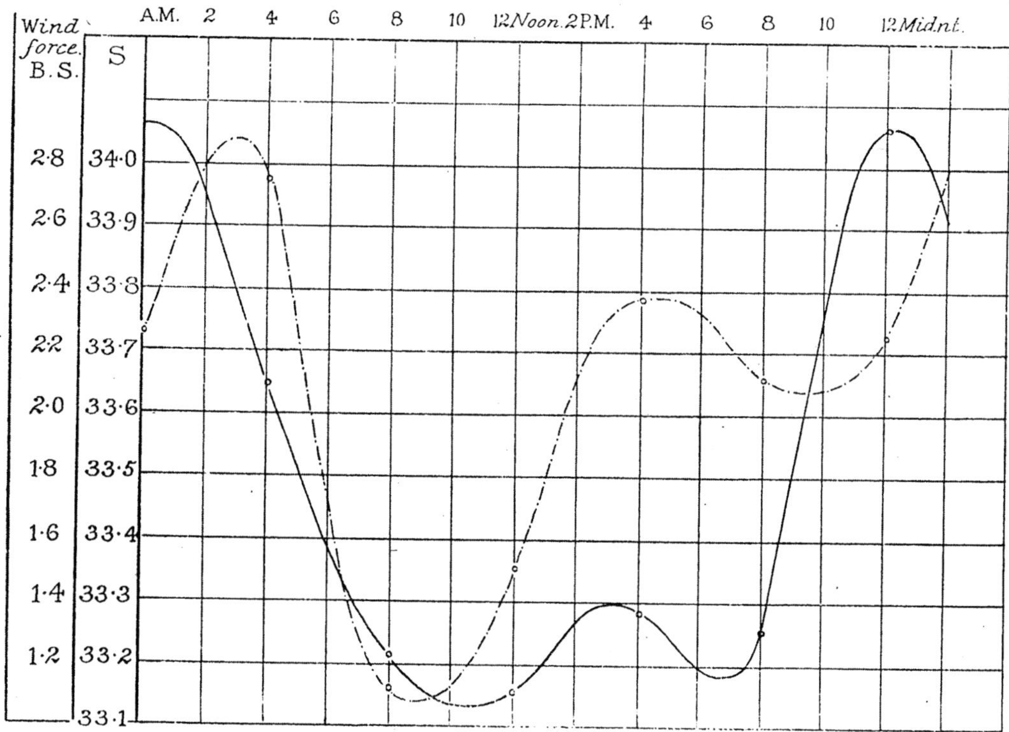

Text-fig. 97. Showing the variation of the surface salinity at different times of the day in the Andaman Sea in November, 1921-22, compared with the wind-force.

by a fall, as was seen in the month of October. It seems probable that a change in the relationships of the water strata has already occurred and that the commencement of the N. E. monsoon rains in this region has caused a dilution of the surface-layer so that it is now less saline than the underlying stratum, a rise in the strength of the wind in consequence now causing the appearance on the surface of more saline water from below, while simultaneously the rate of evaporation will be increased.

Corroborative evidence that this oscillation of the surface salinity does actually exist and, moreover, is not confined to those areas in which the "Investigator" has been employed during the period of my tenure of the appointment of Surgeon-Naturalist, namely in the Laccadive Sea, the Bay of Bengal and the Andaman Sea, is provided by the data that have recently been published by Matthews (1926) for the belt of the Arabian Sea lying between Aden and Bombay. The majority of the observations that have been given by him in Table II at the end of his paper have been taken on board the P. & O. Steamers and unfortunately the samples on which his date are based have, as a rule, been taken at intervals that are too irregular to enable one to calculate the variations except in two series of observations made on board the S.S. "Egypt" while crossing the Arabian Sea, outward bound, on November 5th to 10th and back again, homeward bound, on November 25th to 29th, 1905. During these two trips water-samples were taken at approximately 4-hourly intervals, though usually the early morning and afternoon samples were taken at

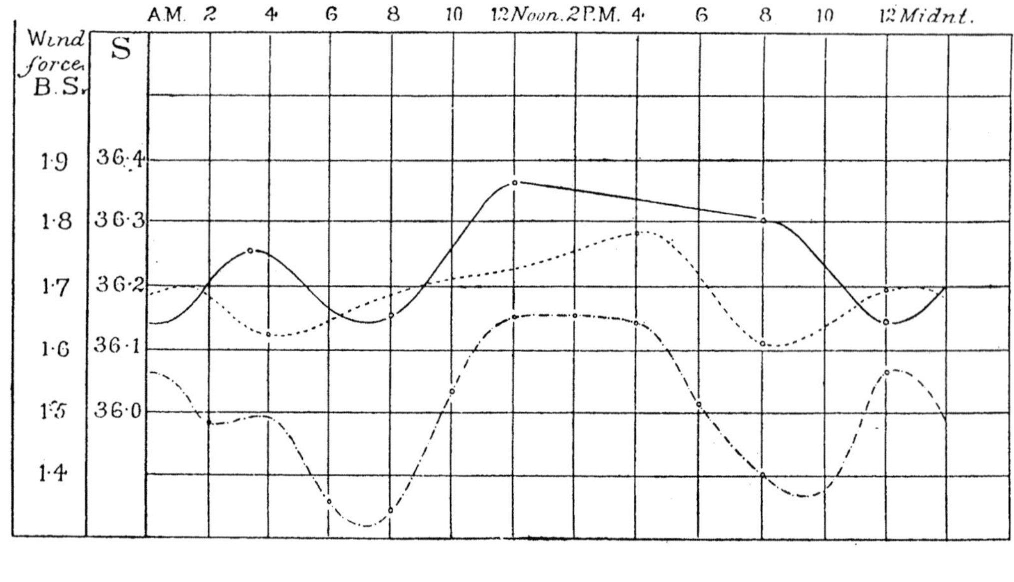

Text-fig. 98. The average salinity of the surface water of the Arabian Sea in November, 1905, as observed by the S.S. "Egypt" (vide Matthews, 1926)

3-30 a.m. and p.m., a few, however, being taken at 4-0 a.m. If we include these latter with the 3-30 a.m. samples we obtain the following average for the two voyages :

	Time of day. A.M. 3-30	A.M. 8	A.M. 12	P.M. 3-30	P.M. 8	P.M. 12
November 5-10, 1905 ..	36·25	36·15	36·36	36·33	36·30	36·14
" 25-29, "	36·12	36·18	36·22	36·28	36·11	36·19

I have plotted the results graphically in Text-figure 98 and it is clear that on both occasions there was a distinct double diurnal oscillation of the salinity of the surface-water, though the times of occurrence of maximum salinity do not quite coincide in the two series. Unfortunately, no data is given regarding the strength of the wind at the time when these samples were taken : I have, therefore, for the purpose of reference and comparison given the average variation in this month as observed by me; the average of all my observations is as follows :

	A.M. 2	4	6	8	10	12 noon	2 P.M.	4	6	8	10	12 Midnt.
Wind force B.S.	1·48	1·49	1·36	1·34	1·53	1·65	1·65	1·64	1·51	1·40	1·37	1·56

The very close agreement in this series of observations between the rise and fall of salinity and the rise and fall of the wind-force is obvious and requires no comment.

December. In this month I again possess only two series of observations, taken,

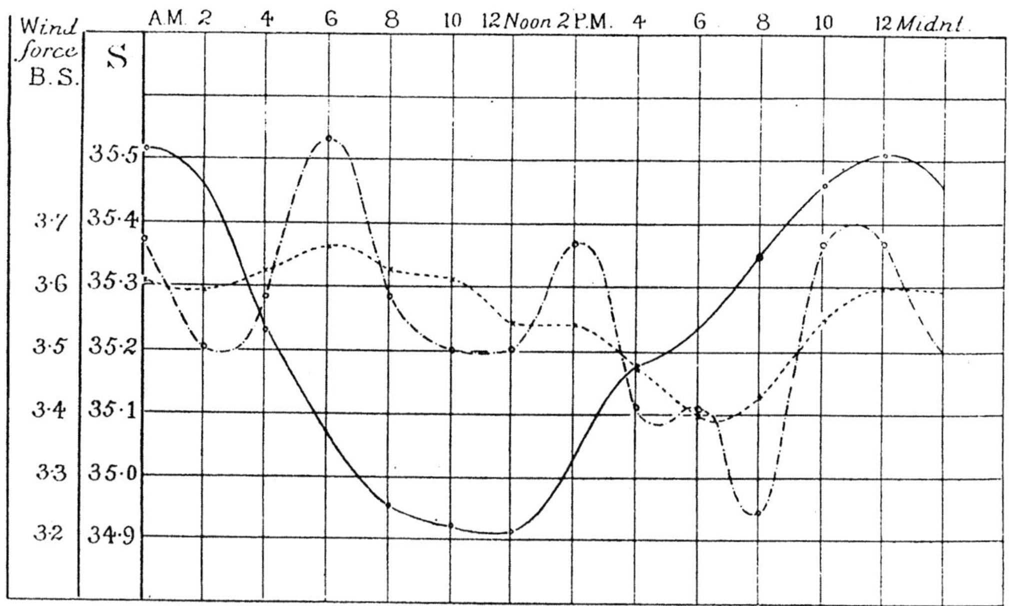

Text-fig. 99. Showing the salinity of the surface water at different times of the day in the Laccadive Sea in December, 1923, compared with the wind-force.

as in the previous month, in the Laccadive Sea and the Andaman Sea respectively. The results of these two series are given below :

		Time of day. A.M. 4	8	10	12	P.M. 4	8	10	12
Laccadive Sea.									
1923	Salinity ..	35·22	34·95	34 92	34·91	35·08	35·36	35·46	35·51
(6½ days)	Wind-force, B.S. ..	*3·58*	*3·58*	*3·90*	*3·58*	*3·42*	*3·25*	*3·67*	*3·67*
Andaman Sea.									
1921–22	Salinity ..	33·13	33·01	..	33·08	33·06	33·28	..	33·54
(4 days)	Wind-force, B.S. ..	*3·33*	*3·05*	..	*3·01*	*3·02*	*3·14*	..	*3·08*
Average Salinity (10½ days)	..	34·42	34·21	..	34·21	34·31	34·56	..	34·76

The variation in the salinity of the surface-water in the Laccadive Sea in 1923 (Text-fig. 99) exhibits a single oscillation having a maximum at about 12 midnight and minimum at 12 noon, which agrees very closely with the type of oscillation shown by the average of all temperature observations throughout the whole season from October to

May (*vide supra* p. 222) and it is possible that this rise and fall is due to the "change over" of the upper levels of the ocean as a result of convection currents set up by the evaporation of the surface-layer during the hot part of the day and the subsequent cooling during the early morning hours from midnight to 6 a.m. On the other hand a second factor may be concerned in this oscillation. At first sight the oscillation of the salinity and the variation in the strength of the wind appear to show no relationship to each other; the wind-force exhibits a quadruple series of oscillations, whereas, as mentioned already, the variation of the salinity follows an almost uniform single rise and fall. If, however, we compute the different strengths of the wind-force at 2-hourly intervals in accordance with the usual formula, $\frac{1a+4b+6c+4d+1e}{16}$, we get the following values :

Time of day ..	2 A.M.	4	6	8	10	12	2 P.M.	4	6	8	10	12
Wind-force, B.S. ..	3·59	3·62	3·66	3·62	3·61	3·54	3·54	3·47	3·40	3·43	3·55	3·60

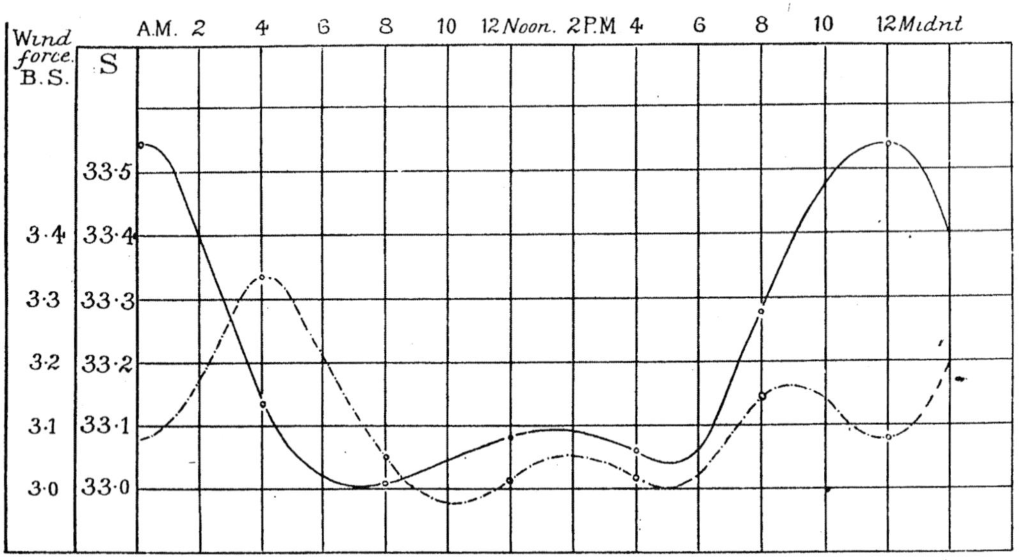

Text-fig. 100. Showing the salinity of the surface water at different times of the day in the Andaman Sea in December, 1921-22, compared with the wind-force.

It is clear that, apart from the minor oscillations, there is a tendency to a single rise and fall in the strength of the wind ; and on comparing this with the variation we find that the two curves tend to alternate with each other, a rise in the wind-force being followed, after a lag of some 6 hours, by a fall in the salinity, and *vice versa* a fall of wind-force is followed by a rise of salinity. A study of the relationship between the surface-salinity and the wind-force in the Laccadive Sea during the months October to December, 1923, inclusive (*vide* Text-figures 92, 96 and 99) indicates that there is towards the latter end of this period a transition in the relationship of the salinity at the surface and at a deeper level ; in October, the salinity and wind-force rise and fall together, in November this state of affairs is by no means as clearly marked and in December they tend to alternate. In November the conditions existing in the upper levels of the Laccadive Sea appear to be such that a layer of low salinity rests on a more saline layer and with an increase in wind the more saline deeper water appears on the surface. I have no record from the Laccadive Sea in January, 1923 but in the Bay of Bengal the local conditions appear to have become reversed there also and we now seem to have a more saline water floating on a stratum of less salinity. In the intervening month of December conditions in the Laccadive Sea appear to be in a critical stage between these two phases, though with a slight tendency towards the later stage, that would probably become well-marked in the following month and is certainly present in the month of February. In the case of the record from the Andaman Sea in 1921-22 (Text-fig. 100) the type of oscillation of the surface-salinity agrees very closely with that found in the same area in the preceding month. The salinity exhibits two maxima, of which by far the greater occurs at or near midnight and the smaller at about 2 p.m. The wind-force on the other hand exhibits a triple oscillation. During the middle of the day from 10 a.m. to 5 p.m. salinity and wind-force tend clearly to rise and fall together, but corresponding to the major oscillation in the surface salinity, between 5 p.m. and 7 a.m., we get a double oscillation in the wind-force with maxima at 9 p.m. and 4 a.m. It is possible that the fall in the wind-force at 12 midnight may be due to error in estimating the strength but here again, if we compute the 4-hourly averages, in accordance with the same formula used above, we find that there is a very clear tendency for the wind-force to rise and fall in a single oscillation that corresponds very closely with the maximum rise and fall of the salinity, and it seems probable that we have here evidence of a similar condition of the surface-layers of the sea as we found indicated in the previous month, namely a dilute stratum lying above a more saline one, so that a rise in the wind-force causes the thinning or even rupture of the upper stratum and the appearance of more saline water on the surface. At the same time the rise in salinity may be due in part to increased evaporation as a result of increased wind, and certainly this factor would seem to be the causative agent in the small rise noted in the middle and hottest part of the day, but this agency is, I think, insufficient to account for the whole of the increase in surface-salinity during the night hours from 6 p.m. to 12 midnight.

January. I have in all three series of observations taken by me in the month of January, the details of which are given below :

		Time of day. A.M. 4	8	10	12	P.M. 4	8	10	12
Bay of Bengal.									
1923	Salinity	34·17	39·92	..	32·85	33·27	33·61	..	33·61
(4 days)	Wind-force, B.S. ..	*2·00*	*2·17*	..	*2·33*	*1·87*	*2·0*	..	*1·87*
1924	Salinity	33·84	33·87	33·90	33·70	33·46	33·90	33·98	33·90
(4 days)	Wind-force, B.S. ..	*2·87*	*2·87*	*2·87*	*3·25*	*2·50*	*3·00*	*3·37*	*3·62*
Andaman Sea.									
1922	Salinity	32·77	33·08	..	32·87	32·95	32·86	..	32·84
(3 days)	Wind-force, B.S. ..	*2·50*	*3·00*	..	*2·50*	*2·50*	*3·50*	..	*25·0*
Average Salinity (11 days)		33·67	33·58	..	33·26	33·25	33·51	..	33·60

In the Bay of Bengal the conditions present in these two years, 1923 and 1924, are the exact opposite of each other. In the earlier year, 1923, (Text-fig. 101) the salinity

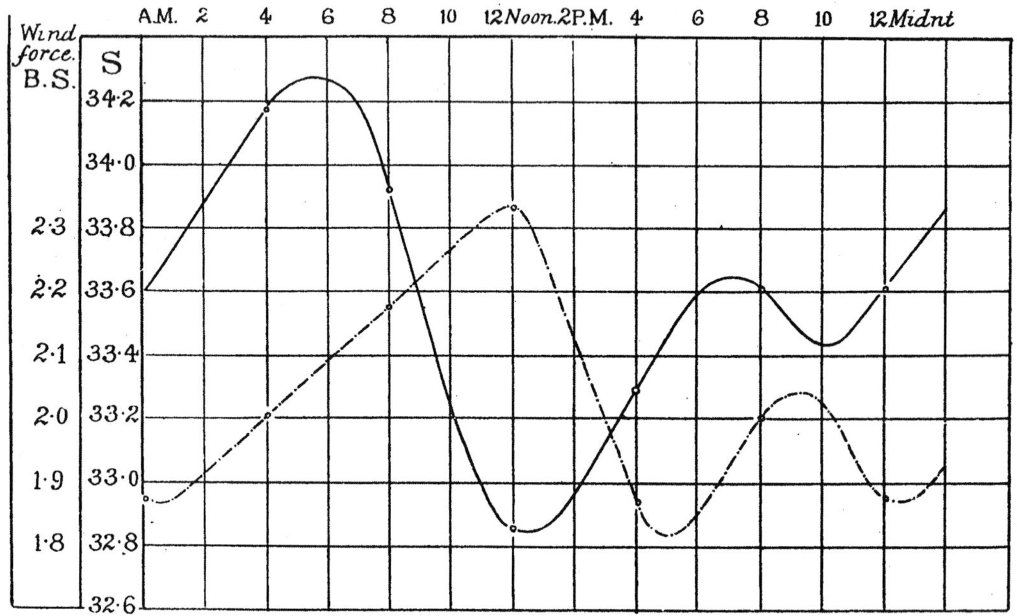

Text-fig. 101. Showing the salinity of the surface water at different times of the day in the Bay of Bengal in January, 1923, compared with the wind-force

of the surface-water clearly oscillates in a manner that is the exact opposite of the variation in the wind-force : a rise in the wind-force is accompanied by a marked fall in the salinity and as the wind drops the salinity again steadily rises. It

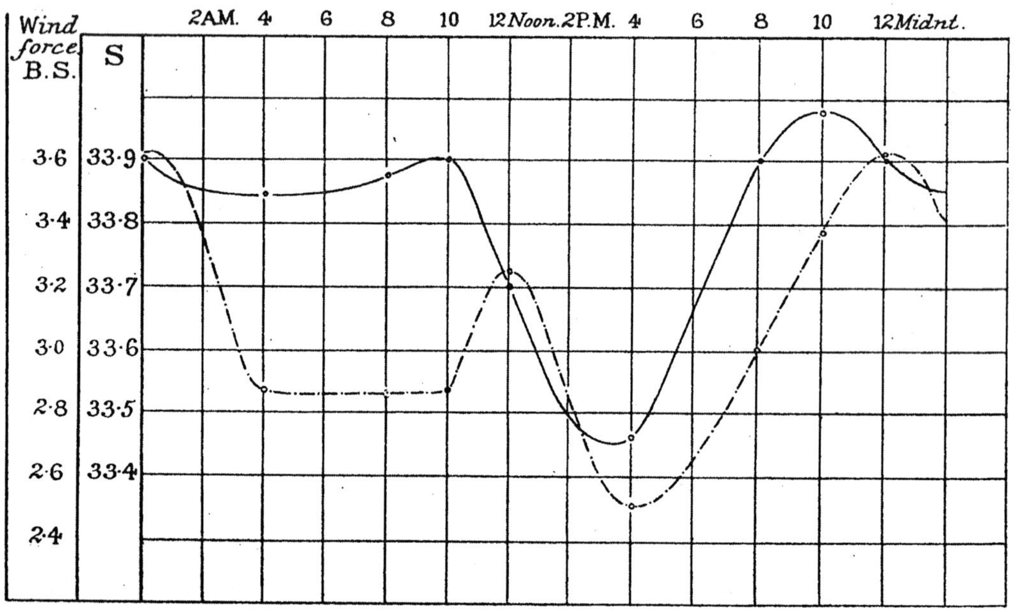

Text-fig. 102. Showing the salinity of the surface water at different times of the day in the Bay of Bengal in January, 1924, compared with the wind-force.

would thus appear probable that in this year there was a layer of surface-water of high salinity reposing on a deeper stratum of lower salinity. In 1924 (Text-fig. 102), however, the condition appears to have been the exact opposite of this; the surface-salinity now varies with the wind-force, though the latter exhibits a slight lag of some two hours in the times of the maxima. In this case, then, it would seem that the explanation may lie either in the effect of increased evaporation owing to increased wind or in the surface-layer having a less salinity than that immediately below. That this latter was in all probability the case is indicated by the rainfall during the months of December and January in the two years: in the first year the rainfall over those parts of India which are most likely to affect the Bay of Bengal (*vide supra* p. 271 *et. seq.*) amounted to 0·472 inches on the average in

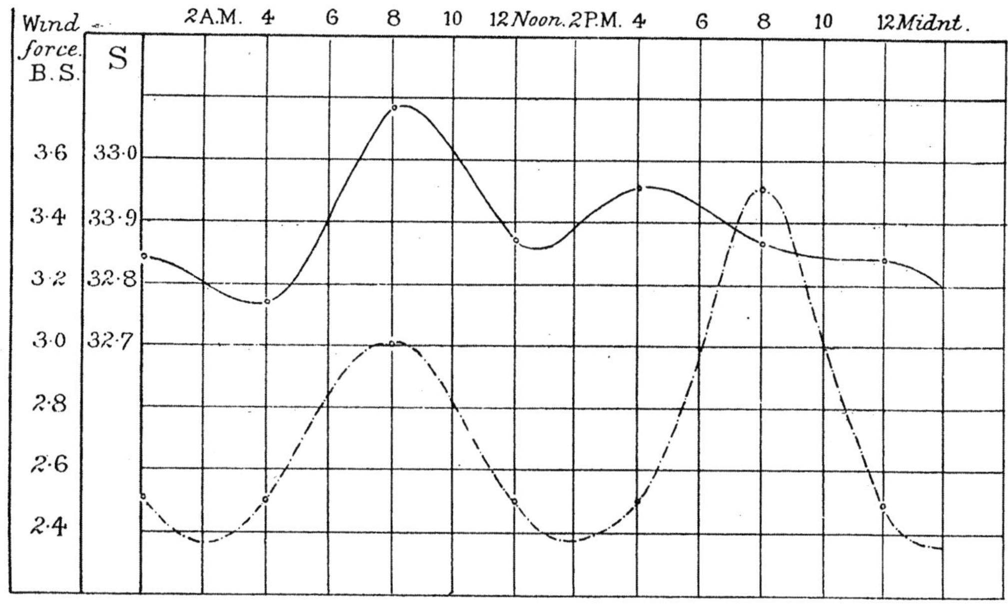

Text-fig. 103. Showing the salinity of the surface water at different times of the day in the Andaman Sea in January, 1922, compared with the wind-force.

December, 1922 and 0·400 inches in January, 1923, the average actual rainfall for the whole area of both months being 0·872 inches; whereas in the following year the average for December, 1923, was 0·560 and for January, 1924, 0·491 or for the whole period 1.051 inches. It seems clear that the upper layer must, therefore, have undergone considerably less dilution during these two months in 1923 than in the corresponding months of 1924 and that this had been insufficient to render the salinity of the surface-water less than that immediately underlying it. In the case of the record from the Andaman Sea in 1922 (Text-fig. 103), there is a distinct tendency for the salinity to rise and fall with the wind force, and this is particularly clearly seen in the daily oscillations between midnight and 4 p.m. Towards the end of the day, however, the correlation between the two is not so well marked, owing to the average salinity at 8 p.m. being lower than one would have expected. On the whole, however, the evidence is in favour of the existence of a condition similar to that in the preceding months (*cf.* Text-figs. 97 and 100).

February. During this month I have been able to make only two series of observations, the data regarding which are given below :

			Time of day.							
			A.M.				P.M.			
			4	8	10	12	4	8	10	12
Laccadive Sea.										
1923	Salinity ..	..	34·51	34·62	..	34·34	34·61	34·68	..	34·55
(6 days)	Wind-force, B.S.	..	1·92	1·70	..	2·21	2·17	2·67	..	1·83
Andaman Sea.										
1922	Salinity ..	..	33·11	33·52	..	32·60	32·97	33·09	..	32·83
(2 days)	Wind-force, B.S.	..	1·25	2·00	..	2·00	1·50	1·25	..	1·00
Average Salinity (8 days)	..	..	34·16	34·34	..	33·90	34·20	34·26	..	34·12

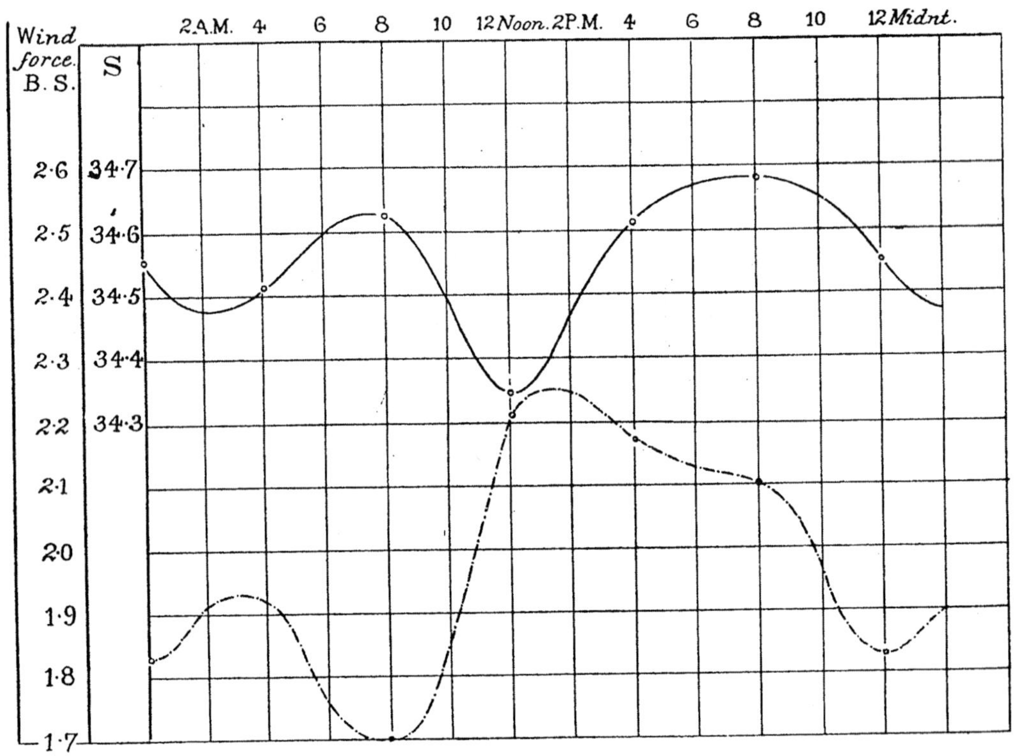

Text-fig. 104. Showing the salinity of the surface water at different times of the day in the Laccadive Sea in February, 1923, compared with the wind-force.

In the Laccadive Sea region (Text-fig. 104) during this month the variation in the surface-salinity tends to follow a double daily oscillation, having a maximum at 8 a.m. and falling to a minimum at or shortly before noon, after which it again steadily rises to 8 p.m., when it shows a second fall to 2 a.m. The strength of the wind also shows a clearly-defined double oscillation : it exhibits a secondary maximum at 3 a.m., after which there follows a fall to 8 a.m. and a very marked rise to 1 p.m. ; the force then slowly falls till midnight, after which it again slightly increases.

A comparison of the curves for the wind-force and the salinity shows that they alternate in a very clear manner, and that the changes in the salinity are the exact opposite of what we should expect, were they due to evaporation. We must, I think, again seek the explanation of this relationship in the different salinities of the water strata, a rise in the wind-force initiating an up-welling of denser and more saline water from below. It is unfortunate that I possess no records in the area for the month of January, but in December as I pointed out (*vide* supra p. 319 and Text-fig. 99) we can detect a tendency for the wind-force and the salinity to alternate with each other and it is clear that in the present month February, this relationship is well-established.

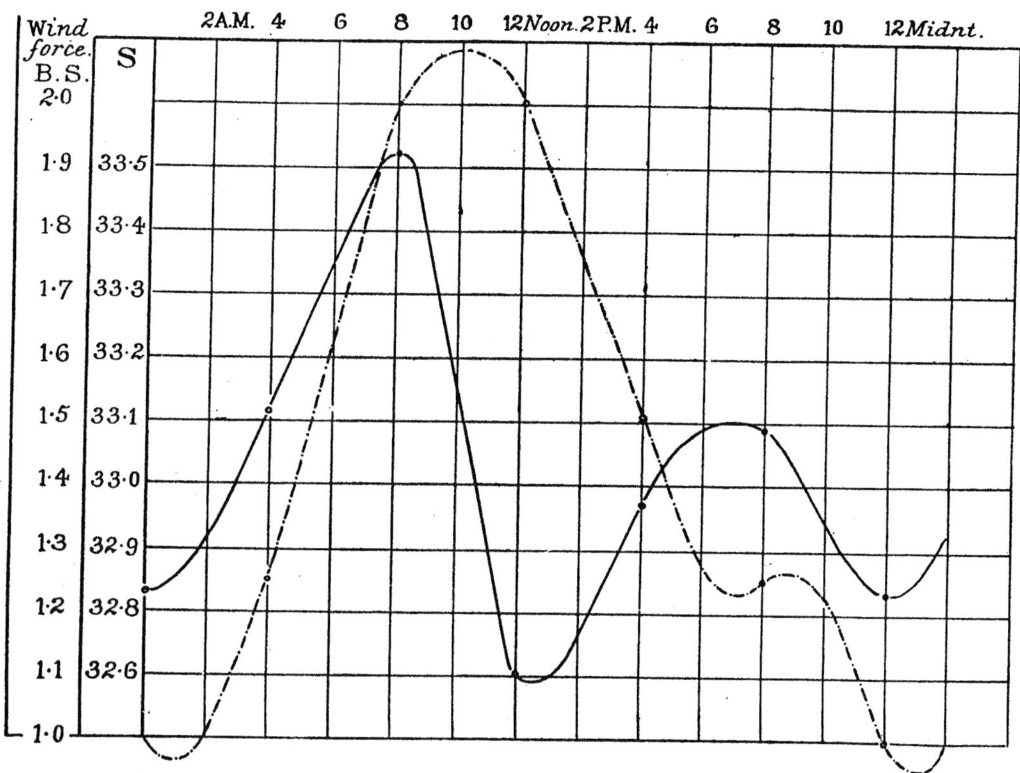

Text-fig. 105. Showing the salinity of the surface water at different times of the day in the Andaman Sea in February, 1922, compared with the wind-force.

In the Andaman Sea region (Text.-fig. 105) the surface salinity and the wind-force exhibit, at first sight, little or no relationship to each other. The salinity shows a clearly marked double oscillation, having a primary maximum at 8 a.m. and a smaller secondary maximum at about 7 p.m., with two minima at about 12 noon and 12 midnight. The wind-force, on the other hand, shows a tendency towards a single oscillation, having a maximum at 10 a.m. and a minimum at 1 a.m.; there is however, evidence of a very small secondary oscillation, having its maximum at 9 p.m. It appears probable that in this case the surface-salinity and wind-force vary in the same direction but the relationship between these two sets of changes, is by no means, a clear one; the rise in salinity in the afternoon, *viz.* from 12 noon to 7 p.m., appears to anticipate the corresponding change in the wind-force by about 7 hours and it is probable that we have here the double effect of the upwelling of more saline water from below under the influence of the wind and the increase in salinity of the surface-water as a result of evaporation during the hottest part of the day.

March. In this month I possess three series of observations, two of which were taken in the Laccadive Sea and the third in the Bay of Bengal. The data of these three series are given below :—

		Time of day.							
		A.M.				P.M.			
		4	8	10	12	4	8	10	12
Laccadive Sea.									
1923	Salinity ..	34·57	34·61	34·60	34·69	34·50	34·58	34·53	34·50
(6 days)	Wind-force, B.S.	1·58	2·40	..	2·20	2·41	2·58	..	1·83
1924	Salinity ..	34·65	34·69	34·67	34·25	34·40	34·53	34·70	34·70
(4 days)	Wind-force, B.S.	1·12	1·60	1·12	2·00	2·00	1·37	1·37	1·37
Bay of Bengal.									
1924	Salinity ..	33·78	33·77	33·57	33·68	33·66	33·70	33·67	33·70
(3.5 days)	Wind-force, B.S.	1·35	1·00	1·17	1·17	1·87	1·62	1·87	1·62
Average Salinity (13.5 days)	..	34·39	34·42	34·35	34·30	34·25	34·34	34·36	34·35

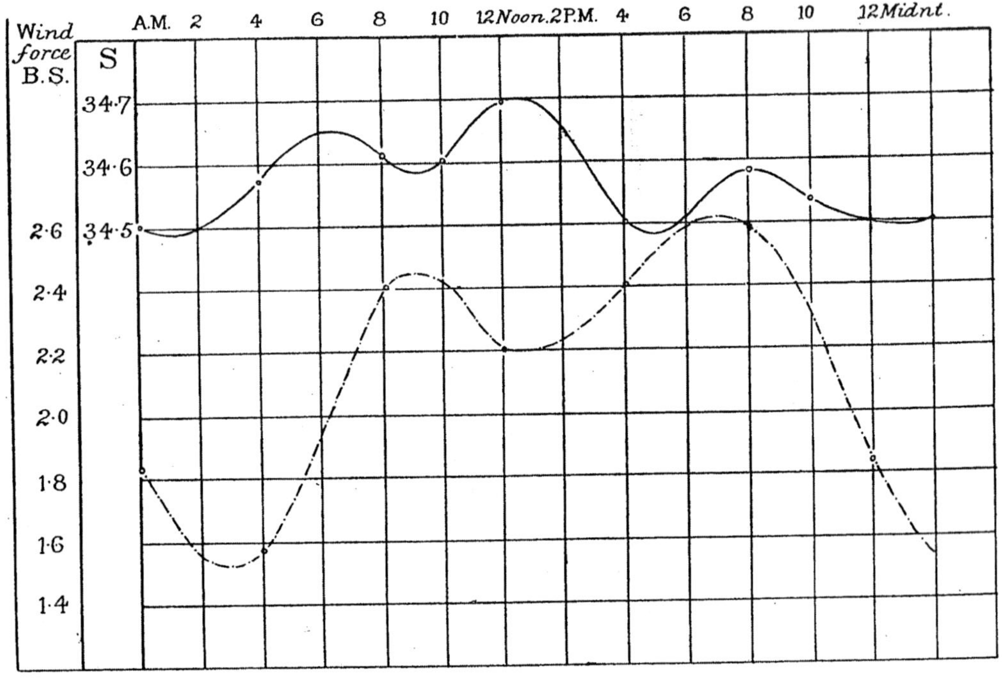

Text-fig. 106. Showing the salinity of the surface water at different times of the day in the Laccadive Sea in March, 1923, compared with the wind-force.

In all three series of observations there is now a very clear tendency for the salinity and the wind-force to alternate with each other. This is least well-shown in the series for the Laccadive Sea in 1923 (Text-fig. 106) : unfortunately, in this year the available data for the wind-force only includes observations at 4-hourly intervals, whereas the data for the surface-salinity includes two additional observations taken at 10 a.m. and 10 P.M. The surface-salinity exhibits a triple oscillation, having maxima at 6 a.m., 1 p.m. and 8 p.m.; the wind-force, however, shows an apparent double oscillation with maxima at 9 a.m. and 7 p.m. The surface-salinity during the day from 6 a.m. to 6 p.m., clearly alternates with the wind-force, but between 8 p.m. and 4 a.m., there is a fall and subsequent rise in the surface-salinity that exhibits a clear correspondence with the oscillation in the wind-force. It is possible that more frequent observations would have revealed such a corresponding triple oscillation in the strength of the wind and this is even rendered probable by a comparison with the data for the other two series, in both of which a subsidiary rise and fall of the wind-force is seen to occur at 10 p.m. and 12 midnight. On the other hand, however, it appears probable from a study of the data for the three months, February—April, 1923

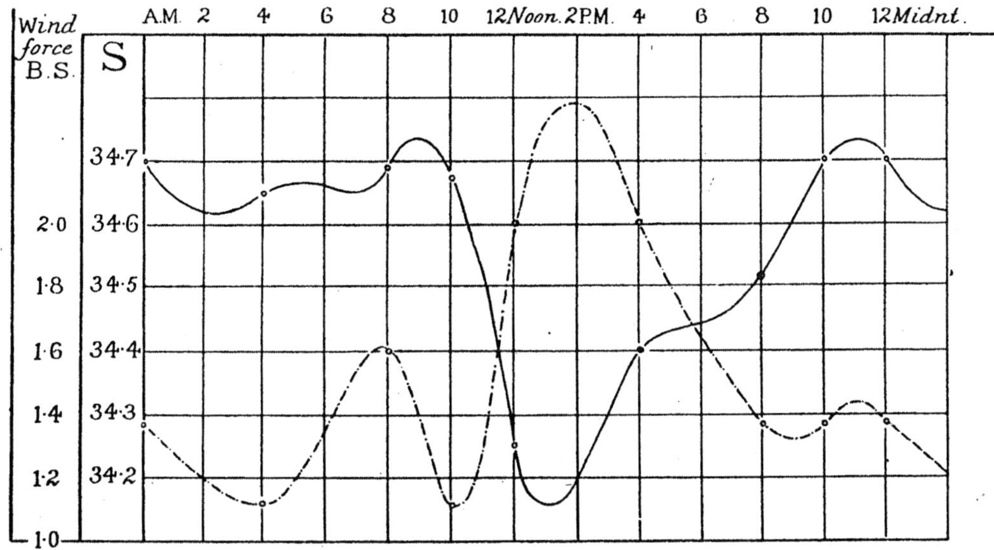

Text-fig. 107. Showing the variation in salinity of the surface water at different times of the day in the Laccadive Sea in March, 1924, compared with the wind-force.

inclusive, that the relationships of the surface-strata of this area have again undergone one of the seasonal changes. In February, 1923, as we have already seen (*vide supra* text-fig. 104) the surface-salinity and wind-force clearly oscillate in opposite directions: in March, 1923, (*vide* text-fig. 105), the same relationship holds good in the main only and by the following month, April 1923, (*vide infra* text-fig. 110), the relationship has become reversed and the wind-force and surface-salinity now oscillate in the same direction. The explanation of the condition present in March is probably to be found in the daily change over of the surface-strata under the influence of convection currents. During the day owing to evaporation the surface-water is probably more saline, though actually, as a result of heating, less dense than the deeper stratum, and in consequence a rise in the wind-force will set up an upwelling of less saline water from below. From 8 p.m. on and throughout the night hours, owing to the change over, the surface-water will be less saline than that below and hence the rise of wind after 3 a.m. causes an upwelling of more saline water. In the data for the Laccadive Sea and the Bay of Bengal in 1924, (Text-figs. 107, 108), the rise and fall of the wind is somewhat irregular and presents a triple or even quadruple oscillation, and in conformity with this we have similar alternating variations in the surface-salinity. It seems probable, therefore, that in this year in both areas of Indian seas, the condition by this month has been reached in which, by evaporation, the surface-layer has been rendered more saline than the deeper water,

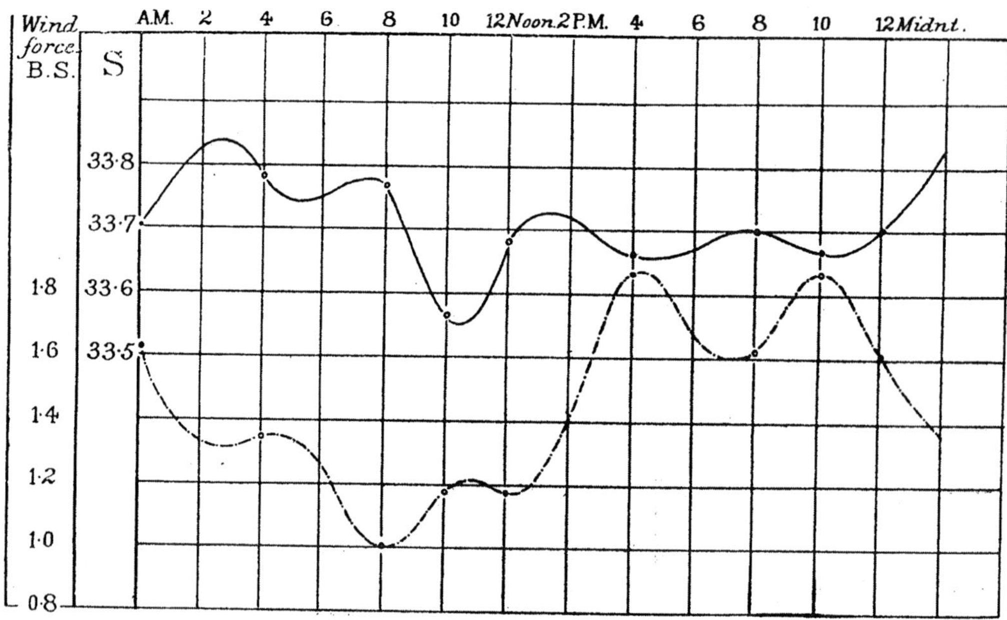

Text-fig. 108. Showing the salinity of the surface water at different times of the day in the Bay of Bengal in March, 1924, compared with the wind-force.

but that owing to a simultaneous rise in the surface-temperature the density of this surface-layer is reduced sufficiently to enable it to float over a less saline but colder and more dense layer. Under these conditions a rise in the wind-force will cause a thinning of the superficial layer and the appearance on the surface of water of less salinity.

April. In the month of April I have been able to make seven series of observations, one in the Andaman Sea, two in the Bay of Bengal and the remainder in the Laccadive Sea.

			Time of day.							
			A.M.				P.M.			
			4	8	10	12	4	8	10	12
Laccadive Sea.										
1914 (7 days)	Salinity	..	34·49	35·33	..	35·09	35·32	35·25	..	34·45
1923	Salinity	..	34·60	34·37	34·49	34·63	34·70	34·50	34·60	34·69
(3 days)	Wind-force,	B.S.	*4·67*	*3·67*	*4·00*	*4·33*	*4·50*	*5·00*	*5·25*	*4·33*
1924	Salinity	..	34·72	34·89	34·87	34·94	34·01	44·82	34·88	34·92
(7 days)	Wind-force,	B.S.	*2·00*	*1·53*	*1·75*	*1·71*	*2·29*	*1·87*	*1·81*	*2·08*
1925	Salinity	..	34·96	34·98	..	34·59	34·93	34·94	..	34·93
(3 days)	Wind-force,	B.S.	*3·00*	*3·00*	..	*3·62*	*3·37*	*3·50*	..	*4·25*
Bay of Bengal.										
1914 (3.5 days)	**Salinity**	..	32·95	32·49	..	32·68	32·73	32·88	..	33·11
1925	**Salinity**	..	33·76	33·48	..	33·88	33·56	34·00	..	34·03
(5·5 days)	**Wind-force,**	B.S.	*2·00*	*1·67*	..	*2·42*	*1·83*	*2·10*	..	*2·50*
Andaman Sea.										
1914 (2 days)	**Salinity**	..	32·58	32·63	..	32·39	32·21	32·07	..	32·82
Average Salinity (31 days)		..	34·17	34·28	..	34·30	34·35	34·36	..	34·29

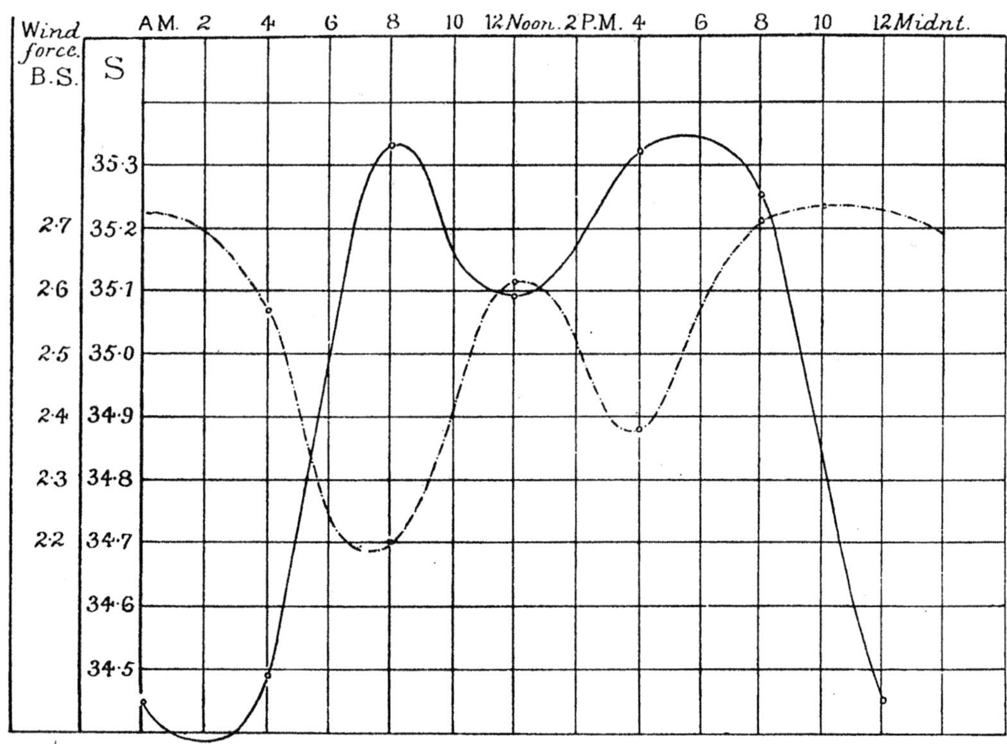

Text-fig. 109. Showing the salinity of the surface water at different times of the day in the Laccadive Sea in April, 1914, compared with the wind-force.

As will be seen from the above data, I have no observations of the strength of the wind in any of these areas in the year 1914, so for the purpose of comparison I have taken the average strength of the wind in all three parts of the Indian seas during the years 1921–25. The figures thus arrived at are as follows:—

Time of day	4 A.M.	8	12 Noon.	4 P.M.	8	12 Midnight.
Wind-force, B.S.	2·57	2·20	2·61	2·38	2·71	2·72

In the Laccadive Sea area the results obtained exhibit very considerable differences in different years. In 1914, (Text-fig. 109), a comparison of the oscillation in the surface-salinity and the average rise and fall of the wind-force shows that both series exhibit a double oscillation during the day and that a rise in the wind-force is accompanied by a fall in the surface-salinity. This relationship is in conformity with the results obtained in the same area in the month of March, 1923, and 1924, (*vide*

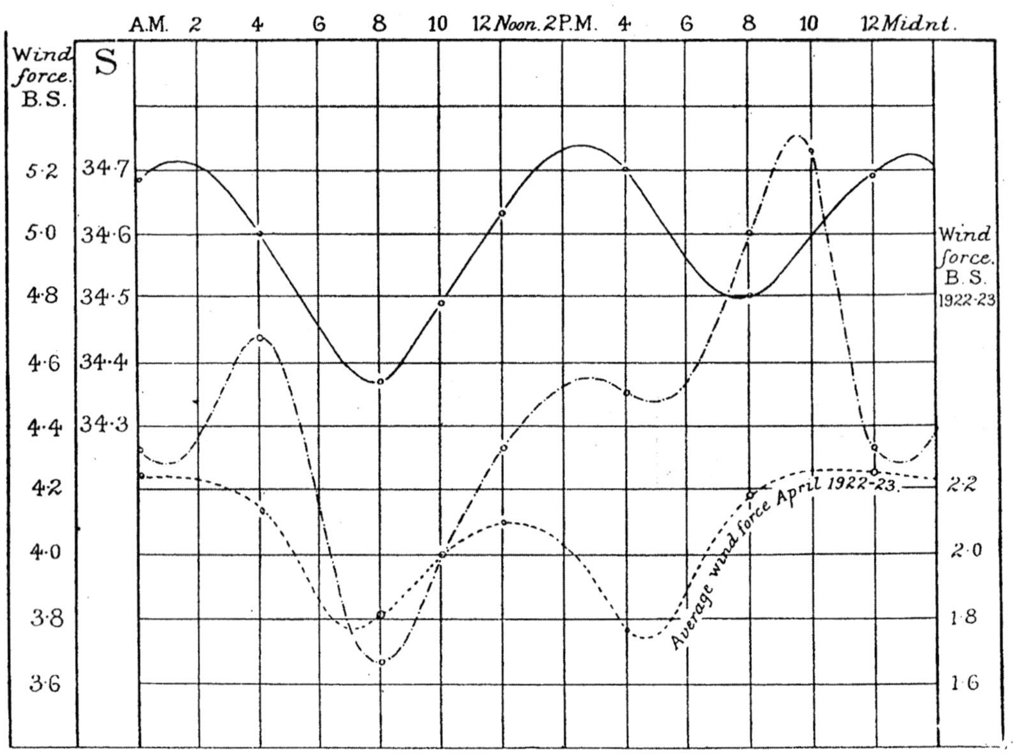

Text-fig. 110. Showing the salinity of the surface water at different times of the day in the Laccadive Sea in April, 1923, compared with the wind-force

Text-figs. 106 and 107). In the case of the data for the same area during April, 1923, (Text-fig. 110), the relationship between the changes in the wind-force and the oscillations in the surface-salinity appears to be of exactly the opposite nature, a rise in the wind-force on this occasion being accompanied by a rise in the salinity. This, however, is by no means as clear as in some of the other series of observations. In 1923 the wind-force exhibits a triple oscillation, whereas the salinity shows only a double one, but the average of all 4-hourly observations in Indian seas taken on board the "Investigator" in this month in the years 1922-23, gives the following data for the wind-force in this month:—

Time of day	4 A.M.	8	12 Noon.	4 P.M.	8	12 Midnight.
Wind-force, B.S.	2·12	1·82	2·09	1·75	2·18	2·25

This clearly shows a double daily oscillation that agrees very closely with the observed variation in the salinity in 1923. In 1924 (Text-fig. 111), the salinity exhibits a triple oscillation, while the wind-force exhibits a quadruple one, and the relationship between the two variations appears to be similar to that seen in this area in March 1923, a rise in the wind-force being accompanied by a fall in the surface-salinity during the day only and the relationship being altered during the night hours: during the morning hours from 4 a.m. to 12 noon the rise and fall of the salinity alternates fairly well with the changes in the wind force, but between 12 noon and 4 a.m. either the salinity-changes exhibit a marked "lag" behind the corresponding wind-variation, amounting to about 3 hours, or else the relationship

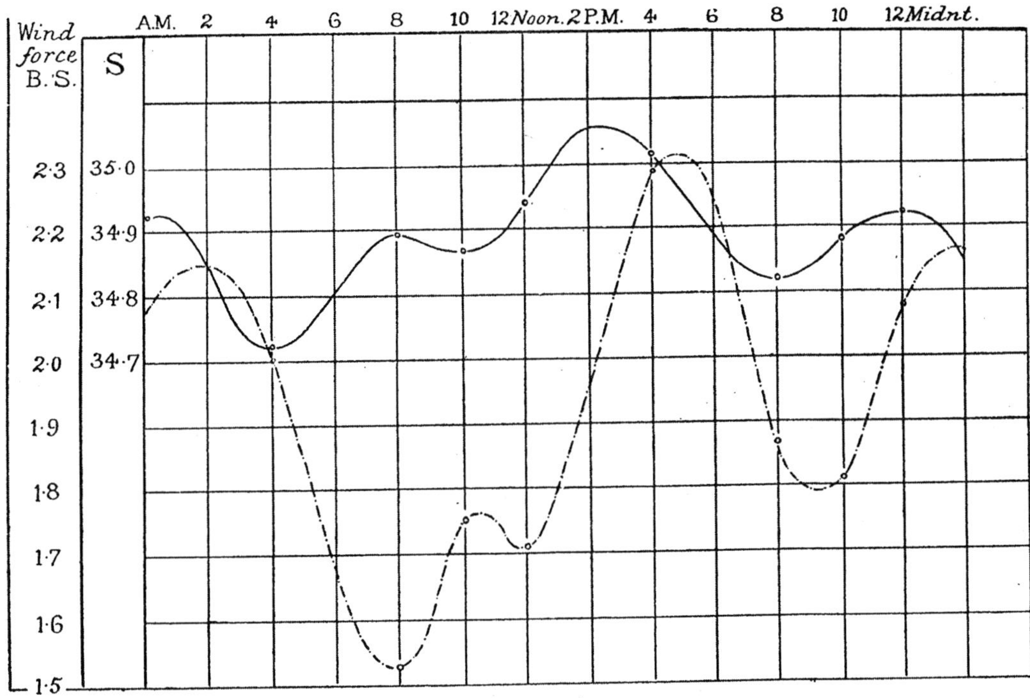

Text-fig. 111. Showing the salinity of the surface water at different times of the day in the Laccadive Sea in April, 1924, compared with the wind-force.

is reversed and the salinity and wind-force vary together, the wind-force exhibiting a lag of some 2-3 hours behind the salinity changes. In the Laccadive Sea in April, 1925, (Text-fig. 112), the salinity and the wind-force exhibit double diurnal oscillations and in this case clearly in the opposite directions, a rise in the wind-force being accompanied by a fall in the salinity. A study of the data for the Laccadive Sea in this and the preceding months indicates that at this season of the year the relationship between the oscillations of salinity and of the wind-force can best be accounted for on the supposition that the surface-layer, as a rule, becomes condensed by evaporation and heated up to an extent sufficient to cause this more saline layer to float on a less saline but colder and, therefore, more dense stratum. In 1923, however, the condition appears to be the reverse of this, the surface-layer being more dilute than the deeper water. In the latter part of the month of April the weather conditions both on the east and west sides of India were decidedly abnormal; according to the Indian Weather Report for that year, "in the last week of the month a temporary advance in the monsoon gave rise to a storm in the neighbourhood of Table Bay, which caused heavy rain in south Burma and died out at sea off the Arakan Coast" and in the same month in the Laccadive Sea high winds were experienced, as can be seen by comparing the average wind-force in this year with that experienced in other years, and there was a considerable amount of rain in the region of the southern Maldive Islands. This abnormal weather commenced in the region of the Maldives on April 11th, when the wind veered from the N.W. with

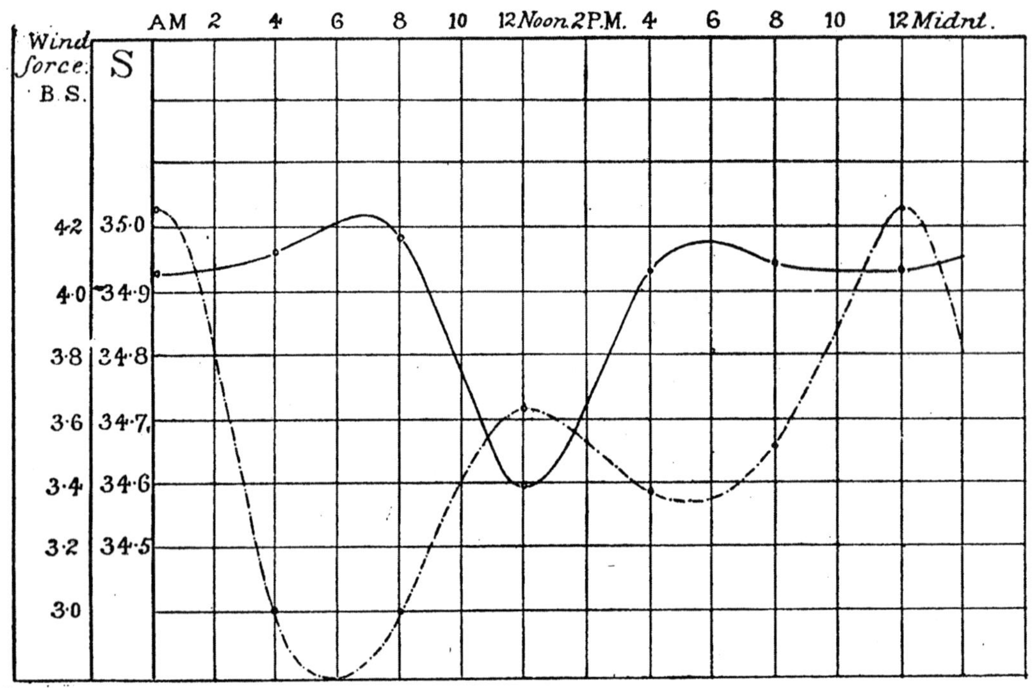

Text-fig. 112. Showing the salinity of the surface-water at different times of the day in Laccadive Sea in April, 1925, compared with the wind-force.

a force 1 to the W.S.W, force 2 to 3; thereafter it blew steadily from the west or the south-west with a fair amount of rain and increased in force to 5 or 5-6 till the 21st, after which the wind moderated to force 2-3. On the 28th the wind again increased in strength, attaining force 5 on the 29th and 30th and causing a strong surface-drift to the eastward. It is probable, therefore, that this abnormal weather had given rise to a dilution of the surface-stratum and had thus rendered it less saline than the deeper layers, so that as early as April there had been a return to a condition of affairs, that is normal in the two monsoon periods. Again in 1924, the Indian Weather Report states that in April "thunderstorms were frequent generally in Lower Burma, Assam, Bengal and the south of the Peninsula," and the rainfall at Bombay was 49% above the normal in this month. Heavy rain was also encountered at sea on April 14th at 4 a.m., when in Lat. 6°27′00″ N: Long. 78°25′00″ E. and it is probable that in this year also there had occurred in this month a certain

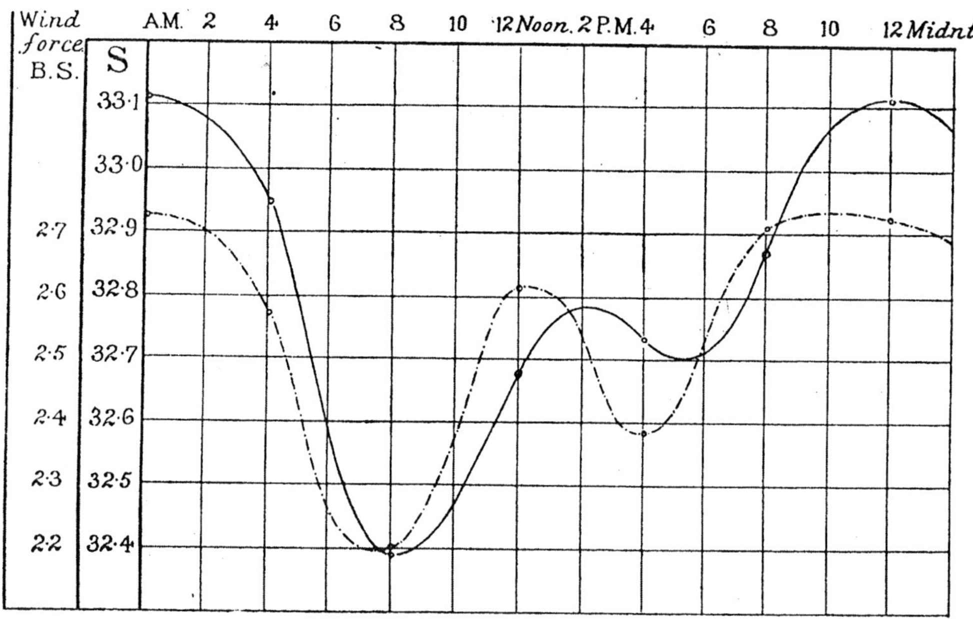

Text-Fig. 113. Showing the salinity of the surface water at different times of the day in the Bay of Bengal in April, 1914, compared with the wind-force.

degree of dilution of the surface-stratum that rendered it less saline than usual and hence diminished the difference between water lying at some depth below and

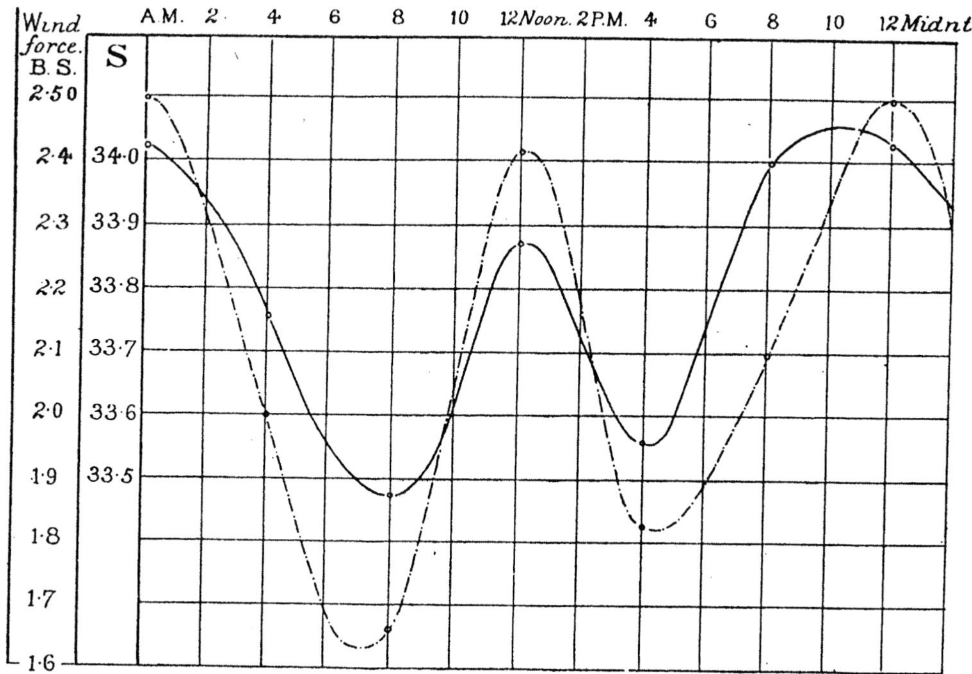

Text-fig. 114. Showing the salinity of the surface water at different times of the day in the Bay of Bengal in April, 1925, compared with the wind-force.

that present on the surface, so that the "normal" relationship between the wind-force and the surface-salinity is less clearly marked than in other years.

In the Bay of Bengal region my observations were taken in 1914 and again in 1925. In the former case I possess only a record of the salinity, whereas in the latter, I have simultaneous records of both salinity and wind-force. These results are shown graphically in Text-figs. 113 and 114. In Text-fig. 113 for the purpose of comparison I have given the average variation of the wind-force in all parts of the Indian Seas in this month. A comparison of the two series reveals a very close agreement between them and in both instances it seems clear that the rise and fall of the salinity coincides with the rise and fall of the wind-force : this is very

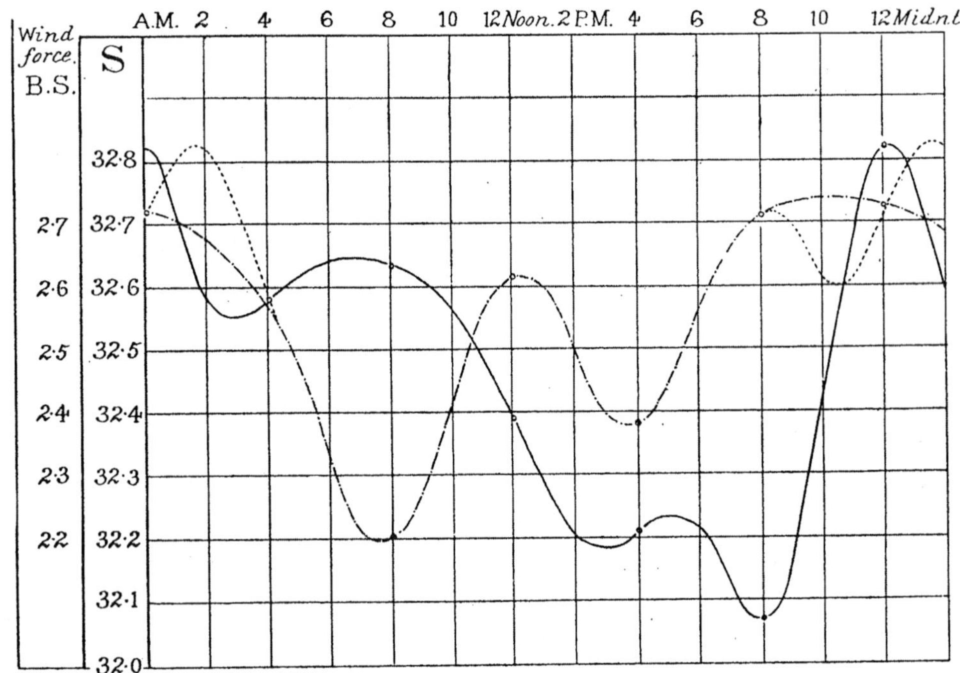

Text-fig. 115. Showing the salinity of the surface water at different times of the day in the Andaman Sea in April 1914, compared with the wind-force.

clearly seen in the data for 1925 (Text-fig. 114). The probable explanation of the agreement between the oscillation in the salinity and that of the wind-force is that at this period of the year the surface-water of the Bay is less saline than that lying immediately below and a rise in the wind-force causes a thinning of this upper layer that is accompanied by a upwelling of more saline water from the deeper levels. At the same time, however, it must be borne in mind that the increased evaporation from the surface as a result of the increased wind-force will tend to increase the corresponding rise in the salinity.

In the single series of observations from the Andaman Sea in April 1914, there appear to be indications that the changes of salinity and the variation of the wind-force (*vide* Text-fig. 115) vary in opposite directions, so that the relationship between them tends to agree with that found in the Laccadive Sea in 1914 : it must, however, be remembered that the observations cover a period of only two days, during which the "Investigator" steamed across the north west part of this area from Rangoon to Manner's Straits and the wind-force, which I have shown, is the average of all observations taken in this month, no observations having been taken simultaneously with the collection of the water samples. I have already pointed out (*vide supra*, p. 86 and Table 12) that in this month we find a well-marked tendency towards a triple oscillation in the wind-force, and the average of the 4-hourly observations would agree just as well with a triple oscillation such as I have indicated by the dotted line in Text-fig. 115 : if this be the true oscillation of the wind-force the manner in which it alternates with the changes in salinity is very striking.

May.

In this month my observations are extremely few as the "Investigator" usually returns to Bombay early in the month. The first series covers only a single day ; while the second was taken during 3½ days at the very beginning of the month and the data of these two series are given below :

		Time of day.							
		A.M.				P.M.			
		4	8	10	12	4	8	10	12
Laccadive Sea.									
1914 (1 day)	Salinity	36·15	36·36	..	36·17	36·26	36·99	..	36·64
1923	Salinity	35·56	35·14	34·97	35·05	35·37	35·17	35·39	35·39
(3.5 days)	Wind-force, B.S. ..	*4·6*	*5·00*	..	*4·7*	*4·5*	*5·0*	..	*5·0*
Average Salinity (4.5 days)	..	35·70	35·41	..	35·30	35·57	35·57	..	35·44

In the first series of obseravations (Text-fig. 116) the salinity exhibits the usual double oscillation, falling steadily from 9 p.m. to 4 a.m., rising slightly till 8 a.m., then falling again till 2 p.m. and finally rising steadily till 9 p.m. For the purpose of comparison I have given the average variation of the wind-force at different times of the day in this month as observed on the "Investigator" (*vide supra*, Table 10, p. 84). In both series we find a double diurnal oscillation and, furthermore, the afternoon rise and fall in each case is greater than the morning variation. One might at first sight be inclined to conclude that in this case the salinity and wind-force alternate with each other, but if we compare the variation in the salinity in Text-fig. 116, with the observed rise and fall of the wind-force in the same area in May 1923 it is at once evident that the two curves almost exactly coincide. It is, therefore, impossible to arrive at any definite conclusion regarding the relationship of the wind-force and the surface-salinity in this year, but it seems probable that the two oscillations varied together, a rise in the wind-force being accompanied by a rise in salinity. In 1923 (Text-fig. 117) in the same region, namely the Laccadive Sea, the salinity again exhibits a clear double oscillation, with an additional slight irregularity at 10 p.m. and this double oscillation alternates with a corresponding variation in the strength of the wind : a rise in the wind-force is clearly accompanied by a fall in the salinity and *vice versa.* It would appear that the abnormal effect of the early burst of the monsoon on the west side of the Laccadive Sea in this year (*vide supra* p. 332 and Text-fig. 112) had passed off or at any rate had not affected the east side, along the coast of India, in which area all the observations were taken during this month, and that the surface-water was here of higher salinity than that below, a rise in the wind-force causing an upwelling of less saline water; but on the other hand it is possible that the variation of the salinity is really positive as regards the wind-force and that the apparent alteration is really due to a very prolonged lag. If this latter view be the correct one, then in this series the major oscillation in the wind-force from 3 p.m. to 3 a.m. corresponds to the major variation in the salinity from 8 p.m. to 10 a.m., that is to say, we have here a lag of 5 hours.

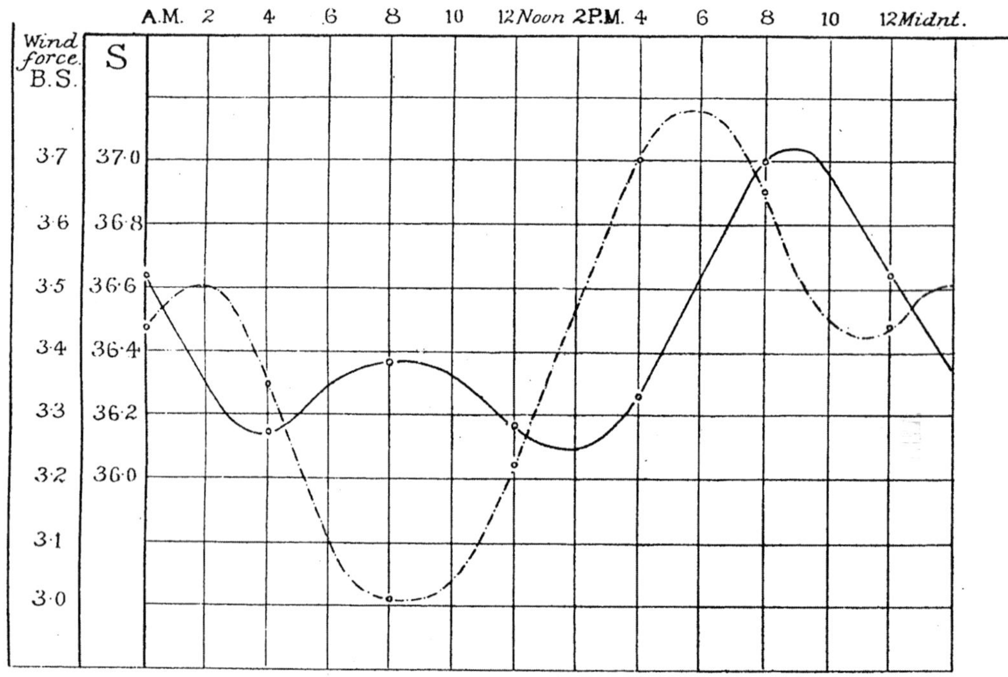

Text-fig. 116. Showing the salinity of the surface water at different times of the day in the Laccadive Sea in May, 1914, compared with the wind-force.

From a consideration of the above data it is, I think, clear that there is conclusive evidence that in all parts of the Indian Seas the surface-salinity is in a constant state of oscillation and that in the vast majority of cases this oscillation is a double one: and, furthermore, it appears more than probable that this variation in the salinity is correlated with the double oscillation that, as I have already shown, is exhibited by the wind-force during the twenty-four hours of the day (*vide supra* p. 79 *et. seq.*). A study of all the available data appears to indicate that this double oscillation of the salinity exhibits two phases at different periods of the year; at one period the salinity rises and falls with the variation in the strength of the wind, while at another it oscillates against the wind-force; and, furthermore, that this oscillation and its two phases are probably dependent on the relative salinity of the surface-water and of the water at some deeper level, which under the influence of the wind wells up towards or actually to the surface.[1] If this be the correct explanation of the observed changes, we should be able to trace a distinct connection between the occurrence of the positive phase, that is a variation with the wind-force, or the negative phase, which goes against the wind-force, and the different seasons of the year. In the wet monsoon months the surface-water will be diluted and be rendered less saline, either directly by rainfall or indirectly by the increased outflow of river-water, and one would, on *a priori* grounds, expect to find that the oscillation of salinity at this season exhibited the positive phase; whereas in the hot dry months, the surface-water will by evaporation become condensed and be more saline than that lying at some little depth

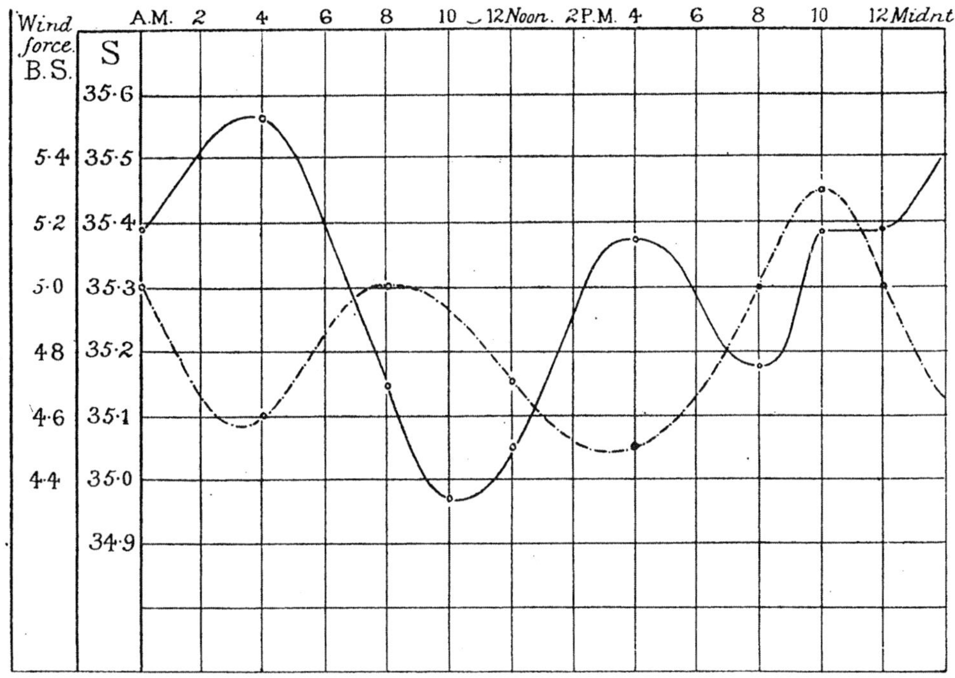

Text-fig. 117. Showing the salinity of the surface water at different times of the day in the Laccadive Sea in May, 1923, compared with the wind-force.

below and we should then expect to find that it was the negative phase that was present. Unfortunately, my data covers only part of the year, from October to May, and therefore includes only one of the wet periods, namely that of the north-east monsoon. In the following table I have given, so far as my data goes, the type of phase that is found to be present in the different months and in each of the three main regions of the Indian Seas. It must be borne in mind that a positive phase will be more easily detected, since in this case the effect of increased evaporation during increased wind-force will of itself tend to produce such a variation and so will emphasise and augment any change that may be due to an upwelling of more

[1] In the following pages where the salinity oscillates in the same direction as the wind-force, *i.e.* a rise in the wind-force being accompanied by a rise in salinity, I have termed the phase "positive," whereas when the salinity falls with a rise in the wind-force I have termed it "negative."

saline water from below; whereas the negative phase will be obscured by the effect of evaporation at the surface and, unless the upwelling of less saline water is marked, we may get no clear indication of it in the variation in the surface-salinity.

I have already (*vide supra*, p. 306 *et seq.*) called attention to the evidence that we at present possess regarding the changes during the course of the year in the salinity of the surface-water and of that lying at some depth below and it now remains to see to what extent the observations made in different months are in conformity with these changes.

Month.	Laccadive Sea.	Bay of Bengal.	Andaman Sea.
October	Positive	Positive	Negative
November	Positive	(? Positive)	Positive
December	Negative	(? Positive)	Positive
January	(Negative ?)	Positive or Negative	Positive
February	Negative	(? Negative)	Positive
March	Negative	Positive	..
April	Usually Negative but in abnormal years may be Positive.	Positive	Negative
May	Negative or Positive.	..	..

Table 80; showing the character, either Positive or Negative, of the phase of oscillation as regards the corresponding oscillation in the wind-force.

In the three regions the changes from the positive to negative phase or *vice versa* appear at first sight to occur at different periods of the year, but if we consider the conditions that are present during the succeeding months in each area separately, it will be found that the changes in the oscillatory phase are in the main in conformity with the seasonal conditions that are present in each area.

In the case of the Laccadive Sea in the months of October and November the oscillatory phase is positive, indicating that the surface-water is more dilute and less saline than the water at some depth below. It would appear probable that we have here in these two months the remains of the effect of the rainfall over the whole Arabian Sea and the western part of the Indian Ocean during the previous south-west monsoon. As this effect passes off the phase in the month of December becomes negative in type and definitely remains so during February and March: in the two succeeding months, April and May, the phase may be either positive or negative depending largely on local meteorological conditions. It would from this appear that the north-east monsoon is in this area without any very appreciable effect on the relationship of the salinity of the surface-water to that of the deeper layers and I have previously pointed out (*vide supra*, p. 57, *et. seq.*) that a similar absence of effect as regards the monthly mean air-temperature on the part of the north-east monsoon can be traced in this area. As I pointed out "it appears then that we can divide the Indian Seas into two areas, lying respectively to the west and east of the Indian Peninsula, in which conditions during the winter months are entirely different..... The differences that are seen to exist....are, I think, correlated with the N.E. Monsoon, the effects of which appear to be felt to a much greater extent on the east side of the Peninsula than on the west." A very similar difference can also be traced in the average humidity of the atmosphere on the two sides of India in each month of the survey season (*vide supra*, p. 94), and I have already remarked that "the data for the Arabian Sea give no indication whatever of any rise in humidity during the period of the N.E. Monsoon....As I have already shown when dealing with the air-temperature in different months on the east and west sides of the Peninsula, the effect of the N.E. monsoon is felt to a much greater extent on the east side, and the "Investigator" series shows in strict accord with this regional difference a very marked increase in humidity (on the east side) from November to December, that continues, though to a less extent, to February and is then succeeded by a fall. This is undoubtedly due to the N.E. Monsoon." It seems clear that the N.E. monsoon has but little effect on the west side of the Indian Peninsula, and the absence of any change during this monsoon season from the negative to the positive phase in the oscillation of the surface-salinity is in strict accord with the meteorological conditions to which I have referred above.

In the case of the Andaman Sea region, with the single exception of the month of October, the oscillation of the surface-salinity exhibits the negative phase throughout the whole series of my observations until the month of April and it seems probable that in the earlier month we still have the dry season condition present in the upper levels of the sea, the surface level being more saline than that immediately below: but in November with the onset of the north-east monsoon this condition is reversed, and from now on to the end of my observations in the month of February the surface-water is diluted either by rainfall or by influx from the rivers, and is, in consequence, less saline than the deeper stratum: in April, however, the condition is again reversed and the resulting oscillation phase is again negative, corresponding to the dry season. My observations in the Bay of Bengal unfortunately do not cover the whole period, but the data that I possess seems to indicate that this area is intermediate between the Laccadive Sea and the Andaman Sea both as regards its geographic position and its reaction to the monsoons. In October, the surface-salinity shows the positive phase that is characteristic of the wet seasons of the year, and thus is in exact agreement with the Laccadive Sea area. Although I have no records for the months of November and December it appears from the January results that this phase has been reversed during the two preceding months and that in certain years the negative or dry-season phase with a high surface-salinity may even persist till January. From February till April the phase is once again positive and this is clearly connected with the effect of the north-east monsoon, as a result of which the surface-water is rendered more dilute; although the direct effect of the monsoon will be over by February, the indirect effect of the increased outflow of river-water and the surface-drift from the Andaman Sea region will maintain a lowered surface-salinity for some considerable time later.

In the case of a positive variation in which the salinity rises with the wind-force we should naturally expect to find that the extent of the oscillations also varies in the same direction, a high wind being accompanied by a high range of oscillation, since in this case both evaporation from the surface and up-welling from below will be at a maximum. In those seasons of the year, however, in which the variation of the salinity is of the negative type, the effect of any up-welling of less saline water from below will to some and possibly to a large extent be counteracted by evaporation, and thus the range of oscillation will be reduced. If in these months we find that there is a correlation between the range of oscillation in the negative direction and the variation of the wind-force then it must, I think, be finally admitted that the up-welling is the more important factor in producing these diurnal change of salinity. Owing to the local differences in the various parts of the Indian Seas, we must, in order to make the comparison, confine ourselves to a single area and, fortunately, I have a good series of negative-phase oscillations occurring in successive months in the Laccadive Sea. In the following table I have given the average range of oscillation of the surface-salinity in the negative phase in each of the months in which it has been obtained in this area and the average wind-force in each month calculated from all observations taken on the "Investigator."

Month.	Wind-force.	Range of Oscillation of Salinity (negative phase.)
December	.. 3·33	0.59.
January	.. 2·26	..
February	.. 2·14	0·34.
March ..	.. 1·70	0·39.
April ..	.. 2·76	0-49.
May ..	.. 4·76	0·71.

It is clear from the above that there is a very close agreement between the strength of the wind and the amount of negative variation in the surface-salinity that can, I think, only be explained on the assumption that it is caused by up-welling from below of water of a low salinity. It seems clear, then, that the general character of this diurnal oscillation in the salinity of the surface-water agrees clearly with what one would expect, were it due, as I believe, to an up-welling of deeper water and, moreover, the results of our analysis of the data regarding the salinity is in close agreement with the conclusions that we have reached from a study of the, temperature-difference between the surface-layer and the supernatant atmosphere. If, now, these changes are respectively due to up-welling from below we should be able to find corroborative evidence of this change in a corresponding alteration of the chemical composition of the surface-water. I have already referred (*vide supra* p. 135) to the difference that I have found to exist between the specific gravity of the surface-water, as found by means of a Buchanan hydrometer and as calculated by means of Knudsen's Tables from the "halogen" content: as I then showed, we can in inshore waters detect a perfectly definite single oscillation in the difference between these two readings and I attributed this difference to the daily "change over," due to convection currents, of the surface-water and water from some depth below. If, now, under the influence of the rise and fall of the wind-force water from below the surface wells up twice a day, in consequence of a thinning of the superficial stratum, to an extent sufficient either to reach the surface itself or by wave action to become mixed with the surface-water, we should be able to find a corresponding change in the relationship between the specific gravity readings taken by the two methods given above, namely by the hydrometer and by titration.

In order to detect any such difference, if present, and to correlate it with the changes noted in the surface-salinity and the wind-force, the 4-hourly samples collected on three successive days, April 28th to 30th, 1925, were pooled in accordance with the time of day when the collection was made. This was necessary in order to obtain sufficient water for examination by the 'Buchanan' hydrometer. Each 4-hourly sample thus obtained was carefully examined, three estimations of the "halogen" content being made by the titration method and three observations of the specific gravity by the 'Buchanan' hydrometer. The average of these series of observations was then taken and the results obtained are given below.

		A.M.			P.M.	
		4	8	12	4	8	12
Ti.	Specific gravity calculated from halogen content.	28·2633	28·4333	28·0067	28·2667	28·3100	28·2933
Hy.	Specific gravity shown by a Buchanan Hydrometer.	28·2882	28·5248	27·9927	28·2863	28 4283	28·4274
	Difference Ti.—Hy.	-0·0249	-0·915	-0·0140	-0·0196	0·1183	-0·1341

I have already shown (*vide supra*, p. 140) that the difference between the specific gravity, as estimated by these two methods, tends to exhibit a single rise and fall in the course of the day, the hydrometer result being higher than the titration result in the middle of the day and, *vice versa*, the titration result being higher at about 6 in the morning. In the present series we see that the hydrometer result is higher than the result given by the titration method except only at 12 noon, but a further study of the differences shows that there is, in this case, not a single oscillation of the difference but a very clear double one. I have plotted these differences against the actual specific gravity as found by the two methods in Text-fig. 118 and it is clear that the difference approaches the zero line or may even become positive at times which correspond exactly with the epochs of minimum salinity: that is to say, that twice daily, namely at, in this instance, 4 A.M. and 12 noon, the surface water possesses a lowered specific gravity and, therefore, a lowered salinity but a relatively increased 'halogen' content. It seems clear that there is at these times an actual change in the percentage chemical composition of the surface-water that cannot be accounted for either by increased evaporation or by changes in the surface water caused by increased activity during the hours of sunlight of the algae that form part of the surface plankton. This change I believe to be due to the up-welling, under the influence of the wind, of water from below the surface. In text-fig. 118 I have also given the average rise and fall of the wind, as observed at 2-hourly intervals when at sea during the months of April, 1924 and 1925, and it is clear that, allowing for minor variations, there is a marked correlation between the wind-force and the changes observed in the sea-water. A rise in the strength of the wind occurs just before midnight and at 12 noon and the minimal strengths are seen to be at 6-30 A.M. and 8 P.M. A rise in the wind-force is accompanied after a short interval, due probably to the inertia of the surface mass of water, by a fall in specific gravity and an increased "halogen" percentage and, *vice versa*, a fall in the wind-force is accompanied by a rise in specific gravity and a fall in the "halogen" content. If this change is due to up-welling during increased wind, the water below the surface must possess a lower specific gravity than the surface-water but a higher "halogen" percentage. I have already (*vide* Pt. III, p. 139 *supra*) called attention to a series of observations taken by me during April, 1925, off the east coast of Ceylon and have shown that

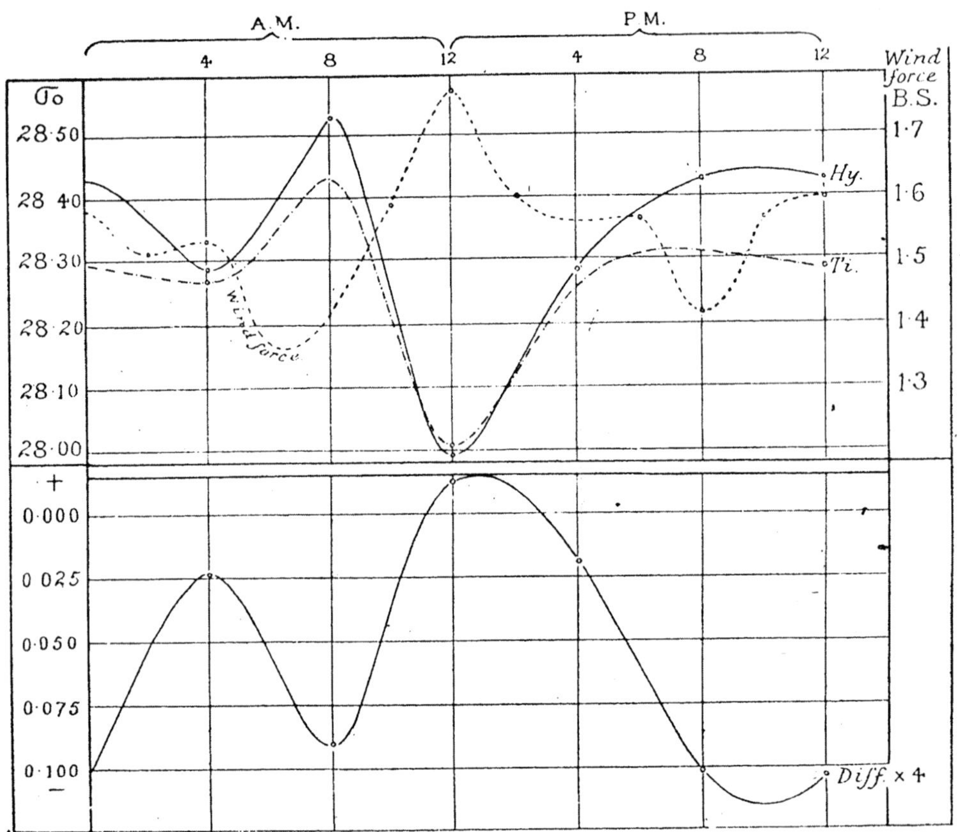

Text-fig 118; Showing the diurnal oscillation in the specific gravity (σ_0) of the surface water in the Laccadive Sea, April 28-30, 1925, as determined by titration and by means of a 'Buchanan' hydrometer

in these waters there is distinct evidence to show that during the daylight hours from 4 A.M. to 8 P.M. there is steady fall in the "halogen" percentage followed by a rapid corresponding rise during the night hours. If now this lowering of the halogen percentage be due, as one may reasonably suspect, to the activity of the planktonic organisms, such an effect will be most marked on the surface, since at this season of the year, owing to the raised average specific gravity and salinity in the water, the bulk of the plankton will be congregated at or near the surface, and it will be less marked in the deeper levels owing to the presence there (1) of less concentrated plankton and (2) a less intensity of light owing to the supernatant water, through which the light has to penetrate.

From the evidence before us it seems reasonably certain that there is each day a double oscillation in the strength of the wind in these Indian regions and that twice a day an increase in the wind-force causes a thinning of the surface-layer of the ocean to an extent that causes an up-welling of water from below either actually to the surface or to such a height that this deeper water becomes mixed with the surface-layer; and that this movement of the water masses produces corresponding effects both on the temperature relationship of the sea and the air and on the surface-salinity and the corresponding relationships of the surface and deeper strata of the ocean.

BIBLIOGRAPHY.

1. Broun, J. A., 1906 .. .. "Taglicher und Jährlicher Gang der Verdunstung in Südindien." Kleinere Mitteilungen. *Meteorologische Zeitschrift*, Vol. XLI, p. 428.

2. Buchan, A., 1895 .. .. "Report on Atmosphere Circulation based on the Observations made on board H.M.S. "Challenger," during the years 1873–1876, and other Meteorological Observations." *Report Sci. Results of H.M.S. "Challenger,"* Physics and Chemistry, Vol. II, London.

3. Dallas, W. L. .. .. "Charts of the Bay of Bengal and adjacent sea north of the Equator, showing the specific gravity, temperature and currents of the sea surface." Published by the Meteorological Department of the Government of India, 1887, Calcutta.

4. Eliot, J., 1898 .. .. "The discussion of the hourly meteorological observations recorded at Trivandrum, during the years 1853–1864." *Indian Meteorological Memoir*, Vol. X, No. I, Calcutta.

5. Harvey, H. W., 1923 .. .. "Hydrographic features of the water in the neighbourhood of Plymouth during the years 1921 and 1922." *Journal Marine Biol. As. Plymouth*, Vol. XIII (New Series), No. I, Plymouth.

6. Hayford, J. F., 1922 .. .. "Effects of winds and of Barometric Pressures on the great Lakes." Carnegie Institution of Washington, Publication No. 317, Washington.

7. Helland Hansen, B. and Nansen, F., 1909 .. .. "The Norwegian Sea." *Report on Norwegian Fishery and Marine Investigations*, Vol. II, No. 2, Christiania.

8. Hjort, J. .. .. .. *Vide* Murray and Hjort.

9. Krummel, O., 1907 .. .. "Handbuch der Ozeanographie" Vol. I, Stutgart.

10. ,, ,, 1911 .. .. "Handbuch der Ozeanographie" Vol. II, Stutgart.

11. Malpas, A. H., 1922 .. .. "Marine Biology." Report of the Government Marine Biologist for 1921. Ceylon Administration Report for 1921. Education, etc.

12. Matthews, D. J., 1926 .. "Physical Oceanography." The Percy Sladen Trust Expedition to the Indian Ocean in 1905, Vol. VIII, No. 7. *Trans. Linn. Soc. London*, Vol. XIX, Pt. I, London.

13. Michaelis, 1923 .. .. "Die wasserbewegung an der Oberfläche des Indischen Ozeans im Januar und Juli." *Veröffentlichungen des Instituts für Meereskunde an der Universitat Berlin.* Neue folge. Geographisch-naturwissenschaftliche Reihe, Heft 8, Berlin.

14. Murray, Sir J., 1898 .. .. 'On the annual Range of Temperature in the surface waters of the Ocean and its relation to other Oceanographical problems." *The Geographical Journal*, Vol. XII, No. 2, p. 113, London.

15. Murray, Sir J. and Hjort, J., 1912 .. "The Depths of the Ocean." London.

16. Pearson, J., 1915 .. .. "Marine Biology." Report of the Government Marine Biologist for 1914.
Ceylon Administration Report for 1914, Education, etc.

17. Pearson, J., 1923 .. .. "Marine Biology." Report of the Government Marine Biologist for 1922.
Ceylon Administration Report for 1922, Education, etc.

18. Pettersson, 1909 .. .. "Gezeitenaehnliche Bewegungen des Tiefen wassers." *Pub. de circonstancc.* No. 47. Conseil Permanent International pour l'exploration de la Mer. Copenhagen.

19. Reichard, A. C., 1913 .. .. "Hydrographische Beobachtungen bei Helgoland in den Jahren 1893–1908."
Wissenschaftliche Meeresuntersuchungen herausgegeben von der Komm. zur wissenschaftlichen untersuchungen der Deutschen Meere in Kiel und der Biologischen Anstalt auf Helgoland, (New Series), Vol. X, Kiel and Leipzig.

20. Ringer, W. E., 1922 .. .. "Temperatuur, zoutgehalte, dichtheid en gasgehalte van hat zeewater."
De Zeeën van Nederlandsch Oost-Indië. Leyden.

21. Schott, G., 1895 .. .. "Die jährliche Temperatureschwankung des Ozeanwassers." *Petermann's Mitteilungen,* 1895, 153–159.

22. Schott, G., 1902 .. .. "Oceanographie und Maritime Meteorologie."
Wissenschaftliche Ergebnisse der Deutschen Tiefsee-Expedition auf dem dampfer "Valdivia," Vol. I, Jena.

23. Weber, Max, 1922 .. .. "Biologie der Zee."
De Zeeën van Nederlandsch Oost-Indië, Leyden.

APPENDIX VIII.

Observations on the Temperature and Salinity of the surface-water of the Bay of Bengal.

Date.	Time.	Position.		Surface Temp.	Air Temp.	Relative Humidity of Atmosphere.	Barometric Pressure.	Cl.	S.	σ_0	σ_t
		Lat. N.	Long. E.	°C	°C						
1914											
April, 18	12 noon	10°. 49′. 30″	91°. 36′. 21″	30·00	..	..	..	18·03	32·58	26.177	19·932
,, ,,	4 p.m.	10 . 36 . 00	91 . 02 . 00	30·56		..	..	17.95	32·43	26·057	19·628
,, ,,	8 p.m.	10 . 19 . 00	90 . 26 . 00	28·89	..	..	..	18·06	32·62	26 212	20·340
,, ,,	12 midnight	10 . 05 . 00	89 . 51 . 00	28·89	..	..	..	18·45	33·33	26·778	20·867
,, 19	4 a.m.	9 . 50 . 00	89 . 16 . 00	28·89	..	..	..	18·33	33·12	26·605	20·706
,, ,,	8 a.m.	9 . 38 . 00	88 . 56 . 00	27·78	..	..	..	18·02	32·55	26·154	20·649
,, ,,	12 noon	9 . 21 . 20	88 . 22 . 20	29·72	..	..	..	18·19	32·86	26·400	20·236
,, ,,	4 p.m.	9 . 06 . 00	87 . 48 . 00	30·00	..	..	..	18·37	33·18	26·664	20·387
,, ,,	8 p.m.	8 . 54 . 00	87 . 16 . 00	28·89	..	..	..	18·30	33·07	26·569	20·673
,, ,,	12 midnight	8 . 38 . 00	86 . 43 . 00	28·33	..	..	..	18·41	33·26	26·719	20·996
,, 20	4 a.m.	8 . 23 . 00	86 . 09 . 00	27·78	..	..	..	18·24	32·95	26·483	20·956
,, ,,	8 a.m.	8 . 10 . 00	85 . 37 . 00	29·44	..	..	..	17·90	32·34	25·985	19·946
,, ,,	12 noon	7 . 55 . 00	84 . 59 . 30	30·28	..	..	..	17·80	32·15	25·831	19·517
,, ,,	4 p.m.	7 . 39 . 00	84 . 26 . 00	30·56	..	..	..	18·04	32·58	26·183	19·747
,, ,,	8 p.m.	7 . 18 . 00	83 . 50 . 00	30·00	..	..	..	18·12	32·74	26·297	20·046
,, ,,	12 midnight	7 . 08 . 00	83 . 21 . 00	28·89	..	..	..	18·13	32·75	26·323	20·544
,, 21	4 a.m.	6 . 53 . 00	82 . 47 . 00	27·78	..	..	..	18·15	32·79	26·352	20·834
,, ,,	8 a.m.	6 . 35 . 00	82 . 12 . 00	29·44	..	..	..	18·03	32·57	26·173	20·120
,, ,,	12 noon	6 . 06 . 15	81 . 25 . 30	29·44	..	..	..	18·33	33·12	26·607	20·523
,, ,,	4 p.m.	5 . 54 . 00	80 . 50 . 00	29·44	..	..	..	18·13	32·75	26·324	20·260
,, ,,	8 p.m.	5 . 50 . 00	80 . 12 . 30	29·44	..	..	..	18·32	33·10	26·595	20·512
1921											
Oct. 18	8 p.m.	5°. 56′. 30″	80°. 57′. 30″	29·8	..	..	30·12	18·99	34·31	27·569	21·296
,, ,,	12 midnight	6 . 05 . 00	81 . 30 . 00	26·0	..	..	29·98	18·77	33·91	27·252	22·238
,, 19	4 a.m.	6 . 19 . 00	82 . 00 . 30	25·5	..	..	29·91	18·99	34·30	27 563	22·682
,, ,,	8 a.m.	6 . 37 . 00	82 . 39 . 00	29·0	..	..	30·00	19·16	34·61	27·817	21·797
,, ,,	12 noon	6 . 51 . 00	83 . 22 . 30	29·9	..	..	29·86	19·10	34·50	27·721	21·403
,, ,,	4 p.m.	7 . 2 . 00	83 . 46 . 00	29·9	28·89	.	29·89	19·59	35·39	28·442	22·075
,, ,,	8 p.m.	7 . 20 . 00	84 . 24 . 00	29·8	28·33	..	29·95	18·84	34·04	27·352	21·095
,, ,,	12 midnight	7 . 40 . 00	85 . 03 . 00	29·0	27·22	..	29·98	18·90	34·14	27·437	21·505
,, 20	4 a.m.	7 . 49 . 00	85 . 25 . 00	27·0	26·67	..	29·91	18·86	34·07	27·377	22·040
,, ,,	8 a.m.	8 . 00 . 00	85 . 46 . 00	28·9	27·22	..	29·96	18·68	33·75	27·116	21·179
,, ,,	12 noon	8 . 11 . 42	86 . 29 . 42	29·8	27·78	..	29·93	18·78	33·93	27·264	21·012
,, ,,	4 p.m.	8 . 28 . 00	86 . 58 . 00	29·0	28 33	..	29·73	18·83	34 02	27·340	21·353
,, ,,	8 p.m.	8 . 48 . 00	87 . 35 . 00	29·0	27·78	..	29·89	18·82	33·99	27·315	21 330
,, 21	4 a.m.	9 . 30 . 00	88 . 50 . 00	27·0	26·11	..	29·70	18·71	33·80	27·164	21·832
,, ,,	8 a.m.	10 . 06 . 20	89 . 19 . 18	29·0	26·67	..	29·91	18·66	33·70	27·082	21·113
,, ,,	12 noon	10 . 10 . 18	89 . 55 . 12	30·0	28·33	..	29·87	18·61	33·62	27·021	20 719
,, ,,	4 p.m.	10 . 22 . 30	90 . 25 . 00	29·1	28·33	..	29·82	18·76	33·90	27·236	21·223
,, ,,	8 p.m.	10 . 37 . 00	91 . 00 . 00	29·0	27·78	..	29·88	18·87	34·09	27·395	21·404
,, ,,	12 midnight	10 . 55 . 30	91 . 37 . 30	29·0	27·78	..	29·90	18·31	33·08	26·581	20·650
,, 22	4 a.m.	11 . 02 . 30	92 . 00 . 30	28·0	25 56	..	29·86	18·25	32 98	26 505	20·910
,, ,,	8 a.m.	11 . 19 . 30	92 . 38 . 00	28·8	22·75	..	29·99	17·92	32·38	26·006	20·180
1922											
Oct. 9	4 p.m.	5 . 48 . 30	80 . 34 . 00	26·22	27·22	..	..	19·31	34·88	28·028	22·895
,, ,,	8 p.m.	5 . 58 . 00	81 . 10 . 00	26·17	26·11	..	..	19·40	35·05	28·159	23·032
,, 10	4 a.m.	6 . 30 . 00	82 . 30 . 00	27·78	27·22	..	..	18·89	34·13	27·430	21·838
,, ,,	8 a.m.	6 . 48 . 30	83 . 07 . 45	27·78	27·78	..	..	19·35	34·96	28·086	22·449
,, ,,	12 noon	6 . 38 . 30	83 . 34 . 30	27·89	27·89	..	..	19·28	34·83	27·993	22·348
,, ,,	4 p.m.	6 . 53 . 30	84 . 01 . 00	28·22	27·50	..	..	19·13	34·57	27·776	22·017
,, ,,	8 p.m.	7 . 19 . 00	84 . 23 . 00	28·22	27·22	..	..	18·98	34·29	27·546	21·802
,, ,,	12 midnight	7 . 34 . 00	85 . 00 . 00	28·06	27·22	..	..	18·96	34·24	27·515	21·826
,, 11	4 a.m.	7 . 54 . 00	85 . 36 . 00	27·33	27·22	..	..	18·79	33·95	27·284	21·847

Date.	Time.	Position.		Surface Temp. °C.	Air Temp. °C.	Relative Humidity of Atmosphere.	Barometric Pressure.	Cl.	S.	σ_0	σ_t
		Lat. N.	Long. E.								
1922.											
Oct. 11	8 a.m.	8°. 10′. 00″	86°. 12′. 00″	27·44	27·78	..	..	19·15	34·60	27·804	22·296
,, ,,	12 noon	8 . 51 . 30	86 . 52 . 00	27·78	27·78	..	..	19·25	34·78	27·949	22·322
,, ,,	4 p.m.	9 . 05 . 00	87 . 50 . 00	28·60	27·50	..	..	19·26	34·79	27·963	22·066
,, ,,	8 p.m.	9 . 22 . 00	87 . 58 . 00	28·50	27·22	..	..	18·79	33·95	27·277	21·460
,, ,,	12 midnight	9 . 40 . 00	88 . 35 . 00	28·33	27·22	..	..	18·60	33·60	27·001	21·258
,, 12	4 a.m.	9 . 53 . 00	89 . 12 30	27·67	27·67	..	..	18·59	33·58	26·981	21·456
,, ,,	8 a.m.	10 . 10 . 00	89 . 48 . 00	28·33	25·56	..	..	18·46	33·34	26·793	21·065
,, ,,	12 noon	10 . 21 . 00	90 . 17 . 30	28·17	26·56	..	..	18·17	32·83	26·384	20·737
,, ,,	4 p.m.	10 . 31 . 00	90 . 44 . 00	28·28	25·78	..	..	17·82	32·19	25·860	20·213
,, ,,	8 p.m.	10 . 41 . 00	91 . 12 . 00	28·00	26·11	..	..	17·88	32·30	25·946	20·386
,, ,,	12 midnight	10 . 53 . 30	91 . 49 . 00	27·78	26·11	..	..	18·03	32·57	26·170	20·664
,, 13	4 a.m.	11 . 11 . 00	92 . 24 . 00	27·44	26·11	..	..	18·11	32·72	26·285	20·880
,, ,,	8 a.m.	11 . 31 . 15	92 . 47 . 00	28·78	27·67	..	..	17·67	31·92	25·654	19·858
1923.											
Jan. 23	4 p.m.	11 . 10 . 30	92 . 19 . 00	27·2	26·78	..	..	18·41	33·26	26·72	21·363
,, ,,	8 p.m.	10 . 52 . 00	91 . 38 . 00	27·1	26·56	..	..	18·46	33·35	26·80	21·469
,, ,,	12 midnight	10 . 36 . 00	91 . 06 . 00	26·67	25·83	..	..	18·49	33·40	26·84	21·643
,, 24	4 a.m.	10 . 22 . 00	90 . 32 . 00	26·67	25·83	..	..	18·64	33·68	27·06	21·813
,, ,,	8 a.m.	10 . 05 . 00	90 . 00 . 00	27·00	26·94	..	..	18·46	33·35	26·80	21·501
,, ,,	12 noon	9 . 38 . 45	89 . 20 . 00	27·2	27·78	..	..	18·40	33·24	26·71	21·353
,, ,,	4 p.m.	9 . 24 . 00	88 . 46 . 00	27·2	26·71	..	..	18·46	33·35	26·80	21·437
,, ,,	8 p.m.	9 . 10 . 00	88 . 11 . 00	27·2	25·67	..	..	18·73	33·84	27·19	21·801
,, ,,	12 midnight	8 . 54 . 00	87 . 35 . 00	26·94	26·11	..	..	18·82	34·00	27·32	22·006
,, 25	4 a.m.	8 . 41 . 00	87 . 01 . 00	26·67	25·67	..	..	18·90	34·14	27·44	22·208
,, ,,	8 a.m.	8 . 26 . 00	86 . 28 . 00	27·50	26·67	..	..	18·95	34·23	27·51	22·003
,, ,,	12 noon	8 . 16 . 30	85 . 46 . 30	27·7	27·5	..	..	18·99	34·31	27·57	21·994
,, ,,	4 p.m.	8 . 00 . 00	85 . 12 . 00	27·7	27·17	..	..	19·08	34·47	27·70	22·116
,, ,,	8 p.m.	7 . 46 . 00	84 . 36 . 00	27·7	26·67	..	..	19·10	34·51	27·73	22·144
,, ,,	12 midnight	7 . 29 . 00	84 . 04 . 00	27·22	26·67	..	..	19·10	34·51	27·73	22·298
,, 26	4 a.m.	7 . 12 . 30	83 . 31 . 00	26·67	25·67	..	..	19·20	34·69	27·87	22·608
,, ,,	8 a.m.	6 . 55 . 30	82 . 58 . 00	26·9	26·67	..	..	18·92	34·18	27·47	22·259
,, 26	12 noon	6°. 32′. 00″	82°. 09′. 00″	27·0	26·11	..	..	17·16	31·00	24·91	19·739
,, ,,	4 p.m.	6 . 6 . 54	81 . 34 . 24	26·9	26·94	..	..	17·71	32·00	25·71	20·517
,, ,,	8 p.m.	5 . 54 . 18	80 . 50 . 18	27·0	26.67	..	..	18·12	32·74	26·30	21·135
,, ,,	12 midnight	5 . 52 . 00	80 . 12 . 30	26·39	25.67	..	..	18·04	32·59	26·19	21·123
1924.											
Jan. 3	8 p.m.	6 . 26 . 54	79 . 47 . 18	27·6	27·22	86·09	29·90	18·72	33·82	27·18	21·66
,, ,,	10 p.m.	6 . 13 . 00	79 . 54 . 30	27·3	27·22	81·74	29·92	18·97	34·27	27·54	22·10
,, ,,	12 midnight	5 . 54 . 42	80 . 06 . 12	27·0	26·67	85·91	29·90	18·81	33·98	27·31	21·98
,, 4	4 a.m.	5 . 47 . 00	80 . 36 . 00	26·0	26·67	77·45	29·84	18·69	33·77	27·13	22·12
,, ,,	8 a.m.	5 . 55 . 30	81 . 06 . 50	26·2	25·78	88·84	29·93	18·58	33·57	26·97	21·91
,, ,,	10 a.m.	6 . 00 . 30	81 . 22 . 00	26·25	25·89	89·15	29·93	18·57	33·55	26·96	21·83
,, ,,	12 noon	6 . 07 . 30	81 . 35 . 30	26·3	24·89	94·68	29·91	18·47	33·37	26·81	21·73
,, ,,	4 p.m.	6 . 22 . 30	82 . 06 . 00	26·8	27·17	85·25	29·82	18·56	33·53	26·94	21·70
,, ,,	8 p.m.	6 . 46 . 00	82 . 43 . 00	27.1	27·06	87·72	29·89	..	..	....	....
,, ,,	10 p.m.	6 . 55 . 00	82 . 58 . 00	27.3	27·22	86·09	29·91	18·71	33·80	27·16	21·74
,, ,,	12 midnight	7 . 05 . 00	83 . 15 . 00	27·5	27·22	84·75	29·89	18·69	33·77	27·13	21·65
,, 5	4 a.m.	7 . 17 . 00	83 . 38 . 00	27·0	27·78	77·70	29·83	18·71	33·80	27·16	21·84
,, ,,	8 a.m.	7 . 27 . 00	84 . 02 . 00	27·0	27·94	84·04	29·89	19·03	34·38	27·63	22·28
,, ,,	10 a.m.	7 . 37 . 00	84 . 16 . 00	27·4	28·22	81·94	29·92	19·21	34·70	27·89	22·39
,, ,,	12 noon	7 . 49 . 00	84 . 37 . 00	28·0	28·28	81·93	29·91	18·72	33·82	27·18	21·53
,, ,,	4 p.m.	8 . 06 . 00	85 . 14 . 00	28·0	26·61	89·05	29·82	18·31	33·08	26·58	20·97
,, ,,	8 p.m.	8 . 22 . 00	85 . 52 . 00	27·9	27·39	86·11	29·91	18·93	34·20	27·48	21·85
,, ,,	10 p.m.	8 . 30 . 00	86 . 10 . 00	27·0	27·22	86·09	29·95	18·79	33·95	27·28	21·95

Date.	Time.	Position.		Surface Temp.	Air Temp.	Relative Humidity of Atmosphere.	Barometric Pressure.	Cl.	S.	σ_0	σ_t
		Lat. N.	Long. E.	°C	°C						
1924.											
Jan. 5	12 midnight	8°. 38′. 00″	86°. 27′. 00″	27·0	26·94	86·11	29·93	18·78	33·93	27·26	21·93
,, 6	4 a.m.	8 . 54 . 00	87 . 02 . 00	26·5	26·94	86·11	29·86	18·68	33·75	27·12	21·96
,, ,,	8 a.m.	9 . 04 . 00	87 . 14 . 00	27·5	27·78	84·55	29·93	18·70	33·78	27·15	21·67
,, ,,	10 a.m.	9 . 17 . 00	87 . 52 . 00	27·7	27·78	85·36	29·94	18·76	33·89	27·23	21·68
,, ,,	12 noon	9 . 34 . 00	88 . 09 . 00	27·8	28·22	82·76	29·91	18·78	33·93	27·26	21·67
,, ,,	4 p.m.	9 . 49 . 00	88 . 46 . 00	27·5	28·06	81·93	29·86	18·67	33·75	27·12	21·64
,, ,,	8 p.m.	10 . 06 . 00	89 . 28 . 00	27·3	27·89	83·22	29·92	18·64	33·68	27·06	21·65
,, ,,	10 p.m.	10 . 14 . 00	89 . 46 . 00	27·0	27·78	81·88	29·97	18·78	33·93	27·26	21·93
,, ,,	12 midnight	10 . 20 . 30	90 . 06 . 00	27·2	27·78	81·88	29·95	18·79	33·95	27·28	21·88
,, 7	4 a.m.	10 . 34 . 00	90 . 40 . 00	27·0	27·50	79·69	29·90	18·83	34·02	27·34	22·01
,, ,,	8 a.m.	10 . 44 . 00	91 . 14 . 00	27·8	27·78	77·70	30·00	18·67	33·73	27·10	21·52
,, ,,	10 a.m.	10 . 51 . 00	91 . 32 . 00	27·4	27·78	81·35	30·01	18·51	33·44	26·87	21·44
,, ,,	12 noon	10 . 55 . 30	91 . 55 . 00	27·9	27·50	80·21	..	18·41	33·26	26·72	21·14
Mar. 4	4 p.m.	11 . 09 . 00	92 . 13 . 00	29·0	29·0	81·32	29·81	18·18	32·84	26·39	20·47
,, ,,	8 p.m.	10 . 52 . 30	91 . 33 . 00	28·0	28·1	81·93	29·87	18·32	33·10	26·59	20·98
,, ,,	10 p.m.	10 . 42 . 00	91 . 12 . 00	27·5	27·8	81·88	29·92	18·50	33·42	26·86	21·40
,, ,,	12 midnight	10 . 36 . 00	90 . 55 . 00	27·5	27·8	81·88	29·91	18·55	33·51	26·93	21·46
,, 5	4 a.m.	10 . 20 . 00	90 . 18 . 00	26·0	27·2	81·74	29·81	18·68	33·75	27·12	22·11
,, ,,	8 a.m,	10 . 02 . 00	89 . 41 . 00	27·8	27·5	83·95	29·90	18·49	33·40	26·84	21·28
,, ,,	10 a.m.	9 . 52 . 00	89 . 24 . 00	28·1	28·1	81·93	29·91	18·47	33·37	26·81	21·16
,, ,,	12 noon	9 . 42 . 00	89 . 07 . 00	28·1	29·3	81·35	29·90	18·44	33·31	26·77	21·12
,, ,,	4 p.m.	9 . 24 . 00	88 . 28 . 00	28·1	27·8	84·03	29·81	18·32	33·10	26·59	20·95
,, ,,	8 p.m.	9 . 08 . 00	87 . 53 . 00	28·0	27·5	84·76	29·88	18·54	33·49	26·91	21·28
,, ,,	10 p.m.	8 . 59 . 00	87 . 34 . 00	27·5	27·2	86·09	29·92	18·66	33·71	27·09	21·61
,, ,,	12 midnight	8 . 51 . 00	87 . 17 . 00	27·5	27·2	86·09	29·91	18·81	33·98	27·31	21·82
,, 6	4 a.m.	8 . 32 . 00	86 . 40 . 00	25·0 (?)	27·2	83·89	29·82	18·79	33·95	27·28	22·57 ?
,, ,,	8 a.m.	8 . 22 . 00	85 . 50 . 00	27·6	27·5	83·95	29·92	18·94	34·22	27·49	21·95
,, ,,	10 a.m.	8 . 13 . 00	85 . 32 . 00	29·0	27·9	80·58	29·93	18·55	33·51	26·93	20·98
,, ,,	12 noon	8 . 05 . 00	85 . 20 . 00	29·2	28·5	81·23	29·89	18·75	33·87	27·22	21·17
,, ,,	4 p.m.	7 . 45 . 00	84 . 43 . 00	28·6	28·3	84·11	29·82	18·89	34·13	27·42	21·56
,, ,,	8 p.m.	7 . 24 . 00	84 . 07 . 12	28·2	28·1	84·59	29·91	18·77	33·91	27·25	21·53
,, ,,	10 p.m.	7 . 14 . 00	83 . 48 . 00	28·0	27·8	81·88	29·96	18·74	33·86	27·20	21·55
,, ,,	12 midnight	7 . 05 . 00	83 . 31 . 00	28·0	27·8	81·88	29·91	18·60	33·60	27·00	21·37
,, 7	4 a.m.	6 . 46 . 00	82 . 52 . 00	27·0	27·8	81·88	29·84	18·64	33·68	27·06	21·74
,, ,,	8 a.m.	6 . 22 . 00	82 . 17 . 00	28·2	27·7	86·99	29·92	18·65	33·69	27·07	21·37
,, ,,	10 a.m.	6 . 18 . 30	81 . 56 . 45	28·2	27·2	83·89	29·94	18·72	33·82	27·18	21·47
,, ,,	12 noon	6 . 10 . 00	81 . 34 . 00	29·4	28·5	75·90	29·86	18·72	33·82	27·18	21·07
,, ,,	4 p.m.	5 . 56 . 30	80 . 58 . 00	28·7	28·8	85·75	29·80	18·67	33·73	27·10	21·23
,, ,,	8 p.m.	5 . 52 . 00	80 . 21 . 30	28·2	28·6	82·03	29·91	18·65	33·69	27·07	21·37
,, ,,	10 p.m.	5 . 56 . 48	80 . 04 . 42	28·0	28·3	81·99	29·93	18·72	33·82	27·18	21·53
,, ,,	12 midnight	6 . 10 . 48	84 . 54 . 42	28·0	28·1	83·88	29·90	18·88	34·11	27·41	21·75
,, 8	4 a.m.	6 . 45 . 00	80 . 46 . 00	27·0	27·5	83·95	29·84	19·03	34·38	27·63	22·28
1925											
April 17	8 a.m.	11°. 03′. 00″	90°. 38′. 00″	29·3	28·61	75·7	30·07	18·62	33·64	27·03	20·96
,, ,,	12 noon	10 . 56 . 00	90 . 00 . 00	30·1	29·72	69·8	30·08	18·48	33·39	26·83	20·51
,, ,,	4 p.m.	10 . 53 . 00	89 . 20 . 00	31·0	30·11	67·8	30·00	18·43	33·30	26·75	20·12
,, ,,	8 p.m.	10 . 51 . 00	88 . 47 . 00	30·1	29·44	69·7	30·07	18·68	33·75	27·12	20·70
,, ,,	12 midnight	10 . 46 . 00	88 . 08 . 00	28·0	29·44	70·4	30·04	18·76	33·89	27·23	21·58
,, 18	4 a.m.	10 . 42 . 00	87 . 30 . 00	29·2	28·89	70·3	30·03	18·46	33·35	26·80	20·78
,, ,,	8 a.m.	10 . 41 . 00	86 . 51 . 00	29·9	29·44	71·6	30·08	18·24	32·95	26·48	20·25
,, ,,	12 noon	10 . 37 . 00	86 . 17 . 00	31·0	30·17	69·5	30·08	18·54	33·49	26·91	20·27
,, ,,	4 p.m.	10 . 34 . 00	85 . 40 . 00	31·0	30·72	66·0	30·00	18·26	32·99	26·51	19·90
,, ,,	8 p.m.	10 . 28 . 00	85 . 00 . 00	31·0	29·72	76·9	30·05	18·41	33·26	26·72	20·13
,, ,,	12 midnight	10 . 24 . 00	84 . 24 . 00	29·0	29·44	78·1	30·05	18·68	33·75	27·12	21·07
,, 19	4 a.m.	10 . 20 . 00	83 . 45 . 00	29·2	28·89	78·0	30·01	18·81	33·98	27·31	21·26

Date.	Time.	Position.		Surface Temp. °C.	Air Temp. °C.	Relative Humidity of Atmosphere.	Barometric. Pressure.	Cl.	S.	σ_0	σ_t
		Lat. N.	Long. E.								
1924. April 19	8 a.m.	10°. 20′. 00″	83°. 06′. 00″	29·4	29·28	77·6	30·07	18·62	33·64	27·03	20·93
,, ,,	12 noon	10 . 21 . 00	82 . 29 . 00	30·2	30·28	75·5	30·05	18 34	33·13	26·62	20·28
,, ,,	4 p.m.	10 . 16 . 00	81 . 52 . 00	30·2	30·00	74·3	29·97	18·39	33·22	26·70	20·35
,, ,,	8 p.m.	10 . 10 . 00	81 . 11 . 00	29·5	29·49	79·4	30·04	18·93	34·20	27·48	21·41
,, ,,	12 midnight	10 . 06 . 00	80 . 31 . 30	28·0	29·72	80·1	30·02	18·62	33·64	27·03	21·39
,, 20	4 a.m.	10 . 03 . 00	80 . 14 . 00	28·2	28·33	82·0	30 00	18·12	32·74	26·30	20·65
,, ,,	8 a.m.	9 . 55 . 00	80 . 05 . 00	29·7	28·50	82·7	30·05	18·12	32 74	26·30	20·6 ;
,, 22	12 noon	9 . 53 . 30	80 . 01 . 30	30·0	29·17	84·2	30·00	19·12	34·54	27·76	21·41
,, ,,	4 p.m.	10 . 04 . 00	80 . 33 . 00	29·8	29·44	82·9	30 05	18·74	33·86	27·20	20·95
,, ,,	8 p.m.	9 . 36 . 00	80 . 59 . 00	29·0	28·94	87·9	30·01	18·94	34·22	27·49	21·49
,, ,,	12 Midnight	9 . 08 . 30	81 . 26 . 00	28·0	27·78	92·9	29·94	18 96	34·25	27·52	21·85
,, 23	4 a.m.	8 . 39 . 00	81 . 51 . 00	27·5	27·78	86·2	30·01	18·95	34·23	27·51	22·00
,, ,,	8 a.m.	8 . 00 . 00	82 . 02 . 00	29·5	29·22	85·2	30·01	19·06	34·43	27·67	21·49
,, ,,	12 noon	7 . 28 . 00	82 . 11 . 00	29·9	29·17	79·6	29·97	18·99	34·31	27·57	21·26
,, ,,	4 p.m.	6 . 50 . 00	82 . 02 . 30	28·8	29·17	82·7	30.03	19·06	34·43	27·67	21·73
,, ,,	8 p.m.	6 . 20 . 00	81 . 52 . 30	28·3	29·06	87·1	29·99	19·14	34·58	27·79	22·00
,, ,,	12 midnight	6 . 00 . 30	81 . 26 . 00	27·0	28·33	86·3	29·88	19·17	34·63	27·83	22·46
,, 24	4 a.m.	5 . 50 . 00	80 . 52 . 00	28·0	26·67	85·9	29·96	19·11	34·52	27·74	22·06
,, ,,	8 a.m.	5 . 49 . 00	80 . 18 . 15	28·2	27·33	92·3	29·95	19·09	34·49	27·71	21·96
,, ,,	12 noon	6 . 14 . 30	79 . 55 . 00	29·0	29·39	79·9	29·93	19·06	34·43	27·67	21·66
,, ,,	4 p.m.	6 . 49 . 12	79 . 46 . 40	29·3	30·00	80·7	29·99	18·87	33·09	27·39	21·20

APPENDIX IX.

Observations on the Temperature and Salinity of the Surface-Water of the Andaman Sea.

Date.	Time.	Position.		Surface Temp. °C.	Air Temp. °C.	Relative Humidity of Atmosphere.	Barometric Pressure.	Cl.	S.	σ_0	σ_t
		Lat. N.	Long. E.								
1914											
April 16	12 noon	16°. 16′. 00″	96°. 15′. 30″	28·90	..	..	..	14·245	25·74	20·680	15·200
,, ,,	4 p.m.	5 . 48 . 30	95 . 56 . 00	31·11	..	..	..	15·33	27·70	22·252	15·571
,, ,,	8 p.m.	15 . 24 . 00	95 . 40 . 15	28·89	..	..	..	17·61	31·82	25·566	19·740
,, ,,	12 midnight	14 . 55 . 00	95 . 19 . 30	27·22	..	..	..	18·29	33·04	26·548	21·196
,, 17	4 a.m.	14 . 26 . 00	94 . 59 . 00	27·22	..	..	..	17·97	32·46	26·084	20·763
,, ,,	8 a.m.	13 . 52 . 40	94 . 34 . 45	28·61	..	..	..	17·91	32·36	26·003	20·238
,, ,,	12 noon	13 . 20 . 45	94 . 15 . 30	30·56	..	..	..	17·93	32·39	26·028	19·603
,, ,,	4 p.m.	12 . 49 . 00	93 . 53 . 30	30·56	..	..	..	17·83	32·21	25·880	19·266
,, ,,	8 p.m.	12 . 18 . 15	93 . 31 . 45	28·89	..	..	..	17·89	32·32	25·970	20·116
,, ,,	12 midnight	11 . 50 . 30	93 . 10 . 00	28·89	..	..	..	18·05	32·61	26·197	20·319
,, 18	4 a.m.	11 . 24 . 00	92 . 45 . 00	27·78	..	..	..	18·11	32·71	26·282	20·767
,, ,,	8 a.m.	11 . 06 . 00	92 . 12 . 45	29·44	..	..	..	18·21	32·90	26·434	20·362
1921											
Oct. 25	12 noon	Port Blair Harbour.		29·2	28·33		29·85	17·20	31·08	24·972	19·084
,, ,,	4 p.m.	11°. 12′. 20″	92°. 52′. 00″	29·0	28·89	..	29·78	17·84	32·23	25·896	20·010
,, ,,	8 p.m.	10 . 35 . 30	92 . 57 . 00	29·0	27·78	..	29·85	17·81	32·18	25·850	19·967
,, ,,	12 midnight	10 . 00 . 00	93 . 04 . 00	28·0	26·67	..	29·85	17·91	32·36	26·002	20·437
,, 26	4 a.m.	9 . 22 . 00	93 . 10 . 00	27·5	27·22	..	29·81	18·53	33·47	26·891	21·426
,, ,,	8 a.m.	8 . 45 . 30	93 . 16 . 00	28·0	27·22	..	..	18·79	33·95	27·282	21·629
Nov. 11	4 p.m.	8 . 10 . 00	93 . 22 . 30	28·5	26·39	..	..	18·71	33·80	27·160	21·341
,, ,,	8 p.m.	8 . 49 . 00	93 . 16 . 00	27·8	26·00	..	..	18·66	33·71	27·094	21·518
,, ,,	12 midnight	9 . 29 . 00	93 . 09 . 00	27·2	25·56		..	18·92	34·18	27·467	22·059
,, 12	4 a.m.	10 . 11 . 00	93 . 02 . 00	27·4	26·11	..	..	18·94	34·22	27·498	22·020
,, ,,	8 a.m.	10 . 52 . 00	92 . 55 . 00	28·0	27·06	..	..	18·04	32·57	26·181	20·604
,, ,,	12 noon	11 . 30 . 35	92 . 48 . 50	28·0	26·67	..	..	17·97	32·47	26·093	20·522
,, 15	4 p.m.	11 . 07 . 00	92 . 53 . 00	29·0	26·94	..	..	18·09	32·68	26·257	20·346
,, ,,	8 p.m.	10 . 25 . 30	93 . 00 . 00	28·0	26·67	..	..	18·21	32·91	26·439	20·881
,, ,,	12 midnight	9 . 44 . 00	93 . 06 . 00	27·5	..	..	..	18·79	33·95	27·278	21·786
,, 16	4 a.m.	9 . 01 . 00	93 . 14 . 00	27·75	..	..	..	18·84	34·04	27·349	21·772
,, ,,	8 a.m.	8 . 31 . 30	93 . 12 . 30	28·0	27·00	..	..	18·74	33·86	27·208	21·560
Dec. 2	12 noon	8 . 09 . 00	93 . 22 . 00	28·0	28·67	..	..	18·52	33·46	26·878	21·252
,, ,,	4 p.m.	8 . 47 . 00	93 . 14 . 00	28·1	28·33	..	..	18·33	33·12	26·608	20·968
,, ,,	8 p.m.	9 . 27 . 00	93 . 05 . 00	28·0	27·22	..	..	18·20	32·88	26·423	20·829
,, ,,	12 midnight	10 . 08 . 30	93 . 05 . 30	28·0	27·22	..	..	18·69	33·77	27·128	21·485
,, 3	4 a.m.	10 . 43 . 00	92 . 57 . 00	27·75	26·39	..	..	18·11	32·71	26·281	20·777
,, ,,	8 a.m.	11 . 28 . 00	92 . 50 . 00	28·1	26·94	..	..	17·87	32·29	25·938	20·345
,, 5	12 noon	Port Blair Harbour.		28·0	28·06	..	..	17·51	31·64	25·417	19·892
,, ,,	4 p.m.	11 . 02 . 30	92 . 53 . 30	28·2	26·94	..	..	17·74	32·05	25·746	20·133
,, ,,	8 p.m.	10 . 24 . 00	92 . 56 . 30	28·1	26·94	..	..	18·06	32·62	26·212	20·600
,, ,,	12 midnight	9 . 46 . 00	93 . 04 . 00	28·0	27·22	..	..	18·56	33·53	26·944	21·314
,, 6	4 a.m.	9 . 02 . 00	93 . 12 . 00	28·5	27·22	..	..	18·35	33·14	26·636	20·863
,, ,,	8 a.m.	8 . 24 . 00	93 . 20 . 00	28·0	26·94	..	..	18·35	33·15	26·643	21·034
,, ,,	12 noon	off W. entrance Nankauri Harbour.		28·1	27·78	..	..	18·32	33·10	26·581	20·938
,, 31	4 p.m.	11°. 12′. 12″	92°. 50′. 24″	28·0	26·66	..	..	18·345	33·14	26·630	21·022
,, ,,	8 p.m.	10 . 34 . 25	93 . 05 . 00	27·9	26·61	..	..	18·47	33·37	26·819	21·231
,, ,,	12 midnight	10 . 06 . 30	93 . 00 . 00	27·0	..	..	..	18·48	33·39	26·831	21·530
1922											
Jan. 1	4 a.m.	9 . 21 . 00	93 . 09 . 00	27·5	..	..	..	18·37	33·19	26·672	21·222
,, ,,	8 a.m.	8 . 49 . 15	93 . 03 . 12	28·0	27·22	..	..	18·245	32·96	26·470	20·873
,, ,,	12 noon	8 . 18 . 30	93 . 18 . 30	28·1	27·78	..	..	18·43	33·30	26·751	21·101
,, 25	12 noon	8 . 32 . 30	93 . 18 . 00	28·2	27·22	..	..	18·45	33·33	26·779	21·095
,, ,,	4 p.m.	9 . 6 . 30	93 . 11 . 00	28·0	28·33	..	..	18·18	32·84	26·387	20·795
,, ,,	8 p.m.	9 . 45 . 00	93 . 03 . 00	27·5	25·67	..	..	18·03	32·57	26·166	20·750

Date.	Time.	Position.		Surface Temp. °C.	Air Temp. °C.	Relative Humidity of Atmosphere.	Barometric Pressure.	Cl.	S.	σ_0	σ_t
		Lat. N.	Long. E.								
1922. Jan. 25	12 midnight	10 . 21 . 00	93 . 04 . 00	27·0	26·67	..	..	18·01	32·55	26·148	20·893
,, 26	4 a.m.	10 . 58 . 00	93 . 06 . 00	27·0	26·67	..	..	17 73	32·03	25·741	20·514
,, ,,	8 a.m.	11 . 32 . 45	92 . 47 . 45	27·9	26·89	..	..	18·27	33·01	26·523	20·955
,, 31	12 noon	11 . 31 . 45	92 . 48 . 15	27·8	27·17	..	..	17·70	31·98	25·691	20·212
,, ,	4 p.m.	10 . 57 . 00	92 . 52 . 00	28·0	27·00	..	..	18·20	32·87	26·414	20·820
,, ,,	8 p.m	10 . 20 . 00	93 . 01 . 00	27·9	26·56	..	..	18·07	32·65	26·227	20·679
,, ,,	12 midnight	9 . 43 . 00	93 . 08 . 00	27·5	26·67	..	..	18·03	32·57	26·167	20·751
Feb. 1	4 a.m.	9 . 05 . 00	93 . 16 . 00	27·5	26·67	..	..	18·32	33·10	26·589	21·145
,, ,,	8 a.m.	8 . 34 . 45	93 . 12 . 30	27·9	27·50	..	..	18·42	33·27	26·734	21·152
,, 25	12 noon	8 . 26 . 45	93 . 17 . 30	29·0	28·00	..	..	18·39	33·22	26·697	20·755
,, ,,	4 p.m.	9 . 05 . 00	93 . 11 . 00	28·6	27·56	..	..	18·31	33·08	26·582	20·780
,, ,,	8 p.m.	9 . 41 . 00	93 . 05 . 00	28·1	26·83	..	..	18·56	33·53	26·942	21·28
,, ,,	12 midnight	10 . 16 . 30	93 . 05 . 00	27·7	26·67	..	..	18·32	33·10	26·586	21·08
,, 26	4 a.m.	10 . 55 . 00	93 . 05 . 00	27·5	26·11	..	..	18·28	33·03	26·530	21·09
,, ,,	8 a.m.	11 . 30 . 12	92 . 49 . 18	28·0	27·33	..	..	18.14	33·78	26·336	20·75
Oct. 17	12 noon	11 . 34 . 18	92 . 47 . 54	28·44	27·66	..	..	17·74	32·05	25·753	20·061
,, ,,	4 p.m.	11 . 03 . 15	92 . 54 . 00	28·44	26·72	..	..	17·87	32·29	25·939	20·234
,, ,,	8 p.m.	10 . 21 . 45	92 . 57 . 30	28·33	26·27	..	..	17·95	32·42	26·052	20·375
,, ,,	12 midnight	9 . 47 . 00	93 . 04 . 30	27·78	26·16	..	..	18·46	33·35	26·798	21·250
,, 18	4 a.m.	9 . 08 . 30	93 . 11 . 00	27·22	26·00	..	..	18 66	33·71	27·091	21·702
1922 Oct. 18	8 a.m.	8°. 32′. 30″	93°. 20′. 00″	27·94	27·83	..	..	18·77	33·90	27·241	21·610
Nov. 9	12 noon	8 . 05 . 00	93 . 24 . 30	27·72	26·16	..	..	18·73	33·84	27·189	21·632
,, ,,	4 p.m.	8 . 42 . 45	93 . 17 . 45	28·06	26·44	..	..	18·80	33·96	27·286	21·613
,, ,,	8 p.m.	9 . 19 . 00	93 . 18 . 00	28·33	26·83	..	..	18·33	33·12	26·606	20·891
,, 10	4 a.m.	10 . 30 . 00	92 . 59 . 00	27·61	25·61	..	..	18·10	32·70	26·274	20·825
Dec. 8	12 noon	8 . 13 . 30	93 . 22 . 00	27·67	27·94	..	..	18·77	33·91	27·252	21·707
,, ,,	4 p.m.	8 . 52 . 00	93 . 14 . 30	27·50	29·50	..	..	18·78	33·93	27·261	21·770
,, ,,	8 p.m.	9 . 27 . 30	93 . 08 . 30	27·33	27·55	..	..	18·96	34·25	27·522	22 069
,, ,,	12 midnight	10 . 05 . 00	93 . 02 . 30	26 94	26·16	..	..	18·53	33·48	26·898	21·613
,, ,,	4 a.m.	10 . 37 . 30	92 . 55 . 30	26·94	25·61	..	..	18·54	33·49	26·912	21·626
,, ,,	8 a.m.	11 . 14 . 30	92 . 53 . 00	27 11	27·00	..	..	18·62	33·64	27·028	21·679
,, 12	12 noon	Port Blair Harbour.		27·83	27·27	..	..	18·72	32·02	25·729	20·238
,, ,,	4 p.m.	11 . 07 . 30	92 . 51 . 00	27·33	26·44	..	..	18·53	33·47	26·893	21·483
,, ,,	8 p.m.	10 . 34 . 00	92 . 55 . 00	27·00	26·16	..	..	18·54	33·50	26·916	21·610
,, ,,	12 midnight	10 . 00 . 00	93 . 01 . 00	26·67	25·61	..	..	18·61	33·61	27·013	21·804
,, 13	4 a.m.	9 . 25 . 00	93 . 09 . 00	26·67	25·61	..	..	18·91	34·17	27·457	22·219
,, ,,	8 a.m.	8 . 43 . 45	93 . 14 . 30	27·22	26·72	..	..	18·52	33·45	26·889	21·514
,, ,,	12 noon	8 . 08 . 30	93 . 23 . 00	27·11	23·66	..	..	18·17	32·83	26·381	21·075

VI. THE TEMPERATURE AND SALINITY OF THE DEEPER WATERS OF THE BAY OF BENGAL AND ANDAMAN SEA.

In the two previous papers of this series, Nos. IV and V, I have dealt with the changes that can be traced in the temperature and salinity of the surface waters of the Bay of Bengal and the Andaman Sea, and in this present contribution I have endeavoured to bring together and collate all the various observations that I have been able to trace regarding the conditions of temperature and salinity that exist in the deeper waters of the same regions and the area to the south and west of Ceylon. As I have previously pointed out, the Bay of Bengal is bounded on its eastern side by the coast of Northern Burma and, to the south of this, by the Andaman-Nicobar ridge, which extends southwards as far as the island of Sumatra; the deep levels of the Bay of Bengal are thus completely cut off from the deeper waters of the Andaman Sea, but it is not possible to consider the various changes that take place in the one area without at the same time considering the changes in the other. The southern boundary of the Bay may be taken as the parallel of Lat. 6° N., but I have included in the following account certain observations taken further to the south and west.

The data at my disposal includes all the observations that have been taken on board the R.I.M.S. 'Investigator' throughout the years 1887 to 1925; and I have also utilised the information given in the 'Annual lists of Deep-sea soundings' published by the Hydrographic Office, Admiralty, Whitehall, London, and certain data taken by the S.S. 'Valdivia', 'Planet', 'Vitiaz', and H.M.S. 'Egeria' and 'Sealark'.

The total number of deep-sea observations of all kinds in this region that I have been able to refer to is 450, and included in this number are 23 sets of serial observations; and I have also given 11 sets taken in the more southerly region between the Equator and Lat. 6° N. The great majority of these observations have already been published, and in the case of those taken by the 'Investigator' many have been printed twice already, once in 'A Naturalist in Indian Seas' (A. Alcock) and again in 'The List of Biological Stations of the R.I.M.S. 'Investigator', published by the Zoological Survey of India; I have, therefore, thought it unnecessary to publish these data once more.

By far the greater number of the temperature observations in the 'Investigator' data have been taken by means of 'Miller-Casella' thermometers of the maximum-minimum type, reading in the Fahrenheit scale; these records I have converted into the Centigrade scale to the nearest decimal place. Since 1921 all deep-sea temperature observations on the 'Investigator' have been taken by means of reversing thermometers reading in the Centigrade scale, manufactured by Negretti and Zambra, London, and tested at the National Physical Research Laboratories, London.

Estimations of the salinity of the deeper waters have only recently been carried out on the 'Investigator'; prior to 1921 not a single observation had, so far as I know, been made by any Surgeon-Naturalist; but during 1921 and 1922 serial observations in the region under consideration were carried out on six occasions while the 'Investigator' was steaming across the southern part of the Bay of Bengal from the south of Ceylon to the Andaman islands. In all these cases the water-samples were taken by means of 'Ekman' reversing water-bottles and the water was stored in spring-stoppered bottles in my laboratory on board for at least twenty-four hours; the samples were then carefully titrated with silver-nitrate solution and compared with standard sea-water of the same temperature, the total salinity being obtained from the 'Halogen' content by means of Knudsen's tables.

General Oceanic Circulation.

During the last few years a great deal of work has been done and a number of interesting papers published regarding the conditions present in the water-masses at different levels in the great oceans, and it is now generally accepted as a proved fact that in all the great oceans a vertical circulation of the water-masses from one level to another is constantly taking place. Schott (1902, p. 165, fig. 33) gave an account of the type of circulation that he believed to be going on in the length of the north and south Atlantic Oceans and he put forward the view that the warm upper layers of the ocean tended to move away from the Equator towards the Poles, while the colder water of the polar regions, after sinking towards the bottom in about Lat. 30° S., moved at first horizontally along the ocean bed and then in the region of the tropics became deflected vertically and made its way towards the surface. Merz and Wüst (1922) later cast doubts upon the existence of this simple bipolar type of circulation and again put forward the older view of Buchanan and Buchan, that was based on the earlier researches of the 'Challenger' and the 'Gazelle'. According to this view the circulation of the water-masses in the Atlantic ocean extends right across both northern and southern regions and is brought about by the combined action of both temperature and salinity. These authors conclude that a movement of the water-masses takes place in the upper levels from the north to the south Atlantic and in the intermediate levels from the south to the north, since the northern area of the Atlantic is warmer than the southern half; in yet deeper levels the water-masses, judging from the difference in salinity, move from north to south, the deep water of the south Atlantic being poor in salt content and that of the north rich; while, finally, in the deepest level of all the Antarctic bottom drift moves slowly from the south polar region towards the equator. In 1926 Drygalski put forward evidence to show that in the Antarctic and south Temperate regions we can clearly recognise these four strata, superposed on one another and all exhibiting characteristic features, and, furthermore, that exactly the same four layers can be detected in the south-west part of the Indian ocean (*vide* Drygalski, 1926(*a*), p. 497 *et seq.*, Pl. VIII). In this southern area we find on the surface a layer of water that can be traced from the south polar region as far north as Lat. 55° S., though owing to climatic and other changes this becomes gradually less marked. Immediately beneath this surface layer lies a mass of polar water that in the main flows towards the north and, as it passes northwards, gradually sinks from a depth of approximately 220 fathoms (400 metres) at the edge of the Antarctic shelf to 490 fathoms (900 metres) in Lat. 45° S. and finally reaches a depth of 710 fathoms (1,300 metres) in the tropical region, where it has a temperature of about 6·6°C. and a salinity of about 34·5 per mille. Beneath this second layer is a third, that Drygalski terms 'tropic' water; this layer, according to him, is moving in a direction from north to south and extends downwards to a depth of some 1,093 fathoms (2,000 metres); this water possesses a temperature about 2°C. warmer and a salinity about 0·2 higher than the supernatant polar water. Finally, there is the true bottom water, that Drygalski believes to be derived by a process of cooling from the 'tropic' water, that again is moving slowly from south to north.

Both Defant (1928) and Möller (1929) in two very interesting papers, in which they deal with the circulation of the water-masses in the western part of the Indian Ocean, have brought forward a mass of evidence to prove that the circulation in this area is also of the four-fold type. For the full details of this evidence I must refer the reader to the original papers; suffice it to say that in the extreme west of the Indian region we find in the region of the equator surface-water having a salinity of 35·5 per mille or over, while water of the same degree of salinity also occurs in Lat. 33° S. In this region the isohalines in a section from south to north exhibit a bend towards Lat. 15° S., at a depth of some 200 fathoms, the northern or equatorial water moving southwards, while the southern water moves towards the north. Between 273 fathoms (500 metres) and 820 fathoms (1,500 metres) depth there is a very clear movement of water, having a salinity between 34·6 and 35·0 per mille, from south to north, reaching even beyond the equator to as far north as Latitude 6°-7° North. At a still greater depth, between 820 fathoms (1,500 metres) and 1,370 fathoms (2,500 metres) a mass of water, having a salinity of approximately 35·0 per mille, is found moving from the area that lies to the north of the equator towards the south; and, finally, there is a bottom stratum of water having a salinity of 34·8 per mille (*vide* Möller, 1929, p. 13, fig. 5) that is moving northwards.

This multiple movement of the water strata is still more clearly seen in the section, given by Defant (1928, p. 479, fig. 25) and Möller (1929, p. 17, fig. 7), through the south-west part of the Indian Ocean in a line extending from Lat. 55° S.; Long. 15° E. to Lat. 8° N.; Long. 77° E., and Möller is able to recognise five different strata namely:—

1. The upper surface water,
2. The subtropical understream,
3. The Antarctic intermediate layer,
4. The north Indian deep stratum, and
5. The Antarctic bottom water.

The surface-water is to a very large extent influenced by the varying conditions of wind force and direction, that determine the direction of the surface currents. A very careful study of these currents and their relationship to the prevailing winds of the Indian oceanic region has been made by Michaelis (1923) and Willimzik (1929). These authors have shown that during the period of the north-east monsoon in December and January there is produced across the whole width of the northern part of the Indian Ocean a convergence zone running from west to east between Lat. 5° and 10° S., along which water coming from the north-east, under the influence of the N.E. monsoon wind, meets the water being driven towards the north-west by the S.E. trade winds. We thus get on the west a mass of water having a salinity of 35·4 per mille, and these two streams, converging together, give rise to the Contra-equatorial current, that in about Lat. 8° S. flows away towards the east. Krümmel (1911, p. 685) had previously pointed out that owing to the action of the earth's rotation the heavier water on the west side of the Arabian Sea flows at first southwards and crosses the equator; it then, in the region of Madagascar and the Seychelles, sinks slowly and at the same time changes its course to the south-east and finally to the east. Under the influence of the south-west monsoon this current is found somewhat to the north of its usual position and at the same time its direction is altered, so that its northern extension sets towards the north-east and can be traced up into the Bay of Bengal. As this subtropical understream flows eastward and gradually sinks, it becomes covered with a layer of less saline water that gradually becomes deeper as we trace the conditions northwards, and at the same time owing to admixture the deeper water becomes somewhat less saline.

As Möller *(loc. cit.*, p. 11) points out there is in the northern part of the Indian Ocean a characteristic difference between the water layers on the east and west. In the east the contrasts of temperature and salinity are smaller than on the west and the salinity of the upper stratum is less than in the west; this is attributed to the more saline character of the deeper layer of water, that has its origin in the region of the Arabian Sea and its offshoots, the Red Sea and Persian Gulf.

Defant (1928, p. 470) points out that in accordance with the temperature curve of the deeper levels of the ocean we can divide the great mass of water into different strata according to whether the temperature gradient changes rapidly or remains nearly constant; thus, as we pass from the surface to the great depths, we first encounter a comparatively thin upper warm stratum, of 100–200 metres (or approximately 50–100 fathoms) depth in which there is only a slight fall of temperature; this he terms the 'Störungszone'. Immediately below this there is a much deeper stratum, of 1,000–1,400 metres depth (or approximately 550–700 fathoms), in which the temperature falls rapidly; and finally beneath this lies a still deeper and thicker stratum, extending from a depth of some 1,400 metres (700 fathoms) to the bottom, in which, again, there is but little change of temperature but in which, unlike the surface stratum, the water is of almost uniformly low temperature. The upper stratum comprises the 'Störungszone' and the zone immediately below it, which he terms the 'Troposphere', and the deeper stratum of constant low temperature he calls the 'Stratosphere'. The upper two zones are the seat of rapid oceanic circulation, whereas in the lower Stratosphere the movements of the water-masses are slow.

A comparison of the depths of the various strata shows clearly that Defant's 'Störungzone' and 'Troposphere' together correspond to the surface-water and the polar water of Drygalski, or to the surface-water, the subtropical understream and the Antarctic intermediate layer of Möller; while the tropic water of Drygalski or the north Indian deep stratum of Möller, and the Antarctic bottom water together constitute Defant's 'Stratosphere'.

The earliest account of the temperatures of the different levels of the ocean in the Indian region is contained in a short paper by Carpenter (1887), in which he gives the results of his observations in the Bay of Bengal. The curve of temperature that he gives is based on a number of observations of bottom temperatures in different parts of the Bay and although, therefore, not strictly comparable with temperatures obtained by 'serial' observations, yet gives a very fair picture of the conditions that exist at the different levels. He shows clearly that from the surface to a depth of about 20 fathoms (37 metres) the temperature shows but little change, but that below this surface layer lies a zone extending downwards for more than the next 100 fathoms (183 metres) in which the temperature falls rapidly from about 26·1°C. to 11·9°C. Then follows a zone that can be traced down to a depth of about 1,200 fathoms (2,195 metres) in which the water temperature falls slowly to 2·2°C.; and finally, in the greatest depth of all there is a bottom stratum in which there is but little change, the actual temperature being in the neighbourhood of only 1·0°C. Eight years later Oldham (1895) gave us a more detailed account of the bottom temperatures in the Laccadive Sea, that lies between the Indian Peninsula and the Maldive and Laccadive Archipelagoes, and he compared the results of his observations with those of Carpenter in the Bay of Bengal. As Oldham remarks, 'The Arabian Sea curve to 100 fathoms depends on so few observations that it is not desirable, up to this depth, to make any comparison with the Bay of Bengal curve. Beyond 100 fathoms, where my observations are numerous, it will be noted that there is a difference of about one or two degrees; the Arabian Sea on the west of Hindustan in October, November and April being that much warmer than the Bay of Bengal. At a depth of 1,300 fathoms the two curves coincide.' This difference between the two areas is undoubtedly to be attributed, as Möller has pointed out, to the differences in the oceanic circulation on the two sides of the Indian Ocean and particularly to the increased effect of the north Indian deep stratum in the western region.

In Tables 81 and 82 I have given in tabular form all the serial observations that have been taken in the region which we are considering, namely the Laccadive Sea, the Bay of Bengal and the Andaman Sea, with the portion of the Indian Ocean that is immediately contiguous. I have given these observations in two separate tables, viz. (*a*) those in which the depths have originally been given in fathoms (Table 81), and (*b*) those in which the depths are in metres (Table 82), and in each case I have given the corresponding depth in the other scale.

Schott (1902, p. 171) has pointed out that 'the Indian vertical circulation is only partially perfect; complete only in the zone of the southern region, but incomplete in the equatorial portion, where it does not reach the surface. The

Position N.	18° 22′	8° 06′ 18″	12° 14′ 06″	12° 06′ 30″	12° 11′ 12″	1° 09′ 00″	6° 50′ 30″	6° 07′ 00″	5° 43′ 00″	5° 48′ 48″	5° 39′ 36″	5° 49′ 30″	6° 01′ 18″	6° 08′ 42″	5° 59′ 30″
Position E.	69° 58′	70° 01′ 30″	70° 54′ 30″	72° 21′ 36″	73° 51′ 30″	74° 44′ 00″	76° 47′ 08″	80° 11′ 00″	80° 25′ 42″	80° 35′ 48″	80° 46′ 42″	80° 54′ 48″	81° 32′ 00″	81° 39′ 12″	81° 42′ 54″
Date ..	—87/88	6-4-94	21-10-93	24-10-92	25-10-92	1-2-23	31-3-94	?-4-89	20-5-09	19-5-09	18-5-09	17-5-09	11-5-09	10-5-09	10-5-09
Ship ..	Investigator.	Egeria	Investigator.	Investigator.	Investigator.	Investigator.	Egeria	Investigator.	?	?	?	?	?	?	?
Depth in Metres / Fms.	°C	°C	°C	°C	C	°C	°C	°C	°C	°C	°C	°C	°C	°C	°C
0 / 0	....	28·61	26·22	27·78	26·67	28·2	28·89	30·4	27·17	27·44	27·56	27·28	27·83	27·28	27·50
46 / 25	....	....	....	26·39(23)	23·33	....	....	26·67	26·05	26·17	26·72	26·22	26·22	25·39	26·89
92 / 50	20·28	....	....	19·28(48)	18·61	....	....	21·00	21·33	22·33	23·00	21·61	21·17	....	23·32
137 / 75	....	....	....	16·50(73)	16·11	....	....	15·56	....	....	....	....	....	18·78	19·11
183 / 100	....	15·72	14·05	15·39	14·83	....	16·17	13·11	15·11	14·95	16·00	14·67	14·11	15·44	15·56
229 / 125	....	....	....	....	....	....	....	11·78	....	....	....	....	....	....	....
274 / 150	....	....	....	....	....	....	....	11·50(154)	....	....	....	....	11·89	13·05	13·44
320 / 175	....	....	....	....	....	....	....	10·44	....	....	....	....	....	....	....
366 / 200	13·44	12·00	11·78	11·95	11·95	11·8	11·17	10·78	10·89	11·39	10·89	10·89	10·56	11·22	12·33
412 / 225	....	....	....	....	....	....	....	11·22	....	....	....	....	....	....	....
457 / 250	....	....	....	....	....	....	....	10·00	....	....	....	....	....	....	....
549 / 300	....	10·28	11·11	10·67	10·56	10·45	10·17	....	....	9·89	....	9·44(320)	....	9·61	11·17
732 / 400	....	9·61	9·33	....	9·28	....	8·61	....	8·50	....	8·56	8·39(420)	....	....	8·74
914 / 500	....	8·00	8·33	8·33	8·33	....	7·89	....	....	7·39	....	7·05(520)	....	....	....
1,097 / 600	....	....	7·28	7·50	7·50	....	....	....	7·11	....	6·39	....	....	....	6·89
1,280 / 700	....	....	6·33	6·22	5·83	....	....	....	....	5·44	....	....	....	....	....
1,463 / 800	....	....	5·22	5·83	5·00	....	....	....	....	....	....	....	....	....	5·17
1,646 / 900	....	....	4·95	5·00(905)	....	....	....	....	....	4·56(878)	....	....	....	....	....
1,829 / 1,000	....	....	3·61	....	3·33	....	3·05(1024)	....	3·28	....	3·44	....	....	....	3·44
Bottom	2·22(1315)	2·11(2340)	....	....	....	....	....	....	....	....	....	....	....	....	....

Position N.		6° 26′ 00″	6° 51′ 00″	6° 38′ 30″	15° 00′ 00″	8° 11′ 42″	8° 51′ 30″	10° 10′ 18″	10° 21′ 00″	6° 16′ 00″	10° 00′ 00″	11° 26′ 00″	
Position E.		81° 51′ 12″	83° 22′ 30″	83° 34′ 30″	85° 00′ 00″	86° 29′ 42″	86° 52′ 00″	89° 55′ 12″	90° 17′ 30″	90° 44′ 00″	91° 07′ 00″	91° 54′ 00″	
Date ..		28-4-09	19-10-21	10-10-22	....	20-10-21	11-10-22	21-10-21	12-10-22	19-4-88	?-4-88	?-4-88	
Ship ..		?	Investigator.	Investigator.	Investigator.	Investigator.	Investigator.	Investigator.	Investigator.	Investigator.	Investigator.	Investigator.	
Depth in Metres	Depth in Fms.	°C	°C	°C	°C	°C	°C	°C	°C	°C	°C	°C	Depth in Fms.
0	0	29·11	29·86	27·89	28·9	29·93	27·78	29·87	27·36	29·06	....	....	0
46	25	27·39	29·0	26·6	28·0	....	....	28·2	26·6	29·0	27·33	....	25
92	50	....	26·97	24·55	25·6	....	23·68	24·56	22·53	25·1	23·22	26·11	50
137	75	....	21·0	20·4	20·72	....	....	19·1	17·6	19·3	16·67	21·50(70)	75
183	100	....	15·55	15·4	16·39	15·87	15·06	14·2	14·67	14·17	14·00	18·22(90) 14·72(110)	100
229	125	....	13·3	13·3	14·00	....	....	12·4	12·7	13·05	12·61	14·44	125
274	150	....	12·2	12·0	12·22	....	....	11·41	11·7	12·11	11·78	12·61	150
320	175	....	11·5	11·2	11·44	....	....	10·9	11·2	11·67	11·22	12·83	175
366	200	11·78	11·02	10·45	10·61	....	10·31	10·5	10·61	11·39	....	11·95	200
412	225	....	....	....	10·28	....	....	....	....	....	....	....	225
457	250	....	....	....	9·83 9·78	....	....	....	....	....	....	....	250
549	300	9·61	....	....	9·33	9·63	....	9·22	....	....	....	....	300
732	400	....	....	....	....	....	....	....	....	....	....	....	400
914	500	7·17	....	7·48	....	7·70	7·62	....	7·37	....	....	....	500
1,097	600	6·22(626)	....	....	....	....	....	....	....	6·05(560)	6·50	....	600
1,280	700	....	....	....	....	....	....	....	....	....	....	....	700
1,463	800	....	....	....	....	....	....	....	....	....	...	5·72(772)	800
1,646	900	....	....	....	....	....	....	....	....	....	3·50(999)	....	900
1,829	1,000	....	....	....	....	....	....	....	....	....	....	....	1,000
Bottom		....	....	....	....	....	....	....	....	....	....	....	Bottom.

Table 81 : Serial Temperature Observations in Indian Waters.
Depth given in Fathoms.

Position N.		1° 52′	6° 36′	2° 30′	6° 23′	3° 32′	7° 43′	6° 07′	1-2° N.	4° 25′
Position E.		74° 45′	75° 57′	76° 4-	05′	88° 20′	88° 45′	92° 03′	96-97 E.	99° 21′
Date ..		30-6-06	31-1-89	18-2-99	4-2-89	17-7-06	10-11-99	31-1-89	....	26-1-09
Ship ..		Planet	Vitiaz	Valdivia	Vitiaz	Planet	Valdivia	Vitiaz	Valdivia	Vitiaz
Depth in		°C	°C	°C	°C	°C	°C	°C	°C	°C
Metres	Fms.									
0	0	29·0	28·2	28·0	26·8	28·3	27·4	28·5	28·0	29·1
25	14	....	28·2	27·7	27·0	....	27·1	28·2	27·6	27·5
50	27	27·5	28·0	27·3	27·8	....	26·5	27·2	27·3	26·9
75	41	....	....	26·7	....	....	25·4	....	27·1	26·3
100	55	22·7	27·4	26·1	27·2	23·4	23·3	19·9	26·6	....
125	68	21·1(120)	....	23·0	....	....	19·5	....	19·2	....
150	82	19·5(140)	....	16·3	....	15·1	16·9	....	15·3	....
175	96	16·8(160)	....	14·0	....	....	15·1	....	12·8	....
200	109	13·6	13·8	13·0	....	13·3	13·9	13·3	12·0	....
225	123	....	....	....	....	....	....	....	....	....
250	137	....	....	....	....	....	....	....	....	....
275	150	....	....	....	....	....	....	....	....	....
300	164	....	....	11·0	....	....	11·3	....	10·0	....
400	219	11·2	10·3	10·2	....	10·3	10·5	10·8	8·7	....
500	273	....	....	9·7	....	....	10·1	....	8·3	....
600	328	....	....	9·2	....	8·7	9·9	....	8·1	....
700	383	....	....	8·3	....	....	9·3	....	7·8	....
800	437	7·9	7·7	7·5	....	7·5	8·7	8·2	6·6	....
900	492	....	....	6·9	....	....	8·1	....	5·9	....
1,000	547	6·6	....	6·1	....	6·3	7·4	....	5·6	....
1,500	820	4·5	....	3·8	..	....	4·6	....	4·4	....
2,000	1,094	....	....	2·5	....	....	3·8	....	3·4	....
3,000	1,640	....	....	2·1	....	....	2·5	....	2·1	....
4,000	2,187	....	....	1·6	....	....	....	....	..	...

Table 82: Serial Temperature Observations in Indian Waters. Depth given in Metres.

explanation lies to hand. In the northern breadth of the Indian ocean the upper surface is not continuously carried away. During the monsoons and the half-yearly changes immediately dependent on them the whole upper circulation is reversed through a complete 180° and there results a to-and-fro movement of the water-masses from west to east and back again in the opposite direction in half-yearly phases, so that no actual removal of water takes place'. Schott (*loc. cit.*, p. 169) has also pointed out that in the region of Lat. 30° S. the conditions, as regards the temperature of the water at different depths, that are present in the Indian Oceau are intermediate between those found to be present in the North and South Atlantic Oceans respectively at a similar distance from the Equator, and that at the Equator itself the temperature in the Indian Ocean is higher than in the Atlantic at the same depths down to 1,000 metres (547 fathoms), but that below this depth it is considerably lower, owing to the greater volume and influence of the Antarctic drift, or, as Möller calls it, the Antarctic intermediate stratum, in the Indian region. It is interesting to take an average of the records of temperature at different depths in the Bay of Bengal in or near Lat. 10° N., and compare the results thus obtained with the figures given by Schott (*loc. cit.*, p. 169) for the equatorial region and Lat. 30° S. in the same ocean. The figures given below for Lat. 10° N. are the average of three sets of serial observations taken by the 'Investigator'.

Depth in		Latitude 30° S.	Equator	Latitude 10° N.
Metres.	Fathoms.	°C	°C	°C
0	0	20·0	27·8	29·0
50	27	19·0	27·0	27·3
100	55	18·9	24·3	22·8
150	82	17·4	17·1	17·2
200	109	15·0	14·5	14·2
400	219	12·5	10·3	10·4
600	328	10·6	8·9	9·0
800	437	8·3	7·5	8·2
1,000	547	5·5	6·2	6·9*
2,000	1,094	2·5	2·7	3·6*

* Average bottom temperature.

Table 83; showing the temperature at different depths on both sides of the Equator in the Indiau Ocean and Bay of Bengal.

The above Table shows that down to a depth of 50 metres (27 fathoms) the temperature of the water is warmer in Lat. 10° N. than it is either at the Equator or in Lat. 30° S., while at 100 metres depth the water is warmer at the Equator than in either area to the north or south. This is almost certainly due to the presence in the equatorial region, and to the immediate northward, of an over-lying stratum of warm water of low salinity and raised temperature, resulting from the dilution of the upper levels by river water, and at a depth of 100 metres to the presence on the Equator of the Contra-equatorial current flowing from west to east. At a depth of 200 metres (109 fathoms) the temperature steadily decreases as one proceeds northwards but below this, down to a depth of 800 metres (437 fathoms), the water in the equatorial zone is colder than the water either to the north or south. In the greatest depth from 800 to 2,000 metres (437 to 1,094 fathoms) the temperature of the water gradually rises as one proceeds from south to north.

In Text-fig. 119, I have given a diagram showing the isothermal lines along Longitude 90° E., passing through the Indian Ocean and the Bay of Bengal. These isotherms are based on the observations taken by the 'Planet', 'Valdivia' and 'Investigator' and in the main the results agree closely with the chart given by Ringer (1922, p. 154, fig. 12). It can be seen that there are indications of a strong and steadily descending current of warm water in the region of Lat. 30° S., that

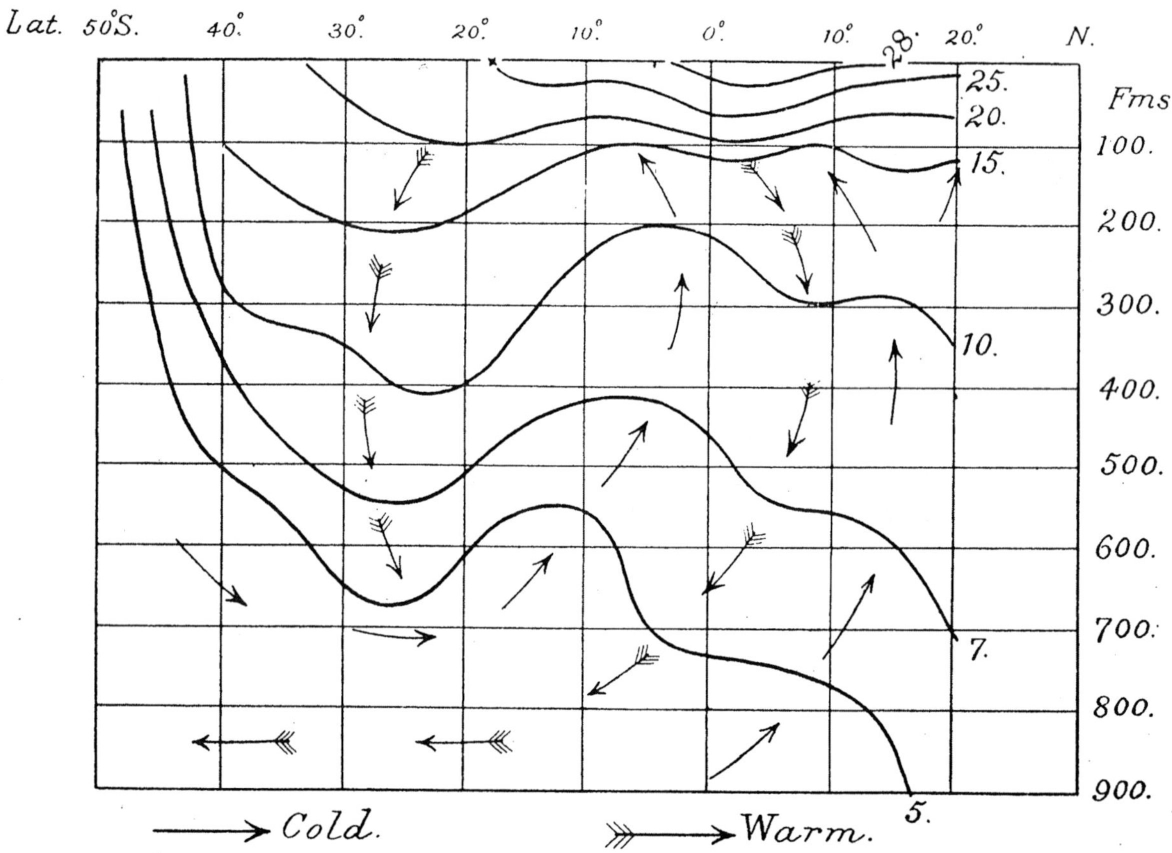

TEXT-FIG. 119.—The isothermal lines in the Indian Ocean along Longitude 90° E.

eventually joins with the cold current of the Antarctic Intermediate layer at a depth of some 1,200 metres (656 fms.) and is then deflected northwards towards the Equatorial region and in about Lat. 15° S. begins to alter its direction and pass upwards towards the surface. In the greatest depths lies the bottom Antarctic drift of cold water that is passing from south to north, and this bottom drift appears to pass beyond the Equator to as far as Lat. 15° N., where it also begins to turn towards the surface. In the region of Lat. 5–10° N. there is some evidence, though not very strong, of a descending stratum of water of a somewhat higher temperature; this presumably is the stratum that Möller refers to as the north Indian deep stratum.

As Schott (1902, p. 149) points out, in order to determine the general features of the oceanic currents, it is permissible to include in a single chart all the observations

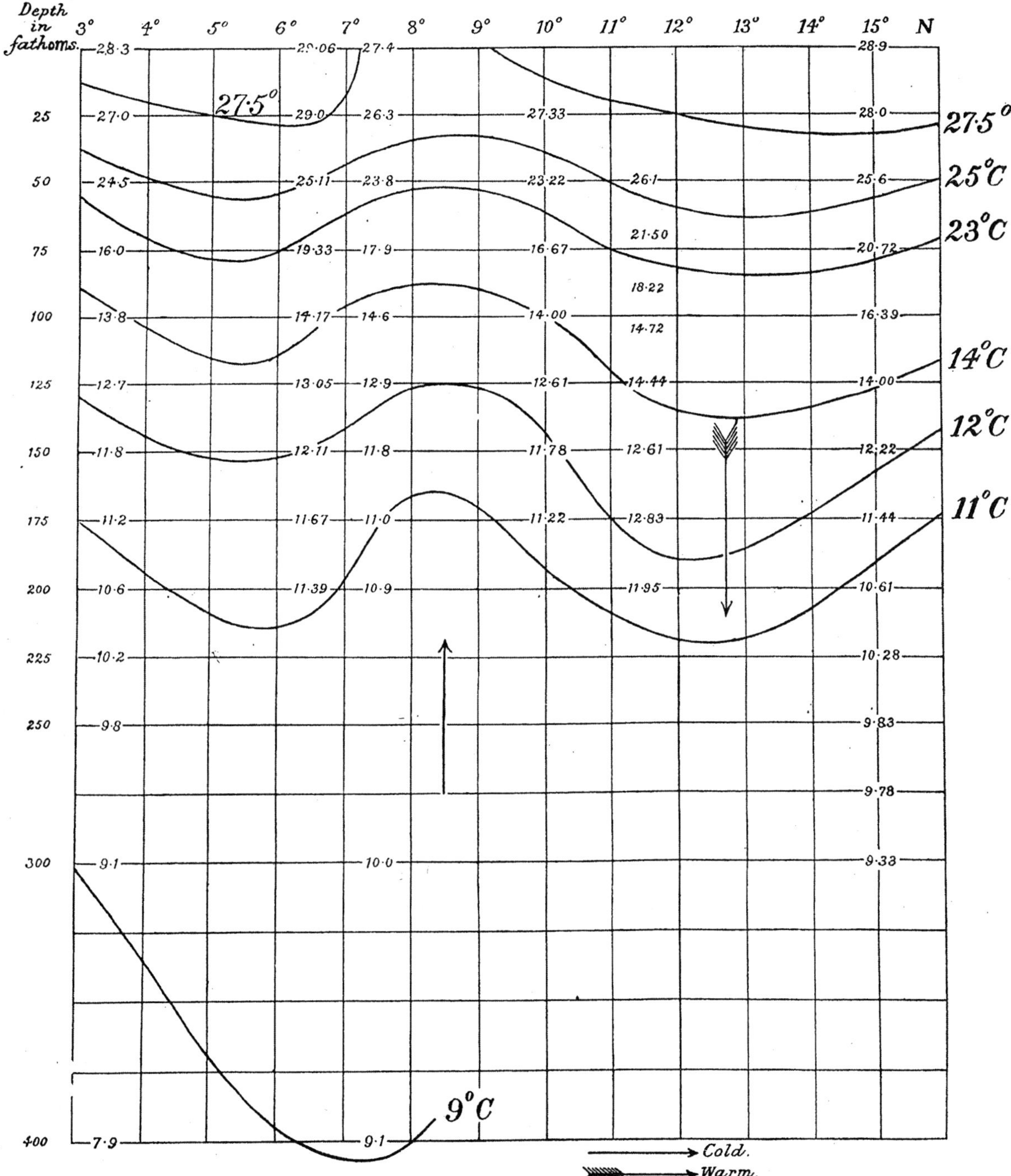

TEXT-FIG. 120.—The isothermal lines in the Bay of Bengal in a north-south direction, roughly along Longitude 90° E.

that have been made in any given area without reference to the month or year in which they were taken; but it must be borne in mind that in so doing we are ignoring any alterations or variations in the circulation that are due to seasonal or other changes. In the main, however, one would expect that such changes would only affect, or at any rate only markedly affect, the upper strata that constitute Defant's 'troposphere' and would produce but little effect on the deeper levels that constitute the 'stratosphere'.

In Text-fig. 120, I have shown the trend of the isothermal lines through the Bay of Bengal along a line that roughly follows the meridian Long. 90° E., the most northerly observation lying, however, somewhat further to the west in about Long. 85° E.

Schott (1902, Pl. XI.) in his chart of the isothermal distribution of the waters of the Indian Ocean at the 100-metres level shows the east and north-east areas of the Bay of Bengal as being colder than the region to the south-west; but a study of the sets of serial observations given above indicates that in the months April–October the opposite is the case and the waters both to the north in Lat. 10–15° N. and to the south in Lat. 3–7° N. are warmer than those in Lat. 7–10°, down to a depth of 200–300 fathoms (366–549 metres), the maximum difference occurring at about 200 fathoms (366 metres). It is probable that Schott's chart of the temperature distribution is correct for the period of the north-east monsoon and that during that season of the year there is an upwelling of cold water from below in the north and east parts of the Bay to counter-balance the drift of the surface-water towards the south-west under the influence of the monsoon wind (*vide infra*, p. 392). On the cessation of this surface drift the upwelling would naturally cease and the temperatures at the different levels tend to become uniform. There must, therefore, be some additional factor that causes this increased temperature in the upper strata down to a depth of some 200 fathoms (183 metres) in the more northerly region. Reference to the charts of the surface currents in the Bay of Bengal at this period of the year, given by myself (*vide supra*, p. 289, Text-fig. 83) and by Möller (1929, p. 42, fig. 22), shows that in the northern and central parts of the Bay there is a meeting of two sets of currents, the one coming from the south-west and south and setting up into the Bay and the other coming from the north-east and east from the Andaman Sea; the meeting of these two streams must almost certainly set up a convergence zone, though possibly a somewhat irregular one, and thus result in the forcing downwards below the surface in Lat. 10–15° N. of some at least of the surface water, and it is probable that this downward movement of the Bay water serves either to augment the north Indian deep current, lying below the Contra-equatorial current, or else to form a deep current, the possible existence of which I shall refer to later (*vide infra*, p. 378), while the other mass of water, lying between Lat. 3–7° N., escapes from the Bay above the Contra-equatorial stream.

Within the limits of the area that we are considering, namely the regions of the Laccadive Sea and Bay of Bengal and the contiguous area of the Indian Ocean, there can be little doubt that we have a series of water-layers, superposed the one over the other and ranging from the deep water of the Antarctic drift to the surface-water that is derived in large part from the influx of fresh water, especially into the Bay of Bengal and the Andaman Sea, from the large rivers of India and Burma. As we have already seen the temperatures at different depths in the line of Long. 90° E. show that even in these northern latitudes as well as in the region of the Equator there is evidence of at least four layers, similar to and continuous with those shown by Drygalski and Möller to be present to the south and west, namely, (*a*) a surface layer of high temperature, and as we shall see (*vide infra*, p. 376), of low salinity, the flow of which is dependent on the changing conditions of the varying monsoon seasons (*vide ante*, p. 280, *et seq.*), being sometimes towards the north-east and at others to the south-west ; (*b*) below this lies a layer of water that is in the main flowing from south to north at a depth of approximately some 600–700 fathoms (1,097–1,280 metres); this (*vide* Drygalski, 1926, p. 530, *et seq.*, Pl. VIII and Möller, 1929, p. 15, fig. 7) is derived from the South Polar region ; (*c*) a third layer, lying between 700 and 1,100 fathoms (1,280 and 2,012 metres), of water of low salinity flowing from north to south ; and (*d*) finally, of the Antarctic bottom drift that is flowing from south to north. At the same time it must be remembered that the movements of these layers are not strictly horizontal, but are part of a rotational movement, that is brought about by a number of influences, and, further, it must be borne in mind that there may be other strata in which the main movement is not from north to south but from east to west or *vice versa*, and which, therefore, would not appear, or only very slightly so, in a section such as that shown in Text-fig. 120.

It is interesting to study the distribution of the isothermal lines in the stretch of water extending from the neighbourhood of the Maldive Archipelago on the west across the southern portion of the Laccadive Sea, past the south of Ceylon and on across the mouth of the Bay of Bengal to the line of the Andaman and Nicobar Islands on the east.

In Text-fig. 121, I have plotted the results obtained from 21 serial observations taken along this line from Long. 73° E. to 92° E. and lying between Lat. 5° and 12° N. There is on the whole a tendency for the isothermal lines to rise steadily as we pass from west to east, but the strata show very distinct vertical oscillations or waves at two points, namely, (1) off the south-east coast of Ceylon between Long. 80° and 82° E., with a maximum oscillation at a depth of about 300 fathoms ; and (2) in the south-east region of the Bay of Bengal between Long. 88°.30′ and 91°.30′ E., with a maximum oscillation at a depth of approximately 450 fathoms. The first series of oscillations is shown clearly in the series of observations taken in this region in 1909 between the 28th of April and the 20th of May. It is interesting to compare these oscillations with those to which Matthews (1926) refers in his report on the observations taken by the 'Sealark' during the Percy Sladen Trust Expedition to the Indian Ocean in 1905. Matthews in his figure (*loc. cit.*, fig. 8.) only traces the oscillations of the temperatures in the upper 150 fathoms, but if we study the changes of level of the isotherms, and especially of the 12° C. isotherm, we see that the maximum oscillation must be occurring at a depth of about 200 fathoms (1,366 metres), or

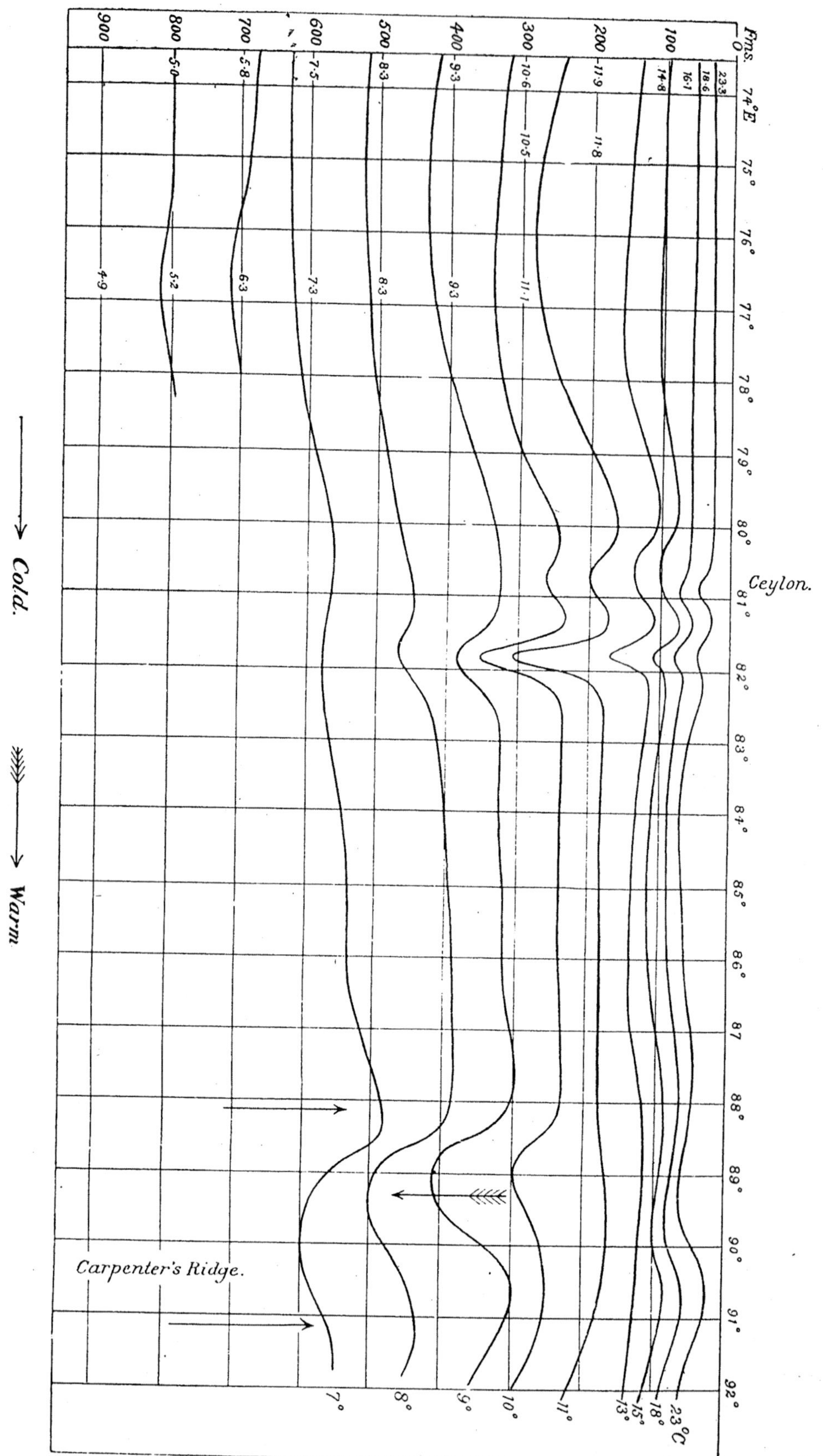

TEXT-FIG. 121.—Showing the isothermal lines from west to east between Latitudes 5° N. and 12° N.

possibly even more, which agrees very closely as regards the level with the results of the observations off the south-east of Ceylon. Matthews remarks that these oscillations cannot be due to seasonal changes at the surface and the cause must be sought in deep currents; and it is to be presumed that this causative current must lie at a depth approximately equal to that at which the greatest oscillation occurs, namely in the vicinity of the 200 to 250 fathom line; a second possible explanation of the oscillations described by Matthews may, however, be found in seasonal changes that cause an alteration in the depth at which the various strata, referred to above, lie and I shall have to return to this subject later, when I am considering the seasonal changes in the temperature of the different levels at different times of the year; but this will not apply to the series of oscillations observed off Ceylon, since these records were all taken at the same period of the year.

The second and eastern series of oscillations, occurring in the region of Long. 88° 30′–91° 30′ E., is in all probability part of the same process to which attention has been called, so far as more southerly latitudes are concerned, by Möller (1929, pp. 24–26). There is a very close agreement between the arrangement of the isothermal lines in the southern part of the Bay of Bengal in an east to west direction in a line approximately extending from Lat. 5° to 12° N., as shown in Text-fig. 121, and that given by Möller in an east to west section across the Indian Ocean in about Lat. 20° S. (*vide* Möller, 1929, p. 26, fig. 14). As Möller points out, in the eastern part of this section the warmer water at about a depth of 400 metres (219 fathoms) can be traced to sink downwards and then flow in the reversed direction from east to west at a depth of about 1,000–1,500 metres (549–820 fathoms) below the cooler water. Similar results have been obtained off the south-west region of Australia. There thus can be little doubt that we have here in the Bay of Bengal indications of the sinking of the warmer water to form the layer that Drygalski terms the 'Tropic' water and Möller the 'north Indian deep current' (Der Nordindische Tiefenstrom).

Our knowledge of the salinity of the water of the different levels in the northern region of the Indian Ocean and especially of the Bay of Bengal is still somewhat scanty, the only exception being the neighbourhood of the Gulf of Aden.

Brennecke (*vide* Krümmel, 1907, p. 342) found from the observations taken by the S.S. 'Planet' between Durban and Ceylon in the months of May and June, 1906, that, as in the Atlantic Ocean, the salinity of the water as we proceed downwards from the surface at first increases to a depth of 25 to 200 metres (14 to 109 fathoms), while below that depth it gradually falls to a minimum in 800 to 1,000 metres (437 to 547 fathoms). Schott (1902, Table 22, Sta. 218) among the observations taken by the 'Valdivia' gives the conditions existing in Lat. 2° 30′ N.; Long. 76° 47′ E., which show a salinity of 35·4 at the surface, and 35·07 at a depth of 25 metres (14 fathoms); below this depth the salinity rises till it is 35·30 at 100 metres (55 fathoms) and from that steadily falls till it is only 34·78 at the bottom in 4,133 metres (2,260 fathoms). In more recent years Schott (1926, p. 426, Pl. XXIX) and Möller (1929, p. 17, fig. 7) have, respectively, given charts showing the trend of the iso-halines in the Indian Ocean, the former on lines running from south to north in the neighbourhood of Long. 60° E. and in a north-easterly direction from the neighbourhood of Mauritius up into the Bay of Bengal; and the latter in a section from south-west to north-east passing through Mauritius and then between the Chagos and Maldive Archipelagoes to the extreme south end of India, or roughly from Lat. 55° S.; Long. 15° E. to Lat. 8° N.; Long. 77° E. This latter section has also been given by Defant (1928, p. 481, fig. 27). In both of these sections there is clear evidence that in the neighbourhood of the Equator there is a current of water, a continuation of the Polar or Antarctic intermediate current, flowing towards the north at a depth of approximately 500 metres (273 fathoms) and that below this we have evidence of the descending Tropic water that at first sinks downwards and then flows southward at a depth of 1,500 metres (820 fathoms). In order to check these results it is important to trace also the general trend of the iso-halines in an east to west direction, and I have given in the accompanying tables 84 and 85 all the data that I have been able to collate regarding the salinity at different depths in the region across the southern end of the Laccadive Sea and the Bay of Bengal, lying to the north of the Equator. In the first of these tables (Table 84) I have given the data obtained by various ships, other than the 'Investigator', that have been engaged in research in these regions and in the second (Table 85) I have given the results of my own observations on the 'Investigator'. In the first of these tables the depths are given in metres and in the second in fathoms. A study of these results shows that there is a considerable range of difference between the various observations and especially between those taken in the west and the east. In the western area, in Long. 74°–79° E. water of over 35·0 salinity is found down to a depth of some 800 metres (437 fathoms), the actual depth showing a steady increase from approximately 600 metres (328 fathoms) in Long. 74° 45′ E. to about 1,000 metres (540 fathoms) in Long. 75° 57′ E., and even to as great a depth as over 2,000 metres (1,094 fathoms) in Long. 76° 47′ E. Further eastward the surface water begins to show a definite trace of dilution, which becomes more and more marked as one passes eastward, and water of a salinity of 35·0 or over is now only found between 200–400 metres depth (109–219 fathoms) in Long. 88° E. and at a depth of 400–800 metres (219–492 fathoms) in Long. 92° E. or 1,500 metres (820 fathoms) in Long. 96°-97° E. At depths greater than these the salinity is found to be between 34·7 and 34·8.

Ship	Planet	Vitiaz	Valdivia	Vitiaz	Planet	Valdivia	Vitiaz	Valdivia	Vitiaz
Lat. N.	1°52′00″	6°56′00″	2°30′00″	6°23′00″	3°32′00″	7°43′00″	6°07′00″	1°-2°	4°43′00″
Long. E.	74 45 00	75 57 00	76 47 00	82 05 00	88 20 00	88 45 00	92 03 00	96°-97°	98 58 00
Metres									
0	35·17	35·14	35·4	33·32	34·20	34·3	32·92	33·9 to 32·9	31·690
25	..	35·14	35·07	34·22	..	..	32·94	..	33·60
50	35·07	35·07	35·19	34·92	..	..	34·20	..	..
75	35·08 (80)	..	..	..	..	..	..	..	34·27
100	35·17	36·45	35·30	34·96	34·99	34·89	34·78	..	..
120	35·19	..	..	..	..	..	..	..	..
140	35·16	..	..	..	..	..	..	33·95	..
150	35·01	..	..	..	34·92	..	..	..	..
200	35·05	35·12	35·25	..	35·03	34·91	34·85	..	..
400	35·05	35·12	..	..	34·99	..	35·08	34·95	..
600	..	..	..	..	35·03	..	..	..	..
800	34·99	35·10	..	..	34·97	..	35·10	..	..
1,000	34·92	..	..	..	34·96	..	..	..	..
1,500	34·81	..	..	..	..	..	..	35·11	..
2,000—3,000.	34·70	..	35·25	..	..	..	..	34·75	..
3,000—4,000.	..	..	..	..	..	34·81	..	34·65	..
4,000—5,000.	..	..	34·78	..	34·74	..	..	..	..
5,000 or over.	..	..	..	..	..	..	..	34·72	..
Date	30. VI. 06	11. II. 89	18. II. 99.	4. II. 89	17. VII. 06	10. XI. 99	31. I. 89	3-5. II. 99	26. I. 89.

Table 84.—Serial observations on Salinity taken by ships other than the 'Investigator' across the Laccadive Sea and the Bay of Bengal.

Lat. N.	..	6° 51′ 00″	6° 38′ 30″	8° 11′ 42″	8° 51′ 30″	10° 10′ 18″	10° 21′ 00″
Long. E.	..	83° 22′ 30″	83° 34′ 30″	86° 29′ 42″	86° 52′ 00″	89° 55′ 12″	90° 17′ 30″
Date	..	19-x-21	10-x-22	20-x-21	11-x-22	21-x-21	12-x-22
Depth in fathoms							
0	..	34·50	34·83	33·93	34·78	33·62	32·83
50	..	35·14	34·89	....	34·99	34·92	34·72
100	..	35·17	34·97	....	34·94	....	34·88
150	..	....	....	....	....	34·97	....
200	..	35·25	35·05	....	34·65	....	....
300	..	....	....	....	....	35·07	....
400	..	34·61	....	....	....	....	....
500	..	....	....	35·09	34·91	....	....

Table 85.—Serial observations on Salinity taken by the 'Investigator'.

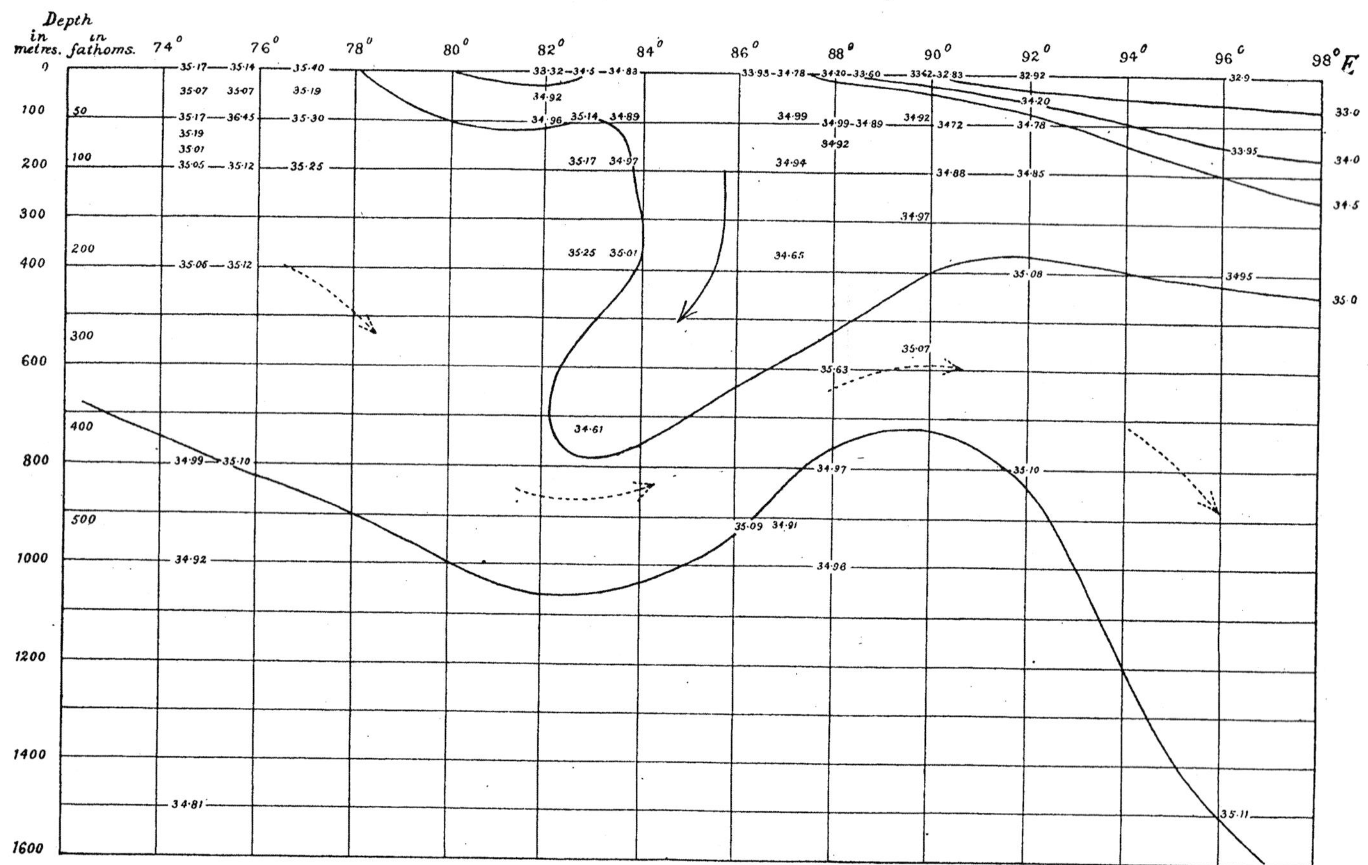

Text-fig. 122.—Showing the isohalines across the Indian Ocean from west to east between Latitudes 1° and 10° N.

In the observations taken on the 'Investigator' we find very similar differences; thus in Long. 83° E. water of a salinity of 35·0 or over occurs between 50 and 250 (approximately) fathoms (274 and 275 metres), the maximum salinity of 35·25 occurring at about 200 fathoms (366 metres). In Long. 86° E. water of this salinity occurs at a depth of 500 fathoms (914 metres) and in Long. 90° E. at 300 fathoms (549 metres).

In Text-fig. 122 I have plotted the results obtained from all observations (*vide* Tables 83 and 84) in a belt from east to west and lying between Lat. 1° and 10° N. of the Equator. We thus get a section that roughly follows a curve, extending from

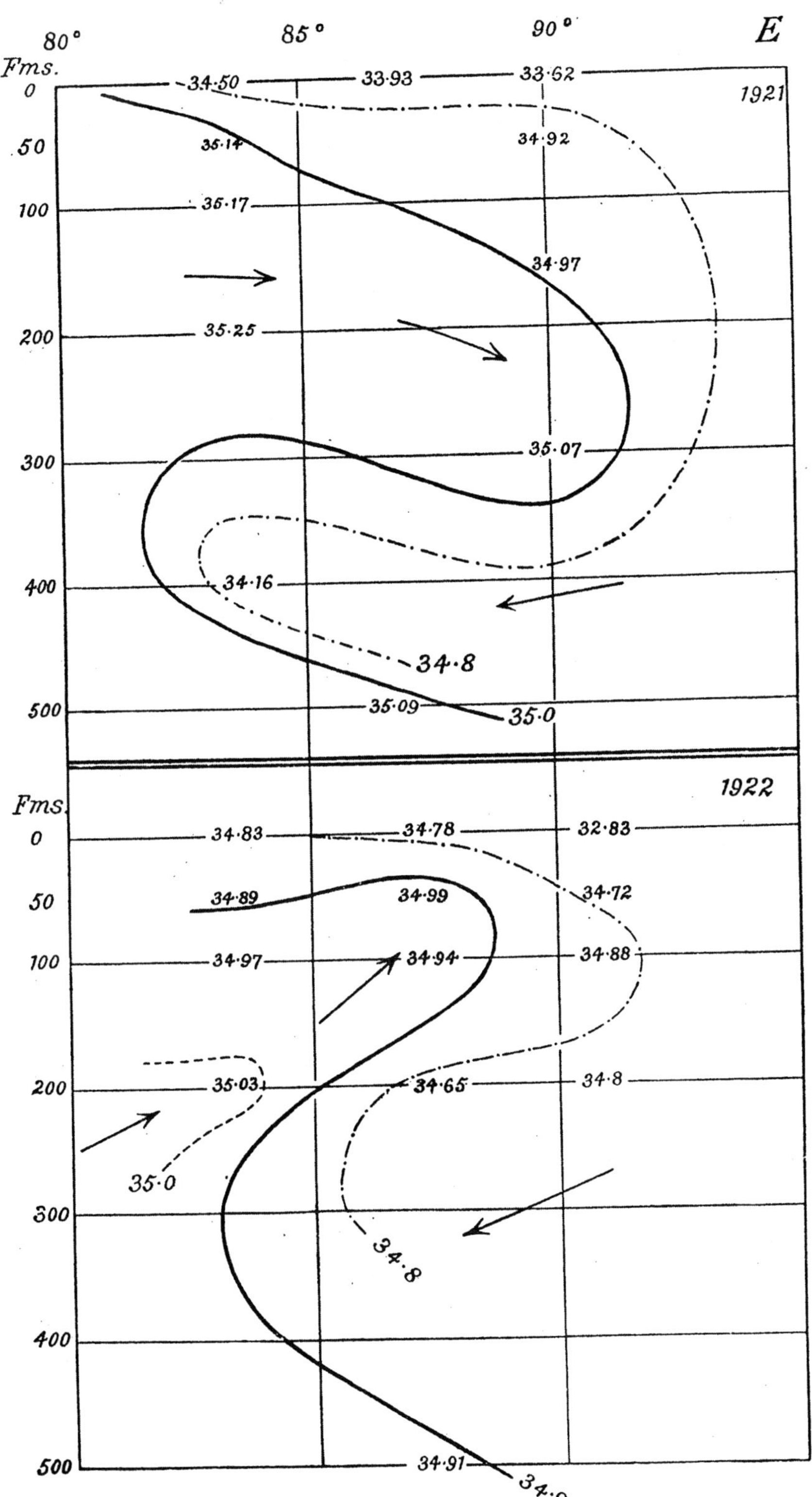

TEXT-FIG. 123.—Showing the isohalines in the southern part of the Bay of Bengal in 1921 and 1922.

approximately Lat. 1° N.; Long. 74° E., passing through Lat. 10° N., Long. 90° E.; and finally terminating in Lat. 1-2° N.; Long. 96° E. It seems clear that there is a stratum of high salinity water, lying at a depth that steadily increases as we pass eastwards from Long. 78° E., at which point it leaves the actual surface, and that this can be traced right across the mouth of the Bay of Bengal. There can be no doubt that this is part of the great Contra-equatorial current that is gradually sinking to reinforce the north Indian deep stratum. The lower limit of this saline stratum lies roughly between 300 and 400 fathoms (164 and 219 metres) depth in Long. 74° E. but sinks to 700 fathoms in Long. 96° E. In the region of Long. 84°–90° E. there is a clear indication that this mass of saline water is being depressed, and possibly split into two streams, by a mass of water of a salinity of under 35·0, that is sinking downwards and appears also to have a distinct trend towards the west.

Comparing the results obtained by the 'Investigator' in the two years 1921 and 1922 (Text-fig. 123), we see that though, on the whole, they conform closely to the results shown above in Text-fig. 122, yet they show an interesting difference in the conditions present in the two years. In 1921 the eastwardly flowing current of high salinity water sinks so that the 35·0 isohaline can be traced to a depth of 175 fathoms (320 metres) in Long. 90° E. and then in Long. 92° E. bends back towards the west at a depth of approximately 300 fathoms (549 metres), finally bending east again in Long. 82-83° E. at a depth of 400 fathoms (732 metres). In 1922, however, the mass of water of less salinity that is dividing the Contra-equatorial current into two, appears from the position of the isohalines to have been considerably increased in volume. There was in this year, as I have previously pointed out (*vide supra*, p. 270 *et seq.*), a very considerable lowering of the salinity of the surface water in the Bay of Bengal in comparison with the conditions that were existing in the previous year, owing apparently to the greatly increased rainfall during the south-west monsoon, and this lowering of salinity can be seen to affect the various levels to a depth as great as 200 fathoms (366 metres) in Long. 83° E., in consequence of which the 35·0 isohaline only extends as far east as Long. 84° E. instead of to Long. 92°-93° E., and the westwardly-moving mass of water of low salinity between 200 and 400 fathoms (366 and 732 metres) is greatly increased.

If now we plot the observations that have been taken in the eastern part of the Bay of Bengal in accordance with their respective positions of Latitude we get the result shown in Text-fig. 124, and here again we can trace the sinking downwards of the less-saline water of the northern region of the Bay in a manner that agrees very closely with the result that we obtained from a study of the temperatures in the same area. There thus seems to be little doubt that, at any rate at certain seasons of the year, there is a descending current of water of increased temperature and low salinity that passes from the northern area in Lat. 7°–10° N. downwards and finally out of the Bay of Bengal, and the only consideration is the direction in which this current flows after it has left the Bay of Bengal region.

In this connection it is extremely interesting to note that at several stations in the Bay of Bengal abnormal temperatures have from time to time been recorded.

At first sight one might be inclined to regard these variations from the normal as due to errors in the thermometer or, less likely, to carelessness in reading the result, and thus be also inclined to disregard these records; but there is a certain degree of agreement between these readings and the presence of the area of water of low salinity and raised temperature that we have just been considering. In Table 86 I have given the details of these abnormal temperatures; in two instances marked* in the table, temperatures actually higher than that of the water either above or below have been obtained; in a third instance the same temperature was recorded at 275 fathoms (504 metres) as at 250 fathoms (457 metres); and in a

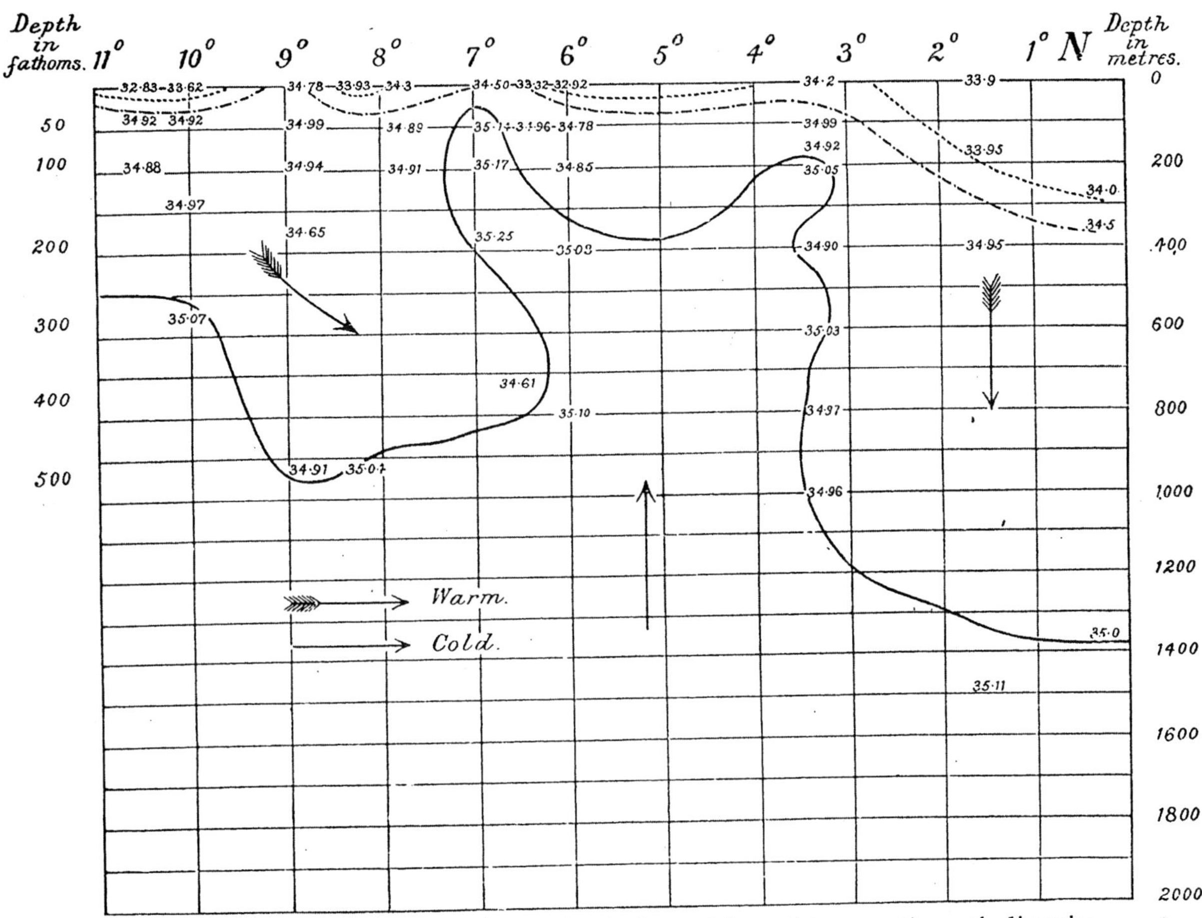

TEXT-FIG. 124.—Showing the isohalines in the Bay of Bengal in a north-south direction.

fourth, a temperature of 11·2°C. (or nearly 2°C. above the normal) was recorded at a depth of 300 fathoms (549 metres). In every case these abnormal temperatures occur at depths between 200 and 400 fathoms (366 and 732 metres) and thus closely correspond to the area of low salinity water that we have just considered. A study of the results obtained by the 'Valdivia' (*vide* Schott, 1902) reveals that the same kind of result was obtained at 'Valdivia' Station 214, Lat. 7° 43′ N.; Long. 88° 45′ E., where a temperature as high as 9·6°C. was recorded at 1,000 metres (547 fathoms). It must, however, be admitted that these abnormally high temperatures have for the most part been recorded in the months of April and May, whereas the records of water of low salinity below the eastwardly moving layer of high salinity water, at a depth of 300–400 fathoms (164–218 metres), were taken in the months of October or February. Were this deep water derived from the surface in the eastern part of the Bay, one would, however, expect to find that its temperature was abnormally high in April and May, the hottest months of the year, and its salinity abnormally low in October and February, following on the dilution of the surface water by the rainfall of the two monsoon seasons. Have we here evidence of a deep current, lying below the great Contra-equatorial current, and possibly to a large extent caused by it? I know of no direct evidence of the existence of such a current, though it would be somewhat surprising if a current of the volume and rapidity of the Contra-equatorial current did not induce a compensating current flowing in the opposite direction, and if that is actually taking place the direction

Long. E. ... Depth in .. Lat. N.		8° 06′ 18″ 70 01 30	12° 14′ 06″ 70 54 30	6° 07′ 00″ 80 11 00	5° 59′ 50″ 81 42 55	15° 00′ 85 00	6° 16′ 90 44	11° 26′ 91 54
fms.	metres	April	October	April	May	April	April	April
		°C	°C	°C	°C	°C	°C	°C
0	0	28·61	26·22	30·40	27·50	28·89	....	29·60
25	46	....	....	26·67	26·90	28·0	29·0	28·6
50	91	....	....	21·00	23·22	25·6	25·11	26·11
75	137	....	....	15·56	19·11	20·72	19·32	20·60
100	183	15·72	14·05	13·11	15·6	16·39	14·17	16·50
125	229	....	....	11·78	14·2	14·0	13·05	14·44
150	274	....	....	11·56	13·2	12·22	12·11	12·61
175	320	....	....	10·44	12·8	11·44	11·67	_12·83*_
200	366	12·00	11·78	_10·78*_	12·3	10·61	_11·39_	_11·95_
225	412	....	....	_11·22*_	....	10·28	....	....
250	457	....	....	10·00	....	9·83	....	....
275	503	....	....	....	....	_9·83_	....	....
300	549	10·28	_11·11_	....	_11·2_	9·33	....	....
400	782	_9·61_	9·33	....	....	....	....	....

Table 86.—showing abnormal temperatures in Indian waters at depths of 209–400 fathoms.

of such a compensating current would be from east to west. In this connection it is interesting to note that Stanley Gardiner (1903, pp. 154, 158, 173–5; 1907–9, pp. 43, 131, 134) has called attention to evidence that certainly appears to point to the existence of deep currents in the region of the Maldive Archipelago and the Seychelles that extend down from the surface to 200 fathoms or more, and in nearly all cases noted by him these surface currents were setting from east to west though in one locality in the Chagos Archipelago and again off the Saya de Malha Banks this westwardly flowing current was, apparently, at depths greater than 160–200 fathoms replaced by a still deeper current flowing in the opposite direction. Finally, a reference to Text-fig. 79 (*vide supra*, p. 284) shows clearly that during the months of September–December, that is to say exactly at that time of the year in which we have obtained proof of the sinking downwards of a mass of less saline water in or about Lat. 10° N., Long. 88° E., there is a well-marked surface current flowing out of the

Andaman Sea basin towards the south-west that in exactly this region meets the surface drift of the Contra-equatorial current and splits it into two parts, one stream passing to the north into the Bay of Bengal and the other continuing eastwards towards the coast of Sumatra. There can thus be little doubt that this surface drift of less saline water towards the south-west is forced downwards beneath the Contra-equatorial current and in all probability continues its course towards the south-west or west at a depth of some 300 fathoms, and that it is this westwardly-flowing current that is responsible for the creation of the oscillations in the isothermal lines that we have seen to be present off the south-east corner of Ceylon (*vide supra*, p. 369).

The next point that we have to consider is the relationship between the temperature of the water at different depths in the Bay of Bengal and in the enclosed area of the Andaman Sea that lies immediately to the east of it. The Andaman Sea is one of that group of enclosed sea-basins that present certain interesting characteristics as regards the temperature of the enclosed water. The Andaman Sea, however, as I have already shown, (*vide supra*, p. 9) is not a single basin but is formed of a group of three; a large one comprising the whole of the central and southern area, and two small subsidiary basins, lying close together in the north-west region. The large basin possesses a depth of 2,200 fathoms (4,023 metres), but the two smaller ones are approximately only 1,100 to 1,200 fathoms (2,012 to 2,195 metres) in depth. On the west side the main basin communicates with the Bay of Bengal through two channels, namely Ten Degree Channel and Great Channel, having respectively depths of approximately 450–500 fathoms (823–914 metres) and 760–800 fathoms (1,390–1,463 metres). The exact depth of the deepest channel between the two areas is not known, owing to insufficient soundings; Carpenter (1889) has given a contour map of the Bay of Bengal and the Andaman Sea and shows the depth of the various channels between them as ranging from 150 to 760 fathoms. A depth of 760 fathoms (1,390 metres) is the deepest sounding that has up to the present time been taken in Great Channel and the trend of the contour lines (*vide* Plate V) shows that it is certainly of less depth than 900 fathoms (1,646 metres). One of the characteristics of such enclosed basins is that the temperature of the water remains practically constant at all depths below that of the deepest channel connecting them with the open ocean. In the case of the Andaman Sea a study of the temperatures in the main basin and those of the waters of the Bay of Bengal shows that the two series agree very fairly closely down to a depth of 800 fathoms and that below this depth the water in the basin shows but little variation, being approximately 5·0°C. whereas the temperature of the water outside the basin, falls with increasing depth to as low as 1·1°C.; this would indicate that the maximum depth of the channel is in the near vicinity of this figure, namely 800 fathoms.

Krümmel (1907, p. 35), in one of the schemes of classification of these enclosed sea basins, divides them according to the difference that exists between the salinity of the water of the enclosed sea and that of the neighbouring ocean; according to this method these seas fall into three groups, namely (1) those in which the salinity is higher than that of the ocean; (2) those in which it is about the same; and (3) those in which it is considerably less. The Andaman Sea belongs to the third of these groups in spite of the fact that the salinity of the surface water of the Bay of Bengal is low, for the combined outflow of freshwater from the great rivers of Burma causes a marked lowering of the surface salinity to an extent that renders it even lower than that of the surface-water of the Bay. It has been pointed out that the currents of water between the open ocean and a neighbouring enclosed sea-basin will primarily depend on the difference in the specific gravity of the water in the two regions, the surface water always tending to move from the area of low specific gravity towards that of higher specific gravity. Conditions in the Andaman Sea are, however, complicated somewhat by the fact that there are several distinct and separate outlets from the sea into the Bay of Bengal on the west and one on the south-east into the enclosed seas of the East Indian Archipelago; furthermore, the main direction of the prevailing winds is completely reversed every six months with the onset of the south-west and north-east monsoons. There is little doubt that the main tendency is for the surface-water to flow in a westerly or south-westerly direction across the Andaman-Nicobar ridge into the Bay of Bengal and although this flow is completely reversed by the effect of the wind during the south-west monsoon during the period June–August, the movement is largely reinforced by the north-east monsoon during the months November–January. During these latter months one would expect to find that the temperature gradients on the two sides of the dividing ridge, that is in the Andaman sea-basin and on the east side of the Bay of Bengal respectively, would show evidence of an inflowing current through the deep channels in order to counterbalance the marked outflow of the surface-water. In the following table I have given the average bottom temperatures at different depths on the two sides of the ridge separating the two areas, and owing to the comparative paucity of observations I have combined the results of the two months, October and November.—

Depth		(I) Bay of Bengal east side °C.	(II) Andaman Sea basin °C.	Difference (I) – (II) °C.
fms.	metres			
200	366	10·9	10·9	0·0
300	549	9·5	9·5	0·0
400	732	8·5	7·7	0·8
500	914	7·3	6·7	0·6
600	1,097	6·5	5·7	0·8
700	1,280	5·6	5·3	0·3
800	1,463	5·0	5·0	0·0
900	1,646	4·3	4·9	–0·6
1,000	1,829	3·8	4·8	–1·0

From these results (*vide* Text-fig. 125) it is clear that at this time of the year the waters of the Bay of Bengal are warmer at all depths from 400 to 700 fathoms (732 to 1,280 metres) than those at the same level in the Andaman Sea.

The evidence at our disposal tends to show that at these depths there is a steady fall in the temperature of the water at the same level as we pass from the west side of the Bay off the south-east corner of Ceylon to Lat. 10° N.; Long. 90° E. and then on through Ten Degree Channel into the Andaman Sea basin; and this, I think, can only be due to the eastwardly flow of the warm tropic water

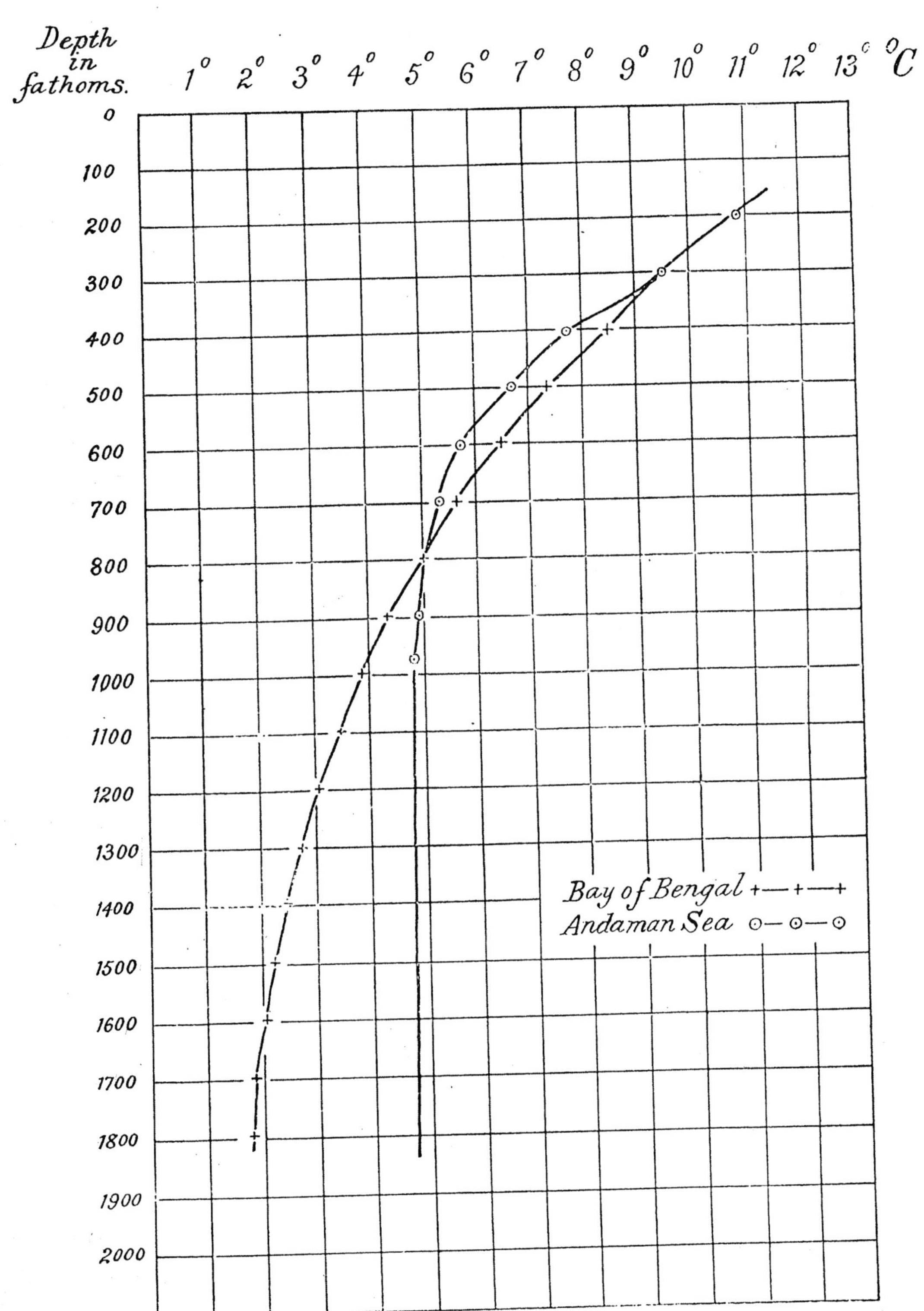

TEXT-FIG. 125.—Showing the Temperature gradients in the Bay of Bengal and the Andaman Sea in the month of October-November.

derived from the Arabian Sea and the descending current in the Bay of Bengal that in this region lies at about this depth, namely 400–700 fathoms.

In the following month, namely December, we can detect the commencement of a marked change in the temperature relationships of the two areas, and in the following table I have given the average results obtained from all available observations. The data for the Bay of Bengal come for the most part from about Lat. 15° N., and are, therefore, rather far removed from the region of the channels connecting the two areas; but up to the present they are all that are available for comparison.

Depth		(I) Bay of Bengal, east side	(II) Andaman Sea basin	Difference (I) – (II)
fms.	metres	°C.	°C.	°C.
200	366	10·9	10·9	0·0
300	549	9·4	9·0	0·4
400	732	7·4	7·7	– 0·3
500	914	6·3	6·9	– 0·6
600	1,097	5·4	6·3	– 0·9
700	1,280	4·6	5·8	– 1·2
800	1,463	3·9	5·4	– 1·5
900	1,646	3·6	5·1	– 1·5
1,000	1,829	3·1	4·8	– 1·7

The first point to note is the change in the temperatures at the different levels in the two areas; in the eastern part of the Bay of Bengal between the depths of 400–800 fathoms (732–1,463 metres) the temperature has risen by approximately 1·0° C. and below this even down to 1,000 fathoms (1,829 metres) there has been a rise, though not quite to so marked an extent; on the other hand in the Andaman Sea basin the temperature of the water is slightly reduced at the 300 fathom (549 metres) level, but at 400 fathoms (732 metres) there has been no change; below this depth from 500 to 900 fathoms (914 to 1,646 metres) the temperatures exhibit a distinct rise that is greatest at 600–700 fathoms (1,097–1,280 metres). Much of this difference is, doubtless, to be attributed to the action of the north-east monsoon that during this month will tend to cause a depression of the water levels on the west side of the Andaman Sea basin, with a consequent rise in the temperature at the deeper levels, while on the west side of the Andaman ridge the action of the wind will tend to cause an upward movement of the deeper water strata to replace the surface-water that is being blown away from the coast, and thus to cause a lowering of the temperature at any given level. At a depth of 900–1,000 fathoms (1,646–1,829 metres) the change in the temperature of the water of the Bay of Bengal is not quite as great as one would expect and is certainly not so great as in the higher levels and I am inclined to attribute this to an outflow at or about this depth of warm water from the Andaman Sea basin through the deep channels into the Bay of Bengal.

During the next two months, January and February, this outflowing deep current appears to increase very considerably in volume. In the following table I have given the average temperatures recorded at different depths in the basin and in the open waters outside :—

Depth fms.	Depth metres	(I) Bay of Bengal, east side °C.	(II) Andaman Sea basin °C.	Difference (I) – (II) °C.
200	366	10·4 ?	10·4	0·0
300	549	9·0 ?	9·0	0·0
400	732	7·9 ?	7·9	0·0
500	914	6·8	7·0	– 0·2
600	1,097	6·1	6·3	– 0·2
700	1,280	5·0	5·7	– 0·7
800	1,463	5·0	5·2	– 0·2
900	1,646	4·9	4·9	0·0
1,000	1,829	4·5	4·9	– 0·3

A study of the temperature gradients, for the month of January, as shown in Text-fig. 126, reveals clearly that we now have two zones of water in the Bay of Bengal, namely at depths of 600–700 fathoms (1,097–1,280 metres) and 700–1,000 fathoms (1,280–1,829 metres) respectively, in which the temperature is warmer than one would expect. Between 500 and 900 fathoms (917 and 1,646 metres) the temperature of the Andaman Sea, as shown in the above Table, is higher than that of the Bay water at the same depth, as one would expect at this season of the year in which the pressure of the north-east monsoon causes the temperature levels on the west side of the basin to be depressed; but in the two zones mentioned above the water of the Bay has a temperature that closely approximates to that of the basin water. This, I think, can only be attributed to outflowing currents of water from the basin to the Bay through the two channels viz., Ten Degree and Great channels, the depths of which correspond very fairly well with the depths of these warm zones.

Unfortunately I have too few records taken during the month of March to enable me to make any comparison between the two areas in this month; but in April the conditions in the Bay appear to be returning to what it was in the month of December. I give the average temperatures in the two areas in the table below:—

Depth fms.	Depth metres	(I) Bay of Bengal, east side °C.	(II) Andaman Sea basin °C.	Difference (I) – (II) °C.
400	732	9·0	9·0	0·0
500	914	7·8	8·0	– 0·2
600	1,097	6·7	6·7	0·0
700	1,280	5·5	5·8	– 0·3
800	1,463	5·1	5·4	– 0·3
900	1,646	4·6	4·9	– 0.3
1,000	1,829	4·0	4·8	– 0.8

From the above it would appear that there is in this month a marked diminution in the outflowing current of warm water through the deep channels. The formation of this outflowing current in December, its marked increase in the months of January-February, and its diminution in April, clearly indicate that it is the result of the north-east monsoon; at first sight one would expect to find that the production of a strong surface-current flowing from north-east to south-west out of the Andaman Sea basin would be accompanied by a deep current flowing, not out of the basin but into it, through the channels connecting it with the Bay; one must, however, bear in mind that during this season of the year there is an enormous volume of water entering the basin from the great rivers of Burma, an influx that will continue for some time

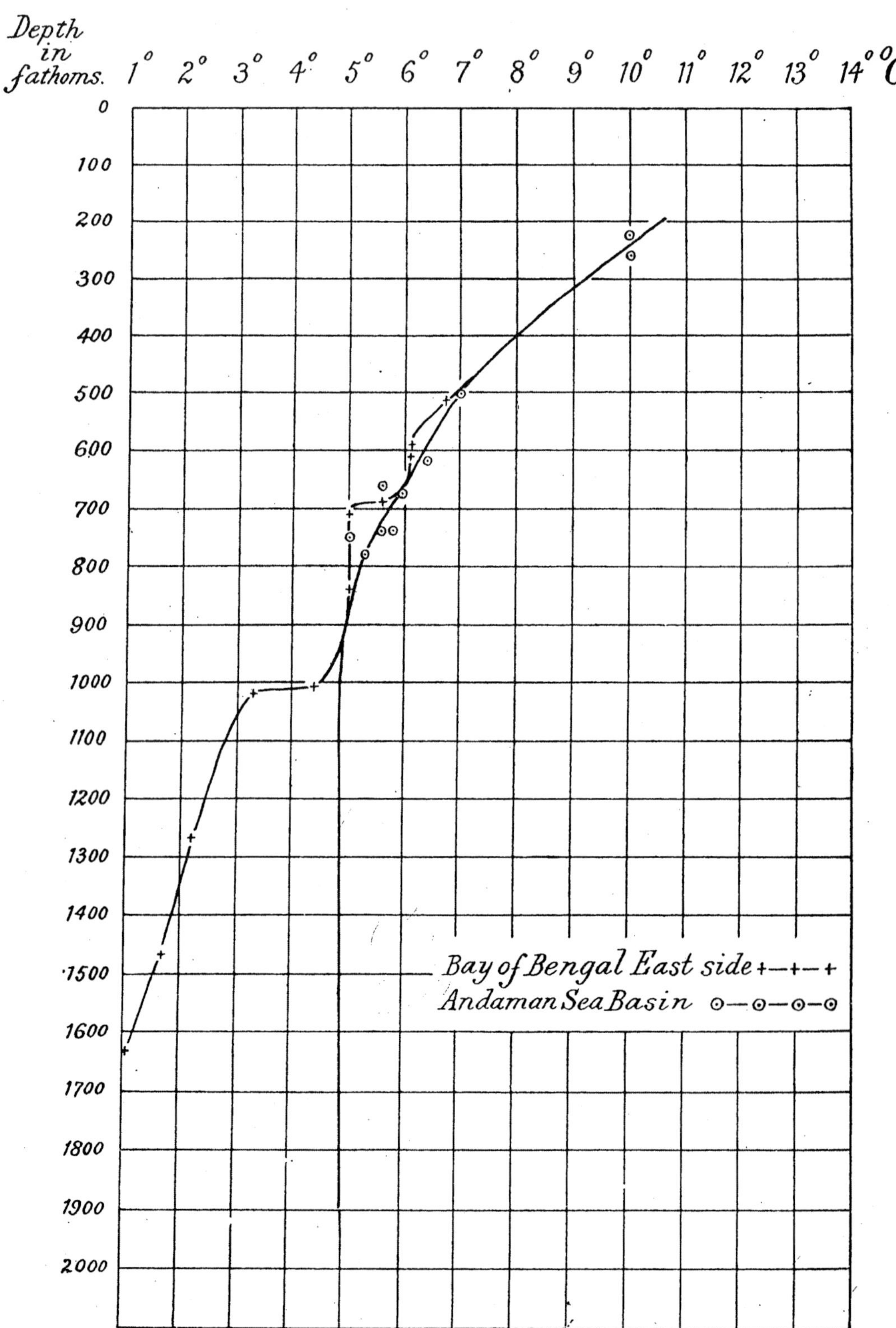

TEXT-FIG. 126.—Showing the Temperature gradient in the Bay of Bengal and the Andaman Sea in the month of January.

after the actual monsoon has died out, and that, in addition, owing to the effect of the north-east monsoon on the waters of the western Pacific ocean and the enclosed seas of the Malay Archipelago, there is a very greatly increased inflow of water through the shallow Straits of Malacca into the south-east part of the Andaman Sea. To compensate for this excessive addition of surface water, it seems that there is an outflow of deep water through both channels, from the Andaman basin to the Bay of Bengal.

The Discontinuity Zone.

Every serial observation that is taken on the temperature of the water at different levels shows clearly that at a depth of about 50 fathoms (91 metres) there is a very rapid drop. The average temperature of the surface-water, as calculated from all the observations given in Table 87, is 28·31°C., and there is, as a rule, but little change in the upper 25 fathoms (46 metres), at which depth the average temperature is 26·75°C. Immediately below this depth, however, there occurs a rapid change and the temperature falls abruptly. The region of this rapid fall of temperature has been termed the 'discontinuity zone'.

Schott (1902, p. 178 *et seq.*) attributes the presence of this zone partly to the alternating effect of the seasons, whereby the surface-water becomes condensed and has a raised temperature during the hot weather, but during the cold season this evaporated water cools and then sinks to find its proper level at a depth of some 50 to 75 fathoms (91 to 137 metres) below the surface; and partly also to the effect of lateral movements of the upper stratum, whereby the increased horizontal movement and the consequent thinning in depth of the surface currents cause the deeper and, therefore, colder and more saline water to approach nearer to the surface. Krümmel (1911, p. 393 *et seq.*), however, refuses to believe that this sudden change in temperature at depths as great as 100 metres (55 fathoms) can in any way be due to the heating up and evaporation of the surface-water. He points out that the true discontinuity layer in other seas occurs at a level that is only a few metres, 15 to 20 (or 8 to 11 fathoms), below the surface and is due to the heating up of the uppermost levels in summer, so as to form a sharply-defined layer, and that such a process must be entirely prevented by both wave-action and surface-currents in the open ocean. 'Only in seas without waves and currents,' he remarks, 'can we expect a sharply marked discontinuity layer'. The formation of a sharply defined belt of water, in which the temperature gradient is as high as 1·75°C. in 25 metres or 3·0°C. in 25 fathoms, is in his opinion due to one of two causes; either (1) to ascending movements of deep strata that carry cold water towards the upper levels, which have been warmed by the sun, or (2) to horizontal movements of large masses of water due to surface-currents, the warm upper layer thus becoming sharply separated from the deeper colder water by a so-called discontinuity zone. In the Bay of Bengal we appear to get such a discontinuity zone that is produced in yet another manner, at least in part; in this case we find that the surface-water is, especially at certain seasons of the year, diluted by rain or river-water and thus, in consequence of its reduced salinity and specific gravity, floats on the top of the more saline water, thus preventing the usual convection currents from making their appearance and so restricting the heating up of the superficial layer by means of the sun's action to the upper and less saline stratum, the discontinuity zone in this case being situated where the two layers of unequal salinity meet.

Schott (1902, p. 178) points out that the average fall of temperature between the surface and a depth of 200 metres is approximately 14°C. or an average fall of 1·75°C. for every 25 metres. He remarks, 'Das Kriterium sowohl dafür, ob überhaupt eine Sprungschicht vorhanden ist, als auch dafür, in welchem Niveau eine vorhandene Sprungschicht im einzelnen Falle liegt, kann auf zweckmässig Weise darin gefunded werden, dass der mittlere Wert des Temperatur-Gradienten von 1·75° pro 25 m. nach oben und unten hin als Schwellenwert gilt'. He fixes 2·0°C. as the temperature variation in a depth of 25 metres that indicates the presence of a discontinuity zone.

In Table 87, I have given all the available observations on the temperature at different levels in the Laccadive Sea and Bay of Bengal and the region to the immediate south of Ceylon. At depths at which no actual temperatures have been recorded, an approximate temperature has been calculated by plotting the series and taking the temperature thus indicated. A study of these observations shows clearly that the average temperature falls rapidly from 28·5°C. at the surface to 13·2°C. at a depth of 125 fathoms, a fall of 15·3°C. This is equivalent to a fall of slightly over 3°C. in each 25 fathoms or 1·65°C. in 25 metres, a result that agrees very closely with Schott's figure of 1·75°C. We may then assume that a discontinuity zone exists at those depths in which the fall of temperature in any interval of 25 fathoms exceeds 3·1°C. and the total extent of the zone is given by taking the difference between the temperatures recorded at intervals of 25 fathoms and noting the depths at which the gradient exceeds this amount.

Schott, from a study of the data before him, concluded that in the Indian Ocean the discontinuity layer lies at a depth below the surface that is intermediate between that found in the Atlantic and Pacific Oceans, and he gives the depths in the three regions as follows:—

In the Atlantic Ocean between 25 and 80 metres depth
" " Indian Ocean " 90 and 140 metres " and
" " Pacific Ocean " 110 and 180 metres "

A closer study of the data from Indian seas appears to me to indicate that, although Schott's figure of 90 to 140 metres depth is the average depth at which the discontinuity zone occurs, it by no means expresses the whole state of affairs. In Table 86 I have given the serial observations and the temperature in each zone of 25 fathoms depth in accordance with the degree of Longitude in which it was taken and the whole series covers a belt from 70° E. to 95° E. If now we divide this belt into intervals of 5° of longitude and calculate the average fall of temper-

Date	Position: Lat. N. ° ′ ″	Position: Long. E. ° ′ ″	0	25	50	75	100	125	150	175	200	225	250	300	400	500	
			Serial Temperatures at depths in fathoms below the surface														
6-iv-94	8 06 18	70 01 30	28·0	25·4	19·8	16·9	15·7	14·7	13·7	12·8	12·0	....	....	10·3	9·6	8·0	Egeria
21-x-93	12 14 06	70 54 30	26·2	23·9	18·6	15·6	14·1	13·3	12·8	12·3	11·8	....	....	11·1	9·3	8·3	Investigator
24-x-92	12 06 30	72 21 36	27·8	26·4	19·3	16·5	15·4	14·4	13·5	12·7	12·0	....	....	10·7	....	8·3	”
25-x-92	12 11 12	73 51 30	26·7	23·3	18·6	16·1	14·8	13·9	13·2	12·6	12·0	....	....	....	....	....	”
30-vi-06	1 52 00	74 45 00	29·0	27·8	28·9	20·3	15·9	12·8	12·2	11·9	11·5	....	....	....	....	....	Planet
31-i-89	6 56 00	75 57 00	28·2	28·1	27·6	24·4	16·8	12·9	11·5	10·9	10·5	....	....	....	....	....	Vitiaz
18-ii-99	2 30 00	76 47 00	28·0	27·2	26·3	18·0	13·6	12·3	11·4	10·8	10·3	....	....	9·4	8·2	6·9	Valdivia
31-iii-94	6 50 30	76 47 18	28·9	24·6	19·2	17·6	16·2	14·8	13·5	12·3	11·2	....	....	10·2	8·6	7·9	Egeria
?-iv-88	6 07 00	80 11 00	30·4	26·7	21·0	15·6	13·1	11·8	11·6	10·4	10·8	11·2	10·0	....	....	....	Investigator
20-v-09	5 43 00	80 25 42	27·1	26·1	21.3	17·3	15·1	13·5	12·4	11·5	10·9	....	....	....	8·5	....	?
19-v-09	5 48 48	80 35 48	27·4	26·2	22·3	17·4	15·0	13·6	12·6	11·9	11·4	....	....	9·9	....	7·4	?
18-v-09	5 39 36	80 46 42	27·6	26·7	23·0	18·6	16·0	14·2	12·7	11·6	10·9	....	....	....	8·6	....	?
17-v-09	5 49 30	80 54 48	27·3	26·2	21·6	16·9	14·7	13·4	12·4	11·5	10·9	..../	....	9·8	8·6	7·4	?
11-v-09	6 01 18	81 32 00	27·8	26·2	21·2	16·5	14·1	12·7	11·9	11·2	10·6	....	....	....	....	....	?
11-v-09	5 55 36	81 35 36	27·8	27·1	20·8	16·6	14·8	....	....	....	....	....	....	....	....	....	?
10-v-09	6 08 24	81 39 12	27·3	25·4	25·4	18·8	15·4	14·2	13·1	12·1	11·2	....	....	9·6	....	....	?
10-v-09	5 59 30	81 42 54	27·5	26·9	23·2	19·1	15·6	14·2	13·4	12·8	12·3	....	....	11·2	8·7	7·8	?
28-iv-09	6 26 00	81 51 12	29·1	27·4	....	....	....	....	....	....	11·8	....	....	9·6	....	7·2	?
19-x-21	6 51 00	83 22 30	29·9	29·0	27·0	22·3	15·6	13·0	11·8	11·2	11·0	....	....	....	....	....	Investigator
10-x-22	6 38 30	83 34 30	27·8	26·8	24·5	20·4	15·4	13·3	12·0	11·2	10·5	....	..	....	....	7·5	”
28-iv-89	15 00 00	85 00 00	28·9	28·0	25·6	20·7	16·4	14·0	12·2	11·4	10·6	10·3	9·8	9·3	....	7·5	”
20-x-21	8 11 42	86 29 42	29·8	28·5	23·5	18·7	15·9	14·3	13·3	12·3	11·8	11·1	10·5	9·6	8·5	7·7	”
11-x-22	8 51 30	86 52 00	27·8	26·6	23·6	18·2	15·1	13·2	12·0	11·1	10·3	....	....	....	....	7·6	”
17-vii-06	3 32 00	88 20 00	28·3	27·0	24·5	16·0	13·8	12·7	11·8	11·2	10·6	10·2	9·8	9·1	7·9	6·8	Planet
10-ii-99	7 43 00	88 45 00	27·4	26·3	23·8	17·9	14·6	12·9	11·8	11·0	10·7	10·4	10·2	10·0	9·1	7·9	Valdivia
21-x-21	10 10 18	89 55 12	30·0	28·4	24·6	19·5	15·4	12·8	11·4	10·9	10·5	....	....	....	9·2	....	Investigator
12-x-22	10 21 00	90 17 30	28·2	26·6	22·5	17·6	14·7	13·0	11·7	11·0	10·6	....	....	....	....	....	”
19-iv-88	6 16 00	90 44 00	29·1	29·0	25·1	19·3	14·2	13·1	12·1	11·7	11·4	....	....	....	....	....	”
?-iv-88	10 00 00	91 07 00	30·4	27·3	23·2	16·7	14·0	12·6	11·8	11·2	....	....	....	....	....	....	”
?-iv-88	11 26 00	91 54 00	29·6	28·6	26·1	20·6	16·5	14·4	12·6	12·8	12·0	....	....	....	....	....	”
31-i-89	6 7 00	92 3 00	28·5	27·4	20·4	17·4	14·5	12·9	12·3	11·8	11·4	10·9	10·5	9·9	8·8	....	Vitiaz
3-5-ii-99	0 46 to 2 12	95 41 to 96 23	28·0	27·4	26·9	17·5	12·5	11·4	10·5	9·8	9·1	8·7	8·3	7·9	6·3	5·5	Valdivia

Table 87.—The Serial Temperatures at intervals of 25 fathoms in Indian waters.

Date	Position		Fall of Temperature in each zone of 25 fathoms depth							
	Lat. N.	Long. E.	0–25	25–50	50–75	75–100	100–125	125–150	150–175	175–200
	° ′ ″	° ′ ″	°C.	°C.	°C.	°C.	°C.	°C.	°C.	°C.
31. I. 89	6 07 00	92 03 00	1·1	*7·0*	3·0	2·9	1·6	0·6	0·5	0·4
18. II. 99	2 30 00	76 47 00	0·8	0·9	*8·3*	*4·4*	1·3	0·9	0·6	0·5
10. II. 99	7 43 00	88 45 00	1·1	2·5	*5·9*	*3·3*	1·7	1·1	0·8	0·3
3-5. II. 99	0 46 to 2 12	95 41 to 96 23	0·6	0·5	*9·4*	*5·0*	1·1	0·9	0·7	0·7
31 III. 94	6 50 30	76 47 18	*4·3*	*5·4*	1·6	1·4	1·4	1·3	1·2	1·1
6. IV. 94	8 06 18	70 01 30	*3·2*	*5·6*	2·9	1·2	1·0	1·0	0·9	0·8
? IV. 88	6 07 00	80 11 00	*3·7*	*5·7*	*5·4*	2·5	1·3	0·2	1·1	—0·4
28. IV. 09	6 26 00	81 51 12	2·7	..	..	..	..	..	..	..
28. IV. 89	15 00 00	85 00 00	0·9	2·4	*4·9*	*4·3*	2·4	1·8	0·8	0·8
19. IV. 88	6 16 00	90 44 00	0·1	*3·9*	*5·8*	*5·1*	1·1	1·0	0·4	0·3
? IV. 88	10 00 00	91 07 00	*3·1*	*4·1*	*6·5*	2·7	1·4	0·8	0·6	..
? IV. 88	11 26 00	91 54 00	1·0	2·5	*5·5*	*4·1*	2·1	1·8	—0·2	0·8
20. V. 09	5 43 00	80 25 42	1·0	*4·8*	*4·0*	2·2	1·6	1·1	0·9	0·6
19. V. 09	5 48 48	80 35 48	1·2	*3·9*	*4·9*	2·4	1·4	1·0	0·7	0·5
18. V. 09	5 39 36	80 46 42	0·9	*3·7*	*4·4*	2·6	1·8	1·5	1·1	0·7
17. V. 09	5 49 30	80 54 48	1·1	*4·6*	*4·7*	2·2	1·3	1·0	0·9	0·6
11. V. 09	6 01 18	81 32 00	1·6	*5·0*	*4·7*	2·4	1·4	0·8	0·7	0·6
11. V. 09	5 55 36	81 35 36	0·7	*6·3*	*4·2*	1·8	..	..	..	..
10. V. 09	6 08 24	81 39 12	1·9	0·0	*6·6*	*3·4*	1·2	1·1	1·0	0·9
10. V. 09	5 59 30	81 42 54	0·6	*3·7*	*4·1*	*3·5*	1·4	0·8	0·6	0·5
17. VII. 06	3 32 00	88 20 00	1·3	2·5	*8·5*	2·2	1·1	0·9	0·6	0·6
21. X. 93	12 14 06	70 54 30	2·3	*5·3*	3·0	1·5	0·8	0·5	0·5	0·5
24. X. 92	12 06 30	72 21 36	1·4	*7·1*	2·8	1·1	1·0	0·9	0·8	0·7
25. X. 92	12 11 12	73 51 30	*3·4*	*4·7*	2·5	1·3	0·9	0·7	0·6	0·6
19. X. 21	6 51 00	83 22 30	0·9	2·0	*4·7*	*6·7*	2·6	1·2	0·6	0·2
10. X. 22	6 38 30	83 34 30	1·0	2·3	*4·1*	*6·0*	2·1	1·3	0·8	0·7
20. X. 21	8 11 42	86 29 42	1·3	*5·0*	*4·8*	2·8	1·6	1·0	1·0	0·5
11. X. 22	8 51 30	86 52 00	1·2	3·0	*5·4*	*3·1*	1·9	1·2	0·9	0·8
21. X. 21	10 10 18	89 55 12	1·6	*3·8*	*5·1*	*4·1*	2·6	1·4	0·5	0·4
12. X. 22	10 21 00	90 17 30	1·6	*4·1*	*4·9*	2·9	1·7	1·3	0·7	0·4

Table 88.—The fall of Temperature in each 25-fathom interval in Indian waters.

ature in each 25 fathoms from the surface downwards we arrive at the following result:—

	Depth in 25 fathom intervals below the surface				
	0–25	25–50	50–75	75–100	100–125
Between Long. E.	°C.	°C.	°C.	°C.	°C.
70°–75°	2·6	5·7	2·8	1·3	0·9
75°–80°	2·6	3·2	4·9	2·9	1·4
80°–85°	1·9	3·3	4·7	5·1	2·0
85°–90°	1·2	3·2	5·8	3·3	1·9
90°–95°	1·3	3·7	5·9	3·8	1·5

These differences in the depth of the zone in the various regions clearly demonstrates that in the Laccadive Sea and round the south end of Ceylon the depth below the surface at which the discontinuity zone lies is comparatively small, but that this depth steadily increases as we proceed eastward. This would appear to bear out Krümmel's view that the zone is caused by the gradual sinking of the warm saline water of the Contra-equatorial current as it flows eastward and meets the less saline surface-water of the Bay of Bengal.

Schott (1902, p. 182 *et seq.*) has pointed out that during the hot season, under the influence of increased evaporation, the upper layer of water will become condensed and that on subsequent cooling this layer will sink downwards, the depth to which it will sink depending on the density of the water at the different levels. He has shown that in the region of the eastern side of the Indian Ocean off the west coast of Sumatra the surface-water, from the effect of evaporation and subsequent convection currents, will sink to a depth of about 110 metres or 60 fathoms; it is interesting to compare his results with those obtained in the region of the Bay of Bengal. The average salinity of the surface-water off the east coast of Ceylon in the month of April, 1924, was 34·29 and the average temperature was 29·45°C. At this period of the year the temperature of the lower layers is as follows:—

At 25	fathoms	26·7°C.	and the corresponding	density would,	therefore,	be	22·30
,, 50	,,	21·0	,,	,,	,,	,,	23·97
,, 75	,,	15·6	,,	,,	,,	,,	25·31
,, 100	,,	13·1	,,	,,	,,	,,	25·84

The average observed density in this area is as follows (*vide* table 37. p. 182),

at 50 fathoms 23·476
,, 100 ,, 25·918

It is clear, then, that in this area water that has been concentrated on the surface to a salinity of 34·29 may on cooling sink to a depth of very nearly 100 fathoms and in the case of the Laccadive Sea, in which the surface water had in the month of April, 1924, a salinity of 34·76, or in the same month in 1914 a salinity of 34·82, might sink to a still greater depth. This is very considerably deeper than Schott suggests. It must be borne in mind that it by no means follows that the sinking of water of raised salinity will be in the vertical direction. We have already seen that the surface-waters of this area are in a continual state of movement and any current that brings more saline water from the western area into the Bay of Bengal will cause convection currents to be set up, which will result in the sinking of the more saline and denser water.

Depth in fathoms below the surface.

Month		0–25 °C.	25–50 °C.	50–75 °C.	75–100 °C.	100–125 °C.	125–150 °C.	150–175 °C.	175–200 °C.
January	..	1·1	*7·0*	*3·0*	2·9	1·6	0·6	0·5	0·4
February	..	0·83	1·30	*7·87*	*4·23*	1·37	0·97	0·70	0·50
March	..	*4·30*	*5·40*	1·60	1·40	1·40	1·30	1·20	1·10
April	..	2·10	*4·03*	*5·17*	*3·32*	1·55	1·10	0·43	0·38
May	..	1·13	*4·00*	*4·70*	2·56	1·44	1·04	0·84	0·63
July	..	1·30	2·50	*8·50*	2·20	1·10	0·90	0·60	0·60
October	..	1·63	*4·14*	*4·14*	*3·28*	1·69	1·06	0·71	0·53

Table 89.—Showing the level and extent of the Discontinuity Zone in Indian waters in different months of the year.

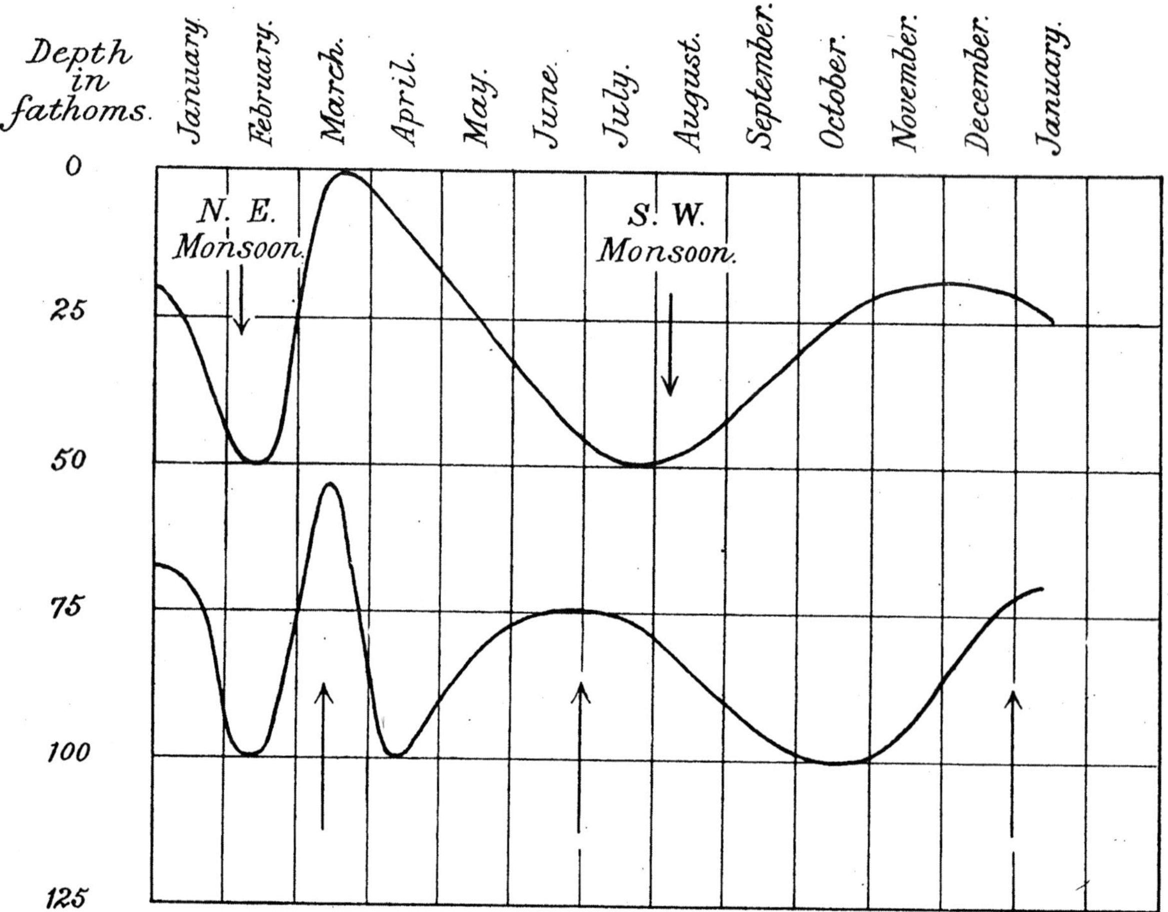

TEXT-FIG. 127.—Showing the upper and lower limits of the Discontinuity Zone throughout the year.

On the other hand a study of the data given above also appears to indicate the possibility that a change of level of the discontinuity zone may be the result of seasonal, as well as local, changes. In the area to the south of Ceylon by far the greater number of observations have been taken during the months of April and May, whereas those in the western area at the mouth of the Laccadive Sea were, with a single exception, taken in October, and in the region of the Bay of Bengal in April or October. I have, therefore, in Table 88 given the available data month by month and in Table 89 the average temperature gradients in each 25 fathoms below the surface in each month; it is unfortunate that there are several months for which we have no data at all and we are, therefore, compelled to base any conclusions on an imperfect series. The average depth of the discontinuity zone has been plotted in text-fig. 127 and this seems to indicate very clearly that there is a marked seasonal change in the depth and also in the vertical extent of this zone. In the month of January the zone lies comparatively near the surface and in the main is confined to a depth between 25 and 50 fathoms (46 and 91 metres), though the close approximation of the amount of the fall of temperature between 50 and 75 fathoms, namely 3·0°C., to the figure that we have taken to indicate the presence of the discontinuity zone, *viz.*, 3·1°C., shows clearly that the zone must extend for some distance below the 50 fathoms 91 metres) level, and in all probability nearly reaches the 75 fathoms (137 metres) mark. A month later in February the data at our disposal indicates that the zone has moved downwards and is now situated at a depth extending between 50 and 100 fathoms (91 and 183 metres). In March the zone lies actually on the surface and extends downwards only to the 50 fathoms (91 metres) level. In April the zone has again increased very considerably in extent and at the same time appears to have sunk to a greater depth, for the evidence before us indicates that it is, as in the month of January, situated between 25 and 75 fathoms (46 and 137 metres). In July the zone has again contracted till it occupies a depth of only some 25 fathoms (46 metres) but this is situated from 50 to 100 fathoms (91 to 183 metres) below the surface. I have no data for the months of either August or September, but in October the zone has greatly increased in extent and now is situated at a depth of 25 to 100 fathoms (46 to 183 metres). One cannot place too much reliance on these figures for in the case of the months of January, March, and August I have only a single observation; nevertheless, it would appear probable that the discontinuity zone exhibits a double variation during the year both as regards the depth at which it lies and also the thickness of the zone itself.

Reference to Text-fig. 127 shows that the upper limit of this zone exhibits two very clear fluctuations during the year, in the latter part of January and February and again during the months of June to September, that is to say during exactly those periods of the year when the two monsoons, with their increased rainfall, are in full swing; and a comparison of the upper level of the zone with the annual variation in the surface temperature in the gulf of Mannar (fig. 36, p. 209 *ante*) reveals a close correspondence. The increased rainfall and lowered surface salinity of the monsoon periods causes a contraction of the discontinuity zone and, owing to the increase in the depth of the upper layer of water of low salinity, forces the zone downwards to greater depths. The lower limit of the zone, however, appears to show a triple oscillation, sinking deeper in the months of February,

April and again in October and rising nearer the surface in January, March and July. It would thus appear that there is some additional factor acting from below that causes this oscillation; any change in the temperature or in the strength of the deeper currents would have the effect of altering the temperature gradient and thus of modifying the discontinuity zone and I shall have to refer to this aspect of the subject later (*vide infra*, p. 411).

Seasonal variation in the temperature of the water at different depths.

The conditions, as regards both temperature and salinity, that exist in the different levels of the Bay of Bengal and neighbouring waters are continually undergoing changes that must be attributed to external causes and especially, so far as the changes in the upper levels are concerned, to the alternating influences of the two monsoon periods. I have already (*vide supra*, pp. 157, 165 and 179) pointed out that the periodic variations in the salinity of the surface-water, experienced in both the Bay of Bengal and the Andaman Sea, indicate that in these waters there is a continual to-and-fro movement or 'seiche' of the deep-water in the basins, and it is clear that such a movement will of itself tend to cause in both the iso-haline and iso-thermal levels at all depths a corresponding oscillation that will be particularly noticeable round the coastal regions and, where the 'seiche' is a bi-nodal one as it appears to be in the Bay of Bengal, at a point midway between the two nodes. Schott (1902, p. 174) has further pointed out that the alternating effect of the two monsoons must to some extent modify the effect that in other oceans is produced by a more or less constant wind and that in consequence the depression of the warmer water on that side of the ocean towards which the wind is blowing, and the compensatory raising of the colder water levels on the side away from which it is blowing, is not so well marked in the Indian Ocean as it is in the Atlantic, though it can still be traced. In addition, it is well known that the mechanical effect of the rotation of the earth causes in the northern hemisphere a depression of the warm upper stratum on the east side of any great ocean, and the same must be true of the Bay of Bengal and the Arabian Sea, so that at one period of the year this effect will be exaggerated by the added influence of the south-west monsoon, while at another period it will be neutralised, either completely or in part, by the influence of the north-east monsoon. There are then a number of different factors that may either all the year round, or at certain stated seasons, affect the level of the isotherms in this area.

In the region around India and to the immediate south we get the great Equatorial system of surface currents that run from east to west or *vice versa* across the whole width of the Indian Ocean. As Krümmel (1911, p. 666) has pointed out, in the Indian Ocean we have during the north-east monsoon a system of currents that closely resembles that of the Atlantic and Pacific Oceans. On the north side we have the great North Equatorial current that flows in the main from east to west, moving from the Andaman Sea past the southern end of India and Ceylon and on towards the Somali coast; a little further to the south between Lat. 2°–5° S. lies the great Contra-equatorial current running from west to east; while further to the south again lies the South Equatorial current, extending between Lat. 10° and 27° S. During the period of the south-west monsoon very considerable changes are brought about in the arrangement of these currents, for the North Equatorial current completely disappears, the whole of the water-masses lying to the north of Lat. 5° S. being in a state of movement towards the east. It is unnecessary for me here to go into details regarding the various lesser changes that take place; this subject has already been treated very fully by Willimzic (1929) and Möller (1929), and I have previously given a somewhat generalised account of the changes in the Indian region in an earlier part of the present publication (*vide supra*, pp. 284, 286, 289 and 292).

The Contra-equatorial current is no mere shallow surface drift, as is indicated by the fact that its effects can be traced down to a considerable depth. Schott (1902, p. 174) has given a table of the temperatures of the water at different depths at the Equator on the two sides of the Indian Ocean, namely off the Somali coast on the west and off the Sumatran coast on the east, and I take the liberty of reproducing these data here:—

Depth		(I) Somali coast °C.	(II) Sumatra coast °C.	Difference (I)−(II) °C.
fms.	metres			
0	0	26·5	28·3	−1·8
27	50	25·8	27·7	−1·9
55	100	23·0	26·6	−3·6
82	150	15·3	15·3	0·0

These temperatures show that at a depth of 100 metres there is on the two sides of the Indian Ocean a difference as great as 3·6°C., the water at that depth being by this amount warmer off the Sumatran coast than off the Somali coast. Further to the north conditions, however, may be exactly the opposite for in 1888-89 two sets of serial observations were taken by the S.S 'Vitiaz' that gave the following results:—

Depth		(I) Lat. N. 6° 56′ Long. E. 75° 57′ °C.	(II) Lat. N. 6° 07′ Long. E. 92° 03′ °C.	Difference (I)−(II) °C.
fms.	metres			
0	0	28·2	28·5	−0·3
14	25	28·2	28·2	0·0
27	50	28·0	27·2	0·8
55	100	27·4	19·9	7·5
109	200	13·8	13·3	0·5
219	400	10·3	10·8	−0·5
437	800	7·7	8·2	−0·5

These observations were taken in the month of January, that is to say at that season of the year in which the north-east monsoon is at its maximum, and the position of the two stations lies in the full strength of the North-Equatorial current; here the water on the west side of the Indian Ocean at a depth of 100 metres (55 fathoms) is as much as 7·5°C. warmer than the water at the same depth on the east side and a difference in the same direction can be traced at the 50-metre (27-fathom) level and even down to a depth of 200 metres (109 fathoms). Below this at depths of 400 to 800 metres (219 to 437 fathoms) the water on the east side is slightly warmer than that on the west. Schott attributes the differences in temperature in the equatorial region to the easterly-flowing current; he remarks, 'The Indian Contra-equatorial current flows in the northern winter to the east and south-east as far as Sumatra; it conveys warm water eastwards (like the Guinea current in the Atlantic Ocean); whereas at its source on the Somali coast and in the region of Zanzibar somewhat colder water pours in from higher latitudes'. He has shown in the chart published (Schott, 1902, Pl. XI) the extent and direction of this flow. At 100 metres (55 fathoms) depth it can be traced across the line of the Maldive Archipelago and then onward in an easterly direction past the south end of India and Ceylon in approximately Lat. 8° N.; it then bends towards the south-east and impinges on the west coast of Sumatra. The differences in the two sets of observations taken by the 'Vitiaz' during the winter months in Lat. 6-7° N. are equally clearly to be attributed to the influence of the North Equatorial current, combined with the effect of the north-east monsoon. Both sets of observations show clearly that these currents are strong drifts that may affect the upper strata of the ocean to a depth as great as 100 metres (55 fathoms) or more, and in consequence we should expect to find that the reversal of the currents under the influence of the alternating monsoons causes on either side of the Bay of Bengal a marked depression of the warm upper stratum at one period of the year and an equally marked upwelling of cold water from below during the opposite season, the effect varying on the two sides in accordance with which monsoon is blowing and, therefore, which coast is to windward and which to leeward. Unfortunately, I have no records of the conditions existing in this area during the south-west monsoon; but I give below such data as I have been able to collate regarding the changes that occur between the months of October to May, both inclusive.

(1) *October.*

In making comparisons between the temperatures at different depths on the two sides of an area such as the Bay of Bengal it is of the greatest importance that, whenever possible, we should compare data taken on the two sides in the same year. We have already seen the extent to which a difference in the strength of the monsoon and the amount of rainfall can affect the salinity of the surface-water (*vide supra*, p. 212) and this difference appears to be by no means confined to the superficial stratum (*vide supra*, p. 376). In the table on next page I have given two sets of serial temperatures from Table 87, taken in close proximity to each other in two consecutive years:—

Depth fms.	Depth metres	(I) Lat. 10° 10′ 18″ N. Long. 89° 55′ 12″ E. 1921 °C.	(II) Lat. 10° 21′ 00″ N. Long. 90° 17′ 30″ E. 1922 °C.	Difference (I) – (II) °C.
0	0	30·0	28·2	1·8
25	46	28·4	26·6	1·8
50	91	24·6	22·5	2·1
75	137	19·5	17·6	1·9
100	183	15·4	14·7	0·7
125	229	12·8	12·7	0·1
150	274	11·4	11·7	–0·3
175	320	10·9	11·2	–0·3
200	366	10·5	10·6	–0·1

It is thus clear that in these two years the water on the east side of the Bay of Bengal was very considerably warmer, down to a depth of over 100 fathoms (183 metres), in 1921 than in 1922 and that this difference is most marked at the 50 fathom level, where it was as great as 2·1°C. An exactly similar state of affairs prevailed in the south-west region of the Bay in these years; if we compare the different temperatures recorded in 1921 and 1922 in this region we find the following differences:—

Depth fms.	Depth metres	(I) Lat. 6° 51′ 00″ N. Long. 83° 22′ 30″ E. 1921 °C.	(II) Lat. 6° 38′ 30″ N. Long. 83° 34′ 30″ E. 1922 °C.	Difference (I) – (II) °C.
0	0	29·8	27·9	1·9
25	46	29·0	26·6	2·4
50	91	27·0	24·6	2·4
75	137	21·0	20·4	0·6
100	183	15.6	15·4	0.2
125	229	13·3	13·3	0·0
150	274	12·2	12·0	0·2
175	320	11·5	11·2	0·3
200	366	11·0	10·5	0·5

Here again we see that in these two years there was a difference in temperature in the same area down to a depth of over 100 fathoms, the water being throughout this depth warmer in 1921 than in 1922; and the maximum difference again occurs at a depth of about 50 fathoms (91 metres) where it was as great as 2·4°C.; at 125 fathoms the temperature is the same but below this from 150–202 fathoms (274 to 366 metres) it is again slightly warmer on the west.

If now we had been comparing the results taken on the south-west side of the Bay in 1921 with those from the east side of the Bay in 1922 we should have found the following difference between the two sets of observations:—

Depth fms.	Depth metres	(I) Lat. 6° 51′ 00″ N. Long. 83° 22′ 30″ E. 1921 °C.	(II) Lat. 10° 21′ 00° N. Long. 90° 17′ 30″ E. 1922 °C.	Difference (I) – (II) °C.
0	0	29·8	27·4	2·4
25	46	29·0	26·6	2·4
50	91	27·0	22·5	4·5
75	137	21·0	17·6	3·4
100	183	15·6	14·7	1·1
125	229	13·3	13·0	0·3
150	274	12·2	11·7	0·5
175	320	11·5	11·2	0·3
200	366	11·0	10·6	0·4

In this comparison the difference in the temperature of the water on the two sides at 50 fathoms (91 metres) depth appears to be as great as 4·5°C. or nearly twice as great as we know actually to have been the case in either year, and at 75 fathoms (137 metres) depth is as great as 3·4°C. Annual differences such as this may thus to a very great extent either exaggerate or diminish or possibly even totally mask any difference that might be due to either surface currents or even seasonal changes.

If now we compare the conditions on the two sides of the Bay in each year we find the following :—

1921 Depth fms.	metres	(I) Lat. N. 6° 51′ 00″ Long. E. 83° 22′ 30″ °C.	(II) Lat. N. 10° 10′ 18″ Long. E. 89° 55′ 12″ °C.	Difference (I) – (II) °C.
0	0	29·9	29·9	0·0
25	46	29·0	28·2	0·8
50	91	27·0	24·6	2·4
75	137	21·0	19·1	1·9
100	183	15·6	14·2	1·4
125	229	13·3	12·4	0·9
150	274	12·2	11·4	0·8
175	320	11·5	10·9	0·6
200	366	11·0	10·5	0·5

1922 Depth fms.	metres	(I) Lat. N. 6° 38′ 30″ Long. E. 83° 34′ 30″ °C.	(II) Lat. N. 10° 21′ 00″ Long. E. 90° 17′ 30″ °C.	Difference (I) – (II) °C.
0	0	27·9	28·2	–0·3
25	46	26·6	26·6	0·0
50	91	24·5	22·5	2·0
75	137	20·4	17·6	2·8
100	183	15·4	14·7	0·7
125	229	13·3	12·7	0·6
150	274	12·0	11·7	0·3
175	320	11·2	11·2	0·0
200	366	10·5	10·6	–0·1

In both years it is clear that at this period of the year the water in the south-west part of the Bay is warmer, with the exception of the actual surface, than that on the east side down to a depth of approximately 150 fathoms or rather more, and that this difference is greatest at a depth of 50 to 75 fathoms. These results are therefore in complete agreement with, though not so marked as, those obtained by the 'Vitiaz' (*vide supra*, p. 393).

(II) *December.*

I have no serial observations taken in this month, but a number of observations of the surface and bottom temperatures at different depths have been taken around the coast and in the following table I have given the averages of all these observations on the two sides of the Bay. While these figures do not conform to the requirements that I have given above, namely, that for the purpose of comparison observations should be taken in the same year, it is, however, probable that the average of a number of such observations taken in different years will at least give an approximately correct picture, since the variations from the normal will tend to neutralise each other.

Depth		Bay of Bengal		Difference
		(I)	(II)	(I) – (II)
fms.	metres	West side	East side	
		°C.	°C.	°C.
0	0	26·6	23·8	2·8
100	183	18·5	17·0	1·5
150	274	14·2	12·1	2·1
200	366	12·1	11·3	0·8
250	457	10·5	10·4	0·1
300	549	9·3	9·1	0·2
400	732	7·8	7·6	0·2

Here again we find that the temperatures on the west side of the Bay, as was the case in the month of October, are warmer than those at the same depth on the east side; but there is this difference between the present data and those that we have been previously considering, that the maximum temperature difference occurs either on the surface itself or at a depth of about 150 fathoms (274 metres).

(III) *January.*

I have already (*vide supra*, p. 393) given the data that were taken by the 'Vitiaz' in this month and we have seen that they show that the water in the western region is warmer than the water at the same depth in the eastern area, especially at a depth of 55 fathoms (100 metres).

For convenience of reference I give the data again below:—

Depth		(I)	(II)	
		Lat. N. 6° 36′	Lat. N. 6° 07′	Difference
fms.	metres	Long. E. 75° 57′	Long. E. 92° 03′	(I) – (II)
0	0	28·2	28·5	–0·3
14	25	28·2	28·2	0·0
27	50	28·0	27·2	0·8
55	100	27·4	19·9	7·5
109	200	13·8	13·3	0·5
219	400	10·3	10·8	–0·5
437	800	7·7	8·2	–0·5

A strong North Equatorial current flowing from east to west would cause the temperature of the water in the upper strata on the west side to be warmer than that on the east at the same depths; but it is extremely interesting to note that in this northern zone of Lat. 6-7° N. in this month there is at a depth of about 300 metres (164 fathoms) and continuing down to below 800 metres (437 fathoms) a very clear reversal of the condition found to be present in the upper stratum, for between these latter depths the water on the east side of the Bay is warmer than that on the west. We have already seen (*vide supra*, p. 377) that at certain seasons of the year there seems to be evidence of a deep current caused by the less saline water in the northern part of the Bay of Bengal sinking downwards and then flowing towards the west or south-west underneath the Contra-equatorial current and it seems possible that the alteration in the temperature relations of the two strata, as indicated above, may be due to the presence in the deeper level of such a current.

(IV) *February.*

Among the data given by Schott (1902, p. 140, Pls. XXI, XXII) we find two other sets of temperature readings on the two sides of the Indian Ocean but more to the south, namely at or near Lat. 2° N., and again at approximately Lat. 3°-4° S. In the first instance the results are as follows:—

Depth metres	Depth fms.	(I) Lat. N. 2° 30′ Long. E. 76° 47′ Date 18-ii-99 °C.	(II) Lat. N, 1-2° Long. E. 96-97° Date 3-5-ii-99 °C.	Difference (I)–(II) °C.
0	0	28·0	28·0	0·0
25	14	27·7	27·6	0·1
50	27	27·3	27·3	0·0
75	41	26·7	27·1	–0·4
100	55	26·1	26·6	–0·5
125	68	23·0	19·2	3·8
150	82	16·3	15·3	1·0
175	96	14·0	12·8	1·2
200	109	13·0	12·0	1·0
300	164	11·0	10·0	1·0
400	219	10·2	8·7	1·5
500	273	9·7	8·3	1·4
600	328	9·2	8·1	1·1
700	383	8·3	7·8	0·5
800	437	7·5	6·6	0·9
900	492	6·9	5·9	1·0
1,000	546	6·1	5·6	0·5

In this series we find that the water exhibits approximately the same temperature at the two stations between the surface and a depth of 50 metres (27 fathoms), but below this the water on the east side at a depth of 75 to 100 metres (41 to 55 fathoms) is slightly warmer than that at the same depth on the west; below this latter depth, however, the condition is reversed and the water at a depth of 125 metres (68 fathoms) to as far down as 1,000 metres (546 fathoms) is appreciably warmer on the west side of the ocean, the greatest difference, amounting to 3·8° C., being found at a depth of 125 metres (68 fathoms). In the second series to the south of the Equator we get the following results:—

Depth metres	Depth fms.	(I) Lat. S. 2° 57′ Long. E. 67° 59′ Date 28-11-99 °C.	(II) Lat. S. 3° 41′ Long. E. 101° 00′ Date 21-1-99 °C.	Difference (I)–(II) °C.
0	0	28·3	27·8	0·5
25	14	28·2	27·0	1·2
50	27	28·0	26·5	1·5
75	41	26·6	26·5	0·1
100	55	21·1	26·4	–5·3
125	68	20·0	19·2	0·8
150	82	18·5	15·1	3·4
175	96	16·8	12·6	4·2
200	109	15·7	11·7	4·0
300	164	11·8	11·0	0·8
400	219	10·1	10·3	–0·2
500	273	9·3	9·4	–0·1
600	328	8·6	8·8	–0·2
700	383	7·9	8·0	–0·1
800	437	7·4	7 1	0·3
900	492	6·9	6·0	0·9
1,000	546	6·2	5·9	0·3

Here again we see that the surface-water extending from 0 to 75 metres (0 to 41 fathoms) depth is warmer on the west side of the ocean, but at 100 metres (55 fathoms) this condition is markedly and suddenly reversed and the water on the west side is now as much as 5·3°C. colder than that on the east; between 125 and 300 metres the water on the west side of the ocean is warmer than that on the east but below this depth the conditions are again reversed and between 400 and 700 metres depth the western area is colder than the east.

(V) *April.*

During this month as many as four different sets of serial observations have been taken by the 'Investigator' and of these two lie on opposite sides of the Bay of Bengal in the neighbourhood of Lat. 6° N. while the other two are from further north and are close together in about Lat. 10°-11° N. I give below the results of the first pair:—

Depth		(I) Lat. N. 6° 07′ Long. E. 80° 11′	(II) Lat. N. 6° 16′ Long. E. 90° 44′	Difference (I) − (II)
fms.	metres	°C.	°C.	°C.
0	0	30·4	29·1	1·3
25	46	26·7	29·0	−2·3
50	91	21·0	25·1	−4·1
75	137	15·6	19·3	−3·7
100	183	13·1	14·2	−1·1
125	229	11·8	13·1	−1·3
150	274	11·6	12·1	−0·5
175	320	10·4	11·7	−1·3
200	366	10·8	11·4	−0·6

In this series the temperature of the water on the eastern side of the Bay of Bengal is, with the single exception of the surface, warmer at all depths down to 200 fathoms (366 metres) than that at the same depth on the west and this difference is most marked at a depth of 50 fathoms (91 metres) though it is still considerable at 75 fathoms (137 metres).

A comparison of the results at different periods of the year clearly reveals that there are very marked changes in the prevailing conditions. Schott (1902, p. 174) states 'Kurzum, während des ganzen Jahres staut sich an der sumatranischen Seite des Oceans das wasser des Oberflächenstromes; daher also ist der relativ wärmste Zone von 50–100 m. Dicke an diese Ostseite des Oceans verlagert, daher ist auch die Ausbildung der Sprungschicht gerade hier sehr intensiv.' While this may be true of the region lying further to the south in the neighbourhood of Lat. 5° S., in which area there is little or no change in the direction of the main surface currents throughout the whole year, it does not hold true for the more northerly region of the Bay of Bengal and the area lying immediately to the south. The northerly area, that we are concerned with, is subject to very great changes in the direction of the surface currents, due to the complete reversal at certain seasons of the year of the main direction of the prevailing wind, and it is of interest to compare the variations in the relative temperatures on the two sides of the area with the monthly current charts for this region given by Möller

(1929, pp. 34-35 and 42-43). In October the Contra-equatorial current occupies a northerly position and the surface-water as far north as Lat. 10° is moving eastward across the southern part of the Bay of Bengal; one would, therefore, expect to find that the water on the east side was warmer at the same depth than that on the west, whereas, as we have seen this is not the case. Möller, however, shows that in the eastern part of the Bay there is in this month a very clear 'convergence' zone, the easterly-moving water of the Contra-equatorial current meeting the westwardly-moving water flowing out of the Andaman Sea, that runs in an S-shaped course from approximately Lat. 5° N.; Long. 95° E. to Lat. 17° N.; Long. 83° E. Along this line surface-water is, as we have already postulated (*vide supra*, p. 377), forced downwards and owing to the land-locked nature of the area is driven towards the south and probably south-west, passing in the southern part of the Bay beneath the upper stratum that is flowing from west to east. We thus have a complete explanation of the conditions of salinity and temperature that we have just been considering (*vide supra*, pp. 367, 377). In the month of December the whole of the surface water of the Bay is moving from east to west and this must result in a depression of the isothermal lines in the western part and a tendency towards the upwelling of colder water from below on the eastern side; we should thus expect to find that the water on the western part of the Bay is warmer at the same depth than that on the eastern portion and this is clearly brought out in the data given above (*vide supra*, p. 396).

In the month of January, as we have seen, in the region of Lat. 6° N. there is clear evidence of the water on the west side of the Indian region being warmer than that at the same depth on the east side down to a depth of about 300 metres (164 fathoms), and, as I have already pointed out, this can equally clearly be attributed to the effect of the North Equatorial current that under the influence of the north-east monsoon is flowing from east to west.

In the month of February the only data at our disposal lies further to the south; we have, however, data for two belts, one at Lat. 2-3° S. and the other at about Lat. 1-2° N. In both these zones we find that the water on the east side is either of the same temperature or a little colder than that on the west down to a depth of 75–100 metres, at which depth there is clear evidence, that is especially strong in the southern belt, that the water on the east side is now considerably warmer than on the west. This condition is the exact opposite of the condition that we found to be present in the previous month, January, in the region more to the north; it seems clear that this difference is due to the fact that the southern data in the month of February are from areas lying within the influence of the Contra-equatorial current and that the water was in movement from west to east, though more markedly so in the southern zone.

Finally, in the month of April the whole of the water in the region under examination is in a state of movement from west to east, the North Equatorial current having been completely obliterated, and it is in this month that we find the conditions in these more northerly areas agreeing exactly with the condition to which Schott called attention in the equatorial zone, the water on the east side of the area being warmer than that on the west at all depths down to about 400 metres.

The Antarctic Bottom Drift

In both the Atlantic and Pacific Oceans the Antarctic Bottom Drift is believed to pass across the Equator to some distance northwards and there is no doubt that a similar condition of affairs is present in the Indian Ocean. Schott (1902, p. 154 and 1926, p. 426, fig. 29) shows this northward extension of the deep current very clearly and he has also pointed out (*loc. cit.*, 1902) that this drift appears to divide into two great tongues one of which lies between Long. 60°—70° E., on the west side of the ocean, and the other between Long. 100°–110° E. on the east side. Matthews (1926, p. 186) has called attention to the low temperatures of the bottom water in certain regions of the western area of the Indian Ocean and he remarks 'the bottom temperature falls from about 4°C. in 1,300 metres (711 fathoms) to 2°C. in 2,400 metres (1,312 fathoms) and then more slowly to about 1·7°C. in 4,000 metres (2,187 fathoms). A reading of 1·1°C. was obtained at 4,459 metres (2,438 fathoms) in 9° 34′ S. Lat. and 52° 26′ E. Long.; this was the greatest depth reached. Temperatures as low as this are found just southward of the line joining Mauritius to the south of Madagascar, and there appears to be free communication here in the deep channel lying between the two islands'. A study of the numerous records of deep temperatures in the north-east part of the Indian Ocean shows that water having a temperature as low as or even lower than 1·1°C. can be traced in the Bay of Bengal as far north as Lat. 9° on the west side of the Bay and to Lat. 12° N. on the east side and I give below a list, compiled from all available sources, of these low readings in various parts of the northern region of the Indian Ocean, and where available, I have also given the name of the ship by which the observation was taken. In order to facilitate reference I have given the depths in both metres and fathoms.

Temperatures of 0·56°C.

Latitude N. ° ′ ″	Longitude E. ° ′ ″	Depth in metres	Depth in fathoms
5 43 00	87 51 00	3,950	2,160
5 48 30	87 35 30	4,023	2,200
5 47 00	87 07 00	4,206	2,300
5 47 00	90 27 00	3,017	1,650
5 48 30	91 28 00	3,929	2,148
5 49 00	91 55 30	4,023	2,200
5 51 00	89 11 30	3,694	2,020
5 52 30	90 12 00	2,816	1,540
5 49 42	88 05 30	3,987	2,180
5 53 00	92 29 00	4,402	2,407
5 53 30	91 12 30	3,758	2,055
5 55 00	91 43 00	3,996	2,185
5 55 42	92 13 30	4,243	2,320
10 46 00	62 36 00	4,515	2,469
11 26 00	60 21 00	4,493	2,457
12 16 30	57 42 30	3,347	1,830

[*Note*:—All these temperatures were taken in the year 1913 and together with certain readings of 1·1°C. taken in the same year form part of a series in which the temperature was read to only the nearest degree Fahrenheit; they, therefore, cannot be taken as exact. The name of the ship which took them is not stated.]

Temperatures of 1·0°C.

Latitude N. ° ′ ″	Longitude E. ° ′ ″	Depth in metres	Depth in fathoms	
8 34 00	91 03 00	3,411	1,865	'Investigator'
11 07 00	89 11 00	3,173	1,735	,,
11 11 00	90 06 00	3,173	1,735	,,
11 15 00	91 16 00	3,365	1,840	,,

Temperatures of 1·1°C.

° ′ ″	° ′ ″			
4 42 00	56 10 00	4,916	2,688	'Stork'
5 37 00	83 44 30	4,098	2,241	?
5 37 42	84 14 42	4,075	2,228	?
5 38 00	84 45 30	4,023	2,200	?
5 39 00	85 16 00	4,060	2,220	?
5 41 00	86 15 30	4,023	2,200	'Investigator'
5 41 00	85 46 00	4,151	2,270	?
5 41 00	86 45 30	4,088	2,235	?
5 43 00	83 59 30	4,075	2,228	?
5 45 00	85 00 30	4,078	2,230	?
5 45 30	89 27 42	3,522	1,926	?
5 45 30	85 31 00	4,060	2,220	?
5 47 00	86 30 30	4,170	2,280	?
5 47 30	86 00 30	4,042	2,210	?
5 52 00	89 42 00	2,991	1,635	?
6 30 30	83 27 42	4,060	2,220	'Investigator'
7 18 00	86 18 30	3,903	2,134	,,
8 03 12	89 35 00	3,658	2,000	,,
8 41 30	91 04 30	3,639	1,990	,,
8 43 00	81 41 45	3,530	1,930	,,
8 45 00	85 29 00	3,694	2,020	,,
9 11 30	81 25 00	3,639	1,990	,,
10 10 30	65 17 00	4,448	2,432	?
10 22 00	64 01 00	4,418	2,416	?
10 38 00	63 10 00	4,459	2,438	?
12 16 30	57 57 00	3,731	2,040	?
20 27 30	67 44 00	3,051	1,668	'Investigator'
24 24 00	61 42 00	2,130	1,165	,,

Note:—Except for the observations taken by the 'Stork' and 'Investigator' those marked with ? belong to the series noted above and were taken in the year 1913.

Temperatures of 1·2°C.

Latitude N. ° ′ ″	Longitude E. ° ′ ″	Depth in metres	Depth in fathoms	
7 16 00	91 07 00	3,174	1,735	'Investigator'
7 25 00	88 16 00	3,722	2,035	?
7 43 00	88 45 00	3,692	2,019	Valdivia
7 58 00	91 47 00	3,974	2,173	,,
10 16 30	88 13 00	3,402	1,860	'Investigator'

Temperatures of 1·3°C.

° ′ ″	° ′ ″			
8 36 00	85 15 00	3,822	2,090	?
9 55 18	89 04 00	3,402	1,860	?
9 20 30	87 50 54	3,566	1,950	?
17 15 45	89 28 00	2,343	1,282	'Investigator'

Temperatures of 1.4°C.

° ′ ″	° ′ ″			
2 30 00	76 47 00	4,133	2,260	'Valdivia'
2 35 00	56 38 00	4,645	2,540	'Stork'
6 25 00	55 10 18	5,289	2,892	,,
9 08 00	91 30 00	3,168	1,732	'Investigator'
10 13 00	81 19 00	3,639	1,990	,,
11 41 00	81 15 45	3,475	1,900	,,
12 24 00	90 08 00	3,052	1,669	,,
12 51 00	88 53 00	3,062	1,674	,,
13 16 00	88 01 30	3,066	1,676	,,
14 14 00	85 58 00	3,025	1,654	,,
15 00 00	85 00 00	3,032	1,658	,,
13 34 00	86 55 00	3,058	1,672	,,
10 22 30	87 43 00	3,411	1,865	,,
23 51 42	65 16 30	3,511	1,920	,,

Temperatures of 1·5°C.

° ′ ″	° ′ ″			
5 56 00	91 05 00	2,908	1,590	'Investigator'
7 46 00	90 25 00	2,969	1,623	,,

Temperatures of 1·6° C.

° ′ ″	° ′ ″			
5 44 00	90 30 00	2,826	1,545	'Investigator'
7 42 00	90 10 00	2,875	1,572	,,
7 38 00	89 48 00	3,019	1,651	,,
7 11 00	78 46 37	2,681	1,466	,,
10 22 30	87 43 00	3,412	1,865	,,
10 58 00	81 19 00	3,603	1,970	,,
15 10 00	92 12 00	2,656	1,452	,,

Temperatures of 1·7°C.

° ′ ″	° ′ ″			
9 21 00	81 22 30	3,685	2,015	'Investigator'
9 34 00	85 43 15	3,652	1,997	,,
11 00 00	87 13 00	3,225	1,763	,,
12 05 00	88 40 15	3,191	1,745	,,
12 20 00	85 08 00	3,297	1,803	,,
13 18 00	91 38 30	3,058	1,672	,,
17 11 30	83 50 00	2,624	1,435	,,
17 43 00	84 23 00	2,469	1,350	,,

In Chart VI, which for purpose of reference I reproduce again here, I have given a more or less generalised plan of what I believe to be the main course of the Antarctic Bottom drift from the South Polar region through the Indian Ocean to the north, and a reference to this shows that there appear to be three main tongues of bottom water, of which the first or most westerly appears to end on the east of Madagascar, while the second and third can be traced respectively into the Arabian Sea on the west of India and the Bay of Bengal on the east. Each of these latter tongues flows at first towards the east, but under the influence of the earth's rotation, combined in the case of the Bay of Bengal current with the obstruction to its path of the Andaman-Nicobar-Sumatra ridge, the streams develop a tendency to sweep round in a left-handed curve and in consequence their course, which is at first in a north-easterly direction, gradually veers towards the north and finally to the north-west, the eastern current sweeping across the mouth of the Bay of Bengal and impinging on the east coast of southern India and Ceylon.

In the region of the Bay of Bengal a study of the bottom temperatures indicates that the major portion of the bottom drift undergoes a further division into two subsidiary streams and the cause of this bifurcation can in all probability be found in the obstruction to its flow that is offered by the rising ground of Carpenter's

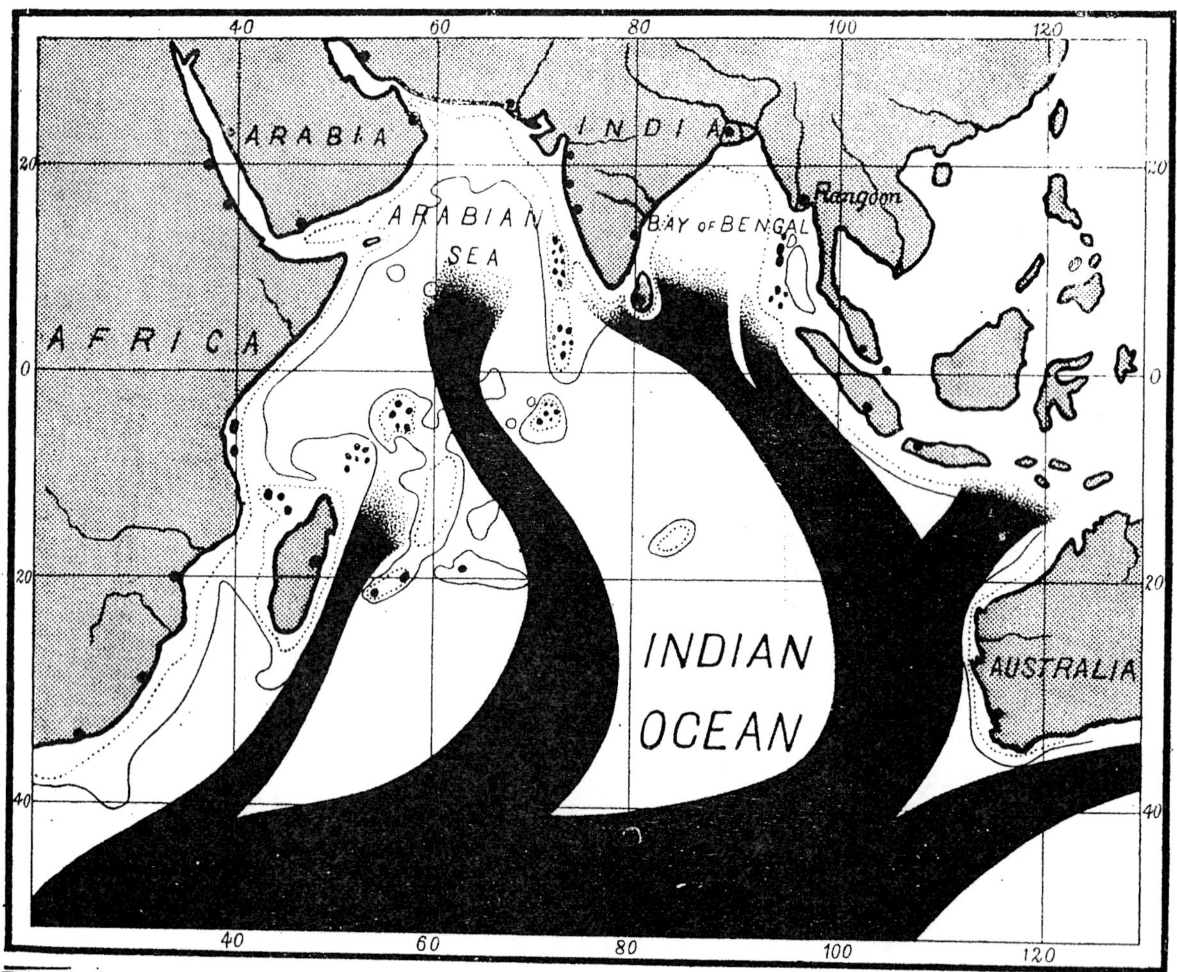

CHART VI.—Chart of the Indian Ocean showing the probable distribution of the bottom-drift of cold Antarctic water.

ridge, which runs towards the south in Long. 90°-91° E. Of the two resulting streams one passes to the northwards between Carpenter's ridge on the west and the Andaman-Nicobar ridge on the east, through the deep channel that is sometimes referred to as the 'Investigator' Deep, where water of a temperature as low as 1·0°C. can be traced on the bottom as far north as Lat. 12° N. The other current flows towards the west and meeting the obstruction of Ceylon and South India appears to again divide into two, one portion continuing westwards across the southern end of Ceylon into the Laccadive Sea, while the other bends towards the north and flows up

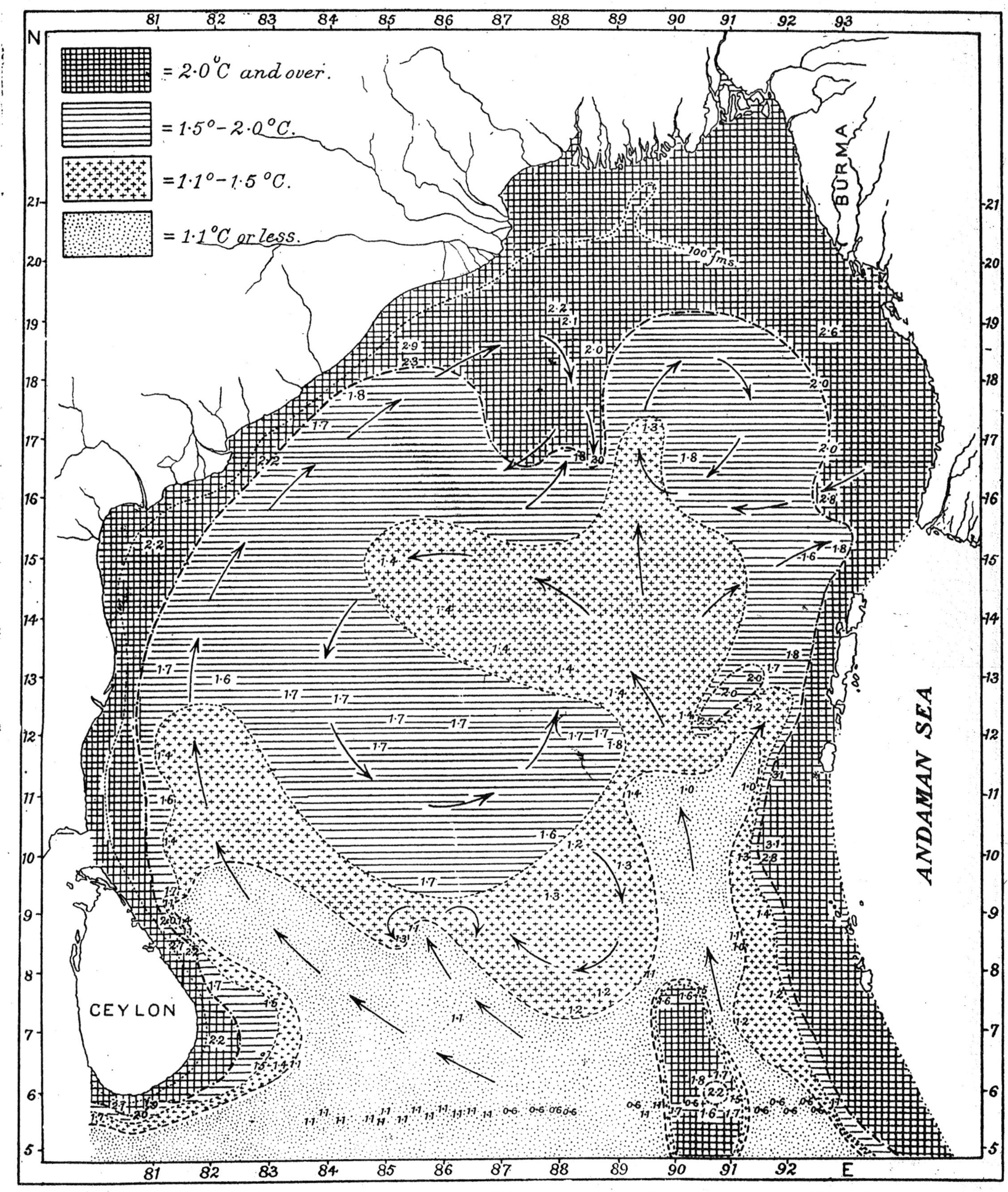

TEXT-FIG. 128.—Showing the probable course of the movement of the water masses at the bottom of the Bay of Bengal.

into the Bay of Bengal along the east coast of India, where water of a temperature of only 1·1°C. has been found at a depth of 1,990 fathoms as far north as Lat. 9° N.

In Text-figure 128, I have plotted the various observations of the bottom temperatures of the Bay of Bengal and I have endeavoured to deduce from these observations the probable course of the flow of the bottom water. It will be seen that the eastern current at first flows towards the north between Carpenter's ridge and the Andaman-Nicobar ridge till opposite the southern end of the Andaman islands; it then in the main gradually bends towards the west and establishes a rotary movement in the centre of the Bay, the centre of the circular movement lying approximately in the neighbourhood of Lat. 13° N. and Long. 86° E. The western branch of the bottom drift, meeting the coast of India and Ceylon appears to turn northward along the coast and finally bending towards the north-east, seems to be flowing straight towards the 'Swatch of no ground' at the mouth of the Gangetic Delta. As a result of the opposite directions of these two great bottom streams in the Bay, the one flowing clock-wise around the western and northern portions and the other flowing counter-clock-wise up the east side and then round the central area, a number of subsidiary vortices appear to be established in different regions of the Bay, as indicated in Text-fig. 128. As I have previously pointed out (*vide supra*, p. 47) Carpenter (1887, p. 231) many years ago insisted that there must be a submarine inflow from the southward into the Bay of Bengal, in order to compensate for the great loss of water from the surface by evaporation, and he even went so far as to suggest that 'although surface disturbance by wind would be unlikely to affect deep currents, yet it is just possible that the great climatic difference between the two monsoons may upset the balance in other ways and so alter the rate of submarine inflow'. To what extent he was correct I shall have occasion to refer to later when considering the changes that can be traced in the deep strata at different seasons of the year.

The last remnant of the most easterly branch of the bottom drift appears to sweep westward to the immediate south of Ceylon. On the west side of India we have a submarine mountain chain that extends from north to south; the more northerly portion of this chain is occupied by the Laccadive Archipelago and lies roughly between Long. 71° 30′ and 73° 30′, while the central part is crowned by the Maldives and lies slightly further to the east between Long. 72° 30′ and 73° 30′ E. Between these two regions in Lat. 8° and 9° N. there is a break in the continuity of the ridge and we get a depth of some 1,250 fathoms, forming Eight and Nine degree channels, separated by the island of Minikoi. To a large extent this ridge cuts off the area between it and peninsular India, that is known as the Laccadive Sea, from the region to the west or the Arabian Sea proper; to the south, however, the Laccadive Sea is widely open to the southern part of the Indian Ocean and in this part of the ridge the depth of water between the atolls of the Maldive Archipelago gradually increases till we reach the gap between the Maldives and the Chagos Archipelago, that crowns the southern end of the mountain chain. Between Latitudes 2° N. and 8° 30′ N. there is below a depth of some 700 fathoms an unbroken barrier, so that any deep current flowing from east to west meets this obstruction and is of necessity deflected from its course.

In Text-fig. 129, I have plotted the results of three serial observations, taken in about Lat. 12° N.; of these the most westerly series lies to the west of the Laccadive Archipelago in Long. 70° 54′ 30″ E., the second lies on the Laccadive ridge and the third is in the Laccadive Sea near the coast of India. It is seen that at a depth of 500 fathoms the temperature recorded is the same at all stations; but that in the

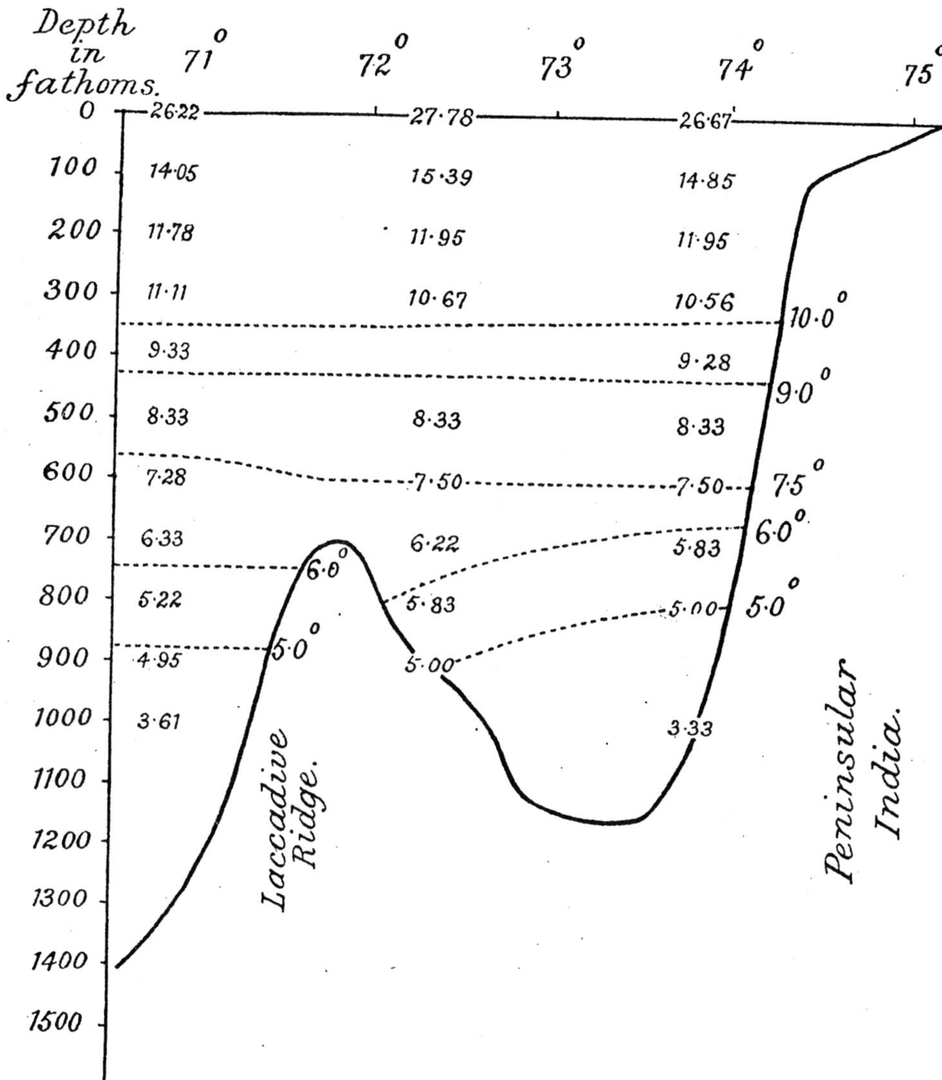

TEXT-FIG. 129.—Showing the isothermals in the eastern part of the Arabian Sea and the Laccadive Sea.

deeper levels the recorded temperatures are slightly higher in the Arabian Sea than in the eastern portion of the Laccadive Sea, thus at a depth of 1,000 fathoms in the Arabian Sea the temperature is 3·61, whereas on the east side of the Laccadive Sea at the same depth it is only 3·33°C. At the same time the isotherms in the Laccadive Sea do not run horizontally but exhibit a distinct upward trend as we pass from west to east across the channel

A temperature as low as 1·56°C. has been recorded at the entrance to the Gulf of Manaar in Lat. 7° 11′ N.; Long. 78° 47′ E. at a depth of 1,466 fathoms (2,681 metres) by the 'Investigator' in 1888-89, while temperatures of 1·67° to 1·95°C. have on several occasions been recorded around Ceylon, as shown in the table below:—

Lat. N. ° ′ ″	Long. E. ° ′ ″	Depth in fms.	Temperature °C.	Ship	Date
8 47 30	70 10 00	2,402	1·67	....	? III 1913
8 21 30	70 56 00	2,118	1·67	....	? III 1913
7 08 00	73 48 12	1,503	1·95	Egeria	1 IV 1894
9 47 30	74 19 42	1,418	1·95	Investigator	29 IV 1892
7 11 00	76 35 30	1,006	1·90	Investigator	5 I 1901
5 43 30	80 05 30	1,500	1·70	Investigator	? I 1913
5 48 15	80 56 00	870	1·90	Investigator	10 I 1901

A study of the bottom temperatures in the Laccadive Sea further indicates that at a given depth the temperature is considerably less on the east side than on the west and in Lat. 10° N., Long. 74° E. at a depth of 1,150 fathoms the water has a temperature of only 2·2°C., whereas this temperature is only reached at a much greater depth in Long. 72° E., the temperature at 1,150 fathoms being 2·89° C. There thus seems to be some evidence that the last part of the Antarctic Bottom drift after passing across the south of Ceylon is deflected by the Maldive ridge to the north and flows northwards into the Laccadive Sea and especially along its east margin.

Seasonal Changes in the Temperature of the Water at Different Levels.

It is now generally recognised that in temperate climates, in which there is a wide range of temperature between winter and summer, there is a regular seasonal variation of temperature and salinity in the upper levels of the ocean and that just as we get succeeding periods of summer and winter at the surface, so we also get corresponding seasons in the deeper levels but at opposite times of the year; in these regions we find that at a depth of about 100 fathoms (183 metres) the warm period occurs in the months of December-January, when the surface water is coldest, and the cold season in August, when the surface water is warmest. In Indian waters it is a matter of some difficulty to decide to what extent the temperatures at the 100 fathoms level at different times of the year may be due to true seasonal changes such as occur in temperate seas or to changes in the currents under the influence of the prevailing winds and the consequent depression or elevation of the isothermal lines along the windward or leeward sea coast. In the following table, I have given the data derived from the average of four serial observations taken off the south-east side of Ceylon in the months of April-May, and two observations in the near vicinity in the month of October:—

Depth in fms.	Depth in metres	(I) Lat. 5°-6° N. Long. 81°-82° E. April-May °C.	(II) 6° N. 83° E. October °C.	Difference (I) – (II) °C.
0	0	27·93	28·88	– 0·95
50	91	22·25	25·96	– 3·71
100	183	15·04	15·55	– 0·51
200	366	11·47	10·74	0·73

It is clear that the water in the month of October is warmer than it is in April-May down to a depth of over 100 fathoms and that the greatest difference is found at or near the 50 fathoms (91 metres) level.

A study of the average temperatures recorded at the bottom in the region of the east coast of India gives a very similar result, as is shown by the temperatures at different depths in the months March-April and again in December :—

Depth fms.	Depth metres	(I) March-April °C.	(II) December °C.	Difference (I)–(II) °C.
0	9	27·8	26·6	1·2
100	183	17·8	18·5	-0·7
150	274	12·7	14·2	-1·5
200	366	11·5	12·1	-0·6
250	457	10·7	10·5	0·2
300	549	10·0	9·3	0·7
400	732	9·0	7·8	1·2

The temperature of the surface-water is naturally higher in the hot-weather months of March and April than in December, which is the period of the north-east monsoon; but from 100 fathoms (183 metres) below the surface down to a depth of nearly 250 fathoms (457 metres) the water is distinctly warmer in December than at the same depth in March-April, the maximum difference occurring at a depth of 150 fathoms (274 metres), at which level it is 1.5°C. At the same time it has to be borne in mind that these differences in temperature may be due to the effect of the north-east monsoon wind in causing a depression of the isothermal strata on the west side of the Bay and a corresponding upwelling of cold water on the east side.

In the Indian region, as I have previously pointed out (*vide supra*, p. 57 *et seq.*) we have an annual double rise and fall of the air temperature, that is more marked on the west side of India than on the east and that we also find a corresponding double variation in the temperature of the surface-water (*vide supra*, p. 142 and p. 208) the two periods of maximum temperature being in April and September respectively. If, therefore, any changes are set up in the deeper levels by convection currents as the result of the heating up and consequent evaporation of the surface-water in these two warm periods, we should equally expect to find that there is evidence of a corresponding double change in the temperature of the deeper water.

The actual data at my disposal is somewhat scanty, but in the following table I have given the average temperatures at the surface and at intermediate depths down to 100 fathoms during the three three-monthly periods of the year in which I have records, namely (I) August–October, the hot weather period following on the cessation of the south-west monsoon, (II) November–January, the coldest period of the year and the period of the north-east monsoon, and (III) February–April, the period of hot weather preceding the onset of the south-west monsoon; unfortunately I have no data for the period May–July.

Depth fms.	Depth metres	(I) August–October °C	(II) November–January °C	(III) February–April °C	Difference (I)–(II) °C	Difference (II)–(III) °C
0	0	29·6	26·9	28·0	2·7	—1·1
20–40	37–73	27·1	23·7	26·0	3·4	—2·3
40–60	73–110	22·2	23·8	24·2	—1·5	—0·4
60–80	110–146	19·7	22·9	23·1	—3·2	—0·2
80–100	146–183	..	19·7	18·1	..	1·6

It thus appears that from the surface down to a depth of 40 fathoms (73 metres) the temperature is high during the period August–October; it then falls during the next period and rises again in the months February–April. Between 40 and 80 fathoms (73 and 146 metres) the temperature is low in August-October, when the surface-water is at its maximum temperature, but rises steadily during the next six months; and finally the temperature at a depth of 80–100 fathoms (146–183 metres) is clearly lower in the months August–October than it is in the next three-monthly period, since in November–January the temperature at this depth is the same as that recorded at a depth of 60–80 fathoms (110–146 metres) in August–October; is apparently at its maximum in November-January and falls again in February–April. It thus appears that in these waters, as in the temperate zone, there is a reversal of the seasons at a depth of about 100 fathoms (183 metres) and that at this depth, that is approximately the depth of the edge of the continental shelf and the commencement of the continental slope, the 'summer' occurs in November–January; and, further, that the seasonal range of temperature at a depth of 60–80 fathoms (110–146 metres) is in the neighbourhood of 3·5°C.

We now come to the consideration of the possible existence of a periodic variation in the temperature of the water in these regions at depths greater than 100 fathoms. It is generally assumed that at depths below this, that is to say below the depth at which seasonal changes, due to convection currents and produced by the alternation of the seasons, can be traced, the temperature in any locality is comparatively constant and especially so in the deep strata; but a reference to the data that I have already given regarding the temperature of the different levels on the eastern side of the Bay of Bengal and in the Andaman Sea in different months (*vide supra*, pp. 379–385) shows clearly that there may be considerable variation from month to month. At the same time one has to recognise that small differences in the records of deep temperatures may be due to (1) error in correctly estimating the depth at which the reading was taken; Carpenter (1887, p. 231) gives a very good description of such an error in a sounding taken in Lat. 19° 34′ N.; Long. 91° 97′ E., in which 'the temperature shown at a cast of 1,400 fathoms, made in 1885,....was far more suitable to a depth of 1,000 fathoms' and a subsequent sounding in the same spot gave a correct reading of 912 fathoms. Errors of less magnitude may also occur unless care is taken to see that the wire of the sounding machine is straight 'up and down' and even then they may still occur owing to the wire being affected by deep horizontal currents, though in modern times with the use of fine wire any error from this last cause will be small; or (2) error in the thermometer.

If, however, in a comparatively large series of observations we find that the average temperatures at different depths show not only a definite variation in the deeper levels at different times of the year, but also corresponding and similar variations in the deeper levels of two adjacent areas, such as the Bay of Bengal and the Andaman Sea, one is justified in assuming that these variations are definite evidence of seasonal or, at least, periodic changes.

I have been carefully through all records of temperatures taken at different times of the year in the above-mentioned areas and by taking the averages for different months I have been able to compile the following gradients :—

Depth		October-November	December	January	February	April
fms.	metres	°C.	°C.	°C.	°C.	°C.
200	366	10·9	11·0	....	....	....
300	549	9·5	9·0	9·0	....	....
400	732	8·5	7·4	7·9	9·4	8·9
500	914	7·3	6·3	6·8	8·1	7·9
600	1,097	6·5	5·4	6·1	7·3	7·0
700	1,280	5·6	4·6	5·0	7·4	6·2
800	1,463	5·0	3·8	5·0	4·9 ?	5·4
900	1,646	4·3	3·6	4·9	5·4	4·6
1,000	1,829	3·8	3·1	4·5	5·2	4·1
1,100	2,012	3·4	2·4	2·7	3·6	3·5
1,200	2,195	3·0	2·0	2·4	3·3	2·8
1,300	2,377	2·7	1·9	2·1	3·1	2·1
1,400	2,560	2·4	1·8	1·8	2·9	1·7
1,500	2,743	2·1	1·7	1·6	2·7	1·6
1,600	2,926	2·0	1·7	1·2	2·4	1·6
1,700	3,109	1·7	1·7	1·1	2·2	1·6
1,800 or over	3,292	1·1 (Oct.) 1·7 (Nov.)	1·7	1·1	1·7	1·6 to 1·3

Table 90; showing the average bottom temperature at different depths and in different months in the Bay of Bengal.

From the above figures, that are based on records of over 100 observations taken in depths ranging from 104 to 2,230 fathoms, there seems to be every reason to conclude that throughout there is a definite seasonal change, the temperature at any depth from 200 to 1,700 fathoms being comparatively high in the months of October and November; but below this, at a depth of 1,800 fathoms and over, the temperature seems to be somewhat greater in the latter month. During the succeeding month, December, the temperature of the water down to a depth of 1,400 fathoms has fallen considerably and appears to have reached its minimum, though in depths greater than this the temperature still continues to fall till January. Between January and February the temperature at all levels rises, and between February and April shows a second clear fall. There thus appears to be in the deep waters of the Bay of Bengal a double fluctuation in the temperature at all depths, having its first maximum at about the end of February and its second probably about the end of September or early in October, though I have no actual evidence of this latter; alternating with these maxima are corresponding minima in December and May or June. It is significant that these two periods of minimum temperature in the deep levels correspond closely with two of the periods in which the discontinuity zone shows an elevation of its lower limit (*vide* Text-fig. 127 p. 390). If now we make a similar analysis of the records of bottom temperature in the Andaman Sea we get the following results :—

Depth fms.	Depth metres	October-November °C.	December °C.
200	366	....	11·9
300	549	9·5	9·4
400	732	7·7	7·7
500	914	6·7	6·9
600	1,097	5·7	6·3
700	1,280	5·3	5·8
800	1,463	5·0	5·3
900	1,646	5·0	4·7
1,000 or over	1,829 or over	5·0	4·7

Depth fms.	Depth metres	January °C.	February °C.	March-April °C.
200	366	10·6	11·0	11·2
300	549	9·0	9·8	9·8
400	732	7·9	8·8	9·0
500	914	7·0	7·8	8·0
600	1,097	6·3	6·6	6·7
700	1,280	5·7	5·9	5·8
800	1,463	5·3	5·3	5·1
900	1,646	5·3	5·3	5·1
1,000 or over	1,829 or over	5·3	5·3	5·1

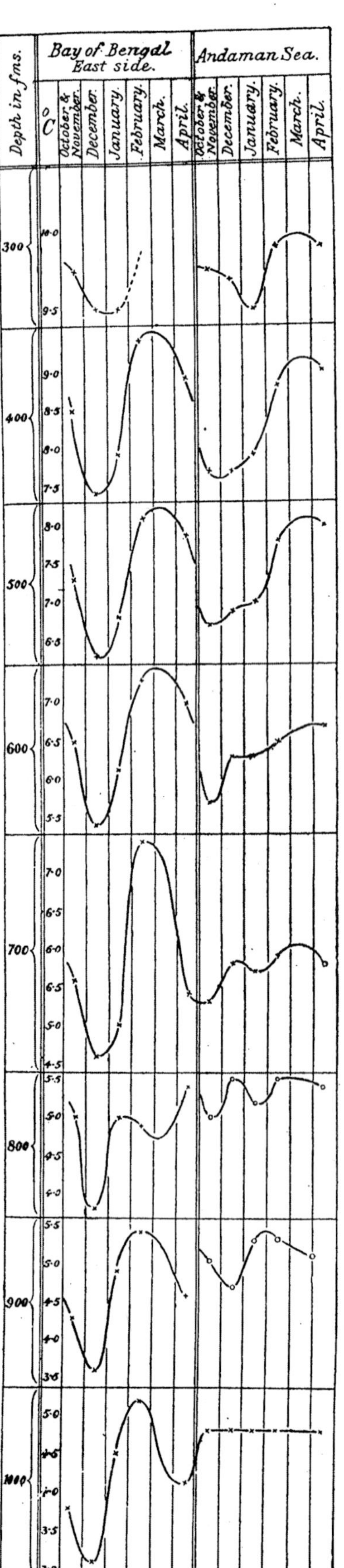

TEXT-FIG. 130.—Seasonal variation in the Temperature of the deep water, from 300–1,000 fathoms, in the Bay of Bengal and Andaman Sea.

For the sake of comparison I have plotted both these sets of temperatures in Text-fig. 130, and it seems to be clearly indicated that in both areas the temperature at any given level tends to follow the same course. At all depths in each area the rise and fall of the temperature follows an almost identical curve; but with this difference, that the changes in the Andaman Sea appear to lag behind those in the Bay of Bengal by an interval of about a month. In the Bay of Bengal the first maximum occurs at about the end of September, or the beginning of October, whereas in the Andaman Sea it is delayed till the end of October or the beginning of November; the minimum occurs in the Bay of Bengal in December but in the Andaman Sea this is not well marked and seems to occur about December or January; and the second maximum occurs in the Bay of Bengal in the end of February or early in March, but in the Andaman Sea at about the middle of this latter month. Exactly the same sequence occurs at all depths in both areas between 400 and 800 fathoms (732 and 1,463 metres) and the explanation seems to lie in changes that are going on in the Bay and are transmitted to the waters of the Andaman Sea basin. A further point to note is that the maximum variation at any one level occurs at or about the 700 fathoms depth and it seems probable that these changes are dependent on corresponding changes, either in volume or in temperature, of the gradually sinking Tropic water, that in this area of the Bay lies at about this depth (*vide supra*, p. 371).

I have already (*vide supra*, p. 369) called attention to the oscillation in the temperature of the water down to a depth of at least 220 fathoms (400 metres) that has been described by Matthews (1926, p. 190) from observations taken by the 'Valdivia' and 'Sealark' in an area extending between Lat. 6-10° S. and Long. 50-57° E. This oscillation Matthews considers to be due to 'deep currents, of which we know so little that it is useless to pursue the question further at present'. If, however, we take the same serial observations and plot them, not according to their geographical positions of Latitude and Longitude, but according to the time of year at which they were taken we get a series that, though incomplete in the months of December and January, gives us some indication of the changes that are going on in these depths during the period from October to March inclusive.

I have plotted the results in Figure 131 and have shown the probable course of the isothermal lines. These certainly appear to indicate that there is a definite oscillation in the temperature at all levels down to at least 400 metres, with maxima in the early part of November or somewhat earlier at the greater depth of 350–400 metres, and again in the latter part of February; between these two maxima there apparently occurs a minimum phase in or about the end of December or early January, but unfortunately we have no actual data for this period. A comparison of these oscillations with those shown to exist in the deep waters of the Bay of Bengal and Andaman Sea reveals a close degree of similarity. Passing from west to east we thus have three adjacent areas in each of which there is some evidence of a periodic variation in the temperature of the water at depths ranging from the surface to 500 metres (0–273 fathoms) or more on the west and from 549 to 1,829 metres (300 to 1,000 fathoms) on the east, and further, that this variation occurs at a progressively later time of year as one proceeds eastward. It seems probable that these changes are due to seasonal alterations in either the temperature or volume, or both, of the Tropic water that gradually sinks in the more northerly region of the Arabian Sea and flows east and south-east to form the north Indian deep stratum. The rise of temperature of the tropic water at depths of 549 to 1,829 metres (300 to 1,000 fathoms) seems to correspond closely with the rise and fall of that of the surface water, as shown in Text-fig. 37 (*vide* p. 211), but as it is clear that some considerable time

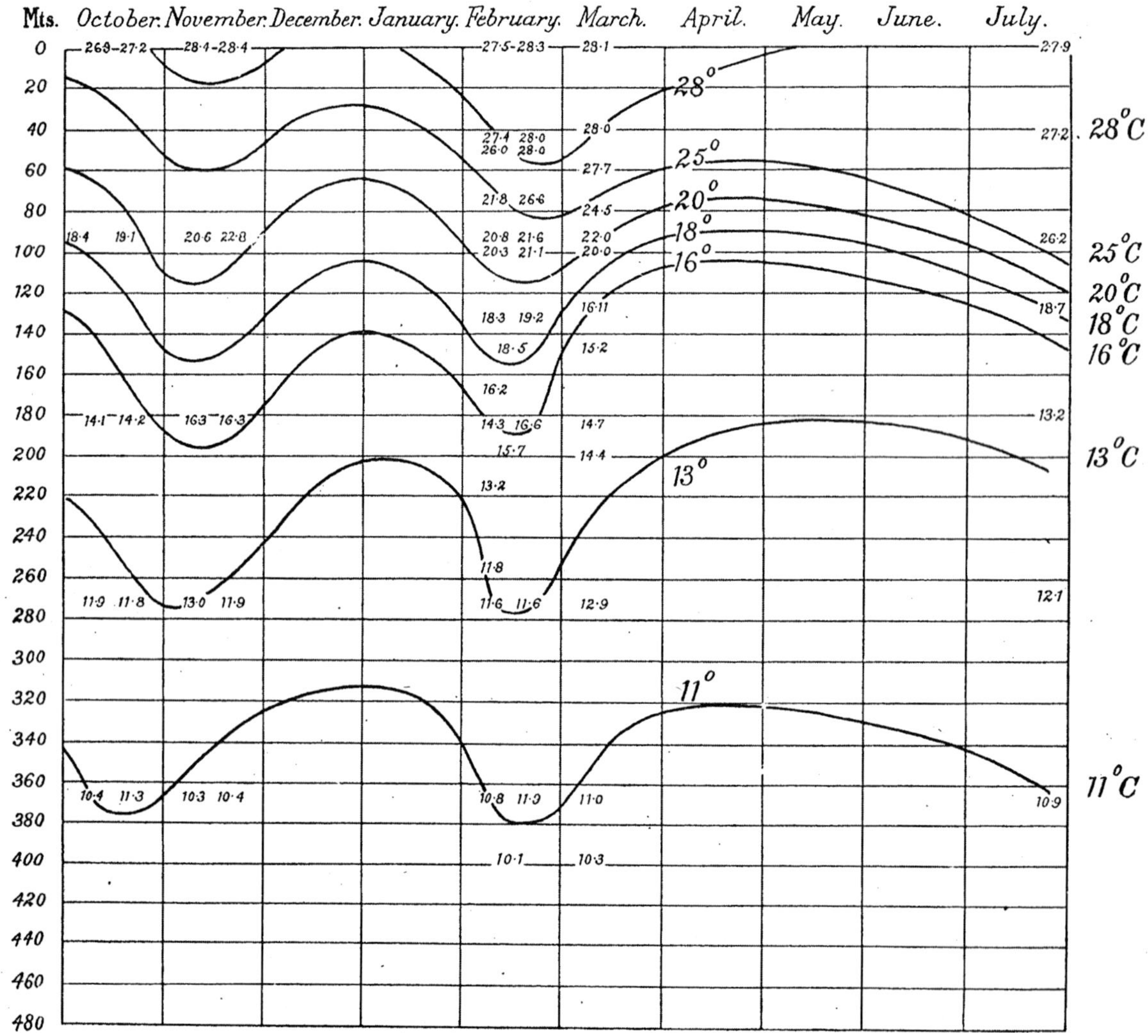

TEXT-FIG. 131.—Evidence of seasonal variation in the Temperature of the deep water of the Arabian Sea and the neighbouring part of the Indian Ocean.

must have elapsed since the water sank below the surface and made its appearance at these great depths, it is equally certain that these changes in the deep levels cannot be attributed to the concurrent surface variation but must be due to changes that have taken place in the surface waters some months previously.

The Bottom Temperature of the Andaman Sea Basin.

Krümmel (1911, p. 494) has pointed out that in an enclosed basin such as the Andaman Sea, if the mean winter temperature over the basin is lower than the temperature of the water of the open ocean at the level of the deepest entrance channel, then the basin will be full of water having this winter temperature; but that if the mean winter temperature is higher than the temperature of the water of the ocean at the level of the channel, then the water of the basin will have the temperature of the water of the open ocean at this depth. It is clear that in tropical regions, such as these with which we are dealing, an enclosed basin should follow the second of these alternatives.

As Krümmel (1907, p. 479) has pointed out, there is a quite considerable range of variation in the temperatures that have been recorded in the deeper waters of the Andaman Sea basin, that is, from depths greater than that of the deepest channel through which the waters of the basin can communicate with those of the Bay of Bengal; he, however, pays but little attention to these variations and assumes that throughout the basin the temperature is uniform and in the vicinity of 4·75°C.

Schott (1902, p. 176) gives the temperature of this basin as 5·4°C. between 1,270 and 1,500 metres (694 to 820 fathoms) and 5·2°C. below this depth down to the bottom. Alcock (1898, p. 9) has, however, recorded a temperature at a depth of 1,159 fathoms (2,120 metres) as low as 4·2°C. There is thus evidence of a considerable range of variation in the deep temperature of this area. One of the characteristics of inland basins such as this is the uniformity of the temperature of the water below the level of the deepest entrance channel and while recognising that the Andaman Sea is probably no exception to this general rule, it is, I think, worth while to attempt to find some explanation of the variations that have been observed, since these variations are in my opinion too great, and, as we shall see, too consistent to be due to mere accident or to careless observation. Any marked variation in the temperature at or near the bottom of such a basin may be attributed to one or other, or possibly to both, of the following causes, namely (1) to a change in the temperature of the water outside the basin or (2) to a heating up of the water through some internal agency in the basin itself. We have already seen that the water of the Bay of Bengal does show a periodic variation in its temperature and this may be sufficient to account for the observed variation in the water of the Andaman Sea; but a further analysis of the data at my disposal seems to indicate the possibility that the second factor may also be concerned.

It is now generally accepted that the deeper water of an inclosed sea must be in a constant state of circulation; only by assuming this, even if no actual proof were forthcoming, can one explain the freedom from stagnation and the presence of living forms at the bottom of such areas, for both these conditions require the presence of oxygen in solution and its constant renewal, together with the removal of carbon dioxide gas. The cause of this circulation is believed to be the heating up of the bottom layers of water, where they are in contact with the earth's surface, and the consequent convection currents that rise upwards in the centre of the basin, countercurrents in the downward direction being produced round the margin. If it be admitted that the ordinary sea bottom is capable of producing this effect and of causing a rise in the temperature of the bottom water, it seems only reasonable to suppose that in an area that is known to be volcanic, and, moreover, still to some degree active, such a heating process on the bottom might be more marked than usual or than in areas that are non-volcanic.

I have previously shown (*vide supra*, p. 9 *et seq.*) that the Andaman Sea is traversed by a volcanic chain that forms a wide sweep and in the southern part of its length is inextricably blended with the non-volcanic continuation of the Aracan Yoma, that forms the main mass of the Andamans and the western part of the Nicobar ridge. The volcanic range, of which Barren Island and Narcondam are the only peaks that rise above sea-level, is still sufficiently active to maintain a hot spring on Barren Island and within recent times there must have been an actual eruption in that part of the range that runs down the east side of the Nicobars, for, as I have already mentioned (*vide supra*, p. 11, footnote) in Lat. 8° 32′ N.; Long. 94° 10′ E. large lumps of Olivine Basalt, that had only very recently been erupted, were brought up in the trawl; the depth from which these masses were obtained was 1,260 fathoms (2,304 metres) and it seems probable that there had very recently been an active submarine eruption that must have had a profound effect on the local temperature of the water. Further to the south the continuation of this volcanic range becomes extremely active in Java and Sumatra and it does not seem beyond the bounds of possibility that this active volcanic area may have a general effect, quite apart from any local effect due to isolated eruptions, in heating up the deeper waters of the Andaman Sea basin.

Assuming that volcanic activity may play a part in causing variations in the temperature of the deep waters of such a basin, this heating up of the bottom stratum might be either local or general throughout the whole length of the range; and in this respect certain observations made by the 'Investigator' are of some interest. In 1897, in Lat. 12° 06′ 30′ N.; Long. 93° 32′ 42″ E., at a depth of 513 fathoms (936 metres) a temperature has been recorded as high as 12·22°C., or approximately 5·0°C. higher than one would expect; this may, of course, be due to error, caused by the thermometer failing to act properly, but on the other hand the sample of the sea-bottom obtained at the same time was also very peculiar; it is described as consisting of 'hard white clay', such a deposit is known from no other locality in the whole basin and the neighbouring deposits are all either green or blue mud. This station lies about mid way between Barren Island and Ritchie's Archipelago to the east of the Andaman islands and due north of Invisible Bank, that is in all probability part of the volcanic range. There is also a small area lying between Lat. 13° 13′ and 13° 35′ N. and Long. 93° 14′ and 93° 44′ E., in which on no less than six occasions at depths ranging from 405 to 1,068 fathoms (822 to 1,954 metres) temperatures have been recorded that were distinctly higher than what one would consider normal for such a depth. The details of these observations are as follows:

Year	Lat. N.	Long. E.	Depth in fathoms	Observed temperature	Normal temperature	Diff.
	° ′ ″	° ′ ″				
1890 ..	13 21 00	93 27 00	922	5·11	4·75	0·36
1890 ..	13 34 15	93 17 15	538	7·17	6·80	0·37
1896 ..	13 27 00	93 14 30	405	8·89	8·15	0·74
1897 ..	13 15 30	93 26 00	498	8·20	7·20	1·00
1903 ..	13 13 30	93 44 00	1,068	5·00	4·75	0·25
1903 ..	13 26 30	93 44 00	935	5·00	4·75	0·25

This area lies about 50 miles to the west of the island of Narcondam, itself an old volcano, and about 10 miles to the east of the station in Lat. 13° 17′ N.; Long. 93° 07′ E., where a number of phosphatic nodules were brought up by the dredge. Tipper (1911, p. 16, footnote) who reported on these nodules found that the phosphatic deposit was surrounding and enclosing the remains of fish; and he suggests that these fish had been killed by volcanic activity, namely, 'by an eruption of the volcanoes, Barren Island and Narcondam, the former of which was active in quite recent times'. It is, however, possible that the centre of this volcanic activity was submarine and lay in the near vicinity of the deposit, namely in the area noted above in which the abnormally high temperature-readings were obtained.

A study of the temperatures recorded from the deeper waters of the Andaman Sea seems to indicate that at depths greater than 800 fathoms (1,463 metres) and thus below the level of the deepest entrance channel, the temperature tends to vary with the proximity to or distance from this volcanic range. The average bottom temperature recorded along the line of the ridge or in its near vicinity is 5·43°C.; it is possible that this average is somewhat too high since in several of the records readings appear to have been taken only to the nearest degree in the Fahrenheit scale, but even after allowing for this and taking all such readings at their lowest possible value, namely half a degree lower than that stated, the average still reaches 5·25°C. In contrast to this the average temperature recorded from stations remote from the ridge gives a figure of only 4·95°C. There thus appears to be an appreciable difference in the temperatures recorded in volcanic and non-volcanic regions of the basin. If this difference be due to the volcanic activity of the range we ought to be able to trace some relationship between the average temperatures in different years and the known activity of the range; in the following table I have grouped together in one series all records from areas that may be suspected to be volcanic and in the second those that have been obtained in regions where no such activity is suspected:

Year	1888-89	1890	1891	1897	1898	1902	1903	1910	1913	1915
Average in volcanic areas	..	5·11°	5·28°	5·25°	5·14°	4·94°	5·00°	....	5·28°	5·28°
Average in non-volcanic areas	4·75°	....	....	....	5·00°	4·78°	4·72°	4·67°	....	....

There seems good reason to suppose that the temperature of the bottom water in the volcanic region is undergoing a periodic rise and fall, a maximum being reached

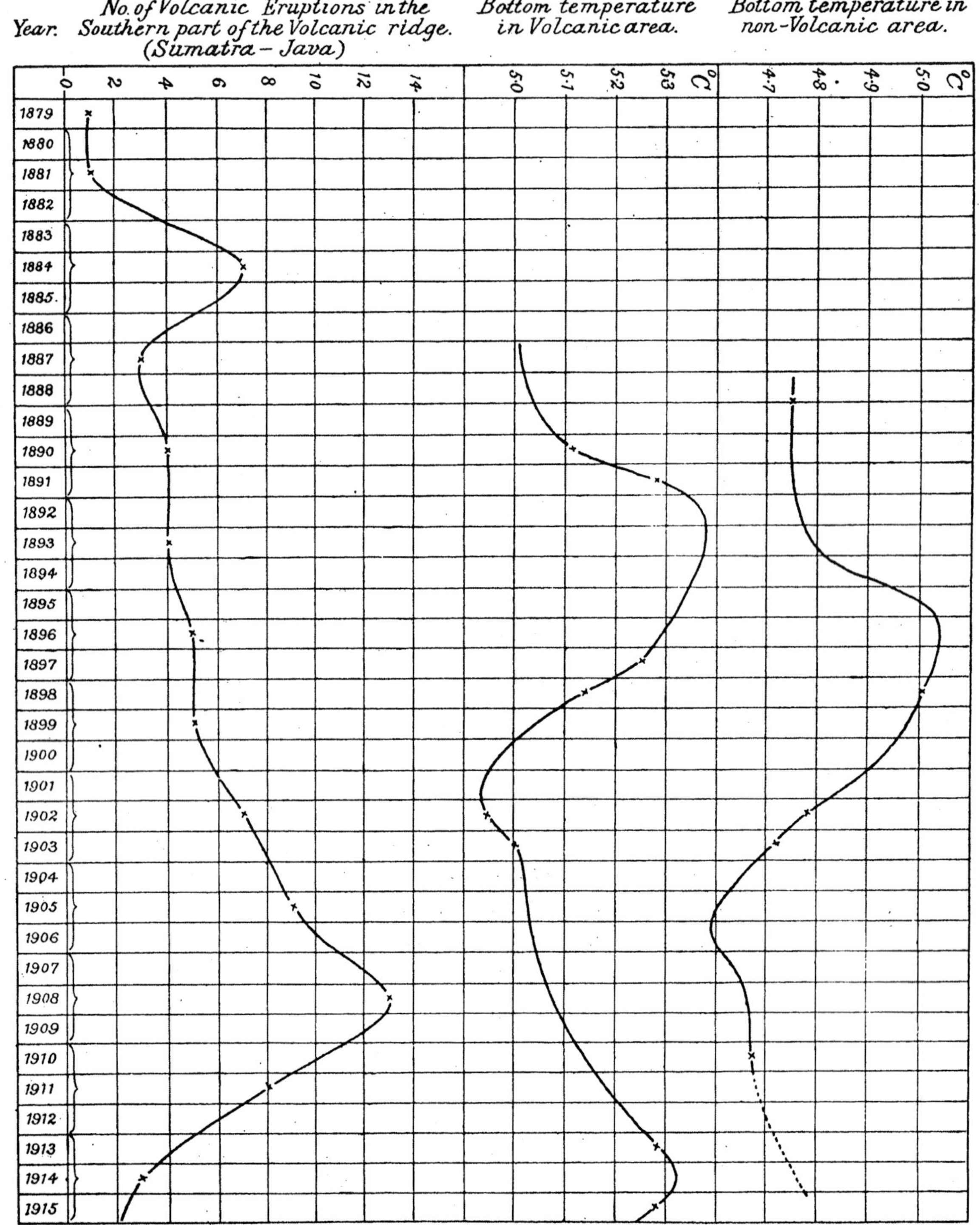

TEXT-FIG. 132.—Showing the degree of volcanic activity in the Sumatra-Java area, and the variation in the bottom temperatures in the Andaman Sea in both volcanic and non-volcanic regions.

about 1891 and again about 1913–15, with a minimum in 1902. In that part of the volcanic range that runs through the Andaman Sea there is now no definitely active volcano, the eruptions of which might be taken as an index of the activity of the whole range; but to the south in Sumatra, Java and Flores there are a number of volcanoes that are frequently in a state of eruption and in an Appendix I have given the years in which these volcanoes have been known to be active within recent times. It is clear that the activity of the range has its main centre in the island of Java, and that this activity gets steadily less and less as we follow the chain towards the north, till we reach Barren island, where there has been no actual eruption since 1852 but where there is still a hot spring. It is now well recognised that there is a distinct periodicity in volcanic outbursts and in Text-fig. 132 I have compiled a graph based on the actual number of volcanic eruptions in the range under consideration for periods of every three years commencing with the three-year period 1877–79 and ending with the period 1913–15. From this it is clear that there was a comparatively small, though clear, increase in volcanic activity during the period 1883–85; this was followed by a decrease in the number of eruptions between 1886 and 1894, after which there was a steady rise until we get the maximum number in the period 1907–09; and finally there was again a decrease lasting at least till 1916–18, beyond which date my data do not extend. In the same text-figure I have plotted the variations in the average temperature of the bottom water in the two areas, volcanic and non-volcanic, of the Andaman Sea basin. The data seems to indicate that these temperatures follow the same general rise and fall but at some considerable time later. The similarity of the curves suggests that the two phenomena are related, and two possible explanations present themselves. In the first it seems possible that the volcanic activity in the region of the island of Java, which involves a great increase in the temperature of the range in that region, gradually causes by conduction along the range a slow but gradual heating up of those portions of the range to the north in the neighbourhood of Barren island, which eventually makes its presence felt by a heating up of the bottom water some ten years later. The water in the bottom regions of the non-volcanic parts of the basin does not show the corresponding rise till some three years later still, but this is, I think, only to be expected; for assuming that in certain areas there is a slow but steady heating up of the bottom water from some such source as a volcanic region, the actual elevation of the temperature will be but small since it will at once set up convection currents in the whole mass, and since the bottom circulation must be comparatively slow, some considerable time, possibly several years, must elapse before this local heating can make its presence felt in the more remote parts of the basin. If this argument is a sound one and the heating up of the water along the volcanic parts of the range is due to increased volcanic activity many hundreds of miles away, one ought to find that there is some agreement, at least, between the interval of time known to elapse between the increased activity in Java and the increased bottom temperature in the region of Barren island on the one hand and the rate at which a rise of temperature will be conducted along a mass of basalt of this length on the other. In this connection I have to acknowledge with thanks the help of Dr. A. M. Heron of the Geological Survey of India, who has called my attention to a paper by Poole (1914), which deals with experiments on

Basalt and Graphite that were designed and carried out to test the conductivity of these substances. Poole (1914) has shown that for temperatures up to 500° or 600°C. the conductivity of the earth's crust may be taken as about 4×10^{-3} without any risk of making a serious error, unless the conductivity is sensibly affected by the large pressure involved. According to Poole the heat flux (Q) in calories per centimeter per second can be calculated from the formula

$$Q = \frac{W}{10 \cdot 14 \times 4 \cdot 18}$$

where W is the product Current × P.D. and 10·14 is the length of the experimental cylinder of basalt. For such a cylinder the value of Q works out to 0·12 calories per cm. per second. The approximate distance between the central area of the volcanic region of Java and Barren island in the Andaman Sea is 1,790 miles, so that if heat can be transmitted along such a volcanic ridge for this distance and if the rate of flux of temperature was at a rate of 1 cm. per second, it would take approximately 9·1 years for the heating up process to affect the region round Barren island. In Text-fig. 132 it will be seen that the first maximum of volcanic activity occurred in 1884 and the corresponding maximum of the bottom temperature in the volcanic region of the Andaman Sea is in or near 1893, an interval of nine years. Dr. J. R. Cotter of the Geological Department, Trinity College, Dublin, to whom I submitted the above possible explanation, does not think it possible that a periodic variation in temperature in Java could reproduce itself at any considerable distance; but he suggests, as an alternative explanation, that there is periodic activity all along the ridge, in which the maxima at successive points occur at successive intervals of time; "Suppose that periodic volcanic activity is due to a gradually increasing stress which finally finds relief in an eruption. This relief of stress throws extra stress on neighbouring regions, and, if there is a line of weakness, the next rupture will probably take place a little further along that line, and the next a little further still. I do not think that the fact that some of the islands are volcanoes which have long been extinct really invalidates this argument. The weak spots may change in course of time, while still remaining on the old fault".

Appendix

A list of eruptions in the volcanoes of the Andaman-Sumatra-Java Range.

(1) *Barren Island* .. .. .. 1852.

(2) *Sumatra.*

Indrapura .. .. .. 1838, 1842.

Goenoeng Dempo .. .. .. 1895, 1908.

Kaba .. .. .. 1875.

Korintji .. .. .. 1908.

Merapi .. .. .. 1861, 1845, 1854, 1855.

(3) *Krakatoa* .. .. .. 1883.

(4) *Java.*

Brama .. .. .. 1906, 1907, 1908, 1909, 1910.

Gedeh .. .. .. 1899, 1909.

Gulung Gung .. .. .. 1868, 1894, 1918.

Guntur .. .. .. 1836, 1847.

Kelut .. .. .. 1848, 1864, 1875, 1891, 1901.

Lamongan .. .. .. 1842, 1847, 1849, 1859, 1869, 1875, 1877, 1883, 1885, 1896, 1898.

Merapi .. .. .. 1846, 1849, 1863, 1865, 1872, 1883, 1894, 1902, 1903, 1904, 1906, 1907, 1909, 1911, 1912.

Merbabu .. .. 1863, 1872.

Pangrano .. .. .. 1840, 1843, 1845, 1847, 1852, 1886.

Raoun .. .. .. 1885, 1903, 1913, 1914.

Semeru .. .. .. 1842, 1844, 1848, 1851, 1857, 1865, 1885, 1886, 1887, 1889, 1890, 1891, 1893, 1895, 1896, 1898, 1899, 1900, 1901, 1903, 1905, 1907, 1908, 1910, 1911.

Sendorol .. .. .. 1882, 1903, 1904, 1906.

Slamat .. .. .. 1845, 1847, 1849, 1904.

Tangkoeban Prahoe .. .. 1896, 1910, 1913.

Tengger .. .. .. 1838, 1841, 1842, 1843, 1844, 1857, 1862, 1868, 1893.

(5) *Bali.*

Batoer .. .. .. 1904, 1905.

(6) *Sumbaya.*

Goenoeng Api .. .. .. 1911.

(7) *Flores.*

Lobetobi .. .. .. 1907, 1909, 1910.

Egon .. .. .. 1907.

The above list has been compiled from the following papers; Malet (1895); Bogolepow (1910); Sapper (1915–18); van Es and Taverne (1924).

Bibliography

Alcock, A., 1898 .. .. 'A summary of the Deep-Sea Zoological work of the Royal Indian Marine Survey Ship Investigator from 1884 to 1897.' *Sci. Mem. Med. Officers Army India.* Part XI. Calcutta.

Bogolepow, M., 1910 .. .. 'Die Periodicität der vulkanischen Erscheinungen auf der Erde.' *Bull. Soc. Imper. des Naturalistes de Moscou.* Vol. XXIII (New Series).

Brennecke, W. and Schott, G., 1922 'Die Atlantische Vertikal Zirculation.' *Zeitschrift der Gesellschaft fur Erdkunde zu Berlin*, No. 7–10.

Brouwer, H. A. .. .. 'Sur les Domes Volcaniques des Volcans actifs de L' Archipel Malais.' *Zeitschrift fur Vulkanologie*, Vol. VI, Pt. 2.

Carpenter, ?A., 1887 .. .. 'Natural History Notes from H.M.'s Indian Marine Survey Steamer 'Investigator'. No. 8. 'The mean Temperature of the Deep Waters of the Bay of Bengal.' *Jour. Asiatic Soc. Bengal*, Vol. LVI, Pt. II.

Defant, ?A., 1928 .. 'Die systematische Erforschung des Weltmeeres.' *Zeitschrift der Gesellschaft fur Erdkunde zu Berlin.* 1828–1928. Sonderband zur hundertjahrfeier der Gesellschaft.

Drygalski, E. V., 1926 .. .. 'Bemerkungen uber die Tiefenströme der Ozean und ihre Beziehungen zur Antarktis.' *Sitzungsberichte Akad. Wissenschaften Munchen.*

Drygalski, E. V., 1926 (*a*) .. 'Ozean und Antarktis. Meeres kundliche forschungen und Ergebnisse der Deutschen Sud-polar—Expedition 1901, 1903.' Vol. VII, Pt. 5.

Gardiner, J. Stanley, 1903 .. 'The Fauna and Geography of the Maldive and Laccadive Archipelagoes.' Vol. I. Cambridge.

Gardiner, J. Stanley, 1907–9 .. 'The Percy Sladen Trust Expedition to the Indian Ocean in 1905.' Vol. I. *Trans. Linn. Soc. London* (2nd series), Vol. XII. Zoology.

van Es, L. J. C., and Taverne, N. J. M., 1924 .. .. 'De Galoenggoeng en Telega Bodas.' *Vulkanologische Mededeelingen*, No. VI.

Krümmel, O., 1907 .. .. 'Handbuch der Ozeanographie,' Vol. I. Stuttgart.

Krümmel, O., 1911 .. .. 'Handbuch der Ozeanographie,' Vol. II. Stuttgart.

Mallet, F. R.? 1895 .. .. 'Bibliography of Barren Island and Narcondam from 1881–1894.' *Rec. Geol. Survey India*, Vol. XXVIII.

Matthews, D. J., 1926 .. The Percy Sladen Trust Expedition to the Indian Ocean in 1905. No. VII; 'Physical Oceanography'. *Trans. Linn. Soc. London* (2nd series), Vol. XIX.

Merz, A., 1922 .. .. 'Temperatureschichtung und vertical-zirculation im Sud-Atlantischen Ozean nach den 'Challenger' und 'Gazelle' Beobachtungen.' *Zeitschrift der Gesellschaft fur Erdkunde zu Berlin*, No. 7–10.

Merz, A., and Wust, G., 1922 .. 'Die Atlantische Vertical-Zirculation.' *Zeitschrift der Gesellschaft fur Erdkunde zu Berlin*, No. 1-2.

Michaelis ? G., 1923 .. .. 'Die Wasserbewegung an der Oberflache des Indischen Ozean im Januar und Juli.'
Veröffentlichungen des Instituts für Meereskunde an der Universität Berlin. Geographisch Naturwissenschaftliche Reihe, Pt. 8.

Moller, L., 1929 .. .. 'Die Zirculation des Indischen Ozean.'
Veröffentlichungen des Instituts für Meereskunde an der Universität Berlin. Geographisch Naturwissenschaftliche Reihe, Pt. 21.

Molengraaf, G. A. F., 1921 .. 'Modern Deep-Sea Research in the East Indian Archipelago.'
The Geographical Journal, Vol, LVII, No. 2. London.

Oldham, C. F., 1895 .. .. Natural History Notes from H.M. Indian Marine Survey Steamer 'Investigator', Series II. No. 18. 'The Topography of the Arabian Sea in the Neighbourhood of the Laccadives.'
Jour. Asiatic Soc. Bengal, Vol. LXIV, No. 1.

Poole, H. H., 1914 .. .. 'On the Thermal Conductivity and Specific Heat of Granite and Basalt at High Temperatures.'
The London, Edinburgh and Dublin Philosophical Magazine and Journal of Science. 6th Series. Vol. XXVII.

Ringer, W. E., 1922 .. .. 'Temperatuur, Zoutgehalte, Dichtheid en Gasgehalte van het Zeewater.'
De Zeen van Nederlandsch Oost-Indie. Leyden.

Sapper, K., 1915–18 .. .. 'Bericht uber die vulkanischen Ereignisse der Jahre,' 1895–1913.
Gerland's Beitrage zur Geophysik. Vol. XIV.

Schott, G., 1902 .. .. 'Oceanographic und Maritime Meteorology.'
Wissenschaftliche Ergebnisse de Deutschen Tiefsee-Expedition auf dem Dampfer 'Valdivia', 1898-1899. Vol. I.

Schott, G., 1926 .. .. 'Die Tiefwasserbewegungen des Indischen Ozeans.'
Ann. der Hydrographie, Vol. LIV, Pt. 12.

Tipper, G. H., 1911 .. .. 'The Geology of the Andaman Islands, with references to the Nicobars.'
Mem. Geol. Survey India, Vol. XXXV, Pt. 4.

Willimzik, M., 1929 .. .. 'Die Strömungen im subtropischen Konvergenzgebiet des Indischen Ozeans.'
Veröffentlichungen des Instituts für Meereskunde an der Universität Berlin. Geographisch Naturwissenschaftliche Reihe. Pt. 14.

VII. THE TOPOGRAPHY AND BOTTOM DEPOSITS OF THE LACCADIVE SEA.*

In the following paper I have attempted to bring together from a variety of sources, including my own observations, the available information regarding the mode of origin and subsequent history of the present character of the west coast of India and of the great chain of islands and atolls that lies to the west and south-west of the Indian peninsula and constitutes the Laccadive and Maldive Archipelagoes. I have also given at the end of the paper a few notes regarding the distribution and general characters of the deep-sea deposits in this region that is termed the Laccadive Sea.

The Laccadive Region.

The Laccadive Archipelago for the most part arises from a single basic plateau, that over the greater part of its extent lies at a depth of about 1,000 fathoms, though at its southern end this increases to 1,100 fathoms. At this south end there are not sufficient soundings to enable one to be certain of the exact run of the contour lines; thus Agassiz regards Suheli Par as being connected with Kalpeni, while he shows Kuvaratti as arising from a separate bank, while Oldham, whose chart I have in the main followed, considers that Kuvaratti and Suheli Par together mark a south-westerly extension of the main plateau and that Kalpeni arises separately. Along the east side of the Archipelago the contours of the sea-bottom are somewhat complicated in the accompanying chart (Chart VII), which I have slightly modified from that given by Oldham (C.F., 1895, pl. I).

The whole of the Laccadive group is seen to be enclosed, and at its north easterly corner connected with the mass of continental India, by the 1,100 fathom line, which Alcock (1902) in his chart of the Indian Seas terms Oldham's Line. Oldham in his original chart showed the presence of an isolated basin, having a depth of over 1,100 fathoms, lying between the Laccadives and the Indian coast; according to him Androth is situated at the extremity of a submarine peninsula that is supposed to jut out from the continental slope of India towards the south-west. He remarks (1895, p. 2) 'in Latitude 11° N. a ridge, on which are situated the Elicalpeni reef and Androth island, projects for fifty miles to the south-west'. If this were indeed the case we should have between the Laccadives and India an isolated basin, cut off both to the north and the south-west by a raised area with a depth of only 1,035 fathoms, and having along its west and southern sides a depth of over 1,150 fathoms.

---

* This paper was written, and received for publication by the Asiatic Society of Bengal, before its Author left India in April, 1933, and prior to his appointment as Leader of the John Murray Oceanographic Expedition.

November, 1934. *General Secretary, A.S.B.*

The deep water of such a basin, if it existed, should show the persistence at all depths below that of its deepest entrance passage of a uniform temperature, that, also, should agree with the temperature present in the water outside the basin at the depth of the connecting channel. In the table below I have given a series of temperature observations in the supposed basin, in the other part of the Laccadive Sea and in the area to the west of the Laccadive Archipelago.

Depth in fathoms.	Supposed basin.	Laccadive Sea between the Maldives and Ceylon.	The Arabian Sea to the west of the Laccadives.
	°C.	°C.	°C.
1,104	..	2·83	..
1,109	..	..	3·22
1,117	2·78	..	..
1,120	..	..	2·72
1,122	2·95	..	..
1,126	2·78	..	..
1,128	..	..	3·05
1,130	2·61	..	..
1,131	..	2·78	..
1,133	..	..	3·22
1,138	2·78	..	..
1,139	3·05	..	3·05
1,140	..	..	2·89
1,145	2·78	..	..
1,152	3·05	..	..
1,154	2·78	..	..
1,155	2·78	..	..
1,160	..	2·20	..
1,161	..	..	2·89
1,179	..	2·22	..

It will be noted that there is, in the records given above, a certain degree of variation in all three sets of observations, and that the temperatures recorded in both the supposed basin and the southern part of the Laccadive Sea are slightly colder than those obtained at corresponding depths in the open waters of the Arabian Sea; there is not sufficient evidence of any great difference between the two areas in the Laccadive Sea and I conclude, therefore, that such an enclosed area, as Oldham postulated, does not exist and that the Elicalpeni Bank is separated from peninsular India by a deep and narrow channel and forms an integral part of the Laccadive Archipelago. It is extremely interesting to note that the position of this bank corresponds very closely with the eastward trend of the hill ranges of the peninsula, where the Western Ghats turn round to be continued into the Eastern Ghats, and it is possible that this feature may have influenced Oldham in reaching the conclusion mentioned above.

The Maldive Region.

Stanley Gardiner has described the Maldives as being situated on a plateau, the depth of which from the few soundings that he made he estimated to be about 200 fathoms; he remarked (1903, p. 153) that 'the trend of the Maldives is such from Kolumadulu to Ihavandifolu that it is almost certain that all must lie on the same plateau' and subsequently he stated that 'it is unlikely that the banks to the south and north (of Kardiva Channel) are separated by any depth greatly in excess of that found in the central basin to the south or in the channels between the atolls. . . . . The various banks then arise as so many plateau from a common plateau, which has a general depth of about 200 fathoms.' Agassiz (1903, p. 9), however, dissented from this view and by his more numerous soundings showed that 'the main chain of the Maldives does not lie alone upon a relatively shallow plateau at a depth of 200 fathoms', but that, like the Laccadives, the atolls are situated on a series of plateau separated from one another by channels of very different depth, many of them being considerably greater than 300 fathoms; a study of Text-fig. 135 (*vide infra*, p. 438) shows clearly that the depth of the intervening channels tends to become very much greater at the two ends of the Archipelago than it is in the middle of the area. Excluding the most southerly peaks on which stand the atoll of Addu and the island Fua-mulaku, we have to go as deep as 1,200 fathoms to find a line that will include the remainder of the Maldive group, or if we still further exclude Minikoi to the north and Haddūmatti to the south, the 800-fathom line will encircle the remainder. It thus appears that we have two basic plateau, for the Laccadive and Maldive archipelagoes respectively, that lie at an average depth of, 1,100 fathoms, the actual depth gradually increasing from 1,000 fathoms at the northern end to 1,200 fathoms at the southern, while the 1,300-fathom line encloses a continuous foundation that at its northern end becomes continuous with the peninsula of India and on either side of which at the southern end there is water of well over 2,000 fathoms.

The General Topography of the West Coast of India.

All along the west coast of India, as around all other continents, there is a well marked continental shelf, extending outwards from the shore line to a depth that varies somewhat in different regions but on the average is about 100 fathoms. By many scientists this continental shelf is regarded as an integral part of the land itself that has become submerged, either by depression of the land area or by elevation of the sea-level. Others, however, consider that the shelf has been built up by the deposition of debris and detritus, derived from the erosion of the continental surface, around the margin of the land area. We shall see that there is a very considerable mass of evidence that points to the occurrence in the past of extensive faulting all along the west coast of India and thus it is possible that the continental shelf may be a submerged part of the Indian peninsula; but on the other hand a study of the shelf in different regions of this west coast (Text-fig. 133) shows certain interesting differences that can be accounted for by differences in oceanic conditions and thus are attributable to variation in the rate and extent of deposition of sand and silt. Along the west coast of India, extending from Karachi in the north to Cape Comorin in the south is a typical shelf that has been termed the 'Bombay shelf'. Krümmel (1907, p. 113) describes this shelf as extending from Karachi to Latitude 12° N. and having an area

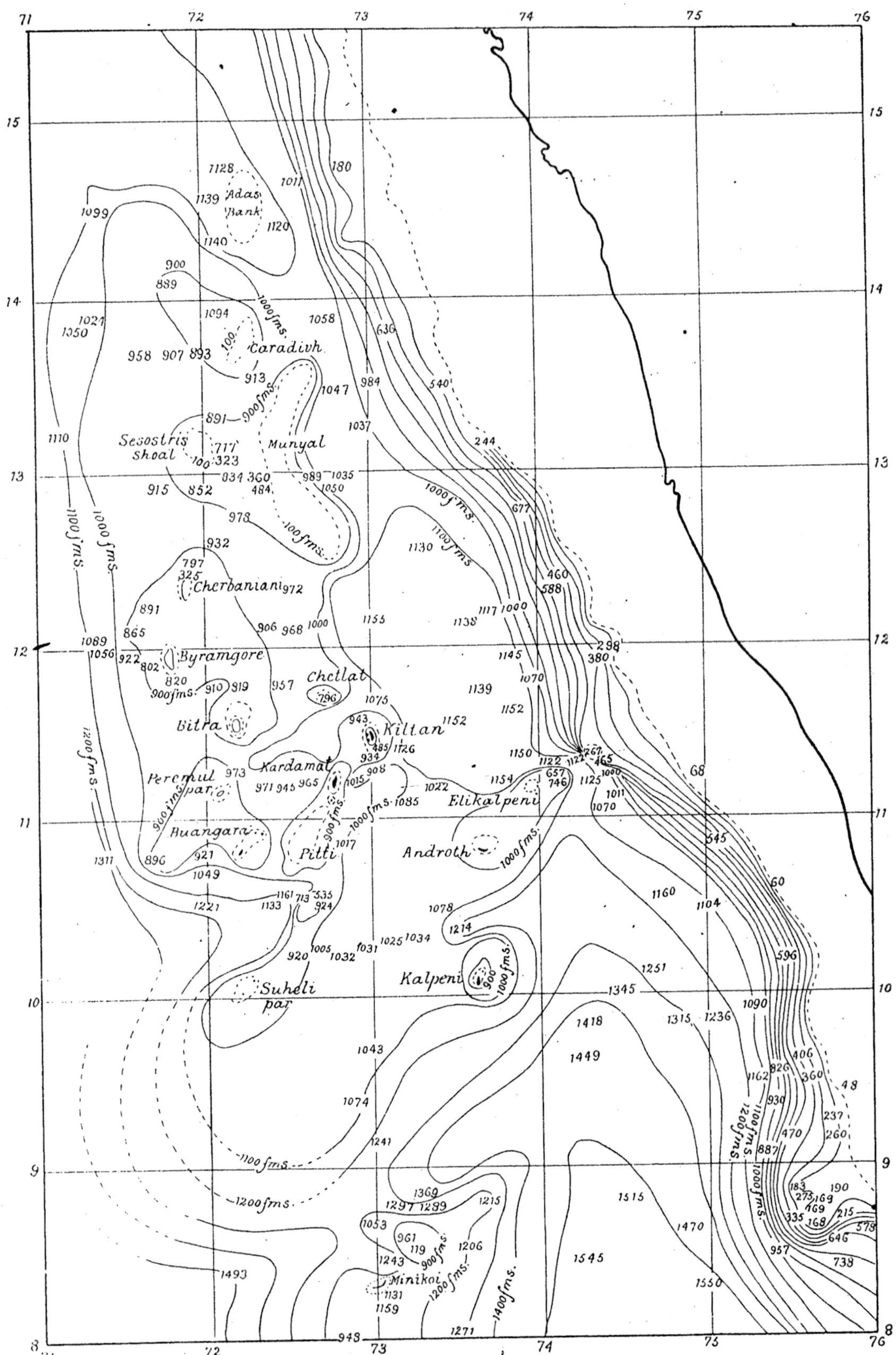

CHART VII.—Chart of the Laccadive Sea from 'Investigator' soundings (modified from Oldham).

of 230,000 square kilometres and a depth ranging from 50–100 metres; but this estimate of the depth gives the average rather than the maximum depth of the edge of the shelf. This shelf is widest at its north end where in Lat. 20° N. it extends out from the coast for a distance of over 120 miles but towards the south it gradually narrows till off Cape Comorin it is only some 30 miles wide. All along this shelf the bottom deposit is terrigenous in character and in the south is largely derived from the small rivers that drain the Western Ghats; whereas the wide area to the north is correlated with the inflow of several large rivers, the chief among them being the Indus and the Narbudda, the former, like the Ganges in the Bay of Bengal, having a deep gully or 'swatch' off its mouth.

Along the Malabar coast, lying to the east of the Laccadive Archipelago, as Alcock (1902, p. 165) has pointed out, the ocean floor slopes very gradually 'to a depth of 100 fathoms but after that, at a distance of 75 to 35 miles from shore, it makes an exceedingly steep descent'; but as we follow this coast to the south we find that the continental shelf and continental slope show very interesting differences in different localities. Opposite Mangalore and down as far as Cochin, the shelf slopes gradually to about 65 fathoms and then drops rapidly away to the 1,000-fathom line and the same condition is again met with in the region of Cape Comorin; but at Quilon the shelf widens considerably and at first slopes gradually to a depth as great as 190 fathoms but then further seaward rises again to a depth of 170 fathoms before it falls steeply away in the continental slope; it seems probable that in this region the formation of a simple continental slope has been prevented by some means or other, such as an oceanic current impinging on the land and preventing the deposition of sand and mud in shallower depths. The Current Charts for the Indian Ocean published by the Koninklijk Nederlandsch Meteorologisch Institut (1927–30) show that in the month of March two strong currents, the one from the north-west and the other from the south, having a velocity of 40–49·9 sea-miles per day, impinge on the west coast of India at about this point (Lat. 10° N.); but in May the current from the south disappears and we now get a strong current setting southwards along this part of the coast; a similar condition is again found in the month of September, though in this latter month the current is less strong and is only of a velocity of some 10–14·9 sea-miles per day. Passing further south to the region off Trivandrum, we there find that

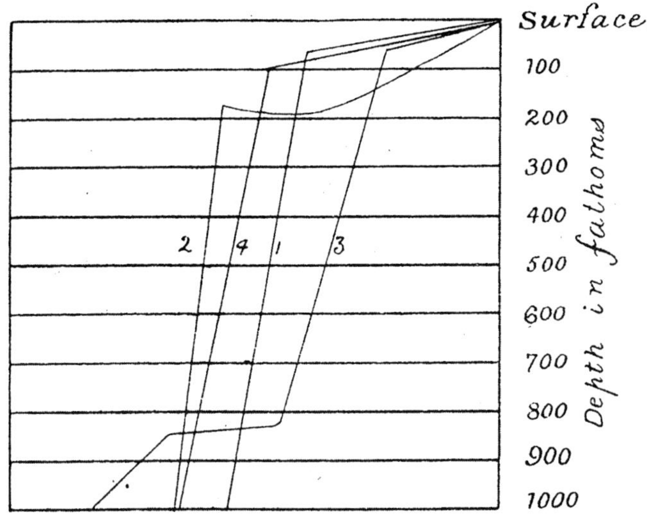

TEXT-FIG. 133.—Section of Continental shelf and slope on W. side of India at—

1. Mangalore—Cochin.
2. Quilon.
3. Trivandrum.
4. Cape Comorin.

the continental shelf is comparatively narrow, being only some 25 miles wide and the continental slope drops steeply from the 50-fathom level down to a depth of some 800 fathoms, at which depth there is situated a wide submarine plateau, to which I shall have occasion to refer later. Around the coasts of Ceylon the continental shelf is narrow and is largely overgrown by coral. As there are no very large rivers in Ceylon and as the land surface is comparatively small, it is not a matter of surprise, if we regard the shelf as being formed by land detritus, that this shelf is narrow. It is interesting to note that along the east coast of this island the shelf exhibits a number of deep indentations, especially in the following localities :—

(1) East of Little Basses Light.
(2) and (3) North and South of Alphee Shoal.
(4) North of Batticaloa Light, and
(5) Off Trincomalee Harbour.

The greater number of these indentations correspond roughly to the mouths of rivers and it might be argued that this furnishes support to the theory that attributes this continental shelf to the submergence of a part of the original land area, these indentations in the shelf representing the now drowned portions of the old river beds; on the other hand, however, since the greater part of this coast is fringed with coral reefs it must be obvious that much of the shelf must have been built up by the formation of a talus slope from the reefs and hence it is exactly in these regions, in which fresh water is being poured into the sea, that coral will be least able to flourish and, therefore, in which the talus slope formation will be slow or even non-existent.

It seems more than probable that the proximity of three plateau, namely those on which the Laccadives and Maldives are situated and the one off the west coast of India, situated so close to each other and at depths that range from only 850 to 1,200 fathoms is more than a mere coincidence; and their existence is particularly interesting in view of the opinion expressed many years ago by Wallace (1895, p. 426) in his review of the distribution of the fauna and flora of the various islands of the western part of the Indian Ocean. Wallace maintained that there could never have been direct land connection between India and Africa, and that all the then known facts were best explained on the assumption that these regions were originally separate, though not widely so, a series of land areas forming a discontinuous chain, and that the extent of these connecting land areas is indicated by the various plateau that are situated at a depth of about 1,000 fathoms and on which the present-day islands rest. Quite recently Scott (1932, p. 139) from his study of the insect fauna of the Seychelles and adjacent islands has reached the conclusion that 'the Seychelles are of ancient continental origin. Whether the former land connections were "bridges" in the older sense, or owed their existence and subsequent disappearance to the displacement of the continental land-masses, is a problem beyond the scope of this paper. But assuming the Seychelles to have been connected with Asia and Africa, their isolation must have been brought about very long ago, or they could hardly exhibit so high a degree of endemism in their fauna and flora. There are slight faunistic indications that connection with Africa may have persisted later than that with Asia, possibly to early Tertiary times, and that the connection with the two continents need not necessarily have included Madagascar.'

As regards the Laccadive and Maldive Archipelagoes there appears to be a mass of evidence that is distinctly in favour of the generally accepted view that there has in the past been extensive subsidences of a very considerable area of land between the present west coast of India and the east coast of Africa, though how much, if any of the present coral reefs were formed during the process it is at present impossible to say, since the evidence lies sunk below the sea-surface. Agassiz (1894, p. 186) in a comparison of the Maldives and the Bermudas remarks 'the islands and islets of the two archipelagoes named may be only the summits of a bank of very irregular outline, upon which corals have established themselves. Subsidence has taken place there to a very considerable extent, as stated by Darwin, but the position of the corals and the shape and distribution of the reefs have only been affected in a limited manner by it, any more than in such localities as the Bermudas and the Bahamas. The depth between Horsburgh Atoll and the southern end of the Mahlos Mahdoo Atoll (over two hundred fathoms) is no more surprising than the great depth of the channel between the Mira por vos Bank and Crooked island Bank, or in the passage between the outlying banks to the south of the Bermudas. Granting that the deep channels have been formed by subsidence, it does not follow that the present distribution and existence of the corals on the summits of the banks is due to the same cause.'

In a recent publication Fox (1931) has given a detailed account of the geological evidence of the existence of the great continent of Gondwana, that is supposed to have at one time connected India and Africa and to have extended on either side to Australia and South America. Readers who wish to obtain further evidence on this subject should consult the above work; suffice it to say here that 'from the evidence brought forward it appears to be practically certain that a vast continental area, relatively permanent, in the upper Palaezoic and lower Mesozoic era, existed in the southern hemisphere and included what are now parts of India, South Africa, South America, Australia and Antarctica. It is possible that these scattered regions were in close association with the South African tract at that time, as claimed by Wegener, and that they have since drifted apart. Although this theory has been criticized and other explanations put forward to explain the geographical features, it has, nevertheless, firmly established the previous existence of Gondwanaland. However extended or compact this continental region may have been, there is little doubt that its shore line varied from time to time. In the Indian region there are evidences of continued variations in the coast line throughout the Mesozoic period until the final break up of Gondwanaland with the eruption of the Deccan lavas at the close of the Cretaceous period.' Pascoe (1927, p. 214) has given a brief but sufficient description of the general condition, as regards the distribution of the land areas, at the end of the Carboniferous and the commencement of the Permian periods. At this time 'we find India forming part of a great southern continent stretching across the Arabian Sea and Indian Ocean, over the site of the Seychelles Islands to Madagascar and South Africa, and thence south-westwards to South America and Antarctica; to the south-east it was united to Australia and may have covered the rest of the Indian Ocean. To the north, girding the greater part of the earth; was a latitudinal sea, the Tethys, of which the Mediterranean is a dwindled relic. The backbone of the Indian end of this old continent of Gondwanaland was the Aravalli Range, the oldest mountain range in India, which at that time must have formed a lofty snow-clad chain comparable to the modern Himalaya.' Still later in geological history a connection is supposed to have existed between India and Madagascar, the so-called continent of 'Lemuria', and this in turn suffered a break up at the beginning of the Tertiary epoch. In Eocene and Miocene times, according to Blanford (1879–87, Vol. I, p. lii) the west coast of India lay much further to the westward than it does at the present time, but in Pleiocene and Pleistocene times the whole area of the Arabian Sea and its coastal borders appears to have suffered great disturbance. Towards the close of the Tertiary Epoch the coast of Baluchistan appears to have assumed its present character by a process of faulting; Blanford (*loc. cit.*, p. lxxi) has called attention to the fact that in this region 'there is a submarine cliff at a distance of about 10 to 20 miles from the shore. This cliff extends from a little west of Cape Monze to the entrance of the Persian Gulf and is about 2,000 feet high, the depth of the sea increasing more or less suddenly from 20 to 30 fathoms down to 300 or 400 fathoms', and, he adds, 'there is a possibility that the line of the submarine cliffs may be a fault'. Associated with this movement there was a rise of land along the coast, resulting in the elevation of the marine beds of the Mekrana group. In western Sind there is evidence of great disturbance about the same time, accompanied by the formation of the Sind mountains, anticlinal ridges running in a north and south direction. A glance at the contour map (Chart VIII) shows that the Khirthar mountains, one of the ranges of the Sind mountains, run in a line from north to south in the northern part of their extent but that as they approach the coast line the range bends towards the south-west and at the present day appears to terminate on the coast at Cape Monze. Oldham (1893, p. 312) remarks, 'a low range of hills, formed of Gaj beds, extends to the south-west past the hot springs at Pir Mangha (Mugger or Manga Pir) to the end of the promontory known as Cape Monze, and the same beds form the low hills east and north-east of Karachi, and furnish the material of which the houses of the town are mostly built. A small island called Churna, in the sea west of Cape Monze also consists of Gaj rocks.' Continuing the line of this range of mountains there is a well marked submarine ridge that rises from a depth of 1,700–1,800 fathoms on each side to a depth of only 1,226 fathoms in the middle of the ridge, while near the extreme western end of the ridge, in about Lat. 22° 18′ N. Long. 62° 20′ E., there is a peak that is covered by only 951 fathoms of water. There can be little doubt that this submarine ridge is a submerged part of the Khirthar mountain range.

In the middle of the Tertiary Epoch the Western Ghats began to assume their present elevation. Oldham (1893, p. 495) states that at this period 'the dry land that stretched westward into the Arabian Sea was depressed, and at the same time that to the east was elevated to form the Western Ghats'. Associated with this movement it appears probable that there was the formation of a great fault along the western coast of India. Fox remarks (1923, p. 126) that 'the presence of a great fault, out at sea parallel to the west coast of India, has long been suspected and it is well known that the hot springs, which occur at intervals along the coastal strip of the Konkan from south of Ratnagiri to north of Bombay, lie on a remarkably straight line indicative of a line of fracture'. La Touche (1919), when describing the submergence of the western part of the island of Bombay, also remarks 'this is not the only instance of a change in the relative level of sea and land having taken place in a region which is perhaps more than usually suspectible to such movements, since the western coast of the Indian Peninsula lies on the edge of a profound depression by which the land formerly connecting India with the African continent has been submerged beneath the waters of the Arabian Sea within a comparatively recent period of geological history'.

Quite recently Schuchert (1932) and Willis (1932) have each contributed very interesting papers on the changes that have taken place in these ancient land connections and have brought forward a mass of evidence derived from numerous sources and from several different sciences in favour of their existence and probable extent. As regards the connection between India and Africa, Willis (1932, p. 939) remarks 'There is biologic evidence of migrations between Africa and India which requires the existence of a land connection at least in Permian time. To define the probable path we may examine the bathymetric map to discover the position of the deeps and to trace the submerged ridges between them. The north-west basin of the Indian Ocean, sometimes called the Arabian Sea, is a typical oceanic basin of great breadth and characteristic depth. The adjoining continents of Africa and India are widely mantled with sediments of Permo-Carboniferous age, which were derived from upraised lands and thus testify to diastrophic movements. The latter according to the thesis of this discussion, are attributed to the activity in the ocean basins that delimit the land masses. . . . . . . The submerged ridge which may be regarded as the trace of the former isthmus between Africa and India is tortuous, but well defined. It runs from Africa east to Madagascar, thence north-east through minor islands to the great arcuate ridge of the Seychelles, which it follows south-eastward for some 500 miles to the broad, barely submerged bank, Saya de Malha. From that plateau it crosses a channel that is 12,000 feet deep to the long swell which supports the Chagos, Maldive and Laquedive islands and which extends northward to the western side of the Indian continental peninsula. The general form of this isthmus is that of the capital letter N.'

There is ample ground then for the belief that the floor of the portion of the Indian Ocean that now forms the deep channel between the Laccadive and Maldive Archipelagoes on the one hand and Peninsular India on the other hand and is known as the Laccadive Sea, as well as a part, at least, of the Arabian Sea has undergone very considerable subsidence; and if any further evidence be necessary it is to be found in the abrupt manner in which the Deccan Trap stops short at the present west coast of India. Oldham (1893, p. 255) remarks 'some faint idea of the extensive area occupied by this formation may be gained from the fact that the railway journey from Bombay to Nagpore, 519 miles long, never leaves the volcanic rocks till it is close to the Nagpore station, and that the traps extend without a break from the sea coast at Bombay to Amarkantak at the head of the Narbada and from near Belgaum to north of Goom. Even this extent, great as it is, by no means represents the whole area originally occupied by the formation, for outliers are found east of Amarkantak as far as Jamira pat in Sarguja, to the south-east a small outcrop occurs close to Rajamahendri, whilst to the westward the series is well developed in Kathiawar and Cutch, and is even believed to be represented, though only by two very thin bands, west of Kotu in Sind.' Glennie (1933) as a result of his studies of the crustal warpings of India has deduced that 'one of the foci of effusion of Deccan trap is centred at Bombay'. It is thus clear that the outflow of the trap reached the line that is now the Bombay coast and from the thickness of the beds in this latter area it is clear that the flow must have proceeded very much further to the west and south-west.

From their study of the Flora of the Maldive Archipelago, Willis and Stanley Gardiner (1901, p. 162) reached the conclusion that 'the flora of the Laccadive and Maldives, as we now find it, may be equally regarded as due to the submergence of the former land areas, or to the appearance of new plants on an area appearing for the first time above the waves, or in other words it is valueless as evidence one way or the other, so far as our present means of interpreting the evidence go. Other evidence seems to render necessary the supposition that where the coral reefs now are there was formerly a great extension of land, but there is no evidence from the present state of the flora for or against this view, nor any certain evidence even of the continual presence of land in the places occupied by the archipelagoes.' Later Stanley Gardiner in two papers (1902, 1906) has postulated an unbroken land connection between India and Africa, that towards the end of the Cretaceous period became narrowed in breadth, partly by subsidence and partly by marine erosion, and that subsequent further marine erosion caused the breaking up of this connecting bridge into a number of separate areas. 'At about the commencement of the Eocene the first straits were formed, probably between the present Saya de Malha, Chagos and Maldive Banks', and as regards these two latter areas he considers that both areas were shaped 'by the cutting down of a land by the action of the seas, and by the subsequent upgrowth of coral reefs, as the currents moderated. Subsidence may have taken a part in perhaps paving the way for this action, but in any case we can have no doubt of the importance of erosion in the past and present in shaping the archipelago.' In the earlier paper (1902) he remarks 'the land may have been the Himalayas of a great continental land joining Ceylon to Madagascar or, as Sir John Murray suggested to me, a series of volcanic erupted masses. What it may be is entirely a matter of theory and a question to which the geological study of the Indian continent has so far yielded no clue. It may be contended and perhaps truly that the separate banks are the remains of some of the peaks; that Kolumadulu or Male perhaps were the sites of mountains, which were cut down 20, 40 or even 60 fathoms, and then built up by the reef organisms to their present level.' In his account of the Maldive Archipelago, Stanley Gardiner (1903, p. 174) reaches the conclusion that 'there was a connection with the other banks towards Madagascar, in fact that these reefs show the positions of the mountains of a great continental land, which once joined Ceylon and Madagascar, but the greater part of which has in past time subsided to great depths and left no trace at the present day. The existence of such a land in the past too is absolutely required to explain the distribution of both animals and plants.'

Schuchert (1932, p. 898), quoting from the works of several Geologists and Palaeontologists remarks that " in the extreme south and south-east of the African continent there occurs, according to Du Toit, a long series of detrital deposits beginning with the Lower Cretaceous (Uitenhage series) and closing with the Upper Cretaceous (Danian). The Uitenhage series (Sundays River beds) is rich in bivalves and ammonites (*Hamites*, *Hoplites* and several forms of *Holcestephanus*). The Trigonias (9 species) are closely related to those of Cutch, India, and likewise of German East Africa (Tanganyika) and Madagascar, showing connections with eastern Tethys and not with the western part of this Mediterranean. Some of this Uitenhage fauna (including *Holcestephanus* in 3 species) recurs in Bolivia, Chile and Argentina, proving a migration route along the southern side of Gondwana. On the other hand, the flora of the Uitenhage is like that of the uppermost Gondwana series of India, indicating continuous land between those places.' Again Clark (The Crinoids of the Indian Ocean, 1917. Ind. Mus., Calcutta) has given a detailed account of the conclusions to be drawn from the known distribution of the Crinoids and he has pointed out that 'Judging from the evidence offered by the recent forms alone the European Crinoids reached the European seas by passage from what is now the Bay of Bengal north of what is now India, or at least Southern India. The Crinoids of south-eastern Africa represent a comparatively young fauna; they must have reached their present habitat by passage south-westward from Ceylon along a more or less complete land bridge since submerged: but few of them have as yet entered the Arabian Sea.'

Davis (1928, pp. 525–532) has recently given a summary of the views that has been put forward regarding the Maldive and Laccadive regions and he has formed the view that these atolls and reefs are perched on the top of fault-blocks, derived from the remains of Gondwanaland, that, as already mentioned, is supposed by geo-

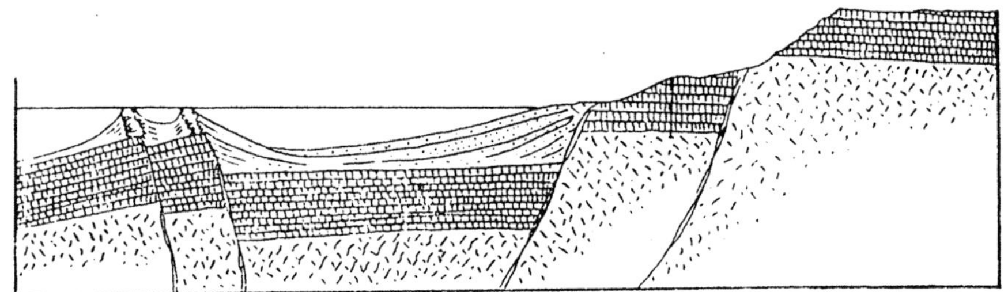

TEXT-FIG. 134.—Hypothetical section from the Maldive atolls to the south-west coast of India, after Davis (1928).

logists to have included Australia, South India, South Africa, part of South America and Antarctica in one continuous area. He assumes 'that the foundation of the

Maldives was a double crested mass, a pair of long narrow fault-block fragments . . . and that this mass either subsided somewhat more slowly than the rest of the vanished continent or perhaps experienced a local upheaval during the general subsidence elsewhere' (Text-fig. 134). Fringing reefs, he thinks, may have been early established on the slopes of this double crested north-south mass, which was later changed by subsidence into a pair of sub-Maldive, fault-block islands and the fringing reef then became converted into a single loop barrier over 400 miles long. 'But it is also probable that that great barrier was drowned at a time of rapid subsidence and replaced by fringing reefs of a new generation in the manner explained by Darwin and that as subsidence continued these fringing in turn grew up as two separate barriers.' These, in their turn, were 'eventually drowned and replaced by some 20 or 30 separate and much smaller barrier reefs corresponding to the separate small islands into which the pair of large islands had by that time been resolved. It is these separate and small barriers that are supposed to have grown up, during further subsidence of their foundations, into the predecessors of the present Maldive atolls.' 'The subsidence during which the predecessor atolls were built up was not long ago reversed into a moderate upheaval; . . . while upheaved, the predecessor reefs were more or less dissected and thus resolved into rings of limestone islets. After the dissection was well advanced, subsidence was for a time resumed at such a rate and to such an amount as to drown any fringing reefs that had been formed on the emerged atoll flanks during the dissection of their crests; and then during a pause in this subsidence each limestone islet came to serve as the centre of a new and minute fringing reef ring. Subsidence then continuing at a more moderate rate, each fringing reef ring grew up for a time as a minute barrier reef, until on the disappearance of the central limestone islet each minute barrier reef became a minute atoll or faro. Little reef patches in the Maldive lagoons may be explained as similarly built up from residual islets formed by the dissection of the emerged limestone in the lagoon floors of the predecessor atolls. The present Maldive lagoon floors would thus represent the lagoon floors of those predecessor atolls, more or less dissected and aggraded.' It will be noted that this theoretical history of the Maldives is entirely in keeping with Darwin's theory of subsiding foundations. Stanley Gardiner (1903, p. 172 *et seq.* and foot-note) on the other hand remarks, 'it will be observed that while rejecting the subsidence theory of Darwin as quite inadequate, I consider that the topographical conditions which made the formation of the coral reefs of this region possible, have probably owed their initiation to the sinking of a great continental land', but he states that he could find no evidence that there had ever been any recent subsidence in this region. There seems but little doubt that in quite recent times certain areas of the western coast of India have undergone slight subsidence; thus in Cutch there was in 1819 a subsidence accompanied by an earthquake, a full account of which has been given by Lyell (1875, p. 98 *et seq.*) and the occurrence of submerged trees, still *in situ*, shows that the island of Bombay has quite recently undergone subsidence, accompanied by tilting (*vide* La Touche, 1919, p. 214 and Fox, 1922, p. 126). All the evidence that we have been considering seems to indicate that originally between

India and Africa lay a connecting mass or series of masses of land forming either a portion of a large continent, namely Gondwanaland, or at a possibly later stage a narrow connecting isthmus, part of the supposed continent of Lemuria, that finally underwent subsidence and was drowned beneath the Indian Ocean, leaving only the topmost peaks as islands. One cannot however ignore the suggestion put forward by Wegener in his work on the origin of Continents and Oceans that India and Africa were at one time part of the same land mass and that they have gradually drifted apart.

The whole process has been summarized by Wait (1931) who points out that 'according to the displacement theory India and Africa were united until the critical rift occurred at the close of the Secondary Age. The lava flows of the Deccan and East Africa would form parts of the same great series of outbursts which occurred in the vicinity of the rift, and would thus be brought into close relationship with each other in space as well as in time. When the rift had completely opened and the drift of India began, the upheaval of the Himalayas may be described as a concertina-like compression of the floor of the Tethys, which fore-shortened India and allow for the ever widening gap between that country and Africa, until the movement was completed.' In order to account for this northward drift of India and the resulting compression of the Himalayas one has to presuppose that the position of the two poles was entirely different then from what it is nowadays and it is supposed that the glaciated areas of Brazil, India, and Australia were grouped around a South Pole that was situated in the centre of South Africa. It is further assumed that as the continents drifted apart detached portions were left behind and formed islands or series of islands, such as the Seychelles in the line of assumed drift of India away from Africa and Madagascar.

A study of the contours of the Laccadive and Maldive regions and of the soundings that have been taken on the two sides, shows clearly that the basis on which these coral reefs and atolls are perched is a magnificent mountain range (Fig. 135), that at the present day rises at its southern end from depth of over 2,000 fathoms and that, if we include the Chagos region, extends through a distance from north to south of nearly 1,500 miles. This great mountain chain runs in a nearly north and south direction and as one would expect in a range of this magnitude, exhibits at certain points along its length breaks in its continuity, some of great extent than others, that are quite comparable to the various passes that one finds in mountain ranges on land. As a result of these breaks, the islands now marking the topmost peaks are divided into geographical groups, each of which is still further subdivided into a series of banks and plateaux. Moresby, who carried out the original survey of this area in 1834–36, speaks of all three groups, namely the Chagos, Maldives and Laccadives, as constituting one great chain. Schott (1902, p. 117) has put forward the view that the Chagos Archipelago and the southern end of the Maldive Archipelago are connected together by a submarine ridge; this view is based on five deep soundings taken by the 'Valdivia' and these certainly indicate the possibility of such a ridge being present, though more soundings are required to enable one to say with certainty that its presence has definitely been proved and to define its limits. Agassiz in 1902 carried out a number of soundings in the Maldive region and these have greatly increased our knowledge of the depths of the channels between the various atolls but all that ~~one can claim~~ with certainty is that the situation of the Chagos plateau in the direct line of continuation of the main chain and the presence at the southern end of the

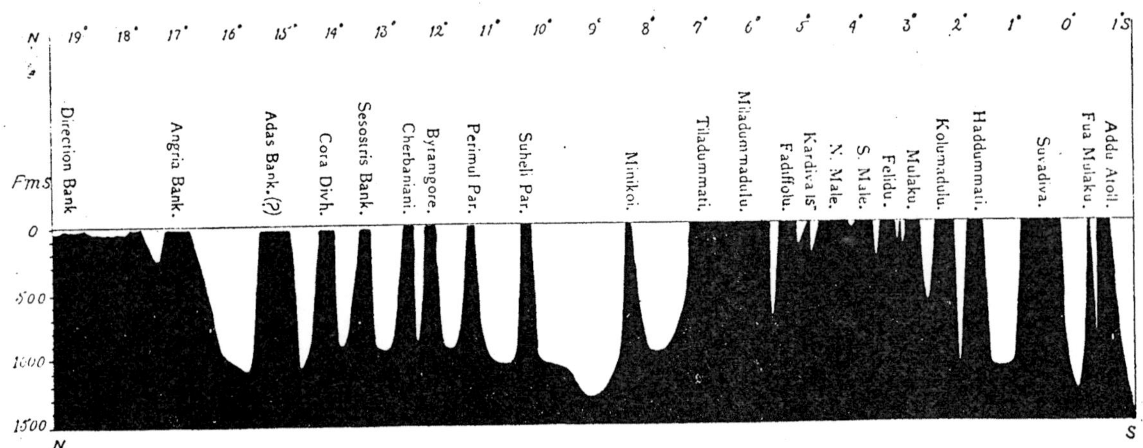

TEXT-FIG. 135.—Plan of Laccadive and Maldive Archipelagoes.

Maldive Archipelago of several outlying islands and atolls, surrounded on all sides by deep water, namely :—

(1) Addu Atoll, that rises from a depth of over 1,500 fathoms on three sides and is separated from

(2) the island of Fua-Mulaku by a depth of over 1,000 fathoms

(3) Suvadiva Atoll, with a depth of 1,282 fathoms on the south and 1,130 fathoms on the north side, and

(4) Haddumatti Atoll, with a depth of 1,100 fathoms between it and the next atoll to the north,

seems to warrant the assumption that they are all parts of one great system.

At the northern end of the great plateau on which the Maldives are situated, the main chain is interrupted by Eight Degree Channel, which has a depth of at least 965 fathoms; but to the north of this the ridge is continued still further, though with a slight trend towards the north-east, through the Atoll of Minikoi as far as 'Investigator' Bank.

The northern part of the chain, on which the Laccadive Archipelago is situated, is separated from the Maldive region by Nine Degree Channel, in which there is a depth of over 1,200 fathoms. The main mass of the Laccadive Archipelago is situated on a single plateau that lies at a depth of about 1,000 fathoms and from this a series of peaks rise up to or nearly to the surface; most of these peaks are situated somewhat to the west of the continuation of the line through the Maldives; and the southern end of the plateau, on which Sukeli par is situated, appears to have been thrust somewhat to the west. On the east side of the main group of the Laccadive lie two other banks, namely Kalapeni (Elikalpeni Bank) and Androth Island, that at first sight might be taken for a continuation of the Maldive ridge, but a study of the deep contours of the sea bottom shows that they are included with the remainder of the Laccadive group within the 1,100-fathom line and are separated from the Maldives by water of considerably greater depth than this, namely about 1,400 fathoms (*vide* Chart VII, p. 428). The Laccadives continue the line of the chain to the north for a further distance of some 250 miles and we then again get a gap, with a depth of water of 1,066 fathoms between Coradivh, the extreme northerly bank of the Laccadives, and peninsular India. It is possible that this gap may be interrupted by Adas Bank; this shoal is reported to have a depth of only 30 fathoms of water on it but its position on the chart is only vaguely defined and, indeed, its actual existence appears to be somewhat doubtful. * At this point the line of the mountain range has approached close to the edge of the continental shelf of the west coast of India and the range might be said to terminate; but it is interesting to note that in the direct continuation of the main chain of submarine peaks and plateaux we find two more banks, situated on the continental shelf itself, namely Angria Bank and Direction Bank. Sowerby (1868) attributes the formation of these two banks to the deposition of sand and mud that is brought down by the rivers that open into the Gulf of Cambay. He believes that 'the deposit is carried down by the ebb tide along the coast to the node of the tide near the Maldive and Laccadive islands. This detritus finds its way gradually down the coast, where it is deposited on Direction and other banks, a portion finding a resting place at the Laccadive and Maldive Islands.' The line connecting these two banks and the Laccadive ridge, if prolonged to the north, runs through the Gulf of Cambay; Oldham (1893, p. 408) remarks that the Gulf is said to be silting up, and there can be little doubt that it was once part of a broad inlet that led to the Ran of Cutch, then an inland sea, and that the remainder of the inlet has been converted into the alluvial plains of Ahmedabad, Broach, Surat and North-west Kathiawar; but this silting up, if the changes be due to that process, is of comparatively recent occurrence, geologically speaking. Whatever view one takes of the formation and mode of origin of the continental shelf, whether one regards it as a 'drowned' portion of the continent or as an accumulation of mud and debris derived by a process of erosion from the land, it seems not improbable that both Angria and Direction Banks have a definite foundation and represent a further continuation towards the north of the Maldive and Laccadive ridge, but that at this point the chain has become obscured by the deposit of silt along the coast of India. I am even inclined to believe that we can find a still more northerly continuation of this great submarine ranges in the Aravalli mountains of Rajputana; the trend of the Aravalli range is in the main from the north-east to the south-west, but at its southermost end the line of the range tends more towards a southerly direction and thus comes into line with the Maldive and Laccadive ridge. To the east of the Gulf of Cambay there is a range of hills known as the Satpura range, that runs in a west to east direction,

* Recent soundings taken by H.M.S. 'Endeavour' show no trace of this bank and it seems probable that there is no such bank.

and it has been suggested that this latter range is a loop of the Aravalli range; this, however, does not in any way invalidate the suggestion that the main line of the Aravalli mountains may be continued to the south. The view that the Aravalli mountains were originally continued much further to the south than is indicated by the remains of this range that we find in India to-day is clearly expressed by Pascoe (1927, p. 217) in the short account that he has given of the geology of India. He writes 'During early Tertiary times the slow rise of the Himalaya along the Tethys coast produced along its southern flank a gulf which extended as far south as the meridian of Lansdowne. At its north-western end it curved southwards to enter the Arabian Sea which was formed about this time in the following way. As a result probably of earth movement, a large block of Gondwanaland west of what is now the Bombay coast was broken off and submerged beneath the waves. The age of the Malabar coast and the Arabian Sea is, therefore, probably early Tertiary; some small coastal deposits near Quilon with early Tertiary fossils confirm this. The straightness of the coast line and its lack of indentation are due to this fracture or faulting and to its comparatively recent age. The same cause brought about the truncation of the Aravalli drainage, and for this reason all the important rivers of Madras and Southern Bombay are easterly flowing and rise within a few miles of the west coast; they are in fact but the lower portions of older rivers which rose in the old Aravalli watershed further west. The disjunction of the submerged part of the continent was assisted by faults in other directions; one of these seems to have coincided with the southern boundary of the Kathiawar Peninsula and to have initiated the Narbada (river), the middle section of which has an unusually straight course.' From the very nature of the case it is impossible to produce any direct evidence of the geological continuity of the Aravalli mountains and the Maldive-Laccadive ridge, and it is probable that during the outflow of Deccan trap the denuded remains of such a southerly continuation of the older mountain chain would to some and possibly to a considerable extent have been buried beneath a layer of Baxer, but there are certain resemblances between the two regions that are, to say the least, suggestive, and appear to be in favour of such a view. True mountain ranges in India are very few and as both Blanford (1879–87, Vol. I, p. vii) and Oldham (1893, p. 4) pointed out, the Aravalli mountains are almost the only true mountain range in the whole of peninsular India; Oldham remarked 'there is not throughout the length and breadth of the Peninsula, with the possible exception of the Aravalli, a single great range of mountains that coincide with a definite axis of elevation'. The work of Oldham (*loc. cit.*) and Heron (1922) has clearly shown that the Aravalli range has been produced by an extensive upthrust, accompanied by great faulting on the south-eastern and eastern sides, and owing to the manner in which the folds of the range have been thrust over towards the south and east the slopes on the east and south-east sides are much steeper than those on the western and north-western side. If, then, the whole extent of the mountain range, including the Aravalli mountains to the north, the Laccadive and Maldive regions and possibly even the Chagos Archipelago to the south, are parts of the same mountain system, one would expect to find the same or similar conditions present throughout the whole chain. A study of the submarine slopes on the two sides of the Laccadive plateaux shows that, as a rule, the eastern side slopes downwards much more rapidly than the western side; according to Alcock (1898, p. 52) 'the bed rises very abruptly to the Plateaux which form the Laccadive islands, the slopes being so abrupt that soundings of a thousand fathoms have several times been taken quite close to shore'. Oldham (C. F., 1895, p. 2) also states that 'the outer slope from the plateaux on which the islands are mostly situated is gradual' and this is clearly indicated in the chart attached to his paper. At the same time he notes that most of the islands are situated on the east side of the lagoons and that large coral boulders are found on the beach on the east and north-east sides of the islands; the explanation that he gives of this latter fact, namely, that it is due to the increased growth of the coral on the south and west sides, consequent on the influence of the south-west monsoon, while the formation of islets and the presence of masses of coral blocks on the east side is due to hurricanes in the Laccadive Sea that lies to the east of the group, may to a certain extent be true, but it does not account satisfactorily for the difference in dip of the two sides of the plateaux; moreoever, further investigation is urgently needed in order to decide whether these boulders and coral masses are, as Oldham thought, fragments of the reef margin that have been hurled upwards and inwards by storms, or whether, on the other hand, they are not parts of a reef that has been elevated above sea-level and owe their present position to the rest of the reef having been eroded down to sea-level.

Such soundings as we possess seem to indicate that the same difference in dip of the two sides is also present in the Maldive region; many more soundings are required before one can state with certainty that this difference is present but a series taken by Stanley Gardiner (1902, p. 285) across the central basin to the south of the atolls of Ari and Male shows the difference in the dip on the two sides very clearly. Brouwer (1918) from his study of the submarine slopes in the region of the Malay Archipelago has suggested that the difference in the slope of the two sides of a submerged mountain chain may be attributed to the effect of a thrusting over of the axis of the chain to one side; in this instance it would be towards the east, thus causing an elevation of the western side and a diminished dip westwards, while the slope on the eastern side would become more steep. If the different slopes in the Laccadive and Maldive regions were due to this cause we should have a clear agreement between this mountain range and the Aravalli mountains to the north. It is interesting to note that in Carboniferous times the Aravalli mountains formed the western boundary of a large sea area. Fermor (1924, p. 15) notes the discovery at Umaria of marine fossils and adds 'this discovery at Umaria suddenly provides evidence of the presence of the sea in Carboniferous times over a portion of what is now Rewa State in the middle of the northern part of the Peninsula, suggesting that in Talchir times the northern Carboniferous sea must have extended at least as far as Umaria . . . . We may perhaps picture this arm or bay of the Carboniferous sea as having been flanked on the west by the Aravalli ranges of Rajputana'.

Whether the Laccadive-Maldive ridge, that I believe to be the continuation of the Aravalli mountains, was completely submerged by the subsidence of the land area to the west of the present coast during the Tertiary epoch there is no means of telling. Blanford (1879–87, Vol. I, p. vii) puts the age of the Aravallis at not later than older Palæozoic, and the range must have undergone many changes since then. All that is now left of the range are the much denuded bases of the original anticlinal folds. Assuming that the Laccadive-Maldive ridge is the continuation of this range, the same or possibly even greater denudation will have taken place, especially in the Laccadive area that lies within the track of the full force of the south-west monsoon. With the sinking of the range subaerial erosion will have been succeeded by marine erosion and the mountains may have thus been converted into a series of plateaux, on which at the close of the glacial epoch coral or coral-forming organisms would have been able to establish themselves, provided that the depth to which erosion had taken place had not exceeded that at which such reef-forming organisms can exist, and allowing for the lowering of water-level that the heavy polar glaciation must have entailed.

Such evidence, then, as I have been able to gather from the geological records of peninsular India points to the conclusion that the Laccadive and Maldive ridge assumed its present condition, except for such changes as marine denudation and coral-formation have subsequently wrought in it, towards the end of the Tertiary epoch. It is interesting to note that a very similar series of changes were simultaneously taking place on the other side of peninsular India to the east of the Bay of Bengal; and a comparison of these two regions reveals a marked degree of resemblance. In both cases a large sea to the west was increased by the subsidence of a large land area and a subsidiary sea became formed and enclosed by a mountain chain that has been in the one case totally and in the other very nearly completely submerged. I give below in parallel columns the main corresponding features in the two areas :—

	The Laccadive and Maldive Archipelagoes.	The Andaman and Nicobar Islands.
1.	The Aravalli mountains.	The Arakan Range.
2.	Adas Bank (?).	Preparis Island.
3.	The Laccadive Archipelago.	The Andaman Islands.
4.	Nine Degree Channel.	Ten Degree Channel.
5.	Minikoi.	Car Nicobar.
6.	The Maldive Archipelago.	The Nicobar Islands.
7.	The Maldive-Chagos gap.	Great Channel.
8.	The Chagos Archipelago.	Sumatra.
9.	The Laccadive Sea.	The Andaman Sea.
10.	Peninsular India, with the Western Ghats.	Burma and the Malay Peninsula, with the Tenasserim range.
11.	Extensive faulting along the Indian coast.	Probable faulting between the Andaman Islands and Barren Islands.

One might even carry the analogy still further. As I have previously pointed out (*vide supra*, p. 16) the eastward slope of the Andaman ridge is much steeper than the westward, and situated within the concavity of the range, on the eastern side, there is a second submarine ridge of volcanic origin running in a north-east direction and separating off a portion of the Andaman Sea as a separate basin. In the Maldive-Laccadive ridge the slopes on the two sides are exactly similar to those of the Andaman ridge, being steeper on the east side, and the trend towards the north-east of the northern region of the Maldive plateau in the neighbourhood of Minikoi and 'Investigator' Bank, together with the line occupied by Kalpeni. Androth and Elicalperi islands of the Laccadive group, provide a further superficial similarity. Schmidt (1932, p. 255, Fig. 198) has put forward the suggestion, based on his experience in the 'Dana', that there is a submarine ridge, running in a north-westerly direction between the Chagos Archipelago and the island of Socotra and separating off a north-eastern part of the Arabian Sea as an isolated basin, but no corresponding ridge can be traced in the Andaman region of the Bay of Bengal.

The subsidence of such an extensive land mass must have been a slow and prolonged process and may well have continued on into the Pleistocene and even the Post-Pleistocene periods; indeed for all that one knows it may even be continuing at the present day. Stanley Gardiner (1903, p. 156) has pointed out that 'a most important point of difference from north to south (of the Maldive chain) lies in the gradual increase of the banks in depth, but whether this be correlated with their position or with the perfecting of the atoll form is not quite clear. The small banks, having relatively a greater amount of reef around them, would naturally be expected to be shallow, as is indeed the case. The large banks admit of a direct comparison, and this shows an irregular increase from Tiladumati to Kolumadulu', and he might have added even to Haddumati and Suvadiva, while Addu at the extreme south of the chain, although only a comparatively small atoll, is very nearly as deep as Haddumati and considerably deeper than the northern atoll of Ihavandifolu, of about the same size. It is possible that this progressive difference is to be attributed to a progressive subsidence of the ridge as we pass from north to south. The fact that such a subsidence has occurred indicates that the area is one of instability of the earth's crust and the steady accumulation of masses of calcium carbonate built up by the corals and coral-forming organisms into extensive reefs may have initiated a later isostatic adjustment, thus causing a further subsidence of the basic mountain chain on which the atolls and reefs are situated. In this connection one observation that was made by the 'Investigator' in 1891 is of some considerable interest. At Station 124 on the 21st of November, 1891, in Lat. 10° 47′ 45″ N.; Long. 72° 40′ 20″ E., from a depth of 703 fathoms the trawl brought up a number of 'Large water-worn fragments of *Reef-coral*' and the bottom is stated to have been hard rock. Unfortunately, no attempt seems to have been made at the time to determine the species of reef-forming coral that were present but so experienced an observer as Alcock can hardly have been mistaken regarding the type of coral that he was dealing with and it seems possible that there is in this locality a submerged reef that can only have reached its present depth by subsidence, though this may have been local, since clearly no reef could have been formed at this depth. Throughout this area the nature of the bottom appears to have been peculiar, and I give below the data from a number of observations taken at about the same time by Alcock.

Date.	Position. Lat. N. ; Long. E.	Depth in fathoms.	Nature of bottom.	Amt. soluble in Acid.	
14-11-91 ..	11° 22′ 20″ ; 72° 51′ 50″	988	Grey Ooze (Coral mud).	75%	A small residence (about 1%) of extremely refractory shells of foraminifera, and about 5% of pumice sand.
17-11-91 ..	11° 30′ 55″ ; 72° 05′ 15″	865	Sand and Grey Ooze (coral mud).	60%	About 30% fine pumice sand.
,, ..	11° 26′ 05″ ; 72° 03′ 15″	944	Grey Ooze (coral mud).	60%	About 40% fine pumice (?) sand.
,, ..	11° 17′ 25″ ; 72′ 05′ 00″	854	Grey Ooze (coral mud).	60%	About 30% pumice (?) sand.
19-11-91 ..	11° 14′ 05″ ; 72° 02′ 20″	707	Grey Ooze (coral mud).	90%	10% fine pumice (?); a few foraminifera, fragments of coral and of large shells.
,, ..	12° 13′ 35″ ; 71′ 58′ 07″	776	Grey Ooze (coral mud).	80%	18% fine pumice (?); about 20% of dry bulk consists of shells of foraminifera.
,, ..	11° 10′ 15″ ; 71° 15′ 25″	719	Grey Ooze (coral mud).		
,, ..	10° 59′ 22″ ; 72° 08′ 00″	882	Grey Ooze (coral mud).	85%	15% fine pumice (?); about 10% of dry bulk consists of shells of foraminifera.
,, ..	10° 53′ 55″ ; 72° 06′ 55″	877	Grey Ooze (coral mud).	87%	13% fine pumice (?); about 5% of dry bulk consists of shells of foraminifera.
20-11-91 ..	10° 47′ 10″ ; 72° 24′ 45″	940	Grey Ooze (coral mud).	91%	9% fine pumice (?); a few foraminifera, fragments of coral and shells of pteropods; spicules of corallines and sponges.
21-11-91 ..	11° 04′ 00″ ; 73° 13′ 45″	1,086	Grey Ooze.	60%	39% fine mud; a few foraminifera; small fragments of coral and pumice.
,, ..	10° 58′ 05″ ; 73° 02′ 50″	1,030	Grey Ooze.	60%	About 40% fine mud; a few foraminifera.

At the present day the great majority of the reefs in the Laccadive and Maldive areas are strewn with water-worn fragments of pumice much of which is of quite recent origin and has, almost certainly, been derived from the great explosion of Krakatra in the early eighties, though other sources appear to be much older. Its presence in this bottom deposit would thus be quite in keeping with the view that we have here a submerged reef.

It seems to me that the evidence that we have been considering is in favour of the view that the general conformation of the basis on which the present-day atolls and reefs of these Archipelagoes are situated is the result of subaerial and marine erosion on a gradually but continually sinking mountain range. Stanley Gardiner, however, as I have already mentioned, has rejected the view that there has been recent subsidence in this region and attributes the general conformation of the Maldives to the effect of submarine erosion of the ridge by deep currents down to a depth of some 200 fathoms. He remarked (1903, p. 173) 'the position of the group, too, would be one eminently favourable to the action of the currents, the plateau rising abruptly and lying right in the middle of the Indian Ocean, fully exposed to the two monsoons. The tidal wave sweeps across the ocean along the lines of latitude, extending from the surface to the greatest depths'. With regard to the possible action of the tidal currents on eroding and moulding the contour of the submarine slopes, a study of the times of high-water at different points around the shores of the Indian Ocean shows that the surface tidal wave swings round a mode that is situated in approximately Lat. 2° S. ; Long. 65° E., passing to the north on the west side and then swinging round the head of the Arabian Sea, and flowing to the south along the west coast of India. Oldham (C. F., 1895, p. 13) remarks that 'the tides are not strong in the Laccadives ; the flood sets to the north-east past the northern reefs and east and south-east in the neighbourhood of the southern ones, that is, in a course of right angles to the length of the reefs'. In the region of the Maldives the tides are also weak and the main tidal flow, so far as the surface currents are concerned, would seem to be in a direction more or less along the line of the ridge and not across it, as suggested by Stanley Gardiner. Two currents, however, do set across the ridge, namely, the north Equatorial and the Contra-equatorial currents. As I have already pointed out (*vide* part VI), these currents vary very considerably both as regards their direction and strength at different seasons of the year in accordance with the changes from the north-east to the south-west monsoons ; indeed, the former seems to disappear completely during the south-west monsoon, the whole of the surface water lying to the north of Lat. 5° S. being during this season in a state of of movement from west to east. To the south of this Latitude the south Equatorial current flows steadily from east to west, but this lies too far south to have any effect on the Maldive region. If there has been any truncation of the ridge from deep erosion by marine currents I should be inclined to attribute it to the effects of the Contra-equatorial and north Equatorial currents rather than to a hypothetical tidal current. Stanley Gardiner (1903, p. 23) has shown that in the Maldives there is a very strong surface current that during the period of his observations appeared to be flowing mainly towards the west, and varying from N. by W. to S. by W., the effects of which in the channel between north and south Male atoll could still be detected down to a depth of 150 fathoms. There is a further possibility that below this surface current lies a still deeper current running in the same general direction, namely, from east to west (*vide supra*, Pt. VI, p. 378), at a depth of about 300 fathoms, but more evidence is required before we can say with certainty that such a deep current is actually present. A study of the action of deep currents in other parts of the world shows that such currents may completely prevent the deposition of ooze on the bottom and thus would keep clear any deep channel such as those between the various atolls of the Maldive Archipelago ; but Stanley Gardiner goes farther than this and attributes these channels to the actual erosive action of these currents. He remarks that 'The general hard bottom can only be explained by the action of currents and on them lies the solution of the question as to the formation of the atolls and banks of the Maldives'. A study of the records of the nature of the bottom does not, however, seem to me to justify the conclusion that the bottom in the region of the Maldive plateaux is, as a rule, hard, for out of some eighty odd soundings that were taken by Agassiz in the Maldives and that range in depth from 100 to 1,547 fathoms, there are only eight that actually record the occurrence of a hard bottom, and, as every oceanographer knows, it is not always safe to trust to the accuracy of such a record. The situation in which these records have been made are as follows :—

Lat. N.	Long. E.	Depth in fathoms.	Position.
2° 33′ 00″	73° 18′ 00″	282	East opening of Kudahurada channel.
2° 37′ 00″	73° 20′ 00″	149	Ditto.
2° 41′ 00″	73° 22′ 00″	403	Ditto.
4° 08′ 00″	73° 31′ 00″	319	Off entrance to Wadu channel.
4° 52′ 30″	73° 24′ 00″	258	S. of Kandu, in east entrance to channel.
4° 85′ 00″	73° 27′ 00″	259	In Wadu channel.
5° 00′ 00″	73° 25′ 00″	312	1 mile E. of Karidu, in east entrance to channel.
5° 33′ 00″	73° 23′ 00″	342	N. of Fadiffolu, in channel between it and Maladu-Malad.

A study of the chart shows that every one of these soundings lies along the line connecting the eastern atolls and moreover the depth is in most cases considerably greater than the 200 fathoms that Stanley Gardiner gave for the supposed depth of the central plateau on which the atolls were perched; so that if this plateau is the result of erosion by deep currents, we must admit that these currents have been active down to a depth of at least 400 fathoms. In every other part of the plateau the bottom is covered by detritus in the shape of sand or coral mud, and it is only at the above spots and such similar spots in the Laccadives, as the one mentioned above where the bottom consisted of hard rock (*vide* table above) that either (1) the original hard bottom still shows above the detritus that is continually being showered down, or (2) the deep currents are of sufficient strength to prevent the deposition of mud and sand. In all probability the latter is the correct explanation, since it is obvious that in other parts of the plateau in the neighbourhood the currents are sufficiently retarded to allow of the deposition of ooze and mud, but it by no means follows from this that the currents are actually strong enough to erode the rocky bases of the atolls. Even if we admit the possibility of deep currents eroding the original ridge down to this depth of 200–400 fathoms, we are still faced with the problem of how banks managed to become established and to extend upwards so nearly to the surface that reef-forming corals were able to establish themselves and grow up into reefs. Clearly this cannot have been done by the accumulation of sediment, for that is entirely contrary to the assumption that there was, and presumably to some extent still is, erosion going on over the plateau, nor can it have been brought about by the establishment of deep-dwelling coral or other bank-forming organisms, such as nullipores, unless we are prepared to admit that since erosion took place there has been a complete change in the general conditions, and that, whereas at first there was sufficient force in the currents to prevent the establishment of such organisms and to erode the rock basis, subsequently, owing to a change in the currents or to a diminution of their strength, organisms, such as *Lithothamnion* or the coral *Lophohelia*, were able to establish themselves and grow up to the level requisite for reef-forming corals to become established in their turn and so give rise to the reefs as we find them to-day. In this connection it is interesting to note that this marine area does seem to be particularly favourable to the growth of this deep-sea coral; Alcock (1898, p. 1) remarked that the work of the 'Investigator' had shown that 'the sea in which corals have been found in the greatest abundance and variety is the narrow basin between the Laccadive and Maldive islands on the west and the Malabar coast on the east. At one spot in this sea, off Elicapeni Bank, at 1,000 fms. we dredged over two hundred specimens of a large new species of *Caryophyllia*; and at another spot, off the Travancore coast, at a depth of about 430 fathoms, Dr. A. R. Anderson, the present Naturalist on the "Investigator" lately dredged "nearly half a ton of living and dead coral . . . . such a haul I have never seen". The corals on this occasion belonged to the genera *Solenosmilia, Lophohelia, Desmophyllum* and *Caryophyllia*,' and a few years later (Alcock, 1902, p. 293) he wrote regarding the deep sea corals of Indian waters, 'some of them, however, are compound branching forms, which must, at times, in certain places grow luxuriantly enough to give rise to veritable submarine reefs'. I, however, agree with Agassiz (1903, p. xviii, foot-note 2) that the fact that *Lophohelia* may occur on deep banks does not necessarily imply or even suggest that it can build up a deep bank from several hundred fathoms below the surface to a height at which reef-forming corals could become established.

I do not think that one is justified in arriving at any conclusion regarding the formation and moulding of the Maldive ridge that will not equally well explain the conditions that are found to be present in the other portions of the range and especially of the Laccadive region; it can hardly be supposed that submarine erosion has caused the formation of the deep channels between the various banks in this northern archipelago any more than the equally deep channels that exist between the outlying atolls at the two ends of the Maldive group. Blanford (*vide* Stanley Gardiner, 1902, p. 299) opposed the view that there had been erosion down to a depth of 200 fathoms, when this was first put forward, and he added 'I would, therefore, suggest that, instead of the bank having been actually formed at a depth of 200 fathoms, below the sea, it was probably formed very much nearer the surface, and that it has sunk, producing precisely the conditions on which Darwin's theory of the formation of coral islands proceeds'. The same view is held by Davis (1928, p. 87) who concludes that 'there is no valid reason for believing that the subsidence of a large land area—Gondwanaland—between India and Madagascar might not have here and there, now and then, gone at such a rate as to permit the formation of fringing and barrier reefs around its sinking slopes and eventually of atolls above its submerged summits'.

THE BOTTOM DEPOSITS OF THE LACCADIVE SEA.

Murray (1889) published a paper on the marine deposits of the Indian and Antarctic Oceans and he remarks that 'in the Bay of Bengal and Arabian Sea, terrigenous matters from the Indus and Ganges are also spread out over a wide extent of the ocean floor and this is always the case where large rivers enter the sea. The colour of the clayey matter of typical oceanic deposits is, as we have seen, usually red or brown, but where land influences prevail it is generally bluish' and in the map attached to this paper he shows the bottom deposits of the Laccadive Sea as being a terrigenous mud in a belt along the west coast of India extending roughly to the foot of the continental slope, while beyond this he shows an area of Globigerina ooze, except where the archipelagoes of the Laccadives, Maldives, and the Chagos are shown as coral mud. Again, in the report of the Deep Sea Deposits obtained by the 'Challenger he gives a very similar map. Collet and Lee (1905) in a map attached to their paper on Glauconite give what they believe to be the areas in all the oceans where deposits of Galuconite occur and it is of interest to see that they make no mention of its occurrence along the Indian coasts.

In the accompanying chart I have plotted all the available data regarding the character of the bottom deposits in the Laccadive Sea region of the Indian waters.

Throughout the whole of the Laccadive Sea coastal region and around the south of Ceylon a study of the bottom deposits reveals the presence of two very distinct areas. Along the continental slopes and extending outwards from the edge of the continental shelf and down the continental slope to a depth of approximately 1,250 fathoms there is an almost uniform deposit of Green mud or Green ooze, with, very occasionally, a sample that has been described as Blue mud. It is generally agreed that Green mud is merely a variety of Blue mud and that it owes its distinctive character to the greater preponderance in the deposit of Glauconite; I have, therefore, treated both types of deposit as one for the purpose of comparison with the deeper and more westerly situated deposit of Grey ooze, that, as a reference to Chart IX shows, occurs throughout the whole of the western part of the Laccadive Sea and throughout the Laccadive and Maldive Archipelagoes. In a certain number of samples there was evidence of stratification, the Grey ooze, and in one instance the Blue mud, being overlaid by a stratum of brown mud or ooze; this difference in colour is in all probability due to chemical changes that have been going on in the deposit, as has been suggested by Murray and Hjort (1912, pp. 174-5): 'thus, in Blue muds it seems to be the rule that the upper portion should be thin and watery and reddish-brown in colour, in striking contrast with the stiff compact blue lower portion, and this is apparently due to the ferric oxide or ferric hydrate being transformed into sulphide and ferrous oxide in the deeper layers'. As Murray and Hjort (1912, p. 162) have pointed out Green and Blue muds are 'found most characteristically on the continental slopes off high and bold coasts where currents from different sources alternate with the season' and although the western coast of Southern India perhaps hardly merits the term high and bold at the present time, owing to the intervention of the low lying coastal plain between the sea and the true Western Ghats, it is possible that in times past the sea actually extended as far as the foot of these Ghats; the currents, by which this coast is washed, are certainly alternating in accordance with the changes brought about by the two monsoon seasons, namely, the south-west and the north-east.

That the green colour of these deposits is due to the presence of Glauconite cannot be doubted and it is of interest to note that this substance is probably present in large quantities in certain areas, especially between Lat. 7° 30′ N. and 10° 10′ N. for in a number of instances the bottom has been stated to consist of 'black sand';

as Murray and Renard (1891, p. 236) have stated, 'the collections of the "Tuscarora" indicate that in depths of 100 to 400 fathoms, off the coasts of California, there are black sands which, if the specimens be in the state in which they were collected, are almost wholly composed of particles of dark green glauconite'. It is probable that these deposits off the coast of India, since they occur at very similar depths, are composed of the same material.

The floor of the Gulf of Manaar is almost entirely composed of Green mud or ooze. All along the west coast of Ceylon and the opposing coast of Southern India there are extensive coral banks, and at first sight one would have expected to find that the sea bottom, like that of the Laccadive Sea, consisted of Grey mud or ooze, derived from the breaking down of the coral reefs; the reason for this difference in the two areas is in all probability to be found in the general set of the strong currents that exist in this region throughout the whole year (*vide* Southwell and Kerkham, 1911). During the north-east monsoon a strong current sets towards the north along the west coast of Ceylon and then bending round sweeps westerly along the southern coast of India; during the south-west monsoon the current, if the monsoon be weak, flows towards the east around the south coast of India and towards the north along the west coast of Ceylon; these two currents combine and pass out towards the north-east through the Pamban Channel into the Bay of Bengal. On the other hand, if the monsoon is strong, the current sets eastwards along the south coast of India and then bending round, runs in a southerly course down the west coast of Ceylon. In either case the result will be to keep the floor of the Gulf of Manaar largely free from coral mud. In one part of the gulf the bottom deposit appears to be of a somewhat unusual character; a sample taken in Lat. 5° 32′ N., Long. 79° 37′ E., from a depth of 675 fathoms, consisted in the main of Green mud and is so listed in the List of Stations of the R.I.M.S. 'Investigator' but mixed with this deposit were a number of stones or nodules. A complete account of this deposit has been given by Jones (1887): 'the nodules are stated to have been associated with sand and mud, which formed a hard calcareous crust at the bottom of the sea. . . . The stones are irregularly rounded and vary in shape from almost spherical to roughly cylindrical with rounded ends. The specimens received varied in size from 1–4 inches in length and $1\frac{1}{4}$—$\frac{3}{4}$ inch in thickness. Externally they are rough and mostly have one or two small excrescences of the size of a pin's head, and a few small pittings of about the same size; the colour is dirty light grey'. A microscopic examination showed that the 'stones' consisted of a number of small aggregates, each of which enclosed the remains of Foraminifera or Radiolaria. The chemical examination of the stones revealed a high percentage of Barium sulphate (approximately 75%), whereas the associated mud contained only Aluminic silicate with small quantities of Calcic carbonate, some iron and a trace of Manganese. Jones concludes that the nodules must have been formed at the bottom of the sea, by a slow segregative action.

CHEMICAL COMPOSITION OF THE BOTTOM DEPOSITS.

As I have already mentioned (*vide supra*, pp. 31 and 36 *et seq.*) Alcock, during his tenure of the appointment of Surgeon-Naturalist on the 'Investigator', carried out a number of analyses of bottom deposits and among these are several from the Laccadive Sea area. In the following tables I have grouped these analyses according to the type of the deposit, namely, Grey mud or ooze, Blue or Green mud or ooze and Globigerina ooze. I have also given the latitude and longitude of the position from which the sample was taken as well as the depth in both fathoms and metres.

A study of these tables shows that there is a fairly close agreement, so far as the percentage content of calcium carbonate is concerned, between the samples of Globigerina ooze and those of Grey mud or ooze; while the samples of Blue or Green mud or ooze are, relatively, poor in calcium carbonate. In the following table, I have given the average calcium carbonate content in both types of deposit, namely, Globigerina ooze and Grey mud or ooze, in descending zones of 500 metres depth.

Depth in metres.	Grey mud or ooze. No. of samples.	Grey mud or ooze. Percentage of $CaCO_3$.	Globigerina ooze. No. of samples	Globigerina ooze. Percentage of $CaCO_3$.
500–1,000	1	40·0	..	..
1,000–1,500	8	49·4	..	..
1,500–2,000	26	60·5	6	57·2
2,000–2,500	13	59·5	4	66·9
2,500–3,000	3	76·7	1	68·3

In both these series there seems to be a clear indication of a progressive increase in the percentage content of calcium carbonate as we proceed downwards. I have already pointed out that in other regions there is evidence of a fall in the carbonate content as we pass from shallow water down to depths of over 1,000 metres (*vide supra*, p. 42), the actual minimum occurring at a depth of about 1,500 metres. In the case of the samples of Green or Blue mud or ooze from the Laccadive Sea we get the following results :—

Depth in metres.	Green or blue mud or ooze. No. of samples.	Green or blue mud or ooze. Percentage of calcium carbonate.
0–500	1	30·0
500–1,000	3	50·0
1,000–1,500	4	37·1
1,500–2,000	4	24·7
2,000–2,500	7	41·4
2,500–3,000	2	52·5
3,000–3,500	3	33·7

In this series we get evidence of this fall in carbonate content of the deposit at a depth of 1,500–2,000 fathoms, at a somewhat deeper level than in the case of the Bay of Bengal samples but not so deep as in the samples from the East Indian Archipelago (*vide* Fig. 4, p. 42 *supra*). Below this depth, as in the other regions, we get a rise of the carbonate content till we reach a depth of 2,500–3,000 metres, below which there is again a falling off as we reach still greater depths. Thus, while the samples of Blue mud follow the usual course as regards the carbonate content, the samples of Grey mud or ooze or the so-called Globigerina ooze show a distinct variance from the normal.

If now we take the samples of Grey mud or ooze from different degrees of latitude we find that there are certain interesting differences; in the following table I have taken, according to their depths, the samples in zones of 3 degrees of latitude, namely, those between 7°–9°, 10°–12°, and 13°–15°:—

	Calcium carbonate content.		
Depth in metres.	7°–9° N.	10°–12° N.	13°–15° N.
500–1,000	..	..	40·0
1,000–1,500	..	51·2	38·5
1,500–2,000	35·0	66·4	54·2
2,000–2,500	66·7	55·0	42·5
2,500–3,000	76·7	..	..

In the case of the samples of Blue or Green mud or ooze we get the following:—

	Calcium carbonate content.			
Depth in metres.	4°–6° N.	7°–9° N.	10°–12° N.	13°–15° N.
0–500	..	..	..	30·0
500–1,000	..	..	55·0	47·5
1,000–1,500	..	21·1	85·0	..
1,500–2,000	..	24·0	25·0	25·0
2,000–2,500	..	50·0	45·4	12·5
2,500–3,000	..	52·5	..	..
3,000–3,500	35·4	..	..	..
3,500–4,000	30·2	..	..	..

A comparison of these two tables indicates that at depths between 2,000 and 2,500 metres there is a distinct tendency for the percentage of calcium carbonate in the bottom samples to decrease as we pass northwards, thus in the case of the Grey ooze at a depth of 2,000–2,500 metres the content steadily diminishes from 66·7% in Lat. 7°–9° N. to 42·5 in Lat. 13°–15° N.; and equally in the samples of Blue or Green mud or ooze at depths between 2,000–2,500 metres the percentage of calcium carbonate decreases from 50·0 in Latitude 7°–9° N. to only 12·5 in Latitude 13°–15° N. Another interesting point brought out by these tables is that in low latitudes, from 7°–9° N. the maximum calcium carbonate content does not occur, as is usually the case, at a depth of 1,500–2,000 metres, but steadily increases below this depth till it reaches as high a figure as 76·7% in a depth of 2,500–3,000 metres, or 1,000 metres deeper than normal.

In order to see whether the same phenomenon occurs in the bottom deposits from the Bay of Bengal I have re-examined the analyses from that region, with the following results. In the case of the deposits that have been classed as Globigerina ooze, we get the following figures for the percentage of calcium carbonate at different depths in different degrees of Latitude :—

	Calcium carbonate content.				
Depth in metres.	7°–9° N.	10°–12° N.	13°–15° N.	16°–18° N.	19°–21° N.
0–500	..	..	..	..	..
500–1,000	..	..	..	..	..
1,000–1,500	..	75·0	..	..	..
1,500–2,000	..	..	..	..	..
2,000–2,500	..	..	..	..	..
2,500–3,000	75·0	..	..	..	..
3,000–3,500	95·0	53·3	..	..	..
3,500–4,000	75·0	..	..	..	..

In the samples that have been classed as mud or ooze the results are as follows :—

	Calcium carbonate content.				
Depth in metres.	7°–9° N.	10°–12° N.	13°–15° N.	16°–18° N.	19°–21° N.
0–500	..	..	12·0	..	..
500–1,000	..	..	..	17·3	..
1,000–1,500	..	..	10·0	trace	..
1,500–2,000	..	..	4·5	trace	..
2,000–2,500	..	..	12·5	7·6	trace
2,500–3,000	..	..	21·6	trace	..
3,000–3,500	36·0	30·0	36·7	..	..
3,500–4,000	24·0	10·0	..	..	..

Here again we see that at a depth of 2,000–2,500 metres there is evidence of a lowering in the calcium carbonate content of the bottom deposit as we pass from lower latitudes, 13°–15° N., in which the content is 12·5 to higher latitudes 19°–21° N. where there is only a trace to be found. In the Bay of Bengal this difference may to a very great extent and possibly wholly be due to the greatly increased amount of terrigenous mud and silt that is being brought down by the great rivers of Bengal but it is doubtful whether any such explanation can account for the similar disappearance of the calcium carbonate in the Laccadive Sea, though, as I have already mentioned (*vide supra*, p. 439), Sowerby considers that the mud and silt, brought down by the rivers that enter the Gulf of Cambay, is swept down the coast by tidal action and is deposited on the sea floor in the region of the Laccadive and Maldive Archipelagoes. A study of the evidence seems to indicate that some other factor is involved. In this connection it is interesting to note that at this depth of 2,000–2,500 metres lies the North Indian Deep Current, that is in the main flowing from north to south and the water of which exhibits a relatively high salinity and temperature and it is possible that at its very origin in the Red Sea and the Persian Gulf and subsequently in its passage towards the south this deep current becomes more and more saturated with calcium carbonate and, therefore, in more southerly latitudes is able to dissolve less and less of this salt from the bottom deposits over which it is flowing. On the other hand, as several observers have pointed out, there is a distinct tendency for calcium carbonate to accumulate to a greater extent in tropical regions than in more northerly latitudes. Murray (1913, p. 215) remarks that 'Generally speaking, organisms secrete Calcium carbonate much more abundantly and rapidly in warm than in cold water. In the Arctic and Antarctic Oceans and in the deep sea, where the temperature approaches the freezing point of fresh water, there are no great accumulations of Calcium carbonate due to secretion by benthonic organisms, and the calcareous shells and skeletons secreted by pelagic organisms are thin and fragile. On the other hand, the most abundant secretion of Calcium carbonate, both by benthonic and planktonic organisms, occurs in tropical and sub-tropical waters'. It may be, therefore, that in the higher latitudes of 19–21° N. we have reached a zone in which there are fewer chalk-secreting organisms both in the plankton and the benthos and hence less calcium carbonate in the bottom deposit.

One of the chief characteristics in which the deposits of Green or Blue mud from the continental slope of the Indian Peninsula differ from the Grey mud or ooze of the more westerly part of the floor of the Laccadive Sea, apart from the lesser percentage of calcium carbonate that we have already considered, is the much greater percentage of silica. In coral rock or sand the silica content is very small; thus in anumber of specimens from different depths in the core obtained by boring in Funafuti atoll (*vide* Judd, p. 369) the insoluble residue 'after submitting them to slow solution in very dilute acid so as to avoid risk of decomposition of basic or hydrous silicates' was extremely small in amount ranging from 0·001 to 0·083 in six samples, and of these residues one gave a percentage of 12·75 per cent. of silica. Stanley Gardiner (1903, pp. 322-323) has shown that in coral rock or sand from the Maldive atolls the percentage of silica varies from 0·013 to 0·076, while in the mud from the lagoon of Suvadiva atoll it ranged from 1·992 to 3·127 per cent. In coral rock from the South sea islands, the percentage varies from 0·72 to 0·789 and in the raised coral rock and limestones of Barbadoes it is as a rule from 0·13 to 3·10, though in one instance a percentage as high as 9·48 was obtained (*vide* Skeats, 1902). In marked contrast to this is the result obtained from analyses of certain bottom deposits in the Mediterranean Sea by Buchanan (1890, p. 131) in which the silicate content ranged from 14·29 to 44·37, the depths from which the samples were obtained ranging from 265 to 1536 fathoms.

Two samples of Green mud from the Laccadive Sea have been analysed by Chatterjea (1927); unfortunately Chatterjea has given to his paper the title 'On three Deep-Sea Deposits from the Bay of Bengal', whereas only one of his samples was from this area, the other two to which I refer being from the west side of India. These two samples were respectively from

(1) Lat. 9° 44′ 00″ N., Long. 75° 29′ 00″ E., 725 fathoms.
(2) Lat. 6° 29′ 05″ N., Long. 78° 34′ 07″ E., 1,430 fathoms.

In the following table I have given the percentage results of analyses of some of the more common ingredients of Glauconitic deposits :—

SiO_2.	Al_2O_3.	Fe_2O_3.	CaO.	MgO.	K_2O.	Na_2O.	Depth fms.	Source.
27·74	13·02	39·93	1·19	4·62	0·95	0·62	155	Challenger.
47·46	1·53	30·83	..	2·41	7·76	..	173	Tuscarora.
50·85	8·92	24·40	1·26	3·13	4·21	0·25	410	Challenger.
51·80	8·67	24·21	1·27	3·04	3·86	0·25	410	,,
55·17	8·12	21·59	1·34	2·83	3·36	0·27	410	,,
56·62	12·54	15·63	1·69	2·49	2·52	0·90	410	,,
50·00	2·26	5·70	trace	1·55	7·50	0·64	725	Investigator.
55·42	1·59	5·22	,,	0·38	4·16	0·24	1,430	,,
49·11	8·03	20·05	0·87	3·18	6·97	0·66	(average of 12)	Heddle.
40·00	13·00	16·81	1·97	1·97	8·21	2·16	..	Hoskins.
52·86	7·08	7·20	trace	2·90	2·23	trace	..	Knerr and Schoenfeld.

For the purpose of comparison I have also given three analyses of Glauconitic rocks from various localities.

It is of some interest to note that in the Glauconitic deposits from the sea bottom there seems to be some evidence that the percentage content of both Iron, in the form of the Ferric oxide, and Magnesium oxide appears to diminish steadily with the increase in the depth from which the example was obtained, while on the whole the silica content increases. The percentage content of the calcium carbonate also seems to increase; unfortunately the two 'Investigator' samples had been treated with acid for the estimation of $CaCO_3$ prior to analysis of the insoluble residue and hence the calcium content is practically nil.

Murray and Renard (1891, p. 387) state that 'Glauconite is almost always accompanied by quartz, orthoclase often kaolinized, white mica, plagioclase, hornblende, magnetite, garnet, epidote, tourmaline, zircon, and fragments of ancient rocks, such as gneiss, mica-schists, chloritic rocks, granite, diabase, etc. In addition to these minerals there seems always to be associated with Glauconite, in modern deposits, a considerable quantity of organic matter, often apparently of a vegetable nature.' In this latter respect the two samples collected by the 'Investigator' were no exception, for Chatterjea states that they contained respectively 31·89 and 31·34 per cent. of material that was lost on ignition and in the case of the specimen from 725 fathoms—example 1—he remarks (*loc. cit.*, p. 25) 'this mud mixed with Glauconite also contains a greenish amorphous matter which in part at least appears to be of an organic nature for it blackens on being heated on a platinum foil, leaving an ash coloured by oxide of iron. This accounts for the high figure for "loss on ignition" as given above'.

With regard to the comparatively high percentage of potash in these Glauconitic deposits Murray and Renard (1891, p. 389) pointed out that 'Glauconite was always associated with terrigenous minerals and in particular with orthoclase more or less kaolinized, and white mica, and with the debris of granite, gneiss, micaschists and other ancient rocks. We cannot fail to be struck with these relations, for it is just these minerals and rocks that must give birth by their decomposition to potassium, derived from the orthoclase and white mica of the gneisses and the granites.'

In this connection it is of some interest to recall the account given by Lomas (1902) of a deposit obtained north of Rameswaram island at the head of the Gulf of Manar in Palk Bay 'where there is an almost complete absence of currents'. This deposit consisted in the main of 'concretions, irregular in form, with here and there a cast of shell along with a few large shells in a fairly fresh but broken condition . . . . The concretions, however, on treatment with a weak acid effervesce strongly and yield a large percentage of fairly coarse sand . . . . Round the periphery of the casts is a fine layer of calcite, which moulds itself into the inequalities of the shell's interior surface; this is succeeded by a darker layer, and then the whole interior is seen to consist of sand grains, quartz, tourmaline, felspar, and zircon, embedded in a mass of secondary calcite . . . . The inorganic fragments dissolved out by acids and fractionated showed a great preponderance of garnets. The heavier portions were pink in colour on this account. Other minerals found were corundum, tourmaline, zircon (inclosed in garnet and free), kyanite, quartz, mica, and felspar.'

Further east Lomas notes the occurrence of Black mud, the analysis of which gave the following result :—

Water and organic matter	16·60 per cent.
Silica	55·00 ,, ,,
Carbonate of Lime	3·50 ,, ,,
Phosphate of Lime	2·25 ,, ,,
Ferric Oxide	4·10 ,, ,,
Alumina	15·80 ,, ,,
Magnesia	2·75 ,, ,,

Finally, off the mouth of Trincomalee Bay in 12 fathoms Foraminifera sand was dredged, the inorganic constituents of which 'include quartz grains, well rounded and ranging up to 3 mm. in diameter, and a black powder much of which could be removed by a magnet. The non-magnetic portion was fractionated by means of the double iodides of mercury and barium, and showed a great number of garnets, corundum, tourmaline, and kyanite in the heavier fractions, while quartz and a few flakes of mica made up the lighter portions.'

Wadia (1919, p. 49) has pointed out that 'the crystalline and gneissic rocks of the Archæan system form an enormous extent of the surface of India. By far the largest part of the Peninsula, the Central and Southern, is occupied by this ancient crystalline complex.' He further points out (*loc. cit.*, p. 50) that 'the constituent minerals of the commoner types of the Archæan gneiss are; orthoclase, oligoclase or microcline, quartz muscovite, biotite, and hornblende with a variable amount of accessory minerals and some secondary or alteration products, like chlorite, tourmaline, epidote and kaolin . . . . Less frequent minerals, and occurring either in the main mass or in the pegmatite veins that cross them, are apatite, epidote, garnets, scapolite, wollastonite, beryl, tourmaline, tremolite, actinolite, jadeite, corundum, sillimanite together with spinels, ilmenite rutile, zircon, graphite iron ore, etc.' There can thus be no doubt that the deposits described by Lomas from Palk Bay and the neighbouring region have been derived from these Archaean rocks and have been carried down into the sea by rivers and streams. Such terrigenous deposits must therefore comprise the greater part of the mud and sand of the continental shelf and slope on both the eastern and western coasts of peninsular India and it is doubtless from this source that much of the glauconite of the deposit in the eastern part of the Laccadive Sea has been derived.

Another possible source of Glauconite is the Deccan Trap that forms the mass of the Western Ghats. According to Oldham (R.D., 1893, p. 257) Glauconite is a characteristic ingredient of this mineral; 'the most striking peculiarity is, perhaps, the great prevalence of Amygdaloid, in which the nodules, chiefly containing zeolite or agate, sometimes form the principle part of the rock. *These nodules are very often coated with Glauconite* (green earth), and the prevalence of this mineral is highly characteristic'. (The italics are mine.)

It is probable that much of the Glauconite of the sea-bottom is deposited from solution, either as the result of decomposition of animal matter or possible as the result of bacterial action but it is also possible that some at least may be the product of sub-aerial erosion and decomposition of the Deccan Trap and has been washed into the sea during the heavy rainfall of the south-west monsoon.

LITERATURE CITED IN THE TEXT.

Agassiz, A., 1894 .. .. .. 'A Reconnaissance of the Bahamas and of the elevated Reefs of Cuba in the Steam Yacht "Wild Duck" January to April, 1893.'
Bull. Mus. Comp. Zool. Harvard, Vol. XXVI.

Agassiz, A., 1903 .. .. .. 'The Coral Reefs of the Maldives.'
Mem. Mus. Comp. Zool. Harvard, Vol. XXIX.

Alcock, A., 1898 .. .. .. 'A Summary of the Deep-Sea Zoological Work of the R.I.M.S. "Investigator" from 1884–1897.'
Sci. Mem. Med. Officers Army India.

Alcock, A., 1902 .. .. .. 'A Naturalist in Indian Seas.' Calcutta.

Blanford, W. T., 1879–87 .. .. See Medlicott and Blanford.

Brouwer, H. A., 1918 .. ... 'On Reef Caps.'
Proc. Royal. Acad. Amsterdam, Vol. XXI.

Buchanan, J. Y., 1890 .. .. 'On the Composition of some Deep-Sea Deposits from the Mediterranean.'
Proc. Royal Soc. Edinburgh, Vol. XVIII.

Chatterjea, N. N., 1927 .. .. 'On three Deep-Sea Deposits from the Bay of Bengal.'
Jour. of the Dept. of Science, Calcutta University, Vol. IX.

Collet, L. W. and Lee, G. W., 1906 .. 'Recherches sur la Glauconie.'
Proc. Royal Soc. Edinburgh, Vol. XXVI.

Davis, W. M., 1928 .. .. 'The Coral Reef Problem.'
American Geographical Society, Special Publication, No. 9.

Fermor, L. L., 1923 .. .. 'General Report of the Geological Survey of India for the year 1921.' *Records Geol. Survey of India*, Vol. LIV.

Fox, C. S., 1923 .. .. .. 'The Occurrence of Bitumen in Bombay Island.' *Records Geol. Survey of India*, Vol. LIV.

Fox, C. S., 1931 .. .. .. 'The Gondwana System and Related Formations.' *Mem. Geol. Survey of India*, Vol. LVIII.

Gardiner, J. S., 1902 .. .. 'The Formation of the Maldives.' *Geographical Journal*, Vol. XIX.

Gardiner, J. S., 1903 .. .. 'The Fauna and Geography of the Maldive and Laccadive Archipelagoes', Vol. 1. Cambridge.

Gardiner, J. S., 1906 .. .. 'The Indian Ocean.' *Geographical Journal*, Vol. XXVIII.

Glennie, E. A., 1933 .. .. 'Crustal Warpings.' *Monthly Notices of the Royal Astronomical Society.* Geophysical Section. Vol. III.

Heron, A. M., 1922 .. .. 'The Gwalior and Vindhyan Systems in South-Eastern Rajputana.' *Mem. Geol. Survey of India*, Vol. XLV.

Jones, E. J., 1887 .. .. 'Natural History Notes from H.M.'s Indian Marine Survey Steamer "Investigator"; No. 5. On Some Nodular Stones obtained by Trawling off Colombo in 675 Fathoms of Water.' *Journ. Asiat. Soc. Bengal*, Vol. LVI.

Judd, J. W., 1904 .. .. 'The Chemical Examination of the Materials from Funafuti.' The Atoll of Funafuti; *Report of the Coral Reef Committee of the Royal Society.*

Krümmel, O., 1907 .. .. 'Handbuch der Ozeanographie', Vol. 1. Stuttgart.

Lyell, C., 1872 .. .. 'The Principles of Geology.'

Lomas, J., 1902 .. .. .. 'On Deposits dredged by Professor Herdman in the Indian Ocean.' *Report of the British Association for the Advancement of Science.* Section D.

Medlicott, and Blanford, H. K., 1879 .. 'A Manual of the Geology of India.' Calcutta.

Murray, J., 1887 .. .. .. 'On some recent Deep-Sea Observations in the Indian Ocean.' *Scottish Geographical Magazine*, Vol. III.

Murray, J., 1889 .. .. .. 'On Marine Deposits in the Indian, Southern, and Antarctic Oceans.' *Scottish Geographical Magazine*, Vol. V.

Murray, Sir J., 1913 .. .. 'The Ocean'. Home University Library of Modern Knowledge, London.

Murray, J., and Hjort, J., 1912 .. 'The Depths of the Ocean.' London.

Murray, J., and Renard, A. F., 1891 .. 'Report on Deep-Sea Deposits.' *Sci. Results of the Voyage of H.M.S. 'Challenger'.*

La Touche, T. D., 1919 .. .. 'Submerged Forests at Bombay.' *Rec. Geol. Survey of India*, Vol. XXXV.

Oldham, C. F., 1895 .. .. 'Topography of the Arabian Sea in the neighbourhood of the Laccadives, ii. The Physical features of some of the Islands.' *Journ. Asiat. Soc. Bengal*, Vol. LXIV, Pt. 2.

Oldham, R. D., 1893 .. .. 'A Manual of the Geology of India.' Calcutta.

Pascoe, E. H., 1927 .. .. 'A Sketch of the Geology of India.' *7th Congress of the Far Eastern Association of Tropical Medicine ; Souvenir Volume of the Indian Empire*, Calcutta.

Schmidt, J., 1932 .. .. 'Dana's Togt Omkring Jorden 1928–1930.' Copenhagen.

Schott, G., 1902 .. .. .. 'Oceanographie und Maritime Meteorology.' Wissenschaftliche Ergebnisse der Deutschen Tief-See Expedition auf dem damfer 'Valdivia', 1898-1899, Vol. I. Jena.

Schuchert, C., 1932 .. . 'Gondwana Land Bridges.' *Bull. Geol. Soc. America*, Vol. XLIII.

Scott, H., 1932 .. .. .. 'Summary and General Conclusions regarding the Insect Fauna of the Seychelles and adjacent Islands.' *Proc. Linn. Soc. London.* May 12th.

Skeats, E. W., 1903 .. 'The Chemical Composition of Limestones from upraised Coral Islands, with notes on their Microscopical Structure.' *Bull. Mus. Comp. Zool. Harvard*, Vol. XLII (Geological Series, Vol. VI).

Southwell, T., and Kerkham, J. C., 1911 .. 'Currents on the Ceylon Pearl Banks, superficial and deep.' *Reports from the Ceylon Marine Biological Laboratory*, No. 20.

Sowerby, W., 1868 .. .. 'Memorandum on the Geological Action on the South Coast of Kattyawar and in the Run of Kutch.' *Trans. Bombay Geographical Soc.*, Vol. XVIII.

Wallace, A. R., 1895 .. .. 'Island Life.' 2nd Edition. London.

Willis, J. C., and Gardiner, S. J., 1901 .. 'The Botany of the Maldive Islands.' *Annals of the Royal Botanic Gardens, Peradeniya*, Vol. I.

Wadia, D. N., 1919 .. .. 'Geology of India.' London.

Willis, B., 1932 .. .. .. 'Isthmian Links.' *Bull. Geol. Soc. America*, Vol. XLIII.

Wegener, A., 1924 .. .. 'The Origin of Continents and Oceans.' (Translated from the third German Edition by J. G. A. Skeol) Methuen & Co., Ltd. London.

Wait, W. E., 1931 .. .. 'The Origin of Continents and Oceans according to the Displacement Theory.' *Journ. Bombay Nat. Hist. Soc.*, Vol. XXXIV.

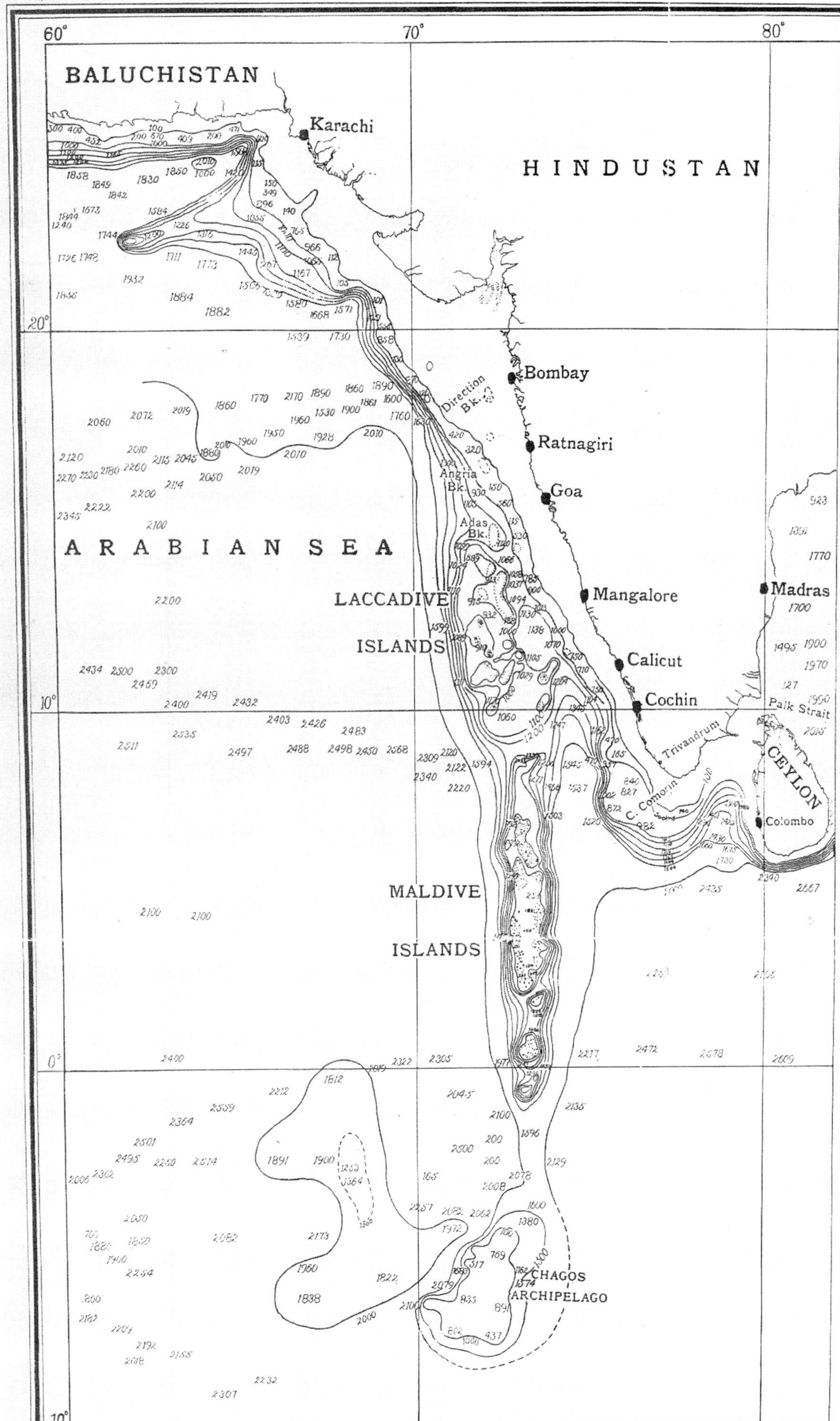

CHART VIII.—Showing the submarine contours to the west and south-west of the Indian Peninsula.

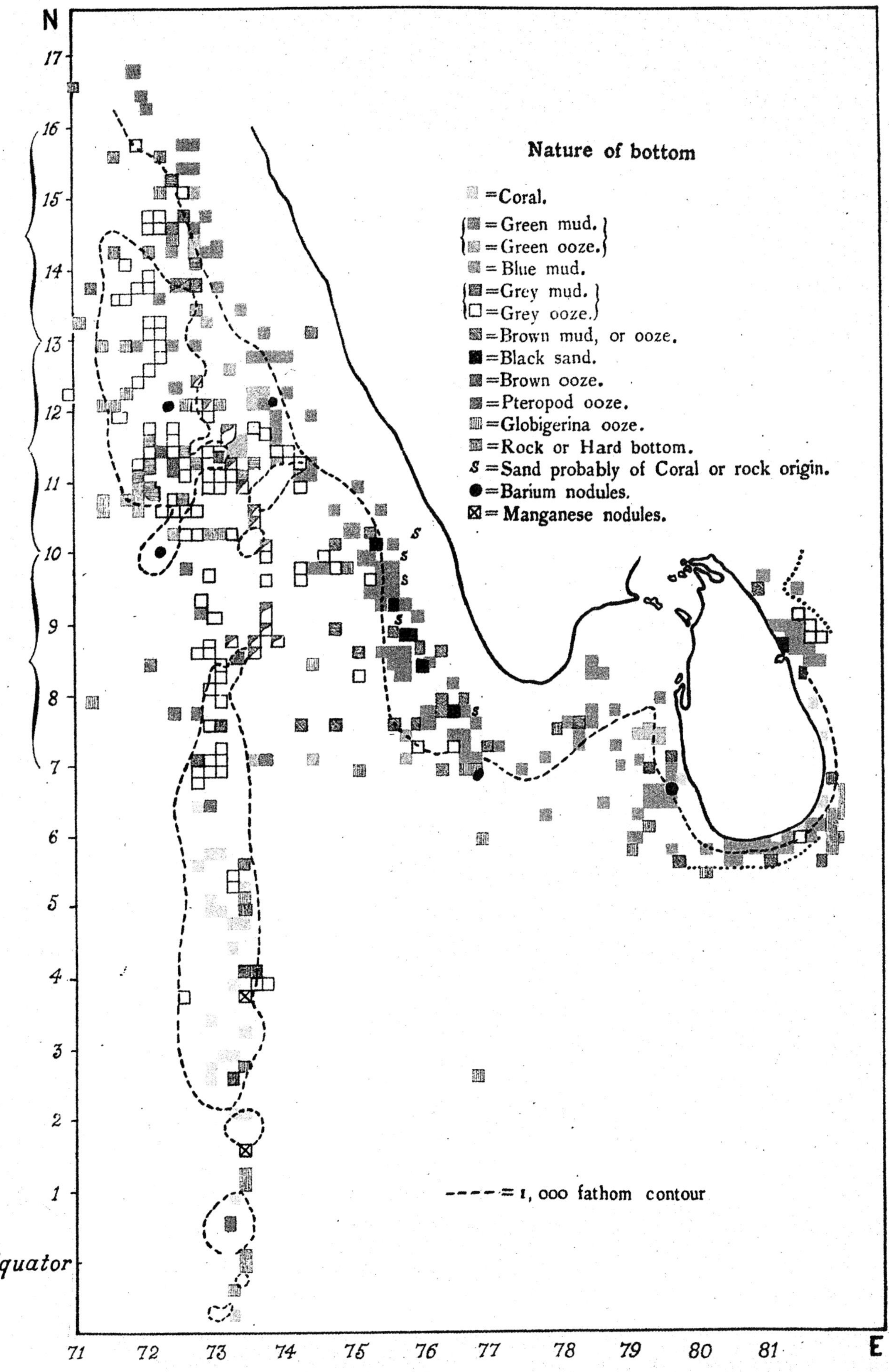
N
17
16
15
14
13
12
11
10
9
8
7
6
5
4
3
2
1
Equator
71
72
73
74
75
76
77
78
79
80
81
E
Nature of bottom
=Coral.
=Green mud.
=Green ooze.
=Blue mud.
=Grey mud.
=Grey ooze.
=Brown mud, or ooze.
=Black sand.
=Brown ooze.
=Pteropod ooze.
=Globigerina ooze.
=Rock or Hard bottom.
S =Sand probably of Coral or rock origin.
=Barium nodules.
= Manganese nodules.
= 1, 000 fathom contour

VIII. STUDIES ON CORAL AND CORAL-FORMATIONS IN INDIAN WATERS.*

In this paper I have attempted to put together a number of observations that I have from time to time, during the course of my service as Surgeon-Naturalist to the Marine Survey of India and since then as Director of the Zoological Survey of India, been able to make on coral growth and the character of coral reefs in various regions of the Indian Seas. I have not attempted to put forward any new theory regarding the mode of origin and formation of these reefs and atolls, but there are certain facts that seem to me to merit attention, especially those that are related to and have a bearing on the changes that appear to have taken place since these atolls were first formed.

Many authors in the past have put forward different views and theories regarding the manner in which atolls originated, and it is around the Laccadive and Maldive region that controversy seems to have reached its highest point. It was a study of the Cocos-Keeling atoll in the Indian Ocean and of the charts and copious notes that had been prepared by Moresby, regarding the above-mentioned archipelagoes, that gave Darwin the idea that Barrier Reefs and Atolls were successively derived from Fringing reefs during the subsidence of the land in a coral-growing area ; it is as well, however, to bear in mind that although this is in the main the view that Darwin put forward, accompanied by a mass of carefully collected evidence, and, in consequence, is usually referred to as Darwin's theory, it by no means expresses the whole of Darwin's views on the subject. Darwin himself remarks that ' a conjecture will perhaps be hazarded that the requisite bases (for the formation of coral atolls) may have been afforded by the accumulation of great banks of sediment that did not quite reach the surface owing to the action of superficial currents aided possibly by the undulatory movements of the sea '. Wood-Jones (1910, pp. 216–218) has pointed out that ' it is often overlooked that Darwin ever admitted any alternative hypothesis to that of subsidence ', and that in justice to Darwin it should be clearly recognized that ' many of the views of Darwin's successors were views that Darwin himself had originated, and the validity of which he had admitted under certain conditions '. Davis (1928) has given a detailed review of the several theories that have been propounded to account for the form and character of the various coral growths found throughout the tropical regions of the three great oceans, and Bonney (*vide* Darwin, 1889, Appendix II) and Hickson (1924) have given summaries of these views. It would be out of place, as well as unnecessary, for me to give a detailed account of them here ; I have, therefore, only given a very brief summary for the benefit of those who have not previously studied this problem.

Following Darwin a theory was put forward by the late Sir John Murray that ascribed the formation of these atolls and barrier reefs to the growth of coral on

* This paper was written, and received for publication by the Asiatic Society of Bengal, before its Author left India in April, 1933, and prior to his appointment as Leader of the John Murray Oceanographic Expedition.
September, 1935. *General Secretary, A.S.B.*

submarine ridges, probably of volcanic origin, that had either been worn down to the requisite level by marine erosion or built up to it by the deposition of ooze, and that these coral patches had subsequently attained their present character by the action on these patches of a process of solution of calcareous material by sea-water. Murray thus adopted as the theoretical base of some reefs an explanation that, as I have already mentioned, had been recognized as a possibility by Darwin himself. Agassiz also followed Murray and even went so far as to remark that in his opinion it was impossible to conceive of the existence of a range of mountains that could have been so completely submerged as not to leave a single trace above water of their original elevated position. In these earlier explanations it was argued that the growth of coral would be more extensive and luxuriant on the margin of such a patch and that, therefore, there would be a natural tendency for the formation in such a patch of a central lagoon and that this would be still further increased by solution. As early as 1898 Stanley Gardiner (1898, p. 485) called attention to the adverse effect that sedimentation had on living corals. He remarked that specimens removed from their natural habitat and fixed between stones in the entrance channel of the atoll of Rotuma in the Pacific 'soon became silted up with sand and were killed... Generally I could not, in the boat channel, get any true coral to grow if placed on the bottom owing to the sand and mud deposited on them.' Wood-Jones elaborated this line of argument and in his work on the Cocos-Keeling atoll (1910, p. 237 *et seq.*) pointed out that the effect of deposition of sand and silt on coral growth will be to kill off the upper colonies and so to give rise to a circular mass of coral that is still flourishing round its margin, whereas the centre of the colony will be occupied by dead coral that is susceptible to the action of sea-water and thus can be dissolved and removed, with the consequent formation of a central depressed area or lagoon. Subsequent observers have also experimented on the adverse effect of silt; thus Mayor (1918) experimented with certain corals on the Murray Island Reef at the northern end of the Great Barrier Reef of Australia and he found that burial in silt rapidly caused the death of a number of species of reef-corals; again more recently (1924) he experimented with live corals placed in a cage near the mouth of a stream in Pago Pago Harbour in Samoa, and he found that the combined effect of lowered salinity and accumulated silt killed off all species with the single exception of a species of *Porites*, that was, however, badly damaged. Edmondson (1928, p. 49–54) has obtained very similar results by burying various species of coral under 4 inches of sand and silt; he found that certain genera, such as *Stephanaria*, *Favia*, *Leptastraea* and *Fungia* appear to be more resistent than others, but that sooner or later such burial caused the death of the coral. Hickson (1924, p. 220) has also called attention to the adverse effect that sedimentation has on the reefs; he remarks 'there seems to be nothing more fatal to the growth of corals than this deposit of silt. The delicate polyps have some power of removing a few light foreign particles that fall upon them, but a continuous shower of grains of sand or mud hinders their power of expansion, interferes with their capacity to capture and ingest food, and by shutting off the light from the canal systems checks the photosynthetic action of the zoochlorellæ. Any change in the set of the tides and currents that drives the silt on to a vigorous part of the reef, or causes stagnation and a fresh deposit of silt elsewhere, may be regarded as among the most destructive of the agents which check the growth of the reefs.' Marshall and Orr as a result of their work on the corals of the Australian Barrier Reef (1931, p. 123 *et seq.*) have shown that sedimentation may affect different species of coral unequally, partly from mechanical reasons, such as the shape of the coral, for in branching species as *Pocillopora* and *Acropora* it is difficult for sediment, especially if coarse grained, to settle on the polyps. 'On the whole the result of the experiments showed that the common types of corals when they are helped by water movements as well as by their own ciliary action, are well able to deal with any ordinary amount of sand falling on them.' On the other hand 'when any circumstance interferes with the permanent level (of the sediment) for more than a few days the corals are killed off up to this new level'. The ability to remove sand and mud varies in different species, '*Favia* and *Fungia* appear to be most capable of dealing with the sediment and *Porites* least so'; and these authors agree that 'continued exposure to sediment has, at least in the laboratory, a harmful effect'. Another factor that must have a profound effect on corals and coral growth especially in the neighbourhood of the mouths of large rivers is the change in the reaction of the river water from alkaline to acid. Ross and Bagchi (1919; 1925) have shown that the water of the Sone and Ganges rivers in India is alkaline for 8 months of the year, but becomes acid during the rainy season from May to August. This change in reaction is attributed to the production of nitric acid in the atmosphere and its solution in the rain, and they have estimated that as much as 2·52 tons of nitric acid are produced per square mile. The effect of this on the river water is to cause a conversion of calcium carbonate into calcium nitrate with the liberation of carbon dioxide gas. As the acid production is in excess, the water of the estuarine region will rapidly dissolve calcium carbonate and to this, as much as to the presence of silt or lowered salinity, may be attributed the absence of all coral from such regions: it would be a matter of considerable interest to investigate the corresponding effect that Tropical thunderstorms have on the acidity of the water in the lagoons of atolls.

In Darwin's day the possibility that the sea-level itself may from time to time have altered had not received much, if any, attention; but he pointed out that in a number of instances there appeared to be evidence in favour of recent elevation in certain coral islands and atolls and he remarked that should this be confirmed 'the question will arise, seeing how immense an area has thus been affected, whether these geologists are not right who believe that the level of the sea is subject to secular changes from astronomical causes'. It is now generally recognized that in the past there have been extensive changes in sea-level and Daly in a series of papers from 1910 has advocated the view that during the last Glacial epoch, owing to the formation of greatly extended ice-caps, a considerable quantity of water was abstracted from the sea, that must of itself have resulted in a fall of level and that, furthermore, the increased masses of ice at the pole would, in addition, have caused a mechanical attraction of water away from the tropical regions and thus have increased the fall of sea-level in the coral-growing areas. The actual extent of this fall he estimated to be about 50 fathoms. This fall of level would have rendered it possible for the forces of nature to effect a wearing down of a number of banks and islands to a level that, on the melting of the ice-caps at the close of the Glacial period, has again, been covered by the gradual rise of the sea-water as it returned to or nearly to its original level, and Daly suggests that it was during this period that coral banks were able to form and to grow up *pari passu* with the rise of sea-level, thus giving rise to the barrier reefs and atolls that we find at the present day. As Davis has pointed out, there is, so far as the end-result is concerned, but little difference between Darwin's theory and that of Daly, since the one postulates a gradual subsidence of land, while the other suggests a slow rise of sea-level.

Finally, it is to Stanley Gardiner that the credit must be given of having realized the importance of a recent slight shift of sea-level, probably only some 4,000 years ago, and of putting forward the view that the actual formation of the islands on the circumference of these atolls must primarily be attributed to this fall of level of the sea-water, rather than, as was supposed by Darwin and most previous writers, to the throwing up of coral and sand by wind- and wave-action alone.

In any given coral-growing area a number of causes must have co-operated in the production of the reefs and atolls and in moulding them into the condition that we find to be present to-day ; some of these causes, such as the fall of sea-level during the glacial period and its subsequent rise were world-wide in their effect, while others such as elevation or subsidence of the land itself may have been purely local.

Evidence of a Fall in Sea-level in the Indian Region in Recent Times.

Darwin (1889, p. 173) and Dana (1875, pp. 284–298) both called attention to the fact that in certain areas of the Pacific Ocean there was evidence of a comparatively recent change in the relative levels of the sea and land ; and although the former noted the possibility that it might be the result of a shift of the sea-level, the latter author considered that there had been no general elevation but that the instances, where elevation of the land had been noticed, were 'so widely scattered that they afford convincing evidence of a cessation in the previous general subsidence'.

In 1898 Stanley Gardiner put forward the view that all the atolls of the Pacific showed signs of having within recent times been raised or that possibly there had been some change of level in the sea. Again in 1902 his observations on the Maldive Atolls convinced him that 'the land has been undoubtedly, by some means or other, raised above the sea and is now everywhere on the larger banks being washed away'. In 1906 (p. 454) he wrote 'starting with the land we are met everywhere (in the Indian Ocean) with evidences of a change of level having been largely responsible for its formation. This is true also of the groups of atolls in the Pacific and seems to be a regular phenomenon in the Indo-Pacific. At the same time we have indubitable evidence of certain islets or coral reefs having been formed by the piling up of reef materials. The change required varies from 10 to about 25 feet.' Agassiz (1903) has also called attention to the fact that in all the great oceans there is evidence of a change of sea- and land-levels. He remarks 'throughout the Pacific and Indian Oceans, and the West Indies the most positive evidence exists of a moderate recent elevation of coral reefs', and he uses this as an argument in favour of his view that coral reefs have arisen on platforms that are rising and not sinking.

A study of the coastal region of the mainland and of the various islands throughout Indian seas reveals the fact that in almost every instance it is possible to obtain evidence that within comparatively recent times there has been an alteration in the relative levels of the sea and land, though the amount of this change seems to have varied to some extent in different regions. In a previous paper (Sewell, 1927) I drew attention to the extent of some of the changes that have been noted to occur in the Indian and the Western Pacific Oceans, and I suggested that there might be evidence in favour of a circular land elevation that had its centre at some point off the north-west of Australia and the maximum elevation in the circle that passes the South Pole, along the east side of Africa, then through the Tien Shan Mountains and Korea, Japan, the coral regions of the west and south-west Pacific, New Zealand and so round to the South Pole again. There is also a considerable mass of evidence, to which Daly (1920, 1926) has drawn attention, that within comparatively recent times there has been a world-wide alteration in the relative levels of sea and land throughout the tropical regions to the extent of some 15–20 feet; and so uniform is this change that it must almost certainly be attributed to a fall of the sea-level rather than to an elevation of the land. As Stanley Gardiner (1930, p. 13, footnote 2) has pointed out, the amount of this eustatic shift varies in different regions with differences of latitude and the neighbouring topography, and he believes that this shift is part of a progressive lowering of the sea-level that is still going on and may possibly, he suggests, be due to a present and actively increasing glacial period for the Southern Hemisphere. Whatever may be the cause, this fall of sea-level is of quite recent occurrence, Daly (1920, 1926) and Edgeworth David (1926) putting it at some 3,500 to 4,000 years ago: the total amount of this fall of sea-level may have been reached in a series of successive stages, but it has had a profound effect on many sea-beaches and especially on coral-reefs; it is, therefore, of some interest to see to what extent we can trace the effects of the fall around the coasts of India and its associated islands. It is in coral-growing regions that evidence of this relative rise of the land is most easily obtainable, owing to the fact that coral, as we know, must have originally been formed below the low-water sea-level, so that if we now find masses of it, still *in situ*, as it grew, but now raised high above the level of low-water, it is a clear indication that a change of level of the sea and land has taken place. Such evidence is, however, by no means limited to these coral-growing areas.

The Cocos-Keeling Atoll.

So far as one can judge from the accounts of this atoll given to us by Darwin, Guppy, and Wood-Jones, there has been but little change in the relative levels of sea and land, though according to Wood-Jones (1910, p. 283) there has been a change of some 3 or 4 feet. He remarks 'between Pulu Tikus and Pulu Pasir these signs of elevation are striking. A notable fact in the disposition is that they occur in series, one layer above the other ; successive land movements being registered by successive steps of breccia, which rise one above the other to a total number of three or four, and to a height of almost as many feet. . . . . . I can imagine no other explanation for the presence of these raised platforms than that afforded by the supposition that the land on which they were laid down has been raised, as a whole, since their formation.' It need hardly be pointed out that a fall of sea-level would produce exactly the same effect, especially if this occurred in a succession of stages of depression, each succeeding phase causing an erosion of the breccia platform.

Coastal Regions of the Bay of Bengal.

I have previously (*vide supra*, p. 13) called attention to the fact that evidence of a recent alteration in the relative levels of the sea and land can be detected in the coral-growing region of the Mergui Archipelago. Around many of the islands in this part of the long coast line of the Bay of Bengal there are extensive coral-reefs that are exposed at low water. One of these is situated on the east side of Paway island and a study of the present conditions appears to indicate that it has recently been relatively elevated. The surface of the reef (Plate II) is dotted over with masses of dead coral, some loose, others firmly attached to the reef-flat, while scattered about are numerous small living colonies ; along the margin of the reef the blocks of dead coral are considerably larger than those nearer the shore, and owing to the sheltered situation of the reef one can, I am convinced, exclude any possibility of their having been thrown up by wave action. It is probable that they are still *in situ* in the original position in which they were formed and have been raised to their present position by a relative elevation. Unfortunately I cannot be absolutely certain on this point, for at the time when I was investigating this region my attention had not been called to this problem and in consequence I did not pay sufficient attention to the character of these coral masses, except to note that they are now dead.

On the opposite side of the Bay of Bengal on the east coast of India we again find evidence of an alteration in the level of the land. Mr. R. Senior White has informed me that in the neighbourhood of Vizagapatam there is a raised sea-beach, that appears to be now some ten feet above the level at which it must have been deposited, and Dr. Srinavasa Rao, of the Zoological Survey of India, who has on several occasions visited this area to study the fauna, has corroborated Mr. Senior White's statement. A little further to the north on the Orissa coast near the Chilka Lake, there is a similar raised beach. Blanford (1872, p. 61), who has called attention to this example, remarks 'near the south-west extremity of the spit there is a considerable deposit of estuarine shells, at a height of 20–30 feet. The Shells found, *Cytherea casta* and *Arca granosa*, have not been found living in the Chilka Lake, and both are estuarine species, not occurring in the sea itself : but the former is now abundant in the estuary connecting the lake with the sea. This deposit appears to afford evidence of a recent elevation of the land.' Wadia (1919, p. 32) points out that 'Oyster-shells discovered lately at Calcutta likewise point to a slight local rise of the eastern coast', or alternatively to a slight fall of sea-level.

The Andaman and Nicobar Islands.

I have previously (*vide supra*, p. 13) called attention to the fact that in both the Andaman and Nicobar Islands we get clear evidence of an apparent fall of sea-level and that the extent of the fall appears to be somewhat greater in the former than in the latter. In both groups of Islands there are remains of raised coral-beaches and these are particularly frequent, though by no means exclusively, along the east coast. Oldham (1885, pp. 144-5) had previously called attention to the occurrence of these structures, and further evidence of an alteration of land- and sea-level can be found in the presence of raised marine conglomerates that were undoubtedly laid down below the level of the sea but are now high above low-water mark and in some cases may be at the level of high-water mark or even higher. As I have previously shown (p. 16 *supra*) the presence of raised coral beaches on the east side of the Andamans and an interrupted barrier reef on the west can be explained by a shifting of the axis of the anticline on which these islands rest: but the presence of raised beaches and wave-cut benches on both east and west sides indicates that there must also have been a relative elevation of the land as a whole. Davis (1928, pp. 495–8) has given us an account of his theories regarding the past history of these islands; but unfortunately his knowledge was very incomplete and he speaks of 'the absence of shore cliffs, which proves that abrasion has not recently acted on the islands, especially on the little islands that rise from detached banks of small area'.

Throughout the islands of Ritchie's Archipelago evidence of a recent eustatic shift of land- and sea-level can be seen in a number of islands. On the west side of Havelock Island, is a sheltered bay, the beach of which is steep and is composed of fine sand. The land at the head of the bay is low-lying, but at each end it rises up in a steep slope, terminating in fine vertical cliffs composed of a white clay, that shows definite lines of horizontal stratification. Immediately above the beach around the centre of the bay the land rises for about two feet; it then drops again and is succeeded by a second rise, so that there is a trough-like depression running along the bay behind the present sea-beach. My impression is that the present beach and the outer lip of the trough are a recent formation, the inner lip of the trough being the old sea-beach. Behind this the land is flat and the soil of this flat area is of a deep rich brown and appears to consist mainly of vegetable mould; scattered about in an extraordinary profusion are sub-fossil shells and fragments of shells, most of which appear to be those of land molluscs, *Cyclophorus* sp., etc. but mixed with these are a few *Pyrazus palustris*, the common inhabitant of mangrove swamps, and the remains of *Trochus*, *Pteroceras*, *Buccinum* and some lamellibranch shells. The obvious conclusion seems to be that this area was originally covered by sea-water but that more recently there has been an alteration of sea- and land-level, that at first caused the area to become dry and that subsequently a new beach was formed; an elevation of about five feet would be sufficient to account for these changes. Gee (1926, p. 219) has noted regarding this island that 'in several parts of the coast a level tract, a few feet above high-water level, extended inwards for a short distance. Occasional pieces of coral and recent shells were met with, suggesting possible relative uplift in recent times. On the other hand in some places these might represent a deposit of fine sand blown up by the monsoons on to the coral reefs which fringe the sea-shore at many points.' Further along the coast towards the south-east a large creek opens into the sea and across the greater part of the mouth of this lies a bar composed of coral and clay-boulders, leaving a narrow deep channel between it and the land. On the west side of the entrance to the creek are some cliffs of clay, and running out from these into the channel are ledges and boulders of hard clay, that are bored through and through by *Pholas* and small Sipunculids. At one point on the reef are the remains of two or three trees, the trunks still standing *in situ*. It is evident that an alteration of the relative land- and sea-level to the extent of about five feet has recently taken place and this agrees well with the condition in the bay to the west ; the presence of the tree trunks, which are not those of mangroves and, therefore, did not grow in the sea, indicates that, since the relative elevation of the land, vegetation had obtained a footing but that a later erosion behind the raised reef has taken place ; at the present time this encroachment of the sea seems still to be in progress, for along the beach to the north of the spit several trees have fallen and at the time of my visit were lying on the beach ; indeed, wherever one goes in this region there seems to be evidence that erosion of the land is steadily proceeding.

At the south end of Henry Lawrence Island (Plate 3) there is a fine cliff of stratified clay and the extreme end of this cliff is much eroded. The cliff runs out in a promontory and in more places than one tunnels have perforated through, while in other places small caves have been formed at or near high-tide mark ; at the extreme end there is a well-marked wave-worn ' bench ', that was eroded when the sea stood at a relatively higher level and this is in its turn now being undercut by erosion at the present day. At the extreme south end of the island the cliff has been completely breached, two isolated masses still standing to show where the cliff originally extended. All around the south end of the island and extending sea-wards for about half a mile is an extensive raised coral-reef, much of which dries at low-water. The surface of the reef is composed entirely of coral-boulders, now dead and eroded, mixed with a fair quantity of sand. At the edge of the reef the coral ends abruptly and the bottom is composed of pure sand, with here and there a few colonies of living coral, mostly of the ' corymbose ' type of *Acropora* and a few alcyonarians. On the west side of the point the beach slopes upward to the edge of the jungle, that commences about one-and-a-half feet above high-water mark. This raised area is then succeeded, as in Havelock Island, by a trough-like depression that runs parallel to the beach and is eight to ten yards wide, inside which the land again rises steeply. This second slope is obviously an old sea-beach ; it is dotted over with shells of *Trochus* and *Pteroceras*, as well as with fragments of *Tridachna*, and its present height is from ten to twelve feet above the present sea-level and five to six feet above the top of the present beach. Gee (1926, p. 221) has noted that ' a short distance up the Kwantung Strait a stretch of loose sand containing recent marine shells and raised about 6 feet above high-water level occurs within the shore, suggesting a relative fall of sea-level within recent times '.

At the north end of Outram Island there is further clear evidence of alteration of level and erosion. For the most part the land terminates in steep cliffs and off the north end are two isolated flat rocks, that still indicate that the land in former times extended at least as far as this, while beyond these rocks the water is shallow for a considerable distance, the bottom consisting of rock, so that the actual point must originally have extended sea-wards for a considerable further distance. All along the shore the cliffs are in places undermined and trees have recently fallen or will soon do so, thus showing that erosion still continues. At the south-east end of the same island considerable erosion is taking place around the point, while extending sea-wards from the foot of the beach there is an extensive raised coral and sand flat that dries at low-water.

On the north-west of Neil Island there is an extensive reef of coral that is exposed at low tide; the shore is composed of large masses of Astræids, that are clearly still in the position in which they grew, and littered all over the beach and to a large extent consolidated with it are fragments of coral that have been torn off the edge of the reef and have been flung inshore. This side of the island is exposed to the full force of the north-east monsoon and this would be quite sufficient to account for the piled-up fragments. In one place these fragments have been piled up into a ridge, closely resembling a boulder zone, that runs parallel to the shore, while behind it mangrove trees have obtained a footing and are now flourishing. The sea-ward edge of the reef slopes gradually down into deep water and is covered with masses of growing coral, conspicuous among which are numerous large examples of *Turbinaria*, and this accounts for the great profusion of fragments of this type of coral on the beach. In-shore there is a strip of sandy beach lying between the reef and the land and all along the coast is littered with the remains of fallen trees, while several more, though at present still standing, are dead and have their roots exposed. Immediately behind the beach the land rises for a foot or two and then sinks again to a muddy, swampy area, behind which is a second raised ridge and swamp; these ridges are frequented by a large burrowing crab and where these have excavated their holes, the debris that has been thrown up consists entirely of coral fragments and broken shells, showing clearly that this part of the island is a raised beach or succession of beaches. In Neill Island Gee (1926, p. 219) has noted that 'the relative uplift of this island appears to have been going on quite recently, for on parts of the coast boulders of recent coral, occurring above high-water mark, extend inland for some distance'.

Again in Wood-Mason Bay at the south-western end of Rutland Island in the South Andaman group and in the Twin Islands off the north-west end of the Bay there are several raised pebble or shingle beaches at a height of six to ten feet above the level of the present beach. Gee (1926, p. 225) also noted that 'the coast also shows evidence of recent relative elevation of the land in the occurrence of a sandy beach about 6 feet above high-tide, and again in the presence of a recent pebble conglomerate similarly raised above the present high-tide level and fringing the coast of the north-western part of the western island'. In one of the small water-pools fed by a small stream at the top of the beach in about the middle of the Bay there is a bank of coral boulders and sandstone blocks ; the presence of this coral is interesting, for nowhere else along the beach in this bay was any dead coral seen. It is probable that this coral represents an earlier formation against which the sand subsequently drifted, the only part of the deposit now exposed being in the bed of the stream that feeds the water-pool, well above the present high-water mark (Plate 12, fig. 1).

Gee has also noted (*loc. cit.*) that in the Cinque Islands 'a raised beach of fine sand with recent shell-fragments about 15 feet above high-water level occurs on the west coast of the southern islands.'

Very similar conditions can be seen in Little Andaman Island. On the west side of the island about a mile or two to the north of Jackson Creek there are some fine cliffs composed of stratified sandstone (*vide* pl. i, fig. 1A) ; at the foot of these cliffs are several caves, some of which are at the present water-level, while others are actually above the level of the present high-tides, and it thus seems clear that they must have been eroded at a time when the sea-level was higher than it is to-day. At the south end of these cliffs there is a well-marked terrace that has been cut by wave action just below the present high-water mark and is of a width of some eight to ten feet ; above this terrace there is a second, some ten feet above the first and now well above high-water. It seems clear that this lower terrace has been cut by the waves in recent years and if, as one supposes, the upper one was cut by the same agency, there must have been a change of level of about ten feet ; such a change would also account for the caves at the base of the cliffs. To a less extent traces of this upper terrace can be seen in the cliffs at the north end of the bay. Further evidence of this eustatic shift is to be seen at the south-east side of the island. Gee (1926, p. 226) records that at the south-west corner of Hut Bay 'unquestionable evidence of the occurrence of beds of recent coral rock *in situ*, above the present sea-level, was noted. Following inland a short distance, a small stream enters from the west. In the bed of this stream boulders of recent coral were abundant, and also occurred in the dense under-growth in the vicinity of the stream. From the lowlying topography of many parts of the island it seems possible that other parts of the coast are composed of similar coral exposed by a recent uplift of the island.'

Turning to the Nicobar Islands we again find clear evidence of a recent change in the relative levels of sea and land. At the northern and north-western side of Camorta Island in the Central group a wide reef, the Perseus Reef, runs out towards the north-west for about three-quarter of a mile. The reef is continued towards the south along the west side of the island but gradually decreases in width. In the edge of the reef are a few narrow passages through which small boats can pass, but there is no deep water inside the reef. The extreme margin of the reef consists almost entirely of coral and nullipore ; the corals are for the most part in large masses that still show the characteristic form in which they grew and in places there are circular or oval patches, raised from six inches to one foot above the general level of the reef, but these are now all dead and water-worn. Scattered over the reef are numerous colonies of living coral and these increase in number near the extreme margin ; they are all small in size and include examples of *Acropora*, of both Stag-horn and 'Corymbose' types, *Millepora*, *Heliopora cœrulea*, *Pocillopora*, *Dendrophyllia*, Astreaceæ, etc., while fragments of dead coral form a substratum between the larger masses that are still *in situ*. In one part of the reef the whole floor is composed of nothing but nullipore, most of it showing the usual pink colour, with here and there patches of Alcyonaceæ. In the southern part of the area the reef-flat is traversed by long inclined ridges that run parallel to the sea-coast and appear to be the remains of hard clay bands, the softer clay of the stratified bands having presumably been washed away. In the main area itself on the land side of the reef lies a very extensive mud- and sand-flat that almost entirely dries at low-water spring-tides; in its outer two-thirds this flat is covered with a profuse growth of weed, but inshore it is of bare sand except at the extreme north end, where even at low-tide the reef is covered by about a foot of water and the weed growth extends right up to the edge of a fringe of mangrove trees. Several creeks extend inwards for a short distance into the island and then turn and run parallel to the beach but separated from it by a belt of dry land covered with vegetation. On the inshore side of these creeks the soil is evidently an old raised marine deposit and consists largely of fragments of coral, shells, etc., with a mixture of small clay boulders. It seems clear that within recent years this area has undergone relative elevation; mangroves have then sprung up along this raised beach on the sea-ward side and subsequently a second beach has been thrown up.

A very similar coral- and mud-reef is to be seen on the east side of Kachal Island at the south end of East Bay. Here the land runs out in a small promontory and terminates in a series of isolated masses of limestone that have been cut off from the main mass of the cliff (Plate 12, fig. 2). Extending sea-wards from the beach there is a wide reef of dead coral and mud that dries at low-water spring-tides; in places on the reef there are small pools that are left by the receding tide and these are occupied by growing colonies of coral, but the greater part of the reef is now composed of dead and water-worn growths that are clearly still in the position in which they lived. That a recent relative elevation of the land has taken place is shown, not merely by the dead coral but also by the presence on some of the limestone masses of the shells of the rock oyster (*Ostrea cucullata*) that are now well above the level of high-water.

I have in a previous paper (Sewell, 1922, p. 984) drawn attention to the presence of a band of raised reef coral in Trinkat Island, lying to the east of Nankauri Harbour, and there is a band of the same material running along the top of the beach on the east side of Kachal Island; it was in the under side of this band that I discovered a colony of the boring barnacle, *Lithotrya nicobarica*, and from the composition of the rock and the presence in it of masses of dead coral and *Tridachna* shells I believe that this was once part of an old coral-reef that was laid down below sea-level. If this be so, it would point to a rise of the land or a fall of sea-level of about six feet.

In all the islands of this group we thus seem to find evidence of a shift of sea-level to an extent that varies from five to ten feet; it is also interesting to note that there are extensive coral reef-flats around most of the islands and that these reefs are best developed around headlands and promontories. At the present day these headlands are all being eroded in spite of the partial protection afforded by the surrounding reefs, and the presence of the reef-flats suggests that the eroded remains of a further extension of the land have served as a basis on which these coral-reefs had been able to grow, but that a further change of sea- and land-level has converted this coral-growth into dead reefs.

Ceylon and Southern India.

As regards the conditions to be found in Ceylon I have been unable to make any personal observations, but Stanley Gardiner (1903, p. 19) records that locally there are 'fringing reefs, mostly in bays along the south and north coasts; above these to the south a loose rock, composed mainly of fragments of coral, often forms broad flats, three to four feet above high tide, now being rapidly washed away wherever they impinge on the beach.' It seems more than probable that these flats must have been laid down below sea-level and have been relatively elevated to their present position.

Foote (1889) and Thurston (1895) have called attention to the presence off the south coast of India of a raised coral-reef that is now situated at a height of about eight feet above the level at which it was originally formed; and in March, 1926, I spent some three weeks in this vicinity, examining the reef and the area around. At the present day there is but little of the actual raised coral still to be seen, but some remains at the north-east corner of Rameswaram Island, near Pamban. As one walks along the shore from the end of the railway-bridge towards the light-house on the north-east point of the island one sees an exposure of a band of shelly conglomerate that lies horizontally; this deposit has been exposed by erosion of the land and there is clear evidence that this erosion is still in progress, for close by there are, or were when I visited the spot, several palms still growing but the roots of which are exposed, and at one place an attempt has been made to prevent further erosion by the construction of a 'bund' but this has in its turn been destroyed. The conglomerate is composed of sand and shells, mixed with a certain proportion of broken coral fragments, and its upper level lies at a height of about six feet above the present high-water mark. About one hundred yards to the north-east of this horizontal stratum one comes across an exposure of sandstone, which is clearly stratified, the strata showing a well-marked dip towards the west or south-west at an angle of approximately 1 in 2 (Pl. 12, fig. 3). Still further along the beach there is a second exposure of sandstone, but in this latter area the exposure shows two zones; the lower zone is composed of strata that show a dip towards the north-east, while above this is a horizontal layer. Above these two exposures is a varying thickness of soft loose sand, the varying depth of the sand being due in all probability to the migration of the sand dunes across the island under the influence of the prevailing winds. The difference of the dip in the two exposures would appear to indicate that the sandstone had been formed beneath a sand ridge and that the dips corresponded to the seaward and landward slopes respectively. The horizontal shelly conglomerate was almost certainly laid down under water in a shallow lagoon or possibly in a back water ; the sub-fossil shells from this latter deposit have been identified for me by Dr. Srinavasa Rao, to whom my thanks are due, and a list of them is given below :—

Cerithium angulifera Sowerby.
Cerithium mankei Deshaye.
Columbella sp.
Columbella versicolor Sowerby.
Euchelus edentulus Ad.
Hemicardium auricula Forskal.
Marginella pacifica Pease, or *M. asselina* Jones.
Meretrix meretrix var. *castanea* (Lamarck).
Murex tenuispina Lamarck.
Nassa nodata Hinds.
Potamides fluviatilis Pot. and Mich.
Potamides (*Tympanotonus*) *cingulatus* Gmelin.
Umbonia vestiaria, Lamarck(=*Rotalla vestiaria*).

On the sea-face and running out in a small spit are several masses of coral, about two to three feet in diameter, that are clearly still *in situ*, though now far above the level at which coral can grow, while other smaller masses are embedded in the exposed face of the reef (Pl. 12, fig. 4 and Pl. 13, fig. 1). The majority of these corals belong to species of a massive character, such as *Porites* or Astraeaceæ, or to a type of brain coral, but in places there are collections of broken fragments of a delicate branching coral of a species of Madrepore, such as one finds to-day growing in sheltered waters. Extending seawards from the shore there is a recent reef-flat that terminates seaward in a fringing reef. From the appearance of the reef and the main masses of coral it seems clear that this raised area once formed the margin of a fringing reef and that the sandstone behind it is the result of consolidation in the beach ; to place the reef at its present position there must have been a change of level between the land and sea of at least seven to eight feet.

Round the point on the east side of the island there is clear evidence of erosion in the shape of exposed roots of Palm trees, and at intervals the old raised reef is exposed ; along the upper part of the beach the outcrop consists entirely of a coral-conglomerate, the upper edge of which is horizontal and is situated at a height of about six to seven feet above the present low-water level. On the seaward side of this horizontal stratum are scattered a number of large, rounded or oval masses of coral of several species ; the conglomerate itself is composed of sand and shells and fragments of a branching Madrepore coral and closely resembles the conglomerate that is being formed at the present day on the inshore side of many reefs. The beach is gradually being eroded and blocks of this conglomerate are being undermined and are in consequence subsiding and breaking away from the rest of the stratum.

Further along the same face of the island towards the town of Rameswaram there is further evidence of the old reef. All along the beach for at least two miles from the

Fisheries Bungalow there is a water-worn outcrop of coral-conglomerate and sandstone, with which are mixed a few blocks of coral and numerous shells; the deposit thus closely resembles that of a reef-flat but at the present time it is six or seven feet above the present sea-level (low tide). About half a mile from the Bungalow there is above this old reef-flat on the landward side a well-marked layer of sandstone that exhibits a clear horizontal stratification; this I consider has been formed from the consolidation of sea- or wind-born sand that lay originally inside the reef. A similar formation is exposed at intervals along the coast, either by a process of natural erosion, or artificially where excavations have been made for the purpose of quarrying the stone for building purposes. The whole length of the exposure is being steadily eroded, large blocks have become undermined and have fallen, while in places 'blow-holes' have formed. From the edge of the small cliff that now forms the seaward face of the exposure the rock slopes gradually upwards to the edge of the cultivated land and in places one can trace parallel lines of low ridges, that closely resemble the successive strata of beach-sandstone in some of the Maldive islands. There can be little doubt that this formation is part of an old fringing-reef, even though actual coral masses are comparatively rare; occasionally one sees a moderate sized block of coral embedded in the sandstone and small fragments of broken Madrepore are common. One large colony was found that must originally have measured seven or eight feet in diameter but is now so eroded that the major part of the mass has disappeared; but with rare exceptions such as this the original coral blocks of the reef face have been eroded and all that is left is the reef-conglomerate. Extending seawards throughout the whole area there is a well-formed and recent reef. In one or two places the line of this old raised reef is broken; one such break occurs just round the first point to the north of the Fisheries Bungalow, and here the deposit is of a red colour and consists for the most part of rounded nodules, each nodule enclosing a Gastropod shell, that has been identified as *Areophanta rugata* Beck, a species that is extremely common all over the sandy region of South India. The interruption of the coral-reef and the occurrence of these subfossil land shells suggest that there was at this point a break in the original reef, probably where a small stream flowed out into the sea, and that these shells had been washed down and had accumulated at this point. This old coral-reef originally extended far beyond the limits of Rameswaram Island, for as Foote has pointed out, there is a raised coral and limestone 'quay-wall' to be seen along the coast of southern India between Muttupettai and the spit opposite Pamban, and also to the east of Kilakarai. I have myself examined this formation in the neighbourhood of the Mandapam spit. On the north side, extending for over half a mile from the end of the spit, there is a fine exposure of limestone and coral-conglomerate; the lower stratum is situated just above the present low-water mark and is a conglomerate, consisting of sand mixed with shells and broken fragments of coral; the upper layer is a pure sandstone, the upper boundary of the deposit being horizontal and the stratum having a dip, amounting in places to 1 in 3 or 1 in 6, away from the sea. It seems probable that the lower stratum was originally part of a reef-flat and that the upper stratum was formed beneath the reverse slope of a sand beach, but that erosion has taken place to such an extent that the greater part including the original reef, most of the reef-flat and the outer part of the sandy beach behind it, has now been eroded away. About three-quarters of a mile further to the north, in the neighbourhood of Mandapam station, one again comes across further evidence of this old reef in the occurrence of an outcrop of shelly conglomerate, but beyond this point, as far as I could see, all traces had disappeared. On the south-west side of the point one finds all along the coast the same outcrop of shelly conglomerate and sandstone; here too erosion has been and still is taking place and blocks of the sandstone are being gradually undermined and are breaking off. The sandstone in this region exhibits a varying dip ; in places the dip runs towards the sea, while in others it is horizontal, and in a few places the inshore part of the exposure shows a 'reverse' slope, the dip of the stratum being away from the sea with a slope of approximately 1 in 1½. In places the top level of the sandstone is actually higher than that of the general surface of the land and between the two beaches on the east and west sides of the spit the whole extent is clearly a marine alluvium, being composed of sand mixed with shells, for the most part belonging to marine species but mixed with others that are those of land snails and especially of *Ariophanta rugata*.

Clearly the old coral-reef has been raised relatively to the sea and we now get exposure of the old reef-flat, while in places, on what was the inshore side of the old reef, there are exposures of the sandstone that was formed beneath the surface of the old sand beach by the gradual percolation of rain-water or sea-spray that dissolved the calcium carbonate from the most superficial layer and redeposited it in a lower stratum. The whole geological formation of this region appears to me to indicate clearly that there has been an alteration in the relative levels of the sea and land to the extent of at least eight feet, and possibly considerably more, for one cannot give any definite figure for the amount of erosion that has occurred on the upper aspect of the reef.

In order to obtain evidence of the time when this reef was formed I made a collection of the various species of coral ; in the specific determination of these I was greatly assisted by Dr. G. Mattai, to whom I am deeply indebted for the help that he so kindly gave me. The various species and their present day distribution are given below :—

Family ASTRÆIDÆ.

Genus *Cyphastræa* Klnzr.

(1) *Cyphastræa microphthalma* (Lam.)

Specimens were obtained from the raised reef at Pamban.

Present distribution : Maldives, Chagos Archipelago, Red Sea, Egmont, Amirante, Saya de Malha, Cocos-Keeling, Australia, New Hebrides (?), etc. ; widely distributed throughout both Indian and Pacific Oceans.

Genus *Favites* Link.

(2) *Favites abdita* (Ell. and Sol.).

Specimens were obtained from the raised reef at Rameswaram.

Present distribution : Banda, Amboina, Tonga Tabu, Fiji, Singapore, Australia, Cocos-Keeling, Mergui Archipelago, Andamans, Ceylon, Maldives, Chagos Archipelago and Red Sea; common throughout both Pacific and Indian Oceans.

Genus *Favia* Oken.

(3) *Favia acropora* (Linn.).

Examples were obtained from the raised reef at Rameswaram.

Present distribution : Solomon Island, Rotuma, Funafuti, Egmont, Amirante, Seychelles and Red Sea. Throughout both Indian and Pacific Oceans.

(4) *Favia doreyensis* M.E. and H.

Taken from the raised reef at Pamban.

Present distribution : New Hebrides (?), Australia, Chagos Archipelago, Maldives, Egmont, Saya de Malha and Red Sea.

(5) *Favia speciosa* (Dana.).

Specimens were obtained from the raised reef both at Mandapam and Pamban.

Present distribution : Solomon Islands, Australia, Cocos-Keeling, Aracan coast of Burma, Ceylon, Chagos Archipelago, Maldives, Egmont, Seychelles and Red Sea. Common throughout Indian seas.

(6) *Favia vallenciennesi* (M.E. and H.).

Specimens taken from the raised reef at Mandapam.

Present distribution : Banda, Padaw Bay, Mergui, Ceylon, Aldabra, Seychelles and Red Sea. Common throughout Indian seas and the Malay region.

Genus *Goniastræa* M.E. and H.

(7) *Goniastræa reteformis* Lamarck.

From the raised reef at Mandapam.

Present distribution : Fiji, Philippine Islands, Great Barrier reef of Australia, Rotuma, Amboina, Ceylon, Singapore, Maldives, Seychelles, Aldabra and Red Sea. Common throughout the Western Pacific and Indian Oceans.

Genus *Symphillia* M.E. and H.

(8) *Symphillia sinuosa* (Q. and G.).

From the reef at Mandapam.

Present distribution : Philippine Islands, Rotuma, Amboina, Singapore, Mergui, Andamans and Maldives.

Family PORITIDÆ.

Genus *Porites* Link.

(9) *Porites nodifera* Klnzr.

Examples taken from the raised reef at Pamban.

Present distribution : Mergui Archipelago and Red Sea.

Family MADREPORIDÆ.

Genus *Turbinaria* Oken.

(10) *Turbinaria crater* (Pallas).

An example from the raised reef at Pamban.

Present distribution : Pacific Ocean, Kermadec Islands, Cape York, Torres Straits, Amboina, Mergui Archipelago, and Ceylon.

Genus *Acropora* Oken.

(11) *Acropora pharaonis* M.E. and H.

One example from the raised reef at Pamban.

Present distribution : Red Sea, west part of Indian Ocean, Cocos-Keeling.

In addition to the above specimens a few corals were collected from the growing reef along the coast ; these included—

Cyphastræa calcidicum Klnzr.
Favia speciosa (Dana).
Goniastræa incrustans Duncan and
Porites nodifera Klnzr.

It is clear, therefore, that this raised reef is ' recent ' so far as its formation in geological times is concerned.

The Western coastal region of India.

Fox (1922) has recorded that in the Island of Bombay ' there are raised terraces of marine sediments 12 feet above sea-level which indicate that large tracts of the Western side of the Island have been recently elevated from beneath the sea ', but as he also records submerged forests on the Eastern side the elevation of the old marine deposits may be due to a tilting of the Island and not a change of sea-level.

Fedden (1885) has given an account of the various changes in level that have occurred in the Kathiawar district of Gujerat. He points out that in this area after the outflow of the Deccan trap ' the lower half of the province was depressed to a certain extent, bringing the southern margins of the traps beneath the sea, and then tertiary strata of miocene age were deposited as a littoral fringe upon the submerged portion of the traps. The south-western part appears to have remained under water longer to admit of later accumulations upon the miocene rocks. After the depression a period of elevation set in ; but at a later date, i.e. in subrecent times, nearly the whole province was again dipped beneath the sea, only the tops of the highest hills probably escaping submersion ; and on its final elevation the receding waters left their mark in the patches and fringes of white miliolite now seen on the sides of the hills and in recesses in the ravines ', these patches lying at a height of some 50–100 feet above the present sea-level. Recent changes of sea-level are also indicated by raised beaches on the south-west coast, as well as by dead coral-reefs that fringe the whole coast facing the Gulf of Cutch from Nawanagar westward, including the islands off the coast ; and still further evidence of this eustatic shift is to be seen in elevated oyster beds in the Malia district. In this connection Wadia (1919, p. 31) remarks ' In some of the creeks of Kathiawar near Porbundar, for example, oyster-shells were found at several places and at levels much above the present height of the tides, while barnacles and *Serpulæ* were found at levels not now reached by the highest tides. In Sind a number of oyster-banks have been seen several feet above high-water mark ', and ' even within historic times the Rann of Cutch was a gulf of the sea, with surrounding coast towns, a few recognizable relics of which yet exist. The gulf was gradually silted up, a process aided, no doubt, by a slow elevation of its floor, and eventually converted into a low-lying tract of land, which at the present day is alternately a dry saline desert for a part of the year, and a shallow swamp for the other part '. While Wadia attributes this change in the Runn of Cutch to, at any rate in part, a slight elevation of the land, it may equally have been due to a slight fall of the sea-level.

The Persian Gulf.

Pilgrim (1911) has called attention to sub-recent changes that have taken place in the sea-level in the Persian Gulf. Evidence of this change is to be seen in several of the islands—

(*a*) *Bahrain Island.*

Pilgrim states that this island subsequent to Eocene times was gradually raised above the sea and much of the formation was denuded as it rose. ' The flat portions of the island bordering the coast, especially to the north of the island, and including the whole area which is under date cultivation, are of sub-recent age '.

(*b*) *Quishm Island* (Kishm Island).

' The whole presents an appearance of having been carved out by the sea, which has within recent times retired from the topmost levels, exactly as has been observed in Bahrain Island. Here too terraces have been left at various levels, in many cases covered with sand, which is probably the remains of a sub-recent beach deposit.'

Unfortunately in neither of these cases does Pilgrim give any indication of the extent to which this change of level has taken place.

(*c*) *Hormuz Island.*

' The level portion of the island to the north consists of a platform of horizontally bedded, sub-recent littoral conglomerate, hardly ever elevated more than 10 or 15 feet above the sea and crowded with recent shells.'

(*d*) *Halul Island.*

' On the south side near the anchorage are cliffs of sub-recent shelly concrete some 15 to 20 feet high and these are found again in places on the western side of the island.'

In all these instances it is probable that we have evidence of a single movement in the relative levels of the sea and land and that this shift is part of the same movement that we have seen in other parts of the Indian coast.

The Coral atolls and islands of the western Indian Ocean.

In all the accounts of these coral atolls and islands there are numerous references to changes in the sea- and land-level. In his account of Minikoi Stanley Gardiner

(1903, p. 36) remarks 'the conglomerate then may be considered to prove conclusively an elevation of the atoll. Its summits are 6 feet above the reef flat (the low-tide level) and allowing 18 feet for the depth at which it was formed, there must have been an alteration of level of at least 24 feet'; but in the southern part of the Maldive Archipelago the extent of elevation appears to be considerably less than this and probably was not more than 5-6 feet.

In the Chagos Archipelago to the south Bourne (1888) reaches the conclusion that there has been elevation of some 4 feet in Diego Garcia, and in both Saloman and Peros Banhos Atolls evidence of this is to be found in masses of coral-rock now standing well above the height at which they can have been formed; though as regards the latter area Stanley Gardiner (1907, p. 39) points out that the greatest height of this rock is 8 feet, which is more than one would expect. Freyer (1911) has called attention to the elevation of Aldabra Atoll and Mauritius, which he estimates to have been in the neighbourhood of some 60 feet; but this is too great to be accounted for by the recent change of sea-level.

The Coral Reefs.

Agassiz (1903*a*, p. 232) from his experience of the Gilbert and Ellice Atolls of the Pacific Ocean concluded that 'in those districts where it is impossible or difficult to observe the structure of the underlying platform, the coral reef question limits itself to the study of the cutting down of the reef platform and of the land rim, of the movements of the mass of material which rests upon the underlying platform, whatever be its nature, the material being supplied by the growth and decay of the corals, of the Nullipores and animals living upon the reefs, either of the sea face or of the lagoon, and to follow the endless combinations resulting from the nature of the reef platform and the local phenomena due to the shape of the land, the tides and winds'.

In every coral-growing area and particularly in oceanic coral reefs two diametrically opposed processes are continually in operation, the one constructive, depending on the growth of the coral and associated plants, such as Nullipores, and the geological formation of conglomerate rocks and sandstones from the coral or other calcareous debris; the other destructive, due to erosion by waves and currents, to the activities of animals that feed upon the coral or of animals and plants (algæ) that bore into it and so render it less solid and more liable to destruction, to adverse conditions of temperature, salinity, etc., and perhaps in some degree to actual solution by the sea-water. In no two given areas will exactly similar conditions be found; even in the same atoll, conditions must be vastly different on the two sides, the one being to windward and the other to leeward. It follows, therefore, that every atoll must be judged independently, and conclusions reached from a study of one atoll or group of atolls may by no means apply equally well to atolls or reefs in some other part of the world, and it is as yet, to my mind, by no means certain, as Stanley Gardiner (1931, p. 131) states, that 'the general resemblance in the topography and in the biology of the builders of the atolls of both oceans (Indian and Pacific) would seem to be so well established that it is safe to apply the phenomena discovered in one ocean to the other '. In the majority of instances this may be, and probably is, justifiable but it does not appear to be of universal application, more especially when one is dealing with the destructive processes that are going on, and Stanley Gardiner himself notes (*loc. cit.* footnote) that ' minor reservations may have to be made in respect to each lagoon, when comparisons are being drawn or deductions from water movements stated '.

Saville Kent (1893, p. 98) put forward the view that the formation of a coral-reef is due primarily to the formation of coral conglomerate and limestone and not, *per se*, to the growth of coral, while the upward extension of the reef depends on the consolidation of dead fragments into a solid basis of limestone on which other corals can become fixed and thrive ; it is in the main only on the upper surface of a coral bed that such a process can go on and, as he points out, this formation only takes place in tropical regions, or at any rate, in regions where tropical conditions exist and where in consequence the rapid evaporation of sea-water causes a deposition of calcium carbonate from solution. ' Temperature, therefore, and not the specific varieties of Madreporaria ' he maintains, ' represents the prime factor in reef construction '. If this be the case, then it is clear that as one passes downwards from the surface to greater depths of water, there will be a more or less rapid falling off in the rate of formation of this conglomerate and limestone and that we shall soon reach the limits of depth at which coral formation can give rise to a definite reef platform, though it must be clearly understood that this may not in any way tally with the actual depth at which so-called reef-forming corals can grow and thrive. That corals of the various reef-forming species can grow at depths of 25 to 30 fathoms is well known and the limiting factor is in all probability decided by the intensity of sunlight at such depths, and is not a result of lowered temperature. But, so far as I am aware, the available evidence is by no means conclusive that a definite coral-reef can be built up from such a depth.

In this connection it is of some interest to study the effect that a marked rise of temperature may have upon the animals living on the reef flat. Mayer (1918) carried out experiments with eight species of corals from the reef at Tortugas, Florida, and he found that exposure for one hour to a temperature of 36·5°C. would suffice to kill 50 per cent. of those examined. Edmondson (1928, p. 19 *et seq.*) has also carried out experiments on the degree to which corals can stand a rise of temperature and he found that ' approximately 70 per cent. of the corals of Waikiki reef used in this experiment are capable of enduring 34°C. for at least one hour. . . . However, if the sea-water is slowly raised to 35°C. about 50 per cent. of the corals are able to endure this temperature for 15 minutes, 36 per cent. survive these conditions for 30 minutes, and 18 per cent. for at least one hour.' Yonge (1931, pp. 154, 166) has shown that one of the results of exposure of corals to a temperature of some 36°C. is to cause the death and disappearance from the coral tissues of the zooxanthellæ and this he attributes entirely to the lowered vitality of the coral polyps and not to any direct effect of the temperature on the zooxanthellæ themselves. The temperature of the water on the reef-flat in one observation that I made was raised from 29·8°C, the temperature of the surface water of the lagoon, to 36·5°C. in the middle of the outer reef-flat. Temperatures of 37·1°C. and 37·8°C. have been recorded in pools in the Great Barrier Reef. It is clear that a temperature of over 36°C. must be extremely harmful to the great majority of corals living on the reef-flat, and if a temperature of 36·5°C. be recorded, as in this instance, the actual surface of the water must have been even hotter. In another part of the same reef it was found that a number of colonies of living coral, of a species of branching *Porites*, had been so affected that all the polyps on the upper surface had been killed off and each dead animal was surrounded and enclosed in a deposit of calcium carbonate. Around the sides of the colony the polyps were still living (Pl. 13, fig. 2). At first sight I was inclined to attribute the death of these upper polyps to the effect of sedimentation, as has been described by Wood-Jones (1910), but a further consideration of the question has lead me to doubt this. Marshall and Orr (1931, p. 131) have pointed out that in the Australian Barrier Reef 'there are often found colonies (of *Porites*) which are flat-topped and dead in the centre. In the moat at Low Isles the tops of all the flat-topped *Porites* occurred at almost the same level, namely just above the constant low-tide level, which indicates that they have been killed by exposure to unfavourable conditions at low tide', and one of these unfavourable conditions and possibly a very important one will be the adverse effect of the raised temperature. The deposition of the calcium carbonate was undoubtedly due to the reaction between the decomposing animal matter in the dead polyp and the calcium sulphate in solution in the sea-water, in the manner described by Murray and Irvine (1889, p. 92 *et seq.*) and such a deposit of calcium carbonate will materially assist in forming a compact reef rock. Stanley Gardiner (1930, p. 7) has noted that 'a similar looking amorphous form of lime often is found on the surfaces of reef coral polyps and sea-weeds, both of which are to some degree dependant on chlorophyll for their nutrition, and all our dredged material of such, especially from the Cargados bank, had to be brushed clean..... we commonly ascribed any death in these organisms to it'. I have seen a similar deposition of calcium carbonate going on in shallow water on the reef-surface at Tor on the Sinai coast of the Gulf of Suez in beds of a species of mussel, *Mytilus variabilis* Krss. 'At the time when I was stationed at Tor in 1916, these beds appeared to be rapidly disappearing and the mussels becoming killed off owing to an accumulation of sand and mud that was covering the beds, and by a simultaneous deposit of a white chalk-like material' (Sewell, 1924, p. 514). Although I attributed the death of these mussels to the deposition of silt and mud it is equally possible that a rise of temperature may have been at least a contributary cause, since the time when I was at Tor was in May and June, one of the hottest times of the year.

Another factor in the building up of the reef and one that is, so far as the evidence goes, particularly important in the formation of the outer part of the reef, is the growth of nullipores. Encrusting *Lithophyllum* and *Lithothamnion* serve to bind all coral growths firmly to the coral rock and at the same time by their own growth add considerably to the deposition of calcareous material; but it also appears probable that this factor plays little or no part in the formation of coral-rock in the inner part of the reef that lies within the boulder zone and that Stanley Gardiner terms the ' boat channel '. Still less does *Lithothamnion* or nullipore enter into the composition of the lagoon margin of the reef. There is some evidence that coral can grow up from comparatively small depths; Darwin described the growth of coral in the Schooner channel in Cocos-Keeling Atoll and mentions that this channel has periodically to be freed from growing coral, and in the Andamans I have myself seen and run aground on a flourishing coral patch that is now covered by only a foot or two of water but where in the chart that was drawn in 1887 a channel is shown having a depth of water of 6 fathoms and I see no reason to doubt the original accuracy of the chart. Incidentally, this would indicate an upward growth of the coral of some 34 feet in 37 years or very nearly a foot a year. Dana (1875), Mayor (1918, 1924) and others have conducted observations that give us some indication of the rates at which individual coral colonies can grow in various parts of the world, and Stanley Gardiner (1903) has also given data dealing with the rate of growth of colonies from Hulule, North Male Atoll in the Maldive Archipelago; but such data prove little or nothing as regards actual reef formation and only indicate the rate at which coral can grow in shallow water, which is naturally rapid. Verstelle (1932) has recently given data regarding the rate of growth of coral reefs in the Dutch East Indian Archipelago and the figures that he gives indicate that the rate of upgrowth in these reefs is more rapid at a depth greater than 5 fathoms and gradually decreases as the depth gets less. The averages of Verstelle's figures are as follows :—

Depth of Reef.	Togian Islands.		Gulf of Tomini.		Kei Islands.	Average.
	Reefs on edge of continental shelf.	Reefs not on edge of continental shelf.	First part.	Second part.		
	cm.	cm.	cm.	cm.	cm.	cm.
Less than 3 metres	4·0	5·7	4·2	3·5	1·7	3·82
3 to 5 metres	13·5	8·3	5·8	7·0	1·4	7·20
More than 5 metres	14·1	15·3	9·0	10·5	5·5	10·88

The maximum increase that Verstelle noted in the height of a reef was at the rate of 41·4 cm. per annum or 16·3 inches, so that the probable rate of upgrowth of 1 foot per annum in the reef in the Andaman Islands to which I have referred is not impossible. Verstelle also noted that certain reefs show evidence of having ' sunk ', or in other words that there is an increase of depth of water over the reef. The average results of this nature that he obtained are as follows :—

Depth of Reef.	Togian Islands.		Gulf of Tomini.		Kei Islands.	Average.
	Reefs on edge of continental shelf.	Reefs not on edge of continental shelf.	First part.	Second part.		
	cm.	cm.	cm.	cm.	cm.	cm.
Less than 3 metres	2·4	3·8	5·4	2·3	3·7	3·52
3 to 5 metres	4·2	8·3	3·2	..	4·1	3·96
More than 5 metres	4·9	5·2	3·6	2·7	2·7	3·82

If, as suggested by Verwey in a postscriptum to Verstelle's paper, this 'increase of reef depth is caused by factors which influence coral growth and not by real sinking' one can only attribute it to the death and subsequent removal either by solution or erosion, of the coral colonies or rock of the reef. It is of some interest to note that the average figures, as given above, indicate that the causative agent may be as effective at depths as great as 5 metres as near the surface : in shallow depths a certain amount of erosion may occur from the scouring action of currents (*vide infra*, p. 512) but it is unlikely that this agency would produce an appreciable result at a depth of 5 metres.

Stanley Gardiner (*loc. cit.*) considers that 'the rate of growth of the whole reef is not probably widely different from that of the individual colonies that mainly serve to build it up' ; but the actual reef formation does not depend on the rate of growth of the corals alone, there is also other factors, namely the filling up of the interspaces by sand, foraminifera, nullipore growth, etc. and this last factor is of equal, if not of greater importance than that of the coral itself. Thus coral growth is not the same as the upward growth of the reef-edge, and it is the growth of the reef-edge that matters, since no coral can retain a position on this part of the reef except in the gullies of the fissure zone or down the sea-ward slope.

The compactness and hardness of a reef may vary very considerably according to local conditions. Nearly every observer has noted that corals grow most rapidly where there is an in-shore current of oceanic water, bringing with it an increased quantity of food-supply ; while *Lithothamnion* and *Lithophyllum* grow best where there are waves and breakers. Hence in areas where we have both these features, as in the Pacific Atolls, the reef will grow better and be more compact on the windward side, and it is not uncommon in this region to find the leeward reef either less developed or even non-existent, this latter state being at least in part due to the smothering of the growing coral by sedimentation ; but in the Laccadive and Maldive regions, where there is an alternation of the two Monsoon seasons and a corresponding change in the surface-currents from north-east to south-west, there is no marked difference between the reefs of the two sides of an atoll, though it has been suggested that the greater rate of growth on the south-west side and the increased number of openings through the reefs on the eastern side of these atolls is attributable to the greater effect of the south-west monsoon.

Whatever view we may hold regarding the foundation on which these coral reefs have arisen and whether we favour the view that the reef has been formed on a submerged volcano or on a bank of detritus or, again, on the eroded remains of land, it seems probable that when once the growth of coral has been established and by secondary processes, such as the geological formation of a conglomerate or the growth of *Lithothamnion* and *Lithophyllum*, has become compacted into a definite reef, such a reef will be able to maintain its position at or near sea-level, even if there be a later subsidence of the land, either relatively or actually, unless this change of level be too rapid. In the vast majority of instances so rapid a change in level is highly unlikely, and with only a gradual movement the amount of formation of coral-rock will be able to equalize the amount of relative subsidence of the reef-surface. On the other hand there seems to be some evidence that coral cannot or, at least, may not grow up to form a reef from any considerable depth ; and there seem to be numerous well-authenticated cases in which no change has taken place in a submerged coral bank for as long as a period of 100 or more years. The Dolphin Reef of Tahiti is believed to have remained at a depth of two-and-a-half fathoms for a period of 67 years. The Macclesfield Bank and Tizzard Bank, as well as others in the Pacific Ocean, appear to have remained stationary and completely submerged since they were discovered ; and certain reefs in the Red Sea have not changed in depth during the last two centuries, so that there has grown up a belief that these latter reefs cannot grow up beyond a certain height, although the local conditions of light, temperature and rate of evaporation of the surface-water would appear to be extremely favourable both to the growth of the coral and to the conversion of this into a solid coral rock and limestone. In Indian waters we find in the Laccadives a number of submerged coral banks, and others are known in the Chagos Archipelago, that is in the main composed of such submerged reefs, the best example being the Great Chagos Bank. As regards the former Stanley Gardiner (1902, p. 277) writes ' of the reefs with land or awash, nine are atolls or preserve traces of having at one time had the ring-shaped form. The rest are mere narrow reefs with a certain amount of land and a greater or less extent of shoal water off them. The three larger submerged banks to the north have depths of 24, 21 and 16 fathoms ; the fourth, Elicalpeni, to the east, having only 6 fathoms . . . . There is little sign of these reefs assuming a ring-shape, the shallowest depths being in the centre of the bank and not on the rim ' ; but in the following year (1903, p. 150) in a general account of the Archipelago he remarks with regard to the more northern banks that ' if they were fairly level, they might conceivably be deemed to be washing away, but they vary in depth to such an extent (Munyal 14 in 34 fathoms), that there can be little reasonable doubt but that they too are in places being built up by corals and other organisms to the sea-level'. On the other hand certain submerged reefs do undoubtedly show the characteristic raised rim of an atoll; examples of this latter type of submerged reef are the reefs to the north of Fiji, to which Stanley Gardiner refers (1931, pp. 22-23), namely Tuscarora, Field Bank and Pasco Bank, and another example is the Great Chagos Bank (*vide* Stanley Gardiner, 1903, p. 17, fig. 4), in which there is a well-marked raised rim, having a depth varying from 5 to 10 fathoms, whereas the greater part of the central area has a depth of 45 to 46 fathoms. Other similar banks in the Chagos are Pitt Bank, Speaker's Bank and Victory Bank, this latter, '4 miles long by 2½ broad, has a perfect rim at 3 feet with 18 feet of water in the centre'. As regards the origin of these rimmed banks there is a general opinion that they must be regarded as drowned atolls, but even so there appears to be some factor which has prevented the upgrowth of the reef to the surface during the interval that has elapsed since their submergence.

An elevation of some 10 to 20 fathoms would convert most, if not all, of these submerged banks into coral atolls or islands and it seems impossible to avoid the conclusion that there is some factor that prevents their upward extension. Guppy (1886, p. 367) as long ago as 1886 expressed doubts whether a coral reef could grow up from a depth greater than some 5 fathoms and he called attention to 'the inability of detached submerged reefs to raise themselves within the construction power of the breakers'. It is possible that a true reef-rock, and consequently a true reef, can only be formed in comparatively shallow water, having a depth that may vary in different localities but that is in some cases as little as two-and-a-half fathoms, because below this depth Lithothamnioneæ cannot flourish sufficiently well to form the reef edge, and the coral growth cannot become compacted and consolidated into a true reef-rock, even though the individual coral colonies may still flourish. Further, Stanley Gardiner (1930, p. 8) has put forward the suggestion that 'the phytophagous corals and plants that build up coral reefs precipitate amorphous carbonate of lime from the super-saturated sea-water owing to the chemical operations of chlorophyll on carbon dioxide and that this material by clinging to their surfaces ultimately compasses their deaths in lagoon conditions beyond five to ten fathoms. In other words the physiological action of the chlorophyll feeding of the important reef building organisms at all depths beyond five to ten fathoms is such as to kill them by a simple chemical reaction in the sea-waters that lave them. If this be so, it is obvious that shoals cannot be built up on lagoon floors from depths beyond ten fathoms'. There seems no reason to limit this action to the lagoon reefs alone and it is equally possible that at a certain depth wave or current action, or the combined action of these agencies and the ciliary mechanism of the coral polyps, may be sufficiently retarded so as to be ineffectual in removing such an amorphous deposit from the surface of the coral or plant, and in consequence coral-growth will be prevented and hence the building up of a coral-reef will be impossible.

In every attempt to trace the changes that have been going on in the various reefs and islands of such coral formations it is of the utmost importance to distinguish most carefully between the various types of rock-formations and to have also a very clear idea as to the conditions under which these rocks have been formed. Among the main types that one can distinguish, a hard and fast line must be drawn between rock that was laid down at the time when the reef was actually growing upwards, and rock that has been formed subsequently during the time in which the reef has been elevated and islands have been formed either by the fall of the sea-level itself or by the piling up of boulders, shingle and sand under the action of wind and waves.

The primary reef-rock must be carefully distinguished from the secondary rocks, and of these latter there are several types that may be met with, their characters differing, according to the degree of coarseness of the coral debris, from a boulder conglomerate, through a shingle conglomerate to a true sandstone. In the case of this latter formation we can distinguish between that which has been laid down beneath the sea and that which has been deposited above low-tide level; and in the latter group various observers have attempted to differentiate between the sandstone that has been deposited, between tide marks and which Stanley Gardiner terms 'beach sandstone', and that which has been deposited at some height above sea-level beneath the surface of the sand and shingle that forms the various islands on the reef flat.

In many cases previous authors have not sufficiently distinguished between these various formations; thus Agassiz (1903*a*), in his detailed account of the atolls of the Pacific, uses the terms coral-breccia, beach-rock conglomerate, or pudding stone for the same rock that forms the basis of the present islands; he considers this to be a secondary formation due to the consolidation of fragments thrown up on the top of the reef flats. In his account of Funafuti, he remarks that 'the coral breccia is made up of rolled recent coral shingle, of pieces of coral, of shells, of sand, quite unlike any breccia we have seen before and partly changed into a hard ringing limestone. Here and there I thought I could detect fragments of the old limestone ledge.' Of this beach rock conglomerate he states (p. 220) that he was unable to satisfy himself of its origin, though as he frequently refers to the fact that it shows evidence of slight elevation, he presumably considers it to have been formed below sea-level. It remained for Stanley Gardiner (1903 and 1931) to emphasise that this rock is a submarine formation and formed part of the reef below the level of the sea, its present position being due to a slight relative elevation.

In attempting to describe a coral reef, and more particularly when attempting to compare one reef with another, it is of the utmost importance that one should be consistent in the use of terms and, further, that so far as possible, allowing for the differences in different reefs due to local or other conditions, the same terms should be used for the corresponding regions. In the following account I have for the most part followed the nomenclature given by Stanley Gardiner (1931), and in the majority of reefs that I have examined one can recognize the following characteristic areas:—

The outer area of the Reef.

1. *The seaward edge,* usually raised by the growth of nullipores, and frequently cut into buttresses and fissures.
2. *The seaward flat.*
3. *The reef-crest and boulder zone*; this may not always be present and in certain cases its position may only be more or less indicated by the presence of a line of coral horses, while in others both a boulder zone and an inner line of coral horses may be present.
4. *The boat channel*; the presence or absence of this channel will depend on the degree of erosion that has taken place on the seaward side of the islands.

The Islands.

5. *The seaward beach* of the islands, where such exist.
6. *The outer rampart* of the islands.
7. *The central depression* of the islands.
8. *The lagoon mound* composed of blown sand and, where present, forming a raised margin on the lagoon side of the islands, corresponding to the piled up outer rampart on the outer face.
9. *The lagoon beach* of the islands that not infrequently shows outcrops of beach sandstone.

The inner area of the Reef.

10. *The lagoon flat,* the extent of which will again depend on the extent to which the islands have been driven inwards or have been eroded by the lagoon waves.
11. *The lagoon edge,* where the reef dips more or less steeply downwards.
12. *The lagoon floor.*

The outer Reef.

The seaward edge of the outer reef, in most of the atolls and fringing reefs that I have examined, exhibits certain differences in accordance with the degree to which it is exposed to the full force of the waves of the open ocean or is more or less protected by neighbouring land or by another atoll. In an exposed situation, open to the full force of the monsoons and especially to the south-west monsoon, the edge of the reef is for the most part composed of *Lithophyllum* ; this strengthens the rock to an extent that enables it to withstand the force of the breakers and at the same time imparts to the reef its characteristic dull-red colour ; Mayor (1924, p. 10) remarks that ' the Lithothamnion itself grows so slowly at tide level that it alone would not form a ridge were it not for the fact that it quickly covers, smooths over and cements upon the reef-flats all corals which have died, thus securing the retention of any accession of limestone which the seaward edge of the reef may have once attained '. The outer area or zone of such a seaward reef is as a rule, in my experience, cut up into buttresses and trenches, through the latter of which the water pours off the reef back to the ocean after each succeeding wave. Differences are, however, met with in different regions ; in the Maldives these buttresses and trenches, especially the latter, are due in my opinion to the scour of the water keeping the trenches open while buttresses grow out, but Crossland (1928*a*) reaches the conclusion that the fissure or trench zone in the Reefs of Tahiti and Moorea is to be attributed to radial faulting in the reef. This, however, appears to be a peculiar local condition and not one generally applicable, and he himself calls attention to the differences that are met with between this part of the reef in these atolls and the characters of the fissure zone in Funafuti. Again, Stephenson, Tandy, and Spender (1931), in their account of Yonge Reef in the Australian Barrier Reef, state that the extreme edge of the reef is composed of an outer raised rim, that is not continuous, and that immediately inside this there is an outer ' moat ', about five feet in depth ; other members of the expedition, however, do not agree with this statement and I have been informed by Dr. Manton that the water over the greater part of this area was shallow and that only in a few patches were holes as deep as 5 feet to be found.

Even in calm weather in the Maldives during the periods of the year when there is little or no trace of any monsoon wind blowing, there is always a heavy surf breaking on the exposed reefs and the effect of the continuous to-and-fro rush of water is clearly demonstrated in the almost total absence of any growing coral colonies in this zone, though such, and in certain cases even apparently very delicate growths, may be found both in some of the fissures or on the seaward face of the reef at a depth of a few fathoms. The ' fissure ' or ' trench ' zone passes into the seaward-flat, which slopes gradually upwards to the reef crest, though few if any of the trenches actually penetrate as far as this.

As one passes inwards from the nullipore zone, one finds that the reef has a comparatively smooth surface, that is, however, thickly dotted over with water-worn, and usually flat, coral fragments, which along the line of the crest may be piled up in a low ridge, running parallel to the reef margin and known as the ' boulder zone '.

The fragments that compose the boulder zone, as Stanley Gardiner (1903, p. 44) has pointed out have a two-fold origin, (*a*) true coral boulders and (*b*) masses of the reef-rock. As he points out, the boulder zone at Minikoi ' has a two-fold origin—boulders cast up on the reefs and masses of the conglomerate, some loose and some attached to the solid platform below ; relatively little consists of masses attached to the reef, the greater part being formed of loose blocks of the old coral rock '. The boulders and coral fragments torn off the reef-edge and cast upon its surface will, of course, originally have been loose, though they may subsequently become consolidated to it. Masses of reef-rock may, however, still be fixed to the reef, but in most cases these masses have been broken down and the old included coral colonies have been set free. Baker (1925, p. 1008) in his account of the reef around Gaua, in the New Hebrides, has described the boulder zone as double. He remarks, ' unlike the usual boulder zone, it is double and consists of two ridges, each about 50 yards wide, separated by a channel of about the same width. The boulder zones project about a foot above the level of the water at low spring-tide, and consist for the most part of rather small pieces of coral, the lower ones being firmly cemented together. There are large boulders here and there. The channel between the boulder zones is a few inches deep at low spring-tides. There is no living coral on the boulder zones nor between them.' I know of no other instance in which the boulder zone exhibits this double character.

Extending inwards from the boulder zone towards the lagoon or the islands, where such still exist, are lines along which these fragments of coral and coral rock are aggregated into definite spits that on the south-west side of the Maldive atolls have a distinct trend towards the north-east, so that they run parallel, more or less, to the main direction of the strongest wind, namely the south-west monsoon. A boulder zone, such as I have just described, is by no means confined to atolls and barrier reefs but may also occur in a fringing reef, provided that general conditions such as heavy seas, etc., are present. Such a boulder zone is present on Krusadai Island at the head of the gulf of Manaar; at the north-east corner of the island there is a well-marked fringing reef, with a clear-cut fissure and buttress zone, and inside this there is a boulder zone, running inwards from which are several spits of coral boulders (Pl. 14, fig. 1). Many larger blocks on the edge of the reef, though now all dead, appear to be *in situ*; but it was impossible to decide this point definitely owing to the degree to which erosion has gone on. The whole reef on this side of the island gives one the impression that it has been raised and is now in process of erosion; there is little or no live coral on the upper surface of the reef but the dead coral masses are encrusted by a few small patches of a red nullipore in the area to the seaward side of the boulder zone. Behind this reef the island is in places clearly undergoing erosion, the beach terminating in a small sand cliff about 2 feet in height.

The term 'Negro-head' has in the past been very loosely applied to any large mass of coral or rock that may be found on the reef-flats and has been used to include not only fragments of coral or portions of the growing face of the reef that have been torn off and flung upwards and inwards on to the reef-flat by the force of the waves, but also large masses of dead coral, that were originally growing *in situ* where they are now seen but have been subsequently killed off by a change of sea-level, and even masses of reef-rock, that still resist erosion and which have also been termed 'Coral horses'.

Agassiz (1898, p. 114) in his description of the Great Barrier Reef of Australia, called attention to the presence of large rock masses on some of the reefs. He remarks, 'The negro-heads occurring on the edge of some of the flats are all composed of beach rock, and remain as monuments of the former extension of the beach rock. ..... The presence of negro-heads, remnants of the former elevated coral reef, and of negro heads, the remnants of former long stretches of elevated coral beach rock conglomerate, give us a ready explanation that their presence on flats where the sea could not have transferred them is due to the disintegration and erosion of the coral reef and of the conglomerate when they extended over the surface of the rock-flats.' On the other hand large masses of coral or coral-rock from the outer face of the reef may be thrown up by wave action, usually the result of cyclonic storms, and occur, as Agassiz emphasizes, over limited areas. Stanley Gardiner (1931, p. 36) also remarks that 'frequently at the base of the sea-ward beach are blocks of limestone not loose but cemented to or part of the reef beneath, sometimes ridges, the "horses" already mentioned, stretching along for many yards. Perhaps there may be one or more lines of "horses" outside each other to seaward on the flat and beyond these solitary sentinels of rock. Where such "negro-heads", as the latter are termed, lie near the reef edge they are usually supposed to have been thrown on to the seaward flat by the waves and thus to have consisted of blocks from the reef edge to seaward. But as often as not they are part of the solid underlying reef.'

It is essential to differentiate between these various classes of coral masses, and in the present paper I have restricted the use of this term 'negro-head' to those masses of coral that have undoubtedly been torn off the reef-face and flung on to the reef flat by wave action. In several instances, as for example on the leeward sides of the reefs of the Great Barrier Reef of Australia are lines and masses of dead coral rock : it is interesting to note that Agassiz (1898, p. 117) in his account of Cairn Reef remarks on the 'mass of negro-heads of all sizes, forming a belt along the western edge of the reef flat', and he says that in South Turtle Reef 'negro-heads are exposed on one part of the edge, rising perhaps two feet above the general line of the flat. *They look to me like fragments of an elevated reef which has been washed away*, leaning only here and there an isolated pinnacle.' (The italics are mine.) Stanley Gardiner (1902, p. 291), although he clearly recognized that such masses of rock in certain of the Maldive Atolls are the remains of an older, 'elevated' reef, applied the term 'negro-head' to them ; 'The reefs to seaward of the rim islands are, for some distance from the land, studded with masses of the rock, which form pinnacles or negro-heads'. Those blocks, that I have examined in the Maldives, though not so large as those found in the islands and atolls of the Pacific Ocean, are, nevertheless, comparable. It is only along the south-west parts of the reefs in the Maldives that I have been able to detect the presence of definite 'negro-heads'. In the Laccadives Oldham (C.F., 1896, p. 12) has called attention to the presence of large blocks of coral thrown up, as he thinks, on the east and north-east sides of the islands, and he attributes their presence on this side and their absence on the south and west to the force of the waves during cyclones in the Laccadive Sea, the position of the islands being such that the strongest winds will under these conditions come from the north and east sides. Stanley Gardiner (1898) experienced the same difficulty in explaining the presence of negro-heads in Funafuti, where they appear to be confined to the north-western and western reefs ; he remarks 'it is remarkable that if there are 'negro-heads' on the leeward reefs there should be none to windward, against which the force of most of the gales is broken. Either they are thrown up in bad cyclones or hurricanes (which always strike to leeward), an almost impossible supposition, or the growth of these masses is firmer to windward'.

In the case of such large coral masses occurring on the leeward side of the reef one is at first sight inclined to believe that these are not true negro-heads but are parts of an older reef that originally grew at a higher level than the present reef and have not yet been entirely destroyed by erosion, or, in other words, that they are of the same nature as 'coral horses'; and yet one cannot, in view of the experience and reliability of many observers, conclude that this is the explanation of these masses in every case.

On Yonge Reef, Three Isles and Low Isles in the Australian Barrier Reef Stephenson (*vide* Stephenson, Tandy, and Spender, 1931, pp. 25, 85) has described a so-called 'Boulder Tract' or 'Boulder Zone' which is composed of moderately large, rounded masses of dead coral that must have, one would imagine, been torn up from the leeward side of the island or reef and have been flung inward on to the reef and this is the explanation given by Steers (1930, p. 15) who states that these 'negro-heads' may occur on the windward side of inner reefs, but reach their maximum, in a case like that of Low Isles, on the leeward side of the reef, the side on which a sudden hurricane from the northward, coming in a direction opposite to that of the trade wind, does material damage to the unwieldy masses of coral accustomed to quiet water; Stephenson (*vide* Stephenson, Tandy, and Spender, 1931, p. 94) gives the same explanation and attributes these boulders on the leeward side of the reefs to 'the result of the action of occasional hurricanes blowing from a direction opposite to that of the prevailing wind and tearing up large masses of coral which have become unusually unwieldy or unstable, growing into exaggerated shapes, as a result of the sheltered conditions under which they usually live'.* These 'negro-heads' must be clearly differentiated from other coral masses that are very similar in appearance and that, as Agassiz suggests, actually grew in the situation in which they are now found but have been killed off by the fall of sea-level that occurred a few thousand years ago and have not yet been completely eroded away. A very similar zone to that on Low Isles in the Australian Barrier Reef, so far as one can judge from the published accounts and photographs given of these latter areas, is to be seen on the north side of Horsburgh Atoll in the Maldives and here also the reef lies on the protected side of the atoll where the force of the waves must to some, and probably to a considerable, extent be reduced. One of the characters of the reef in this situation is the growth of coral colonies on the seaward flat of the reef and the feeble development of nullipores and hence the less consolidated character of the reef-rock to which the living coral colonies are affixed; and it may be that under such conditions of growth even a moderate sea will be able to uproot a living colony or possibly tear off a part of the reef and roll it inwards across the reef-flat, and in support of this view is the fact that in such a so-called 'Boulder Zone' the general character of the coral blocks is different from the blocks seen in the true boulder zone on the exposed face of Addu Atoll, and much more closely resemble masses of dead coral.

Agassiz (1903*a*, p. 227) in his account of Funafuti remarks 'The difference in the width of the outer reef platform of different atolls may be due to the gradual piling up on the sea face of successive rows of shingle and sand beaches. Where this has taken place regularly, an original wide sea face reef may gradually become reduced to a narrow belt'. Such a process is in my view extremely rare and I have seen no example of such action in any Indian reef. That it may occur in the case of certain atolls in the Pacific is shown, for example, in Kwajalong Atoll in the Marshall Islands, where on the leeward side of the atoll, owing to the great movement of sand under the influence of the trade wind, sand dunes have been piled up and have overwhelmed the outer sea beaches but in the vast majority of cases the presence of an exposed basic stratum of reef-rock shows that erosion is taking place. A perusal of the literature reveals a number of instances in which observers have been struck by the evidence of erosion. Agassiz (1898, p. 100) in his account of the Great Barrier

* Since the above was written Kuenen has published (1933, 'Geology of Coral Reefs', The Snellius Expedition, Vol. V, pt. 2) a detailed account of the coral reefs of the East Indias, in which he calls attention to a very similar boulder tract on the East side of Pelokang. He states 'while we might possibly imagine that exceptionally heavy storms occasionally threw corals on to the flat of Pelokang, situated at the edge of the submarine plateau, it is most unlikely that the blocks on the reefs in the centre of the atoll could have been thrown up in the same manner. Moreover, the heaviest gales are experienced here during the west monsoon, but all the boulder ramparts are found on the east side of the flats. . . , the upright position of many of the corals further proves that they grew on the spot.'

Reef of Australia remarks 'Some of the larger patches, such as e Reef, Turtle Reef, Aitch (h) Reef, El (l) and Em (m) Reefs and Eagle Reef, all of which are well isolated reef patches rising from twelve to fourteen fathoms in the channel between the inner edge of the Great Barrier Reef and the mainland. These patches undoubtedly represent the eroded flats of former islands, some of them of considerable size, which occupied the main part of the Coast Channel'......... 'South of Cape Flattery the Low-wooded Isle, Three isles and Two Isles are the only remnants of the numerous but somewhat widely separated islands, which once existed between that Cape and the outer edge of the Barrier Reef, the others having been eroded and changed to the reef flats and patches now existing to the eastward of these islands.' One could quote other instances which Agassiz gives of 'the remnants of former islands which have been reduced to their present level by erosion and denudation', and, throughout, his account of the atolls of the Pacific Ocean is full of instances of the manner in which erosion is steadily going on. Guppy (1889, p. 465) calls attention to the erosion that has taken place in the Button Islands of Cocos-Keeling Atoll and again in South Island. Fryer (1911) in his account of Aldabra states that 'the atoll is losing on every side in its fight with sea and weather; the sea coast is being eaten away; the lagoon is getting larger, the rain is dissolving away the surface of the land, and, to balance all, there is only a slight piling of sand'.

As the islands on the reef-flat are eroded on their seaward faces an ever widening belt of shallow water is formed between the outer margin of the reef or the boulder zone, where such exists, and the seaward beaches. As Mayer (1918, p. 44) has pointed out, the different levels of the *Lithothamnion* ridge and the inner part of the reef-flat is due to the protection afforded by *Lithothamnion* growth in the former area and to erosion in the latter. He remarks in his account of the Murray Island reef 'The Limestone of the fringing reef flat between the lithothamnion ridge and the shore has disappeared, leaving a lagoon about 18 inches deep at low tide, and this despite the growth upon it of one of the most densely clustered coral colonies the writer has ever seen. Were it not for the growth of this coral the reef-flat would have suffered even more serious disintegration, and in hurricane regions where the corals on top of the reef-flats are periodically destroyed, it seems possible that such fringing reef....... might gradually become converted into a barrier reef by the solution of the limestone due to "sand feeders" and to scouring due to currents.' The floor of this area is in its outer part composed of a hard coral rock which, as we pass inwards towards the islands, becomes covered to a greater or less extent by loose sand, though in few or no areas is this deposit a very deep one owing to the strength of the currents that with every rise and fall of the tide sweep across the reef flat. The depth of water in this area is as a rule only one to two feet at low water, though towards the ends of islands, near the passages that run between them and thus connect the inner and outer reef flats, deeper channels may have been scoured out. Sheltered as this area is from the destructive effects of the high seas and yet freely supplied with ever changing water as the tide rises and falls, it forms an almost ideal spot for the growth of coral. Hickson (1924, p. 216) has pointed out that 'the first impression of one coral reef may be that it consists of nothing but huge shrubs of stag's horn Madrepores, of another that it is all palmate Madrepores, of a third that it is all Lithothamnion, although a closer examination shows that many other kinds of coral occur among the prevalent forms'. Mayor (1924, p. 14) and Baker (1925, p. 1009) have called attention to the manner in which, in such an area as we have just considered, the coral growth may be divided into zones, each zone characterized by the profuse growth of a particular species of coral. This condition is clearly seen in certain parts of the reefs in the Maldives and in some of the fringing reefs of the Nicobar Islands. In the reef on the east side of Octavia Bay in Nankauri Harbour (Pl. 14, fig. 2) there is a very clear demarcation into two zones, an outer zone completely covered with a profuse growth of *Sarcophytum*, and an inner zone in which the coral growth is almost completely composed of a branching *Porites*; as Stanley Gardiner (1931, p. 90) has pointed out, 'the visible home of the leathery corals is essentially on the lagoon edges of encircling reefs, on fringing reefs within barriers and on lagoon shoals'. A somewhat similar division of the reef-flat is to be seen in the fringing reef off Reed Point at the north-east corner of Nankauri Island; here the outer zone is composed completely of a branching plate-like growth apparently of *Millepora* sp., while the inner zone is profusely dotted over with a fine growth of more solid colonies of *Heliopora coerulea* (*vide* Pl. 14, fig. 3 and Pl. 15, fig. 1). The outer zone clearly shows the peculiarity, first noted by Darwin, that plate-like forms in such positions grow at right angles to the current, thus presenting their flat faces to it. Such a local distribution of different species of coral is in all probability to be attributed to the fact that corals show a very definite range of adaptability to changes in their external conditions, for information regarding which I would refer the reader to a paper by Edmondson (1928) and papers by Mayor (1918 and 1924). A further factor may be the incompatibility of certain corals, for it seems that one form of coral, when well established, may in some way not as yet understood inhibit the growth of other species.

The Islands on the Reef-Flat.

On the boulder zone or on the lagoon side of it one finds, where such still exist, a series of islands, whose basic structure may differ very considerably; and the question arises, how did these islands come into existence? We have at the present time two views, diametrically opposed to each other, as to the manner in which they originated. According to the older view these islands have all been thrown up by wave-action, in which case they must be regarded as being of the nature of an exaggerated boulder zone. This view has been expressed by Oldham (C.F., 1895, p. 13) regarding the formation of islands in the Laccadives: 'The Seas due to the hurricanes would strike on the eastern and north-eastern sides of the atolls with tremendous force, smashing and tearing the coral boulders off the edge and hurling them on to the centre of the reef; here then would be the foundation for the future island. The currents and tides and ordinary monsoon winds would then be sufficient to complete the remainder of the building-up process'. Wood Jones from his study of the Cocos-Keeling Atoll came to the conclusion that all the land in that atoll has been entirely derived by this process, aided by the deposition of wind-borne sand. He remarks (1910, p. 168) 'The ocean beaches consist essentially of a sloping pile of coral debris piled up on the breccia stratum as a basis', and he also has noted that this basic substratum has been 'elevated' to the extent of some 3-4 feet; but he entirely ignores the part that this raised rocky substratum, which must have been laid down below sea-level, has played in the subsequent formation of the island, nor does he appear to realize that with this 'elevation' of the atoll the breccia must itself have formed an area of dry land, for he adds 'just as the first step in the formation of the barrier consisted of some chance boulder being hurled by storms upon the top of its fellows, *so the island starts to form by some boulder being tossed by the waves upon the platform of the barrier*'. (The italics are mine.) It would appear that this agency, namely the piling up of coral debris, may possibly be responsible for the formation of some, at least, of the smaller islands in the Maldives. In a few places round the margin of an atoll some peculiarity in the conformation of the reef or a local modification of wave-action, as on either side of an entrance channel, may cause rounded and water-worn masses of coral fragments to be piled up to form islands, and examples of such action are to be found in the islands on each side of the single entrance channel to Horsburgh Atoll or on the east side of the south-east entrance of Addu Atoll. This latter island, Mulikadu, in its structure clearly betrays its mode of origin, for the whole of the upper stratum of the island is composed of a series of ridges and furrows that run parallel to each other; with the single exception of the ridge that borders the seaward side of the island, all the ridges run parallel to the lagoon beach, each successive ridge showing where a storm has added a new beach to those that were already in existence. Underlying the island is a basic stratum of raised reef-rock. A very similar formation is seen in Direction Island in Cocos-Keeling Atoll. Here also the island lies at the end of the reef bordering one of the deep entrance channels. Guppy (1889, p. 463) in his description of the island states that 'In the main it has been growing leeward or westward from its eastern portion by the formation of a succession of ridges of reef debris, each ridge thrown up by the waves, and marking a stage in the growth of the island. These ridges are naturally best illustrated in the western or more modern part, where the outermost or most recent of them have not been covered with vegetation'. But these islands I consider to be of secondary formation and to have arisen later than the original islands of the atoll; and this must clearly be the case of the islands at the entrance of Horsburgh Atoll if we adopt the suggestion made to me by Stanley Gardiner that the blackened colour of the water-worn fragments of which these islands are composed is to be attributed to the fact that the coral fragments are not fragments that have been simply torn up off the reef and flung inwards but are fragments that were originally embedded in the reef rock and have been subsequently eroded out and still later have been piled up into the present islands.

The second view of the origin of islands, which was put forward by Stanley Gardiner, supposes that they have been formed by a fall of sea-level that brought the original coral-reef above tide-marks, and that thus there must have been, except for the interruptions caused by the deep entrance channels into the lagoons, a continuous land area all around an atoll. Subsequent changes and especially erosion have given rise to the interrupted chain of islands that are found at the present day, and have also caused the inward drift of these islands across the reef-flat. Stanley Gardiner (1898, p. 428) in his description of the seaward face of some of the islands in Funafuti, calls attention to the character of the rocky basis of the seaward beach and to the presence of pinnacles of rock connected with it, and he concludes that these latter 'are the remains of a part of an old raised reef, which formerly extended for some distance outside the present hurricane beach' and the greater part of which has since been washed away.

In the Report of the Coral-Reef Committee appointed by the Royal Society (1904, p. 71), as a result of their investigations Edgeworth David and Sweet concluded that in the Atoll of Funafuti 'the Breccia has apparently at one time covered the reef proper around perhaps the whole of the rim of the reef platform, though now there are wide stretches of this from which it has been almost completely denuded. Nearly all the islets of the atoll have a foundation of this material or at least a rampart of it along their shore-line facing the ocean. In some cases there is an outer line of extremely dense breccia, marking the most advanced position oceanwards occupied by that material at a time when it formed the foundation for islets which have since been driven inwards over the reef platform towards the shore of the lagoon'.

Stanley Gardiner (1903, pp. 34–6), from his investigations regarding the nature of the rock found in certain atolls and especially of that occurring in certain parts of the atoll of Minikoi in the Maldives, reached the conclusion that in this atoll this rock had formed part of the growing edge of the reef at a time prior to the formation of any reef-flat and that it had subsequently been elevated from its original situation at a depth of about three fathoms to a height of some six feet above the present low-water mark. He lays special emphasis on the absolute necessity of elevation having occurred in an atoll in which such masses of rock are present. This view is very clearly stated in his account of the North Mahlos Bank in the Maldives (1903, p. 171), where he remarks 'that any land has been formed save by this change of level is, as far as I can see, impossible, but many islands in the centre of the lagoon have been washed up as sand banks'. He (1903, p. 149) reached a similar conclusion regarding the Laccadive archipelago, for he remarks that 'practically all land lies on the eastern or leeward side of the reef, the group being completely exposed to the gales of the south-west monsoon, while it must be largely protected by India from those of the north-east. The present contour of the land, throughout the group, would hence seem to be really due, as at Minikoi, to elevation and subsequent erosion, the latter having completely removed all save traces of the land from the western reefs'.

It seems probable that both these views are correct and that most of the present islands have had in reality a double origin.* The first necessity appears to be the

* In a recent paper Kuenen (1933, 'Geology of Coral Reefs,' The Snellius Expedition, Vol. V, pt. 2) has put forward the view that many of the islands on the protected East Indian Reef have been built up 'for many sand cays occur which

formation of the raised basic stratum of coral rock; without this raised part of the old reef the waves will be unable to pile up debris into the necessary seaward ridge, the rush of water across the reef only serving to sweep such debris across the reef into the lagoon. With the fall of sea-level and the emergence of the old reef edge, fragments would be piled up on to it, especially by wave action during heavy storms, and the subsequent destruction of this piled up mass of coral boulders and shingle will be dependent in the main on the gradual destructure of the basic platform by marine erosion acting on its seaward face.

Bourne (1888, p. 443) has pointed out that in the atoll of Diego Garcia in the Chagos he was able to distinguish four different kinds of what he terms coral rock, namely :—

(1) Reef rock, showing a horizontal stratification and a submarine origin;

(2) Boulder rock, exhibiting a stratification dipping downwards towards the sea and formed just above high-tide mark by the consolidation of masses of coral that had been washed across the reef;

(3) Shingle rock of two kinds; the first showing a horizontal stratification as in (1) and only differing from it in its finer and looser texture, with a submarine origin in the more sheltered parts of the lagoon; the second showing a stratification dipping towards the sea and formed above high water mark; and

(4) Sand rock formed above water and showing a stratification dipping towards the sea.

A study of the various islands and their connecting ridges in the Maldives reveals that in those instances in which the original position of the island is still maintained the islands are composed of a solid basic mass of primary coral rock, on to which there has occurred a piling up of numerous water-worn fragments and blocks of coral to form a ridge or outer rampart that may rise to a height of several feet above the present high-water level; and, in the cases that I have investigated, this outer ridge is situated just internal to the present boulder zone of the reef; behind this ridge as a rule sand has accumulated to form the major part of the island. The origin of this rampart-like border has been the subject of considerable discussion; Stanley Gardiner (1903, p. 29), in his description of the atoll of Minikoi, gives the following account of what I take to be a beach of this nature, 'the surface is covered with masses of coral or limestone, some round others flat slabs, weighing up to 1 cwt. or more. To the north of Mou-Rambu they are much larger than to the south, where blown sand has also to some extent intermingled. All the surface masses are pitted and eaten into sharp points by the rain, which rapidly drains through them. No soil has been formed and layer after layer of blocks can be removed, until a firm bed of conglomerate within tidal marks is at last reached'. In this account the use of the term conglomerate is

were built since the present relative levels were reached on reefs that were not laid dry, but were still covered with living corals. Obviously, however, very many and probably most islands have been formed as a result of the emergence of their flats. Without the negative movements the number of islands on reef would be quite small and if no further movements occur their number and extent will in the course of time also undergo considerable reduction'.

likely to be misleading, for, as he explains later, he regarded this rock, as well as the overlying coral blocks, as having originally formed a part of the old reef that had been formed *in situ* in the original reef at some depth below water-level by the silting up of the crevices and the subsequent death and burial of the coral colonies; after elevation to its present height the solution of the cementing material has set the original colonies free once again and these have been eroded by rain and spray to their present shape. That the basic solid rock represents part of the old reef I entirely agree. Unfortunately, he does not state whether the reverse slope of the ridge was continued inwards in a series of spits, so that one is uncertain regarding the exact degree of similarity between this ridge in Minikoi and other similar ridges in Addu Atoll. In many cases, however, there seems to be little doubt that the outer rampart has been thrown up on to a raised reef-platform by wave action. In several islands round Addu Atoll and especially on that part of the reef where erosion by the waves has proceeded least far, as on the east side, we find a structure of this type, and we may take the island of Heratera and especially the south part of it, that in the chart given by Stanley Gardiner (1903, fig. 109, p. 415) is named Putali, as an example. Here the outer ridge is composed of blocks and masses of dead coral of all sizes, that appear to have been thrown up quite irregularly, and all of them are water-worn, so that many are flat and resemble fragments of paving stones; on the inshore side of this ridge the masses form a definite talus slope, running inwards in spits and lines, where the more rounded masses have rolled farther across the surface of the island than those that were of a more flat character. It must not be overlooked that a ridge of this type may, subsequent to its formation, become cemented together into a secondary conglomerate in a manner very similar to the formation of sandstone, as for instance in the island of Hurudu in the North Mahlos Bank, where, as Stanley Gardiner (*loc. cit.*, p. 161) remarks 'along the south and south-east sides (of the island) a coarse beach rock has also been formed of coral and fragments out of the limestone joined together by carbonate of lime deposited from the sea-water'. In such a ridge blown sand may act as a cementing material and by the solution of calcium carbonate from the upper layers by rain water or sea-spray and its redeposition in the lower levels of the stratum the whole may be converted into a solid rock.

In such a secondary conglomerate, composed of fragments of coral that have, since their detachment from the reef, become consolidated together, it is not surprising that some of the fragments may be found to be lying the right way up, and one does occasionally find them thus; but as a rule in such a formation these masses are found lying in all positions. Regarding the formation of such a secondary conglomerate, Stanley Gardiner (1903, p. 34) remarks, 'first the surface, exposed to the tide, has its pores to some extent filled up by sand particles. It further hardens and becomes thoroughly indurated with lime, precipitated from the water, which dries on the surface after each ebb. . . . Behind the surface loose sand becomes more consolidated than where there is rain water alone. The solvent action within the masses, save for a little due to the rain, is inconsiderable and the whole differs very little from the original reef except in its better consolidation and the filling in of its interspaces'.

The distinction between what one may term primary rock and the secondary coral conglomerate, and their recognition, is all important, if one bases the theory of relative elevation of the whole reef on the occurrence of such masses of rock at or above the present sea-level. The occurrence of masses of *secondary* conglomerate on the reef-flat, even though they may project above the level of the present high-water, is *per se* no evidence that any relative alteration of land and sea-level has taken place in recent years ; it is merely a proof that in times past land existed in that area and that denudation and erosion have removed all but the most resistant constituents. The height above sea-level of the secondary conglomerate appears to differ very considerably in different atolls even in the same group ; thus Stanley Gardiner (1903, p. 160) describes a ridge on Kenurus Island in the North Mahlos Group, that has been raised to a height of 13 or 14 feet, and in which the basal eight to nine feet consists of recent coral conglomerate, whereas in Addu Atoll this type of formation is found at a height of only some five feet at the most. On the other hand the height above sea-level at which the primary reef-rock occurs must depend entirely on the extent to which the sea-level has fallen. This primary rock, that was originally part of the old reef, may, subsequent to its relative elevation, have been covered by boulders or coral debris and thus have been enclosed in the mass of an island and protected from erosion. In other localities the upper stratum may consist of shingle or even of sand ; and this sand may equally become consolidated into an æolian limestone. During the subsequent erosion of the island this underlying rock may again become exposed and form the seaward margin, as one sees, for example, in the north beach of the islands of Fehendu and Fuladu in Horsburgh Atoll. An exposed rim of this old reef-rock seems to be of universal or almost universal occurrence. Guppy (1889, p. 262) has called attention to its presence in Cocos-Keeling Atoll, ' such a rock now composes the raised margin of the reef in the wash of the breakers, and in this manner, I imagine this similarly elevated foundation-rock of the islands was largely formed. However, as we approach the lagoon, we find that the conglomerate is composed of much smaller fragments of the same breaker corals, whilst in some places it is mainly made up of the coarse sandstone '. Wood-Jones (1910) pays little or no attention to this rock stratum, though some of his photographs show it quite clearly. Agassiz (1903*a*) frequently directs attention to its presence in the islands of the Pacific, and Steer (1930) has noted its presence in Low Isles in the Australian Barrier Reef.

Reef Rock.

The coral rock of the old reef will differ in its characters according to the situation in which it was laid down ; and one can distinguish between the rock deposited in three different zones of the reef. The rock that was formed in the old reef along the outer zone will present the same characters as the rock that at the present day forms the mass of the ' buttress and trench ' zone of the present reef. The main characteristic of this rock is the growth of *Lithophyllum*, that covers over and protects and finally binds together such corals as are able to exist in the heavy seas to which this zone is exposed. Marshall (1931, p. 68) describes this rock as follows: 'The actual material of this resistant mass on which the waves break is a compact dense white rock in which no structure can be seen even with an ordinary lens.... Microscopic preparations, however, show at once that this dense white rock is mainly composed of the thallus of *Lithothamnion* and of other genera of Algæ. The cell walls have been completely covered with a thick covering of carbonate of lime..... It is always found that some fine sand is embedded in the small crevices between the different organisms...... It is found that carbonate of lime is deposited in the small spaces between these materials and unites the grains and algæ into a solid rock. The deposition of cement takes place so quickly and so strong is the resulting rock, that with a geological hammer it is found to be more difficult to detach fragments from it than from the surface of ordinary hard rocks.'

Inside the lithothamnion zone in a modern reef we find, where such exists, the boulder zone and boat-channel and this is in turn succeeded by the lagoon flat, the character of which changes slowly as we approach the lagoon reef. As the reef became gradually built up, possibly *pari passu* with the slow rise of the sea-level at the close of the glacial period, so the resulting character of the rock will of necessity conform to the character of the reef in which it was being deposited. On the boulder zone the resulting rock will consist of large fragments of coral that were originally torn off the growing face of the reef; these will be cemented into the rock in any position. Further away from the reef edge these fragments will become smaller, forming at first a shingle and then further in still a coarse sand. The dip of the rock in this latter situation will be horizontal or slightly downwards towards the lagoon, and since during its formation it was in all probability covered to at least some extent by growing coral colonies, some of these will have been enclosed and preserved in the conglomerate in the upright position in which they originally stood on the lagoon flat. Darwin himself (1889, p. 156, footnote) put forward the view that 'the corals which lived in the lagoon reefs at each successive level, would be preserved upright and they would consist of many kinds, generally much branched. In this part, however, a very large proportion of the rock, and in some cases nearly all of it, would be formed of sedimentary matter being in an excessively fine or moderately coarse state, with the particles almost blended together. The conglomerate which was formed of rounded pieces of the branched corals on the shores of the lagoon, would differ from that formed on the islets and derived from the outer coast, although both might have been accumulated very near each other. The stratification, taken as a whole, would be horizontal; but the conglomerate beds resting on the exterior reef and the beds of sandstone on the shores of the lagoon and on the external flanks on the reef, would probably be divided (as at Keeling Atoll and at Mauritius) by numerous layers dipping at considerable angles in different directions'. The transition between the conglomerate formed in the old boulder-zone and that formed near the inner margin of the lagoon platform will be gradual and irregular, for in the original condition of the reef, as at the present day, spits of larger fragments will have run inwards from the boulder zone towards the lagoon. Wood Jones (1910) from his study of the Cocos-Keeling Atoll has attempted to draw conclusions regarding the formation of the various types of rock found in such a coral formation; but his account is somewhat difficult to follow and in places appears to be contradictory. He states that inside the boulder-zone lies 'the factory of the breccia' (p. 160). This breccia in his view (*loc. cit.*, p. 154) extends from the outer seaward margin of the reef to the lagoon, forming the Breccia platform, the outer part of which, lying between the seaward margin of the islands and the extreme outer edge of the reef, he designates the 'barrier'. Where islands do not exist the term 'barrier' is used to include the whole extent of the reef surface. 'The outer edge of the barrier is in many places marked by the exposure of large and irregular rock masses at low tide'; these masses he believes to have been derived by wave action from the outer face of the reef and have been torn off and hurled on to the reef. Between this boulder-zone and the outer beaches of the islands lie 'the flat of level cemented rock free of loose fragments, free of jagged edges, and free too of living corals; boulders are washed across and tend to level everything, and only the *Nulliporæ* may flourish. It is here that the finer particles are forced by the spin-drift, as by a sand blast, into every crack and rift of the shore-washed rocks; this is the factory of the breccia'. He attributed the different character of the rock found near the outer edge from that of the lagoon reef solely to the difference in the degree of wave action and the character of the material. 'On the ocean side the fragments are welded together, sand is driven home, and by the cementing action of deposited calcium carbonate the whole is consolidated into a massive mosaic. On the lagoon shore the materials are in a fine state of division, and the force is not nearly so great, and so the product formed is beach sandstone, or beach sandstone in which are embedded coral fragments after the manner of conglomerate'.... 'A fine loose sandstone represents the minimal product of the process—the hard mosaic breccia, the maximal; and all intermediate grades are to be found, *their composition depending entirely upon the wave force that was exercised in their making*.' (The italics are mine.) Later in his account he, however, appears to admit that this is not the whole picture for he states (p. 229) that 'it is easily seen that brecciated rock and sandstone are formed in the lagoon by the actual deposition of calcium carbonate. The deposition of calcium carbonate is a very widespread phenomenon in the lagoon'.

Beach-Sandstone.

Dana (1875, p. 149) called attention to the beach formation of coral-sand rock and mentions that this has been found in a number of the Pacific atolls of the Paumotus group. As he points out, 'the stratified character is always distinct and the layers slope towards the water at the usual small angle, amounting to 5–7 degrees bordering the lagoon and 6–8 degrees on the seaward slope of the land'. He also mentions that this rock is found in the Maldives.

Agassiz (1895, p. 223) clearly differentiated between this beach-sandstone and the basic rock of the islands. He remarks 'the beach rock and the so-called base rock which have been observed at the Bermudas belong, I believe, to two different types. The former, the beach rock, consisting of coral or other sand, is deposited in strata dipping to the sea at a slight inclination, and is characteristic of all coral-reef districts where sand is accumulated along a shelving line of coast. This frequently becomes hardened and changed to a ringing limestone, and is composed usually of rather coarse particles, but not necessarily so. The base rock, considered by some of the writers on the Bermudas to underlie the Æolian hills, I look upon as the modified part of the lower portion of the æolian strata, changed into a hard ringing limestone in which all traces of stratification have often disappeared '.

Agassiz was, however, wrong in his conclusions regarding the origin of the base rock in the Bermudas. Heilpin and Rice both regard this rock as a different and earlier formation, and Vaughan has shown conclusively that it was formed below the sea and not from wind-blown sand.

Stanley Gardiner has also drawn a hard and fast line between the primary coral rock and the secondary beach-sandstone that one sees exposed in places round the margins of the islands on the reef-flat ; he remarks (1903, p. 343) 'the beach-sandstone in the Maldives is everywhere perfectly distinct from the raised rock of the islands, which is a submarine formation, differing radically in its constitution ' and he goes on to remark that 'it cannot be too strongly emphasized that the formation of this sandstone can only take place on the beach and between tide-marks. Its constituents show all the characteristics of such a formation, being much broken, angles rounded and surfaces pitted. Among them may be recognized pieces of coral of all kinds found on the reefs or in the raised rock, but branching species, especially of the genus *Pocillopora*, are by far the most abundant '. He maintained (1903, pp. 341 *et seq.*) that this sandstone is only formed during periods of rest or of actual erosion of a beach ; but he subsequently modified his views on the mode of formation of this rock and he remarks (1931, p. 40), 'The original formation, however, cannot have taken place on the surface of a beach covered with sand washed to and fro by the waves, thus in constant movement, and must have occurred at some distance within the beach.' As he points out, there may be a succession of lines of beach-sandstone and he attributes this to the sea having overlapped an original beach formation. 'All that is wanted to give a fresh line of beach rock is a certain holding up of the washing away of the sand, so as to give time for the evaporation of sea-water between tide marks to make itself felt in the consequent consolidation of the surface sand of the beach . . . . a more general cause may be seen in the solidification of the sand along the line where the fresh-water from the land and the tidal salt-water from the sea meet one another. The former is supersaturated with lime, and this is at once precipitated on the meeting of the salt- and fresh-waters. . . . Here, in the coral sand, the area of precipitation is on a line between tide marks some yards within the beach, and the consequence of the precipitation is a consolidation of a belt of sand into soft rock, which, lying between tide marks, sometimes shows layers and dipping as on a beach.' Personally I do not believe that the process of erosion has any appreciable effect on the initial formation of the beach-sandstone though, if its formation only occurs within the beach, it is of course only during a process of erosion that the sandstone will be exposed. I am by no means fully convinced that one is justified in insisting, as Stanley Gardiner does (1931, p. 43, footnote), on the distinction between this beach-sandstone and the ordinary sandstone that is formed beneath the surface of many, if not all, of the coral islands or beaches in tropical regions.*

Sandstone Formation.

In an account of beach-sandstone Stanley Gardiner (1930, p. 15) remarks that 'at the bases of dunes a sandstone with certain resemblances and dipping more or less as does the sand of the dune was found near Kankasanturi, north of Ceylon, in Farquhar and in Coetivy. In the latter two islands the sand dunes had been blown up on phosphatized coral land, which is very flat and hard, "platin land" as we term it, and in Ceylon they were on raised reefs, all localities above tide level. This dune sandstone was all covered with loose sand, of which a thickness of at least four to six feet seemed usual. We dug down through the top of one dune in Ceylon but only found our formation round the base where its deepest parts were best consolidated. Clearly this sandstone can only be formed by the evaporation of fresh-water, saturated with carbonate of calcium.' The formation of this sandstone is thus not confined to the shores of coral islands or atolls ; but may apparently also be formed along any beach that is fringed by a reef or off which a coral reef is situated, or in any deposit of sand where this contains a moderately high percentage of calcium carbonate. Along the beach at Tuticorin on the south coast of India, where there is an interrupted coral reef lying some little distance out from the shore, on the seaward side of the sand-bar that separates the so-called lagoon from the sea there is an exposure of beach-sandstone, exactly similar to that seen on the beaches of the islands in the Maldives.

Stanley Gardiner (1903, p. 37) in his account of the Atoll of Minikoi remarks, the patch (of coarse sand) at Boni-Kodi forms a soft very friable sandstone on the surface with loose sand beneath, apparently the crowbar finally fetching a hard rock. The amount of consolidation elsewhere is very varied, but at the main high tide level it commonly forms a loose sandstone, increasing gradually in hardness to the low-tide limit. 'In some of my pits I found alternation of harder and softer sandstone or loose sand. It was impracticable to trench across but my pits were sufficiently numerous to warrant one in the belief that there is an irregular series of lines of sand rock, more or less parallel to the lagoon shore with areas of loose sand between.' I doubt the correctness of this latter conclusion and think it more probable that there was a continuous sheet of sandstone, possibly varying to some or even a considerable degree in its hardness, underlying the whole of the loose sand of the island.

I have called attention to the occurrence of a deposit of sandstone behind the margin of the raised reef at Pamban, Rameswaram Island, S. India, and a study of the

---

* Kuenen ('The Geology of Coral Reefs', The Snellius Expedition, Vol. V, Part 2, pp. 86-88) recognizes this distinction between 'Beach-sandstone' and the sandstone formed beneath the sand of coral islands, which he terms 'Cay-sandstone', and he further divides this latter formation into two grades—a harder and a softer—'the hard type is probably formed out of the soft type by evaporation of sea water when it is exposed on the beach. The first hardening is done by percolating rain water.'

accounts of the various reefs and atolls in Indian waters, as well as my own experience, has convinced me that the presence of this rock is of very common, if not of universal, occurrence. Wherever one digs down in the islands of the Maldives or Laccadives one appears to meet an almost identical succession of strata. Moresby (1835, p. 398) records that when digging for fresh-water in one of the islands of 'Malladone-Madone' Atoll the various strata met with were as follows :—'On digging down the sand becomes white like beach sand but more compact ; below three or four feet a soft sandstone is found like particles of beach sand indurated. This sandstone is about two feet thick, below which depth it softens again to sand and fresh-water makes its appearance . . . . . This sandstone which is apparently formed from beach sand and is composed of broken shells and coral, when exposed to the air becomes quite hard and sometimes on the beach looks as if it were vitrified, being sonorous and excessively hard.' Similar

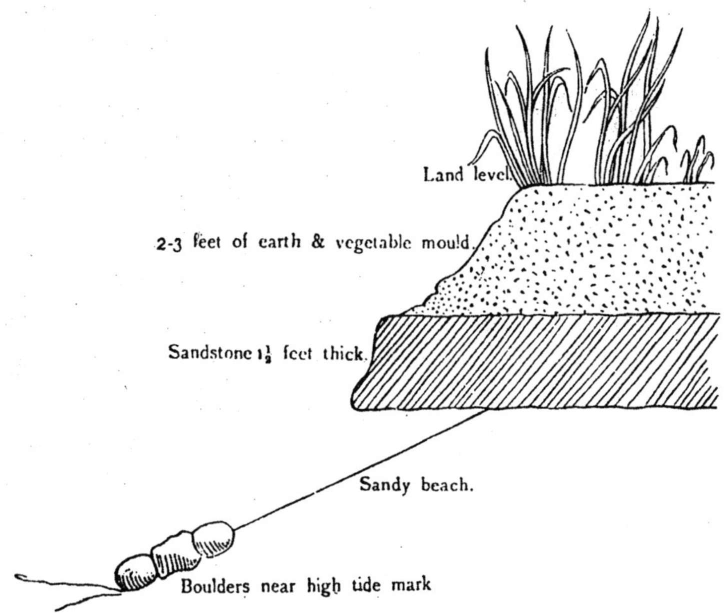

TEXT-FIG. 136.—Exposure of sandstone above beach on N. Side of Fehendu Island, Horsburgh Atoll.

strata were met with when digging a pit in Maradu Island in Addu Atoll; here in the first four feet I encountered at first a layer of earth and the roots of trees, then came a layer of white sand, below which the sand particles had become consolidated to form a layer of sandstone. Under the sandstone was a second layer of sand, and, finally, we came to a mass of coral boulders. Further evidence of the general occurrence of this sandstone formation was found in Fehendu Island, Horsburgh Atoll, where in a cliff on the west side of a bay on the north of the island the whole series of strata were exposed (Pl. 15, fig. 2) ; the same strata could also be seen in a well that the inhabitants had dug near the village at the eastern end of the island, and on the beach close to the village on the north side where the erosion of the beach has undermined this horizontal stratum of sandstone (Text-fig. 136).

As regards the Laccadives Ellis (1924) states that 'several feet under the sand on all the islands, except Minicoy and Kalpeni,* lies a horizontal bed of coral limestone a few inches or perhaps two feet and more in depth. It appears to be continuous with the conglomerate of the reef. There is no indication that any similar bed of limestone is now in process of formation in any position and it may be perhaps that the limestone bed indicates a previous sea bottom at a depth below the action of the waves.' Again as regards the structure of Androth Island, Lat. 10° 48′ N., Long. 73° 57′ E., Ellis remarks (p. 91) 'at the eastern end of the southern face, a conglomerate, which appears to be continuous with that of the reef, rises 5 to 6 feet along the level of the reef-flat and passes under the sand forming the islands; it appears to be continued into the layer of sandstone which everywhere lies beneath the sand'. That a layer of sandstone does underlie some of these islands is thus clear and, indeed, it would seem to be of comparatively common occurrence, since, in his description of Kalpeni (Lat. 10° 7′ N.: Long. 73° 55′ E.) he noted that 'another peculiarity is the absence of the sandstone substratum found in all the other inhabited islands'. Ellis, however, seems to be of the opinion that the sandstone layer is, or at least may be, continuous with the original coral-rock of the old reef. Oldham (C.F., 1896, p. 4) has pointed out that a succession of strata are to be found in some of the islands of the Laccadive Archipelago; thus of Kiltan island he remarks 'the island is formed of coral sand, coral sand rock and broken fragments of coral reef rock, overlying a hard conglomerate composed of fragments of reef corals. The coral sand rock may be seen exposed on the lagoon beach of most of the Laccadive islands; it is of a friable nature, but its surface becomes very hard when exposed to the air.' He thus appears to consider that the beach sandstone is continuous with the sandstone underlying the island.

Bourne (1888,) has given us an account, that is very similar, of the strata in Diego Garcia in the Chagos Archipelago. He there found that there was a double layer of sandstone, or as he calls it coral rock. The strata from above downwards were as follows:—

1. A thin surface layer of sand and mould.
2. A horizontal layer of stratified shingle rock with large coral masses occasionally embedded.
3. Loose coral sand.
4. A second layer of coral rock, the same as 2.
5. A layer of friable sand and
6. Solid rock foundation.

In numerous other localities the existence of beach-sandstone and sandstone or 'tuffe' has been recorded by Stanley Gardiner. Thus in Peros Bahnos Atoll he remarks (1907, p. 39) 'The islands themselves are generally formed of coral rock towards the outside, solid in some places, loose in others, tuffe or consolidated sand-rock in the centre, and loose sand towards the lagoon'. Again, regarding Egmont Atoll he states

---

* This island is named Elikalpeni on the Chart VII, p. 428 *supra*.

(*loc. cit.*, p. 53) that 'The islands themselves are very like those of Salomon, with similarly situated areas of coral-rock, sand and sand-rock (tuffe)'. I have already referred to his account of the formation in Coetivy (*vide supra*, p. 502). In the islands of the Amirante Bank, we again find evidence of this rock formation; in Poivre Island (*loc. cit.*, p. 152) 'the sand has consolidated into rock near the entrances of the bays', and as regards Eagle Island (*loc. cit.*, p. 159) 'the land is entirely formed of sand rock. The shore is sandy with layers of beach sandstone. In most parts it ends in a small sand cliff above, but in some situations in a wall of tuffe rock, 8 to 12 feet high'.

Guppy (1889, p. 402) in his description of Cocos-Keeling Atoll remarks that 'the stratum of conglomerate exposed at the bottom of the wells is usually one or two feet in thickness, and often overlies a bed of semi-consolidated sand of perhaps a similar thickness, underneath which another bed of conglomerate may occur'; probably this upper layer that he terms conglomerate corresponds to the sandstone layer noted in other localities, the deep stratum of conglomerate corresponding to the old reef-rock.

Even on the coast of India at the extreme southern end there is a similar formation of sandstone. In 1926 I was able to pay a visit to Mandapam and the neighbourhood; where the railway leaves the peninsula and crosses the bridge to Rameswaram Island there is a broad spit that is clearly formed of a marine alluvium and is composed of sand mixed with shells, for the most part marine in origin but mixed with others belonging to a species of land Gastropod (*vide supra*, p. 12 *et seq.*). All round this spit and on the other side of the channel on the mainland there is an extensive coral reef and a very fine exposure of sandstone and coral conglomerate. There can be but little doubt that this sandstone, like that on the islands of the atolls and elsewhere, has been formed by percolation of water through a sandy deposit to some depth beneath the surface and that it has subsequently been exposed by erosion along the beach and by removal of the sand, probably by the action of wind, from the upper surface.

A similar formation has been noticed in coral-growing regions in other parts of the world, thus in the Great Barrier Reef of Australia, Jukes, who visited and wrote a report on this great reef in 1842–46 (*vide* Saville Kent, 1893, p. 108) records that the strata in the upper five feet of Raine's Island were as follows:—

		Feet.	Inches.
1.	Good black vegetable mould	..	6
2.	Stone, brown mottled with white, hard and coarse-grained. . .	..	3
3.	Rich moist black soil, like bog-earth ..	1	4
4.	Stone of a light brown colour rather soft but tough and yielding slowly to the pick-axe.	3	..

Below this level he found the rock too hard to allow of his proceeding farther. He came to the conclusion that 'either the stratification and consolidation is the result of a gradual deposition beneath the level of low water, in which case a movement of elevation must have taken place, which in so small a spot seems a doubtful and gratuitous hypothesis; or else the present structure must have been produced in the interior of a mass of loose sand by the infiltration of sea- or rain-water, or some other cause of which we are ignorant . . . After the interior of such a mass of sand has been consolidated the loose exterior may have been washed away and the solid rock exposed.' In the Narrative of the Voyage of the 'Challenger' (Challenger Report, Vol. 1, p. 530) it is clearly stated that this island is composed of blown calcareous sand, which has consolidated towards the centre of the island into a compact limestone. Jukes also makes the interesting observation that in this layer of stone one or two nests of turtles' eggs were subsequently discovered ; as he points out, the presence of these turtles' eggs shows that the consolidation of the sand into stone 'had taken place after it was raised above the sea. It was due, probably, to the infiltration of rain-water percolating through the calcareous sand, that had been gradually piled above high water mark by the combined action of the wind and waves.' It also shows incidentally that the formation had taken place at some depth below the surface of the sand, since turtles dig down and excavate a large and comparatively deep hole in the sand before they deposit their eggs, afterwards covering the eggs up again.

A feature in which this consolidated sandstone may resemble beach sandstone is the stratified nature of the deposit. Mosely (1892, p. 17) called attention to this feature in his account of the sand dunes of Bermuda and he accounts for it as follows :— 'The rain, charged with carbonic acid, percolates through the dunes and taking lime into solution, re-deposits it as a cement, binding the sand grains together. Successive showers of rain, occurring at irregular intervals, some charged more, some less highly, with carbonic acid, and forming each a crest on the surface of the dune of varying thickness, produce a series of very thin, hard layers in the mass of sand, alternating with seams of less consolidated and sometimes quite loose sand. Crusts of consolidated sand are to be observed commonly on the surfaces of fresh sand dunes. These layers or strata of the hardened sand follow in form the contour of the dunes, and thus where these have been perfect domes or mounds dip outwards in all direction, with curved surfaces from a central vertical axis. Such an arrangement is constantly to be seen where sections of the older rocks are exposed.'

Coming now to the consideration of the various ways in which this lithification of the loose sand may take place, and beach sandstone or the sandstone of the island flat may be produced, Dana (1875, p. 122) put forward the view that 'these deposits (of beach-rock) become cemented by being alternately moistened and dried, through the action of the recurring tides and the wash of the sea on the shores. The waters take up some carbonate of lime, and this is deposited and hardens among the particles on the evaporation of the moisture at the retreat of the tides.' In the case of the drift sand rock he (*loc. cit.*, p. 124) considers that 'these sand banks through the agency of infiltrating waters, fresh or salt, become converted into a sand rock, more or less friable, which is frequently oolite'. One of the earliest studies that I have seen is that by Branner (1904) who studied the formation of this rock on the coasts of Brazil. He reached the conclusion that it was formed by the deposition of cementing calcium carbonate derived from four sources, namely :—

(1) Dissolved by rain-water or spray from the beach sands themselves, i.e. carried from the upper layers of the sand and deposited in the lower ones.
(2) From the ocean water after having been derived (through the agency of carbon dioxide) from calcareous organic bodies in the sea.
(3) Brought down from the land by streams.
(4) Dissolved from calcareous beach sands by fresh-water streams cutting behind them and re-deposited while the water is passing seawards through these sands.

While agencies (1), (3) and (4) will act equally upon the sand near the surface of the beach and that below the whole extent of the island, agency (2) will only affect the sand of the sea beach to a depth to which sea-water and spray can penetrate. As Vaugham (1924, p. 325) has pointed out 'the results of the different investigations are accordant, and all agree that the surface layers of the ocean water in tropical and subtropical regions is saturated or even super-saturated with reference to $CaCO_3$ and that any agency that would cause a further concentration of $CaCO_3$ or which would otherwise reduce the capacity of the water to hold $CaCO_3$ in solution, would produce precipitation'. The deposition of the calcium carbonate from solution will be intensified and accelerated in the belt between tide marks by--

(1) The escape of carbon dioxide from the sea-water where the waves break on the beach, and
(2) The escape of carbon dioxide from the sea-water under the influence of a rise of temperature such as would occur while flowing over a shallow reef flat or shelving beach. Branner also suggests that this hardening of the beach-rock may be assisted by the submarine escape of carbon dioxide around volcanic vents, but, as there is no evidence of the Maldive-Laccadive Ridge having a volcanic origin nor of the presence of any volcanic vents in this region, this latter possibility need not detain us.

Again, any increase in the density of the sea-water will hasten the precipitation of calcium carbonate from sea-water, and it is precisely in the tropics and in arid regions that the sea-water is most saline and where there is the greatest tendency for the carbonates to be deposited. In this connection it is interesting to note that in one of the atolls of the Maldives the sea-water flowing over the outer reef flat was found to have had its temperature raised from 29·8°C. to 36·5°C. and in consequence the rate of evaporation had considerably increased, thus causing a rise in specific gravity that on subsequent cooling will cause a rise in density.

In the case of percolation from behind of lake or river-water through the sand beaches, the solution and deposition of calcium carbonate will be greatly assisted by the decay of aquatic plants which charges the water with organic acids. 'As the waters penetrate seaward through the sands of the beach, especially during low tide when the hydrostatic pressure of the water in the pools forces them forwards, they attack the calcareous matter with which they first come in contact, and carry it along in solution until they encounter the sea-water. Here there would be a tendency for the less soluble lime to be precipitated, especially if there were a checking of the movement of the water.' Precipitation would thus take place along the beach line and would be confined to a rather narrow belt, while its length would be determined by the coast-wise length of the pools and of the restraining beaches. While Brànner was impressed by the presence of streams and lakes behind a raised sandy beach and, therefore, lays particular stress on their importance, it is clear that the same process of solution and deposition will occur in any area in which fresh-water is to be found percolating through the sand towards the sea and Crossland (1905) in his account of the Cape Verde Islands gives it as his view that 'Beach Sandstone is formed by the deposition of calcareous cement from fresh-water on meeting the salt.' Field (1919) also put forward a very similar view. He has been quoted by Daly (1924) in full. The only point on which Field differs from other observers is that he introduces the agency of storms; he thinks that during these the sand above sea-level becomes saturated with rain-water, and that quantities of fresh sand are deposited along the beach in certain regions, while other areas will be scoured. This rain-water, percolating through the freshly deposited sand, deposits calcium carbonate, probably as a result of the escape of carbon dioxide, in the interspaces between the sand fragments and thus produces beach rock. He rejects the possibility that the result of putrefaction of organic matter, that has been thrown up by the storm, can have any influence, since on analysis such beach rock is found to be free from any carbonaceous material. Umbgrove (1928) is also of the opinion that coral conglomerate, and presumably beach sandstone, that is found at or near high water mark in such coral islands, is formed not by the evaporation of sea-water, but by fresh-water derived from monsoon showers that is flowing out laterally from the centre of the island towards the shore, where it evaporates on the beach, leaving the lime, which it has dissolved from the centre of the island, to cement the loose fragments of the beach together. Stanley Gardiner in his later contributions to the subject has put forward the view that beach-sandstone is the result of the combined action of both fresh and sea water. He remarks (1930, p. 16) 'Beach sandstone forms between tide levels where the rain water that falls on the land, becoming supersaturated with calcium carbonate in sinking through the soil, meets the salt water, this causing the precipitation of its lime. The consolidation proceeds hence from inland seawards at such a distance behind the sea beach as the sea water penetrates the same. When the overlying sand of the beach has been washed away, these now exposed strata hold up the destruction long enough to be further consolidated by calcium carbonate from the evaporation of the sea-water on their seaward face'; and in the following year (1931, p. 43) he remarks that 'a more general cause may be seen in the solidification of the sand along the line where the fresh-water from the land and the tidal salt water from the sea meet one another. The former is supersaturated with lime, and this is at once precipitated on the meeting of the salt and fresh-water. . . . The area of precipitation is on a line between tide marks some yards within the beach.' Yonge (1930) from his studies on the Great Barrier Reef of Australia also concludes that the cause of the formation of beach sandstone in that area is the result of solution and deposition of calcium carbonate; he remarks 'the summer rains falling on the surface of the cay, drain through it, dissolving as they go all the lime which water can carry in solution; then when the tide is out, the water drains away at the base of the beach. As it emerges it is at once evaporated in the great heat, and leaves behind it the dissolved lime, which is deposited between the particles of sand, finally converting them into the solid rock which now exists. As this process goes on year after year, a series of flat sheets of rock, still with the same gentle upward slope of the original beach, forms round the base of the island.'

These various views presuppose that the only agency in the precipitation of the calcium carbonate is the alteration in the amount of calcium carbonate that can be held in solution; but there are other agencies that have a very marked influence on the production of calcareous mud in coral seas and may equally be a factor in the lithification of the coral sands. Drew, working on the precipitation of calcium carbonate in the sea (1914), produced evidence that this might be due to a specific organism, *Pseudomonas calcis*, and that as a result of this precipitation there might be formed certain kinds of calcareous rock; at the conclusion of his paper Drew remarks 'the investigation can at the most be considered to offer a mere indication of the part played by bacterial growth in the metabolism of the sea'. Later observers, especially Lipman (1924) have cast doubts on the validity of Drew's hypothesis and this author concludes that any denitrifying bacillus may in a suitable medium cause the precipitation of calcium carbonate from solution but that in the sea the main causes are either a chemical interaction, such as was suggested by Murray and Irvine (1905), between calcium sulphate and ammonium carbonate, the latter being produced by the decomposition of any organic matter, or else by a disturbance of equilibrium due to alterations in the quantity of carbon dioxide in solution either by the upwelling of cold water, by changes in the air temperature, the agitation of the water by winds or even the abstraction of this gas by marine plants of the microscopic order.

Vaughan (1924, pp. 321–327) has reviewed the evidence for and against the effect of Drew's organism on calcium precipitation and he concludes that 'bacteria taken from a bottom mud largely composed of chemically precipitated material will precipitate $CaCO_3$ and that the mud contains nutrient material that will support the bacteria. It also appears that the strongly ammonifying *Vibrios* are probably more important agents than the denitrifying *Pseudomonas*.'

Daly (1924) suggests that beach sandstone is only formed in masses of sand that have been suddenly piled up by the action of storms. His conclusions are largely based on observations carried out at the Tortugas Laboratory. Here in both 1910 and again in 1919 a hurricane piled a large quantity of calcareous sand on the side of the Laboratory warf. Much of this sand seems to have been derived from the broad shelf off shore. Less than two years later this sand between tide-marks had already become lithified to a depth reaching a maximum of about 75 cms., resulting in an elongated lens of typical beach-sandstone. The lithification took place under the cover of 25 to 100 cms. of sand, which was not lithified. The explanation that Daly gives of this formation is that ordinary beach sands are comparatively free from organic matter, while shelf-sands are charged with decaying animal and vegetable matter; the bacterial decomposition of this organic matter must cause a precipitation of calcium carbonate from the sea-water, which slowly creeps through the sand because of tidal and other changes of level. He, however, concludes that bacterial action is not sufficient to explain the presence of all the cementing material in beach sandstone and that other factors must come into play. These latter, he concludes in agreement with Branner, are the heating up of the sea-water and its æration by wave action, both of which will cause a precipitation of calcium carbonate. He adds 'the precipitation resulting from the two causes is practically confined to the thin layer of banked up sand, lying between high tide level and a level a little below low tide mark, exactly the position where beach rock occurs. So long as the initially cemented sand keeps its proper relation to the breakers and is, for example, not too deeply covered with imported sand, the cement can thicken and grow more resistent to blows, until the original interstices have become more or less completely filled. On the other hand the condition of shore erosion may lead to the destruction of the new bank before a strong rock can be formed.' If we admit the correctness of the above theory, this would provide a reason for the greater hardness of the beach-sandstone than of the softer sandstone that is found underlying so many of the islands; but its original formation does not appear to be in any material way different from that of the rock that is being formed beneath the raised sand of the islands.

On the vast majority of the islands on the atoll rim in the Indian Archipelagoes or on the coastal region on the main-land there is a thick growth of vegetation and this gives rise to a superficial stratum of vegetable mould, that overlies the calcareous sand (*vide* Pl. 15, fig. 2) and contains quantities of decomposing organic matter, derived in part from the vegetation itself and in part from the various species of animals that frequent the islands, and in this connection it is interesting to note that Steers (1930, p. 12) reports that 'in the Great Barrier Reef in all the cays we visited we did not find one in which there was beach rock and no vegetation' and in only one (Michaelmas Cay) did they find vegetation and no beach rock. In this connection a paper by Hall and Miller (1906) is of considerable interest; these authors have pointed out that the study of the problem of the retention of bases in soils and especially of retention of calcium carbonate has shown that in cultivated soils and presumably therefore also in soils that support a rich tropical vegetation, as do most of these coral islands, 'the process of nitrification, going on in all normal soils, requires some base to combine with the nitrous and nitric acids produced by the oxidation of the ammonia and other nitrogen compounds. In an ordinary way this base is supplied by calcium carbonate' (*loc. cit.*, p. 2). In such soils as one gets in these islands almost the only base is calcium carbonate. These authors go on to remark that 'the destruction of nitrates by bacterial action, with the evolution of the nitrogen as gas, the change commonly known as "denitrification", is always attended by the production of a carbonate of the base with which the nitric acid was combined ..... As also denitrification is most likely to take place in the lower sub-soil where the oxygen of the soil gases has been exhausted, any calcium carbonate reformed in this way would not appear in the analyses set out, which only extend to the depth of 24 inches' (*loc. cit.*, p. 30). In this connection, the work of Ross and Bagchi (1919 and 1925) has indicated that the rain-water of tropical thunderstorms is charged with nitric acid, and this combining with the coral sand will produce calcium nitrate and set free carbon dioxide.

During the monsoon the heavy fall of rain saturates the soil and may in suitable low-lying localities give rise to pools of fresh water. As this water percolates through the soil it becomes charged with organic acids and will when it reaches the lower sandy level dissolve out quantities of calcium carbonate, that will later be deposited at a lower level. Quite apart from the presence of organic acids, the rain-water, being charged with carbon dioxide, will *per se* dissolve a certain amount of calcium carbonate, which from subsequent evaporation or from loss of carbon dioxide will be redeposited at a lower level, as has been pointed out by Skeats (1903, p. 104) who remarks, 'a microscopic study of the limestone also shows that a certain amount of solution must have taken place since the rocks have been upraised. It is noticed that many cavities in the limestone have been more or less filled with a stalagmitic coating of calcite and in certain cases with dolomite. These deposits are no doubt due to the solvent properties of rain-water bearing carbon dioxide in solution, and the precipitation of the dissolved carbonates on the escape of the carbon dioxide.'

Another factor that may have a marked influence on the deposition of calcium carbonate from a state of solution is the presence of algæ that by their ordinary process of photosynthesis can abstract carbon dioxide from solution and thus lower the capacity of water to retain $CaCO_3$ in solution. Such an agency appears to be especially important in the production of Oolite and the work of van Tuyl (1916) and Mawson (1929) seems to indicate that this formation can be traced to the effect of algæ in either secreting carbonate of lime or in utilizing the dissolved CO_2 and thus causing the precipitation of this salt.

For a more complete review of the various agencies that may cause the precipitation of calcium carbonate from a state of solution I would refer the reader to a paper by Johnston and Williamson (1916).

All the evidence that we possess seems to point clearly to the fact that this lithification does *not* take place on the surface but at some depth below, the actual distance varying in different regions, possibly in accordance with local conditions, from 25 to 100 cms., as in the case of the sandstone formation in the sand thrown up at the Tortugas Laboratory, or 4–6 feet in the case of the sand dunes of Ceylon. Because of this limitation of depth the sandstone must of necessity tend to exhibit the same conformation as the actual surface. In my opinion it is only in areas in which erosion is taking place that such a formation will come to view. In the earlier stages of erosion of a beach it is the outer formation, in which the strata exhibit a dip towards the sea owing to their formation beneath the sea-ward slope of the beach, that will first be uncovered, and it is to this portion of the formation that Stanley Gardiner has applied the term 'beach sandstone'. With the exposure of this layer to the action of sea-spray and the mechanical turmoil of the waves, there results a still further deposition of calcareous material, due to causes that we have already discussed, and a consequent increase in the hardness of the stone. This will naturally tend to prevent any further erosion of the seaward face of the island, but should erosion continue we shall eventually reach and expose a layer of cay-sandstone that was formed under the reverse slope of the raised island margin and one, therefore, in which the 'dip' of the sandstone is in the reverse direction, namely away from the sea and towards the land or the centre of the island. Since in a deposit of this nature, i.e. having a reverse slope, the layer of sandstone will rapidly be undermined by wave action, erosion will proceed at a rapid rate and in consequence examples of this type of deposit will not be easy to find; nevertheless that such do occasionally occur is shown in the photograph reproduced in Pl. 12, fig. 3. Finally, still further erosion will expose the horizontal strata of cay-sandstone, formed beneath the flat island or land surface, as shown in text-fig. 136.

Erosion.

Stanley Gardiner (1903, p. 149) has pointed out that in the Laccadive Archipelago 'practically all land lies on the eastern or lee-ward side of the reefs, the group being completely exposed to the gales of the south-west monsoon, while it must be largely protected by India from those of the north-east. The present contour of the land throughout the group would hence seem to be really due, as at Minikoi, to elevation and subsequent erosion, the latter having completely removed all save traces of the land from the western reefs', and with regard to the Maldives (*loc. cit.*, p. 157) he remarks 'as to the land on the rim-reefs of the banks it is noticeable that it lies rather on the east than the west side in all without determination of their position or form . . . It might be pointed out that the strongest and heaviest winds and seas are from the westward rather than the eastward. These might be expected either to pile up coral and so form land, or perhaps if land already exists, to wash it away. The former supposition is untenable in view of the absence of land, and the latter is largely discounted by the fact that Felidu and Mulaku, which must be to a large extent protected on their west sides by the opposite series of banks, are by far the best examples of the absence of land on the western reefs. The presence of land would seem then to have been determined on the formation of the group.' Baker (1925, p. 1017) has, however, called attention to the fact that erosion may go on behind the shelter of a fringing reef; he remarks, 'but when the coral reef has risen to the surface *the battering down of the land does not cease*; on the contrary, it continues and the shore recedes. Tides and currents clear away the ground to a depth of a few feet below low spring tides. The growth of the coral reef outside the original shore-line prevents much submarine erosion, so that a flat platform is left as the shore recedes.' Earlier in this paper I have given examples of this erosion behind coral reefs in the Andaman and Nicobar Islands.

As I have pointed out in a previous paper (Sewell, 1932) whenever we examine the outer beach of the islands in the atolls of the Maldives we find unmistakable evidence that they are being eroded, in some cases only slowly, though in others the process appears to be comparatively rapid ; and it seems clear that, with very few exceptions, these islands are being slowly but none the less surely destroyed. All along the outer shores wave-action and a continuous scouring by currents is gradually breaking down and destroying the old coral-rock rampart, the destruction of these land-masses depending on the force of the waves, the strength of the current and the degree of solidarity that has been reached. In this connection Agassiz (1895, p. 215) has pointed out that ' the solvent action of the salt water cannot be compared in efficiency with the destructive mechanical action of the sea '. In consequence of unequal erosion we find that in places extra hard blocks are left behind, until finally one may get merely a line of such blocks and solid masses, to which the name of ' coral horses ' has been given, to still indicate where the outer shore of an island originally stood. Darwin himself (1889, p. 50) refers to the destruction that was going on in some of the Maldive Atolls : he remarks that ' several islands, and even some of those which are believed to be very old, are now fast wearing away. The work of destruction has, in some instances, been completed in ten years.' In a previous paper I (Sewell, 1932, pp. 458–461) have described several examples of erosion that particularly struck my attention during my visits to two of the Maldive Atolls. Darwin (1889, p. 87) has expressed the opinion that ' the absence of islands on the leeward side of atolls, or, when present, their lesser dimensions composed with those to windward, is a comparatively unimportant fact '. I entirely disagree with such a view. The greater or lesser importance of such a definite difference on the two sides of an atoll depends entirely on the view that one holds regarding the mode of formation of the islands. If these islands have, as Darwin and others have thought, been thrown up by wave action, the absence or the smaller size of islands on the lee side can clearly be correlated with the more sheltered position and the comparative rarity of heavy seas ; but if, on the other hand, the islands have been formed by the fall of sea level and the subsequent erosion of a continuous land area, then their smaller size or absence on the leeward part of the rim of an atoll is entirely contrary to what one would expect.

Since this erosion, as already mentioned, depends very largely on the strength of the wave action and the surface currents along the shore, one would expect in the region of the Maldives and Laccadives that it would be most rapid on the south-western side of the atolls and further, that more erosion would have taken place in those latitudes in which the south-west monsoon is strongest. If now we compare the atolls Diego Garcia (Lat. 10° S.) in the Chagos Archipelago, Addu (Lat. 1° S.), Haddumati (Lat. 2° N.) and Minikoi (Lat. 9° N.) in the Maldives and the atolls and banks of the Laccadives, we find a regular series in which at the southern end there has been but little erosion, whereas this rapidly increases as we proceed further northward. In the Chagos Archipelago, lying well to the south of the equator, we are out of the track of the south-west monsoon and more in the region of the south-east trade winds ; and here we find that the atoll of Diego Garcia is completely surrounded with land, except where the entrance channels into the lagoon form a gap. Other areas in this group have not, however, escaped so lightly. In the case of Solomon Atoll, land is almost entirely absent from the south side and a large part would appear to have been eroded away on the north-east side. This atoll lies to the east side of the archipelago and is protected on the west by Peros Banhos and on the north and north-east sides by Speaker's Bank and Blenheim Rocks. This protection is reflected in the position of the islands that on the south-west, west and north sides abut on the

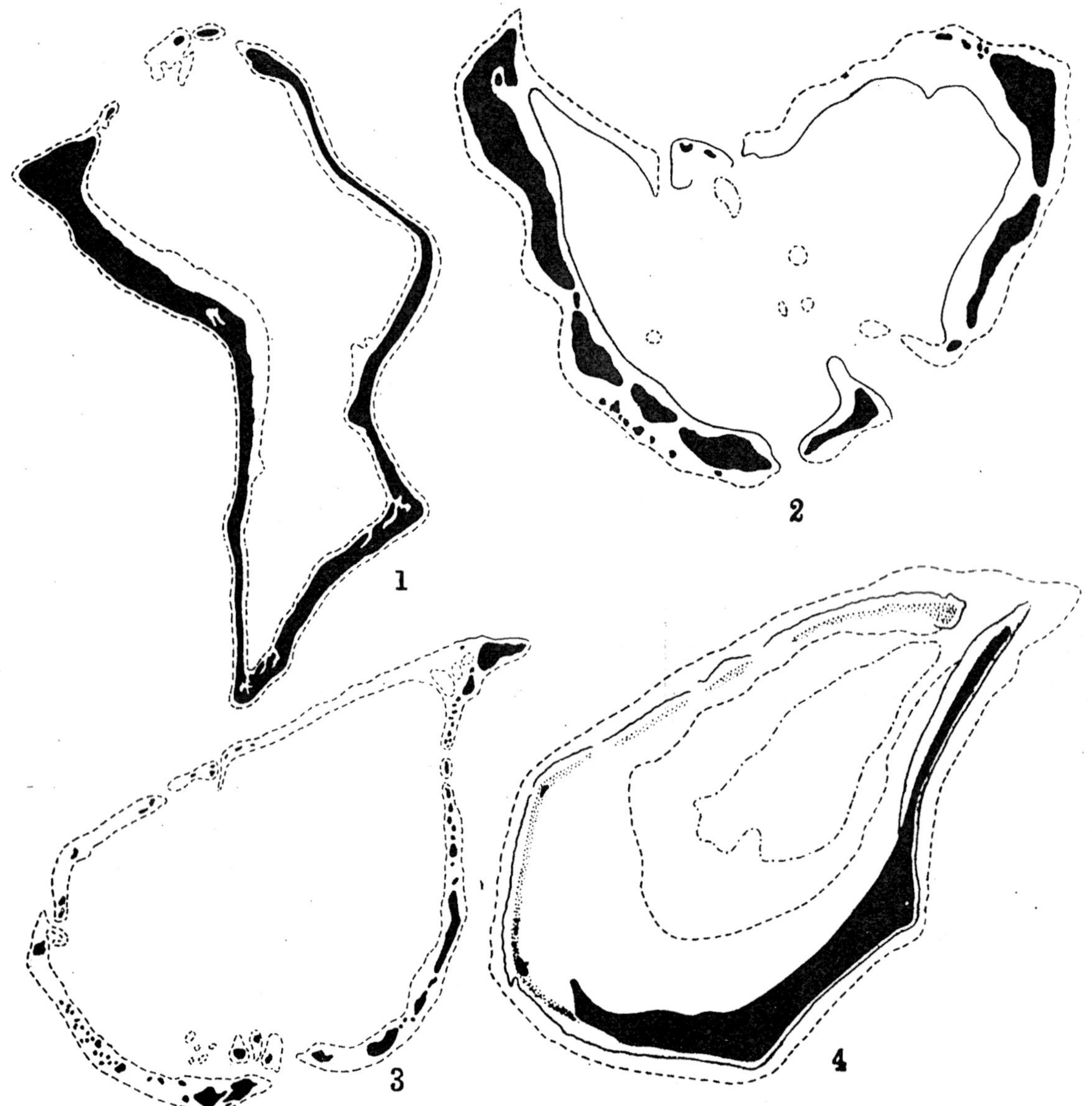

TEXT-FIG. 137.—Charts of some Atolls in the Chagos and Maldive Archipelagoes to show the distribution of land.

1. Chagos Archipelago—Diego Garcia (from Gardiner, 1907).
2. Maldive Archipelago—Addu Atoll (from Gardiner, 1903).
3. Maldive Archipelago—Haddumati (from Gardiner, 1903).
4. Maldive Archipelago—Minikoi (from Gardiner, 1903).

boulder zone and lie close to the present margin of the reef, whereas on the south-east and east they have been driven inwards towards the lagoon. In Peros Banhos Atoll there are several islands on the west and north sides, but between the southern entrance and Ile Yéyé at the north-east corner, are only two small islands, the two Iles Coquillages, and one rock, Coin du Mira : all the rest of the south-east part of the atoll rim is bare of land. Addu Atoll in the Maldives lies slightly to the south of the equator and, therefore, to a considerable extent out of the track of the full force of the south-west monsoon, but here we begin to find evidence of erosion on the northern part of the rim and in the south-west part of the atoll the islands are being driven inwards, a line of small islets and coral horses marking the original limit of the land. Haddumati Atoll, lying about 2° N. of the equator, possesses a large number of islands around the eastern and southern sections of its rim, but as we pass round towards the west and north-west we find that the islands become more and more sparse and scattered, till in the northern part of the western reef all traces of land have completely disappeared. Horsburgh Atoll, lying in Lat. 4° 50′ N., and Minikoi, lying in Lat. 8° 20′ N., are considerably more exposed to the south-west monsoon, though probably not quite far enough north to experience its full force, and here we find that in the former all trace of land has been destroyed on the west and south sides of the atoll, while in the latter all land has disappeared from the west and north-west side of the atoll. Finally, in the Laccadives the majority of the atolls and banks on the west side of the group, where these are exposed to the full force of the south-west wind, show at the present day no trace of land at all, though in a few cases small islands still exist, and it is in the atolls and banks on the east side of the group, where wave action would be to some and possibly a considerable extent reduced, that most of the islands are to be found. One is, I think, justified in adopting the view that the presence of so many islands on the east and south-east sides of the atolls in the Laccadives and Maldives is due, not to their formation in these situations, but to the fact that up to the present time they have managed to escape destruction. While holding this view with regard to the Maldives, one must not, however, overlook the fact that it does not, at any rate at first sight, appear to apply equally well to the atolls of the Pacific Ocean. The atolls of the Pacific that most closely resemble and, therefore, can best be compared with the atolls of the Maldive Archipelago, are those included in the Ellice, Gilbert and Marshall Islands and those of the Paumotus Group. In the case of the Paumotus and in the Gilbert and Ellice Islands the general condition appears to be very similar to that found in the Maldives and Laccadives. In the Paumotus the reefs on the south and south-west sides of the atoll are comparatively free from land ; Agassiz (1903a, p. 91) remarks ' on atolls, which like Takume, run from south-west to the north-east the land rim is mainly on the lee side and only partly on the weather side. On atolls which run in the opposite direction the land is usually best developed on the northern and eastern faces, which may be considered the lee side, the south-west extremity being usually bounded by low bare reef flats with but few islands and islets.' On Rangiroa the whole of the south and south-east reef carries only 4 or 5 small islands ; on the south-west side the gaps between the islands become of great width, some of them being several miles wide. On Tikahau on the south-west and west face there are only five small, widely separated islets on the submerged rim. In Fakarava on the lee side the atoll is fringed by large well-wooded islands, while on the weather side there is a wide reef flat on which small islands are found on the inner edge. Takume is noteworthy for the great number of its islands and islets, which form the land rim on the lee side. As Agassiz points out (1903a, p. 13) ' The cutting down by erosion and denudation and the planing action of the sea are all factors which we still see going on at all points of the Paumutus.' Again (*loc. cit.*, p. 237) Agassiz remarks ' the absence of land rims on the western face of several of the atolls of the Gilbert Islands is a characteristic feature of the group. At Tapeteuea and Tarawa, as at Onoatoa, the eastern faces form a nearly continuous land rim, while the western face is only protected by a sunken reef flat, covered with water of variable depth enclosing very indistinctly marked lagoons. These indistinct lagoons appear to have been formed, at least in the case of Onoatoa and Tapeteuea, from the erosion of wide fringing reef flats once occupying the summit of the shoal areas of the atolls of Onoatoa and Tapeteuea, on the eastern face of which the land rim has been thrown up.' Regarding the Marshall Islands Agassiz (*loc. cit.*, p. 275) states that ' the land rim is especially developed on the side exposed to the raking of the north-east trades. It is somewhat less developed on the faces where the trades strike the shore at a greater angle. The land rim on the southern and western faces is usually bare of islands and islets. . . . . . Wherever the coast line faces the general trend of the trades, the narrow land rim of the Marshall Islands is found ; wherever on the contrary it runs more or less parallel to it, the land rim is limited to a few insignificant islands and islets ; where the outer face of the lagoon is protected during the greater part of the year the land rim is still more insignificant, or becomes a belt awash.' These latter examples are, thus, in marked contrast to the condition found in the Maldive and Laccadive Archipelagoes and the only possible explanation, that I can see, is that in the Pacific the constructive action of the waves in piling up debris from the reef must be greater than their destructive action in eroding and destroying the islands exposed to their full force. Why there should be this difference in the two regions cannot be at present answered ; but I may perhaps be permitted to point out that this difference provides a very good example of the difficulties that may arise when we attempt to apply the phenomena discovered in one ocean to another.

The gradual destruction of the dead coral on the reef and the erosion of the seaward face of the islands give rise to the formation of large quantities of sand and silt, and this is still further augmented by the remains of other marine organisms, such as sponges, Polyzoa, Mollusca, Crustacea and Echinoderma, as well as Foraminifera and the calcareous algæ, such as *Halimeda*. With every rise of the tide this mud and silt is swept across the reef and through the channels between the islands, during which process it acts as a scouring agent producing still further erosion ; it is to a large extent deposited on the lagoon side of the islands or over the floor of the inner part of the lagoon reef, while to some extent it is swept still further inwards into the lagoon and thus becomes deposited over the lagoon floor ; other particles of sand are driven inwards by the wind and are deposited on the level area of the islands or are piled up on the lagoon side of the island as a ridge or ' lagoon mound '. As the islands are eroded on their sea-ward faces and the sand thus formed is deposited on the lagoon side the whole island slowly 'migrates' or drifts across the reef flat towards the lagoon. Under the influence of the prevailing winds and currents, sand and shingle tend to be deposited along certain definite lines and to form spits that run to leeward of the islands. In the case of Addu Atoll under the influence of the north-east monsoon several sand cays and spits have been formed on the lagoon side of the eastern islands with their horns pointing to the west; ultimately such sand spits may coalesce, so as to enclose a shallow area of water and thus give rise to a small pool or lake. Another possible mode of formation of such lakes (*vide* Stanley Gardiner, 1903, p. 417) may be the deposition of sand bars at both ends of a channel running between two islands, thus enclosing a basin between them. Guppy (1889, p. 466) pointed out that this process was going on in the case of a group of small islets in the Cocos-Keeling Atoll on the east side, viz. Pulo Ampang (major and minor) and Pulo Bruko, that stand on the same conglomerate platform. Here the outer margins are becoming joined together by the encroachment of the vegetation on either side of the gaps between the island, while on the inner or lagoon ends sand banks or shoals have been formed, and these are extending laterally, 'so that at no distant date the islands will be united at their inner margins as well as at their outer borders, and two lagoonlets will represent the original gaps between the islands'. The water in such lagoonlets or lakes will originally have been salt, but as a result of rainfall this will in course of time become more or less fresh. With increasing vegetation and the further accumulation of sand, shells, etc., the general level of the island will gradually be raised and such pools will tend to become silted up. A further agent that appears to assist in the ultimate obliteration of such pools is the formation in the water of a flocculent deposit of chalk-like material; in some of the pools on Heratera island in Addu Atoll this formation was going on at such a rate that the whole of the contained water was of a milky appearance and the beds of the pools were covered by this deposit to a depth of some 6 to 9 inches. This deposit would appear to be very similar in character to the flocculent deposit that is found in the shoal waters of Florida and the Bahamas. Agassiz and others have suggested that much of this calcareous deposit must be due to the chemical precipitation of calcium carbonate from its solution in sea-water and is not to be attributed to the mechanical transportation of detritus from the reefs (*vide infra*, p. 526).

As already mentioned, along the inner or lagoon side of an island the sand is often piled up partly by the action of the waves and partly by the wind, to a height greater than that of the general level of the rest of the island, so that it forms a distinct ridge along the inner side that is termed the 'lagoon mound'. A typically perfect island will, therefore, consist of an outer sea-ward ridge of stones and boulders resting on a substratum of the old reef rock, a general level of sand and shingle mixed with shells, etc., the shingle or boulders predominating in its outer half, and an inner rampart or lagoon mound of blown sand. It is in those islands that are still situated near the site of their original formation and opposite which, therefore, there is a wide expanse of the inner lagoon flat, separating them from the deep water of the lagoon, that the inner or lagoon beach is likely to be in a comparatively stable state and it is here that one finds the best examples of the inner ridge of clean sand arising from the level of the reef flat to form a ridge or lagoon mound, frequently some feet above the general level of the island and usually crowned with a growth of scrub and bushes.

One cannot leave the question of the islands and their formation without a reference to the occurrence of pumice. The occurrence of this material on these islands has frequently been recorded. Stanley Gardiner (1906, p. 582) remarks 'one may point out that on the shores of many of the islands there are lines of pumice, which the natives state were washed up about 1885, and would hence have probably owed their origin to the eruption of Krakatoa in 1883'. Many years earlier Alcock (1902, p. 175) recorded the occurrence of a bed of pumice in one of the islands of the Laccadive group and Oldham (C. F., 1896, p. 6) also records this deposit (on Kardamat island); he remarks that 'it is strewn all over the surface and varies in size from a marble to half-a-foot in diameter'. The presence of this line of comparatively recent pumice along the shore or across the level of some of the islands serves to give us a means of estimating the changes that have taken place in the configuration of the islands since 1885 or thereabouts, and in some instances, as for instance in Heratera island in Addu atoll, gives an indication of the amount of inward 'drift' that has taken place in the island since that date.

As the islands migrate towards the lagoon the destructive action of the lagoon waves is able to prevent the addition of further sand and finally erosion commences on this face of the island also. It is not easy to explain why at a certain stage in the history of one of these islands the process of deposition of fresh sand on the lagoon side should cease and be replaced by the action of erosion. There are, I think, two factors involved in this. The first is the proximity of the lagoon beach to the inner margin of the lagoon reef. So long as there is a wide expanse of reef-flat on the lagoon side of an island, wave action will be inhibited, but as this expanse becomes progressively narrowed by the inward 'drift' of an island, the action of the waves will increase and will eventually prevent the further deposition of sand, and may even be sufficiently strong to act up erosion. The second factor is in all probability connected with the condition of the lagoon reef. So long as this is flourishing and is covered with growing coral, it will act as an efficient barrier to the lagoon waves; but with increasing erosion and the deposition of mud or silt, the integrity of the reef-edge is broken up and it no longer can act as an effective barrier, so that the action of the lagoon waves is now felt on the lagoon shores of the islands on the reef-flat. In those islands that have migrated sufficiently far across the reef-flat, we find unmistakable evidence of erosion on their lagoon faces with the exposure of sandstone, that at first is of the beach-sandstone type and slopes towards the lagoon but that at a later stage shows the characters of a sandstone that has been deposited beneath the reverse slope of the lagoon mound. With erosion going on simultaneously on the two sides of an island, its final destruction and disappearance cannot be long delayed.

The Lagoon flat and the Lagoon Edge.

As we pass from the lagoon beach of the islands towards the centre of the lagoon we find that the surface of the lagoon flat slopes gradually downwards and there is simultaneously a steady increase in the quantity of coral that is growing on the surface. At first the lagoon flat is composed almost entirely of sand, that has clearly been derived largely from the destruction of the islands or the coral and rock of the seaward reef and boat channel and that has been swept by the tidal current through the channels between the islands and deposited on the lagoon flat. As one passes further inwards, however, patches of coral begin to make their appearance, for the most part of species of *Acropora*, and these become more and more frequent and eventually coalesce to form definite beds and, finally, to form the reef of the lagoon edge itself, which in many instances possesses a steep and precipitous inner face. Stanley Gardiner (1898, p. 498) states that in the case of the atolls of the Pacific Ocean 'the precipitous character of the reef towards the lagoon and the mushroom-shaped heads of coral are no doubt due to solution. This theory implies that the reef, as fast as it is formed outside, is dissolved away inside, as these barrier and atoll reefs are fairly uniform in breadth. The solution forms the cliffs and undermines the shoals.' It seems to me that far too much is attributed to the effect of solution and too little to the inherent property and peculiarity of the coral growth itself and the effect on this growth of sedimentation of mud and sand and other adverse conditions of temperature, lack of sunlight, etc. Wood Jones (1910) has brought forward evidence in favour of the view that sedimentation is the chief agent in the formation of these atolls and my own observations have led me to accept this agency as one of the possible contributory factors. On the other hand Vaughan (1914, p. 67) has come to the conclusion that as regards the coral reefs of the Pacific, 'atolls are not formed by solution, but by constructional geologic agencies'. Practically every coral reef that I have seen, whether it be inside a lagoon or on the seaward face of a reef, and equally whether it be in a sheltered locality or exposed to heavy seas, provided the depth of water in which the reef is growing is sufficient to allow of its formation, has a vertical or nearly vertical face; and the general appearance of the steep side of a coral reef in any situation that is sheltered is remarkably similar, whether it occurs in an atoll or on the lee side of a barrier reef or even in a fringing reef in a sheltered situation such as occurs around the shores of many of the islands in the Andamans and Nicobars. Mayor (1924, p. 59) in his account of the reefs of Samoa remarks that 'the overhanging sea-ward edge of the flat upper part of the reef is composed chiefly of *Acropora leptocyathus*, which, after it dies, remains bound to the reef by a veneer of lithothamnion. Beneath the overhanging edge there is a precipice or steep slope of from about 1 to 6 fathoms, at the base of which *Acropora* clusters thickly. Thus the reef grows seaward by the formation of an overhanging edge at low tide level, which often breaks off and falling to the bottom of the slope, aids in the upbuilding of the reef.' The description given by Jukes (*vide* Saville Kent, 1893, p. 114) of a part of the Great Barrier Reef of Australia will, with the simple alteration of the names of the different species of coral that he mentions, apply almost equally well to the majority of reefs; 'smooth and rounded masses of *Meandrina* and *Astræa* were contrasted with delicate leaf-like and cup-shaped expansions of *Explanaria* and with an infinite variety of branching *Madreporæ* and *Seriatoporæ*, some with mere finger-shaped projections, others with large branching stems, and others again exhibiting an elegant assemblage of interlacing twigs of the most delicate and exquisite workmanship. Their colours were unrivalled, vivid greens contrasting with more sober browns and yellows, mingled with rich shades of purple, from pale pink to deep blue. Bright red, yellow, and peach-coloured Nulliporæ clothed those masses that were dead, mingled with beautiful pearly flakes of *Eschura* and *Retepora,* the latter looking like lace work in ivory.' Clearly the vertical face of a reef that consists of *living* coral represents the present limit to which the reef has been able to extend. The fringing reefs around the shores of Nankauri Harbour in the Nicobars arise almost vertically from a depth of 15–20 fathoms and extend up to or nearly to the surface and the whole face of the reef appears to be composed of living coral masses. Stanley Gardiner (1903, p. 319) from the appearance of the reefs in Addu atoll concluded that 'on all sides of the lagoon in Addu atoll the reefs appear to have grown out into its basin.' I have made but little attempt to determine the various species of corals that were growing on the reef face in the regions that I have been able to examine, but the general condition of the reefs agrees with the above descriptions and there can be no doubt that the inner vertical face of the lagoon reef is, as a rule, in a state of magnificent vitality ; from sea-level down to a depth of about one fathom and from one end of the reef to the other, practically every available square inch of surface is covered by and is composed of live growing coral, and there does not seem to me to be the slightest foundation for supposing that this part is undergoing decay and solution. Down to this depth (about 1–2 fathoms) the face of the reef in many places has an inclined slope, so that from the top of the reef one is unable to see the character of the lower or deeper area. Below this upper stratum of living corals the lagoon reef very frequently falls steeply and may even be undercut. Stanley Gardiner (1898, p. 123) calls attention to this drop in the reef face and it is to solution that he attributes the presence of 'the small cliff from the edge of the reef to the bottom of the lagoon, which as far as I have seen, is of very general occurrence, and would not be found were the lagoons being gradually filled up by detritus as the subsidence theory requires.' If the vertical face of the lagoon reef were due to the effect of solution, we should, from the general similarity that exists, be forced to conclude that the vertical faces of many fringing reefs were also due to the same cause. That solution or, at any rate, removal of calcareous material does occur is indisputable, but it is probably only dead coral that can thus be acted on and, as Murray and Irvine (1890, p. 95) have pointed out, 'it appears, then, that those portions of a coral or shell that have been recently in direct life-connection with the organism are more soluble than those portions which are older, and have consequently largely passed into the crystalline condition': even as regards this action, Mayor (1924, p. 27), after quoting from the work of Drew (1914) and McClendon (1918), points out that the shallow water over coral reefs is saturated or even supersaturated with lime and in consequence, he remarks, 'it thus appears that solution of calcium carbonate in tropical water is so slight as to be negligible.' On the other hand the analyses of samples of lagoon deposits made by Stanley Gardiner (1903, pp. 322-3) clearly suggest that there is considerable solution of calcareous material from these deposits, since the percentage of silica contained in them is about 50 times as much as is found in the reef rock or sand. It would appear that though the effect of solution may have been considerably overestimated by certain observers, some action at least must be going on in the lagoons of these atolls and in other situations where dead coral or coral rock is exposed to the action of salt water. But other agencies are extremely active, such as fish, holothurians, living algæ, sponges, and molluscs, all of which cause destruction of the dead coral and when these agencies are combined with erosion, due to currents bearing sand and detritus, the lower slopes of the reef get undercut and we have a condition such as that which Paradyce has figured for the coral reefs of the Australian Barrier Reef. Such coral heads were fully described by Dana (1875, p. 110): as he points out 'they are most likely to occur where there are great regions of shallow water extending outward from the barrier, and where tides are not heavy or there is partial protection from them. In some seas, such isolated patches are shaped like a great mushroom—having a narrow trunk or column below, supporting a broad shelf of reef above.' The undercut character of these coral heads is not confined to colonies that are growing at the present sea-level, but it is also clearly seen in a number of localities where the fall of sea-level has put these now dead corals well above the surface. Fryer (1911) has called attention to such undercut rock pinnacles in Aldabra Atoll (*vide* his pl. XXV, fig. 3) and Macfadyen (1930, figs. 1–8) has figured a number of such from the Red Sea. This undercutting action is attributed by Fryer (*loc. cit.*, p. 409) to erosion and to solution, this latter action being due largely to the presence of decaying organic matter in the lagoon mud; while Macfadyen attributes it entirely to solution. Probably both solution and erosion are active agents in its production, the degree to which each is concerned varying in different localities. In living coral the exactly opposite process is going on and calcium salts are abstracted from a state of solution in the water and are built up into the skeleton of the coral.

In my opinion the vertical or precipitous slope of the reef face is, to some extent at least, the natural outcome of the growth of the coral itself and is not due solely to solution or erosion, though there is no doubt that when this steep face has been formed, subsequent events may so act on the reef that solution or erosion may then come into play and by the removal of material from the deeper strata still further increase the steepness of the slope or even undermine the upper living growth. Duerden (1902, p. 438) concluded that 'the presence of zooxanthellæ does not seem to be at all essential to the life of coral polyps, seeing that colourless individuals in the shade flourish apparently as well as those in fully exposed places.' Vaughan (1914*b*), also carried out experiments on this problem and these appear to indicate that certain corals are more susceptible than others, for out of a total of 18 species placed in the observation chamber, after 14 days of confinement without light, one was dead and a second partly so; after 28 days, 3 more species were dead and one partly so; and after 43 days, a further species had died and another was dying round its base. Yonge (1931*a*, p. 160), however, as a result of his work on the corals of the Australian Barrier Reef has reached the conclusion that 'individual reef-building corals at any rate can flourish without contained zooxanthellæ.' So far as I know in none of these series of experiments was any observation made on the rate of growth that was being maintained by the colony either before or during the experiments and it is quite possible that a coral colony might in the absence of zooxanthellæ maintain itself in comparative health and yet not show any appreciable increase in size, while other colonies with zooxanthellæ might show such an increase. There is no doubt that the zooxanthellæ act as very efficient organs of excretion and that the removal of the excretory products of the coral polyps is very largely dependant on the activity of the zooxanthellæ, so that if this activity is interfered with by the absence of light the rate of excretion will show a marked decrease ; and this would *per se* be sufficient to cause a lowered vitality in the coral polyps. There appears to be some evidence of a definite connection between the intensity of the light falling on the coral-colony and the rate of its growth ; in this connection Mayor (1924, p. 23) remarks that 'in many places in Samoa, as over the Tasma Bank, where the water is constantly agitated by the Pacific swell and the bottom is clean, hard and free from silt, the corals at depths of 8·5 fathoms grow only to about one-third the linear dimensions they attain in shallower water, and there are wide spaces between the heads, indicating unfavourable conditions. It looks as if some factor such as light may have a decided influence in determining the growth of coral. Indeed, Vaughan noticed that corals did not grow under the shade of a warf at Tortugas, Florida, although they flourish on those piles of this warf that are exposed to sunlight.' Stanley Gardiner (1902*b*, p. 214) has given an account of certain coral colonies that were growing on the chain of an anchorage buoy in Levuka Harbour, and he notes that 'there is never any coral growth on the buoys, nor on the chains immediately below them. . . . . The absence of corals here is probably due largely to the greater movement of the part near the surface of the water, destroying the colonies as well as causing difficulty of fixation to the larvæ.' In view of the fact that corals can affix themselves and grow on the margins of reefs where the water movements must be greater than in a harbour, this explanation seems to me hardly to bear close examination and I am inclined to think that the absence of the corals from the under part of the buoy and the neighbouring part of the chain is to be attributed to the fact that the buoy floating above on the surface will, to a very great extent, cut off from its own under-surface and the chain immediately below it the rays of light that alone can activate the chlorophyll of the zooxanthellæ. Stanley Gardiner (1898, p. 428) has also called attention to the fact that in certain situations, such as the fissures and clefts in the outer margin of an atoll such as Funafuti, the corals exhibit a distinctly heliotropic type of growth, growing at first outward from the side-walls of the fissure and then turning so as to grow vertically upwards. Wood-Jones (1910, p. 85) has also noted that, 'as a rule, coral zooids and coral colonies tend to grow upwards, and the general form of vegetative growth depends on this fact'.

The most striking character of this deeper part of the reef or coral patch, whether it be part of a barrier reef or a mere mushroom-head, is the absence of growing coral upon it. It is this absence that renders it so liable to solution or erosion by currents. Crossland (1928, p. 593) has called attention to the undercut character of the lagoon reefs in Tahiti; he remarks 'the evidences of lagoon erosion are most striking in the Papete-Papawa area. All the lagoon and shore reefs here, through the greater part of their circumferences, drop off sheer into deep water, almost directly into 10 fathoms or more. The edge takes the remarkable form of a shallow broad shelf, overhanging for six or even ten feet, a structure quite without precedent in the author's experience. Though coral grows abundantly above the shelf, little or none is found upon its edge, or upon the fallen pieces which occur below it in places, except on those shoals which are near the Taunoa Pass; on those near the Papete Pass corals extend a little way down the slopes, when there are slopes, but not in any quantity . . . . . It is clear that the vertical bare rocks below the shelves are the result of erosion.' The absence of growing corals on these undercut vertical walls of the reefs and pinnacles cannot be solely attributed to erosion of the basis; one agency that may prevent the growth of corals in such a situation is in my opinion probably the lack of sunlight due to the overhanging coral growth but other factors may be involved.

With each rise of the tide the water, that sweeps across the reef-flat, on reaching the edge of the lagoon reef to a large extent sinks downwards to the bottom of the lagoon across the face of the reef; this water will have a considerably raised temperature, the extent of this rise being, in all probability, sufficient to produce a deleterious effect on the growing coral, and it will also, during its passage across the reef, have been largely denuded of food-material and oxygen, factors that will still further tend to retard the growth of the coral. In places where erosion of the islands on the reef-flat is going on this water will also be laden with sand and silt, which though possibly not so harmful to the growth of coral as was previously thought, will in any case tend to interfere with the penetration of light to the deeper levels. All these factors taken together may well determine the more rapid growth of those colonies that live nearest to the surface. Where a reef has already attained a steep slope, especially in those cases in which the reef has a north-south direction the upper part of the reef will for at least part of the day cast a shadow over the lower living colonies and, by inhibiting the action of the zooxanthellæ and by retarding the growth of the coral, still further assist in the production of a vertical face.

Along the edge of a reef that possesses a face of this type dead and broken fragments of coral will either be washed up by wave action on to the surface of the reef or more commonly will drop to the bottom and eventually will build up a talus slope; but whereas on the surface of the reef the evaporation of the sea-water and other conditions will expedite the formation of coral-rock and thus assist in the consolidating process, at the bottom of the talus slope the principal action will be one of destruction, for instance by erosion, by the feeding activities of Holothurians, the boring action of Algæ and Sponges, and possibly to some extent by actual solution, all of which will tend to retard or, perhaps, prevent the lateral spreading of the reef. A study of the contours of the bottom of the lagoon in both Addu and Horsburgh atolls clearly indicates the existence of such a talus slope, for the 10-fathom contour line usually lies close to the foot of the lagoon reef and from here the bottom slopes down to an average depth of about 30 fathoms. Owing to the process of solution that this dead material of the talus slope is constantly undergoing, combined with the activity of triturating organisms, such as Holothurians and Sipunculids, the increase in the height of this talus slope will be comparatively slow and this will render equally slow the inward extension of such a lagoon reef, since this must depend in the main on the growth on the vertical face of the reef of such corals as are either directly capable of forming a compact basis for the growth of other corals or else are of such a friable nature that they can be broken off by the action of the lagoon waves and so assist in the formation of the talus slope itself ; branching forms such as the Stag's Horn type of *Acropora*, while able individually to contribute to the height of a reef by assisting in the formation of compact limestone, can add but little to the inward extension of the reef and may even to some extent tend to retard the rate of progress by appropriating to themselves food supplies, using the term in its widest sense, that would otherwise be available for the more solid corals, such as Brain-corals, *Porites* and the Astræaceæ ; they would also assist to only a relatively slight extent in the building up of the talus slope, since they are easily broken up and consumed by the boring and triturating organisms.

If on a reef that possesses a typical vertical inner face, sand and mud, derived from the erosion of the land in its vicinity, be deposited, it will be the corals on the upper surface that will be most affected ; the growth of the corals will be retarded and they may even be completely killed off so that the reef may become broken in its continuity. Wood Jones (1910, p. 123) has laid considerable stress on the importance of the effect of sand and mud deposition in the formation of coral islands and atolls and my experience in the coral regions of the coasts of India and the Maldives has led me to think that he is in the main right ; certainly in most of the reefs that I have examined it is noticeable that in areas where the land is undergoing erosion the character of the reef is affected and study of its structure shows that in most cases the process of destruction is getting the upper hand. While the upper zone of the inner vertical face of the reef may still appear to be in a flourishing condition, the upper surface may consist for the most part of dead coral masses that are undergoing solution and destruction but on which, however, small isolated coral colonies still manage to maintain their existence, though we may find that their growth is stunted or in other ways shows that they are to some extent adversely affected. The destruction of the corals growing on the surface, while those on the vertical face of the reef are but little if at all affected, leads eventually to the reef losing its vertical character, and it acquires a more gradual slope. The process may go even further and the reef may be broken up so that we find only isolated masses of coral, termed coral heads ; a very good example of this latter process appears to be present in the lagoon reef on the north and east sides of Horsburg Atoll ; in the former situation the reef has been broken up into a number of more or less isolated coral heads, whereas in the latter the coral growth appears to indicate that the reef has on more than one occasion been killed off and destroyed. In this atoll (*vide* Stanley Gardiner, 1903, pp. 377–380) practically all traces of land on the south and west sides have already disappeared and extensive erosion is going on in the islands on the north and east sides ; this destruction of the land must have been accompanied by the formation of large quantities of sand and silt which has been swept inwards by each rising tide and has been deposited on the lagoon flat and the growing margin of the lagoon reef. It seems probable that such an extensive process of erosion as this will have proceeded at different rates in different years and that even periods of comparative stability have alternated with periods of rapid erosion, so that the deposition of sand and silt would equally exhibit phases of increased or decreased intensity. With each succeeding phase of destruction of the land, the deposition of sand and silt would result in the partial destruction of the growing reef, leaving behind a series of patches of coral to show where an originally complete and flourishing reef formerly stood, while during each phase of diminished erosion and consequently less deposition the growing edge of the reef would tend to recover and form a further extension inwards towards the centre of the lagoon. Once the land has been completely destroyed, as it has been on the south and west sides, the deposition of sand and mud will cease, and the reef will then be able to extend on the north and east sides and build up a vertical wall such as appears to be typical of reefs in similar localities.

The Lagoon floor.

The character of the lagoon floor is a subject that has given rise to much discussion. According to the Darwinian theory the lagoon floor is formed by debris and detritus from the more rapidly growing, encircling reef: Daly considered that it represents the submerged and eroded remains of the platform on which the encircling atoll reef has grown up: while according to the solution theory of Semper, Murray and others it is the production of solution and erosion acting on an originally continuous coral patch. In the larger lagoons of the Maldives there is in many cases a complex arrangement of shoals and coral masses, constituting what Stanley Gardiner (1903, p. 167) terms a 'jungle' that he believes to be the last remaining traces of a coral reef that has undergone destruction and removal. He remarks 'the general impression in the "jungle" is that the reef-flat is everywhere being driven back and that the whole of its reefs are being removed . . . The last stage in the history of a disappearing reef is its reduction to a mere little surface patch or series of patches, ultimately single coral heads, which sooner or later will themselves topple over, their basal mounds being subsequently reduced to the level of the floor of the bank.' It has frequently been stated that one of the chief characteristics of the floor of a large lagoon is the flatness of its surface; Wharton (1897, p. 390) long ago insisted that 'a flat floor is an invariable characteristic of a large atoll', and this rather sweeping statement would appear to hold good in many instances. On the other hand in a number of smaller atolls the condition is exactly opposite.

Much of the floor of a lagoon is covered with sediment, though the actual proportion of floor thus covered appears to differ very considerably in different atolls and Stanley Gardiner has reached the conclusion (1931, p. 132) that 'lagoons in general are not, on such evidence as exists, to be regarded as areas of sedimentation'. Although in the large lagoons a large area of the lagoon floor may consist of 'hard bottom', in the smaller ones there seems little doubt that the greater part of the floor is covered with a water-borne sediment, that, as Daly has pointed out, may be derived from (1) the erosion of the central island where such exists or has existed, (2) detritus from the encircling reefs, and (3) in part from the remains of non-coral organisms. In the case of the Maldives there are no central islands nor have we any evidence that such ever existed; we may, therefore, pass over this source of sediment; but there is some evidence of a fourth type of sediment that Daly has not mentioned and that is a precipitate of calcium carbonate from the sea-water itself. I have already referred to the formation of a precipitate of this nature in some of the brackish water lakes that have become shut off from the lagoon in Addu atoll (*vide supra*, p. 517) and several observers have also directed attention to its presence in the outflowing water from the lagoon at certain states of the tide during rough weather; thus Stanley Gardiner (1931, p. 135) remarks that 'in rough weather the outflowing water from a lagoon appears milky, and this is due to the suspension in the water of mud formed by the decay of organisms, together with lime precipitated from the supersaturated lagoon water'. I have myself observed this outflow of sediment from a lagoon in the Maldives (*vide infra*, p. 532) but one must not forget that such an outflow is not continuous and only appears to occur at certain states of the tide and under certain weather conditions; at all other times this sediment is being deposited in the lagoon itself. Davis (1923, p. 299), regarding the inwash of detritus from the reef-flat, remarks that 'a more direct indication that the inwash of detritus takes place is found in the occurrence of abundant reef blocks and fragments on the reef-flat, as well as by the frequent slopes of coral sand, on which occasional reef blocks are strewn, in the lagoon waters next inside of the reef flat'.

In addition a further source of sediment in the lagoon, and in certain instances apparently an important source, is to be found in the pelagic organisms that occur in the waters of the lagoon. In this connection Stanley Gardiner (1931, p. 132) remarks, regarding the sand, that 'in protected lagoon pockets it is usually fine, and it may be quite obviously increasing by the accumulation of organically formed limestone particles, in which occasional shells of the animals, which form globigerina or pteropod oozes, may be recognized. Such animals live entirely in the ocean water, so that these shells must be carried in by the tides, but they are not present, as might be expected, in any special abundance immediately behind the encircling reefs, over which the tides rush'. In a foot-note (*loc. cit.*, p. 135) he, however, remarks that 'in Diego Garcia lagoon there was by no means that paucity of suitable plankton-food for corals which we expected', and this was also my experience in the two atolls that I have investigated. Indeed, on one occasion over a litre and a half of a species of Pteropod, *Clio* sp., were taken in a surface tow-net, with a diameter at the mouth of 3 feet, that was left awash at the surface for a period of twelve hours, the organisms being swept into the net by such feeble currents as existed in the lagoon. It is not surprising, therefore, that in some areas the bottom deposit in the lagoon has been described as 'pteropod ooze', and in my experience pteropod shells are of almost universal occurrence in these deposits. Sir John Murray (*vide* Stanley Gardiner, 1906, pp. 584–588) has described in detail a series of deposits from some of the lagoons of the

Maldive Archipelago, including several samples from Suvadiva atoll from depths of over 40 fathoms, which he classes as a true pteropod ooze ; while from other localities samples contained quite appreciable numbers of these delicate shells, though, as Stanley Gardiner points out, they were not present in sufficient numbers to warrant the application of this special term. Out of 70 samples from the channels between the atolls and from the neighbouring waters in the Maldives Agassiz (1903, pp. 29–34) has recorded the occurrence of pteropod shells, in sufficient quantity to merit special mention, in 14 instances ; and it is interesting to note that the number of samples that contain these shells in any quantity steadily decreases as the depth, from which the samples were obtained, increases. The percentage to such pteropod-containing samples at different depths is as follows :—

In samples from	100–200 fathoms,	50%	contain pteropod shells.
,, ,, ,,	200–300 ,,	38·5%	,, ,, ,,
,, ,, ,,	300–500 ,,	33·3%	,, ,, ,,
,, ,, ,,	500–1,000 ,,	33·3%	,, ,, ,,
,, ,, ,,	over 1,000 ,,	0	,, ,, ,,

The gradual disappearance of the pteropod shells from the deeper samples is undoubtedly due to solution, these delicate structures being more easily dissolved than the more solid particles of coral sand. The presence of these shells in so many of the samples of deposit from the lagoon floors seems clearly to point to the fact that the amount of solution that is going on is relatively very slight and it further seems clear that since the lagoon water is so supersaturated with lime that a chemical precipitate is formed, it can have little or no solvent action on the lagoon deposits. The percentage of silicate in the bottom deposit shows an interesting gradation in agreement with the depth of water :

Minikoi	Atoll, Depth 8 fathoms,	percentage of silica	0·193
Horsburgh	,, ,, 23 ,,	,, ,, ,,	0·26
Suvadiva	,, ,, 40-50 ,,	,, ,, ,,	1·992–3·127

These figures seem to indicate that there is considerably less solution of calcareous material in small lagoons with shallow depths than in larger and deeper ones. If this be so, then we are thrown back for an explanation of the removal of detritus from the lagoon by mechanical means and this will depend very largely, if not entirely on the strength of wave-action in stirring up the bottom deposits and on the strength of the currents of water flowing out of the lagoon. With a small lagoon the amount of wave action will be comparatively small since the wind will not have sufficient space in which to raise really effective waves and hence in such atolls we should expect to find that there is evidence of an accumulation of sand and mud in the lagoon and on the lagoon floor.

In the lagoon, as on the reef itself, two diametrically opposed forces are at work and in order to reach any conclusion regarding the formation and the future fate of the lagoon one must bear both factors in mind ; in the one case the process of destruction will tend towards the enlargement of the lagoon at the expense of its surrounding reefs and in the other the process of construction may ultimately cause the filling up of the lagoon and the production of a more or less continuous bank.

Hedley from his studies of the Great Barrier Reef of Australia (1924, p. 62) has reached the conclusion that atolls are in the process of filling up ; he sums up the history of an atoll in the following words, ' first a point in the reef grows upwards till it breaks the surface of the water, then the waves pack round it a mass of drift stones and sand and the islet so formed assumes a crescent shape with the back to windward and the horns to leeward. The waves continue to sweep along further drift matter and the crescent thus grows into a horse-shoe and ultimately into an oval thus enclosing a lagoon. If the process of evolution is continued, the lagoon is filled up and the atoll becomes a solid cay. Finally, seeds drift ashore and a forest clothes the new island '. That this process may happen in the case of small pools or lakes around the margin of an atoll such as one sees in the Maldives has already been shown (*vide supra*, p. 517) but in the case of Addu or Horsburgh atolls there is little or no evidence of any change in the character of the lagoon during the past century.

On the other hand Fryer (1911, p. 409) has come to the conclusion that the lagoon in Aldabra is of more recent formation than the land and has been dissolved out : ' it was probably a low portion of the future atoll '. The lagoon is shallow, and to a great extent is land locked. ' The bottom of the lagoon is rock, but in the centre it is so covered with sand and near the land-rim with mud that it is difficult to determine its nature ; the few observations made pointed to rock of the " champignon " type. The origin of the mud and sand is of great interest . . . Little material is carried in by the tides and there is no doubt that the mud is the product of lagoon erosion.' There can be no doubt that the mud is derived by erosion from the rock of which the atoll is composed mixed with organic debris, and the lagoon appears to be extending.

Before one concludes, however, that the fine, calcareous deposit of these lagoons is derived from the detritus of the coral reefs a careful examination of the deposit is necessary and in this connection it is interesting to note the report on a sample of the bottom mud that I was able to obtain from the central part of the lagoon of Horsburgh atoll in the Maldive Archipelago and that was submitted to the Geological Survey of India for examination. The report is as follows : ' Examination by the petrological microscope shows that the specimen forwarded by you is a crystalline aggregate and not amorphous. Boiling for a few minutes with cobalt nitrate shows that at any rate a good deal is aragonite. The specific gravity is, however, low—2·75. But from the nature of the specimen too much reliance need not be placed on this result. It is probably either a mixture of aragonite and calcite or mostly aragonite. The specimen has been analyzed and found to contain 0.26% SiO_2'. The small percentage of silicate in the deposit forms a marked contrast to the results obtained by Stanley Gardiner in the deposit from Suvadiva Atoll (1903, p. 323).

Vaughan and Matson (1910) made a careful examination of the bottom deposit in the shoal waters of Florida and the Bakamas and reached the conclusion that ' the flocculent, so-called amorphous calcium carbonate is neither of detrital nor organic origin, and, therefore, must be a chemical precipitate'; as already mentioned (*vide supra*, p. 507 *et seq.*) the agency concerned in this precipitation would appear to be, at least in part, the action of denitrifying bacteria. Vaughan (1918, p. 268) sums up the available evidence in the following words, 'it seems to me that all lines of evidence converge and give the same result, which is that in the shoal waters of the tropics ocean water does not dissolve calcium carbonate, but that the contrary process—precipitation by both inorganic and organic (bacterial) agencies—is taking place'.

Other agencies may be the decomposition of animal and vegetable material derived from the organisms of the reef or the plankton, or, as Stanley Gardiner (1930, p. 8) suggests, 'the phytophagus corals and plants that build coral reefs precipitate amorphous carbonate of lime from the supersaturated sea-water owing to the chemical operations of chlorophyll on carbon dioxide'. According to Vaughan (1914, p. 52), Murray and Irvine have shown that calcium carbonate precipitated from sea-water by an alkaline carbonate at a temperature of 34° F. (1·1°C.) is calcite, but that with higher temperatures the character of the precipitate changes to aragonite, so that at 47°F. (8·3°C.) the precipitate is a mixture of calcite and aragonite, and at 80°F. (26·7°C.) or over it consists entirely of aragonite. If this be so then the whole or, at least, the greater part of the precipitate in the lakes on the islands and in the lagoon should be aragonite since the temperature of the water must be in the neighbourhood of 30°C. (86°F.), the air temperature over the Laccadive Sea ranging from a minimum of 25·9C. (78·6 F.) in October to a maximum of 29·7C. (85·5 F.) in the month of April (*vide supra*, Table 1, p. 58); and this is in agreement with the report given above.

Mayor (1924, p. 27) notes that 'Guppy, who studied the Solomon Islands, and Bourne, who describes the Chagos group and Diego Garcia, conclude that the barrier reefs and atolls of these regions are not due to subsidence, but they also maintain that *at present these lagoons are either filling up with silt which is being washed into them or are in a balanced state*, the solution of limestone by the sea-water being offset by the growth of coral and by precipitation of calcium carbonate'. (The italics are mine.) A study of the chart of Diego Garcia shows that the whole of the southern end of the lagoon is covered with only shallow water and is dotted over with numerous shoal patches, water of a depth greater than 10 fathoms being found only in the northern part of the atoll. In the case of the Cocos-Keeling Atoll, Guppy (1889, p. 579), after comparing his soundings with those taken by the Officers of the "Beagle" in 1836, 'came to the conclusion that whilst the general shallowing since 1836 has been slight, the growth of coral patches has been in some places rapid. This is evidently the mode by which the basin of the lagoon will be filled up, and it will be undoubtedly a slow one. In the space of half a century the sandy intervals will shoal but little, whilst the coral patches may rise up considerably'. Wood-Jones (1910, p. 186) in his account of the same atoll notes that 'the shallow southern part of the lagoon extends in the form of a great horse-shoe, skirting the inner shores of the islands and invading at its margins the northern deeper part . . . The bottom of this part is, however, not at all uniform, and deeper hollows are studded about the central portions of the great part of the southern shallows . . . It must not be imagined that the great southern part of the lagoon is by any means devoid of living coral colonies, for dotted about at intervals are some of the most luxuriant of all the coral growths.' In both these atolls, Diego Garcia, and Cocos-Keeling, the shallow end of the lagoon occurs on the windward side. Stanley Gardiner in his account of the atoll of Minikoi (1903, p. 47) remarks that in the lagoon 'the sand ends about 100 yards outside the line, marked 2 fathoms in the chart, opposite and to the north of Minikoi island, the intermediate area generally being covered with coral shoals, intersected by channels with about 2 fathoms of water. These shoals are usually flat on the top, bare or hollowed out in the centre, with perpendicular, or overhanging sides,—especially towards the deeper part of the lagoon—covered with coral growth... The patches decrease in size towards the sand area, on which they seem on the whole to be encroaching. To the south-west and north of the atoll coral patches may occur near the edge of the sand flat, but there is no definite broad line', and he further notes that some of these shoals appeared to have been killed off only a short time previously to his visit; 'the slope from 2 or 2½ fathoms... to the deepest waters of the lagoon is gradual and regular. Coral patches occur everywhere but much more sparingly in the deeper waters and towards the north; indeed the centre of the lagoon is fairly clear.' Here again the shallowing of the lagoon occurs towards the side of the prevailing wind, namely the south-west. In Solomon and Egmont atolls in the Chagos Archipelago, although the main trend of the lagoons is in the line from east-south-east to west-north-west, it is the south-western side of the lagoon that is considerably more shallow than the north-east and, further, is dotted over with shoal patches. In Addu atoll, there is a sandy shelf around the south-western part of the lagoon on which there is a profuse growth of living coral, mostly of the branching Madrepore type, and in Horsburgh atoll, although the deepest part of the lagoon occupies the central area, the bottom inside the reef, except on the southern side from which all trace of land has now disappeared, slopes gradually downwards and this slope is more gradual at the east and west ends of the lagoon. In both these instances the deposition of sand and silt on the lagoon floor is taking place with greatest rapidity in those sides of the lagoon that correspond with the prevailing winds, namely the south-west and north-east. In Aldabra atoll the eastern half of the lagoon is occupied by a wide shoal that dries at low water but in this case the prevailing wind is from the south-east (*vide* Fryer, 1911).

Such shoals in the lagoon are by no means confined to the atolls of the Indian Ocean, but occur also in those of the Pacific. Thus Stanley Gardiner in his account of the atoll of Funafuti (1898, p. 435) states that 'between the patches and knolls of rock outside the lagoon platform, the bottom is generally covered by a sand which is similar in all respects to the sand described by Professor Sollas. It does not extend far into the lagoon off these reefs and the arming of the lead usually comes up clean, or contains a few segments of different species of *Halimeda*. To leeward of all the windward islands, sand is found most abundantly, but only by the main island does it extend more than a couple of hundred yards into the lagoon where, however, it runs out about 1½ miles in a bank with about 17 fathoms of water, the breadth of which a mile out from shore is only about 250 yards.'

A similar deposition of sand and silt may sometimes be observed outside the reef of some of these Pacific atolls; Agassiz (1903*a*, p. 299) in his account of Kwajalong Atoll in the Marshall Group in the Pacific Ocean states that 'the reef-flat slopes gradually into deep water; from the outer edge of the platform we could trace large masses and heads of corals of the same species as those found on the reef-flat itself, only in full activity, growing luxuriantly in from six to seven fathoms of water, and gradually becoming more distant and smaller towards the seventeen fathom line, where the coral boulders have entirely disappeared and are replaced by fine coral sand washed down from the upper part of the reef-flat'. The removal of sand from an atoll in the manner just described is not confined to the action of waves and tides but may also be affected by the wind itself; Agassiz (1903, p. 277) has pointed out that 'the amount of sand which floats off from a lagoon and is blown to sea is very considerable.... The distance to which sand can be carried by the trades is clearly shown by the amount of sand we found in our nets in all the surface hauls in the Marshall Islands when off the lee side of the atolls; sand is floating there in considerable quantities off the east coast of the United States, from Cape Hatteras south shore sand is carried out to sea by the prevailing westerly winds to a very considerable distance.' The action of such deposition of sand may be two-fold, dependent upon the character of the sea floor upon which it is continually settling. If the floor consist of a coral patch there seems little doubt that the effect of the sand and silt will be to gradually kill off the corals so that the patch becomes broken up to a number of isolated coral heads and by the gradual erosion and solution of their supporting pedicles these will be undermined and finally they will overbalance and be destroyed; such an action appears to be taking place in the north-east corner of Addu Atoll and also on the lee side of Low Isles in the Great Barrier Reef (*vide* Stephenson, Tandy, and Spencer, 1931, pp. 61-2) where 'the reef of the anchorage (on the lee side) consist of very irregular masses of dead coral, with sand, or sand and debris, between them'. On the other hand in certain other areas, such as in the lagoon of Addu Atoll, and indeed in many of the smaller atolls these areas of sand deposition appear to form a basis on which large and flourishing colonies and patches of stag's horn Madrepore can become established and can flourish.

The occurrence of these shoals in the lagoons on the leeward side of the windward reefs can only be attributed to one agency, namely the deposition of detritus derived from the reefs themselves under the influence of the increased wind and current action, and the fact that such shoals are not of more frequent occurrence is not a matter for surprise; indeed it would be a matter for surprise if they were, seeing that the effect of waves in the lagoon must be to level the whole floor and obliterate such local irregularities, for, as Davis (1923, p. 299) points out, 'that the lagoon waves and currents are competent to shift detritus on the lagoon floor at times of gales and storms is proved by the turbidity of the lagoon water on such occasions. That a lagoon floor should become nearly level as it is more and more aggraded, whatever the shape of its foundation, is a necessary result of the lifting of detritus by storm waves chiefly from shoals and of its settling from the turbid water everywhere.'

Another of the problems that requires elucidation is that of the origin and maintenance of the deep channels from the lagoon to the open sea. It is frequently stated that these channels occur on the leeward side of the lagoon and are caused and kept open by the outflowing current of water during the ebb-tide that carries with it mud and silt derived from the breaking down of the coral, and that this by its deposition in the channels kills off all coral growth in this region. The solution of the problem is, however, by no means as simple as this. Darwin (1889, p. 157) called attention to the quantity of sediment that may be discharged through the entrance channels and Moresby also noticed the same phenomenon in some of the atolls of the Maldives and Chagos Archipelagces during the change of the monsoons, though why this outflow of sediment should be more evident at this season of the year is not easy to understand. I found a very similar condition present on one occasion off the south-east entrance to Addu atoll; my note on this is as follows:—

> '11-12 A.M. Passed a well-marked line of drift, running roughly parallel to the coast line; numerous large fish (Bonito) jumping in or near the line of the drift. At a short distance inshore from the drift the water was discoloured by sediment, a marked contrast to the outside water that was pure blue. 11-20 A.M. Entering the lagoon and in middle of channel.'

As we were steaming at about eight knots the discolouration of the water commenced at about a mile out. The tide at the time of the occurrence was flooding and had been doing so for over an hour and a half, low tide having been at 9-45 A.M.

In Addu atoll the main entrances are in each case one of a pair situated on the north-west and south-east sides of the atoll. In the northern pair it is the western channel that is the deeper, while in the southern pair it is the eastern. If these channels were kept open by the outflow of sediment one would expect to find that it was the deeper and wider channels that served for the outflow of water from the lagoon but an examination of the currents in the two pairs of channels shows that in the northern pair it is through the western channel that the flood tide sets, whereas in the case of the eastern channel the current runs out of the lagoon at all states of the tide; again in the southern pair it is the eastern channel that carries the flood tide into the lagoon, whereas through the western shallow channel there is a continuous out-going current. Stanley Gardiner (1903, p. 23) has also noted that the current may flow continuously in one direction through the opening of a reef, for he writes 'I also sailed twice up to the passage between Hululefaro and Furenafuri, finding currents of 3·7 and 2·1 knots into the lagoon. Of course all these currents were largely dependent upon the tide but it is a most noticeable feature that the direction remained constant in spite of the latter, which hence only affected the rate.' Agassiz (1903, p. 14) has also noted that in the case of certain atolls of the Paumotus Group in the Pacific Ocean the current may flow continuously into the lagoon through the main pass for some considerable time. He remarks 'at Rangirava we obtained the first intimation of the immense amount of water which flows in and out through the main passes, and according to the strength of the trades, continues sometimes to flow two or three days in one direction'.

A further study of these entrances into Addu and Horsburgh atolls shows that in the case of those entrances through which the current sets into the lagoon there is a marked tendency for the reefs to curve round and be prolonged into the lagoon so as to form a funnel-shaped entrance. Similarly, in Crossland's account of the reefs in Taniti (1928, Text-figs. 6, 7) he figures in every case the exposed upper aspect of the encircling reef as tending to run inwards on each side of the deep channels and this would seem unmistakeably to indicate either that the original reef, which is now undergoing erosion and destruction, took the same course or that the reef is actually spreading inwards, being built up on a growing bank, and that it is not, as suggested by Crossland (1929, p. 603), due to radial faulting across an encircling reef. The production of this inward extension of the reef in those channels through which the current continuously flows inwards to the lagoon is easily explained, since in such channels this inflowing current, assisted by the waves that during storms must sweep through these entrances, must rapidly tend to build up a talus slope of fragments, torn off the reef and piled up on each side of the entrance, and as soon as this reaches the required height corals will become established and, by reason of the favourable situation as regards both oxygen and food-supply, will flourish.

It is difficult to understand why the current should flow, as it undoubtedly does, out of the lagoon through certain channels at all states of the tide; but the whole question of the circulation of water in these lagoons is a most complicated one. Not only does water enter the lagoon through the main entrance channels, but during the flood-tide water passes into the lagoon through subsidiary channels and a great deal finds its way across the reef-flat and between the islands; there is, however, yet another way by which water can, and I have no doubt does, enter the lagoon with every rise of the tide and that is through the reef itself. The main mass of the reef is full of holes and channels, as has clearly been shown by every boring that has been carried out, and it is to this porosity that must be attributed the constant rise and fall of the water that is known to occur in the wells and enclosed lakes on the various islands. Such being the case, there can be no doubt that water could pass from the outside, as the tide rises, through the reef itself into the lagoon, if the level of water in the latter were lower than that of the surrounding sea and that this must be so at least during the flood tide is indicated by the fact that, in my experience, the time of high-water in the lagoon is considerably later, sometimes as much as one hour, than the time of high-water outside the reef.

It is difficult or even impossible to decide to what extent the general character of these entrances is the result of the currents that flow through them or whether on the other hand the fact that the currents are different in the two types of entrance channel, namely shallow and deep, may not be due to the character of the channels themselves. It would seem probable that the shallower channels have been produced secondarily by the destructive effect on coral growth of the sand and silt that is continuously being swept through them; but the deep channels may equally possibly have existed from the very inception of the coral bank and their maintenance be due to the inability of the coral to grow up from this depth to form a reef. Crossland (1929, p. 610) reached the conclusion that in the reefs of Moorea 'all show marked signs of rapid erosion of their sides, and in the shallower passes erosion of their bottoms is clearly seen. The difference between the slopes to the lagoon and ocean on either side of the shallow passes is marked, the former being often the steeper and indicating that the breakdown of the reef-edge, which formed the pass, may have been initiated from the lagoon side.' Another factor in the erosion of the islands and the production of channels from the outer to the inner part of the reef-flat is solution of the surface of the islands by rain-water. Stanley Gardiner in his account of Diego Garcia (1931, p. 139) remarks that 'the land has a series of "sinks" from incipient marshes or bogs, below which the limestone is decaying and in and around which the coconuts are dying, to greater irregular lakes or "barachois" towards the south end. The atoll was surveyed by Moresby in 1837 and the northern half again by Vereker in 1885. More barachois have formed since the first survey and all have enlarged. They are floored by bare rock similar to that of the reef-flat, and low, often overhanging cliffs testify to their circumferential growth. Many open to the lagoon, but none as yet have burst out to seaward.' Fryer has also called attention to similar formations in the island of Aldabra, where 'sinks containing water form in the land below the lowest tide level; they have vertical sides and subsequently acquire openings, by which they become part of the lagoon, (*vide* Stanley Gardiner, 1931, p. 137). In the atolls that I have visited I have seen no such formations, though in the island of Heratera in Addu Atoll there is a shallow pool where the sea has broken through the outer beach and has flooded a low-lying part of the island. A combination of these two factors would produce a shallow channel through an island, and this might rapidly be further eroded by the continual to and fro flow of water with each rise and fall of the tide. In the case of the deeper channels, however, the problem of their formation is not so simple; in certain cases, as in Tahiti, Moorea and Rarotonga, these deep channels can be seen to correspond to the mouths of rivers and therefore can be clearly attributed to the deleterious effect of fresh-water on coral growth and to the submergence of the original river valley; but no such explanation can be put forward for the deep channels in the Maldive atolls unless we are prepared to admit the original presence of a central island that has since been completely eroded and submerged, all trace of its presence having been completely obliterated, or else that there has been radial faulting in the reef, in which case we should be faced with the problem why this faulting should have taken place in the majority of cases on the leeward side of the atoll.

LITERATURE CITED IN THE TEXT.

Agassiz, A., 1894 .. .. 'A Reconnaissance of the Bahamas and of the elevated reefs of Cuba in the Steam Yacht "Wild Duck", January to April, 1893.' *Bull. Mus. Comp. Zool. Harvard*, Vol. XXVI, No. 1.

Agassiz, A., 1895 .. .. 'A visit to the Bermudas in March, 1894.' *Bull. Mus. Comp. Zool. Harvard*, Vol. XXVI, No. 2.

Agassiz, A., 1898 .. .. 'A visit to the Great Barrier Reef of Australia in the steamer "Croydon".' *Bull. Mus. Comp. Zool. Harvard*, Vol. XXVIII, No. 4.

Agassiz, A., 1903 .. .. 'On the formation of barrier reef and of the different types of atolls.' *Proc. Royal Soc. London,* Vol. LXXI.

Agassiz, A., 1903(*a*) .. .. 'Reports on the Scientific Results of the Expedition to the Tropical Pacific.' *Mem. Mus. Comp. Zool. Harvard,* Vol. XXVIII.

Alcock, A., 1902 .. .. 'A Naturalist in Indian Seas,' Calcutta.

Baker, J. R., 1925 .. .. 'A Coral Reef in the New Hebrides.' *Proc. Zool. Soc. London,* 1925, Pt. 2.

Blanford, W. T., 1872 .. .. 'Geology of Orissa.' *Rec. Geol. Survey of India,* Vol. V, Pt. 2, p. 61.

Bourne, G. C., 1888 .. .. 'The Atoll of Diego Garcia and the coral formation of the Indian Ocean.' *Proc. Royal Soc. London,* Vol. XLIII, p. 440.

Branner, J. C., 1904 .. .. 'The Stone Reefs of Brazil, their Geological and Geographical relations, with a chapter on the Coral Reefs.' *Bull. Mus. Comp. Zool. Harvard,* Vol. XLIV, Geological Series, Vol. VII.

Crossland, C., 1905 .. .. 'The Ecology and Deposits of the Cape Verde Marine Fauna.' *Proc. Zool. Soc. London,* 1905, Vol. I.

Crossland, C., 1928 .. .. 'Notes on the Ecology of the Reef-builders of Tahiti.' *Proc. Zool. Soc. London,* 1928, Pt. 2.

Crossland, C., 1928(*a*) .. .. 'The Coral Reefs of Tahiti, Moorea and Rarotonga.' *Journ. Linn. Soc. London,* Vol. XXXVI, Zoology, No. 248.

Daly, R. A., 1920 .. .. 'A recent world-wide sinking of Ocean level.' *Geological Magazine,* Vol. LVII.

Daly, R. A., 1924 .. .. 'The Geology of American Somoa.' *Papers from the Dept. Marine Biol. Carnegie Institute, Washington,* Vol. XIX, No. 6.

Daly, R. A., 1926 .. .. 'Our Mobile Earth.' New York.

Dana, J. D., 1875 .. .. 'Corals and Coral Islands.' London.

Darwin, C., 1889 .. .. 'The Structure and Distribution of Coral Reefs', 3rd Edition. London.

David, Edgeworth, 1926 .. Lecture to the Geological Society. London. 1926.

Davis, W. M., 1923 .. .. 'The Depth of Coral Reef Lagoons.' *Proc. Nat. Acad. Sciences of the United States of America, Washington,* Vol. IX.

Davis, W. M., 1928 .. .. 'The Coral Reef Problem.' *American Geographical Society,* Special Publication, No. 9.

Drew, G. H., 1914 .. .. 'On the Precipitation of calcium in the Sea by Marine Bacteria, and on the action of denitrifying bacteria in Tropical and Temperate Seas.' *Papers from the Tortugas Laboratory, Carnegie Inst. Washington,* Vol. V.

Duenden, I. E., 1902 .. .. 'West Indian Madreporarian Polyps.' *Mem. Nat. Acad. Sci. Washington,* Vol. VIII.

Edmondson, C. H., 1928 .. 'The Ecology of an Hawaiian Coral Reef.' *Bernice P. Bishop Museum Bull.,* 45.

Ellis, R. H., 1924 .. .. 'A short Account of the Laccadive Islands and Minicoy.' Madras.

Fedden, 1885 .. .. 'Geology of Kathiawar Peninsular in Guzerat.' *Mem. Geol. Survey India,* Vol. XXI.

Field, R. M., 1919 .. .. 'Investigations Regarding the Calcium Carbonate Oozes at Tortugas, and the Beach-rock at Loggerhead Key.'
Carnegie Institute, Washington, Year Book, No. 18.

Foote, R. B., 1889 .. .. 'The Coral Reefs of Rameswaram Is.'
Madras Christian College Magazine.
(Extract in *Scott. Geogr. Mag.*, Vol. VI, p. 257, 1890.)

Fryer, J. C. F., 1911 .. .. The Percy Sladen Trust Expedition to the Indian Ocean in 1905; Report No. XIX. 'The Structure and Formation of Aldabra and neighbouring Islands.'
Trans. Linn. Soc. London, 2nd Series, Zoology, Vol. XIV.

Gardiner, J. Stanley, 1898 .. 'The Coral Reefs of Funafuti, Rotuma and Fiji together with some notes on the structure and formation of Coral Reefs in general.'
Proc. Cambridge Philosoph. Soc., Vol. IX, Pt. 8.

Gardiner, J. Stanley, 1902 .. 'The formation of the Maldives.'
Geographical Journal, Vol. XIX.

Gardiner, J. Stanley, 1902(*a*) .. 'The Atoll of Minikoi.'
Proc. Cambridge Philosoph. Soc., Vol. XI.

Gardiner, J. Stanley, 1902(*b*) .. 'On the rate of growth of some Corals from Fiji.'
Proc. Cambridge Phil. Soc., Vol. XI.

Gardiner, J. Stanley, 1903 .. 'The Fauna and Geography of the Maldive and Laccadive Archipelagoes', Cambridge.

Gardiner, J. Stanley, 1906 .. 'The Indian Ocean.'
Geographical Journal, Vol. XXVIII, No. 4 Oct., No. 5 November.

Gardiner, J. Stanley, 1907 .. 'The Percy Sladen Trust Expedition to the Indian Ocean in 1905. No. 1, Description of the Expedition.'
Trans. Linn. Soc. London, 2nd Series, Zoology, Vol. XII.

Gardiner, J. Stanley, 1915 .. 'Submarine slopes.'
The Geographical Journal, March, Vol. XLV.

Gardiner, J. Stanley, 1930 .. 'Studies in Coral Reefs.'
Bull. Mus. Comp. Zool. Harvard, Vol. LXXI, No. 1.

Gardiner, J. Stanley, 1931 .. 'Coral Reefs and Atolls.' London.

Gee, E. R., 1926 .. .. 'The Geology of the Andaman and Nicobar Islands, with special reference to Middle Andaman Island.'
Rec. Geol. Survey India, Vol. LIX, Part 2.

Guppy, H. B., 1886 .. .. 'Notes on the characters and mode of the coral reefs of the Solomon Islands.'
Proc. Royal Soc. Edinburgh, Vol. XIII.

Guppy, H. B., 1889 .. .. 'The Cocos-Keeling Islands.'.
The Scottish Geographical Magazine, Vol. V.

Hall, A. D., and Miller, N. H. J., 1906 .. .. .. 'The Effect of Plant Growth and of Manures upon the Retention of Bases by the Soil.'
Proc. Royal Soc. London; Section B.—Biological Sciences, Vol. LXXVII.

Hedley, C., 1924 .. .. 'The Great Barrier Reef of Australia.'
Journal of the American Museum of Natural History, Vol. XXIV.

Hickson, S. J., 1924 .. .. 'An Introduction to the Study of Corals.' Manchester.

Jones, E. J., 1887 .. .. 'Natural History Notes from H.M.'s Indian Marine Survey Steamer "Investigator", No. 5. On some Nodular Stones obtained by trawling off Colombo in 675 Fathoms of Water.'
Journal, Asiatic Soc. Bengal, Vol. LVI. New Series.

Johnston, J. and Williamson, E. D., 1916 .. .. .. 'The Rôle of Inorganic Agencies in the Deposition of Calcium Carbonate.'
The Journal of Geology, Vol. XXIV, No. 8.

Lipman, C. B., 1924 .. .. 'A Critical and Experimental Study of Drew's bacterial Hypothesis on $CaCO_3$ precipitation in the Sea.'
Papers from the Dept. Marine Biol., Carnegie Institute, Washington, Vol. XIX, No. 8.

Macfadgen, W. A., 1930 .. 'The undercutting of Coral Limestone on the coasts of some islands in the Red Sea.'
The Geographical Journal, Vol. LXXV.

Marshall, P., 1931 .. .. 'Coral Reefs—Rough Water and Calm Water Types.'
Report of the Great Barrier Reef Committee, No. 8.

Marshall, S. M., and Orr, A. P., 1931 'Sedimentation on Low Isles Reef and its relation to coral growth'.
Great Barrier Reef Expedition, 1928-29, Scientific Reports, Vol. I, No. 5. British Museum (Natural History).

Mawson, Sir D., 1929 .. .. 'Some South Australian Algal Limestones in Process of Formation.'
The Quarterly Journ. Geological Soc., Vol. LXXXV.

Mayer, A. G., 1918 .. .. 'Ecology of the Murray Island Coral Reef.'
Papers from the Dept. Marine Biol., Carnegie Institute, Washington, Vol. IX.

Mayer, A. G., 1918(*a*) .. .. 'Toxic effects due to High Temperature.'
Papers from the Dept. Marine Biol., Carnegie Institute, Washington, Vol. XII.

Mayer, A. G., 1924 .. .. 'The Structure and Ecology of Samoan Reefs.'
Papers from the Dept. Marine Biol., Carnegie Institute, Washington, Vol. XIX.

Mayer, A. G., 1924(*a*) .. .. 'Rose Atoll, American Samoa.'
Papers from the Dept. Marine Biol., Carnegie Institute, Washington, Vol. XIX.

Mayer, A. G., 1924(*b*) .. .. 'The Growth rate of Samoan Corals.'
Papers from the Dept. Marine Biol., Carnegie Institute, Washington, Vol. XIX.

McClendon, J. F., 1918 .. 'On changes in the sea and their Relation to Organisms.'
Papers from the Dept. Marine Biol., Carnegie Institute, Washington, Vol. XII, No. XI.

Moseley, H. N., 1892 .. .. 'Notes by a Naturalist. An account of observations made during the voyage of H.M.S. "Challenger".' 2nd Edition, London.

Moresby, 1835 .. .. 'On the Northern Atolls of the Maldives.'
Geographical Journal, Vol. V.

Murray, J. and Irvine, R., 1890 .. 'On Coral Reefs and other Carbonate of Lime Formations in Modern Seas.'
Proc. Royal Soc. Edinburgh, Vol. XVII.

Oldham, C. F., 1896 .. .. (i) 'Topography of the Arabian Sea in the neighbourhood of the Laccadives. (ii) The Physical features of some of the islands.' *Journal of the Asiatic Society of Bengal.* Vol. LXIV, Part ii.

Oldham, R. D., 1885 .. 'Note on the Geology of the Andaman Islands.' *Rec. Geol. Survey India,* Vol. XVIII, Pt. 3.

Pilgrim, G. E., 1911 .. .. 'The Geology of the Persian Gulf and the adjoining portions of Persia and Arabia.' *Mem. Geol. Survey India,* Vol. XXXIV.

Ross, W. C., and Bagchi, K. N., 1919 'The Seasonal Variation in the Reaction and Hardness of River Water in India.' Part 1. *Indian Journ. Med. Research,* Vol. VI.

Ross, W. C., and Bagchi, K. N., 1925 'The Seasonal Variation in the Reaction and Hardness of River Water in India', Part 2. *Indian Journ. Med. Research,* Vol. XII.

Saville Kent, W., 1893 .. 'The Great Barrier Reef of Australia.' London.

Sewell, R. B. S., 1922 .. .. 'A Survey Season in the Nicobar Islands.' *Journal Bombay Nat. Hist. Soc.,* Vol. XXVIII.

Sewell, R. B. S., 1924 .. .. 'Observations on growth in certain Molluscs.' *Rec. Ind. Mus.,* Vol. XXVI, Part VI.

Sewell, R. B. S., 1928 .. 'A study of recent changes of sea-level based largely on a study of coral growths in Indian and Pacific Seas. *Internat. Revue der ges Hydrobiol. u. Hydrographie,* Vol. XX, p. 89.

Sewell, R. B. S., 1932 .. .. 'The Coral Coasts of India.' *The Geographical Journal,* Vol. LXXIX, No. 6.

Skeats, E. W., 1903 .. .. 'The Chemical Composition of Limestones from upraised Coral Islands with notes on their Microscopical Structures.' *Bull. Mus. Comp. Zool. Harvard,* Vol. XLII (Geological Series, Vol. VI).

Steers, J. A., 1930 .. .. 'A Geographical Introduction to the Biological Reports.' *Great Barrier Reef Expedition,* 1928-29. Scientific Reports, Vol. III, No. 1.

Stephenson, T. A., Tandy, G., and Spender, M., 1931 .. .. 'The Structure and Ecology of Low Isles and other reefs.' *Great Barrier Reef Expedition,* 1928-29. Scientific Reports, Vol. III, No. 2.

Thurston, E., 1895 .. .. 'Ramesvaram Island and Fauna of the Gulf of Manaar.' *Madras Government Museum Bulletin,* No. 3.

Tizard, R. N., Moseley, H. N., Buchanan, J. Y., and Murray, J., 1885 .. .. .. Narrative of the Cruise of H.M.S. 'Challenger,' Vol. 1, 2nd Part.

Umbgrove, J. H. F., 1928 .. 'Die Koraalriffen in de Baai van Batavia.' *Dierist van den Mijnbouw in Nederlandsch-India.* Wetenschappelijke Mededeelingen, No. 7.

van Tuyl, F. M., 1916 .. .. 'A contribution to the Oolite problem.' *The Journal of Geology,* Vol. XXIV.

Vaughan, T. W., 1914 .. 'Preliminary Remarks on the Geology of the Bahamas, with special reference to the Origin of the Bahaman and Floridian Oolites.' *Papers from the Tortugas Laboratory of the Carnegie Institute of Washington,* Vol. V, No. 3.

Vaughan, T. W., 1914(*a*) .. 'The Building of the Marquesas and Tortugas atolls and a sketch of the Geologic History of the Florida Reef Tract.' *Papers from the Tortugas Laboratory of the Carnegie Institute of Washington,* Vol. V, No. 14.

Vaughan, T. W., 1914(*b*) .. 'Reef Corals of the Bahamas and of Southern Florida.' *Carnegie Institute, Washington*: Year Book, No. 13.

Vaughan, T. W., 1918 .. 'Some shoal-water Bottom Samples from Murray Island, Australia, and comparisons of them with samples from Florida and the Bahamas.' *Papers from the Dept. Marine Biol., Carnegie Institute, Washington*, Vol. IX.

Vaughan, T. W., 1924 .. 'Oceanography in its relations to other earth sciences.' *Jour. Washington Acad. Sciences*, Vol. XIV, No. 14.

Verstelle, J. Th., 1932 .. .. 'The growth rate at various depths of coral Reefs in the Dutch East Indian Archipelago.' *Treubia*, Vol. XIV.

Wadia, D. N., 1919 .. .. 'Geology of India.' London.

Wharton, W. J. L., 1897 .. 'Foundations of Coral Atolls.' *Nature*, Vol. LV.

Wood-Jones, 1910 .. .. 'Corals and Atolls.' London.

Yonge, C. M., 1930 .. .. 'A year on the Great Barrier Reef.' London.

Yonge, C. M., 1931(*a*) .. .. 'Studies on the Physiology of Corals: IV. The Structure, Distribution and Physiology of the Zooxanthellæ.' *Great Barrier Reef Expedition*, 1928-29 Scientific Reports, Vol. I, No. 6.

Raised Coral Reef on the east side of Paway Island, Mergui Archipelago.

Fig. 1. A bed of dead coral boulders and sandstone blocks in a fresh-water pool at the top of the beach in Wood-Mason Bay, Rutland Island, Andamans.

Fig. 3. Sandstone formation beneath the reverse slope of a sand ridge behind the raised Coral Reef at Pamban, Gulf on Manaar (*Geographical Journ.*, LXXIX, opposite p. 452, 1932).

Fig. 2. Raised Reef on the east side of Kachal Island, Nicobars, showing erosion of Limestone blocks.

Fig. 4. The raised Coral Reef at Pamban, Gulf of Manaar (*Geographical Journ.*, LXXIX, opposite p. 452, 1932).

Fig. 1. Another part of the raised Coral Reef at Pamban, Gulf of Manaar.

Fig. 2. Coral Colonies in the Maldive Archipelago showing the precipitation of amorphous calcium carbonate on the upper surface of the Colonies.

Fig. 1. The Boulder Zone on Krusadai Island, Gulf on Manaar.

Fig. 3. The Outer margin of the Fringing Reef off Reed Point, Nankauri Harbour, Nicobar Islands (*Geographical Journ.*, LXXIX, opposite p. 453, 1932).

Fig. 2. The Fringing Reef in Octavia Bay, Nankauri Harbour, showing the zones of *Sarcophytum* and *Porites*. (Reproduced with kind permission of Messrs. Macmillan & Co.).

Fig. 1. The inner, shore-ward portion of the Fringing Reef off Reed point, Nankauri Harbour, Nicobar Islands, showing extensive growth of *Heliopora coerulea* (*Journ. Bombay Nat. Hist, Soc.*, XXVIII, opposite page 984, 1922)

Fig. 2. Cliff on north side of Fehendu Island, Horsburgh Atoil, showing sandstone formation at a depth of 3 feet.

D